国外优秀数学著作
原版系列

Handbook of Discrete and Computational Geometry, 3e

离散与计算几何手册

[美] 雅各布·E. 古德曼（Jacob E. Goodman）
[美] 约瑟夫·奥罗克（Joseph O'Rourke）
[美] 乔鲍·D. 托特（Csaba D. Tóth）
主编

（上册·第三版）

（英文）

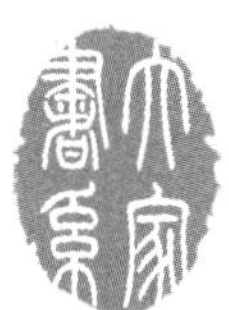

HITP
哈爾濱工業大學出版社
HARBIN INSTITUTE OF TECHNOLOGY PRESS

黑版贸审字 08－2020－202 号

Handbook of Discrete and Computational Geometry，3e/by Jacob E. Goodman，Joseph O'Rourke，Csaba D. Tóth/ISBN：978－1－4987－1139－5
Copyright © 2018 by CRC Press.
Authorized translation from English reprint edition published by CRC Press，part of Taylor & Francis Group LLC；All rights reserved；本书原版由 Tadylor & Francis 出版集团旗下，CRC 出版公司出版，并经其授权出版影印版. 版权所有，侵权必究.
Harbin Institute of Technology Press Ltd is authorized to publish and distribute exclusively the English reprint edition. This edition is authorized for sale throughout Mainland of China. No Part of the publication may be reproduced or distributed by any means，or stored in a database or retrieval system，without the prior written permission of the publisher. 本书英文影印版授权由哈尔滨工业大学出版社独家出版并仅限在中国大陆地区销售. 未经出版者书面许可，不得以任何方式复制或发行本书的任何部分.
Copies of this book sold without a Taylor & Francis sticker on the cover are unauthorized and illegal. 本书封面贴有 Taylor & Francis 公司防伪标签，无标签者不得销售.

图书在版编目(CIP)数据

离散与计算几何手册：第三版＝Handbook of Discrete and Computational Geometry，3e. 上册：英文/(美)雅各布·E. 古德曼(Jacob E. Goodman)，(美)约瑟夫·奥罗克(Joseph O'Rourke)，(美)乔鲍·D. 托特(Csaba D. Tóth)主编. —哈尔滨：哈尔滨工业大学出版社，2023. 1
ISBN 978-7-5767-0652-9

Ⅰ. ①离… Ⅱ. ①雅 … ②约… ③乔… Ⅲ. ①离散数学－计算几何－英文 Ⅳ. ①O18

中国国家版本馆 CIP 数据核字(2023)第 030338 号

LISAN YU JISUAN JIHE SHOUCE：DI-SAN BAN(SHANGCE)

策划编辑 刘培杰 杜莹雪
责任编辑 刘立娟
封面设计 孙茵艾
出版发行 哈尔滨工业大学出版社
社　　址 哈尔滨市南岗区复华四道街 10 号 邮编 150006
传　　真 0451－86414749
网　　址 http://hitpress. hit. edu. cn
印　　刷 哈尔滨市石桥印务有限公司
开　　本 787 mm×1 092 mm 1/16 印张 123 字数 2 371 千字
版　　次 2023 年 1 月第 1 版 2023 年 1 月第 1 次印刷
书　　号 ISBN 978-7-5767-0652-9
定　　价 248.00 元(全 3 册)

(如因印装质量问题影响阅读，我社负责调换)

TABLE OF CONTENTS

PREFACE TO THE FIRST EDITION

While books and journals of high quality have proliferated in discrete and computational geometry during recent years, there has been to date no single reference work fully accessible to the nonspecialist as well as to the specialist, covering all the major aspects of both fields. The *Handbook of Discrete and Computational Geometry* is intended to do exactly that: to make the most important results and methods in these areas of geometry readily accessible to those who use them in their everyday work, both in the academic world—as researchers in mathematics and computer science—and in the professional world—as practitioners in fields as diverse as operations research, molecular biology, and robotics.

A significant part of the growth that discrete mathematics as a whole has experienced in recent years has consisted of a substantial development in discrete geometry. This has been fueled partly by the advent of powerful computers and by the recent explosion of activity in the relatively young field of computational geometry. This synthesis between discrete and computational geometry, in which the methods and insights of each field have stimulated new understanding of the other, lies at the heart of this Handbook.

The phrase "discrete geometry," which at one time stood mainly for the areas of packing, covering, and tiling, has gradually grown to include in addition such areas as combinatorial geometry, convex polytopes, and arrangements of points, lines, planes, circles, and other geometric objects in the plane and in higher dimensions. Similarly, "computational geometry," which referred not long ago to simply the design and analysis of geometric algorithms, has in recent years broadened its scope, and now means the study of geometric problems from a computational point of view, including also computational convexity, computational topology, and questions involving the combinatorial complexity of arrangements and polyhedra. It is clear from this that there is now a significant overlap between these two fields, and in fact this overlap has become one of practice as well, as mathematicians and computer scientists have found themselves working on the same geometric problems and have forged successful collaborations as a result.

At the same time, a growing list of areas in which the results of this work are applicable has been developing. It includes areas as widely divergent as engineering, crystallography, computer-aided design, manufacturing, operations research, geographic information systems, robotics, error-correcting codes, tomography, geometric modeling, computer graphics, combinatorial optimization, computer vision, pattern recognition, and solid modeling.

With this in mind, it has become clear that a handbook encompassing the most important results of discrete and computational geometry would benefit not only the workers in these two fields, or in related areas such as combinatorics, graph theory, geometric probability, and real algebraic geometry, but also the *users* of this body of results, both industrial and academic. This Handbook is designed to fill that role. We believe it will prove an indispensable working tool both for researchers in geometry and geometric computing and for professionals who use geometric tools in their work.

The Handbook covers a broad range of topics in both discrete and computational geometry, as well as in a number of applied areas. These include geometric data structures, polytopes and polyhedra, convex hull and triangulation algorithms, packing and covering, Voronoi diagrams, combinatorial geometric questions, computational convexity, shortest paths and networks, computational real algebraic geometry, geometric arrangements and their complexity, geometric reconstruction problems, randomization and de-randomization techniques, ray shooting, parallel computation in geometry, oriented matroids, computational topology, mathematical programming, motion planning, sphere packing, computer graphics, robotics, crystallography, and many others. A final chapter is devoted to a list of available software. Results are presented in the form of theorems, algorithms, and tables, with every technical term carefully defined in a glossary that precedes the section in which the term is first used. There are numerous examples and figures to illustrate the ideas discussed, as well as a large number of unsolved problems.

The main body of the volume is divided into six parts. The first two, on combinatorial and discrete geometry and on polytopes and polyhedra, deal with fundamental geometric objects such as planar arrangements, lattices, and convex polytopes. The next section, on algorithms and geometric complexity, discusses these basic geometric objects from a computational point of view. The fourth and fifth sections, on data structures and computational techniques, discuss various computational methods that cut across the spectrum of geometric objects, such as randomization and de-randomization, and parallel algorithms in geometry, as well as efficient data structures for searching and for point location. The sixth section, which is the longest in the volume, contains chapters on fourteen applications areas of both discrete and computational geometry, including low-dimensional linear programming, combinatorial optimization, motion planning, robotics, computer graphics, pattern recognition, graph drawing, splines, manufacturing, solid modeling, rigidity of frameworks, scene analysis, error-correcting codes, and crystallography. It concludes with a fifteenth chapter, an up-to-the-minute compilation of available software relating to the various areas covered in the volume. A comprehensive index follows, which includes proper names as well as all of the terms defined in the main body of the Handbook.

A word about references. Because it would have been prohibitive to provide complete references to all of the many thousands of results included in the Handbook, we have to a large extent restricted ourselves to references for either the most important results, or for those too recent to have been included in earlier survey books or articles; for the rest we have provided annotated references to easily accessible surveys of the individual subjects covered in the Handbook, which themselves contain extensive bibliographies. In this way, the reader who wishes to pursue an older result to its source will be able to do so.

On behalf of the sixty-one contributors and ourselves, we would like to express our appreciation to all those whose comments were of great value to the authors of the various chapters: Pankaj K. Agarwal, Boris Aronov, Noga Alon, Saugata Basu, Margaret Bayer, Louis Billera, Martin Blümlinger, Jürgen Bokowski, B.F. Caviness, Bernard Chazelle, Danny Chen, Xiangping Chen, Yi-Jen Chiang, Edmund M. Clarke, Kenneth Clarkson, Robert Connelly, Henry Crapo, Isabel Cruz, Mark de Berg, Jesús de Loera, Giuseppe Di Battista, Michael Drmota, Peter Eades, Jürgen Eckhoff, Noam D. Elkies, Eva Maria Feichtner, Ioannis Fudos, Branko Grünbaum, Dan Halperin, Eszter Hargittai, Ulli Hund, Jürg Hüsler, Peter Johansson, Norman Johnson, Amy Josefczyk, Gil Kalai, Gyula Károlyi, Kevin Klenk, Włodzimierz Kuperberg, Endre Makai, Jr., Jiří Matoušek, Peter McMullen, Hans Melissen, Bengt Nilsson, Michel Pocchiola, Richard Pollack, Jörg Rambau, Jürgen Richter-Gebert, Allen D. Rogers, Marie-Françoise Roy, Egon Schulte, Dana Scott, Jürgen Sellen, Micha Sharir, Peter Shor, Maxim Michailovich Skriganov, Neil J.A. Sloane, Richard

P. Stanley, Géza Tóth, Ioannis Tollis, Laureen Treacy, Alexander Vardy, Gert Vegter, Pamela Vermeer, Siniša Vrećica, Kevin Weiler, Asia Ivić Weiss, Neil White, Chee-Keng Yap, and Günter M. Ziegler.

In addition, we would like to convey our thanks to the editors of CRC Press for having the vision to commission this Handbook as part of their *Discrete Mathematics and Its Applications* series; to the CRC staff, for their help with the various stages of the project; and in particular to Nora Konopka, with whom we found it a pleasure to work from the inception of the volume.

Finally, we want to express our sincere gratitude to our families: Josy, Rachel, and Naomi Goodman, and Marylynn Salmon and Nell and Russell O'Rourke, for their patience and forbearance while we were in the throes of this project.

Jacob E. Goodman
Joseph O'Rourke

PREFACE TO THE SECOND EDITION

This second edition of the Handbook of Discrete and Computational Geometry represents a substantial revision of the first edition, published seven years earlier. The new edition has added over 500 pages, a growth by more than 50%. Each chapter has been thoroughly revised and updated, and we have added thirteen new chapters. The additional room permitted the expansion of the curtailed bibliographies of the first edition, which often required citing other surveys to locate original sources. The new bibliographies make the chapters, in so far as is possible, self-contained. Most chapters have been revised by their original authors, but in a few cases new authors have joined the effort. All together, taking into account the chapters new to this edition, the number of authors has grown from sixty-three to eighty-four.

In the first edition there was one index; now there are two: in addition to the Index of Defined Terms there is also an Index of Cited Authors, which includes everyone referred to by name in either the text or the bibliography of each chapter. The first edition chapter on computational geometry software has been split into two chapters: one on the libraries LEDA and CGAL, the other on additional software. There are five new chapters in the applications section: on algorithms for modeling motion, on surface simplification and 3D-geometry compression, on statistical applications, on Geographic Information Systems and computational cartography, and on biological applications of computational topology. There are new chapters on collision detection and on nearest neighbors in high-dimensional spaces. We have added material on mesh generation, as well as a new chapter on curve and surface reconstruction, and new chapters on embeddings of finite metric spaces, on polygonal linkages, and on geometric graph theory.

All of these new chapters, together with the many new results contained within the Handbook as a whole, attest to the rapid growth in the field since preparation for the first edition began a decade ago. And as before, we have engaged the world's leading experts in each area as our authors.

In addition to the many people who helped with the preparation of the various chapters comprising the first edition, many of whom once again gave invaluable assistance with the present edition, we would also like to thank the following on behalf of both the authors and ourselves: Pankaj Agarwal, David Avis, Michael Baake, David Bremner, Hervé Brönnimann, Christian Buchta, Sergio Cabello,

Yi-Jen Chiang, Mirela Damian, Douglas Dunham, Stefan Felsner, Lukas Finschi, Bernd Gärtner, Ewgenij Gawrilow, Daniel Hug, Ekkehard Köhler, Jeffrey C. Lagarias, Vladimir I. Levenshtein, Casey Mann, Matthias Müller-Hannemann, Rom Pinchasi, Marc E. Pfetsch, Charles Radin, Jorge L. Ramírez Alfonsín, Matthias Reitzner, Jürgen Sellen, Thilo Schröder, Jack Snoeyink, Hellmuth Stachel, Pavel Valtr, Nikolaus Witte, and Chee Yap.

We would also like to express our appreciation to Bob Stern, CRC's Executive Editor, who gave us essentially a free hand in choosing how best to fill the additional 500 pages that were allotted to us for this new edition.

Jacob E. Goodman
Joseph O'Rourke

PREFACE TO THE THIRD EDITION

This third edition of the *Handbook of Discrete and Computational Geometry* appears 20 years after the first edition. A generation of researchers and practitioners have learned their craft using the Handbook. Since the publication of the 2nd edition in 2004, the field has grown substantially. New methods have been developed to solve longstanding open problems, with many successes. Advances in technology demanded new computational paradigms in processing geometric data.

Fifty-eight out of sixty-five chapters from the second edition of the Handbook have been revised and updated, many of them by the original authors, others in collaboration with new co-authors. Ten new chapters have been added. Five of the new chapters are devoted to computational topology and its applications. A new chapter on proximity algorithms gives a comprehensive treatment of relative neighbohood graphs and geometric spanners. New chapters on coresets, sketches, ε-nets, and ε-approximations cover geometric methods to cope with large data. Two new chapters expand on the recent breakthroughs in rigidity theory. We hope the Handbook will remain a one-stop reference book, accessible to both specialists and nonspecialists.

We would like to thank the following, on behalf of the 103 authors and ourselves, for providing valuable feedback on the preliminary version of various chapters: Bernardo Ábrego, Abdo Alfakih, Imre Bárány, Sören Lennart Berg, Mónica Blanco, Prosenjit Bose, Mireille Bousquet-Mélou, Kevin Buchin, Steve Butler, Sergio Cabello, Pablo Camara, Gunnar Carlsson, Cesar Ceballos, Joseph Chan, Danny Chen, Siu-Wing Cheng, Vincent Cohen-Addad, Mónika Csikós, Antoine Deza, Kevin Emmett, Jeff Erickson, Silvia Fernández-Merchant, Moritz Firsching, Florian Frick, Xavier Goaoc, Steven Gortler, Christian Haase, Bill Jackson, Bruno Jartoux, Iyad Kanj, Matya Katz, Balázs Keszegh, Hossein Khiabanian, Csaba Király, Matias Korman, Francis Lazarus, Michael Lesnick, Michael Mandell, Arnaud de Mesmay, Toby Mitchell, Eran Nevo, Patrick van Nieuwenhuizen, Arnau Padrol, Valentin Polishchuk, Hannes Pollehn, Daniel Rosenbloom, Raman Sanyal, Paul Seiferth, Michiel Smid, Yannik Stein, Jean Taylor, Louis Theran, Dimitrios Thilikos, Haitao Wang, Andrew Winslow, Richard Wolff, Ge Xia, and Sakellarios Zairis.

Jacob E. Goodman
Joseph O'Rourke
Csaba D. Tóth

CONTRIBUTORS

Pankaj K. Agarwal
Department of Computer Science
Duke University
Durham, North Carolina, USA
pankaj@cs.duke.edu

John Ralph Alexander, Jr.
Department of Mathematics
University of Illinois
Urbana, Illinois, USA
jralex@math.uiuc.edu

Nina Amenta
Department of Computer Science
University of California, Davis
Davis, California, USA
amenta@cs.udavis.edu

Alexandr Andoni
Department of Computer Science
Columbia University
New York City, New York, USA
aa3815@columbia.edu

Chandrajit L. Bajaj
Department of Computer Science
University of Texas at Austin
Austin, Texas, USA
bajaj@cs.utexas.edu

Gill Barequet
Department of Computer Science
Israel Institute of Technology
Haifa, Israel
barequet@cs.technion.ac.il

Alexander I. Barvinok
Department of Mathematics
University of Michigan
Ann Arbor, Michigan, USA
barvinok@umich.edu

Saugata Basu
Department of Mathematics
Purdue University
West Lafayette, Indiana, USA
sbasu@math.purdue.edu

József Beck
Department of Mathematics
Rutgers University
New Brunswick, New Jersey, USA
jbeck@math.rutgers.edu

Marshall Bern
Protein Metrics
San Carlos, California, USA
bern@proteinmetrics.com

Louis J. Billera
Department of Mathematics
Cornell University
Ithaca, New York, USA
billera@cornell.edu

Anders Björner
Department of Mathematics
Royal Institute of Technology
Stockholm, Sweden
bjorner@math.kth.se

Andrew J. Blumberg
Department of Mathematics
University of Texas at Austin
Austin, Texas, USA
blumberg@math.utexas.edu

Ulrich Brehm
Institut für Geometrie
Technische Universität Dresden
Dresden, Germany
ulrich.brehm@tu-dresden.de

Frédéric Chazal
INRIA Saclay - Ile-de-France
Palaiseau, France
frederic.chazal@inria.fr

William W.L. Chen
Department of Mathematics
Macquarie University
Sydney, New South Wales, Australia
william.chen@mq.edu.au

Otfried Cheong
School of Computing
Korea Advances Inst. Sci. & Tech.
Daejeon, South Korea
otfried@kaist.airpost.net

Éric Colin de Verdière
CNRS, LIGM,
Université Paris-Est
Marne-la-Vallée, France
eric.colindeverdiere@u-pem.fr

Robert Connelly
Department of Mathematics
Cornell University
Ithaca, New York, USA
connelly@math.cornell.edu

Erik D. Demaine
Computer Science and Artificial
Intelligence Laboratory, MIT
Cambridge, Massachusetts, USA
edemaine@mit.edu

Tamal K. Dey
Department of Computer Science
and Engineering
The Ohio State University
Columbus, Ohio, USA
tamaldey@cse.ohio-state.edu

Emilio Di Giacomo
Department of Engineering
Università degli Studi di Perugia
Perugia, Italy
emilio.digiacomo@unipg.it

Yann Disser
Department of Mathematics
Technische Universität Darmstadt
Darmstadt, Germany
disser@mathematik.tu-darmstadt.de

David P. Dobkin
Department of Computer Science
Princeton University
Princeton, New Jersey 08544
dpd@cs.princeton.edu

Martin Dyer
School of Computing
University of Leeds
Leeds, United Kingdom
m.e.dyer@leeds.ac.uk

Herbert Edelsbrunner
Institute of Science and Technology
Klosterneuburg, Austria
edels@ist.ac.at

Gábor Fejes Tóth
Alfréd Rényi Institute of Mathematics
Hungarian Academy of Sciences
Budapest, Hungary
fejes.toth.gabor@renyi.mta.hu

Stefan Felsner
Institut für Mathematik
Technische Universität Berlin
Berlin, Germany
felsner@math.tu-berlin.de

Steven Fortune
Google Inc.
New York City, New York, USA
fortunes@acm.org

Bernd Gärtner
Inst. of Theoret. Computer Science
ETH Zürich
Zürich, Switzerland
gartner@inf.ethz.ch

Solomon W. Golomb
(May 30, 1932—May 1, 2016)

Jacob E. Goodman
Department of Mathematics
City College, CUNY
New York City, New York, USA
jegcc@cunyvm.cuny.edu

Michael T. Goodrich
Department of Computer Science
University of California, Irvine
Irvine, California, USA
goodrich@acm.org

Ronald L. Graham
Computer Science and Engineering
University of California, San Diego
La Jolla, California, USA
rgraham@cs.ucsd.edu

Peter Gritzmann
Department of Mathematics
Technische Universität München
Garching, Germany
gritzman@tum.de

Leonidas J. Guibas
Department of Computer Science
Stanford University
Stanford, California, USA
guibas@cs.stanford.edu

Dan Halperin
School of Computer Science
Tel Aviv University
Tel Aviv, Israel
danha@post.tau.ac.il

Edmund O. Harriss
Department of Mathematical Sciences
University of Arkansas
Fayetteville, Arkansas, USA
eharriss@uark.edu

Martin Henk
Institut für Mathematik
Techniche Universität Berlin
Berlin, Germany
henk@mail.math.tu-berlin.de

Christoph M. Hoffmann
Department of Computer Science
Purdue University
West Lafayette, Indiana, USA
hoffmann@cs.purdue.edu

Michael Hoffmann
Inst. of Theoret. Computer Science
ETH Zürich
Zürich, Switzerland
hoffmann@inf.ethz.ch

Andreas Holmsen
Department of Mathematical Sciences
Korea Advances Inst. of Sci. & Tech.
Daejeon, South Korea
andreash@kaist.edu

Mia Hubert
Department of Mathematics
Katholieke Universiteit Leuven
Leuven, Belgium
mia.hubert@wis.kuleuven.be

Piotr Indyk
Computer Science and Artificial
Intelligence Laboratory, MIT
Cambridge, Massachusetts, USA
indyk@mit.edu

Tibor Jordán
Department of Operations Research
Eötvös Loránd University
Budapest, Hungary
jordan@cs.elte.hu

Michael Joswig
Institut für Matematik
Technische Universität Berlin
Berlin, Germany
joswig@math.tu-berlin.de

Matthew Kahle
Department of Mathematics
The Ohio State University
Columbus, Ohio, USA
mkahle@math.osu.edu

Gil Kalai
Einstein Institute of Mathematics
The Hebrew University of Jerusalem
Jerusalem, Israel
kalai@math.huji.ac.il

Lydia E. Kavraki
Department of Computer Science
Rice University
Houston, Texas, USA
kavraki@rice.edu

Lutz Kettner
NVIDIA Corporation
Berlin, Germany
cg@drkettner.de

Young J. Kim
Computer Graphics Laboratory
Ewha Womans University
Seoul, South Korea
kimy@ewha.ac.kr

David A. Klarner
(October 10, 1940–March 20, 1999)

Victor L. Klee
(September 18, 1925—August 17, 2007)

Patrice Koehl
Department of Computer Science
University of California, Davis
Davis, California, USA
koehl@cs.ucdavis.edu

Marc van Kreveld
Department of Computer Science
Utrecht University
Utrecht, The Netherlands
m.j.vankreveld@uu.nl

Carl W. Lee
Department of Mathematics
University of Kentucky
Lexington, Kentucky, USA
lee@ms.uky.edu

Ming C. Lin
Department of Computer Science
University of North Carolina
Chapel Hill, North Carolina, USA
lin@cs.unc.edu

Giuseppe Liotta
Department of Engineering
Università degli Studi di Perugia
Perugia, Italy
giuseppe.liotta@unipg.it

Benjamin Lorenz
Institut für Mathematik
Technische Universität Berlin
Berlin, Germany
lorenz@math.tu-berlin.de

Dinesh Manocha
Department of Computer Science
University of North Carolina
Chapel Hill, North Carolina, USA
dm@cs.unc.edu

Jiří Matoušek
(March 10, 1963—March 9, 2015)

Nimrod Megiddo
IBM Almaden Research Center
San Jose, California, USA
megiddo@theory.stanford.edu

Bhubaneswar Mishra
Courant Institute
New York University
New York City, New York, USA
mishra@cs.nyu.edu

Joseph S. B. Mitchell
Department of Applied
Mathematics and Statistics
Stony Brook University
Stony Brook, New York, USA
jsbm@ams.stonybrook.edu

Dmitriy Morozov
Lawrence Berkeley National Lab
Berkeley, California, USA.
dmitriy@mrzv.org

David M. Mount
Department of Computer Science
University of Maryland
College Park, Maryland, USA
mount@cs.umd.edu

Ketan Mulmuley
Department of Computer Science
The University of Chicago
Chicago, Illinois, USA
mulmuley@uchicago.edu

Wolfgang Mulzer
Institut für Informatik
Freie Universität Berlin
Berlin, Germany
mulzer@inf.fu-berlin.de

Nabil H. Mustafa
Université Paris-Est,
LIGM, ESIEE Paris
Noisy-le-Grand, France
mustafan@esiee.fr

Stefan Näher
Fachbereich IV - Informatik
Universität Trier
Trier, Germany
naeher@uni-trier.de

Joseph O'Rourke
Department of Computer Science
Smith College
Northampton, Massachusetts, USA
orourke@cs.smith.edu

János Pach
Chair of Combinatorial Geometry
École polytechnique fédérale
Lausanne, Switzerland
janos.pach@epfl.ch

Marco Pellegrini
Istituto di Informatica e Telematica
Consiglio Nazionale dele Ricerche
Pisa, Italy
marco.pellegrini@iit.cnr.it

Jeff M. Phillips
School of Computing
University of Utah
Salt Lake City, Utah, USA
jeffp@cs.utah.edu

Raul Rabadan
Department of Systems Biology
Columbia University
New York City, New York, USA
rr2579@cumc.columbia.edu

Edgar A. Ramos
Escuela de Matematicas
Universidad Nacional de Colombia
Medellin, Colombia
earamos@unal.edu.co

Jürgen Richter-Gebert
Department of Mathematics
Technische Universität München
Garching, Germany
richter@ma.tum.de

Marcel Roeloffzen
Large Graph Project
National Institute of Informatics
Tokyo, Japan
marcel@nii.ac.jp

Peter J. Rousseeuw
Department of Mathematics
Katholieke Universiteit Leuven
Leuven, Belgium
peter.rousseeuw@kuleuven.be

Oren Salzman
Robotics Institute
Carnegie Mellon University
Pittsburgh, Pennsylvania, USA
osalzman@andrew.cmu.edu

Francisco Santos
Department of Mathematics,
Statistics and Computer Science
Universidad de Cantabria
Santander, Spain
francisco.santos@unican.es

Doris Schattschneider
Department of Mathematics
Moravian College
Bethlehem, Pennsylvania, USA
schattschneiderdo@moravian.edu

Rolf Schneider
Mathematisches Institut
Albert-Ludwigs-Universität
Freiburg i. Br., Germany
rolf.schneider@math.uni-freiburg.de

Egon Schulte
Department of Mathematics
Northeastern University
Boston, Massachusetts, USA
schulte@neu.edu

Bernd Schulze
Department of Math. & Stats.
Lancaster University
Lancaster, United Kingdom
b.schulze@lancaster.ac.uk

Raimund Seidel
Theoretical Computer Science
Universität des Saarlandes
Saarbrücken, Germany
rseidel@cs.uni-saarland.de

Marjorie Senechal
Department of Mathematics
Smith College
Northampton, Massachusetts, USA
senechal@math.smith.edu

Vadim Shapiro
Department of Mechanical Engineering
University of Wisconsin, Madison
Madison, Wisconsin, USA
vshapiro@engr.wisc.edu

Micha Sharir
School of Computer Science
Tel Aviv University
Tel Aviv, Israel
michas@tau.ac.il

Vikram Sharma
Institute of Mathematical Sciences
Chennai, Tamil Nadu, India.
vikram@imsc.res.in

Jonathan Shewchuk
Computer Science Division
University of California, Berkeley
Berkeley, California, USA
jrs@berkeley.edu

Anastasios Sidiropoulos
Department of Computer Science
University of Illinois at Chicago
Chicago, Illinois, USA
sidiropo@gmail.com

Nodari Sitchinava
Dept. of Inform.. & Computer Sciences
University of Hawaii, Manoa
Honolulu, Hawaii, USA
nodari@hawaii.edu

Steven S. Skiena
Department of Computer Science
Stony Brook University
Stony Brook, New York, USA
skiena@cs.sunysb.edu

Jack Snoeyink
Department of Computer Science
University of North Carolina
Chapel Hill, North Carolina, USA
snoeyink@cs.unc.edu

Kiril Solovey
School of Computer Science
Tel Aviv University
Tel Aviv, Israel
kirilsol@post.tau.ac.il

Subhash Suri
Department of Computer Science
University of California, Santa Barbara
Santa Barbara, California, USA
suri@cs.ucsb.edu

Roberto Tamassia
Department of Computer Science
Brown University
Providence, Rhode Island, USA
rt@cs.brown.edu

Seth J. Teller
(1964–July 1, 2014)

Csaba D. Tóth
Department of Mathematics
California State University, Northridge
Los Angeles, California, USA
csaba.toth@csun.edu

Godfried T. Toussaint
Computer Science Program
New York University Abu Dhabi
Abu Dhabi, United Arab Emirates
gt@nyu.edu

Kasturi R. Varadarajan
Department of Computer Science
The University of Iowa
Iowa City, Iowa, USA
kasturi-varadarajan@uiowa.edu

Emo Welzl
Inst. of Theoret. Computer Science
ETH Zürich
Zürich, Switzerland
emo@inf.ethz.ch

Rephael Wenger
Department of Computer Science
and Engineering
The Ohio State University
Columbus, Ohio, USA
wenger@cis.ohio-state.edu

Neil L. White
(January 25, 1945–August 11, 2014)

Walter Whiteley
Department of Mathematics
and Statistics
York University
North York, Ontario, Canada
e-mail: whiteley@mathstat.yorku.ca

Chee K. Yap
Courant Institute
New York University
New York City, New York, USA
yap@cs.nyu.edu

Günter M. Ziegler
Institut für Mathematik
Freie Universität Berlin
Berlin, Germany
ziegler@math.fu-berlin.de

Rade Živaljević
Mathematical Institute
Serbian Academy of Sciences and Arts
Belgrade, Serbia
rade@mi.sanu.ac.rs

Part I

COMBINATORIAL AND DISCRETE GEOMETRY

1 FINITE POINT CONFIGURATIONS

János Pach

INTRODUCTION

The study of combinatorial properties of finite point configurations is a vast area of research in geometry, whose origins go back at least to the ancient Greeks. Since it includes virtually all problems starting with "consider a set of n points in space," space limitations impose the necessity of making choices. As a result, we will restrict our attention to Euclidean spaces and will discuss problems that we find particularly important. The chapter is partitioned into incidence problems (Section 1.1), metric problems (Section 1.2), and coloring problems (Section 1.3).

1.1 INCIDENCE PROBLEMS

In this section we will be concerned mainly with the structure of incidences between a finite point configuration P and a set of finitely many lines (or, more generally, k-dimensional flats, spheres, etc.). Sometimes this set consists of all lines connecting the elements of P. The prototype of such a question was raised by Sylvester [Syl93] more than one hundred years ago: Is it true that for any configuration of finitely many points in the plane, not all on a line, there is a line passing through exactly two points? This problem was rediscovered by Erdős [Erd43], and affirmative answers to this question was were given by Gallai and others [Ste44]. Generalizations for circles and conic sections in place of lines were established by Motzkin [Mot51] and Wilson-Wiseman [WW88], respectively.

GLOSSARY

Incidence: A point of configuration P lies on an element of a given collection of lines (k-flats, spheres, etc.).

Simple crossing: A point incident with exactly two elements of a given collection of lines or circles.

Ordinary line: A line passing through exactly two elements of a given point configuration.

Ordinary circle: A circle passing through exactly three elements of a given point configuration.

Ordinary hyperplane: A $(d-1)$-dimensional flat passing through exactly d elements of a point configuration in Euclidean d-space.

Motzkin hyperplane: A hyperplane whose intersection with a given d-dimensional point configuration lies—with the exception of exactly one point—in a $(d-2)$-dimensional flat.

Family of pseudolines: A family of two-way unbounded Jordan curves, any two of which have exactly one point in common, which is a proper crossing.

Family of pseudocircles: A family of closed Jordan curves, any two of which have at most two points in common, at which the two curves properly cross each other.

Regular family of curves: A family Γ of curves in the xy-plane defined in terms of D real parameters satisfying the following properties. There is an integer s such that (a) the dependence of the curves on x, y, and the parameters is algebraic of degree at most s; (b) no two distinct curves of Γ intersect in more than s points; (c) for any D points of the plane, there are at most s curves in Γ passing through all of them.

Degrees of freedom: The smallest number D of real parameters defining a regular family of curves.

Spanning tree: A tree whose vertex set is a given set of points and whose edges are line segments.

Spanning path: A spanning tree that is a polygonal path.

Convex position: P forms the vertex set of a convex polygon or polytope.

k-Set: A k-element subset of P that can be obtained by intersecting P with an open halfspace.

Halving plane: A hyperplane with $\lfloor |P|/2 \rfloor$ points of P on each side.

SYLVESTER-TYPE RESULTS

1. Gallai theorem (dual version): Any set of lines in the plane, not all of which pass through the same point, determines a simple crossing. This holds even for families of pseudolines [KR72].

2. Pinchasi theorem: Any set of at least five pairwise crossing unit circles in the plane determines a simple crossing [Pin01].

   Any sufficiently large set of pairwise crossing pseudocircles in the plane, not all of which pass through the same pair of points, determines an intersection point incident to at most three pseudocircles [ANP$^+$04].

3. Pach-Pinchasi theorem: Given n red and n blue points in the plane, not all on a line, there always exists a bichromatic line containing at most two points of each color [PP00].

   Any finite set of red and blue points contains a monochromatic spanned line, but not always a monochromatic ordinary line [Cha70].

4. Motzkin-Hansen theorem: For any finite set of points in Euclidean d-space, not all of which lie on a hyperplane, there exists a Motzkin hyperplane [Mot51, Han65]. We obtain as a corollary that n points in d-space, not all of which lie on a hyperplane, determine at least n distinct hyperplanes. (A hyperplane is *determined* by a point set P if its intersection with P is not contained in a $(d-2)$-flat.) Putting the points on two skew lines in 3-space shows that the existence of an ordinary hyperplane cannot be guaranteed for $d > 2$.

If $n > 8$ is sufficiently large, then any set of n noncocircular points in the plane determines at least $\binom{n-1}{2}$ distinct circles, and this bound is best possible [Ell67]. The minimum number of ordinary circles determined by n noncocircular points is $\frac{1}{4}n^2 - O(n)$. Here lines are not counted as circles [LMM+17].

5. Csima-Sawyer theorem: Any set of n noncollinear points in the plane determines at least $6n/13$ ordinary lines ($n > 7$). This bound is sharp for $n = 13$ and false for $n = 7$ [KM58, CS93]; see Figure 1.1.1.

   Green-Tao theorem (formerly Motzkin-Dirac conjecture): If n is sufficiently large and even, then the number of ordinary lines is at least $n/2$, and if n is odd, then at least $3\lfloor\frac{n-1}{4}\rfloor$. These bounds cannot be improved [GT13]. In 3-space, any set of n noncoplanar points determines at least $2n/5$ Motzkin hyperplanes [Han80, GS84].

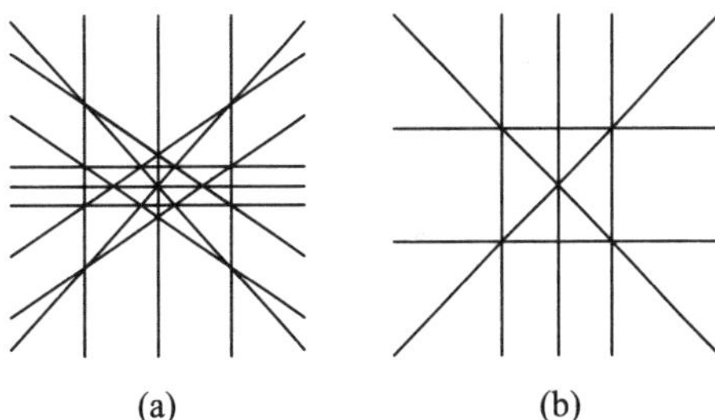

FIGURE 1.1.1
Extremal examples for the (dual) Csima-Sawyer theorem:
(a) 13 *lines (including the line at infinity) determining only* 6 *simple points;*
(b) 7 *lines determining only* 3 *simple points.*

6. Orchard problem [Syl67]: What is the maximum number of collinear triples determined by n points in the plane, no four on a line? There are several constructions showing that this number is at least $n^2/6 - O(n)$, which is asymptotically best possible, cf. [BGS74, FP84]. (See Figure 1.1.2.) Green-Tao theorem [GT13]: If n is sufficiently large, then the precise value of the maximum is $\lfloor\frac{n(n-3)}{6}\rfloor$.

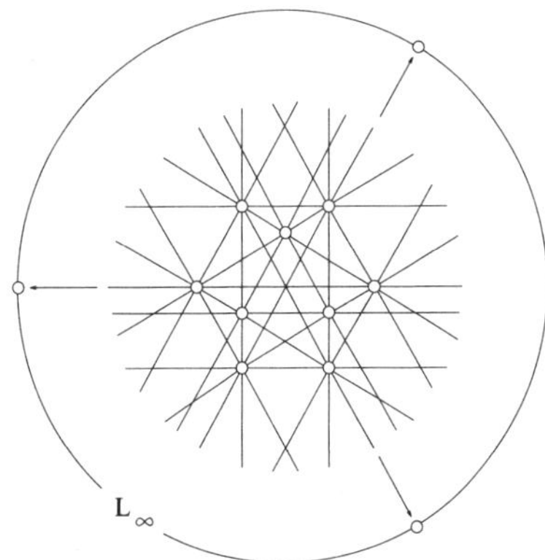

FIGURE 1.1.2
12 *points and* 19 *lines, each passing through exactly* 3 *points.*

7. Dirac's problem [Dir51]: Does there exist a constant c_0 such that any set of n points in the plane, not all on a line, has an element incident to at least $n/2 - c_0$ connecting lines? If true, this result is best possible, as is

shown by the example of n points distributed as evenly as possible on two intersecting lines. (It was believed that, apart from some small examples listed in [Grü72], this statement is true with $c_0 = 0$, until Felsner exhibited an infinite series of configurations, showing that $c_0 \geq 3/2$; see [BMP05, p. 313].) It is known [PW14] that there is a positive constant c such that one can find a point incident to at least cn connecting lines. The best known value of the constant for which this holds is $c = 1/37$. A useful equivalent formulation of this assertion is that any set of n points in the plane, no more than $n - k$ of which are on the same line, determines at least $c'kn$ distinct connecting lines, for a suitable constant $c' > 0$. Note that according to the $d = 2$ special case of the Motzkin-Hansen theorem, due to Erdős (see No. 4 above), for $k = 1$ the number of distinct connecting lines is at least n. For $k = 2$, the corresponding bound is $2n - 4$, $(n \geq 10)$.

8. Ungar's theorem [Ung82]: n noncollinear points in the plane always determine at least $2\lfloor n/2 \rfloor$ lines of different slopes (see Figure 1.1.3); this proves Scott's conjecture. Furthermore, any set of n points in the plane, not all on a line, permits a spanning tree, all of whose $n-1$ edges have different slopes [Jam87]. Pach, Pinchasi, and Sharir [PPS07] proved Scott's conjecture in 3 dimensions: any set of $n \geq 6$ points in $\mathbb{R}^3$, not all of which are on a plane, determine at least $2n - 5$ pairwise nonparallel lines if n is odd, and at least $2n - 7$ if n is even. This bound is tight for every odd n.

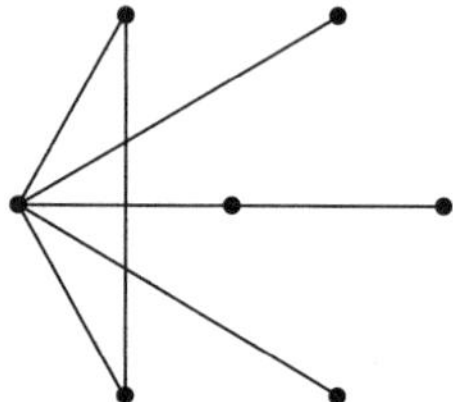

FIGURE 1.1.3
7 points determining 6 distinct slopes.

UPPER BOUNDS ON THE NUMBER OF INCIDENCES

Given a set P of n points and a family Γ of m curves or surfaces, the number of incidences between them can be obtained by summing over all $p \in P$ the number of elements of Γ passing through p. If the elements of Γ are taken from a regular family of curves with D degrees of freedom, the maximum number of incidences between P and Γ is $O(n^{D/(2D-1)}m^{(2D-2)/(2D-1)} + n + m)$. In the most important applications, Γ is a family of straight lines or unit circles in the plane $(D = 2)$, or it consists of circles of arbitrary radii $(D = 3)$. The best upper bounds known for the number of incidences are summarized in Table 1.1.1. It follows from the first line of the table that for any set P of n points in the plane, the number of distinct straight lines containing at least k elements of P is $O(n^2/k^3 + n/k)$, and this bound cannot be improved (Szemerédi-Trotter). In the second half of the table, $\kappa(n,m)$ and $\beta(n,m)$ denote extremely slowly growing functions, which are certainly $o(n^\epsilon m^\epsilon)$ for every $\epsilon > 0$. A family of pseudocircles is *special* if its curves admit a 3-parameter algebraic representation. A collection of spheres in 3-space is said to be in *general position* here if no three of them pass through the same circle [CEG+90, ANP+04].

TABLE 1.1.1 Maximum number of incidences between n points of P and m elements of Γ. [SzT83, CEG+90, NPP+02, KMSS12, Z13]

PT. SET P	FAMILY Γ	BOUND	TIGHT
Planar	lines	$O(n^{2/3}m^{2/3} + n + m)$	yes
Planar	pseudolines	$O(n^{2/3}m^{2/3} + n + m)$	yes
Planar	unit circles	$O(n^{2/3}m^{2/3} + n + m)$	?
Planar	pairwise crossing circles	$O(n^{1/2}m^{5/6} + n^{2/3}m^{2/3} + n + m)$	?
Planar	special pseudocircles	$O(n^{6/11}m^{9/11}\kappa(n,m) + n^{2/3}m^{2/3} + n + m)$	?
Planar	pairwise crossing pseudocircles	$O(n^{2/3}m^{2/3} + n + m^{4/3})$	?
3-dim'l	unit spheres	$O(n^{3/4}m^{3/4} + n + m)$	?
3-dim'l	spheres in gen. position	$O(n^{3/4}m^{3/4} + n + m)$	?
d-dim'l	circles	$O(n^{6/11}m^{9/11}\kappa(n,m) + n^{2/3}m^{2/3} + n + m)$	?

MIXED PROBLEMS

Many problems about finite point configurations involve some notions that cannot be defined in terms of incidences: convex position, midpoint of a segment, etc. Below we list a few questions of this type. They are discussed in this part of the chapter, and not in Section 1.2, which deals with metric questions, because we can disregard most aspects of the Euclidean metrics in their formulation. For example, convex position can be defined by requiring that some sets should lie on one side of certain hyperplanes. This is essentially equivalent to introducing an order along each straight line.

1. Erdős-Klein-Szekeres problem: What is the maximum number of points that can be chosen in the plane so that no three are on a line and no k are in convex position ($k > 3$)? Denoting this number by $c(k)$, it is known that

$$2^{k-2} \leq c(k) \leq 2^{k+o(k)}.$$

The lower bound given in [ES61] is conjectured to be tight, but this has been verified only for $k \leq 6$; [SP06]. The original upper bound, $\binom{2k-4}{k-2}$ in [ES35] was successively improved to $\binom{2k-5}{k-2} - \binom{2k-8}{k-3} + 1$, where the last bound is due to Mojarrad and Vlachos [MV16]. Suk [Suk17] proved that, as k tends to infinity, the lower bound is asymptotically tight.

Let $e(k)$ denote the maximum size of a planar point set P that has no three elements on a line and no k elements that form the vertex set of an "empty" convex polygon, i.e., a convex k-gon whose interior is disjoint from P. We have $e(3) = 2$, $e(4) = 4$, $e(5) = 8$, $e(6) < \infty$, and Horton showed that $e(k)$ is infinite for all $k \geq 7$ [Har78, Hor83, Nic07, Ger08].

2. The number of empty k-gons: Let $H_k^d(n)$ $(n \geq k \geq d+1)$ denote the minimum number of k-tuples that induce an empty convex polytope of k vertices in a set of n points in d-space, no $d+1$ of which lie on a hyperplane. Clearly, $H_2^1(n) = n - 1$ and $H_k^1(n) = 0$ for $k > 2$. For $k = d+1$, we have

$$\frac{1}{d!} \leq \lim_{n\to\infty} H_k^d(n)/n^d \leq \frac{2}{(d-1)!},$$

[Val95]. For $d = 2$, the best estimates known for $H_k^2 = \lim_{n\to\infty} H_k^2(n)/n^2$ are given in [BV04]:

$$1 \le H_3^2 \le 1.62,\ 1/2 \le H_4^2 \le 1.94,\ 0 \le H_5^2 \le 1.02,$$
$$0 \le H_6^2 \le 0.2,\ H_7^2 = H_8^2 = \ldots = 0.$$

3. The number of k-sets [ELSS73]: Let $N_k^d(n)$ denote the maximum number of k-sets in a set of n points in d-space, no $d+1$ of which lie on the same hyperplane. In other words, $N_k^d(n)$ is the maximum number of different ways in which k points of an n-element set can be separated from the others by a hyperplane. It is known that

$$ne^{\Omega(\sqrt{\log k})} \le N_k^2(n) \le O\left(n(k+1)^{1/3}\right),$$

[Tót01, Dey98]. For the number of halving planes, $N^3_{\lfloor n/2\rfloor}(n) = O(n^{5/2})$, and

$$n^{d-1}e^{\Omega(\sqrt{\log n})}) \le N^d_{\lfloor n/2\rfloor}(n) = o(n^d).$$

The most interesting case is $k = \frac{n}{2}$ in the plane, which is the maximum number of distinct ways to cut a set of n points in the plane in half (number of halving lines).

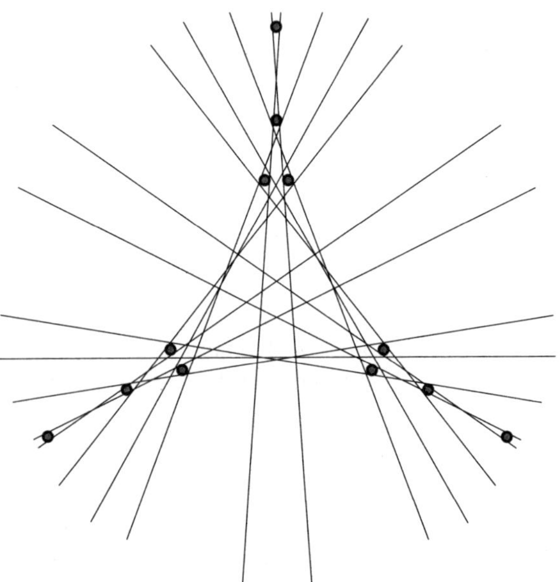

FIGURE 1.1.4
12 points determining 15 combinatorially distinct halving lines.

The maximum number of *at-most-k*-element subsets of a set of n points in d-space, no $d+1$ of which lie on a hyperplane, is $O\left(n^{\lfloor d/2\rfloor}k^{\lceil d/2\rceil}\right)$, and this bound is asymptotically tight [CS89]. In the plane the maximum number of *at-most-k*-element subsets of a set of n points is kn for $k < \frac{n}{2}$, which is reached for convex n-gons [AG86, Pec85].

4. The number of midpoints: Let $M(n)$ denote the minimum number of different midpoints of the $\binom{n}{2}$ line segments determined by n points in convex position

in the plane. One might guess that $M(n) \geq (1-o(1))\binom{n}{2}$, but it was shown in [EFF91] that

$$\binom{n}{2} - \left\lfloor \frac{n(n+1)(1-e^{-1/2})}{4} \right\rfloor \leq M(n) \leq \binom{n}{2} - \left\lfloor \frac{n^2-2n+12}{20} \right\rfloor.$$

5. Midpoint-free subsets: As a partial answer to a question proposed in [BMP05], it was proved by V. Bálint et al.[BBG$^+$95] that if $m(n)$ denotes the largest number m such that every set of n points in the plane has a midpoint-free subset of size m, then

$$\left\lceil \frac{-1+\sqrt{8n+1}}{2} \right\rceil \leq m(n).$$

However, asymptotically, $\Omega(n^{1-\epsilon})n \leq m(n) \leq o(n)$, for every $\epsilon > 0$.

OPEN PROBLEMS

Here we give six problems from the multitude of interesting questions that remain open.

1. Does there exist for every k a number $n = n(k)$ such that in any set P of n points in the plane there are k elements that are either collinear and pairwise see each other? (Two elements of P see each other if the segment connecting them does not pass through any other point of P. [KPW05].

2. Generalized orchard problem (Grünbaum): What is the maximum number $c_k(n)$ of collinear k-tuples determined by n points in the plane, no $k+1$ of which are on a line ($k \geq 3$)? In particular, show that $c_4(n) = o(n^2)$. Improving earlier lower bounds of Grünbaum [Grü76] and Ismailescu [Ism02], Solymosi and Stojaković showed that

$$c_k(n) \geq n^{2-\frac{\gamma_k}{\sqrt{\log n}}},$$

for a suitable $\gamma_k > 0$. [SS13]. For $k=3$, according to the Green-Tao theorem [GT13], we have $c_3(n) = \lfloor \frac{n(n-3)}{6} \rfloor$, provided that n is sufficiently large.

3. Maximum independent subset problem (Erdős): Determine the largest number $\alpha(n)$ such that any set of n points in the plane, no four on a line, has an $\alpha(n)$-element subset with no collinear triples. Füredi [Für91] has shown that $\Omega(\sqrt{n \log n}) \leq \alpha(n) \leq o(n)$.

4. Slope problem (Jamison): Is it true that every set of n points in the plane, not all on a line, permits a spanning path, all of whose $n-1$ edges have different slopes?

5. Empty triangle problem (Bárány): Is it true that every set of n points in the plane, no three on a line, determines at least $t(n)$ empty triangles that share a side, where $t(n)$ is a suitable function tending to infinity?

6. Balanced partition problem (Kupitz): Does there exist an integer k with the property that for every planar point set P, there is a connecting line such that the difference between the number of elements of P on its left side and right side does not exceed k? Some examples due to Alon show that this assertion is not true with $k = 1$. Pinchasi proved that there is a connecting line, for which this difference is $O(\log \log n)$.

1.2 METRIC PROBLEMS

The systematic study of the distribution of the $\binom{n}{2}$ distances determined by n points was initiated by Erdős in 1946 [Erd46]. Given a point configuration $P = \{p_1, p_2, \ldots, p_n\}$, let $g(P)$ denote the number of distinct distances determined by P, and let $f(P)$ denote the number of times that the unit distance occurs between two elements of P. That is, $f(P)$ is the number of pairs $p_i p_j$ $(i<j)$ such that $|p_i - p_j| = 1$. What is the minimum of $g(P)$ and what is the maximum of $f(P)$ over all n-element subsets of Euclidean d-space? These questions have raised deep number-theoretic and combinatorial problems, and have contributed richly to many recent developments in these fields.

GLOSSARY

Unit distance graph: A graph whose vertex set is a given point configuration P, in which two points are connected by an edge if and only if their distance is one.

Diameter: The maximum distance between two points of P.

General position in the plane: No three points of P are on a line, and no four on a circle.

Separated set: The distance between any two elements is at least one.

Nearest neighbor of $\boldsymbol{p \in P}$***:*** A point $q \in P$, whose distance from p is minimum.

Farthest neighbor of $\boldsymbol{p \in P}$***:*** A point $q \in P$, whose distance from p is maximum.

Homothetic sets: Similar sets in parallel position.

REPEATED DISTANCES

Extremal graph theory has played an important role in this area. For example, it is easy to see that the unit distance graph assigned to an n-element planar point set P cannot contain $K_{2,3}$, a complete bipartite graph with 2 and 3 vertices in its classes. Thus, by a well-known graph-theoretic result, $f(P)$, the number of edges in this graph, is at most $O(n^{3/2})$. This bound can be improved to $O(n^{4/3})$ by using more sophisticated combinatorial techniques (apply line 3 of Table 1.1.1 with $m = n$); but we are still far from knowing what the best upper bound is.

In Table 1.2.1, we summarize the best currently known estimates on the maximum number of times the unit distance can occur among n points in the plane, under various restrictions on their position. In the first line of the table—and throughout this chapter—c denotes (unrelated) positive constants. The second and

TABLE 1.2.1 Estimates for the maximum number of unit distances determined by an n-element planar point set P.

POINT SET P	LOWER BOUND	UPPER BOUND	SOURCE
Arbitrary	$n^{1+c/\log\log n}$	$O(n^{4/3})$	[Erd46, SST84]
Separated	$\lfloor 3n-\sqrt{12n-3}\rfloor$	$\lfloor 3n-\sqrt{12n-3}\rfloor$	[Reu72, Har74]
Of diameter 1	n	n	[HP34]
In convex position	$2n-7$	$O(n\log n)$	[EH90, Für90]
No 3 collinear	$\Omega(n\log n)$	$O(n^{4/3})$	Kárteszi
Separated, no 3 coll.	$(2+5/16-o(1))n$	$(2+3/7)n$	[Tót97]

third lines show how many times the minimum distance and the maximum distance, resp., can occur among n arbitrary points in the plane. Table 1.2.2 contains some analogous results in higher dimensions.

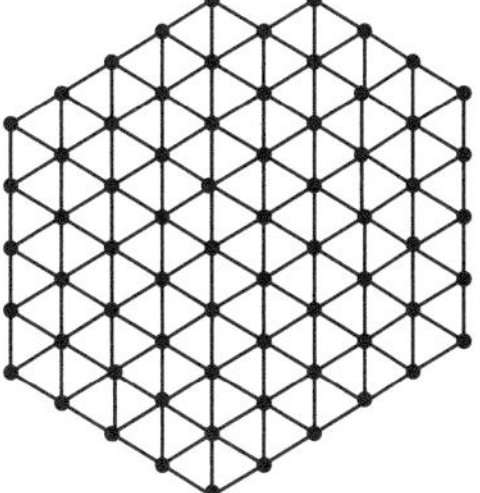

FIGURE 1.2.1
A separated point set with $\lfloor 3n-(12n-3)^{1/2}\rfloor$ *unit distances* ($n=69$). *All such sets have been characterized by Kupitz.*

TABLE 1.2.2 Estimates for the maximum number of unit distances determined by an n-element point set P in d-space.

POINT SET P	LOWER BOUND	UPPER BOUND	SOURCE
$d=3$, arbitrary	$\Omega(n^{4/3}\log\log n)$	$O(n^{3/2})$	[Erd60, KMS12, Zal13]
$d=3$, separated	$6n-O(n^{2/3})$	$6n-\Omega(n^{2/3})$	Newton
$d=3$, diameter 1	$2n-2$	$2n-2$	[Grü56, Hep56]
$d=3$, on sphere (rad. $1/\sqrt{2}$)	$\Omega(n^{4/3})$	$O(n^{4/3})$	[EHP89]
$d=3$, on sphere (rad. $r\neq 1/\sqrt{2}$)	$\Omega(n\log^* n)$	$O(n^{4/3})$	[EHP89]
$d=4$	$\lfloor\frac{n^2}{4}\rfloor+n-1$	$\lfloor\frac{n^2}{4}\rfloor+n$	[Bra97, Wam99]
$d\geq 4$ even, arb.	$\frac{n^2}{2}\left(1-\frac{1}{\lfloor d/2\rfloor}\right)+n-O(d)$	$\frac{n^2}{2}\left(1-\frac{1}{\lfloor d/2\rfloor}\right)+n-\Omega(d)$	[Erd67]
$d>4$ odd, arb.	$\frac{n^2}{2}\left(1-\frac{1}{\lfloor d/2\rfloor}\right)+\Omega(n^{4/3})$	$\frac{n^2}{2}\left(1-\frac{1}{\lfloor d/2\rfloor}\right)+O(n^{4/3})$	[EP90]

The second line of Table 1.2.1 can be extended by showing that the smallest distance cannot occur more than $3n-2k+4$ times between points of an n-element

FIGURE 1.2.2
n points, among which the second-smallest distance occurs $(\frac{24}{7} + o(1))n$ times.

set in the plane whose convex hull has k vertices [Bra92a]. The maximum number of occurrences of the second-smallest and second-largest distance is $(24/7 + o(1))n$ and $3n/2$ (if n is even), respectively [Bra92b, Ves78].

Given any point configuration P, let $\Phi(P)$ denote the sum of the numbers of farthest neighbors for every element $p \in P$. Table 1.2.3 contains tight upper bounds on $\Phi(P)$ in the plane and in 3-space, and asymptotically tight ones for higher dimensions [ES89, Csi96, EP90]. Dumitrescu and Guha [DG04] raised the following related question: given a colored point set in the plane, its *heterocolored diameter* is the largest distance between two elements of different color. Let $\phi_k(n)$ denote the maximum number of times that the heterocolored diameter can occur in a k-colored n-element point set between two points of different color. It is known that $\phi_2(n) = n$, $\phi_3(n)$ and $\phi_4(n) = 3n/2 + O(1)$ and $\phi_k(n) = O(n)$ for every k.

TABLE 1.2.3 Upper bounds on $\Phi(P)$, the total number of farthest neighbors of all points of an n-element set P.

POINT SET P	UPPER BOUND	SOURCE
Planar, n is even	$3n-3$	[ES89, Avi84]
Planar, n is odd	$3n-4$	[ES89, Avi84]
Planar, in convex position	$2n$	[ES89]
3-dimensional, $n \equiv 0 \pmod 2$	$n^2/4 + 3n/2 + 3$	[Csi96, AEP88]
3-dimensional, $n \equiv 1 \pmod 4$	$n^2/4 + 3n/2 + 9/4$	[Csi96, AEP88]
3-dimensional, $n \equiv 3 \pmod 4$	$n^2/4 + 3n/2 + 13/4$	[Csi96, AEP88]
d-dimensional ($d > 3$)	$n^2(1 - 1/\lfloor d/2 \rfloor + o(1))$	

DISTINCT DISTANCES

It is obvious that if all distances between pairs of points of a d-dimensional set P are the same, then $|P| \leq d+1$. If P determines at most g distinct distances, we have that $|P| \leq \binom{d+g}{d}$; see [BBS83]. This implies that if d is fixed and n tends to infinity, then the minimum number of distinct distances determined by n points in d-space is at least $\Omega(n^{1/d})$. Denoting this minimum by $g_d(n)$, for $d \geq 3$ we have the following results:

$$\Omega(n^{2/d-2/(d(d+2))}) \leq g_d(n) \leq O(n^{2/d}).$$

The lower bound is due to Solymosi and Vu [SV08], the upper bound to Erdős. In the plane, Guth and Katz [GK15] nearly proved Erdős's conjecture by showing that $g_2(n) = \Omega(n/\log n)$. Combining the results in [SV08] and [GK15], de Zeeuw showed that

$$g_3(n) = \Omega(n^{3/5}/\log n).$$

In Table 1.2.4, we list some lower and upper bounds on the minimum number of distinct distances determined by an n-element point set P, under various assumptions on its structure.

TABLE 1.2.4 Estimates for the minimum number of distinct distances determined by an n-element point set P in the plane.

POINT SET P	LOWER BOUND	UPPER BOUND	SOURCE
Arbitrary	$\Omega(n/\log n)$	$O(n/\sqrt{\log n})$	[GK15]
In convex position	$\lfloor n/2 \rfloor$	$\lfloor n/2 \rfloor$	[Alt63]
No 3 collinear	$\lceil (n-1)/3 \rceil$	$\lfloor n/2 \rfloor$	Szemerédi [Erd75]
In general position	$\Omega(n)$	$O(n^{1+c/\sqrt{\log n}})$	[EFPR93]

RELATED RESULTS

1. Integer distances: There are arbitrarily large, noncollinear finite point sets in the plane such that all distances determined by them are integers, but there exists no infinite set with this property [AE45].

2. Generic subsets: Any set of n points in the plane contains $\Omega((n/\log n)^{1/3})$ points such that all distances between them are distinct [Cha13]. This bound could perhaps be improved to about $n^{1/2}/(\log n)^{1/4}$, which would be best possible, as is shown by the example of the $\sqrt{n} \times \sqrt{n}$ integer grid. The corresponding quantity in d-dimensional space is $\Omega(n^{1/(3d-3)}(\log n)^{1/3-2/(3d-3)}$. [CFG+15].

3. Borsuk's problem: It was conjectured that every (finite) d-dimensional point set P can be partitioned into $d+1$ parts of smaller diameter. It follows from the results quoted in the third lines of Tables 1.2.1 and 1.2.2 that this is true for $d=2$ and 3. Surprisingly, Kahn and Kalai [KK93] proved that

there exist sets P that cannot be partitioned into fewer than $(1.2)^{\sqrt{d}}$ parts of smaller diameter. In particular, the conjecture is false for $d \geq 64$ [JB14]. On the other hand, it is known that for large d, every d-dimensional set can be partitioned into $(\sqrt{3/2}+o(1))^d$ parts of smaller diameter [Sch88].

4. Nearly equal distances: Two numbers are said to be nearly equal if their difference is at most one. If n is sufficiently large, then the maximum number of times that nearly the same distance occurs among n separated points in the plane is $\lfloor n^2/4 \rfloor$. The maximum number of pairs in a separated set of n points in the plane, whose distance is nearly equal to any one of k arbitrarily chosen numbers, is $\frac{n^2}{2}(1-\frac{1}{k+1}+o(1))$, as n tends to infinity [EMP93].

5. Repeated angles: In an n-element planar point set, the maximum number of noncollinear triples that determine the same angle is $O(n^2 \log n)$, and this bound is asymptotically tight for a dense set of angles (Pach-Sharir). The corresponding maximum in 3-space is at most $O(n^{7/3})$. In 4-space the angle $\pi/2$ can occur $\Omega(n^3)$ times, and all other angles can occur at most $O(n^{5/2+\varepsilon})$ times, for every $\varepsilon > 0$ [Pur88, AS05]. For dimension $d \geq 5$ all angles can occur $\Omega(n^3)$ times.

6. Repeated areas: Let $t_d(n)$ denote the maximum number of triples in an n-element point set in d-space that induce a unit area triangle. It is known that $\Omega(n^2 \log\log n) \leq t_2(n) \leq O(n^{20/9})$, $t_3(n) = O(n^{17/7+\varepsilon})$, for every $\varepsilon > 0$; $t_4(n), t_5(n) = o(n^3)$, and $t_6(n) = \Theta(n^3)$ ([EP71, PS90, RS17, DST09]). Maximum- and minimum-area triangles occur among n points in the plane at most n and at most $\Theta(n^2)$ times [BRS01].

7. Congruent triangles: Let $T_d(n)$ denote the maximum number of triples in an n-element point set in d-space that induce a triangle congruent to a given triangle T. It is known [AS02, ÁFM02] that

$$\begin{aligned} \Omega(n^{1+c/\log\log n}) \leq T_2(n) &\leq O(n^{4/3}), \\ \Omega(n^{4/3}) \leq T_3(n) &\leq O(n^{5/3+\epsilon}), \\ \Omega(n^2) \leq T_4(n) &\leq O(n^{2+\epsilon}), \\ T_5(n) &= \Theta(n^{7/3}), \text{ and} \\ T_d(n) &= \Theta(n^3) \text{ for } d \geq 6. \end{aligned}$$

8. Similar triangles: There exists a positive constant c such that for any triangle T and any $n \geq 3$, there is an n-element point set in the plane with at least cn^2 triples that induce triangles similar to T. For all quadrilaterals Q, whose points, as complex numbers, have an algebraic cross ratio, the maximum number of 4-tuples of an n-element set that induce quadrilaterals similar to Q is $\Theta(n^2)$. For all other quadrilaterals Q, this function is slightly subquadratic. The maximum number of pairwise homothetic triples in a set of n points in the plane is $O(n^{3/2})$, and this bound is asymptotically tight [EE94, LR97]. The number of similar tetrahedra among n points in three-dimensional space is at most $O(n^{2.2})$ [ATT98]. Further variants were studied in [Bra02].

9. Isosceles triangles, unit circles: In the plane, the maximum number of triples that determine an isosceles triangle, is $O(n^{2.137})$ (Pach-Tardos). The maximum number of distinct unit circles passing through at least 3 elements of a planar point set of size n is at least $\Omega(n^{3/2})$ and at most $n^2/3 - O(n)$ [Ele84].

CONJECTURES OF ERDŐS

1. The number of times the unit distance can occur among n points in the plane does not exceed $n^{1+c/\log\log n}$.
2. Any set of n points in the plane determines at least $\Omega(n/\sqrt{\log n})$ distinct distances.
3. Any set of n points in convex position in the plane has a point from which there are at least $\lfloor n/2 \rfloor$ distinct distances.
4. There is an integer $k \geq 4$ such that any finite set in convex position in the plane has a point from which there are no k points at the same distance.
5. Any set of n points in the plane, not all on a line, contains at least $n-2$ triples that determine distinct angles (Corrádi, Erdős, Hajnal).
6. The diameter of any set of n points in the plane with the property that the set of all distances determined by them is separated (on the line) is at least $\Omega(n)$. Perhaps it is at least $n-1$, with equality when the points are collinear.
7. There is no set of n points everywhere dense in the plane such that all distances determined by them are rational (Erdős, Ulam). Shaffaf [Sha15] showed that this conjecture would follow from the Bombieri-Lang conjecture.

1.3 COLORING PROBLEMS

If we partition a space into a small number of parts (i.e., we color its points with a small number of colors), at least one of these parts must contain certain "unavoidable" point configurations. In the simplest case, the configuration consists of a pair of points at a given distance. The prototype of such a question is the Hadwiger-Nelson problem: What is the minimum number of colors needed for coloring the plane so that no two points at unit distance receive the same color? The answer is known to be between 4 and 7.

GLOSSARY

Chromatic number of a graph: The minimum number of colors, $\chi(G)$, needed to color all the vertices of G so that no two vertices of the same color are adjacent.

List-chromatic number of a graph: The minimum number k such that for any assignment of a list of k colors to every vertex of the graph, for each vertex it is possible to choose a single color from its list so that no two vertices adjacent to each other receive the same color.

Chromatic number of a metric space: The chromatic number of the unit distance graph of the space, i.e., the minimum number of colors needed to color all points of the space so that no two points of the same color are at unit distance.

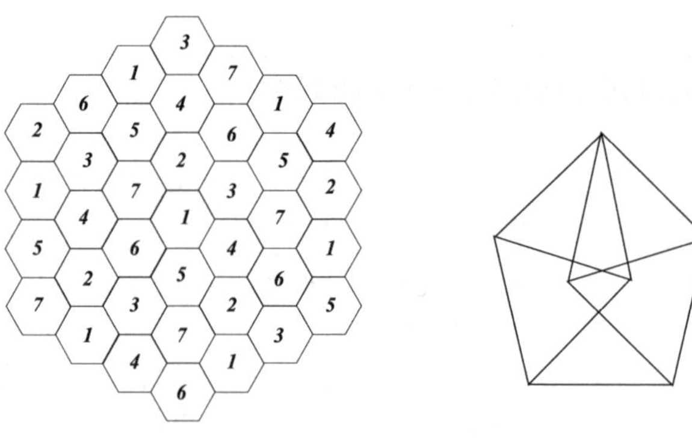

FIGURE 1.3.1
The chromatic number of the plane is (i) *at most* 7 *and* (ii) *at least* 4.

Polychromatic number of metric space: The minimum number of colors, χ, needed to color all points of the space so that for each color class C_i, there is a distance d_i such that no two points of C_i are at distance d_i. A sequence of "forbidden" distances, $(d_1, \ldots, d_\chi)$, is called a ***type*** of the coloring. (The same coloring may have several types.)

Girth of a graph: The length of the shortest cycle in the graph.

A point configuration P is ***k-Ramsey*** in d-space if, for any coloring of the points of d-space with k colors, at least one of the color classes contains a congruent copy of P.

A point configuration P is ***Ramsey*** if, for every k, there exists $d(k)$ such that P is k-Ramsey in $d(k)$-space.

Brick: The vertex set of a right parallelepiped.

FORBIDDEN DISTANCES

Table 1.3.1 contains the best bounds we know for the chromatic numbers of various spaces. All lower bounds can be established by showing that the corresponding unit distance graphs have some *finite* subgraphs of large chromatic number [BE51]. $S^{d-1}(r)$ denotes the sphere of radius r in d-space, where the distance between two points is the length of the chord connecting them.

Next we list several problems and results strongly related to the Hadwiger-Nelson problem (quoted in the introduction to this section).

1. 4-chromatic unit distance graphs of large girth: O'Donnell [O'D00] answered a question of Erdős by exhibiting a series of unit distance graphs in the plane with arbitrary large girths and chromatic number 4.

2. Polychromatic number: Stechkin and Woodall [Woo73] showed that the polychromatic number of the plane is between 4 and 6. It is known that for any $r \in [\sqrt{2}-1, 1/\sqrt{5}]$, there is a coloring of type $(1,1,1,1,1,r)$ [Soi94]. However, the list-chromatic number of the unit distance graph of the plane, which is at least as large as its polychromatic number, is infinite [Alo93].

3. Dense sets realizing no unit distance: The *lower* (resp. *upper*) *density* of an unbounded set in the plane is the lim inf (resp. lim sup) of the ratio of the

TABLE 1.3.1 Estimates for the chromatic numbers of metric spaces.

SPACE	LOWER BOUND	UPPER BOUND	SOURCE
Line	2	2	
Plane	4	7	Nelson, Isbell
Rational points of plane	2	2	[Woo73]
3-space	6	15	[Nec02, Cou02, RT03],
Rational points of 3-space	2	2	Benda, Perles
$S^2(r), \frac{1}{2} \le r \le \frac{\sqrt{3-\sqrt{3}}}{2}$	3	4	[Sim75]
$S^2(r), \frac{\sqrt{3-\sqrt{3}}}{2} \le r \le \frac{1}{\sqrt{3}}$	3	5	Straus
$S^2(r), r \ge \frac{1}{\sqrt{3}}$	4	7	[Sim76]
$S^2\left(\frac{1}{\sqrt{2}}\right)$	4	4	[Sim76]
4-space	9	54	[EIL14, RT03]
Rational points of 4-space	4	4	Benda, Perles
Rational points of 5-space	8	?	[Cib08]
d-space	$(1+o(1))(1.239)^d$	$(3+o(1))^d$	[FW81, Rai00, LR72]
$S^{d-1}(r), r \ge \frac{1}{2}$	d	?	[Lov83]

Lebesgue measure of its intersection with a disk of radius r around the origin to $r^2\pi$, as $r \to \infty$. If these two numbers coincide, their common value is called the ***density*** of the set. Let δ^d denote the maximum density of a planar set, no pair of points of which is at unit distance. Croft [Cro67] and Keleti et al. [KMO16] showed that $0.2293 \le \delta^2 \le 0.2588$.

4. The graph of large distances: Let $G_i(P)$ denote the graph whose vertex set is a finite point set P, with two vertices connected by an edge if and only if their distance is one of the i largest distances determined by P. In the plane, $\chi(G_1(P)) \le 3$ for every P; see Borsuk's problem in the preceding section. It is also known that for any finite planar set, $G_i(P)$ has a vertex with fewer than $3i$ neighbors [ELV89]. Thus, $G_i(P)$ has fewer than $3in$ edges, and its chromatic number is at most $3i$. However, if $n > ci^2$ for a suitable constant $c > 0$, we have $\chi(G_i(P)) \le 7$.

EUCLIDEAN RAMSEY THEORY

According to an old result of Gallai, for any finite d-dimensional point configuration P and for any coloring of d-space with finitely many colors, at least one of the color classes will contain a homothetic copy of P. The corresponding statement is false if, instead of a homothet, we want to find a *translate*, or even a *congruent copy*, of P. Nevertheless, for some special configurations, one can establish interesting positive results, provided that we color a sufficiently high-dimensional space with a sufficiently small number of colors. The Hadwiger-Nelson-type results discussed in the preceding subsection can also be regarded as very special cases of this problem, in which P consists of only two points. The field, known as "Euclidean Ramsey theory," was started by a series of papers by Erdős, Graham, Montgomery, Rothschild, Spencer, and Straus [EGM+73, EGM+75a, EGM+75b].

For details, see Chapter 10 of this Handbook.

OPEN PROBLEMS

1. (Erdős, Simmons) Is it true that the chromatic number of $S^{d-1}(r)$, the sphere of radius r in d-space, is equal to $d+1$, for every $r > 1/2$? In particular, does this hold for $d = 3$ and $r = 1/\sqrt{3}$?

2. (Sachs) What is the minimum number of colors, $\chi(d)$, sufficient to color any system of nonoverlapping unit balls in d-space so that no two balls that are tangent to each other receive the same color? Equivalently, what is the maximum chromatic number of a unit distance graph induced by a d-dimensional separated point set? It is easy to see [JR84] that $\chi(2) = 4$, and we also know that $5 \leq \chi(3) \leq 9$.

3. (Ringel) Does there exist any finite upper bound on the number of colors needed to color any system of (possibly overlapping) disks (of not necessarily equal radii) in the plane so that no two disks that are tangent to each other receive the same color, provided that no three disks touch one another at the same point? If such a number exists, it must be at least 5.

4. (Graham) Is it true that any 3-element point set P that does not induce an equilateral triangle is 2-Ramsey in the plane? This is known to be false for equilateral triangles, and correct for right triangles (Shader). Is every 3-element point set P 3-Ramsey in 3-space? The answer is again in the affirmative for right triangles [BT96].

5. (Bose et al. [BCC+13]) What is the smallest number $c(n)$ such that any set of n lines in the plane, no 3 of which pass through the same point, can be colored by $c(n)$ colors so that no cell of the arrangement is monochromatic? In particular, is it true that $c(n) > n^{\varepsilon}$ for some $\varepsilon > 0$?

1.4 SOURCES AND RELATED MATERIAL

SURVEYS

These surveys discuss and elaborate many of the results cited above.

[PA95, Mat02]: Monographs devoted to combinatorial geometry.

[BMP05]: A representative survey of results and open problems in discrete geometry, originally started by the Moser brothers.

[Pac93]: A collection of essays covering a large area of discrete and computational geometry, mostly of some combinatorial flavor.

[HDK64]: A classical treatise of problems and exercises in combinatorial geometry, complete with solutions.

[KW91]: A collection of beautiful open questions in geometry and number theory, together with some partial answers organized into challenging exercises.

[EP95]: A survey full of original problems raised by the "founding father" of combinatorial geometry.

[JT95]: A collection of more than two hundred unsolved problems about graph colorings, with an extensive list of references to related results.

[Soi09]: A collection of geometric coloring problems with a lot of information about their history.

[Grü72]: A monograph containing many results and conjectures on configurations and arrangements.

RELATED CHAPTERS

Chapter 4: Helly-type theorems and geometric transversals
Chapter 5: Pseudoline arrangements
Chapter 11: Euclidean Ramsey theory
Chapter 13: Geometric discrepancy theory and uniform distribution
Chapter 21: Topological methods in discrete geometry
Chapter 28: Arrangements

REFERENCES

[AE45] N.H. Anning and P. Erdős. Integral distances. *Bull. Amer. Math. Soc.*, 51:598–600, 1945.

[AEP88] D. Avis, P. Erdős, and J. Pach. Repeated distances in space. *Graphs Combin.*, 4:207–217, 1988.

[ÁFM02] B.M. Ábrego and S. Fernández-Merchant. Convex polyhedra in $\mathbb{R}^3$ spanning $\Omega(n^{4/3})$ congruent triangles. *J. Combin. Theory Ser. A*, 98:406–409, 2002.

[AG86] N. Alon and E. Győri. The number of small semispaces of a finite set of points, *J. Combin. Theory Ser. A*, 41:154–157, 1986.

[Alo93] N. Alon. Restricted colorings of graphs. In *Surveys in Combinatorics*, vol. 187 of *London Math. Soc. Lecture Note Ser.*, pages 1–33, Cambridge University Press, 1993.

[Alt63] E. Altman. On a problem of Erdős. *Amer. Math. Monthly*, 70:148–157, 1963.

[ANP$^+$04] P.K. Agarwal, E. Nevo, J. Pach, R. Pinchasi, M. Sharir, and S. Smorodinsky. Lenses in arrangements of pseudo-circles and their applications. *J. ACM*, 51:139–186, 2004.

[AS02] P.K. Agarwal and M. Sharir. The number of congruent simplices in a point set. *Discrete Comput. Geom.*, 28:123–150, 2002.

[AS05] R. Apfelbaum and M. Sharir. Repeated angles in three and four dimensions. *SIAM J. Discrete Math.*, 19:294–300, 2005.

[ATT98] T. Akutsu, H. Tamaki, and T. Tokuyama. Distribution of distances and triangles in a point set and algorithms for computing the largest common point sets. *Discrete Comput. Geom.*, 20:307–331, 1998.

[Avi84] D. Avis. The number of furthest neighbour pairs in a finite planar set. *Amer. Math. Monthly*, 91:417–420, 1984.

[BBG$^+$95] V. Bálint, M. Branická, P. Grešák, I. Hrinko, P. Novotný, and M. Stacho. Several remarks about midpoint-free sets. *Studies of Univ. Transport and Communication, Žilina, Math.-Phys. Ser.*, 10:3–10, 1995.

[BBS83] E. Bannai, E. Bannai, and D. Stanton. An upper bound on the cardinality of an s-distance subset in real Euclidean space II. *Combinatorica*, 3:147–152, 1983.

[BCC$^+$13] P. Bose, J. Cardinal, S. Collette, F. Hurtado, M. Korman, S. Langerman, and P. Taslakian. Coloring and guarding arrangements. *Discrete Math. Theor. Comput. Sci.*, 15:139–154, 2013.

[BE51] N.G. de Bruijn and P. Erdős. A colour problem for infinite graphs and a problem in the theory of relations. *Nederl. Akad. Wetensch. Proc. Ser. A*, 54:371–373, 1951.

[BGS74] S.A. Burr, B. Grünbaum, and N.J.A. Sloane. The orchard problem. *Geom. Dedicata*, 2:397–424, 1974.

[BMP05] P. Brass, W.O.J. Moser, and J. Pach. *Research Problems in Discrete Geometry.* Springer, New York, 2005.

[Bra92a] P. Brass. *Beweis einer Vermutung von Erdős and Pach aus der kombinatorischen Geometrie.* Ph.D. thesis, Dept. Discrete Math., Technische Universität Braunschweig, 1992.

[Bra92b] P. Brass. The maximum number of second smallest distances in finite planar sets. *Discrete Comput. Geom.*, 7:371–379, 1992.

[Bra97] P. Brass. On the maximum number of unit distances among n points in dimension four. In I. Bárány and K. Böröczky, editors, *Intuitive Geometry (Budapest, 1995)*, vol. 6 of *Bolyai Soc. Math. Studies*, pages 277–290, 1997.

[Bra02] P. Brass. Combinatorial geometry problems in pattern recognition. *Discrete Comput. Geom.*, 28:495–510, 2002.

[BRS01] P. Brass, G. Rote and K.J. Swanepoel. Triangles of extremal area or perimeter in a finite planar pointset. *Discrete Comput. Geom.*, 26:51–58, 2001.

[BT96] M. Bóna and G. Tóth. A Ramsey-type problem on right-angled triangles in space. *Discrete Math.*, 150:61–67, 1996.

[BV04] I. Bárány and P. Valtr. Planar point sets with a small number of empty convex polygons. *Studia. Sci. Math. Hungar.*, 41:243–266, 2004.

[CEG$^+$90] K. Clarkson, H. Edelsbrunner, L. Guibas, M. Sharir, and E. Welzl. Combinatorial complexity bounds for arrangements of curves and surfaces. *Discrete Comput. Geom.*, 5:99–160, 1990.

[CFG$^+$15] D. Conlon, J. Fox, W. Gasarch, D.G. Harris, D. Ulrich, and S. Zbarsky. Distinct volume subsets. *SIAM J. Discrete Math.*, 29:472–480, 2015.

[Cha70] G.D. Chakerian. Sylvester's problem on collinear points and a relative *Amer. Math. Monthly*, 77:164–167, 1970.

[Cha13] M. Charalambides. A note on distinct distance subsets. *J. Geom.*, 104:439–442, 2013.

[Cib08] J. Cibulka. On the chromatic number of real and rational spaces. *Geombinatorics*, 18:53–66, 2008.

[Cou02] D. Coulson. A 15-colouring of 3-space omitting distance one. *Discrete Math.*, 256:83–90, 2002.

[Cro67] H.T. Croft. Incidence incidents. *Eureka*, 30:22–26, 1967.

[CS89] K.L. Clarkson and P.W. Shor. Applications of random sampling in computational geometry, II. *Discrete Comput. Geom.*, 4:387–421, 1989.

[CS93] J. Csima and E. Sawyer. There exist $6n/13$ ordinary points. *Discrete Comput. Geom.*, 9:187–202, 1993.

[Csi96] G. Csizmadia. Furthest neighbors in space. *Discrete Math.*, 150:81–88, 1996.

[Dey98] T.K. Dey. Improved bounds for planar k-sets and related problems. *Discrete Comput. Geom.*, 19:373–382, 1998.

[DG04] A. Dumitrescu and S. Guha. Extreme distances in multicolored point sets. *J. Graph Algorithms Appl.*, 8:27–38, 2004.

[Dir51] G.A. Dirac. Collinearity properties of sets of points. *Quart. J. Math. Oxford Ser. (2)*, 2:221–227, 1951.

[DST09] A. Dumitrescu, M. Sharir, and C.D. Tóth. Extremal problems on triangle areas in two and three dimensions. *J. Combin. Theory Ser. A*, 116:1177–1198, 2009.

[EE94] G. Elekes and P. Erdős. Similar configurations and pseudogrids. In K. Böröczky and G. Fejes Tóth, editors, *Intuitive Geometry*, pages 85–104, North-Holland, Amsterdam, 1994.

[EFF91] P. Erdős, P. Fishburn, and Z. Füredi. Midpoints of diagonals of convex n-gons. *SIAM J. Discrete Math.*, 4:329–341, 1991.

[EFPR93] P. Erdős, Z. Füredi, J. Pach, and I.Z. Ruzsa. The grid revisited. *Discrete Math.*, 111:189–196, 1993.

[EGM+73] P. Erdős, R.L. Graham, P. Montgomery, B.L. Rothschild, J. Spencer, and E.G. Straus. Euclidean Ramsey theorems. I. *J. Combin. Theory*, 14:341–363, 1973.

[EGM+75a] P. Erdős, R.L. Graham, P. Montgomery, B.L. Rothschild, J. Spencer, and E.G. Straus. Euclidean Ramsey theorems. II. In A. Hajnal, R. Rado, and V.T. Sós, editors, *Infinite and Finite Sets*, pages 529–558, North-Holland, Amsterdam, 1975.

[EGM+75b] P. Erdős, R.L. Graham, P. Montgomery, B.L. Rothschild, J. Spencer, and E.G. Straus. Euclidean Ramsey theorems. III. In A. Hajnal, R. Rado, and V.T. Sós, editors, *Infinite and Finite Sets*, pages 559–584, North-Holland, Amsterdam, 1975.

[EH90] H. Edelsbrunner and P. Hajnal. A lower bound on the number of unit distances between the points of a convex polygon. *J. Combin. Theory Ser. A*, 55:312–314, 1990.

[EHP89] P. Erdős, D. Hickerson, and J. Pach. A problem of Leo Moser about repeated distances on the sphere. *Amer. Math. Monthly*, 96:569–575, 1989.

[EIL14] G. Exoo, D. Ismailescu, and M. Lim. On the chromatic number of $\mathbb{R}^4$. *Discrete Comput. Geom.*, 52:416–423, 2014.

[Ele84] G. Elekes. n points in the plane determine $n^{3/2}$ unit circles. *Combinatorica*, 4:131, 1984.

[Ell67] P.D.T.A. Elliott. On the number of circles determined by n points. *Acta Math. Acad. Sci. Hungar.*, 18:181–188, 1967.

[ELSS73] P. Erdős, L. Lovász, A. Simmons, and E.G. Straus. Dissection graphs of planar point sets. In G. Srivastava, editor, *A Survey of Combinatorial Theory*, pages 139–149, North-Holland, Amsterdam, 1973.

[ELV89] P. Erdős, L. Lovász, and K. Vesztergombi. Colorings of circles. *Discrete Comput. Geom.*, 4:541–549, 1989.

[EMP93] P. Erdős, E. Makai, and J. Pach. Nearly equal distances in the plane. *Combin. Probab. Comput.*, 2:401–408, 1993.

[EP71] P. Erdős and G. Purdy. Some extremal problems in geometry. *J. Combin. Theory Ser. A*, 10:246–252, 1971.

[EP90] P. Erdős and J. Pach. Variations on the theme of repeated distances. *Combinatorica*, 10:261–269, 1990.

[EP95] P. Erdős and G. Purdy. Extremal problems in combinatorial geometry. In R.L. Graham, M. Grötschel, and L. Lovász, editors, *Handbook of Combinatorics*, pages 809–874, North-Holland, Amsterdam, 1995.

[Erd43] P. Erdős. Problem 4065. *Amer. Math. Monthly*, 50:65, 1943.

[Erd46] P. Erdős. On sets of distances of n points. *Amer. Math. Monthly*, 53:248–250, 1946.

[Erd60] P. Erdős. On sets of distances of n points in Euclidean space. *Magyar Tud. Akad. Mat. Kutató Int. Közl.*, 5:165–169, 1960.

[Erd67] P. Erdős. On some applications of graph theory to geometry. *Canad. J. Math.*, 19:968–971, 1967.

[Erd75] P. Erdős. On some problems of elementary and combinatorial geometry. *Ann. Mat. Pura Appl. Ser. IV*, 103:99–108, 1975.

[ES35] P. Erdős and G. Szekeres. A combinatorial problem in geometry. *Compos. Math.*, 2:463–470, 1935.

[ES61] P. Erdős and G. Szekeres. On some extremum problems in elementary geometry. *Ann. Univ. Sci. Budapest. Eötvös, Sect. Math.*, 3:53–62, 1960/61.

[ES89] H. Edelsbrunner and S.S. Skiena. On the number of furthest-neighbour pairs in a point set. *Amer. Math. Monthly*, 96:614–618, 1989.

[FP84] Z. Füredi and I. Palásti. Arrangements of lines with a large number of triangles. *Proc. Amer. Math. Soc.*, 92:561–566, 1984.

[Für90] Z. Füredi. The maximum number of unit distances in a convex n-gon. *J. Combin. Theory Ser. A*, 55:316–320, 1990.

[Für91] Z. Füredi. Maximal independent subsets in Steiner systems and in planar sets. *SIAM J. Discrete Math.*, 4:196–199, 1991.

[FW81] P. Frankl and R.M. Wilson. Intersection theorems with geometric consequences. *Combinatorica*, 1:357–368, 1981.

[Ger08] T. Gerken. Empty convex hexagons in planar point sets. *Discrete Comput. Geom.*, 39:239–272, 2008.

[GK15] L. Guth and N.H. Katz. On the Erdős distinct distances problem in the plane. *Ann. of Math. (2)*, 181:155–190, 2015.

[Grü56] B. Grünbaum. A proof of Vázsonyi's conjecture. *Bull. Res. Council Israel, Sect. A*, 6:77–78, 1956.

[Grü72] B. Grünbaum. *Arrangements and Spreads*, vol. 10 of *CBMS Regional Conf. Ser. in Math.* Amer. Math. Soc., Providence, 1972.

[Grü76] B. Grünbaum. New views on some old questions of combinatorial geometry. *Colloq. Internaz. Theorie Combin. (Roma, 1973), Tomo I*, 451–468, 1976.

[GS84] B. Grünbaum and G.C. Shephard. Simplicial arrangements in projective 3-space. *Mitt. Math. Sem. Giessen*, 166:49–101, 1984.

[GT13] B. Green and T. Tao. On sets defining few ordinary lines. *Discrete Comput. Geom.*, 50:409–468, 2013.

[Han65] S. Hansen. A generalization of a theorem of Sylvester on lines determined by a finite set. *Math. Scand.*, 16:175–180, 1965.

[Han80] S. Hansen. On configurations in 3-space without elementary planes and on the number of ordinary planes. *Math. Scand.*, 47:181–194, 1980.

[Har74] H. Harborth. Solution to problem 664A. *Elem. Math.*, 29:14–15, 1974.

[Har78] H. Harborth. Konvexe Fünfecke in ebenen Punktmengen. *Elem. Math.*, 34:116–118, 1978.

[HDK64] H. Hadwiger, H. Debrunner, and V. Klee. *Combinatorial Geometry in the Plane.* Holt, Rinehart & Winston, New York, 1964.

[Hep56] A. Heppes. Beweis einer Vermutung von A. Vázsonyi. *Acta Math. Acad. Sci. Hungar.*, 7:463–466, 1956.

[Hor83] J.D. Horton. Sets with no empty 7-gon. *Canad. Math. Bull.*, 26:482–484, 1983.

[HP34] H. Hopf and E. Pannwitz. Aufgabe nr. 167. *Jahresber. Deutsch. Math.-Verein*, 43:114, 1934.

[Ism02] D. Ismailescu. Restricted point configurations with many collinear k-tuplets. *Discrete Comput. Geom.*, 28:571–575, 2002.

[Jam87] R. Jamison. Direction trees. *Discrete Comput. Geom.*, 2:249–254, 1987.

[JB14] T. Jenrich and A.E. Brouwer. A 64-dimensional counterexample to Borsuk's conjecture. *Electron. J. Combin.*, 21:P4.29, 2014.

[JR84] B. Jackson and G. Ringel. Colorings of circles. *Amer. Math. Monthly*, 91:42–49, 1984.

[JT95] T.R. Jensen and B. Toft. *Graph Coloring Problems.* Wiley-Interscience, New York, 1995.

[KK93] J. Kahn and G. Kalai. A counterexample to Borsuk's conjecture. *Bull. Amer. Math. Soc.*, 29:60–62, 1993.

[KMS12] H. Kaplan, J. Matoušek, Z. Safernová, and M. Sharir. Unit distances in three dimensions. *Combin. Probab. Comput.* 21:597–610, 2012.

[KMO16] T. Keleti, M. Matolcsi, F.M. de Oliveira Filho, and I. Ruzsa. Better bounds for planar sets avoiding unit distances. *Discrete Comput. Geom.*, 55:642–661, 2016.

[KM58] L.M. Kelly and W.O.J. Moser. On the number of ordinary lines determined by n points. *Canad. J. Math.*, 10:210–219, 1958.

[KPW05] J. Kára, A. Pór, and D.R. Wood. On the chromatic number of the visibility graph of a set of points in the plane. *Discrete Comput. Geom.*, 34:497–506, 2005.

[KR72] L.M. Kelly and R. Rottenberg. Simple points in pseudoline arrangements. *Pacific. J. Math.*, 40:617–622, 1972.

[KW91] V. Klee and S. Wagon. *Old and New Unsolved Problems in Plane Geometry and Number Theory.* Math. Assoc. Amer., Washington, 1991.

[Lov83] L. Lovász. Self-dual polytopes and the chromatic number of distance graphs on the sphere. *Acta Sci. Math. (Szeged)*, 45:317–323, 1983.

[LR72] D.G. Larman and C.A. Rogers. The realization of distances within sets in Euclidean space. *Mathematika*, 19:1–24, 1972.

[LR97] M. Laczkovich and I.Z. Ruzsa. The number of homothetic subsets. In R.L. Graham and J. Nešetřil, editors, *The Mathematics of Paul Erdős, II*, vol. 14 of *Algorithms and Combinatorics*, pages 294–302, Springer-Verlag, Berlin, 1997.

[LMM$^+$17] A. Lin, M. Makhul, H.N. Mojarrad, J. Schicho, K. Swanepoel, and F. de Zeeuw. On sets defining few ordinary circles. *Discrete Comput. Geom.*, in press, 2017.

[Mat02] J. Matoušek. *Lectures on Discrete Geometry.* Springer-Verlag, New York, 2002.

[Mot51] T. Motzkin. The lines and planes connecting the points of a finite set. *Trans. Amer. Math. Soc.*, 70:451–464, 1951.

[MV16] H.N. Mojarrad and G. Vlachos. An improved bound on the Erdős-Szekeres conjecture. *Discrete Comput. Geom.*, 56:165–180, 2016.

[Nec02] O. Nechushtan. On the space chromatic number. *Discrete Math.*, 256:499–507, 2002.

[Nic07] C.M. Nicolás. The empty hexagon theorem. *Discrete Comput. Geom.*, 38:389–397, 2007.

[O'D00] P. O'Donnell. Arbitrary girth, 4-chromatic unit distance graphs in the plane. II: Graph embedding. *Geombinatorics*, 9:180–193, 2000.

[PA95] J. Pach and P.K. Agarwal. *Combinatorial Geometry.* Wiley, New York, 1995.

[Pac93] J. Pach, editor. *New Trends in Discrete and Computational Geometry.* Springer-Verlag, Berlin, 1993.

[Pec85] G.W. Peck. On 'k-sets' in the plane. *Discrete Math.*, 56:73–74, 1985.

[PP00] J. Pach and R. Pinchasi. Bichromatic lines with few points. *J. Combin. Theory Ser. A*, 90:326–335, 2000.

[PPS07] J. Pach, R. Pinchasi, and M. Sharir. Solution of Scott's problem on the number of directions determined by a point set in 3-space. *Discrete Comput. Geom.*, 38:399–441, 2007.

[PS90] J. Pach and M. Sharir. Repeated angles in the plane and related problems. *J. Combin. Theory Ser. A*, 59:12–22, 1990.

[PW14] M.S. Payne and D.R. Wood. Progress on Dirac's conjecture. *Electron. J. Combin.*, 21:2.12, 2014.

[Pin01] R. Pinchasi. *Problems in Combinatorial Geometry in the Plane.* Ph.D. thesis, Dept. Mathematics, Hebrew Univ., 2001.

[Pur88] G. Purdy. Repeated angles in E_4. *Discrete Comput. Geom*, 3:73–75, 1988.

[Rai00] A.M. Raigorodskii. On the chromatic number of a space. *Russian Math. Surveys*, 55:351–352, 2000.

[Reu72] O. Reutter. Problem 664A. *Elem. Math.*, 27:19, 1972.

[RS17] O.E. Raz and M. Sharir. The number of unit-area triangles in the plane: theme and variations. *Combinatorica*, in press, 2017.

[RT03] R. Radoičić and G. Tóth. Note on the chromatic number of the space. In: *Discrete and Computational Geometry*, vol. 25 of *Algorithms and Combinatorics*, pages 695–698, Springer, Berlin, 2003.

[Sch88] O. Schramm. Illuminating sets of constant width. *Mathematika*, 35:180–199, 1988.

[Sha15] J. Shaffaf. A proof for the Erdős-Ulam problem assuming Bombieri-Lang conjecture. Preprint, `arXiv:1501.00159`, 2015.

[Sim75] G.J. Simmons. Bounds on the chromatic number of the sphere. In *Proc. 6th Southeastern Conf. on Combinatorics, Graph Theory, and Computing, Congr. Numer. 14*, pages 541–548, 1975.

[Sim76] G.J. Simmons. The chromatic number of the sphere. *J. Austral. Math. Soc. Ser. A*, 21:473–480, 1976.

[Soi09] A. Soifer. *The Mathematical Coloring Book.* Springer, New York, 2009.

[Soi94] A. Soifer. Six-realizable set x_6. *Geombinatorics*, 3:140–145, 1994.

[SP06] G. Szekeres and L. Peters. Computer solution to the 17-point Erdős-Szekeres problem. *ANZIAM J.*, 48:151–164, 2006.

[SS13] J. Solymosi and M. Stojaković. Many collinear k-tuples with no $k+1$ collinear points. *Discrete Comput. Geom.*, 50:811-820, 2013.

[SST84] J. Spencer, E. Szemerédi, and W.T. Trotter. Unit distances in the Euclidean plane. In B. Bollobás, editor, *Graph Theory and Combinatorics*, pages 293–303, Academic Press, London, 1984.

[Ste44] R. Steinberg. Solution of problem 4065. *Amer. Math. Monthly*, 51:169–171, 1944. (Also contains a solution by T. Gallai in an editorial remark.)

[Suk17] A. Suk. On the Erdős-Szekeres convex polygon problem. *J. Amer. Math. Soc.*, 30:1047–1053, 2017.

[SV08] J. Solymosi and V.H. Vu. Near optimal bounds for the Erdős distinct distances problem in high dimensions. *Combinatorica*, 28:113–125, 2008.

[Syl67] J.J. Sylvester. Problem 2473. *Educational Times*, 8:104–107, 1867.

[Syl93] J.J. Sylvester. Mathematical question 11851. *Educational Times*, 46:156, 1893.

[Tót97] G. Tóth. The shortest distance among points in general position. *Comput. Geom.*, 8:33–38, 1997.

[Tót01] G. Tóth. Point sets with many k-sets. *Discrete Comput. Geom.*, 26:187–194, 2001.

[Ung82] P. Ungar. $2N$ noncollinear points determine at least $2N$ directions. *J. Combin. Theory Ser. A*, 33:343–347, 1982.

[Val95] P. Valtr. On the minimum number of polygons in planar point sets. *Studia Sci. Math. Hungar.*, 30:155–163, 1995.

[Ves78] K. Vesztergombi. On large distances in planar sets. *Discrete Math.*, 67:191–198, 1978.

[Wam99] P. van Wamelen. The maximum number of unit distances among n points in dimension four. *Beitr. Algebra Geom.*, 40:475–477, 1999.

[Woo73] D.R. Woodall. Distances realized by sets covering the plane. *J. Combin. Theory*, 14:187–200, 1973.

[WW88] P.R. Wilson and J.A. Wiseman. A Sylvester theorem for conic sections. *Discrete Comput. Geom.*, 3:295–305, 1988.

[Zal13] J. Zahl. An improved bound on the number of point-surface incidences in three dimensions. *Contrib. Discrete Math.*, 8:100–121, 2013.

2 PACKING AND COVERING

Gábor Fejes Tóth

INTRODUCTION

The basic problems in the classical theory of packings and coverings, the development of which was strongly influenced by the geometry of numbers and by crystallography, are the determination of the densest packing and the thinnest covering with congruent copies of a given body K. Roughly speaking, the density of an arrangement is the ratio between the total volume of the members of the arrangement and the volume of the whole space. In Section 2.1 we define this notion rigorously and give an account of the known density bounds.

In Section 2.2 we consider packings in, and coverings of, bounded domains. Section 2.3 is devoted to multiple arrangements and their decomposability. In Section 2.4 we make a detour to spherical and hyperbolic spaces. In Section 2.5 we discuss problems concerning the number of neighbors in a packing, while in Section 2.6 we investigate some selected problems concerning lattice arrangements. We close in Section 2.7 with problems concerning packing and covering with sequences of convex sets.

2.1 DENSITY BOUNDS FOR ARRANGEMENTS IN E^d

GLOSSARY

Convex body: A compact convex set with nonempty interior. A convex body in the plane is called a ***convex disk***. The collection of all convex bodies in d-dimensional Euclidean space $\mathbb{E}^d$ is denoted by $\mathcal{K}(\mathbb{E}^d)$. The subfamily of $\mathcal{K}(\mathbb{E}^d)$ consisting of centrally symmetric bodies is denoted by $\mathcal{K}^*(\mathbb{E}^d)$.

Operations on $\mathcal{K}(\mathbb{E}^d)$: For a set A and a real number λ we set $\lambda A = \{x \mid x = \lambda a,\ a \in A\}$. λA is called a ***homothetic copy*** of A. The ***Minkowski sum*** $A + B$ of the sets A and B consists of all points $a + b$, $a \in A$, $b \in B$. The set $A - A = A + (-A)$ is called the ***difference body*** of A. B^d denotes the unit ball centered at the origin, and $A + rB^d$ is called the ***parallel body*** of A at distance r $(r > 0)$. If $A \subset \mathbb{E}^d$ is a convex body with the origin in its interior, then the ***polar body*** A^* of A is $\{x \in \mathbb{E}^d \mid \langle x, a\rangle \le 1 \text{ for all } a \in A\}$.

The ***Hausdorff distance*** between the sets A and B is defined by

$$d(A, B) = \inf\{\varrho \mid A \subset B + \varrho B^d, B \subset A + \varrho B^d\}.$$

Lattice: The set of all integer linear combinations of a particular basis of $\mathbb{E}^d$.

Lattice arrangement: The set of translates of a given set in $\mathbb{E}^d$ by all vectors of a lattice.

Packing: A family of sets whose interiors are mutually disjoint.

Covering: A family of sets whose union is the whole space.

The ***volume*** (Lebesgue measure) of a measurable set A is denoted by $V(A)$. In the case of the plane we use the term ***area*** and the notation $a(A)$.

Density of an arrangement relative to a set: Let $\mathcal{A}$ be an arrangement (a family of sets each having finite volume) and D a set with finite volume. The ***inner density*** $d_{\text{inn}}(\mathcal{A}|D)$, ***outer density*** $d_{\text{out}}(\mathcal{A}|D)$, and ***density*** $d(\mathcal{A}|D)$ of $\mathcal{A}$ relative to D are defined by

$$d_{\text{inn}}(\mathcal{A}|D) = \frac{1}{V(D)} \sum_{A\in\mathcal{A}, A\subset D} V(A),$$

$$d_{\text{out}}(\mathcal{A}|D) = \frac{1}{V(D)} \sum_{A\in\mathcal{A},\, A\cap D\neq\emptyset} V(A),$$

and

$$d(\mathcal{A}|D) = \frac{1}{V(D)} \sum_{A\in\mathcal{A}} V(A\cap D).$$

(If one of the sums on the right side is divergent, then the corresponding density is infinite.)

The ***lower density*** and ***upper density*** of an arrangement $\mathcal{A}$ are given by the limits $d_-(\mathcal{A}) = \liminf_{\lambda\to\infty} d_{\text{inn}}(\mathcal{A}|\lambda B^d)$, $d_+(\mathcal{A}) = \limsup_{\lambda\to\infty} d_{\text{out}}(\mathcal{A}|\lambda B^d)$. If $d_-(\mathcal{A}) = d_+(\mathcal{A})$, then we call the common value the ***density*** of $\mathcal{A}$ and denote it by $d(\mathcal{A})$. It is easily seen that these quantities are independent of the choice of the origin.

The ***packing density*** $\delta(K)$ and ***covering density*** $\vartheta(K)$ of a convex body (or more generally of a measurable set) K are defined by

$$\delta(K) = \sup\{d_+(\mathcal{P}) \mid \mathcal{P} \text{ is a packing of } \mathbb{E}^d \text{ with congruent copies of } K\}$$

and

$$\vartheta(K) = \inf\{d_-(\mathcal{C}) \mid \mathcal{C} \text{ is a covering of } \mathbb{E}^d \text{ with congruent copies of } K\}.$$

The ***translational packing density*** $\delta_T(K)$, ***lattice packing density*** $\delta_L(K)$, ***translational covering density*** $\vartheta_T(K)$, and ***lattice covering density*** $\vartheta_L(K)$ are defined analogously, by taking the supremum and infimum over arrangements consisting of translates of K and over lattice arrangements of K, respectively. It is obvious that in the definitions of $\delta_L(K)$ and $\vartheta_L(K)$ we can take maximum and minimum instead of supremum and infimum. By a theorem of Groemer, the same holds for the translational and for the general packing and covering densities.

Dirichlet cell: Given a set S of points in $\mathbb{E}^d$ such that the distances between the points of S have a positive lower bound, the Dirichlet cell, also known as the ***Voronoi cell***, associated to an element s of S consists of those points of $\mathbb{E}^d$ that are closer to s than to any other element of S.

KNOWN VALUES OF PACKING AND COVERING DENSITIES

Apart from the obvious examples of space fillers, there are only a few specific bodies for which the packing or covering densities have been determined. The bodies for which the packing density is known are given in Table 2.1.1.

TABLE 2.1.1 Bodies K for which $\delta(K)$ is known.

BODY	SOURCE
Circular disk in $\mathbb{E}^2$	[Thu10]
Parallel body of a rectangle	[Fej67]
Intersection of two congruent circular disks	[Fej71]
Centrally symmetric n-gon (algorithm in $O(n)$ time)	[MS90]
Ball in $\mathbb{E}^3$	[Hal05]
Ball in $\mathbb{E}^8$	[Via17]
Ball in $\mathbb{E}^{24}$	[CKM17]
Truncated rhombic dodecahedron in $\mathbb{E}^3$	[Bez94]

We have $\delta(B^2) = \pi/\sqrt{12}$. The longstanding conjecture that $\delta(B^3) = \pi/\sqrt{18}$ has been confirmed by Hales. A packing of balls reaching this density is obtained by placing the centers at the vertices and face-centers of a cubic lattice. We discuss the sphere packing problem in the next section.

For the rest of the bodies in Table 2.1.1, the packing density can be given only by rather complicated formulas. We note that, with appropriate modification of the definition, the packing density of a set with infinite volume can also be defined. A. Bezdek and W. Kuperberg (see [BK91]) showed that the packing density of an infinite circular cylinder is $\pi/\sqrt{12}$, that is, infinite circular cylinders cannot be packed more densely than their base. It is conjectured that the same statement holds for circular cylinders of any finite height.

A theorem of L. Fejes Tóth [Fej50] states that

$$\delta(K) \leq \frac{a(K)}{H(K)} \qquad \text{for } K \in \mathcal{K}(\mathbb{E}^2), \tag{2.1.1}$$

where $H(K)$ denotes the minimum area of a hexagon containing K. This bound is best possible for centrally symmetric disks, and it implies that

$$\delta(K) = \delta_T(K) = \delta_L(K) = \frac{a(K)}{H(K)} \qquad \text{for } K \in \mathcal{K}^*(\mathbb{E}^2).$$

The packing densities of the convex disks in Table 2.1.1 have been determined utilizing this relation.

It is conjectured that an inequality analogous to (2.1.1) holds for coverings, and this is supported by the following weaker result [Fej64]:

Let $h(K)$ denote the maximum area of a hexagon contained in a convex disk K. Let $\mathcal{C}$ be a covering of the plane with congruent copies of K such that no two copies of K cross. Then

$$d_-(\mathcal{C}) \geq \frac{a(K)}{h(K)}.$$

The convex disks A and B ***cross*** if both $A \setminus B$ and $B \setminus A$ are disconnected. As translates of a convex disk do not cross, it follows that

$$\vartheta_T(K) \geq \frac{a(K)}{h(K)} \qquad \text{for } K \in \mathcal{K}(\mathbb{E}^2).$$

Again, this bound is best possible for centrally symmetric disks, and it implies that

$$\vartheta_T(K) = \vartheta_L(K) = \frac{a(K)}{h(K)} \qquad \text{for } K \in \mathcal{K}^*(\mathbb{E}^2). \tag{2.1.2}$$

Based on this, Mount and Silverman gave an algorithm that determines $\vartheta_T(K)$ for a centrally symmetric n-gon in $O(n)$ time. Also the classical result $\vartheta(B^2) = 2\pi/\sqrt{27}$ of Kershner [Ker39] follows from this relation.

The bound $\vartheta_T(K) \leq \frac{a(K)}{h(K)}$ holds without the restriction to non-crossing disks for "fat" disks. A convex disk is r***-fat*** if it is contained in a unit circle and contains a concentric circle of radius r. G. Fejes Tóth [Fej05a] proved the inequality $\vartheta_T(K) \leq \frac{a(K)}{h(K)}$ for 0.933-fat convex disks and, sharpening an earlier result of Heppes [Hep03] also for 0.741-fat ellipses. The algorithm of Mount and Silverman enables us to determine the covering density of centrally symmetric 0.933-fat convex polygons. We note that all regular polygons with at least 10 sides are 0.933-fat, and with a modification of the proof in [Fej05a] it can be shown that the bound $\vartheta_T(K) \leq \frac{a(K)}{h(K)}$ holds also in the case when K is a regular octagon. It follows that if P_n denotes a regular n-gon, then

$$\vartheta(P_{6k}) = \frac{k \sin \frac{\pi}{3k}}{\sin \frac{\pi}{3}}$$

and

$$\vartheta(P_{6k\pm 2}) = \frac{(3k \pm 1) \sin \frac{\pi}{3k\pm 1}}{2 \sin \frac{k\pi}{3k\pm 1} + \sin \frac{(k\pm 1)\pi}{3k\pm 1}}$$

for all $k \geq 1$. The covering density is not known for any convex body other than the space fillers and the examples mentioned above.

The true nature of difficulty in removing the non-crossing condition is shown by an ingenious example by A Bezdek and W. Kuperberg [BK10]. Modifying a pentagonal tile, they constructed convex disks K with the property that in any thinnest covering of the plane with congruent copies of K, crossing pairs occur. The thinnest covering in their construction contains rotated copies of K, so it is not a counterexample for the conjectures that for every convex disk K we have $\vartheta_T(K) \leq \frac{a(K)}{h(K)}$ and $\vartheta_T(K) = \vartheta_L(K)$. The equality $\vartheta_T(K) = \vartheta_L(K)$ was first proved by Januszewski [Jan10] for triangles. Januszewski's result was extended by Sriamorn and Xue [SX15] to a wider class of convex disks containing, besides triangles, all convex quadrilaterals. A *quarter-convex disk* is the affine image of a set of the form $\{(x, y) : 0 \leq x \leq 1, 0 \leq y \leq f(x)\}$ for some positive concave function $f(x)$ defined for $0 \leq x \leq 1$. Sriamorn and Xue proved that $\vartheta_T(K) = \vartheta_L(T)$ for every quarter-convex disk.

One could expect that the restriction to arrangements of translates of a set means a considerable simplification. However, this apparent advantage has not been exploited so far in dimensions greater than 2. On the other hand, the lattice packing density of some special convex bodies in $\mathbb{E}^3$ has been determined; see Table 2.1.2.

TABLE 2.1.2 Bodies $K \in \mathbb{E}^3$ for which $\delta_L(K)$ is known.

BODY	$\delta_L(K)$	SOURCE
$\{x \mid \lvert x\rvert \le 1,\ \lvert x_3\rvert \le \lambda\}$ $(\lambda \le 1)$	$\pi(3-\lambda^2)^{1/2}/6$	[Cha50]
$\{x \mid \lvert x_i\rvert \le 1,\ \lvert x_1+x_2+x_3\rvert \le \lambda\}$	$\begin{cases} \dfrac{9-\lambda^2}{9} & \text{for } 0<\lambda\le\frac{1}{2} \\ \dfrac{9\lambda(9-\lambda^2)}{4(-\lambda^3-3\lambda^2+24\lambda-1)} & \text{for } \frac{1}{2}\le\lambda\le 1 \\ \dfrac{9(\lambda^3-9\lambda^2+27\lambda-3)}{8\lambda(\lambda^2-9\lambda+27)} & \text{for } 1\le\lambda\le 3 \end{cases}$	[Whi51]
$\{x \mid \sqrt{(x_1)^2+(x_2)^2}+\lvert x_3\rvert \le 1\}$	$\pi\sqrt{6}/9 = 0.8550332\ldots$	[Whi48]
Tetrahedron	$18/49 = 0.3673469\ldots$	[Hoy70]
Octahedron	$18/19 = 0.9473684\ldots$	[Min04]
Dodecahedron	$(5+\sqrt{5})/8 = 0.9045084\ldots$	[BH00]
Icosahedron	$0.8363574\ldots$	[BH00]
Cuboctahedron	$45/49 = 0.9183633\ldots$	[BH00]
Icosidodecahedron	$(45+17\sqrt{5})/96 = 0.8647203\ldots$	[BH00]
Rhombic Cuboctahedron	$(16\sqrt{2}-20)/3 = 0.8758056\ldots$	[BH00]
Rhombic Icosidodecahedron	$(768\sqrt{5}-1290)/531 = 0.8047084\ldots$	[BH00]
Truncated Cube	$9(5-3\sqrt{2})/7 = 0.9737476\ldots$	[BH00]
Truncated Dodecahedron	$(25+37\sqrt{5})/120 = 0.8977876\ldots$	[BH00]
Truncated Icosahedron	$0.78498777\ldots$	[BH00]
Truncated Cuboctahedron	$0.8493732\ldots$	[BH00]
Truncated Icosidodecahedron	$(19+10\sqrt{5})/50 = 0.8272135\ldots$	[BH00]
Truncated Tetrahedron	$207/304 = 0.6809210\ldots$	[BH00]
Snub Cube	$0.787699\ldots$	[BH00]
Snub Dodecahedron	$0.7886401\ldots$	[BH00]

All results given in Table 2.1.2 are based on Minkowski's work [Min04] on critical lattices of convex bodies. We emphasize the following special case: Gauss's result that $\delta_L(B^3) = \pi/\sqrt{18}$ is the special case $\lambda = 1$ of Chalk's theorem concerning the frustum of the ball. In [BH00] Betke and Henk gave an efficient algorithm for computing $\delta_L(K)$ for an arbitrary 3-polytope. As an application they calculated the lattice packing densities of all regular and Archimedean polytopes.

Additional bodies can be added to Table 2.1.2 using the following observations.

It has been noticed by Chalk and Rogers [CR48] that the relation $\delta_T(K) = \delta_L(K)$ $(K \in \mathcal{K}(\mathbb{E}^2))$ readily implies that for a cylinder C in $\mathbb{E}^3$ based on a convex disk K we have $\delta_L(C) = \delta_L(K)$. Thus, $\delta_L(C)$ is determined by the lattice packing density of its base.

Next, we recall the observation of Minkowski (see [Rog64, p. 69]) that an arrangement $\mathcal{A}$ of translates of a convex body K is a packing if and only if the arrangement of translates of the body $\frac{1}{2}(K-K)$ by the same vectors is a packing. This implies that, for $K \in \mathcal{K}(\mathbb{E}^d)$,

$$\delta_T(K) = 2^d\delta_T(K-K)\frac{V(K)}{V(K-K)} \quad \text{and} \quad \delta_L(K) = 2^d\delta_L(K-K)\frac{V(K)}{V(K-K)}. \tag{2.1.3}$$

Generally, K is not uniquely determined by $K-K$; e.g., we have $K-K = B^d$ for every $K \subset \mathbb{E}^d$ that is a body of constant width 1, and the determination of $\delta_L(K)$ for such a body is reduced to the determination of $\delta_L(B^d)$, which is established for $d \le 8$ and $d = 24$. We give the known values of $\delta_L(B^d)$ and $\vartheta(B^d)$, in Table 2.1.3.

TABLE 2.1.3 Known values of $\delta_L(B^d)$ and $\vartheta_L(B^d)$.

d	$\delta_L(B^d)$	SOURCE	$\vartheta_L(B^d)$	SOURCE
2	$\frac{\pi}{2\sqrt{3}}$	[Lag73]	$\frac{2\pi}{3\sqrt{3}}$	[Ker39]
3	$\frac{\pi}{\sqrt{18}}$	[Gau]	$\frac{5\sqrt{5}\pi}{24}$	[Bam54]
4	$\frac{\pi^2}{16}$	[KZ72]	$\frac{2\pi^2}{5\sqrt{5}}$	[DR63]
5	$\frac{\pi^2}{15\sqrt{2}}$	[KZ77]	$\frac{245\sqrt{35}\pi^2}{3888\sqrt{3}}$	[RB75]
6	$\frac{\pi^3}{48\sqrt{3}}$	[Bli35]		
7	$\frac{\pi^3}{105}$	[Bli35]		
8	$\frac{\pi^4}{384}$	[Bli35]		
24	$\frac{\pi^{12}}{12!}$	[CK04]		

THE KEPLER CONJECTURE

In 1611 Kepler [Kep87] described the face-centered cubic lattice packing of congruent balls consisting of hexagonal layers. He observed that the packing is built at the same time of square layers. Then he proclaimed that this arrangement is "the tightest possible, so that in no other arrangement could more pallets be stuffed into the same container." This sentence of Kepler has been interpreted by some authors as the conjecture that the density of a packing of congruent balls cannot exceed the density of the face-centered cubic lattice packing, that is, $\pi/\sqrt{18}$. It is doubtful whether Kepler meant this, but attributing the conjecture to him certainly helped in advertising it outside the mathematical community. Today the term "Kepler conjecture" is widely accepted despite the fact that Kepler's statement as quoted above is certainly false if the container is smooth. Schürmann [Sch06] proved that if K is a smooth convex body in d-dimensional space ($d \geq 2$), then there exists a natural number n_0, depending on K, such that the densest packing of $n \geq n_0$ congruent balls in K cannot be part of a lattice arrangement.

Early research concerning Kepler's conjecture concentrated on two easier problems: proving the conjecture for special arrangements and giving upper bounds for $\delta(B^3)$.

We mentioned Gauss's result that $\delta_L(B^3) = \pi/\sqrt{18}$. A stronger result establishing Kepler's conjecture for a restricted class of packings is due to A. Bezdek, W. Kuperberg and Makai [BKM91]. They proved that the conjecture holds for packings consisting of parallel strings of balls. A string of balls is a collection of congruent balls whose centers are collinear and such that each of them touches two others. Before the confirmation of the Kepler conjecture, the best upper bound for $\delta(B^3)$ was given by Muder [Mud93], who proved that $\delta(B^3) \leq 0.773055$.

The first step toward the solution of Kepler's conjecture in its full generality was made in the early 1950s by L. Fejes Tóth (see [Fej72] pp. 174–181). He considered weighted averages of the volumes of Dirichlet cells of a finite collection of balls in a packing. He showed that the Kepler conjecture holds if a particular weighted average of volumes involving not more than 13 cells is greater than or equal the volume of the rhombic dodecahedron circumscribed around a ball (this being the Dirichlet cell of a ball in the face-centered cubic lattice). His argument constitutes a program that, if realizable in principle, reduces Kepler's conjecture to an optimization problem in a finite number of variables. Later, in [Fej64] p. 300, he suggested that with the use of computers $\delta(B^3)$ could be "approximated with great exactitude."

In 1990 W.-Y. Hsiang announced the solution of the Kepler conjecture. His approach is very similar to the program proposed by L. Fejes Tóth. Unfortunately, Hsiang's paper [Hsi93] contains significant gaps, so it cannot be accepted as a proof. Hsiang maintains his claim of having a proof. He gave more detail in [Hsi01]. The mathematical community lost interest in checking those details, however.

About the same time as Hsiang, Tom Hales also attacked the Kepler conjecture. His first attempt [Hal92] was a program based on the Delone subdivision of space, which is dual to the subdivision by Dirichlet cells. He modified his approach in several steps [Hal93, Hal97, Hal98]. His final version, worked out in collaboration with his graduate student Ferguson in [HF06], uses a subdivision that is a hybrid of certain Delone-type tetrahedra and Dirichlet cells. With each ball B in a saturated packing of unit balls, an object, called a *decomposition star*, is associated, consisting of certain tetrahedra having the center of B as a common vertex together with parts of a modified Dirichlet cell of B. A complicated scoring rule is introduced that takes into account the volumes of the different parts of the decomposition star with appropriate weights. The score of a decomposition star in the face-centered cubic lattice is a certain number, which Hales takes to be 8. The key property of the decomposition stars and the scoring rule is that the decomposition star of a ball B, as well as its score, depends only on balls lying in a certain neighborhood of B. From the mathematical point of view, the main step of the proof is the theorem that

> *the Kepler conjecture holds, provided the score of each decomposition star in a saturated packing of unit balls is at most 8.*

The task of proving this, which is an optimization problem in finitely many variables, has been carried out with the aid of computers. As Hales points out, there is hope that in the future such a problem "might eventually become an instance of a general family of optimization problems for which general optimization techniques exist." In the absence of such general techniques, manual procedures had to be used to guide the work of computers.

Computers are used in the proof in several ways. The topological structure of the decomposition stars is described by planar maps. A computer program enumerates around 5000 planar maps that have to be examined as potential counterexamples to the conjecture. Interval arithmetic is used to prove various inequalities. Nonlinear optimization problems are replaced by linear problems that dominate the original ones in order to apply linear programming methods. Even the organization of the few gigabytes of data is a difficult task.

After examining the proof for over two years the team of a dozen referees came to the conclusion that the general framework of the proof is sound, they did not find any error, but they cannot say for certain that everything is correct. In particular they could not check the work done by the computer. The theoretical foundation of the proof was published in the *Annals of Mathematics* [Hal05], and the details, along with historical notes, were given in a series of articles in a special issue of *Discrete and Computational Geometry* ([Hal06a, Hal06b, Hal06c, Hal06d, HF06, HF11].

It is safe to say that the 300-page proof, aided by computer calculations taking months, is one of the most complex proofs in the history of mathematics. Lagarias, who in [Lag02] extracts the common ideas of the programs of L. Fejes Tóth, W.-Y. Hsiang, and Hales and puts them into a general framework, finds that "the Hales–Ferguson proof, assumed correct, is a tour de force of nonlinear optimization."

Disappointed because the referees were unable to certify the correctness of the proof, and realizing that, besides him, probably no human being will ever check all details, Hales launched a project named FLYSPECK designed for a computerized formal verification of his proof. The article [HHM10] by the FLYSPECK team reorganizes the original proof into a more transparent form to provide a greater level of certification of the correctness of the computer code and other details of the proof. The final part of the paper lists errata in the original proof of the Kepler conjecture. The book [Hal12] shows in detail how geometric ideas and elements of proof are arranged and processed in preparation for the formal proof-checking scrutiny.

Marchal [Mar11] proposed an alternate subdivision associated with the packing which is simpler and provides a less complex strategy of proof than that of Hales.

On August 10, 2014 the team of the FLYSPECK project announced the successful completion of the project [Fly14], where they noted that "the formal proof takes the same general approach as the original proof, with modifications in the geometric partition of space that have been suggested by Marchal."

EXISTENCE OF ECONOMICAL ARRANGEMENTS

Table 2.1.4 lists the known bounds establishing the existence of reasonably dense packings and thin coverings. When c appears in a bound without specification, it means a suitable constant characteristic of the specific bound. The proofs of most of these are nonconstructive. For constructive methods yielding slightly weaker bounds, as well as improvements for special convex bodies.

Bound 1 for the packing density of general convex bodies follows by combining Bound 6 with the relation (2.1.3) and the inequality $V(K-K) \leq \binom{2d}{d}V(K)$ of Rogers and Shephard [RS57]). For $d \geq 3$ all methods establishing the existence of dense packings rely on the theory of lattices, thus providing the same lower bounds for $\delta(K)$ and $\delta_T(K)$ as for $\delta_L(K)$.

Better bounds than for general convex bodies are known for balls. Improving earlier results by Ball [Bal92] and Vance [Van11], Venkatesh [Ven13] proved that for any constant $c > \sinh^2(\pi e)/\pi^2 e^3 = 65963.8\ldots$ there is a number $n(c)$ such that for $n > n(c)$ we have $\delta(B^n) \geq \delta_L(B^n) \geq cn2^{-n}$. Moreover, there are infinitely many dimensions n for which $\delta_L(B^n) \geq n \ln\ln n 2^{-n-1}$.

Rogers [Rog59] proved that

$$\vartheta_L(B^n) \leq cn(\log_e n)^{\frac{1}{2}\log_2 2\pi e}.$$

Gritzmann [Gri85] proved a similar bound for a larger class of convex bodies:

$$\vartheta_L(K) \leq cd(\ln d)^{1+\log_2 e}$$

holds for a suitable constant c and for every convex body K in $\mathbb{E}^d$ that has an affine image symmetric about at least $\log_2 \ln d + 4$ coordinate hyperplanes.

TABLE 2.1.4 Bounds establishing the existence of dense packings and thin coverings.

No.	BOUND	SOURCE
	Bounds for general convex bodies in $\mathbb{E}^d$	
1	$\delta_L(K) \geq cd^{3/2}/4^d$ (d large)	[Sch63a, Sch63b]
2	$\vartheta_T(K) \leq d \ln d + d \ln\ln d + 5d$	[Rog57]
3	$\vartheta_L(K) \leq d^{\log_2 \ln d + c}$	[Rog59]
	$\vartheta(K) \leq 3 \quad K \in \mathbb{E}^3$	[Smi00]
	Bounds for centrally symmetric convex bodies in $\mathbb{E}^d$	
4	$\delta_L(K) \geq \zeta(d)/2^{d-1}$	[Hla43]
5	$\delta_L(K) \geq cd/2^d$ (d large)	[Sch63a, Sch63b]
	$\delta_L(K) \geq 0:538\ldots \quad K \in \mathbb{E}^3$	[Smi05]
	Bounds for general convex bodies in $\mathbb{E}^2$	
6	$\delta(K) \geq \sqrt{3}/2 = 0.8660\ldots$	[KK90]
7	$\vartheta(K) \leq 1.2281771\ldots$	[Ism98]
8	$\delta_L(K) \geq 2/3$	[Far50]
9	$\vartheta_L(K) \leq 3/2$	[Far50]
	Bounds for centrally symmetric convex bodies in $\mathbb{E}^2$	
10	$\delta_L(K) \geq 0.892656\ldots$	[Tam70]
11	$\vartheta_L(K) \leq 2\pi/\sqrt{27}$	[Fej72, p. 103]

The determination of the densest packing of congruent regular tetrahedra is mentioned as part of Problem 18 of Hibert's famous problems [Hil00]. In recent years a series of papers was devoted to the construction of dense packing of regular tetrahedra. The presently known best arrangement was constructed by Chen, Engel and Glotzer [CEG10]. It has density $4000/4671 = 0.856347\ldots$. A nice survey on packing regular tetrahedra was written by Lagarias and Zong [LZ02].

UPPER BOUNDS FOR $\delta(K)$ AND LOWER BOUNDS FOR $\vartheta(K)$

The packing density of B^d is not known, except for the cases mentioned in Table 2.1.1. Asymptotically, the best upper bound known for $\delta(B^d)$ is

$$\delta(B^d) \leq 2^{-0.599d+o(d)} \qquad (\text{as } d \to \infty), \tag{2.1.4}$$

given by Kabatiansky and Levenshtein [KL78]. This bound is not obtained directly by the investigation of packings in $\mathbb{E}^d$ but rather through studying the analogous problem in spherical geometry, where the powerful technique of linear programming can be used (see Section 2.4). For low dimensions, Rogers's simplex bound

$$\delta(B^d) \leq \sigma_d \tag{2.1.5}$$

gives a better estimate (see [Rog58]). Here, σ_d is the ratio between the total volume of the sectors of $d+1$ unit balls centered at the vertices of a regular simplex of edge length 2 and the volume of the simplex.

Recently, Rogers's bound has been improved in low dimensions as well. On one hand, K. Bezdek [Bez02] extended the method of Rogers by investigating the surface area of the Voronoi regions, rather than their volume; on the other hand, Cohn and Elkies [CE03, Coh02] developed linear programming bounds that apply directly to sphere packings in $\mathbb{E}^d$. This latter method is very powerful. Using the approach of [CE03] Cohn and Kumar [CK04, CK09] reconfirmed Blichfeldt's result concerning the value of $\delta_L(B^8)$ and determined $\delta_L(B^{24})$. Finally, Viazovska [Via17] succeeded in proving that $\delta(B^8) = \delta_L(B^8)$ and Cohn, Kumar, Miller, Radchenko and Viazovska [CKM17] showed that $\delta(B^)24 = \delta_L(B^{24})$.

Coxeter, Few, and Rogers [CFR59] proved a dual counterpart to Rogers's simplex bound:

$$\vartheta(B^d) \geq \tau_d,$$

where τ_d is the ratio between the total volume of the intersections of $d+1$ unit balls with the regular simplex of edge $\sqrt{2(d+1)/d}$ if their centers lie at the vertices of the simplex, and the volume of the simplex. Asymptotically,

$$\tau_d \sim d/e^{3/2}.$$

In contrast to packings, where there is a sizable gap between the upper bound (2.1.4) and the lower bound (Bound 6 in Table 2.1.4), this bound compares quite favorably with the corresponding Bound 2 in Table 2.1.4.

It is known that there is no tiling of space by regular tetrahedra or octahedra. However, until recently no nontrivial upper bound was known for the packing density of these solids. Gravel, Elser, and Kallus [GEK11] proved upper bounds of $1 - 2.6\ldots \times 10^{-25}$ and $1 - 1.4\ldots \times 10^{-12}$, respectively, for the packing density of the regular tetrahedron and octahedron. According to a result of W. Schmidt (see [Sch61]), we have $\delta(K) < 1$ and $\vartheta(K) > 1$ for every smooth convex body; but the method of proof does not allow one to derive any explicit bound. There is a general upper bound for $\delta(K)$ that is nontrivial (smaller than 1) for a wide class of convex bodies [FK93a]. It is quite reasonable for "longish" bodies. For cylinders in $\mathbb{E}^d$, the bound is asymptotically equal to the Kabatiansky-Levenshtein bound for B^d (as $d \to \infty$). The method yields nontrivial upper bounds also for the packing density of the regular cross-polytope for all dimensions greater than 6 (see [FFV15]).

Zong [Zon14] proved the bound $\delta_T(T) \leq 0.384061$ for the translational packing density of a tetrahedron. De Oliveira Filho and Vallentin [OV] extended the method of Cohn and Elkies to obtain upper bounds for the packing density of convex bodies. In [DGOV17] this extension is used to obtain upper bounds for the translational packing density of superballs and certain Platonic and Archimedean solids in three dimensions. In particular, Zong's upper bound for the translational packing density of a tetrahedron is improved to 0.3745, getting closer to the density $18/49 = 0.3673\ldots$ of the densest lattice packing of tetrahedra.

We note that no nontrivial lower bound is known for $\vartheta(K)$ for any $K \in \mathbb{E}^d$, $d \geq 3$, other than a ball.

REGULARITY OF OPTIMAL ARRANGEMENTS

The packings and coverings attaining the packing and covering densities of a set are, of course, not uniquely determined, but it is a natural question whether there exist among the optimal arrangements some that satisfy certain regularity properties. Of particular interest are those bodies for which the densest packing and/or thinnest covering with congruent copies can be realized by a lattice arrangement.

As mentioned above, $\delta(K) = \delta_L(K)$ for $K \in \mathcal{K}^*(\mathbb{E}^2)$. A plausible interpretation of this result is that the assumption of maximum density creates from a chaotic structure a regular one. Unfortunately, certain results indicate that such bodies are rather exceptional.

Let $\mathcal{L}_p$ and $\mathcal{L}_c$ be the classes of those convex disks $K \in \mathcal{K}(\mathbb{E}^2)$ for which $\delta(K) = \delta_L(K)$ and $\vartheta(K) = \vartheta_L(K)$, respectively. Then, in the topology induced by the Hausdorff metric on $\mathcal{K}(\mathbb{E}^2)$, the sets $\mathcal{L}_p$ and $\mathcal{L}_c$ are nowhere dense [FZ94, Fej95]. It is conjectured that an analogous statement holds also in higher dimensions.

Rogers [Rog64, p. 15] conjectures that for sufficiently large d we have $\delta(B^d) > \delta_L(B^d)$. The following result of A. Bezdek and W. Kuperberg (see [BK91]) supports this conjecture: For $d \geq 3$ there are ellipsoids E in $\mathbb{E}^d$ for which $\delta(E) > \delta_L(E)$. An even more surprising result holds for coverings [FK95]: For $d \geq 3$ every strictly convex body K in $\mathbb{E}^d$ has an affine image K' such that $\vartheta(K') < \vartheta_L(K')$. In particular, there is an ellipsoid E in $\mathbb{E}^3$ for which

$$\vartheta(E) < 1.394 < \frac{3\sqrt{3}}{2}(3 \operatorname{arcsec} 3 - \pi) = \tau_3 \leq \vartheta_T(E) \leq \vartheta_L(E).$$

We note that no example of a convex body K is known for which $\delta_L(K) < \delta_T(K)$ or $\vartheta_L(K) > \vartheta_T(K)$.

Schmitt [Sch88a] constructed a star-shaped prototile for a monohedral tiling in $\mathbb{E}^3$ such that no tiling with its replicas is periodic. It is not known whether a convex body with this property exists; however, with a slight modification of Schmitt's construction, Conway produced a convex prototile that admits only nonperiodic tilings if no mirror-image is allowed (see Section 3.4). Another result of Schmitt's [Sch91] is that there are star-shaped sets in the plane whose densest packing cannot be realized in a periodic arrangement.

2.2 FINITE ARRANGEMENTS

PACKING IN AND COVERING OF A BODY WITH GIVEN SHAPE

What is the size of the smallest square tray that can hold n given glasses? Thue's result gives a bound that is asymptotically sharp as $n \to \infty$; however, for practical reasons, small values of n are of interest.

Generally, for given sets K and C and a positive integer n one can ask for the quantities

$$M_p(K, C, n) = \inf\{\lambda \mid n \text{ congruent copies of } C \text{ can be packed in } \lambda K\}$$

and

$$M_c(K, C, n) = \sup\{\lambda \mid n \text{ congruent copies of } C \text{ can cover } \lambda K\}.$$

Tables 2.2.1–2.2.2 contain the known results of $M_p(K, B^2, n)$ in the cases when K is a circle, square, or regular triangle.

Most of these results were obtained by ad hoc methods. An exception is the case of packing circles in a square. In [GMP94, Pei94] a heuristic algorithm for the determination of $M_p(K, B^2, n)$ and the corresponding optimal arrangements is given in the case where K is the unit square. The algorithm consists of the following steps:

TABLE 2.2.1 Packing of congruent circles in unit squares.

n	$M_p(K, B^2, n)$	SOURCE	n	$M_p(K, B^2, n)$	SOURCE
2	3.414213562...	(elementary)	17	8.532660354...	[GMP94]
3	3.931851653...	(elementary)	18	8.656402355...	[GMP94]
4	4	(elementary)	19	8.907460939...	[GMP94]
5	4.828427125...	(elementary)	20	8.978083353...	[GMP94]
6	5.328201177...	[Gra63, Mel94c]	21	9.358035345...	[NO97]
7	5.732050807...	Schaer (unpublished)	22	9.463845078...	[NO97]
8	5.863703305...	[SM65]	23	9.727406613...	[NO97]
9	6	[Sch95]	24	9.863703306...	[NO97]
10	6.747441523...	[GPW90]	25	10	[Wen87b]
11	7.022509506...	[GMP94]	26	10.37749821...	[NO97]
12	7.144957554...	[GMP94]	27	10.47998305...	[NO97]
13	7.463047839...	[GMP94]	28	10.67548744...	[Mar04]
14	7.732050808...	[Wen87a]	29	10.81512001...	[Mar07]
15	7.863703305...	[GMP94]	30	10.90856381...	[Mar07]
16	8	[Wen83]	36	12	[KW87]

Step 1. Find a good upper bound m for $M_p(K, B^2, n)$. This requires the construction of a reasonably good arrangement, which can be established, e.g., by the Monte Carlo method.

Step 2. Iterate an elimination process on a successively refined grid to restrict possible locations for the centers of a packing of unit circles in mK.

Step 3. Based on the result of Step 2, guess the nerve graph of the packing, then determine the optimal packing with the given graph.

Step 4. Verify that the arrangement obtained in Step 3 is indeed optimal.

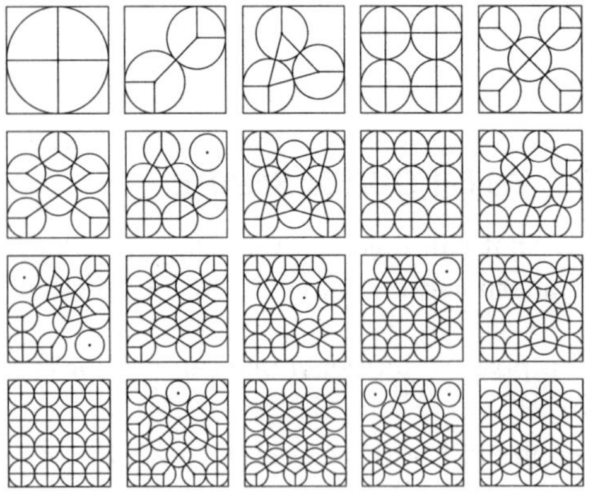

FIGURE 2.2.1
Densest packing of $n \leq 20$ equal circles in a square.

We do not know whether these steps always provide the optimal arrangement in finite time, but in [GMP94, GPW90] the method was implemented successfully for $n \leq 20$. The best arrangements are shown in Figure 2.2.1. Observe that quite often an optimal arrangement can contain a freely movable circle. Using more refined numerical technics Nurmela and Östergård [NO97] and Markót [Mar04, Mar07] solved the cases $21 \leq n \leq 30$ as well.

The sequence $M_p(K, B^2, n)$ seems to be strictly increasing when K is a square

TABLE 2.2.2 Packing of congruent circles in circles (left), and in equilateral triangles T of unit side length (right)

n	$M_p(B^2, B^2, n)$	SORCE	n	$M_p(T, B^2, n)$	SORCE
2	2	(elementary)	2	5.464101615...	(elementary)
3	2.154700538...	(elementary)	3	5.464101615...	(elementary)
4	2.414213562...	(elementary)	4	6.92820323...	[Mel93]
5	2.701301617...	(elementary)	5	7.464101615...	[Mel93]
6	3	(elementary)	6	7.464101615...	[Ole61, Gro66]
7	3	(elementary)	7	8.92820323...	[Mel93]
8	3.304764871...	[Pir69]	8	9.293810046...	[Mel93]
9	3.61312593...	[Pir69]	9	9.464101615...	[Mel93]
10	3.813898249...	[Pir69]	10	9.464101615...	[Ole61, Gro66]
11	3.9238044...	[Mel94a]	11	10.73008794...	[Mel93]
12	4.02960193...	[Fod00]	12	10.92820323...	[Mel94b]
13	4.23606797...	[Fod03a]	$\frac{k(k+1)}{2}$	$2(k+\sqrt{3}-1)$	[Ole61, Gro66]
14	4.32842855...	[Fod03b]			
19	4.86370330...	[Fod99]			

or when K is a circle and $n \geq 7$. In contrast to this, it is conjectured that in the case where K is a regular triangle, we have $M_p(K, B^2, n) = M_p(K, B^2, n-1)$ for all triangular numbers $n = k(k+1)/2$ $(k > 1)$.

The problem of finding the densest packing of n congruent circles in a circle has been considered also in the Minkowski plane. In terms of Euclidean geometry, this is the same as asking for the smallest number $\varrho(n, K)$ such that n mutually disjoint translates of the centrally symmetric convex disk K (the unit circle in the Minkowski metric) can be contained in $\varrho(n, K)K$. Doyle, Lagarias, and Randell [DLR92] solved the problem for all $K \in \mathcal{K}^*(\mathbb{E}^2)$ and $n \leq 7$. There is an n-gon inscribed in K having equal sides in the Minkowski metric (generated by K) and having a vertex at an arbitrary boundary point of K. Let $\alpha(n, K)$ be the maximum Minkowski side-length of such an n-gon. Then we have $\varrho(n, K) = 1 + 2/\alpha(n, K)$ for $2 \leq n \leq 6$ and $\varrho(7, K) = \varrho(6, K) = 3$.

The densest packing of n congruent balls in a cube in $\mathbb{E}^3$ was determined for $n \leq 10$ by Schaer [Sch66a, Sch66b, Sch66c, Sch94] and for $n = 14$ by Joós [Joo09a]). The problem of finding the densest packing of n congruent balls in the regular cross-polytope was solved for $n \leq 7$ by Golser [Gol77]. Böröczky Jr. and Wintsche [BW00] generalized Golser's result to higher dimensions. K. Bezdek [Bez87] solved the problem of packing n congruent balls in a regular tetrahedron, for $n = 5, 8, 9$ and 10. The known values of $M_c(K, B^2, n)$ in the cases when K is a circle, square, or regular triangle are shown in Table 2.2.3.

G. Kuperberg and W. Kuperberg solved the problem of thinnest covering of the cube with k congruent balls for $k = 2$, 3, 4 and 8. Joós [Joo14a, Joo14b] settled the cases $k = 5$ and 6. Joós [Joo08, Joo09b] also proved that the minimum radius of 8 congruent balls that can cover the unit 4-dimensional cube is $\sqrt{5/12}$. The problem of covering the n-dimensional cross-polytope with k congruent balls of minimum radius was studied by Böröczky, Jr., Fábián and Wintsche [BFW06] who found the solution for $k = 2$, n, and $2n$. Remarkably, the case $k = n = 3$ is exceptional, breaking the pattern holding for $k = n \neq 3$.

TABLE 2.2.3 Covering circles, squares, and equilateral triangles with congruent circles.

K	n	$M_c(K, B^2, n)$	SOURCE
B^2	2	2	(elementary)
	3	$2/\sqrt{3}$	(elementary)
	4	$\sqrt{2}$	(elementary)
	5	$1.64100446\ldots$	[Bez84]
	6	$1.7988\ldots$	[Bez84]
	7	2	(elementary)
	8–10	$1+2\cos\frac{2\pi}{n-1}$	[Fej05b]
Unit square	2	$4\sqrt{5}/5$	(elementary)
	3	$\frac{16}{\sqrt{65}}$	[BB85, HM97]
	4	$2\sqrt{2}$	[BB85, HM97]
	5	$3.065975\ldots$	[BB85, HM97]
	6	$\frac{12}{\sqrt{13}}$	[BB85]
	7	$1+\sqrt{7}$	[BB85, HM97]
Regular triangle of side 1	2	2	(elementary)
	3	$2\sqrt{3}$	[BB85, Mel97]
	4	$2+\sqrt{3}$	[BB85, Mel97]
	5	4	[BB85, Mel97]
	6	$3\sqrt{3}$	[BB85, Mel97]
	9	6	[BB85]
	10	$4\sqrt{3}$	[BB85]

SAUSAGE CONJECTURES

Intensive research on another type of finite packing and covering problem has been generated by the sausage conjectures of L. Fejes Tóth and Wills (see [GW93]):

What is the convex body of minimum volume in $\mathbb{E}^d$ that can accommodate k nonoverlapping unit balls? What is the convex body of maximum volume in $\mathbb{E}^d$ that can be covered by k unit balls?

According to the conjectures mentioned above, for $d \geq 5$ the extreme bodies are "sausages" and in the optimal arrangements the centers of the balls are equally spaced on a line segment (Figure 2.2.2).

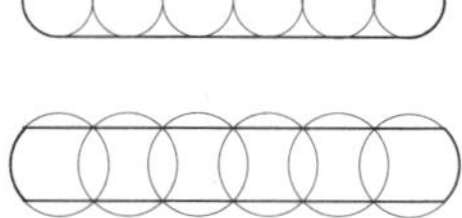

FIGURE 2.2.2
Sausage-like arrangements of circles.

After several partial results supporting these conjectures (see [GW93]) the breakthrough concerning the sausage conjecture for ball packings was achieved by Betke, Henk, and Wills [BHW94]: they proved that the conjecture holds for dimensions $d \geq 13387$. Betke and Henk [BH98] improved the bound on d to $d \geq 42$.

Several generalizations of the problems mentioned above have been considered. Connections of these types of problems to the classical theory of packing and coverings, as well as to crystallography, have been observed. For details we refer to [Bor04].

THE COVERING PROBLEMS OF BORSUK, HADWIGER, and LEVI

In 1933, Borsuk [Bor33] formulated the conjecture that any bounded set in $\mathbb{E}^d$ can be partitioned into $d+1$ subsets of smaller diameter. Borsuk verified the conjecture for $d = 2$, and the three-dimensional case was settled independently by Eggleston, Grünbaum, and Heppes. The Borsuk conjecture was proved for many special cases: for smooth convex bodies by Hadwiger [Had46] and Melzak [Mel67], for sets having the symmetry group of the regular simplex by Rogers [Rog71], and for sets of revolution by Kołodziejczyk [Kol88]. Hadwiger and Melzak's result was generalized by Dekster [Dek93] who proved that the conjecture holds for every convex body for which there exists a direction in which every line tangent to the body contains at least one point of the body's boundary at which the tangent hyperplane is unique.

However, Kahn and Kalai [KK93] showed that Borsuk's conjecture is false in the following very strong sense: Let $b(d)$ denote the smallest integer such that every bounded set in $\mathbb{E}^d$ can be partitioned into $b(d)$ subsets of smaller diameter. Then $b(d) \geq (1.2)^{\sqrt{d}}$ for every sufficiently large value of d. The best known upper bound for $b(d)$ is $b(d) \leq (\sqrt{3/2} + o(1))^n$ due to Schramm [Sch88b]. The lowest dimension known in which the Borsuk conjecture fails is 64 [JB14]. The papers [Rai07] and [Rai08] are excellent surveys on Borsuk's problem.

In the 1950's, Hadwiger and Levi, independently of each other, asked for the smallest integer $h(K)$ such that the convex body K can be covered by $h(K)$ smaller positive homothetic copies of K. Boltjanskiĭ observed that the Hadwiger-Levi covering problem for convex bodies is equivalent to an illumination problem. We say that a boundary point x of the convex body K is ***illuminated from the direction u*** if the ray emanating from x in the direction u intersects the interior of K. Let $i(K)$ be the minimum number of directions from which the boundary of K can be illuminated. Then $h(K) = i(K)$ for every convex body.

It is conjectured that $h(K) \leq 2^d$ for all $K \in \mathcal{K}(\mathbb{E}^d)$ and that equality holds only for parallelotopes. Levi verified the conjecture for the plane, but it is open for $d \geq 3$. Lassak proved Hadwiger's conjecture for centrally symmetric convex bodies in $\mathbb{E}^3$, and K. Bezdek extended Lassak's result to convex polytopes with any affine symmetry. There is a great variety of results confirming the conjecture for special classes of bodies in $\mathbb{E}^n$ by establishing upper bounds for $h(K)$ or $i(K)$ smaller than 2^n. The difficulty of the problem is exposed by the following example of Naszódi [Nas16b]. Clearly, we have $h(B^d) = d+1$. On the other hand, for any $\varepsilon > 0$ there is a centrally symmetric convex body K and a positive constant $c = c(\varepsilon)$ such that K is ε close to B^d and $h(K) \geq c^d$. For literature and further results concerning the Hadwiger-Levi problem, we refer to [Bez93, BMS97, MS99].

2.3 MULTIPLE ARRANGEMENTS

GLOSSARY

k-Fold packing: An arrangement $\mathcal{A}$ such that each point of the space belongs to the interior of at most k members of $\mathcal{A}$.

***k*-Fold covering:** An arrangement $\mathcal{A}$ such that each point of the space belongs to at least k members of $\mathcal{A}$.

Densities: In analogy to the packing and covering densities of a body K, we define the quantities $\delta^k(K)$, $\delta_T^k(K)$, $\delta_L^k(K)$, $\vartheta^k(K)$, $\vartheta_T^k(K)$, and $\vartheta_L^k(K)$ as the suprema of the densities of all k-fold packings and the infima of the densities of all k-fold coverings with congruent copies, translates, and lattice translates of K, respectively.

TABLE 2.3.1 Bounds for k-fold packing and covering densities.

BOUND	SOURCE
$\delta_T^k(K) \geq ck \qquad K \in \mathcal{K}(\mathbb{E}^d)$	[ER62]
$\vartheta_L^k(K) \leq ((k+1)^{1/d}+8d)^d \qquad K \in \mathcal{K}(\mathbb{E}^d)$	[Coh76]
$\delta_L^k(K) \geq k - ck^{2/5} \qquad K \in \mathcal{K}(\mathbb{E}^2)$	[Bol89]
$\vartheta_L^k(K) \leq k + ck^{2/5} \qquad K \in \mathcal{K}(\mathbb{E}^2)$	[Bol89]
$\delta_L^k(P) \geq k - ck^{1/3}$ P convex polygon	[Bol89]
$\vartheta_L^k(P) \leq k + ck^{1/3}$ P convex polygon	[Bol89]
$\delta_L^k(B^d) \leq k - ck^{\frac{d-1}{2d}} \qquad d \not\equiv 1\ (4)$	[Bol79, Bol82]
$\delta_L^k(B^d) \leq k - ck^{\frac{d-3}{2d}} \qquad d \equiv 1\ (4)$	[Bol79, Bol82]
$\vartheta_L^k(B^2) \geq k + ck^{\frac{d-1}{2d}} \qquad d \not\equiv 1\ (4)$	[Bol79, Bol82]
$\vartheta_L^k(B^2) \geq k + ck^{\frac{d-3}{2d}} \qquad d \equiv 1\ (4)$	[Bol79, Bol82]
$\delta^k(B^d) \geq (2k/(k+1))^{d/2}\delta(B^d)$	[Few64]
$\delta_L^k(B^d) \geq (2k/(k+1))^{d/2}\delta_L(B^d)$	[Few64]
$\delta^k(B^d) \leq (1+d^{-1})((d+1)^k-1)(k/(k+1))^{d/2}$	[Few64]
$\delta^2(B^d) \leq \frac{4}{3}(d+2)(\frac{2}{3})^{d/2}$	[Few68]
$\vartheta^k(B^d) \geq ck \qquad c = c_d > 1$	[Fej79]
$\delta^k(B^2) \leq \frac{\pi}{6}\cot\frac{\pi}{6k}$	[Fej76]
$\vartheta^k(B^2) \geq \frac{\pi}{3}\csc\frac{\pi}{3k}$	[Fej76]

The information known about the asymptotic behavior of k-fold packing and covering densities is summarized in Table 2.3.1. There, in the various bounds, different constants appear, all of which are denoted by c. The known values of $\delta_L^k(B^d)$ and $\vartheta_L^k(B^d)$ (for $k \geq 2$) are given in Table 2.3.2.

The k-fold lattice packing density and the k-fold lattice covering density of a triangle T was determined for all k by Sriamorn [Sri16]. We have $\delta_L^k(T) = \frac{2k^2}{2k+1}$ and $\vartheta_L^k(T) = \frac{2k+1}{2}$. Moreover, Sriamorn [Sri14] showed tat $\delta_T^k(T) = \delta_L^k(T)$ and Sriamorn and Wetayawanich [SW15] showed that $\vartheta_T^k(T) = \vartheta_L^k(T)$ for all k.

General methods for the determination of the densest k-fold lattice packings and the thinnest k-fold lattice coverings with circles have been developed by Horváth, Temesvári, and Yakovlev [THY87] and by Temesvári [Tem88], respectively.

These methods reduce both problems to the determination of the optima of finitely many well-defined functions of one variable. The proofs readily provide algorithms for finding the optimal arrangements; however, the authors did not try to implement them. Only the values of $\delta_L^9(B^2)$ and $\vartheta_L^8(B^2)$ have been added in this

TABLE 2.3.2 Known values of $\delta_L^k(B^d)$ and $\vartheta_L^k(B^d)$.

RESULT	SOURCE
$\delta_L^2(B^2) = \frac{\pi}{\sqrt{3}}$	[Hep59]
$\delta_L^3(B^2) = \frac{\sqrt{3}\pi}{2}$	[Hep59]
$\delta_L^4(B^2) = \frac{2\pi}{\sqrt{3}}$	[Hep59]
$\delta_L^5(B^2) = \frac{4\pi}{\sqrt{7}}$	[Blu63]
$\delta_L^6(B^2) = \frac{35\pi}{8\sqrt{6}}$	[Blu63]
$\delta_L^7(B^2) = \frac{8\pi}{\sqrt{15}}$	[Bol76]
$\delta_L^8(B^2) = \frac{3969\pi}{4\sqrt{220-2\sqrt{193}}\sqrt{449+32\sqrt{193}}}$	[Yak83]
$\delta_L^9(B^2) = \frac{25\pi}{2\sqrt{21}}$	[Tem94]
$\delta_L^2(B^3) = \frac{8\pi}{9\sqrt{3}}$	[FK69]
$\vartheta_L^2(B^2) = \frac{4\pi}{3\sqrt{3}}$	[Blu57]
$\vartheta_L^3(B^2) = \frac{\pi\sqrt{27138+2910\sqrt{97}}}{216}$	[Blu57]
$\vartheta_L^4(B^2) = \frac{25\pi}{18}$	[Blu57]
$\vartheta_L^5(B^2) = \frac{32\pi}{7\sqrt{7}}$	[Tem84]
$\vartheta_L^6(B^2) = \frac{98\pi}{27\sqrt{3}}$	[Tem92a]
$\vartheta_L^7(B^2) = 7.672\ldots$	[Tem92b]
$\vartheta_L^8(B^2) = \frac{32\pi}{3\sqrt{15}}$	[Tem94]
$\vartheta_L^2(B^3) = \frac{8\pi}{\sqrt{3}\sqrt{76\sqrt{6}-159}}$	[Few67]

way to the list of values of $\delta_L^k(B^2)$ and $\vartheta_L^k(B^2)$ that had been determined previously by ad hoc methods.

We note that we have $\delta_L^k(B^2) = k\delta_L(B^2)$ for $k \leq 4$ and $\vartheta_L^2(B^2) = 2\vartheta_L(B^2)$. These are the only cases where the extreme multiple arrangements of circles are not better than repeated simple arrangements. These relations have been extended to arbitrary centrally symmetric convex disks by Dumir and Hans-Gill [DH72a, DH72b] and by G. Fejes Tóth (see [Fej84a]). There is a simple reason for the relations $\delta_L^3(K) = 3\delta_L(K)$ and $\delta_L^4(K) = 4\delta_L(K)$ ($K \in \mathcal{K}^*(\mathbb{E}^2)$): Every 3-fold lattice packing of the plane with a centrally symmetric disk is the union of 3 simple lattice packings and every 4-fold packing is the union of two 2-fold packings.

This last observation brings us to the topic of decompositions of multiple arrangements. Our goal here is to find insight into the structure of multiple arrangements by decomposing them into possibly few simple ones. Pach [Pac85] showed that any double packing with positive homothetic copies of a convex disk can be decomposed into 4 simple packings. Further, if $\mathcal{P}$ is a k-fold packing with convex disks such that for some integer L the inradius $r(K)$ and the area $a(K)$ of each member K of $\mathcal{P}$ satisfy the inequality $9\pi^2 kr^2(K)/a(K) \leq L$, then $\mathcal{P}$ can be decomposed into L simple packings.

A set S is *cover-decomposable* if there is an integer $k = k(S)$ such that every locally finite k-fold covering with translates of S can be decomposed into two coverings. Pach [Pac86] proved that centrally symmetric convex polygons are cover-decomposable and conjectured that all convex disks are cover-decomposable. His proof heavily uses the property of central symmetry, and more than two decades elapsed until the cover-decomposability of a non-symmetric polygon, namely of the triangle was proved by Tardos and Tóth [TT07]. Soon after, Pálvölgyi and Tóth [PT10] proved that all convex polygons are cover-decomposable. On the other hand, it turned out [Pál13] that the circle is not cover-decomposable. Moreover, convex sets that have two parallel tangents which both touch it in a point where the boundary has positive finite curvature, are not cover-decomposable. Pálvölgyi [Pál10] showed that concave polygons with no parallel sides are not cover-decomposable. He also proved that polyhedra in 3 and higher dimensions are not cover-decomposable. For details we refer to the survey paper [PPT13] and the web-site `http://www.cs.elte.hu/~dom/covdec/`.

2.4 PROBLEMS IN NONEUCLIDEAN SPACES

Research on packing and covering in spherical and hyperbolic spaces has been concentrated on arrangements of balls. In contrast to spherical geometry, where the finite, combinatorial nature of the problems, as well as applications, have inspired research, investigations in hyperbolic geometry have been hampered by the lack of a reasonable notion of density relative to the whole hyperbolic space.

SPHERICAL SPACE

Let $M(d, \varphi)$ be the maximum number of caps of spherical diameter φ forming a packing on the d-dimensional spherical space $\mathbb{S}^d$, that is, on the boundary of B^{d+1}, and let $m(d, \varphi)$ be the minimum number of caps of spherical diameter φ covering $\mathbb{S}^d$. An upper bound for $M(d, \varphi)$, which is sharp for certain values of d and φ and yields the best estimate known as $d \to \infty$, is the so-called ***linear programming bound*** (see [CS93, pp. 257-266]). It establishes a surprising connection between $M(d, \varphi)$ and the expansion of real polynomials in terms of certain Jacobi polynomials. The Jacobi polynomials, $P_i^{(\alpha,\beta)}(x)$, $i = 0, 1 \ldots, \alpha > -1$, $\beta > -1$, form a complete system of orthogonal polynomials on $[-1, 1]$ with respect to the weight function $(1 - x)^\alpha (1 + x)^\beta$. Set $\alpha = \beta = (d - 1)/2$ and let

$$f(t) = \sum_{i=0}^{k} f_i P_i^{(\alpha,\alpha)}(t)$$

be a real polynomial such that $f_0 > 0$, $f_i \geq 0$ $(i = 1, 2, \ldots, k)$, and $f(t) \leq 0$ for $-1 \leq t \leq \cos\varphi$. Then

$$M(d, \varphi) \leq f(1)/f_0.$$

With the use of appropriate polynomials Kabatiansky and Levenshtein [KL78] obtained the asymptotic bound:

$$\frac{1}{d}\ln M(d,\varphi) \le \frac{1+\sin\varphi}{2\sin\varphi}\ln\frac{1+\sin\varphi}{2\sin\varphi} - \frac{1-\sin\varphi}{2\sin\varphi}\ln\frac{1-\sin\varphi}{2\sin\varphi} + o(1).$$

This implies the simpler bound

$$M(d,\varphi) \le (1-\cos\varphi)^{-d/2}2^{-0.099d+o(d)} \qquad (\text{as } d\to\infty,\ \varphi \le \varphi^* = 62.9974\ldots).$$

Bound (2.1.4) for $\delta(B^d)$ follows in the limiting case when $\varphi \to 0$.

The following is a list of values of d and φ for which the linear programming bound turns out to be exact (see [Lev79]).

$M(2,\arccos 1/\sqrt{5}) = 12$	$M(3,\arccos 1/5) = 120$	$M(4,\arccos 1/5) = 16$
$M(5,\arccos 1/4) = 27$	$M(6,\arccos 1/3) = 56$	$M(7,\pi/3) = 240$
$M(20,\arccos 1/7) = 162$	$M(21,\arccos 1/11) = 100$	$M(21,\arccos 1/6) = 275$
$M(21,\arccos 1/4) = 891$	$M(22,\arccos 1/5) = 552$	$M(22,\arccos 1/3) = 4600$
	$M(23,\pi/3) = 196560$	

For small values of d and specific values of φ the linear programming bound is superseded by the "simplex bound" of Böröczky [Bor78], which is the generalization of Rogers's bound (2.1.5) for ball packings in $\mathbb{S}^d$. The value of $M(d,\varphi)$ has been determined for all d and $\varphi \ge \pi/2$ (see [DH51, Ran55]). We have

$$M(d,\varphi) = i+1 \qquad \text{for } \frac{1}{2}\pi + \arcsin\frac{1}{i+1} < \varphi \le \frac{1}{2}\pi + \arcsin\frac{1}{i}, \qquad i = 1,\ldots,d,$$

$$M(d,\varphi) = d+2 \qquad \text{for } \frac{1}{2}\pi < \varphi \le \frac{1}{2}\pi + \arcsin\frac{1}{d+1},$$

and

$$M(d,\frac{1}{2}\pi) = 2(d+1).$$

Consider a decreasing continuous positive-valued potential function f defined on $(0,4]$. How should N distinct points $\{x_1,x_2,\ldots,x_N\}$ be placed on the unit sphere in n-dimensional space to minimize the potential energy $\sum_{i\neq j} f(\|x_i-x_j\|^2)$. Of special interest are ***completely monotonic*** potential functions f which are infinitely differentiable and satisfy the inequalities $(-1)^k f^{(k)}(x) \ge 0$ for every $k \ge 0$ and every $x \in (0,4]$. Cohn and Kumar [CK07] introduced the notion of universally optimal arrangement. An arrangement of points is ***universally optimal*** if its potential energy is minimal under every completely monotonic potential function.

Cohn and Kumar [CK07] proved universal optimality of the sets of vertices of all regular simplicial polytopes in every dimension, as well as of several other arrangements in dimensions 2–8 and 21–24. These arrangements are listed in Table 1 of their paper and they coincide with the arrangements for which the linear programming bound for $M(d,\varphi)$ is sharp. For the potential function $f(r) = 1/r^s$, the leading term of the potential energy for large s comes from the minimal distance. It follows that for universally optimal arrangements of points the minimal distance is maximal. For, if there were an arrangement with the same number of points but a larger minimal distance, then it would have lower potential energy when s is sufficiently large. With the special choice of $f(r) = 2 - 1/r^2$ and $f(r) = \log(4/r)$, respectively, it also follows that these arrangements maximize the sum, as well as the product, of the distances between pairs of points.

Rogers [Rog63] proved the existence of thin coverings of the sphere by congruent balls. His bound was improved by Böröczky and Wintsche [BW00], Verger-Gaugry [Ver05] and Dumer [Dum07]. The latter author proved that the d-dimensional sphere can be covered by congruent balls with density $\frac{1}{2}d\ln d+\ln d+5d$. In the limiting case when the radius of the balls approaches zero, this improves bound 2 of Table 2.1.4 by a factor of $\frac{1}{2}$. Naszódi [Nas16a] proved the existence of economic coverings of the sphere by congruent copies of a general spherically convex set.

TABLE 2.4.1 Densest packing and thinnest covering with congruent circles on a sphere.

n	a_n	Source	A_n	Source
2	$180°$	(elementary)	$180°$	(elementary)
3	$120°$	(elementary)	$180°$	(elementary)
4	$109.471\ldots°$	[Fej43a]	$141.047\ldots°$	[Fej43b]
5	$90°$	[SW51]	$126.869\ldots°$	[Sch55]
6	$90°$	[Fej43a]	$109.471\ldots°$	[Fej43b]
7	$77.866\ldots°$	[SW51]	$102.053\ldots°$	[Sch55]
8	$74.869\ldots°$	[SW51]		
9	$70.528\ldots°$	[SW51]		
10	$66.316\ldots°$	[Dan86, Har86]	$84.615\ldots°$	[Fej69a]
11	$63.435\ldots°$	[Dan86, Bor83]		
12	$63.435\ldots°$	[Fej43a]	$74.754\ldots°$	[Fej43b]
13	$57.136\ldots°$	[MT12]		
14	$55.670\ldots°$	[MT15]	$69.875\ldots°$	[Fej69a]
24	$43.667\ldots°$	[Rob61]		

Extensive research has been done on circle packings and circle coverings on $\mathbb{S}^2$. Traditionally, here the inverse functions of $M(2,\varphi)$ and $m(2,\varphi)$ are considered. Let a_n be the maximum number such that n caps of spherical diameter a_n can form a packing and let A_n be the minimum number such that n caps of spherical diameter A_n can form a covering on $\mathbb{S}^2$. The known values of a_n and A_n are given in Table 2.4.1. In addition, conjecturally best circle packings and circle coverings for $n\le 130$, as well as good arrangements with icosahedral symmetry for $n\le 55000$, have been constructed [HSS12]. The ad hoc methods of the earlier constructions have recently been replaced by different computer algorithms, but none of them has been shown to give the optimum.

Observe that $a_5=a_6$ and $a_{11}=a_{12}$. Also, $A_2=A_3$. It is conjectured that $a_n>a_{n+1}$ and $A_n>A_{n+1}$ in all other cases.

HYPERBOLIC SPACE

The density of a general arrangement of sets in d-dimensional hyperbolic space $\mathbb{H}^d$ cannot be defined by a limit as in $\mathbb{E}^d$ (see [FK93b]). The main difficulty is that in hyperbolic geometry the volume and the surface area of a ball of radius r are of the same order of magnitude as $r\to\infty$. In the absence of a reasonable definition of density with respect to the whole space, two natural problems arise:

(i) Estimate the density of an arrangement relative to a bounded domain.
(ii) Find substitutes for the notions of densest packing and thinnest covering.

Concerning the first problem, we mention the following result of K. Bezdek (see [Bez84]). Consider a packing of finitely many, but at least two, circles of radius r in the hyperbolic plane $\mathbb{H}^2$. Then the density of the circles relative to the outer parallel domain of radius r of the convex hull of their centers is at most $\pi/\sqrt{12}$.

As a corollary it follows that if at least two congruent circles are packed in a circular domain in $\mathbb{H}^2$, then the density of the packing relative to the domain is at most $\pi/\sqrt{12}$. We note that the density of such a finite packing relative to the convex hull of the circles can be arbitrarily close to 1 as $r \to \infty$. K. Böröczky Jr. [Bor05] proved a dual counterpart to the above-mentioned theorem of K. Bezdek, a corollary of which is that if at least two congruent circles cover a circular domain in $\mathbb{H}^2$, then the density of the covering relative to the domain is at most $2\pi/\sqrt{27}$.

Rogers's simplex bound (2.1.5) for ball packings in $\mathbb{E}^d$ has been extended by Böröczky [Bor78] to $\mathbb{H}^d$ as follows. If balls of radius r are packed in $\mathbb{H}^d$ then the density of each ball relative to its Dirichlet cell is less than or equal to the density of $d+1$ balls of radius r centered at the vertices of a regular simplex of side-length $2r$ relative to this simplex. Of course, we should not interpret this result as a global density bound. The impossibility of such an interpretation is shown by an ingenious example of Böröczky (see [FK93b]). He constructed a packing $\mathcal{P}$ of congruent circles in $\mathbb{H}^2$ and two tilings, $\mathcal{T}_1$ and $\mathcal{T}_2$, both consisting of congruent tiles, such that each tile of $\mathcal{T}_1$, as well as each tile of $\mathcal{T}_2$, contains exactly one circle from $\mathcal{P}$, but such that the tiles of $\mathcal{T}_1$ and $\mathcal{T}_2$ have different areas.

The first notion that has been suggested as a substitute for densest packing and thinnest covering is "solidity." $\mathcal{P}$ is a ***solid packing*** if no finite subset of $\mathcal{P}$ can be rearranged so as to form, together with the rest of $\mathcal{P}$, a packing not congruent to $\mathcal{P}$. Analogously, $\mathcal{C}$ is a ***solid covering*** if no finite subset of $\mathcal{C}$ can be rearranged so as to form, together with the rest of $\mathcal{C}$, a covering not congruent to $\mathcal{C}$. Obviously, in $\mathbb{E}^d$ a solid packing with congruent copies of a body K has density $\delta(K)$, and a solid covering with congruent copies of K has density $\vartheta(K)$. This justifies the use of solidity as a natural substitute for "densest packing" and "thinnest covering" in hyperbolic space.

The tiling with Schläfli symbol $\{p, 3\}$ (see Chapters 18 or 20 of this Handbook) has regular p-gonal faces such that at each vertex of the tiling three faces meet. There exists such a tiling for each $p \geq 2$: for $p \leq 5$ on the sphere, for $p \geq 7$ on the hyperbolic plane, while for $p = 6$ we have the well-known hexagonal tiling in Euclidean plane. The incircles of such a tiling form a solid packing and the circumcircles form a solid covering. In addition, several packings and coverings by incongruent circles, including the incircles and the circumcircles of certain trihedral Archimedean tilings have been confirmed to be solid (see [Fej68, Fej74, Hep92, Flo00, Flo01, FH00]).

Other substitutes for the notion of densest packing and thinnest covering have been proposed in [FKK98] and [Kup00]. A packing $\mathcal{P}$ with congruent copies of a body K is ***completely saturated*** if no finite subset of $\mathcal{P}$ can be replaced by a greater number of congruent copies of K that, together with the rest of $\mathcal{P}$, form a packing. Analogously, a covering $\mathcal{C}$ with congruent copies of K is ***completely reduced*** if no finite subset of $\mathcal{C}$ can be replaced by a smaller number of congruent copies of K that, together with the rest of $\mathcal{C}$, form a covering. While there are convex bodies that do not admit a solid packing or solid covering, it has been shown [Bow03, FKK98] that each body in $\mathbb{E}^d$ or $\mathbb{H}^d$ admits a completely saturated packing and a completely reduced covering. However, the following rather counterintuitive result of Bowen makes it doubtful whether complete saturatedness and complete reducedness are good substitutes for the notions of densest packing and thinnest covering in hyperbolic space. For any $\varepsilon > 0$ there is a body K in $\mathbb{H}^d$ that admits a tiling and at the same time a completely saturated packing $\mathcal{P}$ with the following property. For every point p in $\mathbb{H}^d$, the limit

$$\lim_{\lambda\to\infty} \frac{1}{V(B_\lambda(p))} \sum_{P\in\mathcal{P}} V(P \cap (B_\lambda(p))$$

exists, is independent of p, and is less than ε. Here $V(\cdot)$ denotes the volume in $\mathbb{H}^d$ and $B_\lambda(p)$ denotes the ball of radius λ centered at p.

Bowen and Radin [BR03, BR04] proposed a probabilistic approach to analyze the efficiency of packings in hyperbolic geometry. Their approach can be sketched as follows.

Instead of studying individual arrangements, one considers the space Σ_K consisting of all saturated packings of $\mathbb{H}^d$ by congruent copies of K. A suitable metric on Σ_K is introduced that makes Σ_K compact and makes the natural action of the group $\mathcal{G}^d$ of rigid motions of $\mathbb{H}^d$ on Σ_K continuous. We consider Borel probability measures on Σ_K that are invariant under $\mathcal{G}^d$. For such an invariant measure μ the ***density*** $d(\mu)$ of μ is defined as $d(\mu) = \mu(A)$, where A is the set of packings $\mathcal{P} \in \Sigma_K$ for which the origin of $\mathbb{H}^d$ is contained in some member of $\mathcal{P}$. It follows easily from the invariance of μ that this definition is independent of the choice of the origin. The connection of density of measures to density of packings is established by the following theorem.

If μ is an ergodic invariant Borel probability measure on Σ_K, then—with the exception of a set of μ-measure zero—for every packing $\mathcal{P} \in \Sigma_K$, and for all $p \in \mathbb{H}^d$,

$$\lim_{\lambda\to\infty} \frac{1}{V(B_\lambda(p))} \sum_{P\in\mathcal{P}} V(P \cap (B_\lambda(p)) = d(\mu). \tag{2.4.1}$$

(A measure μ is ergodic if it cannot be expressed as the positive linear combination of two invariant measures.)

The ***packing density*** $\delta(K)$ of K can now be defined as the supremum of $d(\mu)$ for all ergodic invariant measures on Σ_K. A packing $\mathcal{P} \in \Sigma_K$ is ***optimally dense*** if there is an ergodic invariant measure μ such that the orbit of $\mathcal{P}$ under $\mathcal{G}^d$ is dense in the support of μ and, for all $p \in \mathbb{H}^d$, (2.4.1) holds.

It is shown in [BR03] and [BR04] that there exists an ergodic invariant measure μ with $d(\mu) = \delta(K)$ and a subset of the support of μ of full μ-measure of optimally dense packings. Bowen and Radin prove several results justifying that this is a workable notion of optimal density and optimally dense packings. In particular, the definitions carry over without any change to $\mathbb{E}^d$, and there they coincide with the usual notions. The advantage of this probabilistic approach is that it neglects pathological packings such as the example by Böröczky. As for packings of balls, it is shown in [BR03] that there are only countably many radii for which there exists an optimally dense packing of balls of the given radius that is periodic.

2.5 NEIGHBORS

GLOSSARY

Neighbors: Two members of a packing whose closures intersect.

Newton number $N(K)$ of a convex body K: The maximum number of neighbors of K in all packings with congruent copies of K.

Hadwiger number $H(K)$ of a convex body K: The maximum number of neighbors of K in all packings with translates of K.

n-neighbor packing: A packing in which each member has exactly n neighbors.

n^+-neighbor packing: A packing in which each member has at least n neighbors.

Table 2.5.1 contains the results known about Newton numbers and Hadwiger numbers. Cheong and Lee [CL07] showed that, surprisingly, there are non-convex Jordan regions with arbitrary large Hadwiger number.

It seems that the maximum number of neighbors of one body in a lattice packing with congruent copies of K is considerably smaller than $H(K)$. While $H(B^d)$ is of exponential order of magnitude, the highest known number of neighbors in a lattice packing with B^d occurs in the Barnes-Wall lattice and is $c^{O(\log d)}$ [CS93]. Moreover, Gruber [Gru86] showed that, in the sense of Baire categories, most convex bodies in $\mathbb{E}^d$ have no more than $2d^2$ neighbors in their densest lattice packing. Talata [Tal98b] gave examples of convex bodies in $\mathbb{E}^d$ for which the difference between the Hadwiger number and the maximum number of neighbors in a lattice packing is 2^{d-1}. Alon [Alo97] constructed a finite ball packing in $\mathbb{E}^d$ in which each ball has $c^{O(\sqrt{d})}$ neighbors.

A problem related to the determination of the Hadwiger number concerns the maximum number $C(K)$ of mutually nonoverlapping translates of a set K that have a common point. No more than four nonoverlapping translates of a topological disk in the plane can share a point [BKK95], while for $d \geq 3$ there are starlike bodies in $\mathbb{E}^d$ for which $C(K)$ is arbitrarily large.

For a given convex body K, let $M(K)$ denote the maximum natural number with the property that an $M(K)$-neighbor packing with finitely many congruent copies of K exists. For $n \leq M(K)$, let $L(n, K)$ denote the minimum cardinality, and, for $n > M(K)$, let $\lambda(n, K)$ denote the minimum density, of an n-neighbor packing with congruent copies of K. The quantities $M_T(K)$, $M^+(K)$, $M_T^+(K)$, $L_T(n, K)$, $L^+(n, K)$, $L_T^+(n, K)$, $\lambda_T(n, K)$, $\lambda^+(n, K)$, and $\lambda_T^+(n, K)$ are defined analogously.

Österreicher and Linhart [OL82] showed that for a smooth convex disk K we have $L(2, K) \geq 3$, $L(3, K) \geq 6$, $L(4, K) \geq 8$, and $L(5, K) \geq 16$. All of these inequalities are sharp. We have $M_T^+(K) = 3$ for all convex disks, and there exists a 4-neighbor packing of density 0 with translates of any convex disk. There exists a 5-neighbor packing of density 0 with translates of a parallelogram, but Makai [Mak85] proved that $\lambda_T^+(5, K) \geq 3/7$ and $\lambda_T^+(6, K) \geq 1/2$ for every $K \in \mathcal{K}(\mathbb{E}^2)$ that is not a parallelogram, and that $\lambda_T^+(5, K) \geq 9/14$ and $\lambda_T^+(6, K) \geq 3/4$ for every $K \in \mathcal{K}^*(\mathbb{E}^2)$ that is not a parallelogram. The case of equality characterizes triangles and affinely regular hexagons, respectively. According to a result of Chvátal [Chv83], $\lambda_T^+(6, P) = 11/15$ for a parallelogram P.

TABLE 2.5.1 Newton and Hadwiger numbers.

BODY K	RESULT	SOURCE
B^3	$N(K) = 12$	[SW53]
B^4	$N(K) = 24$	[Mus08]
B^8	$N(K) = 240$	[Lev79, OS79]
B^{24}	$N(K) = 196560$	[Lev79, OS79]
Regular triangle	$N(K) = 12$	[Bor71]
Square	$N(K) = 8$	[You39, Bor71, KLL95]
Regular pentagon	$N(K) = 6$	[ZX02]
Regular n-gon for $n \geq 6$	$N(K) = 6$	[Bor71, Zha98]
Isosceles triangle with base angle $\pi/6$	$N(K) = 21$	[Weg92]
Convex disk of diameter d and width w	$N(K) \leq (4 + 2\pi)\frac{d}{w} + \frac{w}{d} + 2$	[Fej69b]
Parallelotope in $\mathbb{E}^d$	$H(K) = 3^d - 1$	[Had46]
Tetrahedron	$H(K) = 18$	[Tal99a]
Octahedron	$H(K) = 18$	[LZ99]
Rhombic dodecahedron	$H(K) = 18$	[LZ99]
Starlike region in $\mathbb{E}^2$	$H(K) \leq 35$	[Lan11]
Centrally symmetric starlike region in $\mathbb{E}^2$	$H(K) \leq 12$	[Lan09]
Convex body in $\mathbb{E}^d$	$H(K) \leq 3^d - 1$	[Had46]
Convex body in $\mathbb{E}^d$	$H(K) \geq 2^{cd},\ c > 0$	[Tal98a]
Simplex in $\mathbb{E}^d$	$H(K) \geq 1.13488^{d-o(d)}$	[Tal00]
Unit ball in L_p norm	$2^{n+o(1)}$ as $p \to \infty$	[Xu07]
Set in $\mathbb{E}^d$ with $\text{int}\,(K - K) \neq \emptyset$	$H(K) \geq d^2 + d$	[Smi75]

A construction of Wegner (see [FK93c]) shows that $M(B^3) \geq 6$ and $L(6, B^3) \leq 240$, while Kertész [Ker94] proved that $M(B^3) \leq 8$. It is an open problem whether an n-neighbor or n^+-neighbor packing of finitely many congruent balls exists for $n = 7$ and $n = 8$.

The long-standing conjecture of L. Fejes Tóth [Fej69c] that a 12-neighbor packing of congruent balls consist of parallel hexagonal layers was recently confirmed independently by Hales [Hal13] and by Böröczky and Szabó [BS15]. Both proofs heavily depend on the use of computers. Hales uses the technique he developed for the proof of the Kepler conjecture, while the proof by Böröczky and Szabó relies on the computer-aided solution of the "strong thirteen spheres problem" by Musin and Tarasov.

Harborth, Szabó and Ujváry-Menyhárt [HSU02] constructed finite n-neighbor packings of incongruent balls in $\mathbb{E}^3$ for all $n \leq 12$ except for 11. The question whether a finite 11-neighbor packing of balls exists remains open. Since the smallest ball in a finite ball-packing has at most twelve neighbors, there is no finite n-neighbor packing for $n \geq 12$. On the other hand, the average number of neighbors in a finite packing of balls can be greater than 12. G. Kuperberg and Schramm [KS94] constructed a finite packing of balls in which the average number of neighbors is $666/53 = 12.566$ and showed that the average number of neighbors in every finite packing of balls is at most $8 + 4\sqrt{3} = 14.928$.

For 6^+-neighbor packings with (not necessarily equal) circles, the following nice theorem of Bárány, Füredi, and Pach [BFP84]) holds:

In a 6^+-neighbor packing with circles, either all circles are congruent or arbitrarily small circles occur.

2.6 SELECTED PROBLEMS ON LATTICE ARRANGEMENTS

In this section we discuss, from the vast literature on lattices, some special problems concerning arrangements of convex bodies in which the restriction to lattice arrangements is automatically imposed by the nature of the problem.

GLOSSARY

Point-trapping arrangement: An arrangement $\mathcal{A}$ such that every component of the complement of the union of the members of $\mathcal{A}$ is bounded.

Connected arrangement: An arrangement $\mathcal{A}$ such that the union of the members of $\mathcal{A}$ is connected.

j-impassable arrangement: An arrangement $\mathcal{A}$ such that every j-dimensional flat intersects the interior of a member of $\mathcal{A}$.

Obviously, a point-trapping arrangement of congruent copies of a body can be arbitrarily thin. On the other hand, Bárány, Böröczky, Makai, and Pach showed that the density of a point-trapping *lattice* arrangement of any convex body in $\mathbb{E}^d$ is greater than or equal to 1/2. For $d \geq 3$, equality is attained only in the "checkerboard" arrangement of parallelotopes (see [BBM86]).

Bleicher [Ble75] showed that the minimum density of a point-trapping lattice of unit balls in $\mathbb{E}^3$ is equal to

$$32\sqrt{(7142 + 1802\sqrt{17})^{-1}} = 0.265\ldots .$$

The extreme lattice is generated by three vectors of length $\frac{1}{2}\sqrt{7+\sqrt{17}}$, any two of which make an angle of $\arccos \frac{\sqrt{17}-1}{8} = 67.021\ldots^\circ$.

For a convex body K, let $c(K)$ denote the minimum density of a connected lattice arrangement of congruent copies of K. According to Groemer [Gro66],

$$\frac{1}{d!} \leq c(K) \leq \frac{\pi^{d/2}}{2^d\Gamma(1+d/2)} \qquad \text{for } K \in \mathcal{K}^d.$$

The lower bound is attained when K is a simplex or cross-polytope, and the upper bound is attained for a ball.

For a given convex body K in $\mathbb{E}^d$, let $\varrho_j(K)$ denote the infimum of the densities of all j-impassable lattice arrangements of copies of K. Obviously, $\varrho_0(K) = \vartheta_L(K)$. Let $\widehat{K} = (K-K)^*$ denote the polar body of the difference body of K. Between $\varrho_{d-1}(K)$ and $\delta_L(\widehat{K})$ Makai [Mak78], and independently also Kanan and Lovász [KL88], found the following surprising connection:

$$\varrho_{d-1}(K)\delta_L(\widehat{K}) = 2^d V(K)V(\widehat{K}).$$

Little is known about $\varrho_j(K)$ for $0 < j < d-1$. The value of $\varrho_1(B^3)$ has been determined by Bambah and Woods [BW94]. We have

$$\varrho_1(B^3) = 9\pi/32 = 0.8835\ldots .$$

An extreme lattice is generated by the vectors $\frac{4}{3}(1,1,0)$, $\frac{4}{3}(0,1,1)$, and $\frac{4}{3}(1,0,1)$.

2.7 PACKING AND COVERING WITH SEQUENCES OF CONVEX BODIES

In this section we consider the following problem: Given a convex set K and a sequence $\{C_i\}$ of convex bodies in $\mathbb{E}^d$, is it possible to find rigid motions σ_i such that $\{\sigma_i C_i\}$ covers K, or forms a packing in K? If there are such motions σ_i, then we say that the sequence $\{C_i\}$ permits an ***isometric covering*** of K, or an ***isometric packing*** in K, respectively. If there are not only rigid motions but even translations τ_i so that $\{\tau_i C_i\}$ is a covering of K, or a packing in K, then we say that $\{C_i\}$ permits a ***translative covering*** of K, or a ***translative packing*** in K, respectively.

First we consider translative packings and coverings of cubes by sequences of boxes. By a ***box*** we mean an orthogonal parallelotope whose sides are parallel to the coordinate axes. We let $I^d(s)$ denote a cube of side length s in $\mathbb{E}^d$.

Groemer (see [Gro85]) proved that a sequence $\{C_i\}$ of boxes whose edge lengths are at most 1 permits a translative covering of $I^d(s)$ if

$$\sum_i V(C_i) \geq (s+1)^d - 1,$$

and that it permits a translative packing in $I^d(s)$ if

$$\sum_i V(C_i) \leq (s-1)^d - \frac{s-1}{s-2}((s-1)^{d-2} - 1).$$

Slightly stronger conditions (see [Las97]) guarantee even the existence of on-line algorithms for the determination of the translations τ_i. This means that the determination of τ_i is based only on C_i and the previously fixed sets $\tau_i C_i$.

We recall (see [Las97]) that to any convex body K in $\mathbb{E}^d$ there exist two boxes, say Q_1 and Q_2, with $V(Q_1) \geq 2d^{-d}V(K)$ and $V(Q_2) \leq d!V(K)$, such that $Q_1 \subset K \subset Q_2$. It follows immediately that if $\{C_i\}$ is a sequence of convex bodies in $\mathbb{E}^d$ whose diameters are at most 1 and

$$\sum_i V(C_i) \geq \frac{1}{2}d^d((s+1)^d - 1),$$

then $\{C_i\}$ permits an isometric covering of $I^d(s)$; and that if

$$\sum_i V(C_i) \leq \frac{1}{d!}\left((s-1)^d - \frac{s-1}{s-2}((s-1)^{d-2} - 1)\right),$$

then it permits an isometric packing in $I^d(s)$.

The sequence $\{C_i\}$ of convex bodies is ***bounded*** if the set of the diameters of the bodies is bounded. As further consequences of the results above we mention the following. If $\{C_i\}$ is a bounded sequence of convex bodies such that $\sum V(C_i) = \infty$, then it permits an isometric covering of $\mathbb{E}^d$ with density $\frac{1}{2}d^d$ and an isometric packing in $\mathbb{E}^d$ with density $\frac{1}{d!}$. Moreover, if all the sets C_i are boxes, then $\{C_i\}$ permits a translative covering of $\mathbb{E}^d$ and a translative packing in $\mathbb{E}^d$ with density 1.

In $\mathbb{E}^2$, any bounded sequence $\{C_i\}$ of convex disks with $\sum a(C_i) = \infty$ permits even a translative packing and covering with density $\frac{1}{2}$ and 2, respectively. It is an open problem whether for $d > 2$ any bounded sequence $\{C_i\}$ of convex bodies in $\mathbb{E}^d$ with $\sum V(C_i) = \infty$ permits a translative covering. If the sequence $\{C_i\}$ is unbounded, then the condition $\sum V(C_i) = \infty$ no longer suffices for $\{C_i\}$ to permit even an isometric covering of the space. For example, if C_i is the rectangle of side lengths i and $\frac{1}{i^2}$, then $\sum a(C_i) = \infty$ but $\{C_i\}$ does not permit an isometric covering of $\mathbb{E}^2$. There is a simple reason for this, which brings us to one of the most interesting topics of this subject, namely Tarski's plank problem.

A ***plank*** is a region between two parallel hyperplanes. Tarski conjectured that if a convex body of minimum width w is covered by a collection of planks in $\mathbb{E}^d$, then the sum of the widths of the planks is at least w. Tarski's conjecture was first proved by Bang. Bang's theorem immediately implies that the sequence of rectangles above does not permit an isometric covering of $\mathbb{E}^2$, not even of $(\frac{\pi^2}{12} + \epsilon)B^2$.

In his paper, Bang asked whether his theorem can be generalized so that the width of each plank is measured relative to the width of the convex body being covered, in the direction normal to the plank. Bang's problem has been solved for centrally symmetric bodies by Ball [Bal91]. This case has a particularly appealing formulation in terms of normed spaces: If the unit ball in a Banach space is covered by a countable collection of planks, then the total width of the planks is at least 2.

The paper of Groemer [Gro85] gives a nice account on problems about packing and covering with sequences of convex bodies.

2.8 SOURCES AND RELATED MATERIAL

SURVEYS

The monographs [Bor04, Fej72, Rog64, Zon99] are devoted solely to packing and covering; also the books [CS93, CFG91, EGH89, Fej64, Gru07, GL87, BMP05, PA95] contain results relevant to this chapter. Additional material and bibliography can be found in the following surveys: [Bar69, Fej83, Fej84b, Fej99, FK93b, FK93c, FK01, Flo87, Flo02, GW93, Gro85, Gru79].

RELATED CHAPTERS

Chapter 3: Tilings
Chapter 7: Lattice points and lattice polytopes
Chapter 13: Geometric discrepancy theory and uniform distribution
Chapter 18: Symmetry of polytopes and polyhedra
Chapter 20: Polyhedral maps
Chapter 64: Crystals, periodic and aperiodic

REFERENCES

[Alo97] N. Alon. Packings with large minimum kissing numbers. *Discrete Math.*, 175:249–251, 1997.

[Bal91] K. Ball. The plank problem for symmetric bodies. *Invent. Math.*, 104:535–543, 1991.

[Bal92] K. Ball. A lower bound for the optimal density of lattice packings. *Internat. Math. Res. Notices*, 1992:217–221, 1992.

[Bam54] R.P. Bambah. On lattice coverings by spheres. *Proc. Nat. Inst. Sci. India*, 20:25–52, 1954.

[Bar69] E.P. Baranovskiĭ. Packings, coverings, partitionings and certain other distributions in spaces of constant curvature. (Russian) *Itogi Nauki—Ser. Mat. (Algebra, Topologiya, Geometriya)*, 14:189–225, 1969. Translated in *Progr. Math.*, 9:209–253, 1971.

[BB85] A. Bezdek and K. Bezdek. Über einige dünnste Kreisüberdeckungen konvexer Bereiche durch endliche Anzahl von kongruenten Kreisen. *Beitr. Algebra Geom.*, 19:159–168, 1985.

[BBM86] K. Böröczky, I. Bárány, E. Jr. Makai, and J. Pach. Maximal volume enclosed by plates and proof of the chessboard conjecture. *Discrete Math.*, 60:101–120, 1986.

[Bez02] K. Bezdek. Improving Rogers' upper bound for the density of unit ball packings via estimating the surface area of Voronoi cells from below in Euclidean d-space for all $d \geq 8$. *Discrete Comput. Geom.*, 28:75–106, 2002.

[Bez84] K. Bezdek. Ausfüllungen in der hyperbolischen Ebene durch endliche Anzahl kongruenter Kreise. *Ann. Univ. Sci. Budapest. Eötvös Sect. Math.*, 27:113–124, 1984

[Bez87] K. Bezdek. Densest packing of small number of congruent spheres in polyhedra. *Ann. Univ. Sci. Budapest. Eötvös Sect. Math.*, 30:177–194, 1987.

[Bez93] K. Bezdek. Hadwiger-Levi's covering problem revisited. In J. Pach, editor, *New Trends in Discrete and Computational Geometry*, pages 199–233. Springer, New York, 1993.

[Bez94] A. Bezdek. A remark on the packing density in the 3-space. In K. Böröczky and G. Fejes Tóth, editors, *Intuitive Geometry*, vol. 63 of *Colloq. Math. Soc. János Bolyai*, pages 17–22. North-Holland, Amsterdam, 1994.

[BFP84] I. Bárány, Z. Füredi, and J. Pach. Discrete convex functions and proof of the six circle conjecture of Fejes Tóth. *Canad. J. Math.*, 36:569–576, 1984.

[BFW06] K. Böröczky Jr., I. Fábián, and G. Wintsche. Covering the crosspolytope by equal balls. *Period. Math. Hungar.*, 53:103–113, 2006.

[BH00] U. Betke and M. Henk. Densest lattice packings of 3-polytopes. *Comput. Geom.*, 16:157–186, 2000.

[BH98] U. Betke and M. Henk. Finite packings of spheres. *Discrete Comput. Geom.*, 19:197–227, 1998.

[BHW94] U. Betke, M. Henk, and J.M. Wills. Finite and infinite packings. *J. Reine Angew. Math.*, 453:165–191, 1994.

[BK10] A. Bezdek and W. Kuperberg. Unavoidable crossings in a thinnest plane covering with congruent convex disks. *Discrete Comput. Geom.*, 43:187–208, 2010.

[BK91] A. Bezdek and W. Kuperberg. Packing Euclidean space with congruent cylinders and with congruent ellipsoids. In *Applied Geometry and Discrete Mathematics*, vol. 4 of *DIMACS Ser. Discrete Math. Theoret. Comp. Sci.*, pages 71–80, AMS, Providence, 1991.

[BKK95] A. Bezdek, K. Kuperberg, and W. Kuperberg. Mutually contiguous and concurrent translates of a plane disk. *Duke Math. J.*, 78:19–31, 1995.

[BKM91] A. Bezdek, W. Kuperberg, and E. Makai. Jr. Maximum density space packings with parallel strings of balls. *Discrete Comput. Geom.*, 6:277–283, 1991.

[Ble75] M.N. Bleicher. The thinnest three dimensional point lattice trapping a sphere. *Studia Sci. Math. Hungar.*, 10:157–170, 1975.

[Bli35] H.F. Blichfeldt. The minimum values of positive quadratic forms in six, seven and eight variables. *Math. Z.*, 39:1–15, 1935.

[Blu57] W.J. Blundon. Multiple covering of the plane by circles. *Mathematika*, 4:7–16, 1957.

[Blu63] W.J. Blundon. Multiple packing of circles in the plane. *J. London Math. Soc.*, 38:176–182, 1963.

[BMP05] P. Brass, W. Moser, and J. Pach. *Research Problems in Discrete Geometry.* Springer, New York, 2005.

[BMS97] V. Boltjanski, H. Martini, and P.S. Soltan. *Excursions into Combinatorial Geometry.* Springer, Berlin, 1997.

[Bol76] U. Bolle. *Mehfache Kreisanordnungen in der euklidischen Ebene, Dissertation.* Dortmund, 1976.

[Bol79] U. Bolle. Dichteabschätzungen für mehrfache gitterförmige Kugelanordnungen im $\mathbb{R}^m$. *Studia Sci. Math. Hungar.*, 14:51–68, 1979.

[Bol82] U. Bolle. Dichteabschätzungen für mehrfache gitterförmige Kugelanordnungen im $\mathbb{R}^m$ II. *Studia Sci. Math. Hungar.*, 17:429–444, 1982.

[Bol89] U. Bolle. On the density of multiple packings and coverings of convex discs. *Studia Sci. Math. Hungar.*, 24:119–126, 1989.

[Bor04] K. Böröczky Jr. *Finite packing and covering.* Cambridge Univ. Press, Cambridge, 2004.

[Bor05] K. Böröczky Jr. Finite coverings in the hyperbolic plane. *Discrete Comput. Geom.*, 33:165–180, 2005.

[Bor33] K. Borsuk. Drei Sätze über die n-dimensionale euklidische Sphäre. *Fund. Math.*, 20:177–190, 1933.

[Bor71] K. Böröczky. Über die Newtonsche Zahl regulärer Vielecke. *Period. Math. Hungar.*, 1:113–119, 1971.

[Bor78] K. Böröczky. Packing of spheres in spaces of constant curvature. *Acta Math. Acad. Sci. Hungar.*, 32:243–261, 1978.

[Bor83] K. Böröczky. The problem of Tammes for $n = 11$. *Studia Sci. Math. Hungar.*, 18:165–171, 1983.

[Bow03] L. Bowen. On the existence of completely saturated packings and completely reduced coverings. *Geom. Dedicata*, 98:211–226, 2003.

[BR03] L. Bowen and C. Radin. Densest packing of equal spheres in hyperbolic space. *Discrete Comput. Geom.*, 29:23–39, 2003.

[BR04] L. Bowen and C. Radin. Optimally dense packings of hyperbolic space. *Geometriae Dedicata*, 104:37–59, 2004.

[BS15] K. Böröczky and L. Szabó. 12-neighbour packings of unit balls in $\mathbb{E}^3$. *Acta Math. Hungar.*, 146:421–448, 2015.

[BW00] K. Böröczky Jr. and G. Wintsche. Sphere packings in the regular crosspolytope. *Ann. Univ. Sci. Budapest. Eötvös Sect. Math.*, 43:151–157, 2000.

[BW94] R.P. Bambah and A.C. Woods. On a problem of G. Fejes Tóth. *Proc. Indian Acad. Sci. Math. Sci.*, 104:137–156, 1994.

[CE03] H. Cohn and N. Elkies. New upper bounds on sphere packings I. *Ann. of Math.*, 157:689–714, 2003.

[CEG10] E.R. Chen, M. Engel, and S.C. Glotzer. Dense crystalline dimer packings of regular tetrahedra. *Discrete Comput. Geom.*, 44:253–280, 2010.

[CFG91] H.T. Croft, K.J. Falconer, and R.K. Guy. *Unsolved Problems in Geometry.* Springer, New York, 1991.

[Cha50] J.H.H. Chalk. On the frustrum of a sphere. *Ann. of Math.*, 52:199–216, 1950.

[Chv75] V. Chvátal. On a conjecture of Fejes Tóth. *Period. Math. Hungar.*, 6:357–362, 1975.

[CK04] H. Cohn and A. Kumar. The densest lattice in twenty-four dimensions. *Electronic Research Announcements of the AMS*, 10:58–67, 2004.

[CK07] H. Cohn and A. Kumar. Universally optimal distribution of points on spheres. *J. AMS*, 20:99–148, 2007.

[CK09] H. Cohn and A. Kumar. Optimality and uniqueness of the Leech lattice among lattices. *Ann. of Math.*, 170:1003–1050, 2009.

[CKM17] H. Cohn, A. Kumar, S.D. Miller, D. Radchenko, and M. Viazovska. The sphere packing problem in dimension 24. *Ann. of Math.*, 185:1017–1033, 2017.

[CL07] O. Cheong and M. Lee. The Hadwiger number of Jordan regions is unbounded. *Discrete Comput. Geom.*, 37:497–501, 2007.

[Coh02] H. Cohn. New upper bounds on sphere packings II. *Geometry & Topology*, 6:329–353, 2002.

[Coh76] M.J. Cohn. Multiple lattice coverings of space. *Proc. London Math. Soc.*, 32:117–132, 1976.

[CFR59] H.S.M. Coxeter, L. Few, and C.A. Rogers. Covering space with equal spheres. *Mathematika*, 6:147–157, 1959.

[CR48] J.H.H. Chalk and C.A. Rogers. The critical determinant of a convex cylinder. *J. London Math. Soc.*, 23:178–187, 1948.

[CS93] J.H. Conway and N.J.A. Sloane. *Sphere Packings, Lattices and Groups*, 2nd edition. Springer, New York, 1993.

[Dan86] L. Danzer. Finite point-sets on S^2 with minimum distance as large as possible. *Discrete Math.*, 60:3–66, 1986.

[Dek93] B.V. Dekster. The Borsuk conjecture holds for convex bodies with a belt of regular points. *Geom. Dedicata*, 45:301–306, 1993.

[DGOV17] M. Dostert, C. Guzmán, F.M. de Oliveira Filho, and F. Vallentin. New upper bounds for the density of translative packings of three-dimensional convex bodies with tetrahedral symmetry. *Discrete Comput. Geom.*, 58:449–481, 2017

[DH51] H. Davenport and Gy. Hajós. Problem 35. *Matematikai Lapok*, 2:63, 1951.

[DH72a] V.C. Dumir and R.J. Hans-Gill. Lattice double coverings in the plane. *Indian J. Pure Appl. Math.*, 3:466–480, 1972.

[DH72b] V.C. Dumir and R.J. Hans-Gill. Lattice double packings in the plane. *Indian J. Pure Appl. Math.*, 3:481–487, 1972.

[DLR92] P.G. Doyle, J.C. Lagarias, and D. Randall. Self-packing of centrally symmetric convex discs in $\mathbb{R}^2$. *Discrete Comput. Geom.*, 8:171–189, 1992.

[DR63] B.N. Delone and S.S. Ryškov. Solution of the problem on the least dense lattice covering of a 4-dimensional space by equal spheres. (Russian) *Dokl. Akad. Nauk SSSR*, 152:523–524, 1963.

[Dum07] I. Dumer. Covering spheres with spheres. *Discrete Comput. Geom.*, 38:665–679, 2007.

[EGH89] P. Erdős, P.M. Gruber, and J. Hammer. *Lattice Points.* vol. 39 of *Pitman Monographs.* Longman Scientific/John Wiley, New York, 1989.

[ER62] P. Erdős and C.A. Rogers. Covering space with convex bodies. *Acta Arith.*, 7:281–285, 1961/1962.

[Far50] I. Fáry. Sur la densité des réseaux de domaines convexes. *Bull. Soc. Math. France*, 78:152–161, 1950.

[Fej43a] L. Fejes Tóth. Über eine Abschätzung des kürzesten Abstandes zweier Punkte eines auf einer Kugelfläche liegenden Punktsystems. *Jber. Deutsch. Math. Verein.*, 53:66–68, 1943.

[Fej43b] L. Fejes Tóth. Covering the sphere with congruent caps. (Hungarian) *Mat. Fiz. Lapok*, 50:40–46, 1943.

[Fej50] L. Fejes Tóth. Some packing and covering theorems. *Acta Sci. Math. Szeged*, Leopoldo Fejer et Frederico Riesz LXX annos natis dedicatus, Pars A, 12:62–67, 1950.

[Fej64] L. Fejes Tóth. *Regular Figures.* Pergamon Press, Oxford, 1964.

[Fej67] L. Fejes Tóth. On the arrangement of houses in a housing estate. *Studia Sci. Math. Hungar.*, 2:37–42, 1967.

[Fej68] L. Fejes Tóth. Solid circle-packings and circle-coverings. *Studia Sci. Math. Hungar.*, 3:401–409, 1968.

[Fej69a] G. Fejes Tóth. Kreisüberdeckungen der Sphäre. *Studia Sci. Math. Hungar.*, 4:225–247, 1969.

[Fej69b] L. Fejes Tóth. Scheibenpackungen konstanter Nachbarnzahl. *Acta Math. Acad. Sci. Hungar.*, 20:375–381, 1969.

[Fej69c] L. Fejes Tóth. Remarks on a theorem of R. M. Robinson. *Studia Sci. Math. Hungar.*, 4:441–445, 1969.

[Fej71] L. Fejes Tóth. The densest packing of lenses in the plane. (Hungarian) *Mat. Lapok*, 22:209–213, 1972.

[Fej72] L. Fejes Tóth. *Lagerungen in der Ebene auf der Kugel und im Raum*, 2nd edition. Springer, Berlin, 1972.

[Fej74] G. Fejes Tóth. Solid sets of circles. *Stud. Sci. Math. Hungar.*, 9:101–109, 1974

[Fej76] G. Fejes Tóth. Multiple packing and covering of the plane with circles. *Acta Math. Acad. Sci. Hungar.*, 27:135–140, 1976.

[Fej79] G. Fejes Tóth. Multiple packing and covering of spheres. *Acta Math. Acad. Sci. Hungar.*, 34:165–176, 1979.

[Fej83] G. Fejes Tóth. New results in the theory of packing and covering. In P.M. Gruber and J.M. Wills, editors, *Convexity and Its Applications*, pages 318–359, Birkhäuser, Basel, 1983.

[Fej84a] G. Fejes Tóth. Multiple lattice packings of symmetric convex domains in the plane. *J. London Math. Soc.*, (2)29:556–561, 1984.

[Fej84b] L. Fejes Tóth. Density bounds for packing and covering with convex discs. *Expo. Math.*, 2:131–153, 1984.

[Fej95] G. Fejes Tóth. Densest packings of typical convex sets are not lattice-like. *Discrete Comput. Geom.*, 14:1–8, 1995.

[Fej99] G. Fejes Tóth. Recent progress on packing and covering. In B. Chazelle, J.E. Goodman, and R. Pollack, editors, *Advances in Discrete and Computational Geometry*, pages 145–162, AMS, Providence, 1999.

[Fej05a] G. Fejes Tóth. Covering with fat convex discs. *Discrete Comput. Geom.*, 37:129–141, 2005.

[Fej05b] G. Fejes Tóth. Covering a circle by eight, nine, or ten congruent circles. In J.E. Goodman, J. Pach, and E. Welzl, editors, *Combinatorial and Computational Geometry*, vol. 52 of *MSRI Publ.*, pages 359–374, Cambridge University Press, 2005.

[Few64] L. Few. Multiple packing of spheres. *J. London Math. Soc.*, 39:51–54, 1964.

[Few67] L. Few. Double covering with spheres. *Mathematika*, 14:207–214, 1967.

[Few68] L. Few. Double packing of spheres: A new upper bound. *Mathematika*, 15:88–92, 1968.

[FFV15] G. Fejes Tóth, F. Fodor and V. Vígh. The packing density of the n-dimensional cross-polytope. *Discrete Comput. Geom.*, 54:182–194, 2015.

[FH00] A. Florian and A. Heppes. Solid coverings of the Euclidean plane with incongruent circles. *Discrete Comput. Geom.*, 23:225–245, 2000.

[FK01] G. Fejes Tóth and W. Kuperberg. Sphere packing. In *Encyclopedia of Phiscal Sciences and Technology*, 3rd edition, vol. 15, pages 657–665, Academic Press, New York, 2001.

[FK69] L. Few and P. Kanagasabapathy. The double packing of spheres. *J. London Math. Soc.*, 44:141–146, 1969.

[FK93a] G. Fejes Tóth and W. Kuperberg. Blichfeldt's density bound revisited. *Math. Ann.*, 295:721–727, 1993.

[FK93b] G. Fejes Tóth and W. Kuperberg. Packing and covering with convex sets. In P.M. Gruber and J.M. Wills, editors, *Handbook of Convex Geometry*, pages 799–860, North-Holland, Amsterdam, 1993.

[FK93c] G. Fejes Tóth and W. Kuperberg. A survey of recent results in the theory of packing and covering. In J. Pach, editor, *New Trends in Discrete and Computational Geometry*, pages 251–279, Springer, New York, 1993.

[FK95] G. Fejes Tóth and W. Kuperberg. Thin non-lattice covering with an affine image of a strictly convex body. *Mathematika*, 42:239–250, 1995.

[FKK98] G. Fejes Tóth, G. Kuperberg, and W. Kuperberg. Highly saturated packings and reduced coverings. *Monatsh. Math.*, 125:127–145, 1998.

[Flo00] A. Florian. An infinite set of solid packings on the sphere. *Österreich. Akad. Wiss. Math.-Natur. Kl. Sitzungsber. Abt. II*, 209:67–79, 2000.

[Flo01] A. Florian. Packing of incongruent circles on a sphere. *Monatsh. Math.*, 133:111–129, 2001.

[Flo02] A. Florian. Some recent results in discrete geometry. *Rend. Circ. Mat. Palermo (2) Suppl. No. 70, part I*, 297–309, 2002.

[Flo87] A. Florian. Packing and covering with convex discs. In K. Böröczky and G. Fejes Tóth, editors, *Intuitive Geometry (Siófok, 1985)*, vol. 48 of *Colloq. Math. Soc. János Bolyai*, pages 191–207, North-Holland, Amsterdam, 1987.

[Fly14] Flyspeck. `http://code.google.com/p/flyspeck/wiki/AnnouncingCompletion`, 2014.

[Fod00] F. Fodor. The densest packing of 12 congruent circles in a circle. *Beitr. Algebra Geom.*, 41:401–409, 2000.

[Fod03a] F. Fodor. The densest packing of 13 congruent circles in a circle. *Beitr. Algebra Geom.*, 44:21–69, 2003.

[Fod03b] F. Fodor. The densest packing of 14 congruent circles in a circle. *Stud. Univ. Žilina Math. Ser.*, 16:25–34, 2003.

[Fod99] F. Fodor. The densest packing of 19 congruent circles in a circle. *Geom. Dedicata*, 74:139–145, 1999.

[FZ94] G. Fejes Tóth and T. Zamfirescu. For most convex discs thinnest covering is not lattice-like. In K. Böröczky and G. Fejes Tóth, editors, *Intuitive Geometry*, vol. 63 of *Colloq. Math. Soc. János Bolyai*, pages 105–108, North-Holland, Amsterdam, 1994.

[Gau] C.F. Gauss. Untersuchungen über die Eigenschaften der positiven ternären quadratischen Formen von Ludwig August Seeber, *Göttingische gelehrte Anzeigen*, July 9, 1831. Reprinted in: *Werke*, vol. 2, Königliche Gesellschaft der Wissenschaften, Göttingen, 1863, 188–196 and *J. Reine Angew. Math.* 20:312–320, 1840.

[GEK11] S. Gravel, V. Elser, and Y. Kallus. Upper bound on the packing density of regular tetrahedra and octahedra. *Discrete Comput. Geom.*, 46:799–818, 2011.

[GL87] P.M. Gruber and C.G. Lekkerkerker. *Geometry of Numbers.* Elsevier, North-Holland, Amsterdam, 1987.

[GMP94] C. de Groot, M. Monagan, R. Peikert, and D. Würtz. Packing circles in a square: Review and new results. In P. Kall, editor, *System Modeling and Optimization, Proceedings of the 15th IFIP Conference, Zürich, 1991*, vol. 180 of *Lecture Notes in Control and Information Services*, pages 45–54, 1994.

[Gol77] G. Golser. Dichteste Kugelpackungen im Oktaeder. *Studia Sci. Math. Hungar.*, 12:337–343, 1977.

[GPW90] C. de Groot, R. Peikert, and D. Würtz. The optimal packing of ten equal circles in a square. *IPS Research Report 90-12*, ETH Zürich, 1990.

[Gra63] R.L. Graham. Personal communication, 1963.

[Gri85] P. Gritzmann. Lattice covering of space with symmetric convex bodies. *Mathematika*, 32:311–315, 1985.

[Gro66] H. Groemer. Zusammenhängende Lagerungen konvexer Körper. *Math. Z.*, 94:66–78, 1966.

[Gro85] H. Groemer. Coverings and packings by sequences of convex sets. In J.E. Goodman, E. Lutwak, J. Malkevitch, and R. Pollack, editors, *Discrete Geometry and Convexity*, vol. 440 of *Annals of the New York Academy of Sciences*, pages 262–278, 1985.

[Gru07] P.M. Gruber. *Convex and Discrete Geometry.* Springer, Berlin, 2007.

[Gru79] P.M. Gruber. Geometry of numbers. In J. Tölke and J.M. Wills, editors, *Contributions to Geometry*, Proc. Geom. Symp. (Siegen, 1978), pages 186–225, Birkhäuser, Basel, 1979.

[Gru86] P.M. Gruber. Typical convex bodies have surprisingly few neighbours in densest lattice packings. *Studia Sci. Math. Hungar.*, 21:163–173, 1986.

[GW93] P. Gritzmann and J.M. Wills. Finite packing and covering. In P.M. Gruber and J.M. Wills, editors, *Handbook of Convex Geometry*, pages 861–897, North-Holland, Amsterdam, 1993.

[Had46] H. Hadwiger. Mitteilung betreffend meine Note: Überdeckung einer Menge durch Mengen kleineren Durchmessers. *Comment. Math. Helv.*, 19:72–73, 1946.

[Hal05] T.C. Hales. A proof of the Kepler conjecture. *Ann. of Math.*, 162:1065–1185, 2005.

[Hal06a] T.C. Hales. Historical overview of the Kepler conjecture. *Discrete Comput. Geom.*, 36:5–20, 2006.

[Hal06b] T.C. Hales. Sphere packings, III. Extremal cases. *Discrete Comput. Geom.*, 36:71–110, 2006.

[Hal06c] T.C. Hales. Sphere packings, IV. Detailed bounds. *Discrete Comput. Geom.*, 36:111–166, 2006.

[Hal06d] T.C. Hales. Sphere packings, VI. Tame graphs and linear programs. *Discrete Comput. Geom.*, 36:205–265, 2006.

[Hal12] T.C. Hales. *Dense sphere packings: A blueprint for formal proofs.* vol. 400 of *London Math. Soc. Lecture Note Series*, Cambridge Univ. Press, Cambridge, 2012.

[Hal13] T.C. Hales. The strong dodecahedral conjecture and Fejes Tóth's conjecture on sphere packings with kissing number twelve. In *Discrete geometry and optimization*, vol. 69 of *Fields Inst. Commun.*, pages 121–132, Springer, New York, 2013.

[Hal92] T.C. Hales. The sphere packing problem. *J. Comput. Appl. Math.*, 44:41–76, 1992.

[Hal93] T.C. Hales. Remarks on the density of sphere packings in three dimensions. *Combinatorica*, 13:181–187, 1993.

[Hal97] T.C. Hales. Sphere packings I. *Discrete Comput. Geom.*, 17:1–51, 1997.

[Hal98] T.C. Hales. Sphere packings II. *Discrete Comput. Geom.*, 18:135–149, 1998.

[Har86] L. Hárs. The Tammes problem for $n = 10$. *Studia Sci. Math. Hungar.*, 21:439–451, 1986.

[Hep03] A. Heppes. Covering the plane with fat ellipses without non-crossing assumption. *Disicrete Comput. Geom.*, 29:477–481, 2003.

[Hep59] A. Heppes. Mehrfache gitterförmige Kreislagerungen in der Ebene. *Acta Math. Acad. Sci. Hungar.*, 10:141–148, 1959.

[Hep92] A. Heppes. Solid circle-packings in the Euclidean plane. *Discrete Comput. Geom.*, 7:29–43, 1992.

[HF06] T.C. Hales and S.P. Ferguson. A formulation of the Kepler conjecture. *Discrete Comput. Geom.*, 36:21–69, 2006.

[HF11] T.C. Hales and S.P. Ferguson. *The Kepler conjecture. The Hales-Ferguson proof. Including papers reprinted from Discrete Comput. Geom. 36 (2006).* Edited by J.C. Lagarias, Springer, New York, 2011.

[HHM10] T.C. Hales, J. Harrison, S. McLaughlin, T. Nipkow, S. Obua, and R. Zumkeller. A revision of the proof of the Kepler conjecture. *Discrete Comput. Geom.*, 44:1–34, 2010.

[Hil00] D. Hilbert. Mathematical problems. *Bull. AMS*, 37:407–436, 2000. Reprinted from *Bull. AMS* 8:437–479, 1902.

[Hla43] E. Hlawka. Zur Geometrie der Zahlen. *Math. Z.*, 49:285–312, 1943.

[HM97] A. Heppes and J.B.M. Melissen. Covering a rectangle with equal circles. *Period. Math. Hungar.*, 34:63–79, 1997.

[Hoy70] D.J. Hoylman. The densest lattice packing of tetrahedra. *Bull. AMS*, 76:135–137, 1970.

[Hsi01] W.-Y. Hsiang. *Least action principle of crystal formation of dense packing type and Kepler's conjecture.* Vol. 3 of *Nankai Tracts in Mathematics*, World Scientific, Singapore, 2001.

[Hsi93] W.-Y. Hsiang. On the sphere packing problem and the proof of Kepler's conjecture. *Internat. J. Math.*, 93:739–831, 1993.

[HSS12] R.J. Hardin, N.J.A. Sloane, and W.D. Smith. *Tables of Spherical Codes with Icosahedral Symmetry.* Published electronically at `http://NeilSloane.com/icosahedral.codes/`, 2012.

[HSU02] H. Harborth, L. Szabó, and Z. Ujváry-Menyhárt. Regular sphere packings. *Arch. Math.*, 78:81–89, 2002.

[Ism98] D. Ismailescu. Covering the plane with copies of a convex disc. *Discrete Comput. Geom.*, 20:251–263, 1998.

[Jan10] J. Januszewski. Covering the plane with translates of a triangle. *Discrete Comput. Geom.*, 43:167–178, 2010.

[JB14] T. Jenrich and A.E. Brouwer. A 64-dimensional counterexample to Borsuk's conjecture. *Electron. J. Combin.*, 21:#4, 2014.

[Joo08] A. Joós. Covering the unit cube by equal balls. *Beiträge Algebra Geom.*, 49:599–605, 2008.

[Joo09a] A. Joós. On the packing of fourteen congruent spheres in a cube. *Geom. Dedicata*, 140:49–80, 2009.

[Joo09b] A. Joós. Erratum to: A. Joós: Covering the unit cube by equal balls. *Beiträge Algebra Geom.*, 50:603–605, 2009.

[Joo14a] A. Joós. Covering the k-skeleton of the 3-dimensional unit cube by five balls. *Beiträge Algebra Geom.*, 55:393–414, 2014.

[Joo14b] A. Joós. Covering the k-skeleton of the 3-dimensional unit cube by six balls. *Discrete Math.*, 336:85–95, 2014.

[Kep87] J. Kepler. Vom sechseckigen Schnee. Strena seu de Nive sexangula. Translated from Latin and with an introduction and notes by Dorothea Goetz. *Ostwalds Klassiker der Exakten Wissenschaften*, 273. Akademische Verlagsgesellschaft Geest & Portig K.-G., Leipzig, 1987.

[Ker39] R.B. Kershner. The number of circles covering a set. *Amer. J. Math.*, 61:665–671, 1939.

[Ker94] G. Kertész. Nine points on the hemisphere. In K. Böröczky and G. Fejes Tóth, editors, *Intuitive Geometry*, vol. 63 of *Colloq. Math. Soc. János Bolyai*, pages 189–196, North-Holland, Amsterdam, 1994.

[KK90] G. Kuperberg and W. Kuperberg. Double-lattice packings of convex bodies in the plane. *Discrete Comput. Geom.*, 5:389–397, 1990.

[KK93] J. Kahn and G. Kalai. A counterexample to Borsuk's conjecture. *Bull. AMS*, 29:60–62, 1993.

[KL78] G.A. Kabatiansky and V.I. Levenshtein. Bounds for packings on the sphere and in space. (Russian) *Problemy Peredači Informacii*, 14:3–25, 1978. English translation in *Probl. Inf. Transm.*, 14:1–17, 1978.

[KL88] R. Kannan and L. Lovász. Covering minima and lattice-point-free convex bodies. *Ann. of Math.*, 128:577–602, 1988.

[KLL95] M.S. Klamkin, T. Lewis and A. Liu. The kissing number of the square. *Math. Mag.*, 68:128–133, 1995.

[Kol88] D. Kołodziejczyk. Some remarks on the Borsuk conjecture. *Comment. Math. Prace Mat.*, 28:77–86, 1988.

[KS94] G. Kuperberg and O. Schramm. Average kissing numbers for non-congruent sphere packings. *Math. Res. Lett.*, 1:339–344, 1994.

[Kup00] G. Kuperberg. Notions of densness. *Geometry & Topology*, 4:277–292, 2000.

[KW87] K. Kirchner and G. Wengerodt. Die dichteste Packung von 36 Kreisen in einem Quadrat. *Beitr. Algebra Geom.*, 25:147–159, 1987.

[KZ72] A. Korkine and G. Zolotareff. Sur les formes quadratiques positives quaternaires. *Math. Ann.*, 5:581–583, 1872.

[KZ77] A. Korkine and G. Zolotareff. Sur les formes quadratiques positives. *Math. Ann.*, 11:242–292, 1877.

[Lag02] J.C. Lagarias. Bounds for local density of sphere packings and the Kepler conjecture. *Discrete Comput. Geom.*, 27:165–193, 2002.

[Lag73] J.C. Lagrange. Recherches d'arithmétique. *Nouv. Mém. Acad. Roy. Belles Letters Berlin*, 265–312, 1773. Vol. 3 of *Œuvres complète*, pages 693–758, Gauthier-Villars, Paris, 1869,

[Lan09] Zs. Lángi. On the Hadwiger numbers of centrally symmetric starlike disks. *Beitr. Algebra Geom.*, 50:249–257, 2009.

[Lan11] Zs. Lángi. On the Hadwiger numbers of starlike disks. *European J. Combin.*, 32:1203–1212, 2011.

[Las97] M. Lassak. A survey of algorithms for on-line packing and covering by sequences of convex bodies. In I. Bárány and K. Böröczky, editors, *Intuitive Geometry*, vol. 6 of *Bolyai Soc. Math. Studies*, pages 129–157. János Bolyai Math. Soc., Budapest, 1997.

[Lev79] V.I. Levenšteĭn On bounds for packings in n-dimensional Euclidean space. (Russian) *Dokl. Akad. Nauk SSSR*, 245:1299–1303, 1979. English translation in *Soviet Math. Dokl.*, 20:417–421, 1979.

[LZ02] J.C. Lagarias and C. Zong. Mysteries in packing regular tetrahedra. *Notices AMS*, 58:1540–1549, 2012.

[LZ99] D.G. Larman and C. Zong. On the kissing number of some special convex bodies. *Discrete Comput. Geom.*, 21:233–242, 1999.

[Mak78] E. Makai, Jr. On the thinnest nonseparable lattice of convex bodies. *Studia Sci. Math. Hungar.*, 13:19–27, 1978

[Mak85] E. Makai, Jr. Five-neighbour packing of convex plates. *Intuitive geometry (Siófok, 1985)* 373–381, Vol. 48 of *Colloq. Math. Soc. János Bolyai*, North-Holland, Amsterdam, 1987.

[Mar04] M.C. Markót. Optimal packing of 28 equal circles in a unit square—the first reliable solution. *Numer. Algorithms*, 37:253–261, 2004.

[Mar07] M.Cs. Markót. Interval methods for verifying structural optimality of circle packing configurations in the unit square. *J. Comput. Appl. Math.*, 199:353–357, 2007.

[Mar11] C. Marchal. Study of the Kepler's conjecture: the problem of the closest packing. *Math. Z.*, 267:737–765, 2011.

[Mel67] Z.A. Melzak. A note on the Borsuk conjecture. *Canad. Math. Bull.*, 10:1–3, 1967.

[Mel93] J.B.M. Melissen. Densest packings of congruent circles in an equilateral triangle. *Amer. Math. Monthly*, 100:816–825, 1993.

[Mel94a] J.B.M. Melissen. Densest packings of eleven congruent circles in a circle. *Geom. Dedicata*, 50:15–25, 1994.

[Mel94b] J.B.M. Melissen. Optimal packings of eleven equal circles in an equilateral triangle. *Acta Math. Hungar.*, 65:389–393, 1994.

[Mel94c] J.B.M. Melissen. Densest packing of six equal circles in a square. *Elem. Math.*, 49:27–31, 1994.

[Mel97] J.B.M. Melissen. Loosest circle coverings of an equilateral triangle. *Math. Mag.*, 70:119–125, 1997.

[Min04] H. Minkowski. Dichteste gitterförmige Lagerung kongruenter Körper. *Nachr. Akad. Wiss. Göttingen Math.-Phys. Kl.*, 311–355, 1904. *Gesammelte Abhandlungen, Vol. II*, pages 3–42, Leipzig, 1911.

[MS90] D.M. Mount and R. Silverman. Packing and covering the plane with translates of a convex polygon. *J. Algorithms*, 11:564–580, 1990.

[MS99] H. Martini and V. Soltan. Combinatorial problems on the illumination of convex bodies. *Aequationes Math.*, 57:121–152, 1999.

[MT12] O.R. Musin and A.S. Tarasov. The strong thirteen spheres problem. *Discrete Comput. Geom.*, 48:128–141, 2012.

[MT15] O.R. Musin and A.S. Tarasov. The Tammes problem for $N = 14$. *Exp. Math.*, 24:460–468, 2015.

[Mud93] D.J. Muder. A new bound on the local density of sphere packings. *Discrete Comput. Geom.*, 10:351–375, 1993.

[Mus08] O.R. Musin. The kissing number in four dimensions. *Ann. of Math.*, 168:1–32, 2008.

[Nas16a] M. Naszódi. On some covering problems in geometry. *Proc. Amer. Math. Soc.*, 144:3555–3562, 2016.

[Nas16b] M. Naszódi A spiky ball. *Mathematika*, 62:630–636, 2016.

[NO97] K.J. Nurmela, P.R.J. Östergård. Packing up to 50 circles in a square. *Discrete Comput. Geom.*, 18:111–120, 1997.

[OL82] F. Österreicher and J. Linhart. Packungen kongruenter Stäbchen mit konstanter Nachbarnzahl. *Elem. Math.*, 37:5–16, 1982.

[Ole61] N. Oler. A finite packing problem. *Canad. Math. Bull.*, 4:153–155, 1961.

[OS79] A.M. Odlyzko and N.J.A. Sloane. New bounds on the number of unit spheres that can touch a unit sphere in n dimensions. *J. Combin. Theory Ser. A*, 26:210–214, 1979.

[OV] F.M. de Oliveira Filho and F. Vallentin. Computing upper bounds for packing densities of congruent copies of a convex body. Preprint, `arXiv:1308.4893`, 2013.

[PA95] J. Pach and P.K. Agarwal. *Combinatorial Geometry.* John Wiley, New York, 1995.

[Pac85] J. Pach. A note on plane covering. In *Diskrete Geometrie, 3. Kolloq.*, Salzburg, pages 207–216, 1985.

[Pac86] J. Pach. Covering the plane with convex polygons. *Discrete Comput. Geom.*, 1:73–81, 1986.

[Pál10] D. Pálvölgyi. Indecomposable coverings with concave polygons. *Discrete Comput. Geom.*, 44:577–588, 2010.

[Pál13] D. Pálvölgyi. Indecomposable coverings with unit discs. Preprint, `arXiv:1310.6900v1`, 2013.

[Pei94] R. Peikert. Dichteste Packung von gleichen Kreisen in einem Quadrat. *Elem. Math.*, 49:16–26, 1994.

[Pir69] U. Pirl. Der Mindestabstand von n in der Einheitskreisscheibe gelegenen Punkten. *Math. Nachr.*, 40:111–124, 1969.

[PPT13] J. Pach, D. Pálvölgyi, and G. Tóth. Survey on decomposition of multiple coverings. In I. Bárány, K.J. Böröczky, G. Fejes Tóth, and J. Pach, editors, *Geometry—Intuitive, Discrete, and Convex*, vol. 24 of *Bolyai Society Mathematical Studies*, pages 219–257, Springer, 2013.

[PT10] D. Pálvölgyi and G. Tóth. Convex polygons are cover-decomposable. *Discrete Comput. Geom.*, 43:483–496, 2010.

[Rai07] A.M. Raigorodskii. Around the Borsuk conjecture. (Russian) *Sovrem. Mat. Fundam. Napravl.*, 23:147–164, 2007. Translated in *J. Math. Sci. (N.Y.)*, 154:604–623, 2008.

[Rai08] A.M. Raigorodskii. Three lectures on the Borsuk partition problem. In N. Young and Y. Choi, editors, *Surveys in Contemporary Mathematics*, vol. 347 of *London Math. Soc. Lecture Note Ser.*, pages 202–247, Cambridge Univ. Press, Cambridge, 2008.

[Ran55] R.A. Rankin. The closest packing of spherical caps in n dimensions. *Proc. Glasgow Math. Assoc.*, 2:139–144, 1955.

[RB75] S.S. Ryškov and E.P. Baranovskiĭ. Solution of the problem of the least dense lattice covering of five-dimensional space by equal spheres. (Russian) *Dokl. Akad. Nauk SSSR*, 222:9–42, 1975.

[Rob61] R.M. Robinson. Arrangements of 24 points on a sphere. *Math. Ann.*, 144:17–48, 1961.

[Rog57] C.A. Rogers. A note on coverings. *Mathematika*, 4:1–6, 1957.

[Rog58] C.A. Rogers. The packing of equal spheres. *Proc. London Math. Soc.*, 8:609–620, 1958.

[Rog59] C.A. Rogers. Lattice coverings of space. *Mathematika*, 6:33–39, 1959.

[Rog63] C.A. Rogers. Covering a sphere with spheres. *Mathematika*, 10:157–164, 1963.

[Rog64] C.A. Rogers. *Packing and Covering.* Cambridge Univ. Press, Cambridge, 1964.

[Rog71] C.A. Rogers. Symmetrical sets of constant width and their partitions. *Mathematika*, 18:105–111, 1971.

[RS57] C.A. Rogers and G.C. Shephard. The difference body of a convex body. *Arch. Math.*, 8:220–233, 1957.

[Sch06] A. Schürmann. On packing spheres into containers. About Kepler's finite sphere packing problem. *Documenta Math.*, 11:393–406, 2006.

[Sch55] K. Schütte. Überdeckungen der Kugel mit höchstens acht Kreisen. *Math. Ann.*, 129:181–186, 1955.

[Sch61] W.M. Schmidt. Zur Lagerung kongruenter Körper im Raum. *Monatsh. Math.*, 65:54–158, 1961.

[Sch63a] W.M. Schmidt. On the Minkowski-Hlawka theorem. *Illinois J. Math.*, 7:18–23, 1963.

[Sch63b] W.M. Schmidt. Correction to my paper, "On the Minkowski-Hlawka theorem." *Illinois J. Math.*, 7:714, 1963.

[Sch66a] J. Schaer. On the densest packing of spheres in a cube. *Canad. Math. Bull.*, 9:265–270, 1966.

[Sch66b] J. Schaer. The densest packing of five spheres in a cube. *Canad. Math. Bull.*, 9:271–274, 1966.

[Sch66c] J. Schaer. The densest packing of six spheres in a cube. *Canad. Math. Bull.*, 9:275–280, 1966.

[Sch88a] P. Schmitt. An aperiodic prototile in space. Preprint, University of Vienna, 1988.

[Sch88b] O. Schramm. Illuminating sets of constant width. *Mathematika*, 35:180–189, 1988.

[Sch91] P. Schmitt. Disks with special properties of densest packings. *Discrete Comput. Geom.*, 6:181–190, 1991.

[Sch94] J. Schaer. The densest packing of ten congruent spheres in a cube. In K. Böröczky and G. Fejes Tóth, editors, *Intuitive Geometry*, volume 63 of *Colloq. Math. Soc. János Bolyai*, pages 403–424. North-Holland, Amsterdam, 1994.

[Sch95] J. Schaer. The densest packing of 9 circles in a square. *Canad. Math. Bull.*, 8:273–277, 1965.

[SM65] J. Schaer and A. Meir. On a geometric extremum problem. *Canad. Math. Bull.*, 8:21–27, 1965.

[Smi00] E.H. Smith. A bound on the ratio between the packing and covering densities of a convex body. *Discrete Comput. Geom.*, 23:325–331, 2000.

[Smi05] E.H. Smith. A new packing density bound in 3-space. *Discrete Comput. Geom.*, 34:537–544, 2005.

[Smi75] M.J. Smith. Packing translates of a compact set in Euclidean space. *Bull. London Math. Soc.*, 7:129–131, 1975.

[Sri14] K. Sriamorn. On the multiple packing densities of triangles. *Discrete Comput. Geom.*, 55:228–242, 2016.

[Sri16] K. Sriamorn. Multiple lattice packings and coverings of the plane with triangles. *Discrete Comput. Geom.*, 55:228–242, 2016.

[SX15] K. Sriamorn and F. Xue. On the covering densities of quarter-convex disks. *Discrete Comput. Geom.*, 54:246–258, 2015.

[SW51] K. Schütte and B.L. van der Waerden. Auf welcher Kugel haben 5, 6, 7, 8 oder 9 Punkte mit Mindestabstand Eins Platz? *Math. Ann.*, 123:96–124, 1951.

[SW53] K. Schütte and B.L. van der Waerden. Das Problem der dreizehn Kugeln. *Math. Ann.*, 125:25–334, 1953.

[SW15] K. Sriamorn and A. Wetayawanich. On the multiple covering densities of triangles. *Discrete Comput. Geom.*, 54:717–727, 2015.

[Tal00] I. Talata. A lower bound for the translative kissing numbers of simplices. *Combinatorica*, 20:281–293, 2000.

[Tal98a] I. Talata. Exponential lower bound for the translative kissing numbers of d-dimensional convex bodies. *Discrete Comput. Geom.*, 19:447–455, 1998.

[Tal98b] I. Talata. On a lemma of Minkowski. *Period. Math. Hungar.*, 32:199–207, 1998.

[Tal99a] I. Talata. The translative kissing number of tetrahedra is 18. *Discrete Comput. Geom.*, 22:231–248, 1999.

[Tam70] P. Tammela. An estimate of the critical determinant of a two-dimensional convex symmetric domain. (Russian) *Izv. Vysš. Učebn. Zaved. Matematika*, 103:103–107, 1970.

[Tem84] Á. Temesvári. Die dünnste gitterförmige 5-fache Kreisüberdeckung der Ebene. *Studia Sci. Math. Hungar.*, 19:285–298, 1984.

[Tem88] Á. Temesvári. Eine Methode zur Bestimmung der dünnsten gitterförmigen k-fachen Kreisüberdeckungen. *Studia Sci. Math. Hungar.*, 23:23–35, 1988.

[Tem92a] Á. Temesvári. Die dünnste gitterförmige 6-fache Kreisüberdeckung. *Berzsenyi Dániel Tanárk. Föisk. Tud. Közl.*, 8:93–112, 1992.

[Tem92b] Á. Temesvári. Die dünnste gitterförmige 7-fache Kreisüberdeckung. *Berzsenyi Dániel Tanárk. Föisk. Tud. Közl.*, 8:113–125, 1992.

[Tem94] Á. Temesvári. Die dichteste gitterförmige 9-fache Kreispackung. *Rad. Hrvatske Akad. Znan. Umj. Mat.*, 11:95–110, 1994.

[Thu10] A. Thue. Über die dichteste Zusammenstellung von kongruenten Kreisen in einer Ebene. Definitionen und Theoreme. *Christiania Vid. Selsk. Skr.*, 1:3–9, 1910.

[THY87] Á.H. Temesvári, J. Horváth, and N.N. Yakovlev. A method for finding the densest lattice k-fold packing of circles. (Russian) *Mat. Zametki*, 41:625–636, 1987. English translation *Mathematical Notes*, 41:349–355, 1987.

[TT07] G. Tardos and G. Tóth. Multiple coverings of the plane with triangles. *Discrete Comput. Geom.*, 38:443–450, 2007.

[Xu07] L. Xu. A note on the kissing numbers of superballs. *Discrete Comput. Geom.*, 37:485–491, 2007.

[Yak83] N.N. Yakovlev. The densest lattice 8-packing on a plane. (Russian) *Vestnik Moskov. Univ. Ser. I Mat. Mekh.*, 5:8–16, 1983.

[You39] J.W.T. Youngs. A lemma on squares. *Amer. Math. Monthly*, 46:20–22, 1939.

[Van11] S. Vance. Improved sphere packing lower bounds from Hurwitz lattices. *Adv. Math.*, 227:2144–2156, 2011.

[Ven13] A. Venkatesh. A note on sphere packings in high dimension. *Int. Math. Res. Not.*, 2013:1628–1642, 2013.

[Ver05] J.-L. Verger-Gaugry. Covering a ball with smaller equal balls in $\mathbb{R}^n$. *Discrete Comput. Geom.*, 33:143–155, 2005.

[Via17] M. Viazovska. The sphere packing problem in dimension 8. *Ann. of Math.*, 185:991–1015, 2017.

[Weg92] G. Wegner. Relative Newton numbers. *Monatsh. Math.*, 114:149–160, 1992.

[Wen83] G. Wengerodt. Die dichteste Packung von 16 Kreisen in einem Quadrat. *Beitr. Algebra Geom.*, 16:173–190, 1983.

[Wen87a] G. Wengerodt. Die dichteste Packung von 14 Kreisen in einem Quadrat. *Beitr. Algebra Geom.*, 25:25–46, 1987.

[Wen87b] G. Wengerodt. Die dichteste Packung von 25 Kreisen in einem Quadrat. *Ann. Univ. Sci. Budapest. Eötvös Sect. Math.*, 30:3–15, 1987.

[Whi48] J.V. Whitworth. On the densest packing of sections of a cube. *Ann. Mat. Pura Appl.*, 27:29–37, 1948.

[Whi51] J.V. Whitworth. The critical lattices of the double cone. *Proc. London Math. Soc.* 53:422–443, 1951.

[Zha98] L. Zhao. The kissing number of the regular polygon. *Discrete Math.*, 188:293–296, 1998.

[Zon99] C. Zong. *Sphere Packings.* Springer, New York, 1999.

[Zon14] C. Zong. On the translative packing densities of tetrahedra and cuboctahedra. *Adv. Math.*, 260:130–190, 2014.

[ZX02] L. Zhao and J. Xu. The kissing number of the regular pentagon. *Discrete Math.*, 252:293–298, 2002.

3 TILINGS

Edmund Harriss, Doris Schattschneider, and Marjorie Senechal

INTRODUCTION

Tilings of surfaces and packings of space have interested artisans and manufacturers throughout history; they are a means of artistic expression and lend economy and strength to modular constructions. Today scientists and mathematicians study tilings because they pose interesting mathematical questions and provide mathematical models for such diverse fields as the molecular anatomy of crystals, cell packings of viruses, n-dimensional algebraic codes, "nearest neighbor" regions for a set of discrete points, meshes for computational geometry, CW-complexes in topology, the self-assembly of nano-structures, and the study of aperiodic order.

The world of tilings is too vast to discuss in a chapter, or even in a gargantuan book. Even such basic questions as: *What bodies can tile space? In what ways do they tile?* are intractable unless the tiles and tilings are subject to constraints, and even then the subject is unmanageably large.

In this chapter, due to space limitations, we restrict ourselves, for the most part, to tilings of unbounded spaces. Be aware that this is a severe restriction. Tilings of the sphere and torus, for example, are also subtle and important.

In Section 3.1 we present some general results that are fundamental to the subject as a whole. Section 3.2 addresses tilings with congruent tiles. In Section 3.3 we discuss the classical subject of periodic tilings, which continues to be an active field of research. Section 3.4 concerns nonperiodic and aperiodic tilings. We conclude with a very brief description of some kinds of tilings not considered here.

3.1 GENERAL CONSIDERATIONS

In this section we define terms that will be used throughout the chapter and state some basic results. Taken together, these results state that although there is no algorithm for deciding which bodies can tile, there are criteria for deciding the question in certain cases. We can obtain some quantitative information about their tilings in particularly well-behaved cases.

Like many ideas that seem simple in life, it is complicated to give a precise mathematical definition of the notion of tiling. In particular this relates to the hard problem of defining the general notion of shape. Beyond the simple examples such as squares and circles there are shapes with holes, or fractal boundary, the latter allowing a shape of finite volume to have an infinite number of holes. As many of these strange examples turn up as examples of tilings, the definition needs to include them. We therefore define a *body* to be a compact subset of a manifold $S \subset \mathbb{E}^n$ that is the closure of its interior. A tiling is then a division of S into a countable (finite or infinite) number of bodies.

GLOSSARY

Body: A compact subset (of a manifold $S \subset \mathbb{E}^n$) that is the closure of its (nonempty) interior.

Tiling (of S): A decomposition of S into a countable number of n-dimensional bodies whose interiors are pairwise disjoint. In this context, the bodies are also called n-cells and are the tiles of the tiling (see below). Synonyms: ***tessellation***, ***parquetry*** (when $n = 2$), ***honeycomb*** (for $n \geq 2$).

Tile: A body that is an n-cell of one or more tilings of S. To say that a body ***tiles*** a region $R \subseteq S$ means that R can be covered exactly by congruent copies of the body without gaps or overlaps.

Locally finite tiling: Every n-ball of finite radius in S meets only finitely many tiles of the tiling.

Prototile set (for a tiling $\mathcal{T}$ of S): A minimal subset of tiles in $\mathcal{T}$ such that each tile in the tiling $\mathcal{T}$ is the congruent copy of one of those in the prototile set. The tiles in the set are called prototiles and the prototile set is said to admit $\mathcal{T}$.

k-face (of a tiling): An intersection of at least $n - k + 1$ tiles of the tiling that is not contained in a j-face for $j < k$. (The 0-faces are the ***vertices*** and 1-faces the ***edges***; the $(n-1)$-faces are simply called the ***faces*** of the tiling.)

Cluster and Patch (in a tiling $\mathcal{T}$): The set of bodies in a tiling $\mathcal{T}$ that intersect a compact subset of $K \subset S$ is a *cluster*. The set is a *patch* if K can be chosen to be convex. See Figure 3.1.1.

FIGURE 3.1.1
Three clusters in a tiling of the plane by "chairs." Two of these clusters are patches.

Normal tiling: A tiling in which (i) each prototile is homeomorphic to an n-ball, and (ii) the prototiles are uniformly bounded (there exist $r > 0$ and $R > 0$ such that each prototile contains a ball of radius r and is contained in a ball of radius R). It is technically convenient to include a third condition: (iii) the intersection of every pair of tiles is a connected set. (A normal tiling is necessarily locally finite.)

Face-to-face tiling (by polytopes): A tiling in which the faces of the tiling are also the $(n-1)$-dimensional faces of the polytopes. (A face-to-face tiling by convex polytopes is also k-face-to-k-face for $0 \leq k \leq n-1$.) In dimension 2, this is an ***edge-to-edge*** tiling by polygons, and in dimension 3, a face-to-face tiling by polyhedra.

Dual tiling: Two tilings $\mathcal{T}$ and $\mathcal{T}^*$ are dual if there is an incidence-reversing bijection between the k-faces of $\mathcal{T}$ and the $(n-k)$-faces of $\mathcal{T}^*$ (see Figure 3.1.2).

Voronoi (Dirichlet) tiling: A tiling whose tiles are the Voronoi cells of a discrete set Λ of points in S. The Voronoi cell of a point $p \in \Lambda$ is the set of all points in S that are at least as close to p as to any other point in Λ (see Chapter 27).

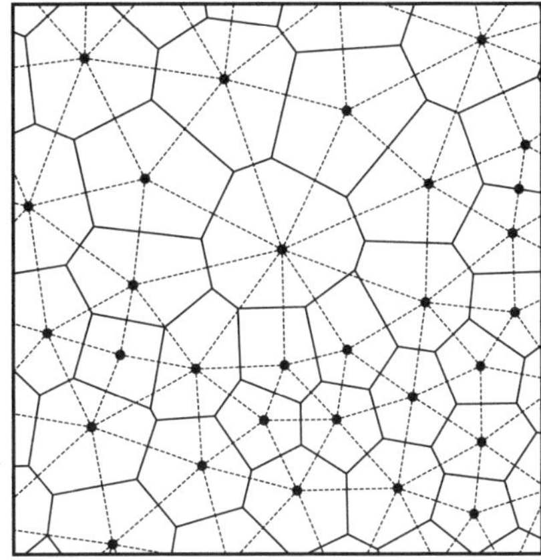

FIGURE 3.1.2
A Voronoi tiling (solid lines) and its Delaunay dual (dashed lines).

Delaunay (or Delone) tiling: A face-to-face tiling by convex circumscribable polytopes (i.e., the vertices of each polytope lie on a sphere).

Isometry: A distance-preserving self-map of S.

Symmetry group (of a tiling): The set of isometries of S that map the tiling to itself.

MAIN RESULTS

1. **The Undecidability Theorem.** There is no algorithm for deciding whether or not an arbitrary set of bodies admits a tiling of $\mathbb{E}^2$ [Ber66]. This was initially proved by Berger in the context of Wang tiles, squares with colored edges (the colors can be replaced by shaped edges to give a result for tiles). Berger's result showed that any attempted algorithm would not work (would run forever giving no answer) for some set of Wang tiles. This set can be arbitrarily large. The result was improved by Ollinger [Oll09] to show that any attempted algorithm could be broken by a set of at most 5 polyominoes.
2. **The Extension Theorem (for $\mathbb{E}^n$).** Let A be any finite set of bodies, each homeomorphic to a closed n-ball. If A tiles regions that contain arbitrarily large n-balls, then A admits a tiling of $\mathbb{E}^n$. (These regions need not be nested, nor need any of the tilings of the regions be extendable!) The proof for $n = 2$ in [GS87] extends to $\mathbb{E}^n$ with minor changes.
3. **The Normality Lemma (for $\mathbb{E}^n$).** In a normal tiling, the ratio of the number of tiles that meet the boundary of a spherical patch to the number of tiles in the patch tends to zero as the radius of the patch tends to infinity. In fact, a stronger statement can be made: For $s \in S$ let $t(r, s)$ be the number of tiles in

the spherical patch $P(r,s)$. Then, in a normal tiling, for every $x > 0$,

$$\lim_{r\to\infty} \frac{t(r+x,s) - t(r,s)}{t(r,s)} = 0.$$

The proof for $n = 2$ in [GS87] extends to $\mathbb{E}^n$ with minor changes.

4. **Euler's Theorem for tilings of $\mathbb{E}^2$.** Let $\mathcal{T}$ be a normal tiling of $\mathbb{E}^2$, and let $t(r,s)$, $e(r,s)$, and $v(r,s)$ be the numbers of tiles, edges, and vertices, respectively, in the circular patch $P(r,s)$. Then if one of the limits $e(\mathcal{T}) = \lim_{r\to\infty} e(r,s)/t(r,s)$ or $v(\mathcal{T}) = \lim_{r\to\infty} v(r,s)/t(r,s)$ exists, so does the other, and $v(\mathcal{T}) - e(\mathcal{T}) + 1 = 0$. Like Euler's Theorem for Planar Maps, on which the proof of this theorem is based, this result can be extended in various ways [GS87].
5. **Voronoi and Delaunay Duals.** Every Voronoi tiling has a Delaunay dual and conversely (see Figure 3.1.2) [Vor09].

OPEN PROBLEM

1. Is there an algorithm to decide whether any set of at most four bodies admits a tiling in $\mathbb{E}^n$? The question can also be asked for other spaces. The tiling problem for the hyperbolic plane was shown to be undecidable independently by Kari and Morgenstern [Kar07, Mar08].

3.2 TILINGS BY ONE TILE

To say that a body tiles $\mathbb{E}^n$ usually means that there is a tiling all of whose tiles are congruent copies of this body. The artist M.C. Escher has demonstrated how intricate such tiles can be even when $n = 2$. But in higher dimensions the simplest tiles—for example, cubes—can produce surprises, as the counterexample to Keller's conjecture attests (see below).

GLOSSARY

Monohedral tiling: A tiling with a single prototile.

r-morphic tile: A prototile that admits exactly r distinct (non-superposable) monohedral tilings. Figure 3.2.1 shows a 3-morphic tile and its three tilings, and Figure 3.2.2 shows a 1-morphic tile and its tiling.

k-rep tile: A body for which k copies can be assembled into a larger, similar body. (Or, equivalently, a body that can be partitioned into k congruent bodies, each similar to the original.) More formally, a k-rep tile is a compact set A_1 in S with nonempty interior such that there are sets $A_2, \ldots, A_k$ congruent to A_1 that satisfy

$$\text{Int}\, A_i \cap \text{Int}\, A_j = \emptyset$$

for all $i \neq j$ and $A_1 \cup .... \cup A_k = g(A_1)$, where g is a similarity mapping.

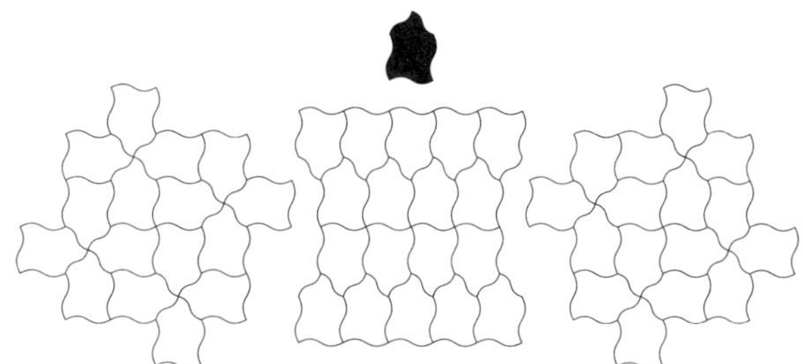

FIGURE 3.2.1
A trimorphic tile.

Transitive action: A group G is said to act transitively on a nonempty set $\{A_1, A_2, \ldots\}$ if the set is an orbit for G. (That is, for every pair A_i, A_j of elements of the set, there is a $g_{ij} \in G$ such that $g_{ij}A_i = A_j$.)

Isohedral (tiling): A tiling whose symmetry group acts transitively on its tiles.

Anisohedral tile: A prototile that admits monohedral tilings but no isohedral tilings. In Figure 3.2.2, the prototile admits a unique nonisohedral tiling; the black tiles are each surrounded differently, from which it follows that no isometry can map one to the other (while mapping the tiling to itself). This tiling is periodic, however; see Section 3.3.

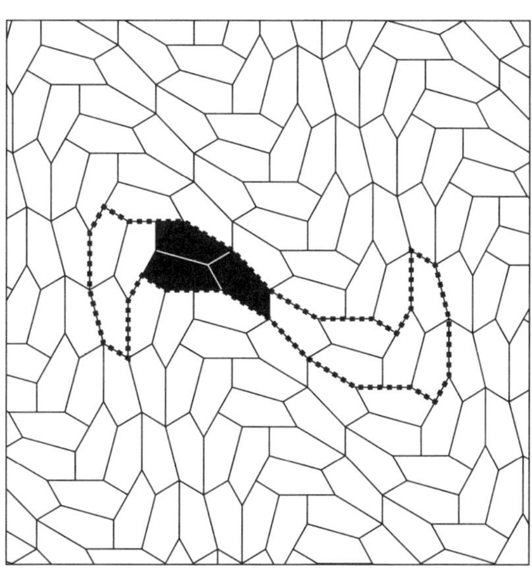

FIGURE 3.2.2
An anisohedral tile. This tile was the fifteenth type of pentagon found to tile the plane [MMVD17].

Corona (of a tile P in a tiling $\mathcal{T}$): Define $C^0(P) = P$. Then $C^k(P)$, the kth corona of P, is the set of all tiles $Q \in \mathcal{T}$ for which there exists a path of tiles $P = P_0, P_1, \ldots, P_m = Q$ with $m \leq k$ in which $P_i \cap P_{i+1} \neq \emptyset$, $i = 0, 1, \ldots, m-1$.

Lattice: The group of integral linear combinations of n linearly independent vectors in S. A point orbit of a lattice, often called a ***point lattice***, is a particular case of a ***regular system of points*** (see Chapter 64).

Translation tiling: A monohedral tiling of S in which every tile is a translate of a fixed prototile. See Figure 3.2.3.

Lattice tiling: A monohedral tiling on whose tiles a lattice of translation vectors acts transitively. Figure 3.2.3 is not a lattice tiling since it is invariant by multiples of just one vector.

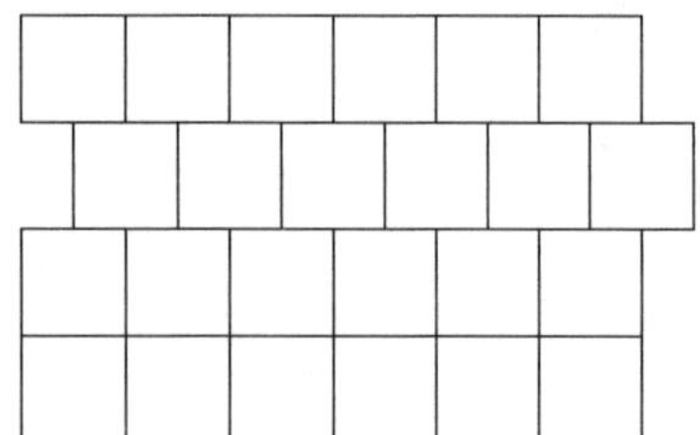

FIGURE 3.2.3
This translation non-lattice tiling is nonperiodic but not aperiodic.

n-parallelotope: A convex n-polytope that tiles $\mathbb{E}^n$ by translation.

Belt (of an n-parallelotope): A maximal subset of parallel $(n-2)$-faces of a parallelotope in $\mathbb{E}^n$. The number of $(n-2)$-faces in a belt is its length.

Center of symmetry (for a set A in $\mathbb{E}^n$***):*** A point $a \in A$ such that A is invariant under the mapping $x \mapsto 2a - x$; the mapping is called a ***central inversion*** and an object that has a center of symmetry is said to be ***centrosymmetric***.

Stereohedron: A convex polytope that is the prototile of an isohedral tiling. A Voronoi cell of a regular system of points is a stereohedron.

Linear expansive map: A linear transformation all of whose eigenvalues have modulus greater than one.

MAIN RESULTS

1. **The Local Theorem**. Let $\mathcal{T}$ be a monohedral tiling of S, and for $P \in \mathcal{T}$, let $S_i(P)$ be the subgroup of the symmetry group of P that leaves invariant $C^i(P)$, the ith corona of P. $\mathcal{T}$ is isohedral if and only if there exists an integer $k > 0$ for which the following two conditions hold: (a) for all $P \in \mathcal{T}$, $S_{k-1}(P) = S_k(P)$ and (b) for every pair of tiles P, P' in $\mathcal{T}$, there exists an isometry γ such that $\gamma(P) = P'$ and $\gamma(C^k(P)) = C^k(P')$. In particular, if P is asymmetric, then $\mathcal{T}$ is isohedral if and only if condition (b) holds for $k = 1$ [DS98].
2. A convex polytope is a parallelotope if and only if it is centrosymmetric, its faces are centrosymmetric, and its belts have lengths four or six. First proved by Venkov, this theorem was rediscovered independently by McMullen [Ven54, McM80]. There are two combinatorial types of parallelotopes in $\mathbb{E}^2$ and five in $\mathbb{E}^3$.
3. The number $|F|$ of faces of a convex parallelotope in $\mathbb{E}^n$ satisfies Minkowski's inequality, $2n \le |F| \le 2(2^n - 1)$. Both upper and lower bounds are realized in every dimension [Min97].
4. The number of faces of an n-dimensional stereohedron in $\mathbb{E}^n$ is bounded. In fact, if a is the number of translation classes of the stereohedron in an isohedral tiling, then the number of faces is at most the Delaunay bound $2^n(1 + a) - 2$ [Del61].
5. Anisohedral tiles exist in $\mathbb{E}^n$ for every $n \ge 2$ [GS80]. (The first example, given for $n = 3$ by Reinhardt [Rei28], was the solution to part of Hilbert's 18th problem.) H. Heesch gave the first example for $n = 2$ [Hee35] and R. Kershner the first convex examples [Ker68].
6. Every n-parallelotope admits a lattice tiling. However, for $n \ge 3$, there are nonconvex tiles that tile by translation but do not admit lattice tilings [SS94].

7. A lattice tiling of $\mathbb{E}^n$ by unit cubes must have a pair of cubes sharing a whole face [Min07, Haj42]. However, a famous conjecture of Keller, which stated that for every n, any tiling of $\mathbb{E}^n$ by congruent cubes must contain at least one pair of cubes sharing a whole face, is false: for $n \geq 8$, there are translation tilings by unit cubes in which no two cubes share a whole face [LS92].
8. Every linear expansive map that transforms the lattice $\mathbb{Z}^n$ of integer vectors into itself defines a family of k-rep tiles; these tiles, which usually have fractal boundaries, admit lattice tilings [Ban91].

OPEN PROBLEMS

1. Which convex n-polytopes in $\mathbb{E}^n$ are prototiles for monohedral tilings of $\mathbb{E}^n$? This is unsolved for all $n \geq 2$ (see [GS87, Sch78b, MMVD17] for the case $n = 2$). In July 2017, M. Rao announced that the list of 15 convex pentagon types that tile is complete; the proof had not been verified at the time of writing [Rao17]. The most recent addition to the list is shown in Figure 3.2.2. For higher dimensions, little is known; it is not even known which tetrahedra tile $\mathbb{E}^3$ [GS80, Sen81].
2. **Voronoi's conjecture:** Every convex parallelotope in $\mathbb{E}^n$ is affinely equivalent to the Voronoi cell of a lattice in $\mathbb{E}^n$. The conjecture has been proved for $n \leq 4$, for the case when the parallelotope is a zonotope, and for certain other special cases [Erd99, Mag15].

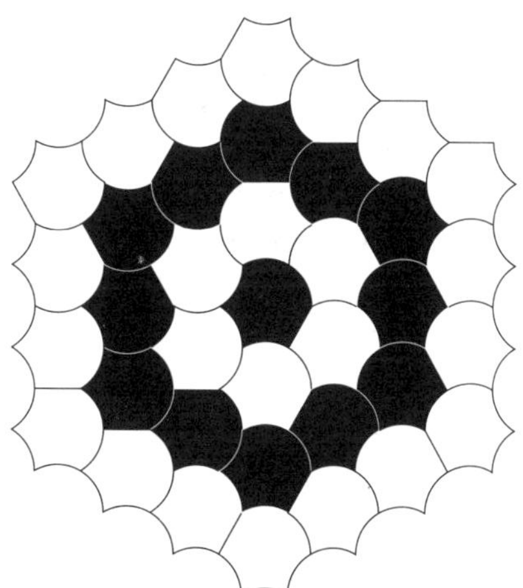

FIGURE 3.2.4
A 3-corona tile. (It cannot be surrounded by a fourth corona.) 4-corona and 5-corona tiles also exist.

3. **Heesch's Problem.** Is there an integer k_n, depending only on the dimension n of the space S, such that if a body A can be completely surrounded k_n times by tiles congruent to A, then A is a prototile for a monohedral tiling of S? (A is completely surrounded once if A, together with congruent copies that have nonempty intersection with A, tile a patch containing A in its interior.) When $S = \mathbb{E}^2$, $k_2 > 5$. The body shown in Figure 3.2.4 can be completely surrounded three times but not four. William Rex Marshall and, independently, Casey Mann, found 4-corona tiles, and Mann 5-corona tiles [Man04]. Michael DeWeese has found hexagons with generalized matching rules having Heesch numbers 1 through 9 and 11 [MT15]. This problem is unsolved for all $n \geq 2$.

4. **Keller's conjecture** is true for $n \leq 6$ and false for $n \geq 8$ (see Result 7 above). The case $n = 7$ is still open.
5. Find a good upper bound for the number of faces of an n-dimensional stereohedron. Delaunay's bound, stated above, is evidently much too high; for example, it gives 390 as the bound in $\mathbb{E}^3$, while the maximal known number of faces of a three-dimensional stereohedron (found by P. Engel [Eng81]) is 38.
6. For monohedral (face-to-face) tilings by convex polytopes there is an integer k_n, depending only on the dimension n of S, that is an upper bound for the constant k in the Local Theorem [DS98]. Find the value of this k_n. For the Euclidean plane $\mathbb{E}^2$ it is known that $k_2 = 1$ (convexity of the tiles is not necessary) [SD98], but for the hyperbolic plane, $k_2 \geq 2$ [Mak92]. For $\mathbb{E}^3$, it is known that $2 \leq k_3 \leq 5$.

3.3 PERIODIC TILINGS

Periodic tilings have been studied intensely, in part because their applications range from ornamental design to crystallography, and in part because many techniques (algebraic, geometric, and combinatorial) are available for studying them.

GLOSSARY

Periodic tiling of $\mathbb{E}^n$***:*** A tiling, not necessarily monohedral, whose symmetry group contains an n-dimensional lattice. This definition can be adapted to include "subperiodic" tilings (those whose symmetry groups contain $1 \leq k < n$ linearly independent vectors) and tilings of other spaces (for example, cylinders). Tilings in Figures 3.2.1, 3.2.2, 3.3.1, and 3.3.3 are periodic.

Fundamental domain (generating region) for a periodic tiling: In the general case of a group acting on a space, a fundamental domain is a subset of the space containing exactly one point from each orbit. It generally requires that the subset be connected and have some restrictions on its boundary. In the case of tilings a fundamental domain can be a (usually connected) minimal subset of the set of tiles that generates the whole tiling under the symmetry group. For example, a fundamental domain may be a tile (Figure 3.2.1), a subset of a single tile (Figure 3.3.1), or a subset of tiles (three shaded tiles in Figure 3.2.2).

Lattice unit (or translation unit) for a periodic tiling: A (usually connected) minimal region of the tiling that generates the whole tiling under the translation subgroup of the symmetry group. A lattice unit can be a single tile or contain several tiles. Figure 3.3.1 has a 3-tile lattice unit; Figure 3.2.2 has a 12-tile lattice unit (outlined).

Orbifold (of a tiling of S***):*** An orbifold is a generalization of a manifold to allow singularities. They are usually formed by folding up a space by a discrete symmetry group, and they are therefore a powerful tool to study periodic patterns. For the plane this is described in [CBGS08].

k-isohedral (tiling): A tiling whose tiles belong to k transitivity classes under the action of its symmetry group. Isohedral means 1-isohedral (Figures 3.3.1 and 3.3.3). The tiling in Figure 3.2.2 is 3-isohedral.

Equitransitive (tiling by polytopes): A tiling in which each combinatorial class of tiles forms a single transitivity class under the action of the symmetry group of the tiling.

k-isogonal (tiling): A tiling whose vertices belong to k transitivity classes under the action of its symmetry group. Isogonal means 1-isogonal.

k-uniform (tiling of a 2-dimensional surface): A k-isogonal tiling by regular polygons.

Uniform (tiling for n > 2): An isogonal tiling with congruent edges and uniform faces.

Flag of a tiling (of S): An ordered $(n+1)$-tuple $(X_0, X_1, ..., X_n)$, with X_n a tile and X_k a k-face for $0 \leq k \leq n-1$, in which $X_{i-1} \subset X_i$ for $i = 1, \ldots, n$.

Regular tiling (of S): A tiling $\mathcal{T}$ whose symmetry group is transitive on the flags of $\mathcal{T}$. (For $n > 2$, these are also called regular honeycombs.) See Figure 3.3.3.

k-colored tiling: A tiling in which each tile has a single color, and k different colors are used. Unlike the case of map colorings, in a colored tiling adjacent tiles may have the same color.

Perfectly k-colored tiling: A k-colored tiling for which each element of the symmetry group G of the uncolored tiling effects a permutation of the colors. The ordered pair (G, Π), where Π is the corresponding permutation group, is called a k-color symmetry group.

CLASSIFICATION OF PERIODIC TILINGS

There is a variety of notations for classifying the different "types" of tilings and tiles. Far from being merely names by which to distinguish types, these notations tell us the investigators' point of view and the questions they ask. Notation may tell us the global symmetries of the tiling, or how each tile is surrounded, or the topology of its orbifold. Notation makes possible the computer implementation of investigations of combinatorial questions about tilings.

Periodic tilings are classified by symmetry groups and, sometimes, by their skeletons (of vertices, edges, ..., $(n-1)$-faces). The groups are known as ***crystallographic groups***; up to isomorphism, there are 17 in $\mathbb{E}^2$, 219 in $\mathbb{E}^3$, and 4894 in $\mathbb{E}^4$. For $\mathbb{E}^2$ and $\mathbb{E}^3$, the most common notation for the groups has been that of the International Union of Crystallography (IUCr) [Hah83]. This is cross-referenced to earlier notations in [Sch78a]. Recently developed notations include Delaney-Dress symbols [Dre87] and orbifold notation for $n = 2$ [Con92, CH02] and for $n = 3$ [CDHT01, CBGS08].

GLOSSARY

International symbol (for periodic tilings of $\mathbb{E}^2$ and $\mathbb{E}^3$): Encodes lattice type and particular symmetries of the tiling. In Figure 3.3.1, the lattice unit diagram at the right encodes the symmetries of the tiling and the IUCr symbol p31m indicates that the highest-order rotation symmetry in the tiling is 3-fold, that there is no mirror normal to the edge of the lattice unit, and that there is a mirror at 60° to the edge of the lattice unit. These symbols are augmented to denote symmetry groups of perfectly 2-colored tilings.

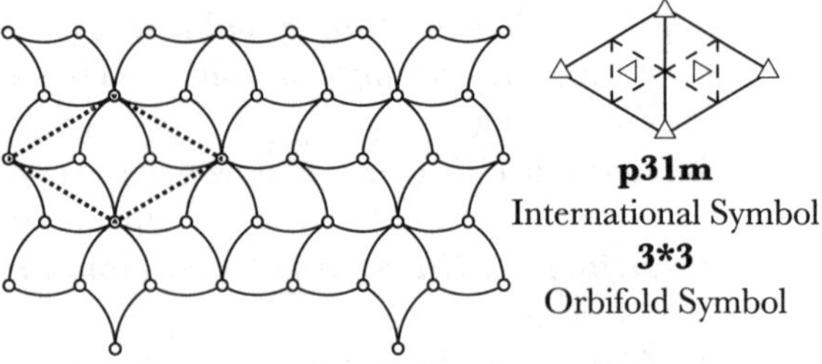

FIGURE 3.3.1
An isohedral tiling with standard IUCr lattice unit in dotted outline; a half-leaf is a fundamental domain. The classification symbols are for the symmetry group of the tiling.

Delaney-Dress symbol (for tilings of Euclidean, hyperbolic, or spherical space of any dimension): Associates an edge-colored and vertex-labeled graph derived from a ***chamber system*** (a formal barycentric subdivision) of the tiling. In Figure 3.3.2, the nodes of the graph represent distinct triangles A, B, C, D in the chamber system, and colored edges (dashed, thick, or thin) indicate their adjacency relations. Numbers on the nodes of the graph show the degree of the tile that contains that triangle and the degree of the vertex of the tiling that is also a vertex of that triangle.

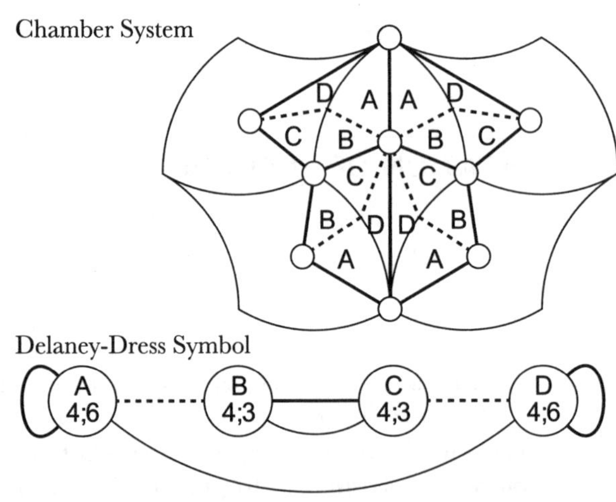

FIGURE 3.3.2
A chamber system of the tiling in Figure 3.3.1 determines the graph that is its Delaney-Dress symbol.

Orbifold notation (for symmetry groups of tilings of 2-dimensional surfaces of constant curvature): Encodes properties of the orbifold induced by the symmetry group of a periodic tiling of the Euclidean plane or hyperbolic plane, or a finite tiling of the surface of a sphere; introduced by Conway and Thurston. In Figure 3.3.1, the first 3 in the orbifold symbol 3*3 for the symmetry group of the tiling indicates there is a 3-fold rotation center (gyration point) that becomes a cone point in the orbifold, while *3 indicates that the boundary of the orbifold is a mirror with a corner where three mirrors intersect.

See Table 3.3.1 for the IUCr and orbifold notations for $\mathbb{E}^2$.

Isohedral tilings of $\mathbb{E}^2$ fall into 11 combinatorial classes, typified by the Laves nets (Figure 3.3.3). The Laves net for the tiling in Figure 3.3.1 is [3.6.3.6]; this gives the vertex degree sequence for each tile. In an isohedral tiling, every tile is surrounded in the same way. Grünbaum and Shephard provide an incidence

TABLE 3.3.1 IUCr and orbifold notations for the 17 symmetry groups of periodic tilings of $\mathbb{E}^2$.

IUCr	ORBIFOLD	IUCr	ORBIFOLD
p1	o or o1	p3	333
pg	×× or 1××	p31m	3*3
cm	*× or 1*×	p3m1	*333
pm	** or 1**	p4	442
p2	2222	p4g	4*2
pgg	22×	p4m	*442
pmg	22*	p6	632
cmm	2*22	p6m	*632
pmm	*2222		

FIGURE 3.3.3
The 11 Laves nets. The three regular tilings of $\mathbb{E}^2$ are at the top of the illustration.

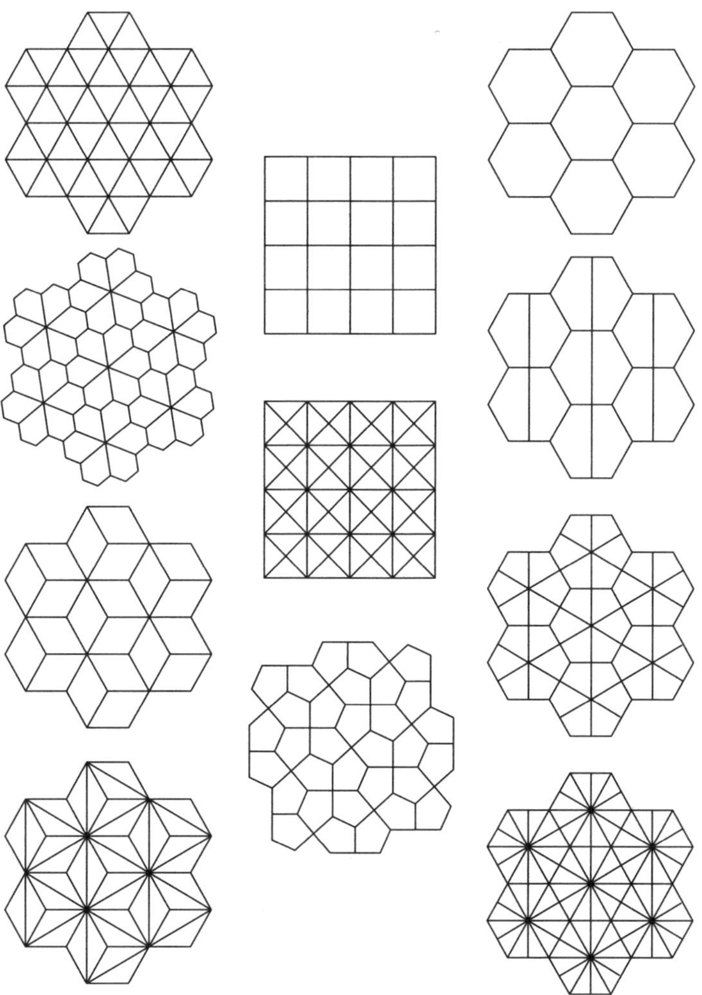

symbol for each isohedral type by labeling and orienting the edges of each tile [GS79]. Figure 3.3.4 gives the incidence symbol for the tiling in Figure 3.3.1. The tile symbol $a^+a^-b^+b^-$ records the cycle of edges of a tile and their orientations with respect to the (arrowed) first edge (+ indicates the same, − indicates opposite orientation). The adjacency symbol b^-a^- records for each different letter edge of a single tile, beginning with the first, the edge it abuts in the adjacent tile and their relative orientations (now − indicates same, + opposite).

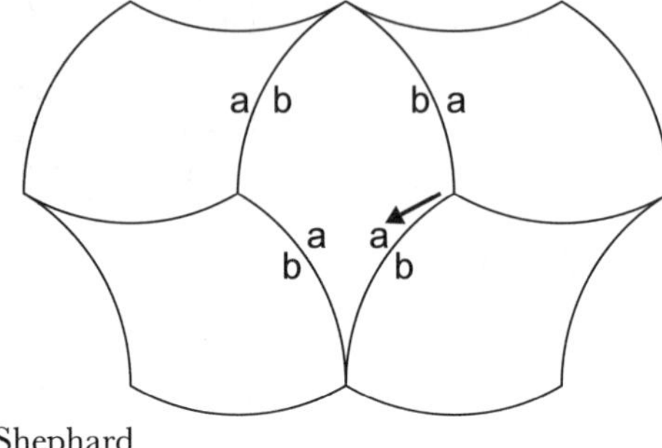

FIGURE 3.3.4
Labeling and orienting the edges of the isohedral tiling in Figure 3.3.1 determines its Grünbaum-Shephard incidence symbol.

These symbols can be augmented to adjacency symbols to denote k-color symmetry groups. Earlier, Heesch devised signatures for the 28 types of tiles that could be fundamental domains of isohedral tilings without reflection symmetry [HK63]; this signature system was extended in [BW94].

MAIN RESULTS

1. If a finite prototile set of polygons admits an edge-to-edge tiling of the plane that has translational symmetry, then the prototile set also admits a periodic tiling [GS87].
2. The number of symmetry groups of periodic tilings in $\mathbb{E}^n$ is finite (this is a famous theorem of Bieberbach [Bie10] that partially solved Hilbert's 18th problem: see also Chapter 64); the number of symmetry groups of corresponding tilings in hyperbolic n-space, for $n = 2$ and $n = 3$, is infinite.
3. Using their classification by incidence symbols, Grünbaum and Shephard proved there are 81 classes of isohedral tilings of $\mathbb{E}^2$, 93 classes if the tiles are ***marked*** (that is, they have decorative markings to break the symmetry of the tile shape) [GS77]. There is an infinite number of classes of isohedral tilings of $\mathbb{E}^n$, $n > 2$.
4. Every k-isohedral tiling of the Euclidean plane, hyperbolic plane, or sphere can be obtained from a $(k-1)$-isohedral tiling by a process of ***splitting*** (splitting an asymmetric prototile) and ***gluing*** (amalgamating two or more equivalent asymmetric tiles adjacent in the tiling into one new tile) [Hus93]; there are 1270 classes of normal 2-isohedral tilings and 48,231 classes of normal 3-isohedral tilings of $\mathbb{E}^2$.
5. Classifying isogonal tilings in a manner analogous to isohedral ones, Grünbaum and Shephard have shown [GS78] that there are 91 classes of normal isogonal tilings of $\mathbb{E}^2$ (93 classes if the tiles are marked).

6. For every k, the number of k-uniform tilings of $\mathbb{E}^2$ is finite. There are 11 uniform tilings of $\mathbb{E}^2$ (also called ***Archimedean***, or ***semiregular***), of which 3 are regular. The Laves nets in Figure 3.3.3 are duals of these 11 uniform tilings [GS87]. There are 28 uniform tilings of $\mathbb{E}^3$ [Grü94] and 20 2-uniform tilings of $\mathbb{E}^2$ [Krö69]; see also [GS87]. In the hyperbolic plane, uniform tilings with vertex valence 3 and 4 have been classified [GS79].
7. There are finitely many regular tilings of $\mathbb{E}^n$ (three for $n = 2$, one for $n = 3$, three for $n = 4$, and one for each $n > 4$) [Cox63]. There are infinitely many normal regular tilings of the hyperbolic plane, four of hyperbolic 3-space, five of hyperbolic 4-space, and none of hyperbolic n-space if $n > 4$ [Sch83, Cox54].
8. If two orbifold symbols for a tiling of the Euclidean or hyperbolic plane are the same except for the numerical values of their digits, which may differ by a permutation of the natural numbers, then the number of k-isohedral tilings for each of these orbifold types is the same [BH96].
9. There is a one-to-one correspondence between perfect k-colorings of a tiling whose symmetry group G acts on it freely and transitively, and the subgroups of index k of G. See [Sen79].

OPEN PROBLEMS

1. Conjecture: Every convex pentagon that tiles $\mathbb{E}^2$ admits a k-isohedral tiling for some $k \leq 3$. Michaël Rao announced a proof of this conjecture in July 2017 depending on an exhaustive computer search; the proof had not been verified at the time of writing [Rao17].
2. Enumerate the uniform tilings of $\mathbb{E}^n$ for $n > 3$.
3. Delaney-Dress symbols and orbifold notations have made progress possible on the classification of k-isohedral tilings in all three 2-dimensional spaces of constant curvature; extend this work to higher-dimensional spaces.

3.4 NONPERIODIC TILINGS

Nonperiodic tilings are found everywhere in nature, from cracked glazes to biological tissues to crystals. In a remarkable number of cases, such tilings exhibit strong regularities. For example, many nonperiodic tilings repeat on increasingly larger scales. An even larger class of tilings are those called repetitive, in which every bounded configuration appearing anywhere in the tiling is repeated infinitely many times throughout it.

Aperiodic prototype sets are particularly interesting. They were first introduced to prove the Undecidability Theorem (Section 3.1). Later, after Penrose found pairs of aperiodic prototiles (see Figure 3.4.1), they became popular in recreational mathematical circles.

The deep mathematical properties of Penrose tilings were first studied by Penrose, Conway, de Bruijn, and others. After the discovery of quasicrystals in 1984, aperiodic tilings became the focus of intense research. We present only the basic ideas of this rapidly developing subject here.

FIGURE 3.4.1
Patches of Penrose tilings of the plane. On the left the tiling by kites and darts, but note that these two unmarked tiles can tile the plane periodically. This is fixed on the right, where the matching rules that ensure nonperiodicity are enforced by the shapes of the edges. As a result the two shapes from the right-hand tiling constitute an aperiodic prototile set.

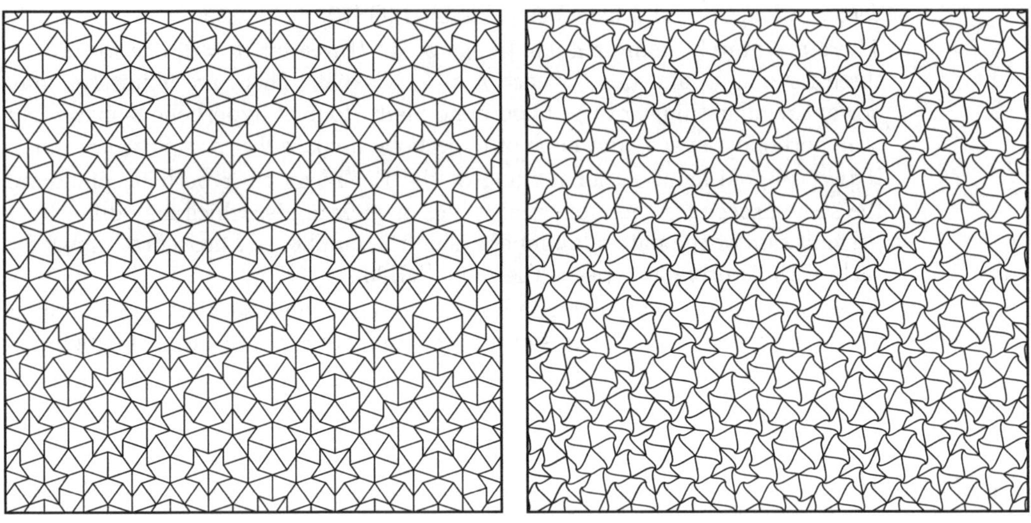

GLOSSARY

Nonperiodic tiling: A tiling with no translation symmetry.

Aperiodic tiling: A nonperiodic tiling that does not contain arbitrarily large periodic patches.

Aperiodic prototile set: A prototile set that admits only aperiodic tilings; see Figure 3.4.1.

Relatively dense configuration: A configuration C of tiles in a tiling for which there exists a radius r_C such that every ball of radius r_C in the tiling contains a copy of C.

Repetitive: A tiling in which every bounded configuration of tiles is relatively dense in the tiling.

Local isomorphism class: A family of tilings such that every bounded configuration of tiles that appears in any of them appears in all of the others. (For example, the uncountably many Penrose tilings with the same prototile set form a single local isomorphism class.)

Matching rules: A list of rules for fitting together the prototiles of a given prototile set.

Mutually locally derivable tilings: Two tilings are mutually locally derivable if the tiles in either tiling can, through a process of decomposition into smaller tiles, or regrouping with adjacent tiles, or a combination of both processes, form the tiles of the other (see Figure 3.4.2).

The first set of aperiodic prototiles numbered 20,426. After Penrose found his set of two, the question naturally arose, does there exist an "einstein," that is, a

FIGURE 3.4.2
The Penrose tilings by kites and darts and by rhombs are mutually locally derivable.

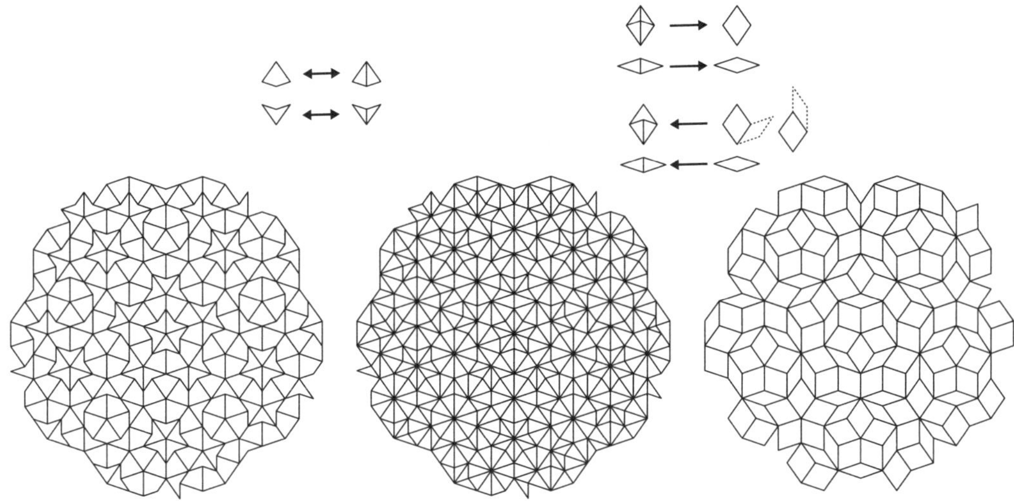

single tile that tiles only nonperiodically? Interesting progress was made by Socolar and Taylor [ST11], who found an einstein with several connected components. Rao's announcement of the classification of convex monohedral tiles (Section 3.2) implies that einsteins cannot be convex.

The next two sections discuss the best-known families of nonperiodic tilings.

3.4.1 SUBSTITUTION TILINGS

GLOSSARY

Substitution tiling: A tiling whose tiles can be composed into larger tiles, called ***level-one*** tiles, whose level-one tiles can be composed into level-two tiles, and so on *ad infinitum*. In some cases it is necessary to partition the original tiles before composition.

Self-similar tiling: A substitution tiling for which the larger tiles are copies of the prototiles (all enlarged by a constant expansion factor λ). k-rep tiles are the special case when there is just one prototile (Figure 3.1.1).

Unique composition property A substitution tiling has the unique composition property if j-level tiles can be composed into $(j+1)$-level tiles in only one way $(j=0,1,\ldots)$.

Inflation rule (for a substitution tiling): The equations $T_i' = m_{i1}T_1 \cup \ldots \cup m_{ik}T_k$, $i = 1, \ldots, k$, that describe the numbers m_{ij} of each prototile T_j in the next higher level prototile T_i'. These equations define a linear map whose matrix has i,j entry m_{ij}.

Pisot number: A Pisot number is a real algebraic integer greater than 1 such that all its Galois conjugates are less than 1 in absolute value. A Pisot number is called a unit if its inverse is also an algebraic integer.

MAIN RESULTS

1. Tilings with the unique composition property are nonperiodic (the proof given in [GS87] for $n = 2$ extends immediately to all n). Conversely, nonperiodic self-similar tilings have the unique composition property [Sol98].
2. Mutual local derivability is an equivalence relation on the set of all tilings. The existence or nonexistence of hierarchical structure and matching rules is a class property [KSB93].
3. The prototile set of every substitution tiling can be equipped with matching rules that force the hierarchical structure [GS98].

3.4.2 PROJECTED TILINGS

The essence of the "cut-and-project" method for constructing tilings is visible in the simplest example. Let L be the integer lattice in $\mathbb{E}^2$ and $\mathcal{T}$ the tiling of $\mathbb{E}^2$ by squares whose vertices are the points of L. Let $\mathcal{V}$ be the dual Voronoi tiling—in this example, the Voronoi tiles are squares centered at the points of L. Let E be any line in $\mathbb{E}^2$ of slope α, where $0 < \alpha < \pi/2$, and let F be a "face" of $\mathcal{V}$ of dimension k, $k \in \{0, 1, 2\}$. If $E \cap F \neq \emptyset$, project the corresponding $2-k$ face of $\mathcal{T}$. (Thus, if E cuts a face of $\mathcal{V}$ of dimension 2, project the point $x \in L$ at its center onto E; if E cuts the edge of $\mathcal{V}$ joining two faces with centers x, y, project the edge of $\mathcal{T}$ with endpoints x and y.) This gives a tiling of E by line segments which are projections of a "staircase" whose runs and rises are consecutive edges of $\mathcal{T}$. The tiling is nonperiodic if and only if α is irrational. Note that the construction does not guarantee that the tiling possesses substitution or matching rules, though for certain choices of L and α it may.

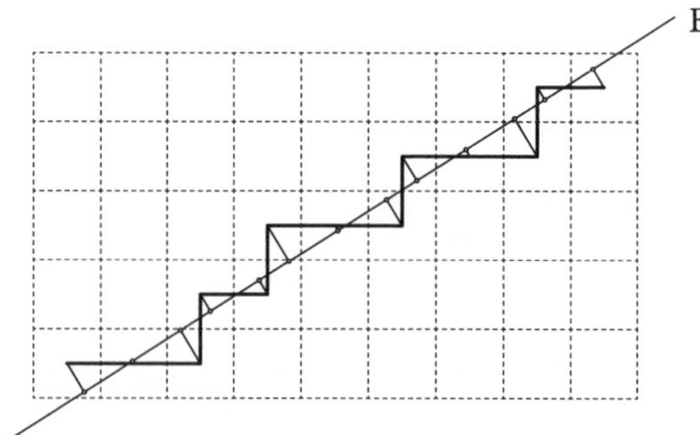

FIGURE 3.4.3
A lattice L with its Voronoi tiling. We connect successive lattice points whose cells are cut by the line E to create a "staircase." Projecting the staircase onto the line, we get a tiling by line segments of two lengths (the projections of the horizontal and vertical line segments, respectively).

GLOSSARY

Canonical projection method for tilings: Let L be a lattice in $\mathbb{E}^n$, $\mathcal{V}$ its Voronoi tiling, and $\mathcal{T}$ the dual Delaunay tiling. Let E be a translate of a subset of $\mathbb{E}^n$ of dimensions $m < n$, and let F be a "face" of $\mathcal{V}$ of dimension k, $k \in \{0, \ldots, n\}$. If $E \cap F \neq \emptyset$, project the corresponding $n - k$ face of $\mathcal{T}$ onto E. Thus, if E cuts a face of $\mathcal{V}$, project the point $x \in L$ at its center onto E, and so forth.

Cut-and-project method for tilings: A more general projection method of which the canonical is a special case (see below).

MAIN RESULTS

- A canonically projected tiling is nonperiodic if and only if $|E \cap L| \leq 1$.
- The canonical projection method is equivalent to the following: Let L be a lattice in $\mathbb{E}^n$ with Voronoi tiling $\mathcal{V}$, and let $\mathcal{D}$ be the dual Delaunay tiling. Let E be a translate of a k-dimensional subspace and $E^\perp$ its the orthogonal complement, and $\mathcal{V}^\perp$ be the projection of $\mathcal{V}$ onto $E^\perp$. The elements of $\mathcal{V}$ that are projected onto E are those for which the dual element projects into $\mathcal{V}^\perp$, $\mathcal{V}^\perp$ is the *window* of the projection.
- The canonical projection method is a special case of the more general cut-and-project method [BG13], in which the window and possibly E are modified in any of several possible ways (e.g., the window may be larger, smaller, discontinuous, fractal).
- In [Har04], Harriss determined which canonically projected tilings admit a substitution rule, and gives a method for constructing any substitution rule that generates the tiling.

FIGURE 3.4.4
An Ammann-Beenker tiling; the relative frequencies of the two marked vertex stars are the (relative) areas of the marked regions in the octagon.

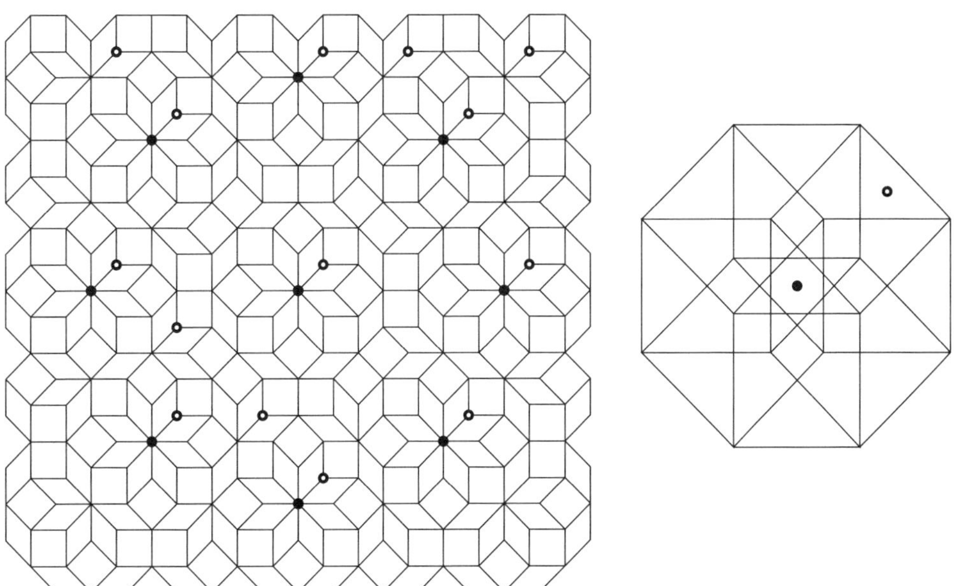

Some of the best known projected nonperiodic tilings are listed in Table 3.4.1 (see [Sen96]). In all these cases, the lattice L is the standard integer lattice and the window Ω is a projection of a hypercube (the Voronoi cell of L). The subspace E is translated so that it does not intersect any faces of $\mathcal{V}$ of dimension less than $n-k$ (thus only a subset of the faces of the Delaunay tiling $\mathcal{D}$ of dimensions $0, 1, \ldots, k$ will be projected). All of these tilings are substitution tilings.

TABLE 3.4.1 Canonically projected nonperiodic tilings.

TILING FAMILY	L	E
Fibonacci tiling	I_2	line with slope $1/\tau$ $(\tau = (1+\sqrt{5})/2)$
Ammann-Beenker tiling	I_4	plane stable under 8-fold rotation
Danzer's Tetrahedra tiling	D_6	plane stable under 5-fold rotation
Danzer 3D tiling	I_6	3-space stable under icosahedral rotation group

The relative frequencies of the vertex configurations of a canonically projected tiling are determined by the window: they are the ratios of volumes of the intersections of the projected faces of $V(0)$ [Sen96]).

OPEN PROBLEMS

The Pisot conjecture states that, for any $1-d$ substitution with a unit Pisot scaling factor, there is a window that makes it a projection tiling. This difficult conjecture can be formulated in many ways [AH14, KLS15]. The Rauzy fractals (see Figure 3.4.5) show how complex the windows can get (especially the right-hand window).

FIGURE 3.4.5
Rauzy fractals, named for their discoverer [Rau82], give the projection windows for three-letter substitution rules, in this case $(a \to ab, b \to ac, c \to a)$, $(a \to bc, b \to c, c \to a)$, $(a \to ab, b \to c, c \to a)$. *The windows can become very complicated, as seen on the right. This gives a sense of the complexity of the Pisot conjecture, as it states that these shapes will always be the closure of their interior.*

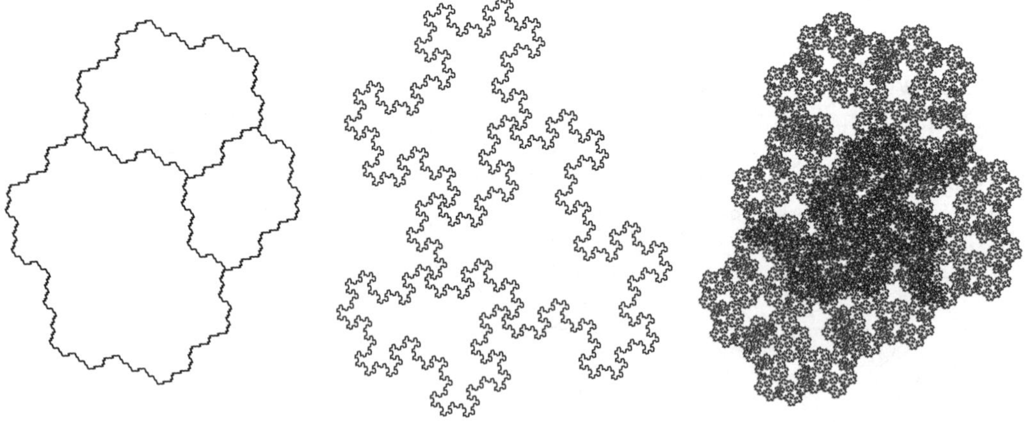

3.5 OTHER TILINGS, OTHER METHODS

There is a vast literature on tilings (or dissections) of bounded regions (such as rectangles and boxes, spheres, polygons, and polytopes). This and much of the recreational literature focuses on tilings whose prototiles are of a particular type, such as rectangles, clusters of n-cubes (polyominoes—see Chapter 14—and polycubes) or n-simplices (polyiamonds in $\mathbb{E}^2$), or tilings by recognizable animate figures. In the search for new ways to produce tiles and tilings, both mathematicians (such as P.A. MacMahon [Mac21]) and amateurs (such as M.C. Escher [Sch90]) have contributed to the subject. Recently the search for new shapes that tile a given bounded region S has produced knotted tiles, toroidal tiles, and twisted tiles. Kuperberg and Adams have shown that for any given knot K, there is a monohedral tiling of $\mathbb{E}^3$ (or of hyperbolic 3-space, or of spherical 3-space) whose prototile is a solid torus that is knotted as K. Also, Adams has shown that, given any polyhedral submanifold M with one boundary component in $\mathbb{E}^n$, a monohedral tiling of $\mathbb{E}^n$ can be constructed whose prototile has the same topological type as M [Ada95].

Other directions of research seek to broaden the definition of prototile set: in new contexts, the tiles in a tiling may be homothetic or topological (rather than congruent) images of tiles in a prototile set. A tiling of $\mathbb{E}^n$ by polytopes in which every tile is combinatorially isomorphic to a fixed convex n-polytope (the combinatorial prototile) is said to be ***monotypic***. It has been shown that in $\mathbb{E}^2$, there exist monotypic face-to-face tilings by convex n-gons for all $n \geq 3$; in $\mathbb{E}^3$, every convex 3-polytope is the combinatorial prototile of a monotypic tiling [Sch84a]. Many (but not all) classes of convex 3-polytopes admit monotypic face-to-face tilings [DGS83, Sch84b].

Dynamical systems theory is perhaps the most powerful tool used to study tilings today, but it takes us beyond the scope of this chapter and we do not discuss it here. For more on this and further references see [BG13, KLS15, Sad08].

3.6 SOURCES AND RELATED MATERIALS

SURVEYS

The following surveys are useful, in addition to the references below.

[GS87]: The definitive, comprehensive treatise on tilings of $\mathbb{E}^2$, state of the art as of the mid-1980's. All subsequent work (in any dimension) has taken this as its starting point for terminology, notation, and basic results. The Main Results of our Section 3.1 can be found here.

[BG13]: A survey of the rapidly growing field of aperiodic order, of which tilings are a major part.

The *Bielefeld Tilings Encyclopedia*, http://tilings.math.uni-bielefeld.de/ gives details of all substitution tilings found in the literature.

[PF02]: A useful general reference for substitutions.

[Moo97]: The proceedings of the NATO Advanced Study Institute on the Mathematics of Aperiodic Order, held in Waterloo, Canada in August 1995.

[Sch93]: A survey of tiling theory especially useful for its accounts of monotypic and other kinds of tilings more general than those discussed in this chapter.

[Sen96]: Chapters 5–8 form an introduction to the theory of aperiodic tilings.

[SS94]: This book is especially useful for its account of tilings in $\mathbb{E}^n$ by clusters of cubes.

RELATED CHAPTERS

Chapter 14: Polyominoes
Chapter 27: Voronoi diagrams and Delaunay triangulations
Chapter 64: Crystals, periodic and aperiodic

REFERENCES

[Ada95] C. Adams. Tilings of space by knotted tiles. *Math. Intelligencer*, 17:41–51, 1995.

[AH14] P. Arnoux and E. Harriss. What is ... a Rauzy Fractal? *Notices Amer. Math. Soc.*, 61:768–770, 2014.

[Ban91] C. Bandt. Self-similar sets 5. Integer matrices and fractal tilings of $\mathbb{R}^n$. *Proc. Amer. Math. Soc.*, 112:549–562, 1991.

[Ber66] R. Berger. The undecidability of the domino problem. *Mem. Amer. Math. Soc.*, 66:1–72, 1966.

[BG13] M. Baake and U. Grimm. *Aperiodic Order: a Mathematical Invitation.* Cambridge University Press, 2013.

[BH96] L. Balke and D.H. Huson. Two-dimensional groups, orbifolds and tilings. *Geom. Dedicata*, 60:89–106, 1996.

[Bie10] L. Bieberbach. Über die Bewegungsgruppen der euklidischen Räume (Erste Abh.). *Math. Ann.*, 70:297–336, 1910.

[BW94] H.-G. Bigalke and H. Wippermann. *Reguläre Parkettierungen.* B.I. Wissenschaftsverlag, Mannheim, 1994.

[CBGS08] J.H. Conway, H. Burgiel, and C. Goodman-Strauss. *The Symmetries of Things.* A.K. Peters/CRC Press, 2008.

[CDHT01] J.H. Conway, O. Delgado Friedrichs, D.H. Huson, and W.P. Thurston. Three-dimensional orbifolds and space groups. *Beiträge Algebra Geom.*, 42:475–507, 2001.

[CH02] J.H. Conway and D.H. Huson. The orbifold notation for two-dimensional groups. *Structural Chemistry*, 13:247–257, 2002.

[Con92] J.H. Conway. The orbifold notation for surface groups. In M. Liebeck and J. Saxl, editors, *Groups, Combinatorics and Geometry*, pages 438–447, Cambridge University Press, 1992.

[Cox54] H.S.M. Coxeter. Regular honeycombs in hyperbolic space. In *Proc. Internat. Congress Math.*, volume III, Nordhoff, Groningen and North-Holland, Amsterdam, 1954, pages 155–169. Reprinted in *Twelve Geometric Essays*, S. Illinois Univ. Press, Carbondale, 1968, and *The Beauty of Geometry: Twelve Essays*, Dover, Mineola, 1999.

[Cox63] H.S.M. Coxeter. *Regular Polytopes,* second edition. Macmillan, New York, 1963. Reprinted by Dover, New York, 1973.

[Del61] B.N. Delone. Proof of the fundamental theorem in the theory of stereohedra. *Dokl. Akad. Nauk SSSR*, 138:1270–1272, 1961. English translation in *Soviet Math.,* 2:812–815, 1961.

[DGS83] L. Danzer, B. Grünbaum, and G.C. Shephard. Does every type of polyhedron tile three-space? *Structural Topology,* 8:3–14, 1983.

[Dre87] A.W.M. Dress. Presentations of discrete groups, acting on simply connected manifolds. *Adv. Math.*, 63:196–212, 1987.

[DS98] N. Dolbilin and D. Schattschneider. The local theorem for tilings. In J. Patera, editor, *Quasicrystals and Discrete Geometry*, vol, 10 of *Fields Inst. Monogr.*, pages 193–199, AMS, Providence, 1998.

[Eng81] P. Engel. Über Wirkungsbereichsteilungen von kubischer Symmetrie, *Z. Kristallogr.*, 154:199–215, 1981.

[Erd99] R. Erdahl. Zonotopes, dicings, and Voronois conjecture on parallelohedra. *European J. Combin.*, 20:527–549, 1999.

[Grü94] B. Grünbaum. Uniform tilings of 3-space. *Geombinatorics*, 4:49–56, 1994.

[GS77] B. Grünbaum and G.C. Shephard. The eighty-one types of isohedral tilings in the plane. *Math. Proc. Cambridge Phil. Soc.,* 82:177–196, 1977.

[GS78] B. Grünbaum and G.C. Shephard. The ninety-one types of isogonal tilings in the plane. *Trans. Amer. Math. Soc.*, 242:335–353, 1978 and 249:446, 1979.

[GS79] B. Grünbaum and G.C. Shephard. Incidence symbols and their applications. In D.K. Ray-Chaudhuri, editor, *Relations Between Combinatorics and Other Parts of Mathematics*, vol. 34 of *Proc. Sympos. Pure Math.*, pages 199–244. AMS, Providence, 1979.

[GS80] B. Grünbaum and G.C. Shephard. Tilings with congruent tiles. *Bull. Amer. Math. Soc.*, 3:951–973, 1980.

[GS87] B. Grünbaum and G.C. Shephard. *Tilings and Patterns.* Freeman, New York, 1987. Second edition, Dover, New York, 2016.

[GS98] C. Goodman-Strauss. Matching rules and substitution tilings. *Ann. Math.,* 147:181–223, 1998.

[Hah83] T. Hahn, editor. *International Tables for Crystallography*, volume A. *Space Group Symmetry.* Reidel, Dordrecht, 1983.

[Haj42] G. Hajós. Über einfache und mehrfache Bedeckung des n-dimensionalen Raumes mit einem Würfelgitter. *Math Z.*, 47:427–467, 1942.

[Har04] E.O. Harriss. *On Canonical Substitution Tilings.* PhD thesis, Imperial College, London, 2004.

[Hee35] H. Heesch. Aufbau der Ebene aus kongruenten Bereichen. *Nachr. Ges. Wiss. Göttingen, New Ser.,* 1:115–117, 1935.

[HK63] H. Heesch and O. Kienzle. *Flächenschluss. System der Formen lückenlos aneinanderschliessender Flachteile.* Springer-Verlag, Berlin, 1963.

[Hus93] D.H. Huson. The generation and classification of tile-k-transitive tilings of the Euclidean plane, the sphere, and the hyperbolic plane. *Geom. Dedicata*, 47:269–296, 1993.

[Kar07] J. Kari. The tiling problem revisited. In *Proc. 5th Conf. Machines, Computations and Universality*, vol. 4664 of *LNCS*, pages 72–79, Springer, Berlin, 2007.

[Ker68] R.B. Kershner. On paving the plane. *Amer. Math. Monthly*, 75:839–844, 1968.

[KLS15] J. Kellendonk, D. Lenz, and J. Savinien, editors. *Mathematics of Aperiodic Order*. Vol. 309 of *Progress in Math.*, Springer, Berlin, 2015.

[Krö69] O. Krötenheerdt. Die homogenen Mosaike n-ter Ordnung in der euklidischen Ebene, I. *Wiss. Z. Martin-Luther-Univ. Halle-Wittenberg Math.-Natur. Reihe*, 18:273–290, 1969.

[KSB93] R. Klitzing, M. Schlottmann, and M. Baake. Perfect matching rules for undecorated triangular tilings with 10-, 12-, and 8-fold symmetry. *Internat. J. Modern Phys.*, 7:1453–1473, 1993.

[LS92] J.C. Lagarias and P.W. Shor. Keller's cube-tiling conjecture is false in high dimensions. *Bull. Amer. Math. Soc.*, 27:279–283, 1992.

[Mac21] P.A. MacMahon. *New Mathematical Pastimes.* Cambridge University Press, 1921. Reprinted by Forgotten Books, London, 2017

[Mag15] A. Magazinov. Voronoi's conjecture for extensions of Voronoi parallelohedra. *Moscow J. Combin. Number Theory*, 5:3, 2015.

[Mak92] V.S. Makarov. On a nonregular partition of n-dimensional Lobachevsky space by congruent polytopes. *Discrete Geometry and Topology, Proc. Steklov. Inst. Math,* 4:103–106, 1992.

[Man04] C. Mann. Heesch's Tiling Problem. *Amer. Math. Monthly*, 111:509–517, 2004.

[Mar08] M. Margenstern. The domino problem of the hyperbolic plane is undecidable. *Theoret. Comput. Sci.*, 407:29–84, 2008.

[McM80] P. McMullen. Convex bodies which tile space by translation. *Mathematika,* 27:113–121, 1980; 28:191, 1981.

[Min07] H. Minkowski. *Diophantische Approximationen.* Teubner, Leipzig, 1907; reprinted by Chelsea, New York, 1957.

[Min97] H. Minkowski. Allgemeine Lehrsätze über die konvexen Polyeder. *Nachr. Ges. Wiss. Göttingen. Math-Phys. Kl.*, 198–219, 1897. In *Gesammelte Abhandlungen von Hermann Minkowski*, reprint, Chelsea, New York, 1967.

[MMVD17] C. Mann, J. McLoud-Mann, and D. Von Derau. Convex pentagons that admit i-block transitive tilings. *Geom. Dedicata*, in press, 2017.

[Moo97] R.V. Moody, editor. *Mathematics of Long Range Aperiodic Order.* Vol. 489 of *NATO Science Series C*, Kluwer, Dordrecht, 1997.

[MT16] C. Mann and B.C. Thomas. Heesch numbers of edge-marked polyforms. *Exp. Math.*, 25: 281–294, 2016.

[Oll09] N. Ollinger. Tiling the plane with a fixed number of polyominoes. In *Proc. 3rd Conf. Language Automata Theory Appl.*, vol. 5457 of *LNCS*, pages 638–647, Springer, Berlin, 2009.

[PF02] N. Pytheas Fogg. *Substitutions in Dynamics, Arithmetics and Combinatorics.* Vol. 1794 of *Lecture Notes in Math.*, Springer, Berlin, 2002.

[Rao17] M. Rao. Exhaustive search of convex pentagons which tile the plane. Preprint, `arXiv:1708.00274`, 2017.

[Rau82] G. Rauzy. Nombres algébriques et substitutions. *Bull. Soc. Math. France*, 110:147–178, 1982.

[Rei28] K. Reinhardt. Zur Zerlegung der euklidischen Räume durch kongruente Würfel. *Sitzungsber. Preuss. Akad. Wiss. Berlin*, 150–155, 1928.

[Sad08] L. Sadun. *Topology of Tiling Spaces.* Vol. 46 of *University Lecture Series*, AMS, Providence, 2008.

[Sch78a] D. Schattschneider. The plane symmetry groups: their recognition and notation. *Amer. Math. Monthly*, 85:439–450, 1978.

[Sch78b] D. Schattschneider. Tiling the plane with congruent pentagons. *Math. Mag.*, 51: 29-44, 1978. Reprinted with afterword in *The Harmony of the World: 75 Years of Mathematics Magazine*, pages 175–190, MAA, Washington, 2007.

[Sch83] V. Schlegel. Theorie der homogen zusammengesetzen Raumgebilde. *Verh. (= Nova Acte) Kaiserl. Leop.-Carol. Deutsch. Akad. Naturforscher*, 44:343–459, 1883.

[Sch84a] E. Schulte. Tiling three-space by combinatorially equivalent convex polytopes. *Proc. London Math. Soc.*, 49:128–140, 1984.

[Sch84b] E. Schulte. Nontiles and nonfacets for Euclidean space, spherical complexes and convex polytopes. *J. Reine Angew. Math.*, 352:161–183, 1984.

[Sch90] D. Schattschneider. *Visions of Symmetry. Notebooks, Periodic Drawings, and Related Work of M.C. Escher.* Freeman, New York, 1990. New edition with afterword, *M. C. Escher, Visions of Symmetry*, Harry Abrams, New York, 2004.

[Sch93] E. Schulte. Tilings. In P.M. Gruber and J.M. Wills, editors, *Handbook of Convex Geometry*, volume B, pages 899–932, North Holland, Amsterdam, 1993.

[SD98] D. Schattschneider and N. Dolbilin. One corona is enough for the Euclidean plane. In J. Patera, editor, *Quasicrystals and Geometry*, vol. 10 of *Fields Inst. Monogr.*, pages 207–246, AMS, Providence, 1998.

[Sen79] M. Senechal. Color groups. *Discrete Applied Math.*, 1:51–73, 1979.

[Sen81] M. Senechal. Which tetrahedra fill space? *Math. Mag.*, 54:227–243, 1981.

[Sen96] M. Senechal. *Quasicrystals and Geometry.* Cambridge University Press, corrected paperback edition, 1996.

[Sol98] B. Solomyak. Nonperiodicity implies unique composition for self-similar translationally finite tilings. *Discrete Comput. Geom.*, 20:265–279, 1998.

[SS94] S. Stein and S. Szabó. *Algebra and Tiling: Homomorphisms in the Service of Geometry.* Vol. 25 of *Carus Math. Monographs*, MAA, Washington, 1994.

[ST11] J.E.S. Socolar and J.M. Taylor. An aperiodic hexagonal tile. *J. Combin. Theory Ser. A*, 118:2207–2231, 2011.

[Ven54] B.A. Venkov. On a class of Euclidean polyhedra. *Vestnik Leningrad. Univ. Ser. Mat. Fiz. Khim.*, 9:11–31, 1954.

[Vor09] G. Voronoï. Nouvelles applications des paramètres continus à la théorie des formes quadratiques II. *J. Reine Angew. Math.*, 136:67–181, 1909.

4 HELLY-TYPE THEOREMS AND GEOMETRIC TRANSVERSALS

Andreas Holmsen and Rephael Wenger

INTRODUCTION

Let $\mathcal{F}$ be a family of convex sets in $\mathbb{R}^d$. A *geometric transversal* is an affine subspace that intersects every member of $\mathcal{F}$. More specifically, for a given integer $0 \leq k < d$, a k-dimensional affine subspace that intersects every member of $\mathcal{F}$ is called a *k-transversal* to $\mathcal{F}$. Typically, we are interested in necessary and sufficient conditions that guarantee the existence of a k-transversal to a family of convex sets in $\mathbb{R}^d$, and furthermore, to describe the space of all k-transversals to the given family. Not much is known for general k and d, and results deal mostly with the cases $k = 0$, 1, or $d - 1$.

Helly's theorem gives necessary and sufficient conditions for the members of a family of convex sets to have a point in common, or in other words, a 0-transversal. Section 4.1 is devoted to some of the generalizations and variations related to Helly's theorem. In the study of k-transversals, there is a clear distinction between the cases $k = 0$ and $k > 0$, and Section 4.2 is devoted to results and questions dealing with the latter case.

NOTATION

Most of the notation and terminology is quite standard. This chapter deals mainly with families of subsets of the d-dimensional real linear space $\mathbb{R}^d$. In certain cases we consider $\mathbb{R}^d$ to be equipped with a metric, in which case it is the usual Euclidean metric. The convex hull of a set $X \subset \mathbb{R}^d$ is denoted by $\operatorname{conv} X$. The cardinality of a set X is denoted by $|X|$. If $\mathcal{F}$ is a family of sets, then $\bigcap \mathcal{F}$ denotes the intersection and $\bigcup \mathcal{F}$ the union of all members of $\mathcal{F}$. If $\mathcal{F}$ is a family of subsets of $\mathbb{R}^d$, then $\operatorname{conv} \mathcal{F}$ denotes the convex hull of $\bigcup \mathcal{F}$.

4.1 HELLY'S THEOREM AND ITS VARIATIONS

One of the most fundamental results in combinatorial geometry is Helly's classical theorem on the intersection of convex sets.

THEOREM 4.1.1 *Helly's Theorem* [Hel23]

Let $\mathcal{F}$ be a family of convex sets in $\mathbb{R}^d$, and suppose that $\mathcal{F}$ is finite or at least one member of $\mathcal{F}$ is compact. If every $d + 1$ or fewer members of $\mathcal{F}$ have a point in common, then there is a point in common to every member of $\mathcal{F}$.

Helly's theorem has given rise to numerous generalizations and variations. A typical *"Helly-type"* theorem has the form:

> If every m or fewer members of a family of objects have property $\mathcal{P}$, then the entire family has property $\mathcal{P}$.

In many cases the extension to infinite families can be dealt with by standard compactness arguments, so for simplicity we deal only with finite families. Under this assumption it is usually not any restriction to assume that every member of the family is open, or that every member of the family is closed, so we will usually choose the form which is the most convenient for the statement at hand. The reader should also be aware that results of this kind are by no means restricted to combinatorial geometry, and Helly-type theorems appear also in graph theory, combinatorics, and related areas; but here we will focus mainly on geometric results.

Let us briefly describe the kind of variations of Helly's theorem we will be treating. The first variation deals with *replacing the convex sets by other objects.* For instance, we may replace convex sets by a more general class of subsets of $\mathbb{R}^d$. In the other direction, one might obtain more structure by specializing the convex sets to homothets, or translates, of a fixed convex set.

The second variation deals with *changing the local condition.* Instead of asking for every $d+1$ or fewer members of $\mathcal{F}$ to have a point in common, one may ask for something less. As an example, we may suppose that among any $d+2$ members of $\mathcal{F}$ *some* $d+1$ have a point in common. This direction has uncovered several deep results concerning the intersection patterns of convex sets. Another direction would be to strengthen the local condition: What if we assume that any $d+1$ or fewer members have an intersection that is at least 1-dimensional, or has volume at least 1? Do any of these assumptions lead to a stronger conclusion? If not, what if we assume the same for the intersection of any $100d$ or fewer members?

This brings us to the third and final variation, which deals with *changing the conclusion.* For instance, our goal may not only be that the intersection is nonempty, but that its diameter is at least some $\delta > 0$. Or that the intersection contains a point with integer coordinates, or more generally, a point from some given set M.

In the last decade we have seen an enormous activity in this area and we have tried to emphasize some of the recent developments. Unfortunately, it is not possible to include every result related to Helly's theorem, and it has been necessary to make some subjective choices. For instance, we have chosen not to discuss Helly-type results related to spherical convexity, and we do not treat abstract convexity spaces in any detail.

GLOSSARY

Helly-number: The Helly-number of a family of sets $\mathcal{C}$ is the smallest integer h such that if $\mathcal{F}$ is a finite subfamily of $\mathcal{C}$ and any h or fewer members of $\mathcal{F}$ have nonempty intersection, then $\bigcap \mathcal{F} \neq \emptyset$. In the case when such a number does not exist, we say that the Helly-number is unbounded, or simply write $h(\mathcal{C}) = \infty$.

Fractional Helly-number: A family of sets $\mathcal{F}$ has fractional Helly-number k if for every $\alpha > 0$ there exists a $\beta > 0$ such that for any finite subfamily $\mathcal{G} \subset \mathcal{F}$ where at least $\alpha\binom{|\mathcal{G}|}{k}$ of the k-member subfamilies of $\mathcal{G}$ have nonempty intersection, there exists a point common to at least $\beta|\mathcal{G}|$ members of $\mathcal{G}$.

Piercing number: The piercing number of a family of sets $\mathcal{F}$ is the smallest integer k for which it is possible to partition $\mathcal{F}$ into subfamilies $\mathcal{F}_1, \ldots, \mathcal{F}_k$ such that $\bigcap \mathcal{F}_i \neq \emptyset$ for every $1 \leq i \leq k$.

Homology cell: A topological space X is a homology cell if it is nonempty and its (singular reduced) homology groups vanish in every dimension. In particular, nonempty convex sets are homology cells.

Good cover: A family of open subsets of $\mathbb{R}^d$ is a good cover if any finite intersection of its members is empty or a homology cell.

Convex lattice set: A convex lattice set in $\mathbb{R}^d$ is a subset $S \subset \mathbb{R}^d$ such that $S = K \cap \mathbb{Z}^d$ for some convex set $K \subset \mathbb{R}^d$.

Diameter: The diameter of a point set $X \subset \mathbb{R}^d$ is the supremum of the distances between pairs of points in X.

Width: The width of a closed convex set $X \subset \mathbb{R}^d$ is the smallest distance between parallel supporting hyperplanes of X.

GENERALIZATIONS TO NONCONVEX SETS

Helly's theorem asserts that the Helly-number of the family of all convex sets in $\mathbb{R}^d$ equals $d+1$. One of the basic problems related to Helly's theorem has been to understand to which extent the assumption of convexity can be weakened.

Early generalizations of Helly's theorem involve families of homology cells in $\mathbb{R}^d$ such that the intersection of any $d+1$ or fewer members is also a homology cell [Hel30, AH35], or that the intersection of any $1 \leq j \leq d$ is j-acyclic [Deb70] (which means that all the homology groups up to dimension j vanish). Montejano showed that these conditions can be relaxed even further.

THEOREM 4.1.2 [Mon14]

Let $\mathcal{F}$ be a family of open subsets of $\mathbb{R}^d$ such that the intersection of any j members of $\mathcal{F}$ has vanishing (singular reduced) homology in dimension $d-j$, for every $1 \leq j \leq d$. Then $h(\mathcal{F}) \leq d+1$.

In fact, Theorem 4.1.2 holds for families of open subsets of any topological space X for which the homology of any open subset vanishes in dimensions greater or equal to d.

In 1961, Grünbaum and Motzkin [GM61] conjectured a Helly-type theorem for families of *disjoint unions* of sets. Their conjecture (now a theorem) was formulated in a rather abstract combinatorial setting, and here we discuss some of the related results in increasing order of generality. The basic idea is that if $\mathcal{C}$ is a family of sets with a bounded Helly-number, and $\mathcal{F}$ is a finite family of sets such that the intersection of any subfamily of $\mathcal{F}$ can be expressed as a disjoint union of a bounded number of members of $\mathcal{C}$, then the Helly-number of $\mathcal{F}$ is also bounded.

THEOREM 4.1.3 [Ame96]

Let $\mathcal{C}$ be the family of compact convex sets in $\mathbb{R}^d$, and let $\mathcal{F}$ be a finite family of sets in $\mathbb{R}^d$ such that the intersection of any subfamily of $\mathcal{F}$ can be expressed as the union of at most k pairwise disjoint members of $\mathcal{C}$. Then $h(\mathcal{F}) \leq k(d+1)$.

Note that Theorem 4.1.3 reduces to Helly's theorem when $k=1$. Amenta gave a short and elegant proof of a more general result set in the abstract combinatorial framework of "Generalized Linear Programming," from which Theorem 4.1.3 is a simple consequence (see Chapter 49). Examples show that the upper bound $k(d+1)$ cannot be reduced.

Recently, Kalai and Meshulam obtained a topological generalization of Theorem 4.1.3. A family $\mathcal{C}$ of open subsets of $\mathbb{R}^d$ is called a *good cover* if any finite intersection of members of $\mathcal{C}$ is empty or a homology cell. Note that the family of all open convex subsets of $\mathbb{R}^d$ is a good cover.

THEOREM 4.1.4 [KM08]

Let $\mathcal{C}$ be a good cover in $\mathbb{R}^d$, and let $\mathcal{F}$ be a finite family of sets in $\mathbb{R}^d$ such the intersection of any subfamily of $\mathcal{F}$ can be expressed as a union of at most k pairwise disjoint members of $\mathcal{C}$. Then $h(\mathcal{F}) \leq k(d+1)$.

Let us make a remark concerning "topological Helly-type theorems." Let $\mathcal{F}$ be a finite family of subsets of some ground set. The *nerve* $N(\mathcal{F})$ of $\mathcal{F}$ is the simplicial complex whose vertex set is $\mathcal{F}$ and whose simplices are all subfamilies $\mathcal{G} \subset \mathcal{F}$ such that $\bigcap \mathcal{G} \neq \emptyset$. If $\mathcal{F}$ is a family of convex sets in $\mathbb{R}^d$, then the nerve $N(\mathcal{F})$ is *d-collapsible.* This implies that every induced subcomplex of $N(\mathcal{F})$ has vanishing homology in all dimensions greater or equal to d, a property often called *d-Leray.* In the case when $\mathcal{F}$ is a good cover in $\mathbb{R}^d$, the *Nerve theorem* from algebraic topology implies that $\bigcup \mathcal{F}$ is topologically equivalent to $N(\mathcal{F})$ (on the level of homology), which again implies that $N(\mathcal{F})$ is d-Leray. These important tools allow us to transfer combinatorial properties regarding the intersection patterns of convex sets, or good covers, in $\mathbb{R}^d$ into properties of simplicial complexes. The proof of Theorem 4.1.4 is based on simplicial homology and the computation of certain spectral sequences. In fact, Kalai and Meshulam established a more general topological result concerning the Leray-numbers of "dimension preserving" projections of simplicial complexes, from which Theorem 4.1.4 is deduced via the Nerve theorem. See the survey [Tan13] for more information regarding simplicial complexes arising from the intersection patterns of geometric objects.

A remarkable feature of the proofs of Theorems 4.1.3 and 4.1.4 is that they are set in quite abstract frameworks, and as we remarked earlier, the original motivating conjecture of Grünbaum and Motzkin [GM61] was formulated in the general combinatorial setting of *abstract convexity spaces*, which includes Theorems 4.1.3 and 4.1.4 as special cases. The general conjecture was treated by Morris [Mor73], but his proof is extremely involved, and the correctness of some of his arguments have been open to debate [EN09]. However, the skepticism has been put to rest after Eckhoff and Nischke [EN09] revisited Morris's approach and provided a complete proof of the Grünbam–Motzkin conjecture.

To describe this abstract setting we must introduce a few definitions. Let $\mathcal{C}$ be a family of subsets of some ground set. We say that $\mathcal{C}$ is *intersectional* if for any finite subfamily $\mathcal{F} \subset \mathcal{C}$, the intersection $\bigcap \mathcal{F}$ is empty or belongs to $\mathcal{C}$. Furthermore, we say that $\mathcal{C}$ is *nonadditive* if for any finite subfamily $\mathcal{F} \subset \mathcal{C}$ consisting of at least two nonempty pairwise disjoint members of $\mathcal{C}$, the union $\bigcup \mathcal{F}$ does not belong to $\mathcal{C}$. For example, the family of all compact convex sets in $\mathbb{R}^d$ is an intersectional and nonadditive family with Helly-number $d+1$. The same is also true for any good cover in $\mathbb{R}^d$ if we include all possible intersections as well. Thus the following theorem, conjectured by Grünbaum and Motzkin, is a common generalization of both Theorems 4.1.3 and 4.1.4.

THEOREM 4.1.5 [EN09]

Let $\mathcal{C}$ be an intersectional and nonadditive family with Helly-number $h(\mathcal{C})$, and let $\mathcal{F}$ be a family of sets such that the intersection of any k or fewer members of $\mathcal{F}$ can be expressed as a union of at most k pairwise disjoint members of $\mathcal{C}$. Then $h(\mathcal{F}) \leq k \cdot h(\mathcal{C})$.

Before ending the discussion of the Grünbaum–Motzkin conjecture let us mention a related result. For a locally arcwise connected topological space Γ, let d_Γ denote the smallest integer such that every open subset of Γ has trivial rational homology in dimension d_Γ and higher. A family $\mathcal{F}$ of open subsets of Γ is called *acyclic* if for any nonempty subfamily $\mathcal{G} \subset \mathcal{F}$, the intersection of the members of $\mathcal{G}$ has trivial rational homology in dimensions greater than zero. This means that the intersections of members of $\mathcal{F}$ need not be connected, but each connected component of such an intersection is a homology cell.

THEOREM 4.1.6 [CVGG14]

Let $\mathcal{F}$ be a finite acyclic family of open subsets of a locally arcwise connected topological space Γ, and suppose that for any subfamily $\mathcal{G} \subset \mathcal{F}$, the intersection $\bigcap \mathcal{G}$ has at most r connected components. Then $h(\mathcal{F}) \leq r(d_\Gamma + 1)$.

It should be noted that Theorem 4.1.6 includes both Theorems 4.1.3 and 4.1.4, but does not seem to follow from Theorem 4.1.5 since it does not require $\mathcal{F}$ to be intersectional. Thus, Theorems 4.1.5 and 4.1.6 appear to be distinct generalizations of Theorem 4.1.4, and their proofs differ significantly. While the proof of Theorem 4.1.5 is combinatorial and uses elementary methods, the proof of Theorem 4.1.6 is based on the homology of simplicial posets and introduces the concept of the *multinerve* (which generalizes the nerve complex). It would be interesting to find a common generalization of Theorems 4.1.5 and 4.1.6. A generalization of Theorem 4.1.6 is also given in [CVGG14] which implies bounds on Helly-numbers for higher dimensional transversals.

The results related to the Grünbaum–Motzkin conjecture deal with families of sets that are built up as *disjoint* unions of a bounded number of members from a sufficiently "nice" class of sets for which the Helly-number is known (with an exception of Theorem 4.1.6). If we do not require the unions to be disjoint, it is still possible to bound the Helly-number.

THEOREM 4.1.7 [AK95, Mat97]

For any integers $k \geq 1$ and $d \geq 1$ there exists an integer $c = c(k, d)$ such that the following holds. Let $\mathcal{F}$ be a finite family of subsets of $\mathbb{R}^d$ such that the intersection of any subfamily of $\mathcal{F}$ can be expressed as the union of at most k convex sets (not necessarily disjoint). Then $h(\mathcal{F}) \leq c$.

Alon and Kalai [AK95] obtain Theorem 4.1.7 as a consequence of a more general theorem. On the other hand, their method is more complicated than Matoušek's proof [Mat97], which is based mostly on elementary methods. Both proof methods yield rather poor numerical estimates on the Helly-numbers $c(k, d)$. For instance, for unions of convex sets in the plane, Matoušek's proof gives $c(2, 2) \leq 20$, $c(2, 3) \leq 90$, $c(2, 4) \leq 231$, etc. It would be interesting to obtain sharper bounds.

We conclude this discussion with a recent interesting Helly-type result due to Goaoc et al., which generalizes Theorem 4.1.7. For a topological space X, let $\tilde{\beta}_i(X)$ denote the i-th reduced Betti-number of X with coefficients in $\mathbb{Z}_2$.

THEOREM 4.1.8 [GP+15]

For any integers $b \geq 0$ and $d \geq 1$ there exists an integer $c = c(b,d)$ such that the following holds. Let $\mathcal{F}$ be a finite family of subsets of $\mathbb{R}^d$ such that for any proper subfamily $\mathcal{G} \subset \mathcal{F}$ we have $\tilde{\beta}_i(\bigcap G) \leq b$ for all $0 \leq i \leq \lceil d/2 \rceil - 1$. Then $h(F) \leq c$.

The proof of Theorem 4.1.8 relies on a general principle for obtaining Helly-type theorems from nonembeddability results for certain simplicial complexes, combined with an application of Ramsey's theorem. This generalizes previous work by Matoušek [Mat97] who established Theorem 4.1.8 in the special case $b = 0$ (using homotopy instead of homology). As a result, the bounds obtained on $c(b,d)$ are enormous and probably very far from the truth. However, examples show that Theorem 4.1.8 is sharp in the sense that *all* the (reduced) Betti numbers $\tilde{\beta}_i$ with $0 \leq i \leq \lceil d/2 \rceil - 1$ must be bounded in order to obtain a finite Helly-number.

INTERSECTIONS IN MORE THAN A POINT

Here we discuss some generalizations of Helly's theorem that apply to families of convex sets, but strengthen both the hypothesis and the conclusion of the theorem. The typical goal here is to guarantee that the sets intersect in more than a single point. The first such result is due to Vincensini [Vin39] and Klee [Kle53] and is a direct consequence of Helly's theorem.

THEOREM 4.1.9 [Vin39, Kle53]

Let $\mathcal{F}$ be a finite family of convex sets in $\mathbb{R}^d$ and let B be some convex set in $\mathbb{R}^d$. If for any $d+1$ or fewer members of $\mathcal{F}$ there exists a translate of B contained in their common intersection, then there exists a translate of B contained in $\bigcap \mathcal{F}$.

Note that we obtain Helly's theorem when K is a single point. In Theorem 4.1.9, one may also replace the words "contained in" by "containing" or "intersecting." The following related result was obtained by De Santis [San57], who derived it from a generalization of Radon's Theorem. Note that it reduces to Helly's theorem for $k = 0$.

THEOREM 4.1.10 [San57]

Let $\mathcal{F}$ be a finite family of convex sets in $\mathbb{R}^d$. If any $d-k+1$ or fewer members of $\mathcal{F}$ contain a k-flat in common, then there exists a k-flat contained in $\bigcap \mathcal{F}$.

One may also interpret the conclusion of Helly's theorem as saying that the intersection of the sets is at least 0-dimensional. Similarly, if $\mathcal{F}$ is a finite family of convex sets in $\mathbb{R}^d$, then Theorem 4.1.9 implies that if the intersection of every $d+1$ or fewer members is d-dimensional then $\bigcap \mathcal{F}$ is d-dimensional. For arbitrary $1 \leq k < d$, Grünbaum [Grü62] showed that if the intersection of every $2n-k$ or fewer sets is at least k-dimensional, then $\bigcap \mathcal{F}$ is at least k-dimensional. It turns out that the threshold $2n-k$ is not optimal, and the exact Helly-numbers for the dimension were obtained by Katchalski around a decade later.

THEOREM 4.1.11 [Kat71]

Let $\mathcal{F}$ be a finite family of convex sets in $\mathbb{R}^d$. Let $\psi(0,d) = d+1$ and $\psi(k,d) =$ $\max(d+1, 2(d-k+1))$ for $1 \leq k \leq d$. If the intersection of any $\psi(k,d)$ or fewer members of $\mathcal{F}$ has dimension at least k, then $\bigcap \mathcal{F}$ has dimension at least k.

Another direction in which Helly's theorem has been generalized is with a "quantitative" conclusion in mind. Loosely speaking, this means that we want the intersection of a family of convex sets to be "large" in some metrical sense. The first such result was noted by Buchman and Valentine [BV82]. For a nonempty closed convex set K in $\mathbb{R}^d$, the *width* of K is defined as the smallest possible distance between two supporting hyperplanes of K.

THEOREM 4.1.12 [BV82]

Let $\mathcal{F}$ be a finite family of closed convex sets in $\mathbb{R}^d$. If the intersection of any $d+1$ or fewer members of $\mathcal{F}$ has width at least w, then $\bigcap \mathcal{F}$ has width at least w.

Notice that Theorem 4.1.12 follows directly from Theorem 4.1.9 by observing that a convex set has width at least ω if and only if it contains a translate of every segment of length ω. Replacing "width" by "diameter" we get the following result due to Bárány Katchalski, and Pach.

THEOREM 4.1.13 [BKP84]

Let $\mathcal{F}$ be a finite family convex sets in $\mathbb{R}^d$. If the intersection of any $2d$ or fewer members of $\mathcal{F}$ has diameter at least 1, then $\bigcap \mathcal{F}$ has diameter at least $d^{-2d}/2$.

The proof of Theorem 4.1.13 is less obvious than the proof of Theorem 4.1.12 outlined above, and relies on a quantitative version of Steinitz' theorem. Examples show that the Helly-number $2d$ in Theorem 4.1.13 is best possible, but it is conjectured that the bound on the diameter can be improved.

CONJECTURE 4.1.14 [BKP84]

Let $\mathcal{F}$ be a finite family convex sets in $\mathbb{R}^d$. If the intersection of any $2d$ or fewer members of $\mathcal{F}$ has diameter at least 1, then $\bigcap \mathcal{F}$ has diameter at least $c_1 d^{-1/2}$ for some absolute constant $c_1 > 0$.

Bárány et al. [BKP84] also proved a Helly-type theorem for the volume, and conjectured a lower bound for the volume of the intersection given that the intersection of any $2d$ or fewer members has volume at least 1. A proof of this conjecture, using John's decomposition of the identity and the Dvoretzky–Rogers lemma, was recently given by Naszódi.

THEOREM 4.1.15 [Nas16]

Let $\mathcal{F}$ be a finite family of convex sets in $\mathbb{R}^d$. If the intersection of any $2d$ or fewer members of $\mathcal{F}$ has volume at least 1, then $\bigcap \mathcal{F}$ has volume at least d^{-cd} for some absolute constant $c > 0$.

The current best estimate for the constant c in Theorem 4.1.15 is given in [Bra17].

Before ending the discussion on quantitative Helly theorems, we mention some extensions of Theorems 4.1.13 and 4.1.15 that answer a question by Kalai and Linial.

THEOREM 4.1.16 [DLL$^+$15a]

For every $d \in \mathbb{N}$ and $\epsilon > 0$ there exists integers $n_1 = n_1(d, \epsilon)$ and $n_2 = n_2(d, \epsilon)$ such that the following hold. Let $\mathcal{F}$ be a finite family of convex sets in $\mathbb{R}^d$.

1. *If the intersection of any $n_1 d$ or fewer members of $\mathcal{F}$ has diameter at least 1,*

then $\bigcap \mathcal{F}$ has diameter at least $1-\epsilon$.

2. *If the intersection of any $n_2 d$ or fewer members of $\mathcal{F}$ has volume at least 1, then $\bigcap \mathcal{F}$ has volume at least $1-\epsilon$.*

These results witness a trade-off between the Helly-numbers and the lower bounds on the diameter and volume of $\bigcap \mathcal{F}$. The numbers $n_1(d,\epsilon)$ and $n_2(d,\epsilon)$ are related to the minimal number of facets needed to approximate an arbitrary convex body in $\mathbb{R}^d$ by a polytope. Moreover, for any fixed d, the numbers $n_1(d,\epsilon)$ and $n_2(d,\epsilon)$ are in $\Theta(\epsilon^{-(d-1)/2})$ [DLL+15a, Sob16]. See also [LS09] for other quantitative versions of Helly's theorem.

WEAKENING THE LOCAL CONDITION

For a finite family $\mathcal{F}$ of convex sets in $\mathbb{R}^d$, let $f_k(\mathcal{F})$ denote the number of subfamilies of $\mathcal{F}$ of size $k+1$ with nonempty intersection. In particular, $f_0(\mathcal{F})$ denotes the number of (nonempty) members of $\mathcal{F}$. Helly's theorem states that if $f_d(\mathcal{F})$ equals $\binom{|\mathcal{F}|}{d+1}$, then there is a point in common to all the members of $\mathcal{F}$. What can be said if $f_d(\mathcal{F})$ is some value less than $\binom{|\mathcal{F}|}{d+1}$? This question was considered by Katchalski and Liu [KL79] who showed that if "almost every" subfamily of size $d+1$ have nonempty intersection, then there is a point in common to "almost every" member of $\mathcal{F}$. The precise meaning of "almost every" can be viewed as a consequence of "the upper bound theorem" for families of convex sets, and gives us the following.

THEOREM 4.1.17 [Kal84]

Let $\mathcal{F}$ be a finite family of $n \geq d+1$ convex sets in $\mathbb{R}^d$. For any $0 \leq \beta \leq 1$, if $f_d(\mathcal{F}) > (1-(1-\beta)^{d+1})\binom{n}{d+1}$, then some $\lfloor \beta n \rfloor + 1$ members of $\mathcal{F}$ have a point in common.

This result is commonly known as the *fractional Helly theorem* and has many important applications in discrete geometry (most notably its role in the (p,q)-problem discussed below). The upper bound theorem from which it is derived was discovered by Kalai [Kal84], and independently by Eckhoff [Eck85], and gives optimal upper bounds for the numbers $f_d(\mathcal{K}), \ldots, f_{d+r-1}(\mathcal{K})$ in terms of $f_0(\mathcal{F})$, provided $f_{d+r}(\mathcal{K}) = 0$. Here we state the particular instance that implies Theorem 4.1.17.

THEOREM 4.1.18 [Kal84, Eck85]

Let $\mathcal{F}$ be a finite family of $n \geq d+1$ convex sets in $\mathbb{R}^d$. For any $0 \leq r \leq n-d-1$, if $f_d(\mathcal{F}) > \binom{n}{d+1} - \binom{n-r}{d+1}$, then some $d+r+1$ members of $\mathcal{F}$ have a point in common.

The lower bound on $f_d(\mathcal{F})$ given in Theorem 4.1.18 cannot be reduced, which can be seen by considering r copies of $\mathbb{R}^d$ and $n-r$ hyperplanes in general position. The two proofs of Theorem 4.1.18 are quite different, but they both use the Nerve theorem and deal with the more general setting of *d-collapsible* simplicial complexes. It is known that the result extends to general *d-Leray* complexes as well [AKM+02]. Yet another proof of Theorem 4.1.18 was given by Alon and Kalai [AK85].

Due to its importance, several generalizations of the fractional Helly theorem have been considered. Let $\mathcal{F}$ be an arbitrary set system. We say that $\mathcal{F}$ has *fractional Helly-number* k if for every $\alpha > 0$ there exists a $\beta > 0$ such that for any finite subfamily $\mathcal{G} \subset \mathcal{F}$ where at least $\alpha\binom{|\mathcal{G}|}{k}$ of the k-member subfamilies of $\mathcal{G}$

have nonempty intersection, there exists a point common to at least $\beta|\mathcal{G}|$ members of $\mathcal{G}$. We say that $\mathcal{F}$ has the *fractional Helly-property* if it has a finite fractional Helly-number. An important remark is that a set system may have the fractional Helly-property even though its Helly-number is unbounded.

Matoušek [Mat04] showed that any set system with bounded VC-dimension has the fractional Helly-property. More precisely, he shows that if the dual shatter function of $\mathcal{F}$ is bounded by $o(m^k)$, then $\mathcal{F}$ has fractional Helly-number k. For related geometric variants see also [Pin15]. Another noteworthy example is given by Rolnick and Soberón [RS17], who develop a framework to show that families of convex sets in $\mathbb{R}^d$ have the fractional Helly-property with respect to an abstract quantitative function defined on the family of all convex sets in $\mathbb{R}^d$. As a consequence, they obtain fractional analogues of Theorem 4.1.16 with respect to volume and surface area. A similar result has also been shown for the diameter [Sob16].

Another direction that has received a great deal of attention involves *piercing problems* for families of convex sets. Let $\mathcal{F}$ be a family of sets and suppose $\mathcal{F}$ can be partitioned into subfamilies $\mathcal{F}_1, \ldots, \mathcal{F}_k$ such that $\bigcap \mathcal{F}_i \neq \emptyset$ for every $1 \leq i \leq k$. The smallest number k for which this is possible is the *piercing number* of $\mathcal{F}$ and will be denoted by $\tau(\mathcal{F})$.

One early generalization of Helly's theorem involving the piercing number was considered by Hadwiger and Debrunner [HD57], where the hypothesis that every $d+1$ members of $\mathcal{F}$ have a point in common was replaced by the hypothesis that among any p members of $\mathcal{F}$ some q have a point in common, where $p \geq q \geq d+1$. For certain values of p and q, they obtained the following result.

THEOREM 4.1.19 [HD57]

Let $\mathcal{F}$ be a finite family of at least p convex sets in $\mathbb{R}^d$. If among any p members of $\mathcal{F}$ some q have a point in common, where $p \geq q \geq d+1$ and $p(d-1) < (q-1)d$, then $\tau(\mathcal{F}) \leq p-q+1$.

Examples show that the value $p-q+1$ is tight. Notice that Theorem 4.1.19 reduces to Helly's theorem when $p = q = d+1$. For general values of p and q, not covered by Theorem 4.1.19, even the existence of a bounded piercing number remained unknown for a long time, and this became known as the *Hadwiger-Debrunner (p,q)-problem.* (Notice that repeated applications of Theorem 4.1.17 only give an upper bound of $\log|\mathcal{F}|$.) The (p,q)-problem was finally resolved by Alon and Kleitman, and is now referred to as the *(p,q)-theorem.*

THEOREM 4.1.20 [AK92]

For any integers $p \geq q \geq d+1$, there exists an integer $c = c(p,q,d)$ such that the following holds. Let $\mathcal{F}$ be a finite family of at least p convex sets in $\mathbb{R}^d$. If among any p members of $\mathcal{F}$ some q have a point in common, then $\tau(\mathcal{F}) \leq c$.

The proof of the (p,q)-theorem combines several tools from discrete geometry, and the most prominent roles are played by the fractional Helly theorem (Theorem 4.1.17) and the weak ϵ-net theorem for convex sets. It is of considerable interest to obtain better bounds on $c(p,q,d)$ (the only exact values known are the ones covered by Theorem 4.1.19). For the first open case of convex sets in the plane, the best known bounds are $3 \leq c(4,3,2) \leq 13$ [KGT99]. See [Eck03] for a survey on the (p,q)-problem, and also [KT08] and [Mül13].

It should be mentioned that the (p,q)-theorem holds in much more general settings. For instance, it holds for families of sets that are unions of convex sets as well as for good covers in $\mathbb{R}^d$ [AK95, AKM$^+$02]. In fact, a very general combinatorial framework was established in [AKM$^+$02] that provides bounds on the piercing number for general set systems that have the fractional Helly-property.

Recently, several other variations of the (p,q)-theorem have been considered. This includes a fractional version [BFM$^+$14] as well as quantitative versions [RS17]. As a sample of these results we mention the following theorem due to Soberón.

THEOREM 4.1.21 [Sob16]

For any integers $p \geq q \geq 2d > 1$ and real $0 < \epsilon < 1$ there exists an integer $c = c(p,q,d,\epsilon)$ such that the following holds. Let $\mathcal{F}$ be a finite family of at least p convex sets in $\mathbb{R}^d$, each of diameter at least 1. If among any p members of $\mathcal{F}$ some q have intersection of diameter at least 1, then there are c segments $S_1, \ldots, S_c$ each of length at least $1-\epsilon$ such that every member of $\mathcal{F}$ contains at least one of the S_i.

Piercing problems have also been studied for restricted classes of convex sets, in which case more precise (or even exact) bounds are known. For the special case of homothets, the intersection of every two members of $\mathcal{F}$ suffices to guarantee a bounded piercing number.

THEOREM 4.1.22 [Grü59]

For any integer $d \geq 1$ there exists an integer $c = c(d)$ such that the following holds. If $\mathcal{F}$ is a finite family of homothets of a convex set in $\mathbb{R}^d$ and any two members of $\mathcal{F}$ intersect, then $\tau(\mathcal{F}) \leq c$.

The special case when $\mathcal{F}$ is a family of circular disks in $\mathbb{R}^2$ was a question raised by Gallai, and answered by Danzer in 1956 (but not published until 1986).

THEOREM 4.1.23 [Dan86]

Let $\mathcal{F}$ be a finite family of circular disks in $\mathbb{R}^2$. If any two members of $\mathcal{F}$ intersect, then $\tau(\mathcal{F}) \leq 4$.

An example consisting of 10 disks shows that the number 4 cannot be reduced, and it is known that for 9 disks in the plane that pairwise intersect the piercing number is at most 3. For the even more restricted case when $\mathcal{F}$ is a family of pairwise intersecting *unit* disks in $\mathbb{R}^2$ it is known that $\tau(\mathcal{F}) \leq 3$. This is a special case of a conjecture of Grünbaum that stated that $\tau(\mathcal{F}) \leq 3$ for any family $\mathcal{F}$ of pairwise intersecting translates of a compact convex set $K \subset \mathbb{R}^2$. The conjecture was confirmed in full generality by Karasev.

THEOREM 4.1.24 [Kar00]

Let $\mathcal{F}$ be a finite family of translates of a compact convex set $K \subset \mathbb{R}^2$. If any two members of $\mathcal{F}$ intersect, then $\tau(\mathcal{F}) \leq 3$.

Karasev [Kar08] extended the methods from the proof of Theorem 4.1.24 to obtain higher dimensional analogues, where "pairwise intersecting" is replaced by the property that any d members have a point in common. For any family $\mathcal{F}$ of Euclidean balls in $\mathbb{R}^d$ whose radii differ by no more than a factor of d, he shows that if every d members of $\mathcal{F}$ have a point in common, then $\tau(F) \leq d+1$. Using this result he gives an upper bound on the piercing number for families of Euclidean balls without any size constraint.

THEOREM 4.1.25 [Kar08]

Let $\mathcal{F}$ be a family of Euclidean balls in $\mathbb{R}^d$ such that any d members of $\mathcal{F}$ have a point in common. Then $\tau(\mathcal{F}) \leq 4(d+1)$ for $d \leq 4$ and $\tau(\mathcal{F}) \leq 3(d+1)$ for $d \geq 5$.

The bounds in Theorem 4.1.25 are probably not tight, for instance when $d = 2$ it only gives $\tau(\mathcal{F}) \leq 12$ (whereas Theorem 4.1.23 shows that $\tau(\mathcal{F}) \leq 4$), and it would be interesting to obtain sharper bounds. By the same method Karasev also shows that for families $\mathcal{F}$ of positive homothets of a simplex where every d members of $\mathcal{F}$ have a point in common, we have $\tau(\mathcal{F}) \leq d+1$ (which is not known to be tight for $d > 2$).

We conclude this discussion with a conjecture due to Katchalski and Nashtir.

CONJECTURE 4.1.26 [KN99]

There exists a constant k_0 such that $\tau(\mathcal{F}) \leq 3$ for every finite family $\mathcal{F}$ of convex sets in $\mathbb{R}^2$ where $\tau(\mathcal{G}) = 2$ for every subfamily $\mathcal{G} \subset \mathcal{F}$ with $|\mathcal{G}| \leq k_0$.

COLORFUL VERSIONS

The following remarkable generalization of Helly's theorem was discovered by Lovász and described by Bárány.

THEOREM 4.1.27 [Bár82]

Let $\mathcal{F}_1, \ldots, \mathcal{F}_{d+1}$ be finite families of convex sets in $\mathbb{R}^d$. If $\bigcap_{i=1}^{d+1} K_i \neq \emptyset$ for each choice of $K_i \in \mathcal{F}_i$, then $\bigcap \mathcal{F}_i \neq \emptyset$ for some $1 \leq i \leq d+1$.

This is commonly known as the *colorful Helly theorem.* Notice that it reduces to Helly's theorem by setting $\mathcal{F}_1 = \cdots = \mathcal{F}_{d+1}$. The dual version, known as the *Colorful Carathéodory theorem,* was discovered by Bárány, and has several important applications in discrete geometry. For example, it plays key roles in Sarkaria's proof of Tverberg's theorem [Sar92, BO97] and in the proof of the existence of weak ϵ-nets for the family of convex sets in $\mathbb{R}^d$ [ABFK92].

Recently, Kalai and Meshulam gave a far-reaching topological generalization of Theorem 4.1.27. Recall that a *matroid* $\mathcal{M}$ defined on a finite set E is uniquely determined by its *rank function* ρ, and that a subset $A \subset E$ is *independent* in $\mathcal{M}$ if and only if $\rho(A) = |A|$.

THEOREM 4.1.28 [KM05]

Let $\mathcal{F}$ be a good cover in $\mathbb{R}^d$ and let $\mathcal{M}$ be a matroid with rank function ρ defined on the members of $\mathcal{F}$. If every subfamily of $\mathcal{F}$ that is independent in $\mathcal{M}$ has a point in common, then there exists a subfamily $\mathcal{G} \subset \mathcal{F}$ such that $\rho(\mathcal{F} \setminus \mathcal{G}) \leq d$ and $\bigcap \mathcal{G} \neq \emptyset$.

This reduces to Theorem 4.1.27 in the case when $\mathcal{F} = \mathcal{F}_1 \cup \ldots \cup \mathcal{F}_{d+1}$ is a family of convex sets in $\mathbb{R}^d$ and $\mathcal{M}$ is the partition matroid induced by the $\mathcal{F}_i$. The proof of Theorem 4.1.28 is based on simplicial homology and, in fact, Kalai and Meshulam prove an even more general result concerning arbitrary d-Leray complexes. Further extensions of Theorem 4.1.28 were obtained in [Hol16]. See also [Flø11] for an algebraic formulation of Theorem 4.1.27.

Quantitative versions of the colorful Helly theorem have also been obtained.

For instance, a colorful version of Theorem 4.1.16 for the volume was proved in [DLL+15a], with an essentially different proof given in [RS17]. Here is a colorful version of Theorem 4.1.16 for the diameter.

THEOREM 4.1.29 [Sob16]

For every integer $d \geq 1$ and real $0 < \epsilon < 1$ there exists an integer $n = n(d, \epsilon)$ such that the following holds. Let $\mathcal{F}_1, \ldots, \mathcal{F}_n$ be finite families of convex sets in $\mathbb{R}^d$, and suppose $\bigcap_{i=1}^n K_i$ has diameter 1 for every choice of $K_i \in \mathcal{F}_i$. Then $\bigcap \mathcal{F}_i$ has diameter at least $1 - \epsilon$ for some $1 \leq i \leq n$. Moreover, for any fixed d, $n(d, \epsilon) = \Theta(\epsilon^{-(d-1)/2})$.

There are also colorful versions of the (p, q)-theorem. One such theorem was proved by Bárány and Matoušek [BM03], while another variation was obtained in [BFM+14].

THEOREM 4.1.30 [BFM+14]

For any integers $p \geq q \geq d + 1$ there exists an integer $c = c(p, q, d)$ such that the following holds. Let $\mathcal{F}_1, \ldots, \mathcal{F}_p$ be finite families of convex sets in $\mathbb{R}^d$. Suppose for any choice $K_1 \in \mathcal{F}_1, \ldots, K_p \in \mathcal{F}_p$ some q of the K_i have a point in common. Then there are at least $q - d$ of the families for which $\tau(\mathcal{F}_i) \leq c$.

We conclude with a conjecture regarding a colorful generalization of Theorem 4.1.24.

CONJECTURE 4.1.31 [JMS15]

Let K be a compact convex set in the plane and let $\mathcal{F}_1, \ldots, \mathcal{F}_n$ be nonempty families of translates of K, with $n \geq 2$. Suppose $K_i \cap K_j \neq \emptyset$ for every choice $K_i \in \mathcal{F}_i$, $K_j \in \mathcal{F}_j$, and $i \neq j$. Then there exists $m \in \{1, 2, \ldots, n\}$ such that $\tau(\bigcup_{i \neq m} \mathcal{F}_i) \leq 3$.

HELLY THEOREMS FOR CONVEX LATTICE SETS

Let $\mathbb{Z}^d$ denote the integer lattice in $\mathbb{R}^d$. A *convex lattice set* in $\mathbb{R}^d$ is a subset $S \subset \mathbb{R}^d$ such that $S = K \cap \mathbb{Z}^d$ for some convex set $K \subset \mathbb{R}^d$. Note that the convex hull operator in $\mathbb{R}^d$ equips $\mathbb{Z}^d$ with the structure of a general *convexity space*, but the specific structure of convex lattice sets has drawn particular interest due to its connection with integer linear programming, geometry of numbers, and crystallographic lattices. The family of all subsets of $\{0, 1\}^d$ of size $2^d - 1$ shows that the Helly-number for convex lattice sets in $\mathbb{R}^d$ is at least 2^d, and the following theorem, due to Doignon [Doi73], shows that this is indeed the correct Helly-number.

THEOREM 4.1.32 [Doi73]

Let $C_{\mathbb{Z}^d}$ be the family of convex lattice sets in $\mathbb{R}^d$. Then $h(C_{\mathbb{Z}^d}) = 2^d$.

This theorem was rediscovered independently by Bell [Bel77] and Scarf [Sca77]. Recently, Aliev et al. [ABD+16, ADL14] obtained the following quantitative version of Theorem 4.1.32.

THEOREM 4.1.33 [ADL14]

For any integers $d \geq 1$ and $k \geq 1$ there exists a constant $c = c(d, k)$ such that the following holds. Let $\mathcal{F}$ be a finite family of convex lattice sets in $\mathbb{R}^d$. If the *intersection of every c or fewer members of $\mathcal{F}$ contains at least k points, then $\bigcap \mathcal{F}$ contains at least k points.*

This result was initially proved in [ABD+16]. The bounds on $c(d,k)$ were improved in [ADL14] and in subsequent work [CHZ15, AGP+17] it was shown that $c(d,k)$ is in $\Theta(k^{(d-1)/(d+1)})$.

The fractional version of Theorem 4.1.32 has also been considered and it was shown in [AKM+02] that the family of convex lattice sets in $\mathbb{R}^d$ has the fractional Helly-property with the fractional Helly-number 2^d. This was later improved by Bárány and Matoušek.

THEOREM 4.1.34 [BM03]

For any integer $d \geq 1$ and real number $0 < \alpha < 1$ there exists a real number $\beta = \beta(d, \alpha) > 0$ such that the following holds. Let $\mathcal{F}$ be a finite family of convex lattice sets in $\mathbb{R}^d$ and suppose there are at least $\alpha\binom{|\mathcal{F}|}{d+1}$ subfamilies of size $d+1$ that have nonempty intersection. Then there is a point contained in at least $\beta|\mathcal{F}|$ members of $\mathcal{F}$.

This result shows that the large Helly-number of Theorem 4.1.32 can be regarded as a "local anomaly" and that the relevant number for other, more global Helly-type properties is only $d+1$. For instance, by the tools developed in [AKM+02], it can be shown that Theorem 4.1.34 implies a (p, q)-theorem for convex lattice sets [BM03]. The proof of Theorem 4.1.34 uses surprisingly little geometry of $\mathbb{Z}^d$ and relies mostly on tools from extremal combinatorics. These methods also yield interesting colorful Helly-type theorems for convex lattice sets [AW12, BM03].

We now describe a common generalization of Theorem 4.1.32 and Helly's theorem, which was stated by Hoffman [Hof79], and rediscovered by Averkov and Weismantel [AW12]. Let M be a subset of $\mathbb{R}^d$. A subset $S \subset M$ is called *M-convex* if $S = M \cap C$ for some convex set $C \subset \mathbb{R}^d$. Thus, the convex hull operator in $\mathbb{R}^d$ equips M with the structure of a general convexity space.

Averkov and Weismantel proved the following Helly-type theorem for mixed integer spaces, that is, sets of the form $M = \mathbb{R}^m \times \mathbb{Z}^n$.

THEOREM 4.1.35 [Hof79, AW12]

Let $\mathcal{C}_{\mathbb{R}^m \times \mathbb{Z}^n}$ be the family of all $(\mathbb{R}^m \times \mathbb{Z}^n)$-convex sets in $\mathbb{R}^{m+n}$. Then $h(\mathcal{C}_{\mathbb{R}^m \times \mathbb{Z}^n}) = (m+1)2^n$.

In fact, Averkov and Weismantel deduce Theorem 4.1.35 from more general inequalities concerning the Helly-numbers of spaces of the form $\mathbb{R}^m \times M$ and $M \times \mathbb{Z}^n$. They also establish fractional and colorful Helly-type theorems for mixed integer spaces, and they obtain the following generalization of Theorem 4.1.34.

THEOREM 4.1.36 [AW12]

Let M be a nonempty closed subset of $\mathbb{R}^d$ and let $\mathcal{C}_M$ be the family of all M-convex sets in $\mathbb{R}^d$. If $h(\mathcal{C}_M)$ is finite, then $\mathcal{C}_M$ has the fractional Helly-property with fractional Helly-number $d+1$.

Further generalizations and variations of Helly's theorem for M-convex sets have been established in [Ave13, Hal09, DLL+15b].

PROBLEM 4.1.37 [DLL+15b]

Let $\mathbb{P}$ denote the set of prime numbers, and let $\mathcal{C}_{\mathbb{P}\times\mathbb{P}}$ denote the family of all $(\mathbb{P}\times\mathbb{P})$-convex sets in $\mathbb{R}^2$. Does $\mathcal{C}_{\mathbb{P}\times\mathbb{P}}$ have a finite Helly-number ?

It is known that $h(\mathcal{C}_{\mathbb{P}\times\mathbb{P}}) \geq 14$, and it is conjectured that the Helly-number is unbounded [DLL+15b].

4.2 GEOMETRIC TRANSVERSALS

GLOSSARY

Transversal: A k-transversal to a family $\mathcal{F}$ of convex sets is an affine subspace of dimension k that intersects every member of $\mathcal{F}$.

Line transversal: A 1-transversal to a family of convex sets in $\mathbb{R}^d$.

Hyperplane transversal: A $(d-1)$-transversal to a family of convex sets in $\mathbb{R}^d$.

Separated: A family $\mathcal{F}$ of convex sets is k-separated if no $k+2$ members of $\mathcal{F}$ have a k-transversal.

Ordering: A k-ordering of a family $\mathcal{F} = \{K_1, \ldots, K_n\}$ of convex sets is a family of orientations of $(k+1)$-tuples of $\mathcal{F}$ defined by a mapping $\chi : \mathcal{A}^{k+1} \to \{-1, 0, 1\}$ corresponding to the orientations of some family of points $X = \{x_1, \ldots, x_n\}$ in $\mathbb{R}^k$. The orientation of $(a_{i_0}, a_{i_1}, \ldots, a_{i_k})$ is the orientation of the corresponding points $(x_{i_0}, x_{i_1}, \ldots, x_{i_k})$, i.e.,

$$\operatorname{sgn}\det\begin{pmatrix} 1 & x^1_{i_0} & \cdots & x^k_{i_0} \\ \vdots & \vdots & \ddots & \vdots \\ 1 & x^1_{i_k} & \cdots & x^k_{i_k} \end{pmatrix}.$$

Geometric permutation: A geometric permutation of a $(k-1)$-separated family $\mathcal{F}$ of convex sets in $\mathbb{R}^d$ is the pair of k-orderings induced by some k-transversal of $\mathcal{F}$.

Perhaps due to the fact that the space of all affine k-flats in $\mathbb{R}^d$ is no longer contractible when $k > 0$, there is a clear distinction between the cases $k = 0$ and $k > 0$ in the study of k-transversals. In 1935, Vincensini asked whether Helly's theorem can be generalized to k-transversals for arbitrary $k < d$. In other words, is there a number $m = m(k, d)$ such that for any family $\mathcal{F}$ of convex sets in $\mathbb{R}^d$, if every m or fewer members of $\mathcal{F}$ have a k-transversal, then $\mathcal{F}$ has a k-transversal? The answer to Vincensini's question is no, and as Santaló pointed out, even the number $m(1, 2)$ does not exists in general, that is, there is no Helly-type theorem for *line transversals* to convex sets in the plane.

It is evident that to get a "Helly-type theorem with transversals" one needs to impose additional conditions on the shapes and/or the relative positions of the sets of the family. Helly-type theorems for transversals to families of a restricted class of convex sets in $\mathbb{R}^d$ are closely related to the combinatorial complexity of the *space of transversals*, and a bound on the Helly-number can often be deduced from the topological Helly theorem or one of its variants.

As before, we mainly restrict our attention to finite families of compact convex sets, unless stated otherwise. In most cases, this causes no loss in generality and replacing compact sets by open sets is usually straightforward.

HADWIGER-TYPE THEOREMS

In 1957, Hadwiger [Had57] gave the first necessary and sufficient conditions for the existence of a line transversal to a finite family $\mathcal{F}$ of pairwise disjoint convex sets in $\mathbb{R}^2$. The basic observation is that if a line L intersects every member of $\mathcal{F}$, then L induces a linear ordering of $\mathcal{F}$; this is simply the order in which L meets the members of $\mathcal{F}$ (as it is traversed in one of its two opposite directions). In particular, this implies that there exists a linear ordering of $\mathcal{F}$ such that any *three* members of $\mathcal{F}$ can be intersected by a directed line consistently with the ordering. Hadwiger's transversal theorem asserts that this necessary condition is also sufficient.

THEOREM 4.2.1 *Hadwiger's Transversal Theorem* [Had57]

Let $\mathcal{F}$ be a finite family of pairwise disjoint convex sets in $\mathbb{R}^2$. If there exists a linear ordering of $\mathcal{F}$ such that every three members of $\mathcal{F}$ can be intersected by a directed line in the given order, then $\mathcal{F}$ has a line transversal.

An interesting remark is that even though the conditions of Theorem 4.2.1 guarantee the existence of a line transversal to $\mathcal{F}$, we are not guaranteed a line that intersects the members of $\mathcal{F}$ in the given order. To obtain this stronger conclusion we need to impose the condition that every *four* members can be intersected by a directed line in the given order, which is a theorem discovered by Wenger [Wen90c], and independently by Tverberg [Tve91].

Hadwiger's transversal theorem has been generalized to higher dimensions and gives necessary and sufficient conditions for the existence of a *hyperplane* transversal to a family of compact convex sets in $\mathbb{R}^d$. Partial results in this direction were obtained by Katchalski [Kat80], Goodman and Pollack [GP88], and Wenger [Wen90c], before Pollack and Wenger [PW90] found a common generalization of these previous results.

Let $\mathcal{F}$ be a finite family of compact convex sets in $\mathbb{R}^d$ and let P be a subset of $\mathbb{R}^k$. We say that $\mathcal{F}$ *separates consistently* with P if there exists a map $\varphi : \mathcal{F} \to P$ such that for any two of subfamilies $\mathcal{F}_1$ and $\mathcal{F}_2$ of $\mathcal{F}$ we have

$$\operatorname{conv}\mathcal{F}_1 \cap \operatorname{conv}\mathcal{F}_2 = \emptyset \;\Rightarrow\; \operatorname{conv}\varphi(\mathcal{F}_1) \cap \operatorname{conv}\varphi(\mathcal{F}_2) = \emptyset.$$

Note that, if $k < d$ and $\mathcal{F}$ separates consistently with a set $P \subset \mathbb{R}^k$, then every $k+2$ members of $\mathcal{F}$ have a k-transversal. (This follows from Radon's theorem.) Moreover, if $\mathcal{F}$ has a hyperplane transversal, then $\mathcal{F}$ separates consistently with a set $P \subset \mathbb{R}^{d-1}$; simply choose one point from each member of $\mathcal{F}$ contained in the hyperplane transversal.

THEOREM 4.2.2 [PW90]

A family $\mathcal{F}$ of compact convex sets in $\mathbb{R}^d$ has a hyperplane transversal if and only if $\mathcal{F}$ separates consistently with a set $P \subset \mathbb{R}^{d-1}$.

When $\mathcal{F}$ is a family of pairwise disjoint sets in the plane, the separation condition of Theorem 4.2.2 is equivalent to the ordering condition in Hadwiger's theorem. The proof of Theorem 4.2.2 uses the Borsuk-Ulam theorem and Kirchberger's theorem, and it was generalized by Anderson and Wenger [AW96] who showed that the point set P can be replaced by an acyclic oriented matroid of rank at most d.

Theorem 4.2.2 was strengthened further by Arocha et al. [ABM$^+$02] to include a description of the topological structure of the space of hyperplane transversals [ABM$^+$02].

Let $\mathcal{G}_{d-1}^d$ denote the space of all affine hyperplanes in $\mathbb{R}^d$. Note that $\mathcal{G}_{d-1}^d$ is retractible to the space of hyperplanes passing through the origin, and therefore homotopy equivalent to $\mathbb{RP}^{d-1}$. For a family $\mathcal{F}$ of convex sets in $\mathbb{R}^d$, let $\mathcal{T}_{d-1}^d(\mathcal{F})$ denote the subspace of $\mathcal{G}_{d-1}^d$ of all hyperplane transversals to $\mathcal{F}$. We say that $\mathcal{F}$ has a *virtual k-transversal* if the homomorphism induced by the inclusion, $H_{d-1-k}(\mathcal{T}_{d-1}^d(\mathcal{F})) \to H_{d-1-k}(\mathcal{G}_{d-1}^d)$, is nonzero. In particular, if L is a k-transversal to $\mathcal{F}$, then the set of all hyperplanes containing L shows that $\mathcal{F}$ has a virtual k-transversal. Thus, the property of having a virtual k-transversal can be interpreted as saying that there are "as many" hyperplane transversals as if there exists a k-transversal.

THEOREM 4.2.3 [ABM$^+$02]

Let $\mathcal{F}$ be a finite family of compact convex sets in $\mathbb{R}^d$ and let P be a set of points in $\mathbb{R}^k$, for some $0 \leq k < d$. If $\mathcal{F}$ separates consistently with P, then $\mathcal{F}$ has a virtual k-transversal.

The proof of Theorem 4.2.3 follows the same ideas as the proof of Theorem 4.2.2, and uses Alexander duality to obtain the stronger conclusion.

There are currently no known conditions, similar in spirit of Hadwiger's transversal theorem, which guarantee the existence of a k-transversal to a family of compact convex sets in $\mathbb{R}^d$ for $0 < k < d-1$. Already in the simplest case, line transversals in $\mathbb{R}^3$, examples show that there is no direct analogue of Hadwiger's theorem. In particular, for any integer $n \geq 3$ there is a family $\mathcal{F}$ of pairwise disjoint convex sets in $\mathbb{R}^3$ and a linear ordering of $\mathcal{F}$ such that any $n-2$ members of $\mathcal{F}$ are met by a directed line consistent with the ordering, yet $\mathcal{F}$ has no line transversal [GPW93]. It is even possible to restrict the members of $\mathcal{F}$ to be pairwise disjoint translates of a fixed compact convex set in $\mathbb{R}^3$ [HM04]. In view of these examples, the following result is all the more remarkable.

THEOREM 4.2.4 [BGP08]

Let $\mathcal{F}$ be a finite family of pairwise disjoint Euclidean balls in $\mathbb{R}^d$. If there exists a linear ordering of $\mathcal{F}$ such that every $2d$ or fewer members of $\mathcal{F}$ can be intersected by a directed line in the given order, then $\mathcal{F}$ has a line transversal.

The special case when the members of $\mathcal{F}$ are congruent was established in [CGH$^+$08], and it was shown by Cheong et al. [CGH12] that the number $2d$ in Theorem 4.2.4 is nearly optimal, in particular it cannot be reduced to $2d-2$, and it is conjectured that the correct number should be $2d-1$. The proof of Theorem 4.2.4 uses a homotopy argument due to Klee, which was also used by Hadwiger in his proof of Theorem 4.2.1. Essentially, one continuously contracts the members of $\mathcal{F}$ until the hypothesis is about to fail. In the planar case it is easily seen that the "limiting configuration" consists of precisely three members of $\mathcal{F}$ that "pin" a unique line, but in dimensions greater than two the situation is more difficult to analyze.

We conclude our discussion of Hadwiger's transversal theorem with a colorful generalization due to Arocha et al.

THEOREM 4.2.5 [ABM08]

Let $\mathcal{F}_1$, $\mathcal{F}_2$, and $\mathcal{F}_3$ be nonempty finite families of compact convex sets in the plane. Suppose $\mathcal{F}_1 \cup \mathcal{F}_2 \cup \mathcal{F}_3$ admits a linear ordering $\prec$ such that $B \cap \mathrm{conv}(A \cup C) \neq \emptyset$ for any choice $A \prec B \prec C$ of sets belonging to distinct $\mathcal{F}_i$. Then one of the $\mathcal{F}_i$ has a line transversal.

Note that when $\mathcal{F}_1 = \mathcal{F}_2 = \mathcal{F}_3$ we recover the planar case of Theorem 4.2.2. The proof of Theorem 4.2.5 combines the topological arguments from [Wen90c] with some additional combinatorial arguments. Arocha et al. conjectured that Theorem 4.2.5 holds in every dimension, and some partial results (requiring roughly d^2, rather than $d+1$, color classes) were established in [HR16], but the general conjecture remains open.

THE SPACE OF TRANSVERSALS

Given a family $\mathcal{F}$ of convex sets in $\mathbb{R}^d$, let $\mathcal{T}_k^d(\mathcal{F})$ denote the space of all k-transversals of $\mathcal{F}$. The space of *point* transversals, that is $\mathcal{T}_0^d$, has a relatively simple structure, since the intersection of convex sets is convex. For $k \geq 1$, the space $\mathcal{T}_k^d(\mathcal{F})$ can be much more complicated; it need not be connected, and each connected component may have nontrivial topology. The "combinatorial complexity" of $\mathcal{T}_k^d(\mathcal{F})$ can be measured in various ways. These give rise to problems that are interesting in their own right, but are also closely related to Helly-type transversal theorems. It is usual to witness a drop in the complexity of $\mathcal{T}_k^d(\mathcal{F})$ as one restricts the family $\mathcal{F}$ to more specialized classes of convex sets, and typically, one gets a Helly-type theorem when the complexity of $\mathcal{T}_k^d(\mathcal{F})$ is universally bounded over all families $\mathcal{F}$ within the given class.

If $\mathcal{F}$ is a family of pairwise disjoint convex sets, then a directed line that intersects every member of $\mathcal{F}$ induces a well-defined order on $\mathcal{F}$. Thus an undirected line transversal to $\mathcal{F}$ induces a pair of opposite linear orderings or "permutations" on $\mathcal{F}$. More generally, a family $\mathcal{F}$ of convex sets is $(k-1)$-*separated* if no $k+1$ members have a $(k-1)$-transversal. An oriented k-transversal H intersects a $(k-1)$-separated family $\mathcal{F} = \{K_1, \ldots, K_n\}$ of convex sets in a well-defined k-ordering. The orientation of $(K_{i_0}, K_{i_1}, \ldots, K_{i_k})$ is the orientation in H of any corresponding set of points $(x_{i_0}, x_{i_1}, \ldots, x_{i_k})$, where $x_{i_j} \in K_{i_j} \cap H$. An unoriented k-transversal to a $(k-1)$-separated family $\mathcal{F}$ of convex sets induces a pair of k-orderings on $\mathcal{F}$, consisting of the two k-orderings on $\mathcal{F}$ induced by the two orientations of the k-transversal. We identify each such pair of k-orderings and call this a *geometric permutation* of $\mathcal{F}$.

If $\mathcal{F}$ is $(k-1)$-separated, then two k-transversals that induce different geometric permutations on $\mathcal{F}$ must necessarily belong to different connected components of $\mathcal{T}_k^d(\mathcal{F})$. The converse is also true for hyperplane transversals.

THEOREM 4.2.6 [Wen90b]

Let $\mathcal{F}$ be a $(d-2)$-separated family of compact convex sets in $\mathbb{R}^d$. Two hyperplane transversals induce the same geometric permutation on $\mathcal{F}$ if and only if they lie in the same connected component of $\mathcal{T}_{d-1}^d(\mathcal{F})$.

It is not hard to show that if $|\mathcal{F}| \geq d$, then each connected component of $\mathcal{T}_{d-1}^d(\mathcal{F})$ is contractible. For k-transversals with $0 < k < d-1$, the situation is not so pleasant. There are constructions of families $\mathcal{F}$ of pairwise disjoint translates in

$\mathbb{R}^3$ where $\mathcal{T}_1^3(\mathcal{F})$ has arbitrarily many connected components, each corresponding to the same geometric permutation [HM04]. However, for the specific case of disjoint balls an analogous result holds for line transversals.

THEOREM 4.2.7 [BGP08]

Let $\mathcal{F}$ be a family of pairwise disjoint Euclidean balls in $\mathbb{R}^d$. Two line transversals induce the same geometric permutation on $\mathcal{F}$ if and only if they lie in the same connected component of $\mathcal{T}_1^d(\mathcal{F})$.

In fact, it can also be shown that each connected component of $\mathcal{T}_1^d(\mathcal{F})$ is contractible when $|\mathcal{F}| \geq 2$. It would be interesting to know if similar results hold for families of Euclidean balls in $\mathbb{R}^d$ with respect to k-transversals for $1 < k < d-1$.

A natural problem that arises is to bound the number of distinct geometric permutations that $\mathcal{T}_k^d(\mathcal{F})$ induces on $\mathcal{F}$ in terms of the size of $\mathcal{F}$. Let $g_k^d(n)$ denote the maximum number of geometric permutations over all $(k-1)$-separated families of n convex sets in $\mathbb{R}^d$. It is not hard to show that $g_1^2(n)$ is linear in n, and in this case the precise bound is known. In general, the following is known about $g_k^d(n)$:

THEOREM 4.2.8

1. $g_1^2(n) = 2n-2$ [ES90].
2. $g_1^d(n) = \Omega(n^{d-1})$ [KLL92].
3. $g_{d-1}^d(n) = O(n^{d-1})$ [Wen90a].
4. $g_k^d(n) = O(n^{k(k+1)(d-k)})$ for fixed k and d [GPW96].

One of the longest-standing conjectures concerning geometric permutations is that the lower bound in part 2 of Theorem 4.2.8 is tight, that is, $g_1^d(n) = \Theta(n^{d-1})$. For nearly 20 years, the best known upper bound was $g_1^d(n) = O(n^{2d-2})$ [Wen90a], until Rubin et al. improved the bound using the Clarkson-Shor probabilistic analysis and a charging scheme technique developed by Tagansky.

THEOREM 4.2.9 [RKS12]

The maximum number of geometric permutations induced by the line transversals to a family of n pairwise disjoint convex sets in $\mathbb{R}^d$ is $O(n^{2d-3}\log n)$.

By further specializing the shape of the convex sets of the family, it is possible to obtain sharper bounds on the number of geometric permutations induced by line transversals. Smorodinsky et al. [SMS00], showed that any family $\mathcal{F}$ of n pairwise disjoint Euclidean balls in $\mathbb{R}^d$ has a "separating set" of size linear in n. In other words, there exists a family $\mathcal{H}$ of hyperplanes in $\mathbb{R}^d$, with $|\mathcal{H}| = O(n)$, such that every pair of members in $\mathcal{F}$ can be separated by some hyperplane in $\mathcal{H}$. This "separation lemma" is specific to families of "fat" convex objects [KV01] and does not generalize to arbitrary families of convex sets. As a consequence we have the following.

THEOREM 4.2.10 [SMS00]

The maximum number of geometric permutations induced by the line transversals to a family of n pairwise disjoint Euclidean balls in $\mathbb{R}^d$ is $\Theta(n^{d-1})$.

This theorem shows that the conjectured bound for $g_1^d(n)$ is tight for the case of families of pairwise disjoint Euclidean balls in $\mathbb{R}^d$. See also [AS05] for other partial results.

CONJECTURE 4.2.11

The maximum number of geometric permutations induced by the line transversals to a family of n pairwise disjoint convex sets in $\mathbb{R}^d$ is $\Theta(n^{d-1})$.

For families of pairwise disjoint translates in the plane, the bounds on the maximum number of geometric permutations can be reduced even further.

THEOREM 4.2.12 [KLL87, KLL92]

The line transversals to a family of pairwise disjoint translates of a compact convex set in $\mathbb{R}^2$ induce at most three geometric permutations.

The possible patterns of geometric permutations induced on families of pairwise disjoint translates in the plane have also been studied. This work was initiated by Katchalski [Kat86] in connection with a conjecture of Grünbaum, and a complete characterization is given in [AHK$^+$03]. For families of n pairwise disjoint translates in $\mathbb{R}^3$, it is known that the maximum number of geometric permutations is $\Omega(n)$ [AK05]. Again, the situation changes drastically when we restrict our attention to the Euclidean ball.

THEOREM 4.2.13 [HXC01, KSZ03, CGN05, NCG$^+$16]

The line transversals to a family of pairwise disjoint translates of the Euclidean ball in $\mathbb{R}^d$ induce at most three distinct geometric permutations. Furthermore, if the family has size at least 7, then there are at most two distinct geometric permutations.

The planar version of this result was proved by Smorodinsky et al. [SMS00], and improved in [AHK$^+$03] where it was shown that a family of at least *four* pairwise disjoint unit disks in the plane can have at most two distinct geometric permutations.

CONJECTURE 4.2.14 [NCG$^+$16]

The line transversals to a family of at least four disjoint unit balls in $\mathbb{R}^3$ induce at most two geometric permutations.

We now discuss a different type of "combinatorial complexity" of the space of transversals. If the members of $\mathcal{F}$ are closed, then the boundary of $\mathcal{T}_k^d(\mathcal{F})$ consists of k-flats that support one or more members of $\mathcal{F}$. This boundary can be partitioned into subspaces of k-flats that support the same subfamily of $\mathcal{F}$. Each of these subspaces can be further partitioned into connected components. The *combinatorial complexity* of $\mathcal{T}_k^d(\mathcal{F})$ is the number of such connected components.

Even in $\mathbb{R}^2$, the boundaries of two convex sets can intersect in an arbitrarily large number of points and may therefore have an arbitrarily large number of common supporting lines. Thus the space of line transversals to two convex sets in $\mathbb{R}^2$ can have arbitrarily large combinatorial complexity. However, if $\mathcal{F}$ consists of pairwise disjoint convex sets in $\mathbb{R}^2$ or, more generally, suitably separated convex sets in $\mathbb{R}^d$, then the complexity is bounded. If the convex sets have constant description complexity, then again the transversal space complexity is bounded. Finally, if the sets are convex polytopes, then the transversal space is bounded by the total number of polytope faces. Table 4.2.1 gives bounds on the transversal space complexity for various families of sets.

The function $\alpha(n)$ is the very slowly growing inverse of the Ackermann function. The function $\lambda_s(n)$ is the maximum length of an (n,s) Davenport-Schinzel sequence,

TABLE 4.2.1 Bounds on $\mathcal{T}_k^d(\mathcal{F})$.

FAMILY $\mathcal{F}$	k	d	COMPLEXITY OF $\mathcal{T}_k^d(\mathcal{F})$	SOURCE
$(d-2)$-separated family of n compact and strictly convex sets	$d-1$	d	$O(n^{d-1})$	[CGP+94]
n connected sets such that any two sets have at most s common supporting lines	1	2	$O(\lambda_s(n))$	[AB87]
n convex sets with const. description complexity	1	3	$O(n^{3+\epsilon})$ for any $\epsilon > 0$	[KS03]
n convex sets with const. description complexity	2	3	$O(n^{2+\epsilon})$ for any $\epsilon > 0$	[ASS96]
n convex sets with const. description complexity	3	4	$O(n^{3+\epsilon})$ for any $\epsilon > 0$	[KS03]
n line segments	$d-1$	d	$O(n^{d-1})$	[PS89]
Convex polytopes with a total of n_f faces	$d-1$	d	$O(n_f^{d-1}\alpha(n_f))$	[PS89]
Convex polytopes with a total of n_f faces	1	3	$O(n_f^{3+\epsilon})$ for any $\epsilon > 0$	[Aga94]
n $(d-1)$-balls	$d-1$	d	$O(n^{\lceil d/2 \rceil})$	[HII+93]

which equals $n\alpha(n)^{O(\alpha(n)^{s-3})}$. Note that $\lambda_s(n) \in O(n^{1+\epsilon})$ for any $\epsilon > 0$. The asymptotic bounds on the worst case complexity of hyperplane transversals ($k = d-1$) to line segments and convex polytopes are tight. There are examples of families $\mathcal{F}$ of convex polytopes where the complexity of $\mathcal{T}_1^3(\mathcal{F})$ is $\Omega(n_f^3)$.

As may be expected, the time to construct a representation of $\mathcal{T}_k^d(\mathcal{F})$ is directly related to the complexity of $\mathcal{T}_k^d(\mathcal{F})$. Most algorithms use upper and lower envelopes to represent and construct $\mathcal{T}_k^d(\mathcal{F})$. Table 4.2.2 gives known bounds on the worst case time to construct a representation of the space $\mathcal{T}_k^d(\mathcal{F})$ for various families of convex sets. All sets are assumed to be compact. As noted, for $\mathcal{T}_1^3(\mathcal{F})$ and $\mathcal{T}_3^4(\mathcal{F})$, the bound is for expected running time, not worst case time.

TABLE 4.2.2 Algorithms to construct $\mathcal{T}_k^d(\mathcal{F})$.

FAMILY $\mathcal{F}$	k	d	TIME COMPLEXITY	SOURCE
$(d-2)$-separated family of n strictly convex sets with constant description complexity	$d-1$	d	$O(n^{d-1}\log^2(n))$	[CGP+94]
n convex sets with const. description complexity s.t. any two sets have at most s common supporting lines	1	2	$O(\lambda_s(n)\log n)$	[AB87]
n convex sets with const. description complexity	1	3	$O(n^{3+\epsilon})$ $\forall\, \epsilon > 0$ (exp'd.)	[KS03]
n convex sets with const. description complexity	2	3	$O(n^{2+\epsilon})$ $\forall\, \epsilon > 0$	[ASS96]
n convex sets with const. description complexity	3	4	$O(n^{3+\epsilon})$ $\forall\, \epsilon > 0$ (exp'd.)	[KS03]
Convex polygons with a total of n_f faces	1	2	$\Theta(n_f\log(n_f))$	[Her89]
Convex polytopes with a total of n_f faces	1	3	$O(n_f^{3+\epsilon})$ $\forall\, \epsilon > 0$	[PS92]
Convex polytopes with a total of n_f faces	2	3	$\Theta(n_f^2\alpha(n_f))$	[EGS89]
Convex polytopes with a total of n_f faces	$d-1$	d	$O(n_f^d)$, $d > 3$	[PS89]
n $(d-1)$-balls	$d-1$	d	$O(n^{\lceil d/2 \rceil+1})$	[HII+93]
n convex homothets	1	2	$O(n\log(n))$	[Ede85]
n pairwise disjoint translates of a convex set with constant description complexity	1	2	$O(n)$	[EW89]

The model of computation used in the lower bound for the time to construct $\mathcal{T}_1^2(\mathcal{F})$ is an algebraic decision tree. In the worst case, $\mathcal{T}_2^3(\mathcal{F})$ may have $\Omega(n_f^2\alpha(n_f))$ complexity, which is a lower bound for constructing $\mathcal{T}_2^3(\mathcal{F})$. Similarly, $\mathcal{T}_1^3(\mathcal{F})$ may have $\Omega(n_f^3)$ complexity, giving an $\Omega(n_f^3)$ lower bound for the time to construct it.

HELLY-TYPE THEOREMS

As remarked earlier, there is no Helly-type theorem for general families of convex sets with respect to k-transversals for $k \geq 1$. Therefore, one direction of research in geometric transversal theory has focused on finding restricted classes of families of convex sets for which there is a Helly-type theorem. The earliest such results were established by Santaló.

THEOREM 4.2.15 [San40]

Let $\mathcal{F}$ be a family of parallelotopes in $\mathbb{R}^d$ with edges parallel to the coordinate axes.

1. *If any $2^{d-1}(d+1)$ or fewer members of $\mathcal{F}$ have a hyperplane transversal, then $\mathcal{F}$ has a hyperplane transversal.*
2. *If any $2^{d-1}(2d-1)$ or fewer members of $\mathcal{F}$ have a line transversal, then $\mathcal{F}$ has a line transversal.*

It is known that the Helly-number $2^{d-1}(d+1)$ for hyperplane transversals is tight, but for the case of line transversals the correct Helly-number seems to be unknown. Grünbaum [Grü64] generalized Theorem 4.2.15 to families of polytopes whose vertex cones are "related" to those of a fixed polytope in $\mathbb{R}^d$.

Let us remark that Theorem 4.2.15 can be deduced from Kalai and Meshulam's topological version of Amenta's theorem (Theorem 4.1.4). To see this, note that it is no loss in generality to assume that any subfamily of $\mathcal{F}$ that admits a transversal, admits a transversal that is not orthogonal to any of the coordinate axes. We can then partition the set of all hyperplanes (or lines) that are not orthogonal to any coordinate axis into 2^{d-1} distinct "direction classes." The next step is to show that the space of transversals (to any given subfamily) within a fixed direction class is a contractible set. Now Theorem 4.2.15 follows from Theorem 4.1.4 since the space of hyperplanes in $\mathbb{R}^d$ is d-dimensional and the space of lines in $\mathbb{R}^d$ is $(2d-2)$-dimensional.

In most cases, when a certain class of families of convex sets admits a Helly-type theorem for k-transversals, it is possible to show that the Helly-number is bounded by applying Theorem 4.1.8. This reduces the problem of bounding the Helly-number to the problem of bounding the complexity of the space of transversals. The resulting upper bound on the Helly-number will in general be very large, and often more direct arguments are needed to obtain sharper bounds.

Let us illustrate how Theorem 4.1.8 can be used to prove an upper bound on the Helly-number. A finite family $\mathcal{F}$ of compact convex sets in $\mathbb{R}^d$ is *ϵ-scattered* if, for every $0 < j < d$, any j of the sets can be separated from any other $d-j$ of the sets by a hyperplane whose distance is more than $\epsilon D(\mathcal{F})/2$ away from all d of the sets, where $D(\mathcal{F})$ is the largest diameter of any member of $\mathcal{F}$. It can be shown that for every $\epsilon > 0$ there exists a constant $C_d(\epsilon)$ such that the space of hyperplane transversals of an ϵ-scattered family induces at most $C_d(\epsilon)$ distinct geometric permutations. By Theorem 4.2.6, this implies that the space of hyperplane transversals has at most

$C_d(\epsilon)$ connected components. Moreover, each connected component is contractible if $|\mathcal{F}| \geq d$, and if $|\mathcal{F}| = k < d$, then the space of hyperplane transversals is homotopy equivalent to $\mathbb{RP}^{d-k}$. For each member $K \in \mathcal{F}$, let $H_K = \mathcal{T}^d_{d-1}(\{K\})$, that is, the set of hyperplanes that intersect K. Note that the family $\{H_K\}_{K\in\mathcal{F}}$ can be parameterized as a family of compact subsets of $\mathbb{R}^{2d}$ and that the Betti-numbers of the intersection of any subfamily can be bounded by some absolute constant. Therefore, Theorem 4.1.8 implies that there is a finite Helly-number (depending on ϵ and d) for hyperplane transversals to ϵ-scattered families of convex sets in $\mathbb{R}^d$.

By a direct geometric argument, the Helly-number can be reduced drastically under the assumption that the family $\mathcal{F}$ is sufficiently large.

THEOREM 4.2.16 [AGPW01]

For any integer $d > 1$ and real $\epsilon > 0$ there exists a constant $N = N_d(\epsilon)$, such that the following holds. If $\mathcal{F}$ is an ϵ-scattered family of at least N compact convex sets in $\mathbb{R}^d$ and every $2d+2$ members of $\mathcal{F}$ have a hyperplane transversal, then $\mathcal{F}$ has a hyperplane transversal.

Goodman and Pollack conjectured that there is a bounded Helly-number for plane transversals to 1-separated families translates of the Euclidean ball in $\mathbb{R}^3$. A counter-example to this conjecture was given in [Hol07], and can be seen as a consequence of the fact that there exists a 1-separated family $\mathcal{F}$ of unit balls in $\mathbb{R}^3$ whose plane transversals induce a linear number of geometric permutations on $\mathcal{F}$. This example illustrates that in order to obtain a bounded Helly-number for hyperplane transversals it is necessary to assume that the family is ϵ-scattered.

From the previous discussion it is clear that there is a bounded Helly-number for *line transversals* to families of disjoint unit balls in $\mathbb{R}^d$. This follows from Theorems 4.2.7, 4.2.13, and 4.1.8 as in the argument above. By a more direct argument based on Theorem 4.2.4, the Helly-number can be reduced even further.

THEOREM 4.2.17 [CGH+08]

Let $\mathcal{F}$ be a family of pairwise disjoint unit balls in $\mathbb{R}^d$. If any $4d-1$ or fewer members of $\mathcal{F}$ have a line transversal, then $\mathcal{F}$ has a line transversal.

The planar case of Theorem 4.2.17 was proven by Danzer [Dan57] with the optimal Helly-number five, who also conjectured that the Helly-number was bounded for arbitrary dimension. For $d \geq 6$ the Helly-number can be further reduced to $4d-2$ [CVGG14]. A lower bound construction [CGH12] shows that the bound is tight up to a factor of two.

Motivated by Danzer's Helly-theorem for line transversals to unit disks in the plane, Grünbaum [Grü58] proved a Helly-theorem for line transversals to families of pairwise disjoint translates of a parallelogram in the plane, again with Helly-number five. The same result for arbitrary families of pairwise disjoint translates was proved by Tverberg.

THEOREM 4.2.18 [Tve89]

Let $\mathcal{F}$ be a family of pairwise disjoint translates of a compact convex set in $\mathbb{R}^2$. If every five or fewer members of $\mathcal{F}$ have a line transversal, then $\mathcal{F}$ has a line transversal.

The fact that there is a bounded Helly-number follows from Theorems 4.2.6, 4.2.12, and 4.1.8, but obtaining the optimal Helly-number is significantly more difficult. Tverberg's proof involves a detailed analysis of the patterns of the possible geometric permutations of families of translates, and it would be interesting to find a simpler proof. Several generalizations of Theorems 4.2.18 are known. For instance, the condition that the members of $\mathcal{F}$ are pairwise disjoint can be weakened and Robinson [Rob97] showed that for every $j > 0$ there exists a number $c(j)$ such that if $\mathcal{F}$ is a family of translates of a compact convex set in $\mathbb{R}^2$ such that the intersection of any j members of $\mathcal{F}$ is empty and such that every $c(j)$ or fewer members of $\mathcal{F}$ have a line transversal, then $\mathcal{F}$ has a line transversal. See also [BBC$^+$06] for further variations.

For families of pairwise disjoint translates in $\mathbb{R}^3$ there is no universal bound on the Helly-number with respect to line transversals. For any integer $n > 2$, there exists a family $\mathcal{F}$ of n pairwise disjoint translates of a compact convex set such that every $n-1$ members of $\mathcal{A}$ have a line transversal, but $\mathcal{F}$ does not have a line transversal [HM04]. This specific example shows that there can be arbitrarily many connected components in the space of transversals that all correspond to the same geometric permutation, and again illustrates that there is no Helly-theorem when the complexity of the space of transversals is unbounded.

PARTIAL TRANSVERSALS AND GALLAI-TYPE PROBLEMS

Even though there is no pure Helly-type theorem for line transversals to general families of convex sets in the plane, there is still much structure to a family of convex sets that satisfy the local Helly-property with respect to line transversals. In particular, families of convex sets in the plane admit a so-called *Gallai-type* theorem. In other words, for a family $\mathcal{F}$ of convex sets in the plane, if every k or fewer members of $\mathcal{F}$ have a line transversal, then there exists a small number of lines whose union intersects every member of $\mathcal{F}$. The natural problem is to determine the smallest number of lines that will suffice, and Eckhoff has determined nearly optimal bounds for all values of $k \geq 3$.

THEOREM 4.2.19 [Eck73, Eck93a]

Let $\mathcal{F}$ be a finite family of convex sets in $\mathbb{R}^2$.

1. *If every four or fewer members of $\mathcal{F}$ have a line transversal, then there are two lines whose union intersects every member of $\mathcal{F}$.*
2. *If every three or fewer members of $\mathcal{F}$ have a line transversal, then there are four lines whose union intersects every member of $\mathcal{F}$.*

The bound in part 1 is obviously tight, and the lines can actually be chosen to be orthogonal. The proof of part 1 is relatively simple and follows by choosing a minimal element of a suitably chosen partial order on the set of orthogonal pairs of lines in the plane. The proof of part 2 is more difficult and is a refinement of an argument due to Kramer [Kra74]. The idea is to consider a pair of disjoint members of $\mathcal{F}$ that are extremal in the sense that their separating tangents form the smallest angle among all disjoint pairs in the family. Using this pair he defines four candidate lines, and proceeds to show that through a series of rotations and translations of these lines one can reach a position in which their union meets every member of $\mathcal{F}$.

CONJECTURE 4.2.20 [Eck93a]

If every three or fewer members of $\mathcal{F}$ have a line transversal, then there are three lines whose union intersects every member of $\mathcal{F}$.

A different question one can consider is to look for the largest *partial transversal.* This was considered by Katchalski, who in 1978 conjectured that if $\mathcal{F}$ is a finite family of convex sets in the plane such that every three members of $\mathcal{F}$ have a line transversal, then there is a line that intersects at least $\frac{2}{3}|\mathcal{F}|$ members of $\mathcal{F}$. Katchalski and Liu considered the more general problem, when every k or fewer members have a line transversal.

THEOREM 4.2.21 [KL80a]

For any integer $k \geq 3$ there exists a real number $\rho = \rho(k)$ such that the following holds. Let $\mathcal{F}$ be a family of compact convex sets in the plane. If every k or fewer members of $\mathcal{F}$ have a line transversal, then there is a subfamily of $\mathcal{F}$ of size at least $\rho|\mathcal{F}|$ that has a transversal. Moreover, $\rho(k) \to 1$ as $k \to \infty$.

In view of the absence of a Helly-type theorem for line transversals, Theorem 4.2.21 is quite remarkable. It is an interesting problem to determine the optimal values for $\rho(k)$, especially for small values of k. The current best bounds for $\rho(k)$ are given in [Hol10]: the upper bound is $\rho(k) \leq 1 - \frac{1}{k-1}$, disproving Katchalski's conjecture for $\rho(3)$, and a lower bound is of order $1 - \frac{\log k}{k}$. For small values of k, the known bounds are $1/3 \leq \rho(3) \leq 1/2$, $1/2 \leq \rho(4) \leq 2/3$, $1/2 \leq \rho(5) \leq 3/4$, etc.

Alon and Kalai [AK95] showed that the family of all convex sets in $\mathbb{R}^d$ has fractional Helly-number $d+1$ with respect to hyperplane transversals. In particular, for every $\alpha > 0$ there exists a $\beta > 0$ (which depends only on α and d) such that if $\mathcal{F}$ is a finite family of convex sets in $\mathbb{R}^d$ such that $\alpha\binom{|\mathcal{F}|}{d+1}$ of the $(d+1)$-tuples of $\mathcal{F}$ have a hyperplane transversal, then there exists a hyperplane that intersects at least $\beta|\mathcal{F}|$ members of $\mathcal{F}$. By the techniques developed in [AKM$^+$02] this implies the following (p,q)-theorem for hyperplane transversals.

THEOREM 4.2.22 [AK95]

For any integers $p \geq q \geq d+1$ there exists an integer $c = c(p,q,d)$ such that the following holds. If $\mathcal{F}$ is a finite family of at least p convex sets in $\mathbb{R}^d$ and out of every p members of $\mathcal{F}$ there are some q that have a hyperplane transversal, then there are c hyperplanes whose union intersects every member of $\mathcal{F}$.

The theorem above can not be extended to k-transversals for $0 < k < d-1$. In particular, for every $k \geq 3$, Alon et al. [AKM$^+$02] construct a family of convex sets in $\mathbb{R}^3$ in which every k members have a line transversal, but no $k+4$ members have a line transversal.

Problems concerning partial transversals have also been studied for families of pairwise disjoint translates in the plane. Katchalski and Lewis [KL80b] showed that there is a universal constant C such that if $\mathcal{F}$ is a family of pairwise disjoint translates of a compact convex set K in the plane such that every three members of $\mathcal{F}$ have a line transversal, then there is a line that intersects all but at most C members of $\mathcal{F}$. The original upper bound given in [KL80b] was $C \leq 603$, but they conjectured that $C = 2$. Bezdek [Bez94] gave an example of $n \geq 6$ pairwise disjoint congruent disks where every 3 have a line transversal, but no line meets more than $n-2$ of the disks, showing that the Katchalski–Lewis conjecture is best possible.

Heppes [Hep07] proved that the Katchalski–Lewis conjecture holds for the disk.

THEOREM 4.2.23 [Hep07]

Let $\mathcal{F}$ be a family of pairwise disjoint congruent disks in the plane. If every three members of $\mathcal{F}$ have a line transversal, then there is a line that intersects all but at most two members of $\mathcal{F}$.

It turns out that the conjectured shape-independent upper bound $C = 2$ does not hold in general. It was shown in [Hol03] that, for any $n \geq 12$, there exists a family $\mathcal{F}$ of n pairwise disjoint translates of a parallelogram where every three members of $\mathcal{F}$ have a line transversal, but no line meets more than $n - 4$ of the members of $\mathcal{F}$. For families of disjoint translates of a general convex set K, the best known upper bound is $C \leq 22$ [Hol03].

In view of Theorem 4.2.18 and Theorem 4.2.23 it is natural to ask about families of disjoint translates in the plane where every four or fewer members have a line transversal. This problem was originally investigated by Katchalski and Lewis [KL82] for families of disjoint translates of a parallelogram who showed that there exists a line that intersects all but at most two members of the family. In a series of papers, Bisztriczky et al. [BFO05a, BFO05b, BFO08] investigated the same problem for families of disjoint congruent disks, and obtained the following result.

THEOREM 4.2.24 [BFO08]

Let $\mathcal{F}$ be a family of pairwise disjoint congruent disks in the plane. If every four members of $\mathcal{F}$ have a line transversal, then there is a line that intersects all but at most one member of $\mathcal{F}$.

The result is best possible, as can be seen from a family of five nearly touching unit disks with centers placed at the vertices of a regular pentagon. It is not known whether Theorem 4.2.24 generalizes to families of disjoint translates of a compact convex set in the plane, but combining Theorem 4.2.18 with an elementary combinatorial argument shows that if a counter-example exists, then there exists one consisting of at most 12 translates.

We conclude this section with another transversal problem concerning congruent disks in the plane. Let D denote a closed disk centered at the origin in $\mathbb{R}^2$, let $\mathcal{F} = \{x_i + D\}$ be a finite family of translates of D, and let $\lambda\mathcal{F} = \{x_i + \lambda D\}$ denote the family where each disk has been inflated by a factor of $\lambda > 0$ about its center.

CONJECTURE 4.2.25 *Dol'nikov, Eckhoff*

If every three members of $\mathcal{F} = \{x_i + D\}$ have a line transversal, then the family $\lambda\mathcal{F} = \{x_i + \lambda D\}$ has a line transversal for $\lambda = \frac{1+\sqrt{5}}{2}$.

It is easily seen the factor $\lambda = \frac{1+\sqrt{5}}{2}$ is best possible by centering the disks at the vertices of a regular pentagon. Heppes [Hep05] has shown that $\lambda \leq 1.65$, under the additional assumption that the members of $\mathcal{F}$ are pairwise disjoint. If we instead assume that every *four* members of $\mathcal{F}$ have a line transversal, Jerónimo [Jer07] has shown that $\lambda = \frac{1+\sqrt{5}}{2}$ is the optimal inflation factor.

4.3 SOURCES AND RELATED MATERIAL

SURVEYS

[DGK63]: The classical survey of Helly's theorem and related results.

[Eck93b]: A survey of Helly's theorem and related results, updating the material in [DGK63].

[GPW93]: A survey of geometric transversal theory.

[SA95]: Contains applications of Davenport-Schinzel sequences and upper and lower envelopes to geometric transversals.

[Mat02]: A textbook covering many aspects of discrete geometry including the fractional Helly theorem and the (p, q)-problem.

[Eck03]: A survey on the Hadwiger–Debrunner (p, q)-problem.

[Tan13]: A survey on intersection patterns of convex sets from the viewpoint of nerve complexes.

[ADLS17]: A recent survey on Helly's theorem focusing on the developments since [Eck93b].

RELATED CHAPTERS

Chapter 2: Packing and covering
Chapter 3: Tilings
Chapter 6: Oriented matroids
Chapter 17: Face numbers of polytopes and complexes
Chapter 21: Topological methods in discrete geometry
Chapter 28: Arrangements
Chapter 41: Ray shooting and lines in space
Chapter 49: Linear programming

REFERENCES

[AB87] M. Atallah and C. Bajaj. Efficient algorithms for common transversals. *Inform. Process. Lett.*, 25:87–91, 1987.

[ABD+16] I. Aliev, R. Bassett, J.A. De Loera, and Q. Louveaux. A quantitative Doignon-Bell-Scarf theorem. *Combinatorica*, online first, 2016.

[ABFK92] N. Alon, I. Bárány, Z. Füredi, and D.J. Kleitman. Point selections and weak e-nets for convex hulls. *Combin. Probab. Comput.*, 1:189–200, 1992.

[ABM08] J.L. Arocha, J. Bracho, and L. Montejano. A colorful theorem on transversal lines to plane convex sets. *Combinatorica*, 28:379–384, 2008.

[ABM+02] J.L. Arocha, J. Bracho, L. Montejano, D. Oliveros, and R. Strausz. Separoids, their categories and a Hadwiger-type theorem for transversals. *Discrete Comput. Geom.*, 27:377–385, 2002.

[ADL14] I. Aliev, J.A. De Loera, and Q. Louveaux. Integer programs with prescribed number of solutions and a weighted version of Doignon-Bell-Scarf's theorem. In *Proc. 17th Conf. Integer Progr. Combin. Opt.*, vol. 8494 of *LNCS*, pages 37–51, Springer, Berlin, 2014.

[ADLS17] N. Amenta, J.A. De Loera, and P. Soberón. Helly's theorem: New variations and applications. in H.A. Harrington, M. Omar, and M. Wright, editors, *Algebraic and Geometric Methods in Discrete Mathematics*, vol. 685 of *Contemp. Math.*, pages 55–96, Amer. Math. Soc., Providence, 2017.

[Aga94] P.K. Agarwal. On stabbing lines for polyhedra in 3D. *Comput. Geom.*, 4:177–189, 1994.

[AGP$^+$17] G. Averkov, B. González Merino, I. Paschke, M. Schymura, and S. Weltge. Tight bounds on discrete quantitative Helly numbers. *Adv. Appl. Math.*, 89:76–101, 2017.

[AGPW01] B. Aronov, J.E. Goodman, R. Pollack, and R. Wenger. A Helly-type theorem for hyperplane transversals to well-separated convex sets. *Discrete Comput. Geom.*, 25:507–517, 2001.

[AH35] P. Alexandroff and H. Hopf. *Topologie I*, vol. 45 of *Grundlehren der Math.* Springer-Verlag, Berlin, 1935.

[AHK$^+$03] A. Asinowski, A. Holmsen, M. Katchalski, and H. Tverberg. Geometric permutations of large families of translates. In: Aronov et al., editors, *Discrete and Computational Geometry: The Goodman-Pollack Festschrift*, vol. 25 of *Algorithms and Combinatorics*, pages 157–176, Spinger, Heidelberg, 2003.

[AK85] N. Alon and G. Kalai. A simple proof of the upper bound theorem. *European J. Combin.* 6:211–214, 1985.

[AK92] N. Alon and D. Kleitman. Piercing convex sets and the Hadwiger–Debrunner (p, q)-problem. *Adv. Math.*, 96:103–112, 1992.

[AK95] N. Alon and G. Kalai. Bounding the piercing number. *Discrete Comput. Geom.*, 13:245–256, 1995.

[AK05] A. Asinowski and M. Katchalski. The maximal number of geometric permutations for n disjoint translates of a convex set in $\mathbb{R}^3$ is $\Omega(n)$. *Discrete Comput. Geom.*, 35:473–480, 2006.

[AKM$^+$02] N. Alon, G. Kalai, J. Matoušek, and R. Meshulam. Transversal numbers for hypergraphs arising in geometry. *Adv. Appl. Math.*, 29:79–101, 2002.

[Ame96] N. Amenta. A short proof of an interesting Helly-type theorem. *Discrete Comput. Geom.*, 15:423–427, 1996.

[AS05] B. Aronov and S. Smorodinsky. On geometric permutations induced by lines transversal through a fixed point. *Discrete Comput. Geom.*, 34:285–294, 2005.

[ASS96] P.K. Agarwal, O. Schwarzkopf, and M. Sharir. The overlay of lower envelopes and its applications. *Discrete Comput. Geom.*, 15:1–13, 1996.

[Ave13] G. Averkov. On maximal S-free sets and the Helly number for the family of S-convex sets. *SIAM J. Discrete Math.*, 27:1610–1624, 2013.

[AW96] L. Anderson and R. Wenger. Oriented matroids and hyperplane transversals. *Adv. Math.*, 119:117–125, 1996.

[AW12] G. Averkov and R. Weismantel. Transversal numbers over subsets of linear spaces. *Adv. Geom.*, 12:19–28, 2012.

[Bár82] I. Bárány. A generalization of Carathéodory's theorem. *Discrete Math.*, 40:141–152, 1982.

[BBC$^+$06] K. Bezdek, T. Bisztriczky, B. Csikós, and A. Heppes. On the transversal Helly numbers of disjoint and overlapping disks. *Arch. Math.*, 87:86–96, 2006.

[Bel77] D.E. Bell. A theorem concerning the integer lattice. *Stud. Appl. Math.*, 56:187–188, 1977.

[Bez94] A. Bezdek. On the transversal conjecture of Katchalski and Lewis. In G. Fejes Tóth, editor, *Intuitive Geometry*, vol. 63 of *Colloquia Math. Soc. János Bolyai*, pages 23–25, North-Holland, Amsterdam, 1991.

[BFM$^+$14] I. Bárány, F. Fodor, L. Montejano, and D. Oliveros. Colourful and fractional (p,q)-theorems. *Discrete Comput. Geom.*, 51:628–642, 2014.

[BFO05a] T. Bisztriczky, F. Fodor, and D. Oliveros. Large transversals to small families of unit disks. *Acta. Math. Hungar.*, 106:273–279, 2005.

[BFO05b] T. Bisztriczky, F. Fodor, and D. Oliveros. A transversal property to families of eight or nine unit disks. *Bol. Soc. Mat. Mex.*, 3:59–73, 2005.

[BFO08] T. Bisztriczky, F. Fodor, and D. Oliveros. The $T(4)$ property of families of unit disks. *Isr. J. Math*, 168:239–252, 2008.

[BGP08] C. Borcea, X. Goaoc, and S. Petitjean. Line transversals to disjoint balls. *Discrete Comput. Geom.*, 39:158–173, 2008.

[BKP84] I. Bárány, M. Katchalski, and J. Pach. Helly's theorem with volumes. *Amer. Math. Monthly*, 91:362–365, 1984.

[BM03] I. Bárány and J. Matoušek. A fractional Helly theorem for convex lattice sets. *Adv. Math.*, 174:227–235, 2003.

[BO97] I. Bárány and S. Onn. Carathéodory's theorem, colorful and applicable. In I. Bárány and K. Böröczky, editors, *Intuitive Geometry*, vol. 6 of *Bolyai Soc. Math. Stud.*, pages 11–21, János Bolyai Math. Soc., Budapest, 1997.

[Bra17] S. Brazitikos. Brascamp-Lieb inequality and quantitative versions of Helly's theorem. *Mathematika.*, 63:272–291, 2017.

[BV82] E.O. Buchman and F.A. Valentine. Any new Helly numbers? *Amer. Math. Monthly*, 89:370–375, 1982.

[CGH12] O. Cheong, X. Goaoc, and A. Holmsen. Lower bounds to Helly numbers of line transversals to disjoint congruent balls. *Israel J. Math.*, 190:213–228, 2012.

[CGH$^+$08] O. Cheong, X. Goaoc, A. Holmsen, and S. Petitjean. Helly-type theorems for line transversals to disjoint unit balls. *Discrete Comput. Geom.*, 39:194–212, 2008.

[CGN05] O. Cheong, X. Goaoc, and H.-S. Na. Geometric permutations of disjoint unit spheres. *Comput. Geom.*, 30:253–270, 2005.

[CGP$^+$94] S.E. Cappell, J.E. Goodman, J. Pach, R. Pollack, M. Sharir, and R. Wenger. Common tangents and common transversals. *Adv. Math.*, 106:198–215, 1994.

[CHZ15] S.R. Chestnut, R. Hildebrand, and R. Zenklusen. Sublinear bounds for a quantitative Doignon-Bell-Scarf theorem. Preprint, `arXiv:1512.07126`, 2015.

[CVGG14] É. Colin de Verdière, G. Ginot and X. Goaoc. Helly numbers of acyclic families. *Adv. Math.*, 253:163–193, 2014.

[Dan57] L. Danzer. Über ein Problem aus der kombinatorischen Geometrie. *Arch. Math.*, 8:347–351, 1957

[Dan86] L. Danzer. Zur Lösung des Gallaischen Problems über Kreisscheiben in der euklidischen Ebene. *Studia Sci. Math. Hungar.*, 21:111–134, 1986.

[Deb70] H. Debrunner. Helly type theorems derived from basic singular homology. *Amer. Math. Monthly*, 77:375–380, 1970.

[DGK63] L. Danzer, B. Grünbaum, and V. Klee. Helly's theorem and its relatives. In *Convexity*, vol. 7 of *Proc. Sympos. Pure Math.*, pages 101–180, AMS, Providence, 1963.

[DLL$^+$15a] J.A. De Loera, R.N. La Haye, D. Rolnick, and P. Soberón. Quantitative Tverberg, Helly, & Carathéodory theorems. Preprint, `arXiv:1503.06116`, 2015.

[DLL$^+$15b] J.A. De Loera, R.N. La Haye, D. Oliveros, and E. Roldán-Pensado. Helly numbers of algebraic subsets of $\mathbb{R}^d$. Preprint, `arxiv:1508.02380`, 2015. To appear in *Adv. Geom.*

[Doi73] J.P. Doignon. Convexity in cristallographical lattices. *J. Geometry*, 3:71–85, 1973).

[Eck03] J. Eckhoff. A survey on the Hadwiger–Debrunner (p, q)-problem. In B. Aronov et al., editors, *Discrete and Computational Geometry: The Goodman-Pollack Festschrift*, vol. 25 of *Algorithms and Combinatorics*, pages 347–377, Springer, Berlin, 2003.

[Eck73] J. Eckhoff. Transversalenprobleme in der Ebene. *Arch. Math.*, 24:195–202, 1973.

[Eck85] J. Eckhoff. An upper bound theorem for families of convex sets. *Geom. Dedicata*, 19:217–227, 1985.

[Eck93a] J. Eckhoff. A Gallai-type transversal problem in the plane. *Discrete Comput. Geom.*, 9:203–214, 1993.

[Eck93b] J. Eckhoff. Helly, Radon and Carathéodory type theorems. In P.M. Gruber and J.M. Wills, editors, *Handbook of Convex Geometry*, pages 389–448, North-Holland, Amsterdam, 1993.

[Ede85] H. Edelsbrunner. Finding transversals for sets of simple geometric figures. *Theoret. Comput. Sci.*, 35:55–69, 1985.

[EGS89] H. Edelsbrunner, L.J. Guibas, and M. Sharir. The upper envelope of piecewise linear functions: algorithms and applications. *Discrete Comput. Geom.*, 4:311–336, 1989.

[EN09] J. Eckhoff and K. Nischke. Morris's pigeonhole principle and the Helly theorem for unions of convex sets. *Bull. Lond. Math. Soc.*, 41:577–588, 2009.

[ES90] H. Edelsbrunner and M. Sharir. The maximum number of ways to stab n convex non-intersecting sets in the plane is $2n - 2$. *Discrete Comput. Geom.*, 5:35–42, 1990.

[EW89] P. Egyed and R. Wenger. Stabbing pairwise-disjoint translates in linear time. In *Proc. 5th Sympos. Comput. Geom.*, pages 364–369, ACM Press, 1989.

[Flø11] G. Fløystad. The colorful Helly theorem and colorful resolutions of ideals. *J. Pure Appl. Algebra*, 215:1255–1262, 2011.

[GM61] B. Grünbaum and T.S. Motzkin. On components of families of sets. *Proc. Amer. Math. Soc.*, 12:607-613, 1961.

[GP88] J.E. Goodman and R. Pollack. Hadwiger's transversal theorem in higher dimensions. *J. Amer. Math. Soc.* 1:301–309, 1988.

[GP$^+$15] X. Goaoc, P. Paták, Z. Safernová, M. Tancer, and U. Wagner. Bounding Helly numbers via Betti numbers. In *Proc. 31st Sympos. Comput. Geom.*, vol. 34 of *LIPIcs*, pages 507–521, Schloss Dagstuhl, 2015.

[GPW93] J.E. Goodman, R. Pollack, and R. Wenger. Geometric transversal theory. In J. Pach, editor, *New Trends in Discrete and Computational Geometry*, vol. 10 of *Algorithms and Combinatorics*, pages 163–198, Springer, Berlin, 1993.

[GPW96] J.E. Goodman, R. Pollack, and R. Wenger. Bounding the number of geometric permutations induced by k-transversals. *J. Combin. Theory Ser. A*, 75:187–197, 1996.

[Grü58] B. Grünbaum. On common transversals. *Arch. Math.*, 9:465–469, 1958.

[Grü59] B. Grünbaum. On intersections of similar sets. *Portugal Math.*, 18:155–164, 1959.

[Grü62] B. Grünbaum. The dimension of intersections of convex sets. *Pacific J. Math.*, 12:197–202, 1962.

[Grü64] B. Grünbaum. Common secants for families of polyhedra. *Arch. Math.*, 15:76–80, 1964.

[Had57] H. Hadwiger. Über eibereiche mit gemeinsamer treffgeraden. *Portugal Math.*, 6:23–29, 1957.

[Hal09] N. Halman. Discrete and lexicographic Helly-type theorems. *Discrete Comput. Geom.*, 39:690–719, 2008.

[HD57] H. Hadwiger and H. Debrunner. Über eine Variante zum Helly'schen Satz. *Arch. Math.*, 8:309–313, 1957.

[Hel23] E. Helly. Über Mengen konvexer Körper mit gemeinschaftlichen Punkten. *Jahresber. Deutsch. Math.-Verein.*, 32:175–176, 1923.

[Hel30] E. Helly. Über Systeme abgeschlossener Mengen mit gemeinschaftlichen Punkten. *Monatsh. Math.*, 37:281–302, 1930.

[Hep05] A. Heppes. New upper bound on the transversal width of $T(3)$-families of discs. *Discrete Comput. Geom.*, 34:463–474, 2005.

[Hep07] A. Heppes. Proof of the Katchalski–Lewis transversal conjecture for $T(3)$-families of congruent discs. *Discrete Comput. Geom.*, 38:289–304, 2007.

[Her89] J. Hershberger. Finding the upper envelope of n line segments in $O(n \log n)$ time. *Inform. Process. Lett.*, 33:169–174, 1989.

[HII$^+$93] M.E. Houle, H. Imai, K. Imai, J.-M. Robert, and P. Yamamoto. Orthogonal weighted linear L_1 and L_∞ approximation and applications. *Discrete Appl. Math.*, 43:217–232, 1993.

[HM04] A. Holmsen and J. Matoušek. No Helly theorem for stabbing translates by lines in $\mathbb{R}^3$. *Discrete Comput. Geom.*, 31:405–410, 2004.

[Hof79] A.J. Hoffman. Binding constraints and Helly numbers. In *2nd International Conference on Combinatorial Mathematics*, vol. 319 of *Annals New York Acad. Sci*, pages 284–288, New York, 1979.

[Hol03] A. Holmsen. New bounds on the Katchalski-Lewis transversal problem. *Discrete Comput. Geom.*, 29:395–408, 2003.

[Hol07] A. Holmsen. The Katchalski-Lewis transversal problem in $\mathbb{R}^d$. *Discrete Comput. Geom.*, 37:341–349, 2007.

[Hol10] A.F. Holmsen. New results for $T(k)$-families in the plane. *Mathematika*, 56:86–92, 2010.

[Hol16] A.F. Holmsen. The intersection of a matroid and an oriented matroid. *Adv. Math.*, 290:1–14, 2016.

[HR16] A.F. Holmsen and E. Roldán-Pensado. The colored Hadwiger transversal theorem in $\mathbb{R}^d$. *Combinatorica*, 36:417–429, 2016.

[HXC01] Y. Huang, J. Xu, and D.Z. Chen. Geometric permutations of high dimensional spheres. In *Proc. 12th ACM-SIAM Sympos. Discrete Algorithms*, pages 244–245, 2001.

[Jer07] J. Jerónimo-Castro. Line transversals to translates of unit discs. *Discrete Comput. Geom.*, 37:409–417, 2007.

[JMS15] J. Jerónimo-Castro, A. Magazinov, and P. Soberón. On a problem by Dol'nikov. *Discrete Math.*, 338:1577–1585, 2015.

[Kal84] G. Kalai. Intersection patterns of convex sets. *Israel J. Math.*, 48:161–174, 1984.

[Kar00] R.N. Karasev. Transversals for families of translates of a two-dimensional convex compact set. *Discrete Comput. Geom.*, 24:345–353, 2000.

[Kar08] R.N. Karasev. Piercing families of convex sets with the d-intersection property in $\mathbb{R}^d$. *Discrete Comput. Geom.*, 39:766–777, 2008.

[Kat71] M. Katchalski. The dimension of intersections of convex sets. *Israel J. Math.*, 10:465–470, 1971.

[Kat80] M. Katchalski. Thin sets and common transversals. *J. Geom.*, 14:103–107, 1980.

[Kat86] M. Katchalski. A conjecture of Grünbaum on common transversals. *Math. Scand.*, 59:192–198, 1986.

[KGT99] D.J. Kleitman, A. Gyárfás, and G. Tóth. Convex sets in the plane with three of every four meeting. *Combinatorica*, 21:221–232, 2001.

[KL79] M. Katchalski and A. Liu. A problem of geometry in $\mathbb{R}^n$. *Proc. Amar. Math. Soc.*, 75:284–288, 1979.

[KL80a] M. Katchalski and A. Liu. Symmetric twins and common transversals. *Pacific J. Math*, 86:513–515, 1980.

[KL80b] M. Katchalski and T. Lewis. Cutting families of convex sets. *Proc. Amer. Math. Soc.*, 79:457–461, 1980.

[KL82] M. Katchalski and T. Lewis. Cutting rectangles in the plane. *Discrete Math.*, 42:67–71, 1982.

[Kle53] V. Klee. The critical set of a convex body. *Amer. J. Math.*, 75:178–188, 1953.

[KLL87] M. Katchalski, T. Lewis, and A. Liu. Geometric permutations of disjoint translates of convex sets. *Discrete Math.*, 65:249–259, 1987.

[KLL92] M. Katchalski, T. Lewis, and A. Liu. The different ways of stabbing disjoint convex sets. *Discrete Comput. Geom.*, 7:197–206, 1992.

[KM05] G. Kalai and R. Meshulam. A topological colorful Helly theorem. *Adv. Math.*, 191:305–311, 2005.

[KM08] G. Kalai and R. Meshulam. Leray numbers of projections and a topological Helly-type theorem. *J. Topol.*, 1:551–556, 2008.

[KN99] M. Katchalski and D. Nashtir. A Helly type conjecture. *Discrete Comput. Geom.*, 21:37–43, 1999.

[Kra74] D. Kramer. Transversalenprobleme vom Hellyschen und Gallaischen Typ. Dissertation, Universität Dortmund, 1974.

[KS03] V. Koltun and M. Sharir. The partition technique for overlays of envelopes. *SIAM J. Comput.*, 32:841–863, 2003.

[KSZ03] M. Katchalski, S. Suri, and Y. Zhou. A constant bound for geometric permutations of disjoint unit balls. *Discrete Comput. Geom.*, 29:161–173, 2003.

[KT08] J. Kynčl and M. Tancer. The maximum piercing number for some classes of convex sets with the $(4,3)$-property. *Electron. J. Combin.* 15:27, 2008

[KV01] M.J. Katz and K.R. Varadarajan. A tight bound on the number of geometric permutations of convex fat objects in $\Re^d$. *Discrete Comput. Geom.*, 26:543–548, 2001.

[LS09] M. Langberg and L.J. Schulman. Contraction and expansion of convex sets. *Discrete Comput. Geom.* 42:594–614, 2009.

[Mat97] J. Matoušek. A Helly-type theorem for unions of convex sets. *Discrete Comput. Geom.*, 18:1–12, 1997.

[Mat02] J. Matoušek. *Lectures on Discrete Geometry.* Vol. 212 of *Graduate Texts in Math.*, Springer, New York, 2002.

[Mat04] J. Matoušek. Bounded VC-dimension implies a fractional Helly theorem. *Discrete Comput. Geom.*, 31:251–255, 2004.

[Mon14] L. Montejano. A new topological Helly theorem and some transversal results. *Discrete Comput. Geom.*, 52:390–398, 2014.

[Mor73] H.C. Morris. *Two pigeon hole principles and unions of convexly disjoint sets.* Ph.D. thesis, Caltech, Pasadena, 1973.

[Mül13] T. Müller. A counterexample to a conjecture of Grünbaum on piercing convex sets in the plane. *Discrete Math.*, 313:2868–271, 2013.

[Nas16] M. Naszódi. Proof of a conjecture of Bárány, Katchalski and Pach. *Discrete Comput. Geom.* 55:243–248, 2016.

[NCG+16] J.-S. Haa, O. Cheong, X. Goaoc, and J. Yang. Geometric permutations of non-overlapping unit balls revisited. *Comput. Geom.*, 53:36–50, 2016.

[Pin15] R. Pinchasi. A note on smaller fractional Helly numbers. *Discrete Comput. Geom.*, 54:663–668, 2015.

[PS89] J. Pach and M. Sharir. The upper envelope of piecewise linear functions and the boundary of a region enclosed by convex plates: combinatorial analysis. *Discrete Comput. Geom.*, 4:291–309, 1989.

[PS92] M. Pellegrini and P.W. Shor. Finding stabbing lines in 3-space. *Discrete Comput. Geom.*, 8:191–208, 1992.

[PW90] R. Pollack and R. Wenger. Necessary and sufficient conditions for hyperplane transversals. *Combinatorica*, 10:307–311, 1990.

[RKS12] N. Rubin, H. Kaplan, and M. Sharir. Improved bounds for geometric permutations. *SIAM J. Comput*, 41:367–390, 2012.

[Rob97] J.-M. Robert. Geometric orderings of intersecting translates and their applications. *Comput. Geom.*, 7:59–72, 1997.

[RS17] D. Rolnick and P. Soberón. Quantitative (p, q) theorems in combinatorial geometry. *Discrete Math.*, 340:2516–2527, 2017.

[SA95] M. Sharir and P.K. Agarwal. *Davenport-Schinzel Sequences and Their Geometric Applications.* Cambridge University Press, 1995.

[San40] L.A. Santaló. Un teorema sóbre conjuntos de paralelepipedos de aristas paralelas. *Publ. Inst. Mat. Univ. Nac. Litoral*, 2:49–60, 1940.

[San57] R. De Santis. A generalization of Helly's theorem. *Proc. Amer. Math. Soc.*, 8:336–340, 1957.

[Sar92] K.S. Sarkaria. Tverberg's theorem via number fields. *Isreal. J. Math.*, 79:317–320, 1992.

[Sca77] H.E. Scarf. An observation on the structure of production sets with indivisibilities. *Proc. Nat. Acad. Sci. U.S.A.*, 74:3637–3641, 1977.

[SMS00] S. Smorodinsky, J.S.B. Mitchell, and M. Sharir. Sharp bounds on geometric permutations for pairwise disjoint balls in $\Re^d$. *Discrete Comput. Geom.*, 23:247–259, 2000.

[Sob16] P. Soberón. Helly-type theorems for the diameter. *Bull. London Math. Soc.* 48:577–588, 2016.

[Tan13] M. Tancer. Intersection patterns of convex sets via simplicial complexes, a survey. In J. Pach, editor, *Thirty Essays on Geometric Graph Theory*, pages 521–540, Springer, New York, 2013.

[Tve89] H. Tverberg. Proof of Grünbaum's conjecture on common transversals for translates. *Discrete Comput. Geom.*, 4:191–203, 1989.

[Tve91] H. Tverberg. On geometric permutations and the Katchalski-Lewis conjecture on partial transversals for translates. In J.E. Goodman, R. Pollack, and W. Steiger, editors, *Discrete and Computational Geometry*, vol. 6 of *DIMACS Ser. Discrete Math. Theor. Comp. Sci.*, pages 351–361, AMS, Providence, 1991.

[Vin39] P. Vincensini. Sur une extension d'un théorème de M. J. Radon sur les ensembles de corps convexes. *Bull. Soc. Math. France*, 67:115–119, 1939.

[Wen90a] R. Wenger. Upper bounds on geometric permutations for convex sets. *Discrete Comput. Geom.*, 5:27–33, 1990.

[Wen90b] R. Wenger. Geometric permutations and connected components. Technical Report TR-90-50, DIMACS, Piscataway, 1990.

[Wen90c] R. Wenger. A generalization of Hadwiger's transversal theorem to intersecting sets. *Discrete Comput. Geom.*, 5:383–388, 1990.

5 PSEUDOLINE ARRANGEMENTS

Stefan Felsner and Jacob E. Goodman

INTRODUCTION

Pseudoline arrangements generalize in a natural way arrangements of straight lines, discarding the straightness aspect, but preserving their basic topological and combinatorial properties. Elementary and intuitive in nature, at the same time, by the Folkman-Lawrence topological representation theorem (see Chapter 6), they provide a concrete geometric model for oriented matroids of rank 3.

After their explicit description by Levi in the 1920's, and the subsequent development of the theory by Ringel in the 1950's, the major impetus was given in the 1970's by Grünbaum's monograph *Arrangements and Spreads*, in which a number of results were collected and a great many problems and conjectures posed about arrangements of both lines and pseudolines. The connection with oriented matroids discovered several years later led to further work. The theory is by now very well developed, with many combinatorial and topological results and connections to other areas as for example algebraic combinatorics, as well as a large number of applications in computational geometry. In comparison to arrangements of lines arrangements of pseudolines have the advantage that they are more general and allow for a purely combinatorial treatment.

Section 5.1 is devoted to the basic properties of pseudoline arrangements, and Section 5.2 to related structures, such as arrangements of straight lines, configurations (and generalized configurations) of points, and allowable sequences of permutations. (We do not discuss the connection with oriented matroids, however; that is included in Chapter 6.) In Section 5.3 we discuss the stretchability problem. Section 5.4 summarizes some combinatorial results known about line and pseudoline arrangements, in particular problems related to the cell structure of arrangements. Section 5.5 deals with results of a topological nature and Section 5.6 with issues of combinatorial and computational complexity. Section 5.8 with several applications, including sweeping arrangements and pseudotriangulations.

Unless otherwise noted, we work in the real projective plane $\mathbf{P}^2$.

5.1 BASIC PROPERTIES

GLOSSARY

Arrangement of lines: A labeled set of lines not all passing through the same point (the latter is called a ***pencil***).

Pseudoline: A simple closed curve whose removal does not disconnect $\mathbf{P}^2$.

Arrangement of pseudolines: A labeled set of pseudolines not a pencil, every pair meeting no more than once (hence exactly once and crossing).

Isomorphic arrangements: Two arrangements such that the mapping induced by their labelings is an isomorphism of the cell complexes into which they partition $\mathbf{P}^2$. (Isomorphism classes of pseudoline arrangements correspond to reorientation classes of oriented matroids of rank 3; see Chapter 6.)

Stretchable: A pseudoline arrangement isomorphic to an arrangement of straight lines. Figure 5.1.1 illustrates what was once believed to be an arrangement of straight lines, but which was later proven not to be stretchable. We will see in Section 5.6 that most pseudoline arrangements, in fact, are not stretchable.

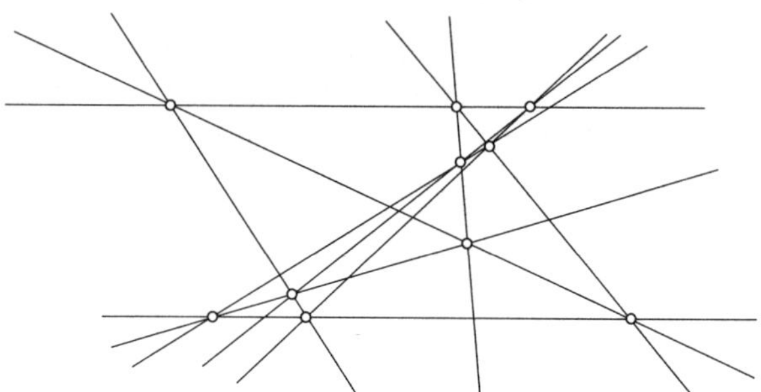

FIGURE 5.1.1
An arrangement of 10 *pseudolines, each containing* 3 *triple points; the arrangement is nonstretchable.*

Vertex: The intersection of two or more pseudolines in an arrangement.

Simple arrangement: An arrangement (of lines or pseudolines) in which there is no vertex where three or more pseudolines meet.

Euclidean arrangement of pseudolines: An arrangement of x-monotone curves in the Euclidean plane, every pair meeting exactly once and crossing there. In this case the canonical labeling of the pseudolines is using $1,\ldots,n$ in upward order on the left and in downward order on the right.

Wiring diagram: A Euclidean arrangement of pseudolines consisting of piecewise linear "wires." The wires (pseudolines) are horizontal except for small neighborhoods of their crossings with other wires; see Figure 5.1.2 for an example.

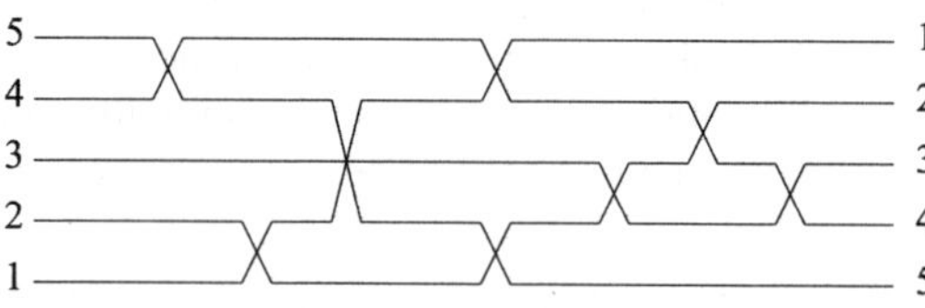

FIGURE 5.1.2
A wiring diagram.

A fundamental tool in working with arrangements of pseudolines, which takes the place of the fact that two points determine a line, is the following.

THEOREM 5.1.1 *Levi Enlargement Lemma* [Lev26]

If $\mathcal{A} = \{L_1,\ldots,L_n\}$ is an arrangement of pseudolines and $p,q \in \mathbf{P}^2$ are two distinct points not on the same member of $\mathcal{A}$, there is a pseudoline L passing through p and q such that $\mathcal{A} \cup \{L\}$ is an arrangement.

Theorem 5.1.1 has been shown by Goodman and Pollack [GP81b] not to extend to arrangements of pseudohyperplanes. It has, however, been extended in [SH91] to the case of "2-intersecting curves" with three given points. But it does not extend to k-intersecting curves with $k+1$ given points for $k>2$.

The Levi Enlargement Lemma is used to prove generalizations of a number of convexity results on arrangements of straight lines, duals of statements perhaps better known in the setting of configurations of points: Helly's theorem, Radon's theorem, Carathéodory's theorem, Kirchberger's theorem, the Hahn-Banach theorem, the Krein-Milman theorem, and Tverberg's generalization of Radon's theorem (cf. Chapter 4). To state two of these we need another definition. If $\mathcal{A}$ is an arrangement of pseudolines and p is a point not contained in any member of $\mathcal{A}$, $L \in \mathcal{A}$ is in the ***p-convex hull*** of $\mathcal{B} \subset \mathcal{A}$ if every path from p to a point of L meets some member of $\mathcal{B}$.

THEOREM 5.1.2 *Helly's theorem for pseudoline arrangements* [GP82a]

If $\mathcal{A}_1, \ldots, \mathcal{A}_n$ are subsets of an arrangement $\mathcal{A}$ of pseudolines, and p is a point not on any pseudoline of $\mathcal{A}$ such that, for any i, j, k, $\mathcal{A}$ contains a pseudoline in the p-convex hull of each of $\mathcal{A}_i, \mathcal{A}_j, \mathcal{A}_k$, then there is an extension $\mathcal{A}'$ of $\mathcal{A}$ containing a pseudoline lying in the p-convex hull of each of $\mathcal{A}_1, \ldots, \mathcal{A}_n$.

THEOREM 5.1.3 *Tverberg's theorem for pseudoline arrangements* [Rou88b]

If $\mathcal{A} = \{L_1, \ldots, L_n\}$ is a pseudoline arrangement with $n \geq 3m-2$, and p is a point not on any member of $\mathcal{A}$, then $\mathcal{A}$ can be partitioned into subarrangements $\mathcal{A}_1, \ldots, \mathcal{A}_m$ and extended to an arrangement $\mathcal{A}'$ containing a pseudoline lying in the p-convex hull of $\mathcal{A}_i$ for every $i = 1, \ldots, m$.

Some of these convexity theorems, but not all, extend to higher dimensional arrangements; see [BLS+99, Sections 9.2,10.4] and Section 26.3 of this Handbook.

Planar graphs admit straight line drawings, from this it follows that the pseudolines in an arrangement may be drawn as polygonal lines, with bends only at vertices. Eppstein [Epp14] investigates such drawings on small grids. Related is the following by now classical representation, which will be discussed further in Section 5.2.

THEOREM 5.1.4 [Goo80]

Every arrangement of pseudolines is isomorphic to a wiring diagram.

Theorem 5.1.4 is used in proving the following duality theorem, which extends to the setting of pseudolines the fundamental duality theorem between lines and points in the projective plane.

THEOREM 5.1.5 [Goo80]

Given an arrangement $\mathcal{A}$ of pseudolines and a set $\mathcal{S}$ of points in $\mathbf{P}^2$, there is a point set $\hat{\mathcal{A}}$ and a pseudoline arrangement $\hat{\mathcal{S}}$ so that a point $p \in \mathcal{S}$ lies on a pseudoline $L \in \mathcal{A}$ if and only if the dual point $\hat{L}$ lies on the dual pseudoline $\hat{p}$.

THEOREM 5.1.6 [AS05]

For Euclidean arrangements, the result of Theorem 5.1.5 *holds with the additional property that the duality preserves above-below relationships as well.*

5.2 RELATED STRUCTURES

GLOSSARY

Circular sequence of permutations: A doubly infinite sequence of permutations of $1, \ldots, n$ associated with an arrangement $\mathcal{A}$ of lines $L_1, \ldots, L_n$ by sweeping a directed line across $\mathcal{A}$; see Figure 5.2.3 and the corresponding sequence below.

Local equivalence: Two circular sequences of permutations are locally equivalent if, for each index i, the order in which it switches with the remaining indices is either the same or opposite.

Local sequence of unordered switches: In a Euclidean arrangement (wiring diagram), the permutation α_i given by the order in which the remaining pseudolines cross the ith pseudoline of the arrangement. In Figure 5.1.2, for example, α_1 is $(2, \{3, 5\}, 4)$.

Configuration of points: A (labeled) family $\mathcal{S} = \{p_1, \ldots, p_n\}$ of points, not all collinear, in $\mathbf{P}^2$.

Order type of a configuration $\mathcal{S}$: The mapping that assigns to each ordered triple i, j, k in $\{1, \ldots, n\}$ the orientation of the triple (p_i, p_j, p_k).

Combinatorial equivalence: Configurations $\mathcal{S}$ and $\mathcal{S}'$ are combinatorially equivalent if the set of permutations of $1, \ldots, n$ obtained by projecting $\mathcal{S}$ onto every line in general position agrees with the corresponding set for $\mathcal{S}'$.

Generalized configuration: A finite set of points in $\mathbf{P}^2$, together with an arrangement of pseudolines such that each pseudoline contains at least two of the points and for each pair of points there is a connecting pseudoline. Also called a ***pseudoconfiguration***. For an example see Figure 5.2.1.

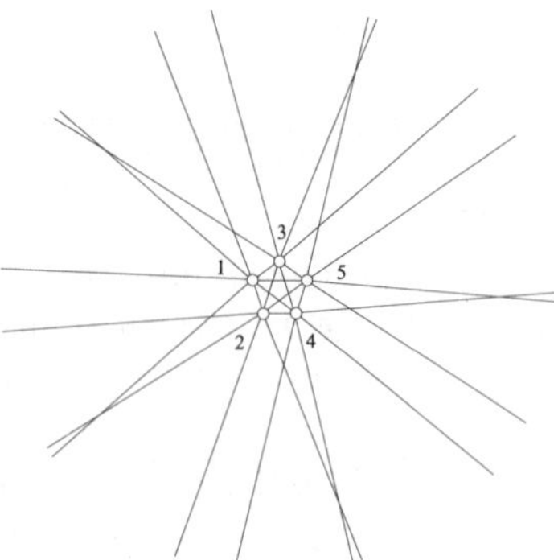

FIGURE 5.2.1
A generalized configuration of 5 points, this example is known as the ***bad pentagon***.

Allowable sequence of permutations: A doubly infinite sequence of permutations of $1, \ldots, n$ satisfying the three conditions of Theorem 5.2.1. It follows from those conditions that the sequence is periodic of length $\leq n(n-1)$, and that its period has length $n(n-1)$ if and only if the sequence is ***simple***, i.e., each move consists of the switch of a single pair of indices.

ARRANGEMENTS OF STRAIGHT LINES

Much of the work on pseudoline arrangements has been motivated by problems involving straight line arrangements. In some cases the question has been whether known results in the case of lines really depended on the *straightness* of the lines; for many (but not all) combinatorial results the answer has turned out to be negative. In other cases, generalization to pseudolines (or, equivalently, reformulations in terms of allowable sequences of permutations) has permitted the solution of a more general problem where none was known previously in the straight case. Finally, pseudolines have turned out to be more useful than lines for certain algorithmic applications; this will be discussed in Section 5.8.

For arrangements of straight lines, there is a rich history of combinatorial results, some of which will be summarized in Section 5.4. Much of this is discussed in [Grü72].

Line arrangements are often classified by isomorphism type. For (unlabeled) arrangements of five lines, for example, Figure 5.2.2 illustrates the four possible isomorphism types, only one of which is simple.

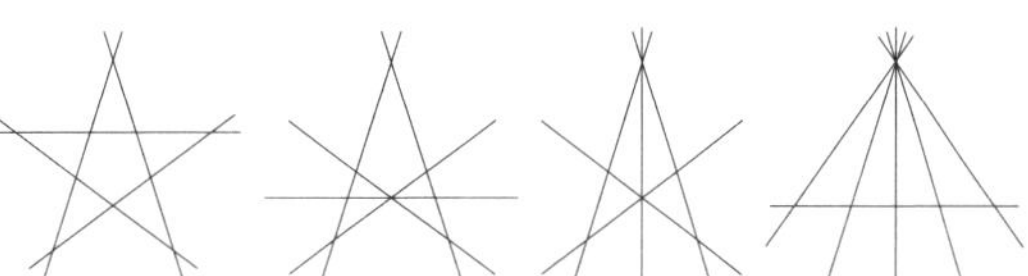

FIGURE 5.2.2
The 4 isomorphism types of arrangements of 5 lines.

There is a second classification of line arrangements, which has proven quite useful for certain problems. If a distinguished point not on any line of the arrangement is chosen to play the part of the "vertical point at infinity," we can think of the arrangement $\mathcal{A}$ as an arrangement of nonvertical lines in the Euclidean plane, and of P_∞ as the "upward direction." Rotating a directed line through P_∞ then amounts to sweeping a directed vertical line through $\mathcal{A}$ from left to right (say). We can then note the order in which this directed line cuts the lines of $\mathcal{A}$, and we arrive at a periodic sequence of permutations of $1, \ldots, n$, known as the ***circular sequence of permutations*** belonging to $\mathcal{A}$ (depending on the choice of P_∞ and the direction of rotation). This sequence is actually doubly infinite, since the rotation of the directed line through P_∞ can be continued in both directions. For the arrangement in Figure 5.2.3, for example, the circular sequence is

$$\mathcal{A}: \quad \ldots 12345 \xrightarrow{12,45} 21354 \xrightarrow{135} 25314 \xrightarrow{25,14} 52341 \xrightarrow{234} 54321 \ldots.$$

We have indicated the "moves" between consecutive permutations.

THEOREM 5.2.1 [GP84]

A circular sequence of permutations arising from a line arrangement has the following properties:

(i) *The move from each permutation to the next consists of the reversal of one or more nonoverlapping adjacent substrings;*

(ii) *After a move in which i and j switch, they do not switch again until every other pair has switched;*

(iii) $1, \ldots, n$ *do not all switch simultaneously with each other.*

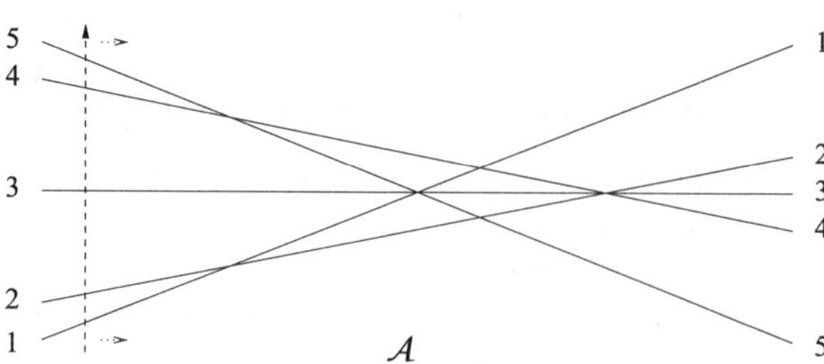

FIGURE 5.2.3
An arrangement of 5 *lines.*

If two line arrangements are isomorphic, they may have different circular sequences, depending on the choice of P_∞ (and the direction of rotation). We do have, however,

THEOREM 5.2.2 [GP84]

If $\mathcal{A}$ and $\mathcal{A}'$ are arrangements of lines in $\mathbf{P}^2$, and Σ and Σ' are any circular sequences of permutations corresponding to $\mathcal{A}$ and $\mathcal{A}'$, then $\mathcal{A}$ and $\mathcal{A}'$ are isomorphic if and only if Σ and Σ' are locally equivalent.

CONFIGURATIONS OF POINTS

Under projective duality, arrangements of lines in $\mathbf{P}^2$ correspond to configurations of points. Some questions seem more natural in this setting of points, however, such as the Sylvester-Erdős problem about the existence of a line with only two points (ordinary line), and Scott's question whether n noncollinear points always determine at least $2\lfloor n/2\rfloor$ directions. These problems are discussed in Chapter 1 of this Handbook.

Corresponding to the classification of line arrangements by isomorphism type, it turns out that the "dual" classification of point configurations is by order type.

THEOREM 5.2.3 [GP84]

If $\mathcal{A}$ and $\mathcal{A}'$ are arrangements of lines in $\mathbf{P}^2$ and $\mathcal{S}$ and $\mathcal{S}'$ the point sets dual to them, then $\mathcal{A}$ and $\mathcal{A}'$ are isomorphic if and only if $\mathcal{S}$ and $\mathcal{S}'$ have the same (or opposite) order types.

From a configuration of points one also derives a circular sequence of permutations in a natural way, by projecting the points onto a rotating line. The sequence for the arrangement in Figure 5.2.3 comes from the configuration in Figure 5.2.4 in this way. Circular sequences yield a finer classification than order type; the order types of two point sets may be identical while their circular sequences are different.

It follows from projective duality that

THEOREM 5.2.4 [GP82b]

A sequence of permutations is realizable as the circular sequence of a set of points if and only if it is realizable as the sequence of an arrangement of lines.

The circular sequence of a point configuration can be reconstructed from the *set* of permutations obtained by projecting it in the directions that are not spanned by the points. The corresponding result in higher dimensions is useful (there the

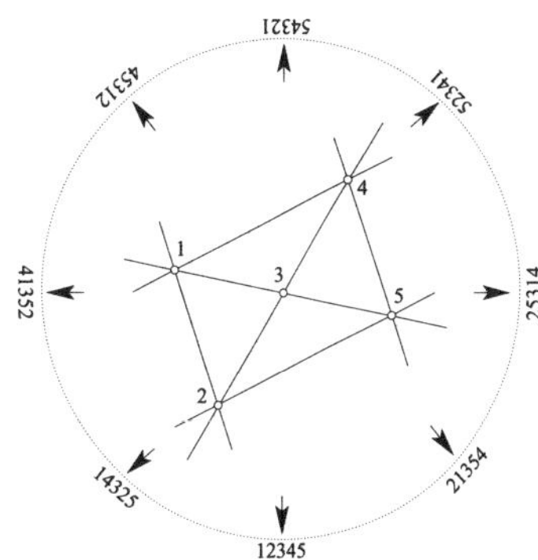

FIGURE 5.2.4
A configuration of 5 *points and its circular sequence.*

circular sequence generalizes to a somewhat unwieldy cell decomposition of a sphere with a permutation associated with every cell), since it means that all one really needs to know is the *set* of permutations; how they fit together can then be determined. Chapter 1 of this Handbook is concerned with results and some unsolved problems on point configurations.

GENERALIZED CONFIGURATIONS

Just as pseudoline arrangements generalize arrangements of straight lines, generalized configurations provide a generalization of configurations of points.

For example, a circular sequence for the generalized configuration in Figure 5.2.1, which is determined by the cyclic order in which the connecting pseudolines meet a distinguished pseudoline (in this case the "pseudoline at infinity"), is

$$\ldots 12345 \overset{34}{-} 12435 \overset{12}{-} 21435 \overset{14}{-} 24135 \overset{35}{-} 24513 \overset{24}{-} 42513 \overset{25}{-} 45213 \overset{13}{-} 45231 \overset{23}{-} 45321 \overset{45}{-} 54321 \ldots$$

Another generalization of a point configuration is given by an abstraction of the order type. Intuitively an ***abstract order type*** prescribes an orientation clockwise, collinear, or counterclockwise, for triples of elements (points). The concept can be formalized by defining a (projective) abstract order type as a reorientation class of oriented matroids of rank 3. From the Folkman-Lawrence topological representation theorem it follows that abstract order types and pseudoline arrangements are the same. When working with abstract order types it is convenient to choose a line at infinity to get a Euclidean arrangement. In the Euclidean arrangement each triple of lines forms a subarrangement equivalent to one of the three arrangements shown in Figure 5.2.5; they tell whether the orientation of the corresponding triple of 'points' is clockwise, collinear or counterclockwise.

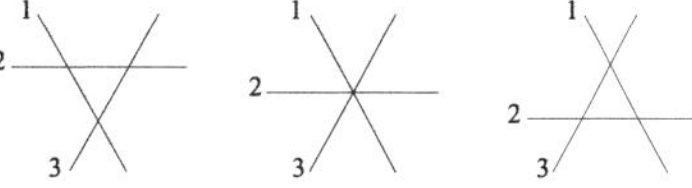

FIGURE 5.2.5
The Euclidean arrangements of 3 lines.

Knuth [Knu92] proposed an axiomatization of (Euclidean) abstract order types in general position (no collinearities) with five axioms. He calls a set together with a ternary predicate obeying the axioms a ***CC system***.

As in the case of circular sequences and order types we observe that generalized configurations yield a finer classification than abstract order types.

ALLOWABLE SEQUENCES

An allowable sequence of permutations is a combinatorial abstraction of the circular sequence of permutations associated with an arrangement of lines or a configuration of points. We can define, in a natural way, a number of geometric concepts for allowable sequences, such as *collinearity*, *betweenness*, *orientation*, *extreme point*, *convex hull*, *semispace*, *convex n-gon*, *parallel*, etc. [GP80a]. Not all allowable sequences are realizable, however, the smallest example being the sequence corresponding to Figure 5.2.1. A realization of this sequence would have to be a drawing of the bad pentagon of Figure 5.2.1 with straight lines, and it is not hard to prove that this is impossible; a proof of the nonrealizability of a larger class of allowable sequences can be found in [GP80a].

Allowable sequences provide a means of rephrasing many geometric problems about point configurations or line arrangements in combinatorial terms. For example, Scott's conjecture on the minimum number of directions determined by n lines has the simple statement: "Every allowable sequence of permutations of $1,\dots,n$ has at least $2\lfloor n/2 \rfloor$ moves in a half-period." It was proved in this more general form by Ungar [Ung82], and the proof of the original Scott conjecture follows as a corollary; see also [Jam85], [BLS+99, Section 1.11], and [AZ99, Chapter 9].

The Erdős-Szekeres problem (see Chapter 1 of this Handbook) looks as follows in this more general combinatorial formulation:

PROBLEM 5.2.5 *Generalized Erdős-Szekeres Problem* [GP81a]

What is the minimum n such that for every simple allowable sequence Σ on $1,\dots,n$, there are k indices with the property that each occurs before the other $k-1$ in some term of Σ?

Allowable sequences arise from Euclidean pseudoline arrangements by sweeping a line across from left to right, just as with an arrangement of straight lines, and they arise as well from generalized configurations just as from configurations of points. In fact, the following theorem is just a restatement of Theorem 5.1.5.

THEOREM 5.2.6 [GP84]

Every allowable sequence of permutations can be realized both by an arrangement of pseudolines and by a generalized configuration of points.

Half-periods of allowable sequences correspond to Euclidean arrangements; they can be assumed to start with the identity permutation and to end with the reverse of the identity. If the sequence is simple it is the same as a ***reduced decomposition*** of the reverse of the identity in the Coxeter group of type A (symmetric group). Theorem 5.6.1 below was obtained in this context.

WIRING DIAGRAMS

Wiring diagrams provide the simplest "geometric" realizations of allowable sequences. To realize the sequence

$$\mathcal{A}: \quad \ldots 12345 \xrightarrow{45} 12354 \xrightarrow{12} 21354 \xrightarrow{135} 25314 \xrightarrow{25,14} 52341 \xrightarrow{23} 53241 \xrightarrow{24} 53421 \xrightarrow{34} 54321 \ldots$$

for example, simply start with horizontal "wires" labeled $1, \ldots, n$ in (say) increasing order from bottom to top, and, for each move in the sequence, let the corresponding wires cross. This gives the wiring diagram of Figure 5.1.2, and at the end the wires have all reversed order. (It is then easy to extend the curves in both directions to the "line at infinity," thereby arriving at a pseudoline arrangement in $\mathbf{P}^2$.)

We have the following isotopy theorem for wiring diagrams.

THEOREM 5.2.7 [GP85]

If two wiring diagrams numbered $1, \ldots, n$ in order are isomorphic as labeled pseudoline arrangements, then one can be deformed continuously to the other (or to its reflection) through wiring diagrams isomorphic as pseudoline arrangements.

Two arrangements are related by a ***triangle-flip*** if one is obtained from the other by changing the orientation of a triangular face, i.e., moving one of the three pseudolines that form the face across the intersection of two others.

THEOREM 5.2.8 [Rin57]

Any two simple wiring diagrams numbered $1, \ldots, n$ in order can be obtained from each other with a sequence of triangle-flips.

This result has well-known counterparts in the terminology of *mutations* for oriented matroids and *Coxeter relations* for reduced decompositions, see [BLS$^+$99, Section 6.4].

If $\mathcal{A}$ and $\mathcal{A}'$ are simple wiring diagrams and there are exactly t triples of lines such that the orientations of the induced subarrangement in $\mathcal{A}$ and $\mathcal{A}'$ differ, then it may require more than t triangle-flips to get from $\mathcal{A}$ to $\mathcal{A}'$; an example is given in [FZ01].

Wiring diagrams have also been considered in the bi-colored setting. Let L be a simple wiring diagram consisting of n blue and n red pseudolines, and call a vertex P *balanced* if P is the intersection of a blue and a red pseudoline such that the number of blue pseudolines strictly above P equals the number of red pseudolines strictly above P (and hence the same holds for those strictly below P as well).

THEOREM 5.2.9 [PP01]

A simple wiring diagram consisting of n blue and n red pseudolines has at least n balanced vertices, and this result is tight.

LOCAL SEQUENCES AND CLUSTERS OF STARS

The following theorem (proved independently by Streinu and by Felsner and Weil) solves the "cluster of stars" problem posed in [GP84]; we state it here in terms of local sequences of wiring diagrams, as in [FW01].

THEOREM 5.2.10 [Str97, FW01]

A set $(\alpha_i)_{i=1,\ldots,n}$ with each α_i a permutation of $\{1, \ldots, i-1, i+1, \ldots, n\}$, is the set of local sequences of unordered switches of a simple wiring diagram if and only if for all $i < j < k$ the pairs $\{i,j\}$, $\{i,k\}$, $\{j,k\}$ appear all in natural order or all in inverted order in α_k, α_j, α_i (resp.).

HIGHER DIMENSIONS

Just as isomorphism classes of pseudoline arrangements correspond to oriented matroids of rank 3, the corresponding fact holds for higher-dimensional arrangements, known as arrangements of pseudohyperplanes: they correspond to oriented matroids of rank $d+1$ (see Theorem 6.2.4 in Chapter 6 of this Handbook).

It turns out, however, that in higher dimensions, generalized configurations of points are (surprisingly) more restrictive than such oriented matroids; thus it is only in the plane that "projective duality" works fully in this generalized setting; see [BLS+99, Section 5.3].

5.3 STRETCHABILITY

STRETCHABLE AND NONSTRETCHABLE ARRANGEMENTS

Stretchability can be described in either combinatorial or topological terms:

THEOREM 5.3.1 [BLS+99, Section 6.3]

Given an arrangement $\mathcal{A}$ or pseudolines in $\mathbf{P}^2$, the following are equivalent.

(i) *The cell decomposition induced by $\mathcal{A}$ is isomorphic to that induced by some arrangement of straight lines;*

(ii) *Some homeomorphism of $\mathbf{P}^2$ to itself maps every $L_i \in \mathcal{A}$ to a straight line.*

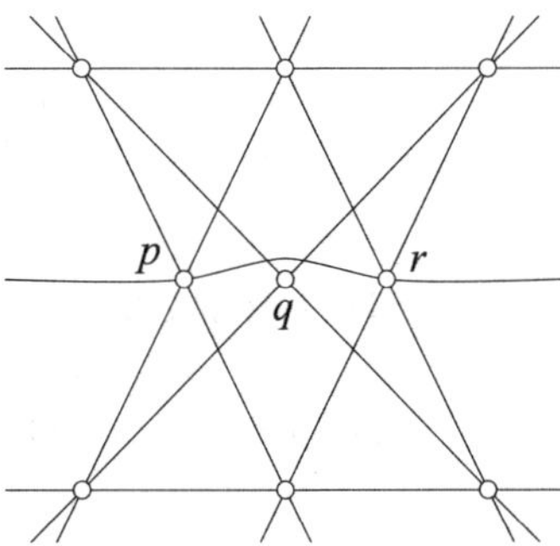

FIGURE 5.3.1
An arrangement that violates the theorem of Pappus.

Among the first examples observed of a nonstretchable arrangement of pseudolines was the non-Pappus arrangement of 9 pseudolines constructed by Levi: see Figure 5.3.1. Since Pappus's theorem says that points p, q, and r must be collinear if the pseudolines are straight, the arrangement in Figure 5.3.1 is clearly nonstretchable. A second example, involving 10 pseudolines, can be constructed similarly by violating Desargues's theorem.

Ringel showed how to convert the non-Pappus arrangement into a *simple* arrangement that was still nonstretchable. A symmetric drawing of it is shown in Figure 5.3.2.

Using allowable sequences, Goodman and Pollack proved the conjecture of

FIGURE 5.3.2
A simple nonstretchable arrangement of 9 pseudolines.

Grünbaum that the non-Pappus arrangement has the smallest size possible for a nonstretchable arrangement:

THEOREM 5.3.2 [GP80b]

Every arrangement of 8 or fewer pseudolines is stretchable.

In addition, Richter-Gebert proved that the non-Pappus arrangement is unique among simple arrangements of the same size.

THEOREM 5.3.3 [Ric89]

Every simple arrangement of 9 pseudolines is stretchable, with the exception of the simple non-Pappus arrangement.

The "bad pentagon" of Figure 5.2.1, with extra points inserted to "pin down" the intersections of the sides and corresponding diagonals, provides another example of a nonstretchable arrangement. The bad pentagon was generalized in [GP80a] yielding an infinite family of nonstretchable arrangements that were proved, by Bokowski and Sturmfels [BS89a], to be "minor-minimal." This shows that stretchability of simple arrangements cannot be guaranteed by the exclusion of a finite number of "forbidden" subarrangements. A similar example was found by Haiman and Kahn; see [BLS$^+$99, Section 8.3].

As for arrangements of more than 8 pseudolines, we have

THEOREM 5.3.4 [GPWZ94]

Let $\mathcal{A}$ be an arrangement of n pseudolines. If some face of $\mathcal{A}$ is bounded by at least $n-1$ pseudolines, then $\mathcal{A}$ is stretchable.

Finally, Shor shows in [Sho91] that even if a stretchable pseudoline arrangement has a symmetry, it may be impossible to realize this symmetry in any stretching.

THEOREM 5.3.5 [Sho91]

There exists a stretchable, simple pseudoline arrangement with a combinatorial symmetry such that no isomorphic arrangement of straight lines has the same combinatorial symmetry.

COMPUTATIONAL ASPECTS

Along with the Universality Theorem (Theorem 5.5.7 below), Mnëv proved that the problem of determining whether a given arrangement of n pseudolines is stretchable is NP-hard, in fact as hard as the problem of solving general systems of polynomial equations and inequalities over $\mathbb{R}$ (cf. Chapter 37 of this Handbook):

THEOREM 5.3.6 [Mnë85, Mnë88]

The stretchability problem for pseudoline arrangements is polynomially equivalent to the "existential theory of the reals" decision problem.

Shor [Sho91] presents a more compact proof of the NP-hardness result, by encoding a "monotone 3-SAT" formula in a family of suitably modified Pappus and Desargues configurations that turn out to be stretchable if and only if the corresponding formula is satisfiable. (See also [Ric96a].)

The following result provides an upper bound for the realizability problem.

THEOREM 5.3.7 [BLS+99, Sections 8.4, A.5]

The stretchability problem for pseudoline arrangements can be decided in singly exponential time and polynomial space in the Turing machine model of complexity. The number of arithmetic operations needed is bounded above by $2^{4n \log n + O(n)}$.

The NP-hardness does not mean, however, that it is pointless to look for algorithms to determine stretchability, particularly in special cases. Indeed, a good deal of work has been done on this problem by Bokowski, in collaboration with Guedes de Oliveira, Pock, Richter-Gebert, Scharnbacher, and Sturmfels. Four main algorithmic methods have been developed to test for the realizability (or nonrealizability) of an oriented matroid, i.e., in the rank 3 case, the stretchability (respectively nonstretchability) of a pseudoline arrangement:

(i) The ***inequality reduction method:*** this attempts to find a relatively small system of inequalities that still carries all the information about a given oriented matroid.

(ii) The ***solvability sequence method:*** this attempts to find an elimination order with special properties for the coordinates in a potential realization of an order type.

(iii) The ***final polynomial method:*** this attempts to find a bracket polynomial (cf. Chapter 60) whose existence will imply the *non*realizability of an order type.

(iv) Bokowski's's ***rubber-band method:*** an elementary heuristic that has proven surprisingly effective in finding realizations [Bok08].

Not every realizable order type has a solvability sequence, but it turns out that every nonrealizable one does have a final polynomial, and an algorithm due to Lombardi [Lom90] can be used to find one. Fukuda et al. [FMM13] have refined the techniques and used them successfully for the enumeration of realizable oriented matroids with small parameters.

All of these methods extend to higher dimensions. For details about the first three, see [BS89b].

DISTINCTIONS BETWEEN LINES AND PSEUDOLINES

The asymptotic number of pseudoline arrangements is strictly larger than the number of line arrangements; see Table 5.6.2 below. Nevertheless, we know only very few combinatorial properties or algorithmic problems that can be used to distinguish between line and pseudoline arrangements.

Combinatorial properties of this type are given by the number of triangles in nonsimple arrangements.

The first such example was given by Roudneff in [Rou88a]. He proved the following statement that had been conjectured of Grünbaum: An arrangement of n lines with only n triangles is simple. However, there exist nonsimple arrangements of n pseudolines with only n triangles. An example of such an arrangement is obtained by "collapsing" the central triangle in Figure 5.3.2.

Felsner and Kriegel [FK99] describe examples of nonsimple Euclidean pseudoline arrangements with only $2n/3$ triangles; see Figure 5.3.3. A theorem of Shannon [Sha79] (see also [Fel04, Thm. 5.18]), however, asserts that a Euclidean arrangement of n lines has at least $n-2$ triangles. Hence the examples from [FK99] are nonstretchable.

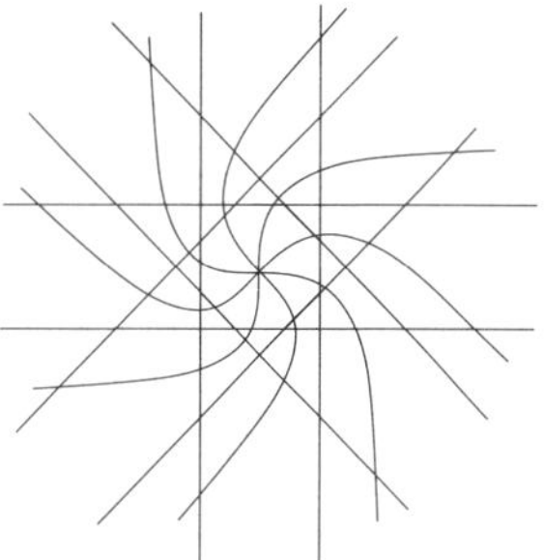

FIGURE 5.3.3
A nonsimple Euclidean arrangement of 12 pseudolines with 8 triangles.

Recently applications of the polynomial method to problems in incidence geometry have led to some breakthroughs [Gut16] The improved bounds apply only to line arrangements. This may open the door to a range of additional properties that distinguish between lines and pseudolines. Candidates could be the number of ordinary points or the constant for the strong Dirac conjecture; see [LPS14].

Another candidate problem where pseudolines and lines may behave differently is the maximum length of x-monotone paths, see Theorem 5.4.14.

Steiger and Streinu [SS94] consider the problem of x-sorting line or pseudoline intersections, i.e., determining the order of the x-coordinates of the intersections of the lines or pseudolines in a Euclidean arrangement. They prove that in comparison-based sorting the vertices of a simple arrangement of n lines can be x-sorted with $O(n^2)$ comparisons while the x-sorting of vertices of a simple arrangement of n pseudolines requires at least $\Omega(n^2 \log n)$ comparisons. The statement for pseudolines is a corollary of Theorem 5.6.1, i.e., based on the number of possible x-sortings.

5.4 COMBINATORIAL RESULTS

In this section we survey combinatorial results. This includes several results that update Grünbaum's comprehensive 1972 survey [Grü72]. For a discussion of *levels in arrangements* (dually, *k-sets*), see Chapters 28 and 1, respectively. Erdős and Purdy [EP95] survey related material with an emphasis on extremal problems.

GLOSSARY

Simplicial arrangement: An arrangement of lines or pseudolines in which every cell is a triangle.

Near-pencil: An arrangement with all but one line (or pseudoline) concurrent.

Projectively unique: A line arrangement $\mathcal{A}$ with the property that every isomorphic line arrangement is the image of $\mathcal{A}$ under a projective transformation.

RELATIONS AMONG NUMBERS OF VERTICES, EDGES, AND FACES

THEOREM 5.4.1 *Euler*

If $f_i(\mathcal{A})$ is the number of faces of dimension i in the cell decomposition of $\mathbf{P}^2$ induced by an arrangement $\mathcal{A}$, then $f_0(\mathcal{A}) - f_1(\mathcal{A}) + f_2(\mathcal{A}) = 1$.

In addition to ***Euler's formula***, the following inequalities are satisfied for arbitrary pseudoline arrangements (here, $n(\mathcal{A})$ is the number of pseudolines in the arrangement $\mathcal{A}$). We state the results for projective arrangements. Similar inequalities for Euclidean arrangements can easily be derived.

THEOREM 5.4.2 [Grü72, SE88]

(i) $1+f_0(\mathcal{A}) \leq f_2(\mathcal{A}) \leq 2f_0(\mathcal{A})-2$, *with equality on the left for precisely the simple arrangements, and on the right for precisely the simplicial arrangements;*

(ii) $n(\mathcal{A}) \leq f_0(\mathcal{A}) \leq \binom{n(\mathcal{A})}{2}$, *with equality on the left for precisely the near-pencils, and on the right for precisely the simple arrangements;*

(iii) *For $n \gg 0$, every f_0 satisfying $n^{3/2} \leq f_0 \leq \binom{n}{2}$, with the exceptions of $\binom{n}{2} - 3$ and $\binom{n}{2} - 1$, is the number of vertices of some arrangement of n pseudolines (in fact, of straight lines);*

(iv) $2n(\mathcal{A}) - 2 \leq f_2(\mathcal{A}) \leq \binom{n(\mathcal{A})}{2} + 1$, *with equality on the left for precisely the near-pencils, and on the right for precisely the simple arrangements;*

(v) $f_2(\mathcal{A}) \geq 3n(\mathcal{A}) - 6$ *if $\mathcal{A}$ is not a near-pencil.*

There are gaps in the possible values for $f_2(\mathcal{A})$, as shown by Theorem 5.4.3. This proves a conjecture of Grünbaum which was generalized by Purdy, it refines Theorem 5.4.2(iv).

THEOREM 5.4.3 [Mar93]

There exists an arrangement $\mathcal{A}$ of n pseudolines with $f_2(\mathcal{A}) = f$ if and only if, for some integer k with $1 \le k \le n-2$, we have $(n-k)(k+1) + \binom{k}{2} - \min(n-k, \binom{k}{2}) \le f \le (n-k)(k+1) + \binom{k}{2}$. Moreover, if $\mathcal{A}$ exists, it can be chosen to consist of straight lines.

Finally, the following result (proved in the more general setting of geometric lattices) gives a complete set of inequalities for the flag vectors $(n(\mathcal{A}), f_0(\mathcal{A}), i(\mathcal{A}))$. The component $i(\mathcal{A})$ is the number of vertex–pseudoline incidences in the arrangement $\mathcal{A}$.

THEOREM 5.4.4 [Nym04]

The closed convex set generated by all flag vectors of arrangements of pseudolines is characterized by the following set of inequalities: $n \ge 3$, $f_0 \ge n$, $i \ge 2f_0$, $i \le 3f_0 - 3$, and $(k-1)i - kn - (2k-3)f_0 + \binom{k+1}{2} \ge 0$, for all $k \ge 3$. Moreover, this set of inequalities is minimal.

THE NUMBER OF CELLS OF DIFFERENT SIZES

It is easy to see by induction that a *simple* arrangement of more than 3 pseudolines must have at least one nontriangular cell. This observation leads to many questions about numbers of cells of different types in both simple and nonsimple arrangements, some of which have not yet been answered satisfactorily.

On the minimum number of triangles, we have a classical result of Levi.

THEOREM 5.4.5 [Lev26]

In any arrangement of $n \ge 3$ pseudolines, every pseudoline borders at least 3 triangles. Hence every arrangement of n pseudolines determines at least n triangles.

This minimum is achieved by the "cyclic arrangements" of lines generated by regular polygons, as in Figure 5.4.1.

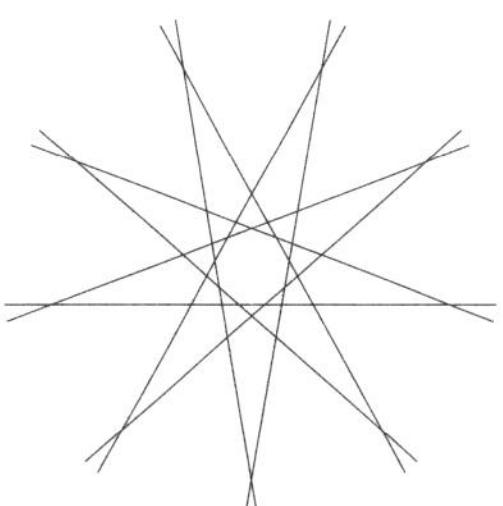

FIGURE 5.4.1
A cyclic arrangement of 9 lines.

Grünbaum [Grü72] asked for the maximum number of triangles and provided the upper bound of $\lfloor n(n-1)/3 \rfloor$ for the simple case. Harborth [Har85] introduced the doubling method to construct infinite families of simple arrangements of pseudolines attaining the upper bound. Roudneff [Rou96] proved that the upper bound also holds for nonsimple arrangements. Forge and Ramírez Alfonsín [FR98] constructed infinite families of line arrangements attaining the upper bound.

Already Grünbaum knew that his upper bound could not be attained for all values of n. Building on results from [BBL08], Blanc proved

THEOREM 5.4.6 [Bla11]

If $\mathcal{A}$ is a simple arrangement of n pseudolines, with $n \geq 4$ and $p_3(\mathcal{A})$ is the number of triangles in $\mathcal{A}$, then

$$p_3(\mathcal{A}) \leq \begin{cases} (n(n-1))/3 & \text{if } n \equiv 0,4 \pmod 6 \\ (n(n-2)-2)/3 & \text{if } n \equiv 1 \pmod 6 \\ (n(n-1)-5)/3 & \text{if } n \equiv 2 \pmod 6 \\ (n(n-2))/3 & \text{if } n \equiv 3,5 \pmod 6 \end{cases}$$

Furthermore, each of these bounds is tight for infinitely many integers n, and for $n \not\equiv 2 \pmod 6$ a family of straight line arrangements attaining the bound is known.

For arrangements in the Euclidean plane $\mathbb{R}^2$, on the other hand, we have

THEOREM 5.4.7 [FK99]

(i) *Every simple arrangement of n pseudolines in $\mathbb{R}^2$ contains at least $n-2$ triangles, with equality achieved for all $n \geq 3$.*
(ii) *Every arrangement of n pseudolines in $\mathbb{R}^2$ contains at least $2n/3$ triangles, with equality achieved for all $n \equiv 0 (\mod 3)$.*

Regarding the maximum number of triangles in simple Euclidean arrangements of n pseudolines Blanc proves a theorem similar to Theorem 5.4.6 that improves on the upper bound of $n(n-2)/3$ depending on the residue modulo six.

The following result disproved a conjecture of Grünbaum:

THEOREM 5.4.8 [LRS89]

There is a simple arrangement of straight lines containing no two adjacent triangles.

The proof involved finding a pseudoline arrangement with this property, then showing (algebraically, using Bokowski's "inequality reduction method"—see Section 5.3) that the arrangement, which consists of 12 pseudolines, is stretchable.

The following general problem was posed by Grünbaum [Grü72].

PROBLEM 5.4.9

What is the maximum number of k-sided cells in an arrangement of n pseudolines, for $k > 3$?

Some results about quadrilaterals have been obtained.

THEOREM 5.4.10 [Grü72, Rou87, FR01]

(i) *Every arrangement of $n \geq 5$ pseudolines contains at most $n(n-3)/2$ quadrilaterals. For straight line arrangements, this bound is achieved by a unique arrangement for each n.*
(ii) *A pseudoline arrangement containing $n(n-3)/2$ quadrilaterals must be simple.*

There are infinitely many simple pseudoline arrangements with no quadrilaterals, contrary to what was once believed. The following result implies, however, that there must be many quadrilaterals or pentagons in *every* simple arrangement.

THEOREM 5.4.11 [Rou87]

Every pseudoline in a simple arrangement of $n > 3$ pseudolines borders at least 3 quadrilaterals or pentagons. Hence, if p_4 is the number of quadrilaterals and p_5 the number of pentagons in a simple arrangement, we have $4p_4 + 5p_5 \geq 3n$.

Leaños et al. [LLM+07] study simple Euclidean arrangements of pseudolines without faces of degree ≥ 5. They show that they have exactly $n-2$ triangles and $(n-2)(n-3)/2$ quadrilaterals. Moreover, all these arrangements are stretchable.

SIMPLICIAL ARRANGEMENTS

Simplicial arrangements were first studied by Melchior [Mel41]. Grünbaum [Grü71] listed 90 "sporadic" examples of simplicial arrangements together with the following three infinite families:

THEOREM 5.4.12 [Grü72]

Each of the following arrangements is simplicial:

(i) *the near-pencil of n lines;*

(ii) *the sides of a regular n-gon, together with its n axes of symmetry;*

(iii) *the arrangement in* (ii), *together with the line at infinity, for n even.*

In [Grü09] Grünbaum presented his collection from 1971 with corrections and in a more user-friendly way. Subsequently, Cuntz [Cun12] did an exhaustive computer enumeration of all simplicial arrangements with up to 27 pseudolines. As a result he found four additional sporadic examples.

On the other hand, additional infinite families of (nonstretchable) simplicial arrangements of pseudolines are known, which are constructible from regular polygons by extending sides, diagonals, and axes of symmetry and modifying the resulting arrangement appropriately. More recently Berman [Ber08] elaborated an idea of Eppstein to generate simplicial arrangements of pseudolines with rotational symmetry. The key idea is to collect segments contributed by the pseudolines of an orbit in a wedge of symmetry, they can be interpreted as (pseudo)light-beams between two mirrors. The method, also described in [Grü13], was used by Lund et al. [LPS14] to construct counterexamples for the strong Dirac conjecture. Figure 5.4.2 shows an example.

FIGURE 5.4.2
A simplicial arrangement of 25 *pseudolines (including line at infinity). Each pseudoline is incident to* ≤ 10 *vertices.*

One of the most important problems on arrangements is the following.

PROBLEM 5.4.13 [Grü72]

Classify all simplicial arrangements of pseudolines. Which of these are stretchable? In particular, are there any infinite families of simplicial line arrangements besides the three of Theorem 5.4.12*?*

It has apparently not been disproved that every (pseudo)line arrangement is a subarrangement of a simplicial (pseudo)line arrangement.

PATHS IN PSEUDOLINE ARRANGEMENTS

A ***monotone path*** π of a Euclidean arrangement $\mathcal{A}$ is a collection of edges and vertices of $\mathcal{A}$ such that each vertical line contains exactly one point of π. The length of π is defined as the number of vertices where π changes from one support line to another. Edelsbrunner and Guibas [EG89] have shown that the maximum length of a monotone path in an arrangement of n pseudolines can be computed via a topological sweep. They also asked for the maximum length λ_n of a monotone path where the maximum is taken over all arrangements of n lines. Matoušek [Mat91] proved that λ_n is $\Omega(n^{5/3})$. This was raised to $\Omega(n^{7/4})$ by Radoičić and Tóth [RT03], they also showed $\lambda_n \leq 5n^2/12$.

THEOREM 5.4.14 [BRS$^+$05]

The maximum length λ_n of a monotone path in an arrangement of n lines is $\Omega(n^2/2^{c\sqrt{\log n}})$, for some constant $c > 0$.

Dumitrescu [Dum05] studied arrangements with a restricted number of slopes. One of his results is that if there are at most 5 slopes, then the length of a monotone path is $O(n^{5/3})$. Together with the construction of Matoušek we thus know that the maximum length of a monotone path in arrangements with at most 5 slopes is in $\Theta(n^{5/3})$.

Matoušek [Mat91] showed that in the less restrictive setting of arrangements of pseudolines there are arrangements with n pseudolines which have a monotone path of length $\Omega(n^2/\log n)$.

For related results on k-levels in arrangements, see Chapter 28.

A ***dual path*** of an arrangement is a path in the dual of the arrangement, i.e., a sequence of cells such that any two consecutive cells of the sequence share a boundary edge. A dual path in an arrangement with lines colored red and blue is an *alternating dual path* if the path alternatingly crosses red and blue lines. Improving on work by Aichholzer et al. [ACH$^+$14], the following has been obtained by Hoffmann, Kleist, and Miltzow.

THEOREM 5.4.15 [HKM15]

Every arrangement of n lines has a dual path of length $n^2/3 - O(n)$ and there are arrangements with a dual path of length $n^2/3 + O(n)$.

Every bicolored arrangement has an alternating dual path of length $\Omega(n)$ and there are arrangements with $3k$ red and $2k$ blue lines where every alternating dual path has length at most $14k$.

COMPLEXITY OF SETS OF CELLS IN AN ARRANGEMENT

The ***zone*** of a line ℓ in $\mathcal{A}$ is the collection of all faces that have a segment of ℓ on their boundary. The complexity of the zone of ℓ is the sum of the degrees of all faces in the zone of ℓ.

THEOREM 5.4.16 *Zone Theorem* [BEPY91, Pin11]

The sum of the numbers of sides in all the cells of an arrangement of $n+1$ pseudolines that are supported by one of the pseudolines is at most $19n/2 - 3$; this bound is tight.

For general sets of faces, on the other hand, Canham proved

THEOREM 5.4.17 [Can69]

If $F_1, \ldots, F_k$ are any k distinct faces of an arrangement of n pseudolines, then $\sum_{i=1}^{k} p(F_i) \leq n + 2k(k-1)$, where $p(F)$ is the number of sides of a face F. This is tight for $2k(k-1) \leq n$.

For $2k(k-1) > n$, this was improved by Clarkson et al. to the following result, with simpler proofs later found by Székely and by Dey and Pach; the tightness follows from a result of Szemerédi and Trotter, proved independently by Edelsbrunner and Welzl.

THEOREM 5.4.18 [ST83, EW86, CEG$^+$90, Szé97, DP98]

The total number of sides in any k distinct cells of an arrangement of n pseudolines is $O(k^{2/3}n^{2/3} + n)$. This bound is (asymptotically) tight in the worst case.

There are a number of results of this kind for arrangements of objects in the plane and in higher dimensions; see Chapter 28, as well as [CEG$^+$90] and [PS09].

5.5 TOPOLOGICAL PROPERTIES

GLOSSARY

Topological projective plane: $\mathbf{P}^2$, with a distinguished family $\mathcal{L}$ of pseudolines (its "lines"), is a topological projective plane if, for each $p, q \in \mathbf{P}^2$, exactly one $L_{p,q} \in \mathcal{L}$ passes through p and q, with $L_{p,q}$ varying continuously with p and q.

Isomorphism of topological projective planes: A homeomorphism that maps "lines" to "lines."

Universal topological projective plane: One containing an isomorphic copy of every pseudoline arrangement.

Topological sweep: If $\mathcal{A}$ is a pseudoline arrangement in the Euclidean plane and $L \in \mathcal{A}$, a topological sweep of $\mathcal{A}$ "starting at L" is a continuous family of pseudolines including L, each compatible with $\mathcal{A}$, which forms a partition of the plane.

Basic semialgebraic set: The set of solutions to a finite number of polynomial equations and strict polynomial inequalities in $\mathbb{R}^d$. (This term is sometimes used even if the inequalities are not necessarily strict.)

GRAPH AND HYPERGRAPH PROPERTIES

An arrangement of pseudolines can be interpreted as a graph drawn crossing free in the projective plane. Hence, it is natural to study graph-theoretic properties of arrangements.

THEOREM 5.5.1 [FHNS00]

The graph of a simple projective arrangement of $n \geq 4$ pseudolines is 4-connected.

Using wiring diagrams, the same authors prove

THEOREM 5.5.2 [FHNS00]

Every projective arrangement with an odd number of pseudolines can be decomposed into two edge-disjoint Hamiltonian paths (plus two unused edges), and the decomposition can be found efficiently.

Harborth and Möller [HM02] identified a projective arrangements of 6 lines that admits no decomposition into two Hamiltonian cycles.

Bose et al. [BCC+13] study the maximum size of independent sets and vertex cover as well as the chromatic numbers of three hypergraphs defined by arrangements. They consider the hypergraphs given by the line-cell, vertex-cell and cell-zone incidences, respectively.

THEOREM 5.5.3 [BCC+13, APP+14]

The lines of every arrangement of n pseudolines in the plane can be colored with $O(\sqrt{n/\log n})$ colors so that no face of the arrangement is monochromatic. There are arrangements requiring at least $\Omega(\frac{\log n}{\log\log n})$ colors.

Let P be a set of points in the plane and let $\alpha(P)$ be the maximum size of a subset S of P such that S is in general position. Let $\alpha_4(n)$ be the minimum of $\alpha(P)$ taken over all sets P of n points in the plane with no four points on a line. Any improvement of the upper bound in the above theorem would imply an improvement on the known lower bound for $\alpha_4(n)$. The study of $\alpha_4(n)$ was initiated by Erdős, see [APP+14] for details.

EMBEDDING IN LARGER STRUCTURES

In [Grü72], Grünbaum asked a number of questions about extending pseudoline arrangements to more complex structures. The strongest result known about such extendibility is the following, which extends results of Goodman, Pollack, Wenger, and Zamfirescu [GPWZ94].

THEOREM 5.5.4 [GPW96]

There exist uncountably many pairwise nonisomorphic universal topological projective planes.

In particular, this implies the following statements, all of which had been conjectured in [Grü72].

(i) Every pseudoline arrangement can be extended to a topological projective plane.

(ii) There exists a universal topological projective plane.

(iii) There are nonisomorphic topological projective planes such that every arrangement in each is isomorphic to some arrangement in the other.

Theorem 5.5.4 also implies the following result, established earlier by Snoeyink and Hershberger (and implicitly by Edmonds, Fukuda, and Mandel; see [BLS$^+$99, Section 10.5]).

THEOREM 5.5.5 *Sweeping Theorem* [SH91]

A pseudoline arrangement $\mathcal{A}$ in the Euclidean plane can be swept by a pseudoline, starting at any $L \in \mathcal{A}$.

PROBLEM 5.5.6 [Grü72]

Which arrangements are present (up to isomorphism) in every *topological projective plane?*

MOVING FROM ONE ARRANGEMENT TO ANOTHER

In [Rin56], Ringel asked whether an arrangement $\mathcal{A}$ of straight lines could always be moved continuously to a given isomorphic arrangement $\mathcal{A}'$ (or to its reflection) so that all intermediate arrangements remained isomorphic. This question, which became known as the "isotopy problem" for arrangements, was eventually solved by Mnëv, and (independently, since news of Mnëv's results had not yet reached the West) by White in the nonsimple case, then by Jaggi and Mani-Levitska in the simple case [BLS$^+$99]. Mnëv's results are, however, by far stronger.

THEOREM 5.5.7 *Mnëv's Universality Theorem* [Mnë85]

If V is any basic semialgebraic set defined over $\mathbb{Q}$, there is a configuration $\mathcal{S}$ of points in the plane such that the space of all configurations of the same order type as $\mathcal{S}$ is stably equivalent to V. If V is open in some $\mathbb{R}^n$, then there is a simple configuration $\mathcal{S}$ with this property.

From this it follows that the space of line arrangements isomorphic to a given one may have the homotopy type of *any* semialgebraic variety, and in particular may be disconnected, which gives a (very strongly) negative answer to the isotopy question. For a further generalization of Theorem 5.5.7, see [Ric96a].

The line arrangement of smallest size known for which the isotopy conjecture fails consists of 14 lines in general position and was found by Suvorov [Suv88]; see also [Ric96b]. Special cases where the isotopy conjecture *does* hold include:

(i) every arrangement of 9 or fewer lines in general position [Ric89], and

(ii) an arrangement of n lines containing a cell bounded by at least $n-1$ of them.

There are also results of a more combinatorial nature about the possibility of transforming one pseudoline arrangement to another. Ringel [Rin56, Rin57] proved

THEOREM 5.5.8 *Ringel's Homotopy Theorem*

If $\mathcal{A}$ and $\mathcal{A}'$ are simple arrangements of pseudolines, then $\mathcal{A}$ can be transformed to $\mathcal{A}'$ by a finite sequence of steps each consisting of moving one pseudoline continuously across the intersection of two others (triangle flip). If $\mathcal{A}$ and $\mathcal{A}'$ are simple arrangements of lines, this can be done within the space of line arrangements.

The second part of Theorem 5.5.8 has been generalized by Roudneff and Sturmfels [RS88] to arrangements of planes; the first half is still open in higher dimensions.

Ringel also observed that the isotopy property does hold for pseudoline arrangements.

THEOREM 5.5.9 [Rin56]

If $\mathcal{A}$ and $\mathcal{A}'$ are isomorphic arrangements of pseudolines, then $\mathcal{A}$ can be deformed continuously to $\mathcal{A}'$ through isomorphic arrangements.

Ringel did not provide a proof of this observation, but one method of proving it is via Theorem 5.2.7, together with the following isotopy result.

THEOREM 5.5.10 [GP84]

Every arrangement of pseudolines can be continuously deformed (through isomorphic arrangements) to a wiring diagram.

5.6 COMPLEXITY ISSUES

THE NUMBER OF ARRANGEMENTS

Various exact values, as well as bounds, are known for the number of equivalence classes of the structures discussed in this chapter. Early work in this direction is documented in [Grü72, GP80a, Ric89, Knu92, Fel97, BLS+99]. Table 5.6.1 shows the values for $n \leq 11$, additional values are known for rows 3, 6, and 7.

TABLE 5.6.1 Exact numbers known for low n.

	3	4	5	6	7	8	9	10	11	
Arr's of n lines	1	2	4	17	143	4890	460779			[FMM13]
Simple arr's of n lines	1	1	1	4	11	135	4381	312114	41693377	[AK07]
Simplicial arr's of n lines	1	1	1	2	2	2	2	4	2	[Grü09]
Arr's of n pseudolines	1	2	4	17	143	4890	461053	95052532		[Fin]
Simple arr's of n p'lines	1	1	1	4	11	135	4382	312356	41848591	[AK07]
Simplicial arr's of n p'lines	1	1	1	2	2	2	2	4	2	[Cun12]
Simple Eucl. arr's of n p'lines	2	8	62	908	24698	1232944	112018190	18410581880	5449192389984	[KSYM11]
Simple Eucl. config's	1	2	3	16	135	3315	158817	14309547	2334512907	[AAK01]
Simple Eucl. gen'd config's	1	2	3	16	135	3315	158830	14320182	2343203071	[AAK01]

The only exact formula known for arbitrary n follows from Stanley's formula:

THEOREM 5.6.1 [Sta84]

The number of simple allowable sequences on $1, \ldots, n$ containing the permutation $123 \ldots n$ is

$$\frac{\binom{n}{2}!}{1^{n-1}3^{n-2}5^{n-3}\cdots(2n-3)^{1}}.$$

For n arbitrary, Table 5.6.2 indicates the known asymptotic bounds.

TABLE 5.6.2 Asymptotic bounds for large n (all logarithms are base 2).

EQUIVALENCE CLASS	LOWER BOUND	UPPER BOUND	
Isom classes of simple (labeled) arr's of n p'lines	$2^{.1887n^2}$	$2^{.6571n^2}$	[FV11]
Isom classes of (labeled) arr's of n p'lines	"	$2^{1.0850n^2}$	[BLS+99, p. 270]
Order types of (labeled) n pt configs (simple or not)	$2^{4n\log n+\Omega(n)}$	$2^{4n\log n+O(n)}$	[GP93, p. 122]
Comb'l equiv classes of (labeled) n pt configs	$2^{7n\log n}$	$2^{8n\log n}$	[GP93, p. 123]

CONJECTURE 5.6.2 [Knu92]

The number of isomorphism classes of simple pseudoline arrangements is $2^{\binom{n}{2}+o(n^2)}$.

Consider simple Euclidean arrangements with $n+1$ pseudolines. Removing pseudoline $n+1$ from such an arrangement yields an arrangement of n pseudolines, the *derived* arrangement. Call $\mathcal{A}$ and $\mathcal{A}'$ equivalent if they have the same derived arrangement. Let γ_n be the maximum size of an equivalence class of arrangements with $n+1$ pseudolines. Dually, γ_n is the maximum number of extensions of an arrangement of n pseudolines. Knuth proved that $\gamma_n \leq 3^n$ and suggested that $\gamma_n \leq n2^n$ might be true. This inequality would imply Conjecture 5.6.2. In the context of social choice theory, $\gamma_n \leq n2^n$ was also conjectured by Fishburn and by Galambos and Rainer. This conjecture was disproved by Ondřej Bílka in 2010. The current bounds on γ_n are $2.076^n \leq \gamma_n \leq 4n\, 2.487^n$, see [FV11].

HOW MUCH SPACE IS NEEDED TO SPECIFY AN ARRANGEMENT?

Allowable sequences of $1,\ldots,n$ can be encoded with $O(n^2\log n)$ bits. This can be done via the *balanced tableaux* of Edelman and Greene or by encoding the ends of the interval that is reversed between consecutive permutations with $\leq 2\log n$ bits.

Given their asymptotic number, wiring diagrams should be encodable with $O(n^2)$ bits. The next theorem shows that this is indeed possible.

THEOREM 5.6.3 [Fel97]

Given a wiring diagram $\mathcal{A} = \{L_1,\ldots,L_n\}$, *let* $t^i_j = 1$ *if the* j*th crossing along* L_i *is with* L_k *for* $k > i$, 0 *otherwise. Then the mapping that associates to each wiring diagram* $\mathcal{A}$ *the binary* $n \times (n-1)$ *matrix* (t^i_j) *is injective.*

The number of stretchable pseudoline arrangements is much smaller than the total number, which suggests that it could be possible to encode these more compactly. The following result of Goodman, Pollack, and Sturmfels (stated here for the dual case of point configurations) shows, however, that the "naive" encoding, by coordinates of an integral representative, is doomed to be inefficient.

THEOREM 5.6.4 [GPS89]

For each configuration $\mathcal{S}$ *of points* (x_i, y_i), $i = 1,\ldots,n$, *in the integer grid* $\mathbb{Z}^2$, *let*

$$\nu(\mathcal{S}) = \min\max\{|x_1|,\ldots,|x_n|,|y_1|,\ldots,|y_n|\},$$

the minimum being taken over all configurations $\mathcal{S}'$ *of the same order type as* $\mathcal{S}$, *and* *let* $\nu^*(n) = \max \nu(\mathcal{S})$ *over all* n*-point configurations. Then, for some* $c_1, c_2 > 0$,

$$2^{2^{c_1 n}} \leq \nu^*(n) \leq 2^{2^{c_2 n}}.$$

5.7 APPLICATIONS

Planar arrangements of lines and pseudolines, as well as point configurations, arise in many problems of computational geometry. Here we describe several such applications involving pseudolines in particular.

GLOSSARY

Pseudoline graph: Given a Euclidean pseudoline arrangement Γ and a subset E of its vertices, the graph $G = (\Gamma, E)$ whose vertices are the members of Γ, with two vertices joined by an edge whenever the intersection of the corresponding pseudolines belongs to E.

Extendible set of pseudosegments: A set of Jordan arcs, each chosen from a different pseudoline belonging to a simple Euclidean arrangement.

TOPOLOGICAL SWEEP

The original idea behind what has come to be known as *topologically sweeping an arrangement* was applied, by Edelsbrunner and Guibas, to the case of an arrangement of straight lines. In order to construct the arrangement, rather than using a line to sweep it, they used a pseudoline, and achieved a saving of a factor of $\log n$ in the time required, while keeping the storage linear.

THEOREM 5.7.1 [EG89]

The cell complex of an arrangement of n *lines in the plane can be computed in* $O(n^2)$ *time and* $O(n)$ *working space by sweeping a pseudoline across it.*

This result can be applied to a number of problems, and results in an improvement of known bounds on each: minimum area triangle spanned by points, visibility graph of segments, and (in higher dimensions) enumerating faces of a hyperplane arrangement and testing for degeneracies in a point configuration.

The idea of a topological sweep was then generalized, by Snoeyink and Hershberger, to sweeping a pseudoline across an arrangement of *pseudolines*; they prove the possibility of such a sweep (Theorem 5.5.5), and show that it can be performed in the same time and space as in Theorem 5.7.1. They also apply this result to finding a short Boolean formula for a polygon with curved edges.

The topological sweep method was also used by Chazelle and Edelsbrunner [CE92] to report all k-segment intersections in an arrangement of n line segments in (optimal) $O(n \log n + k)$ time, and has been generalized to higher dimensions.

APPLICATIONS OF DUALITY

Theorem 5.1.6, and the algorithm used to compute the dual arrangement, are used by Agarwal and Sharir to compute incidences between points and pseudolines and to compute a subset of faces in a pseudoline arrangement [AS05]. An additional application is due to Sharir and Smorodinsky. Define a ***diamond*** in a Euclidean arrangement as two pairs $\{l_1, l_2\}, \{l_3, l_4\}$ of pseudolines such that the intersection of one pair lies above each member of the second and the intersection of the other pair below each member of the first.

THEOREM 5.7.2 [SS03]

Let Γ be a simple Euclidean pseudoline arrangement, E a subset of vertices of Γ, and $G = (\Gamma, E)$ the corresponding pseudoline graph. Then there is a drawing of G in the plane, with the edges constituting an extendible set of pseudosegments, such that for any two edges e, e' of G, e and e' form a diamond if and only if their corresponding drawings cross.

Conversely, for any graph $G = (V, E)$ drawn in the plane with its edges constituting an extendible set of pseudosegments, there is a simple Euclidean arrangement Γ of pseudolines and a one-to-one mapping ϕ from V onto Γ with each edge $uv \in E$ mapped to the vertex $\phi(u) \cap \phi(v)$ of Γ, such that two edges in E cross if and only if their images are two vertices of Γ forming a diamond.

This can then be used to provide a simple proof of the Tamaki-Tokuyama theorem:

THEOREM 5.7.3 [TT03]

Let Γ and G be as in Theorem 5.7.2. *If Γ is diamond-free, then G is planar, and hence $|E| \leq 3n - 6$.*

PSEUDOTRIANGULATIONS

A ***pseudotriangle*** is a simple polygon with exactly three convex vertices, and a ***pseudotriangulation*** is a tiling of a planar region into pseudotriangles. Pseudotriangulations first appeared as subdivisions of a polygon obtained by adding a collection of noncrossing geodesic paths between vertices of the polygon. The name pseudotriangulation was coined by Pocchiola and Vegter [PV94]. They discovered them when studying the visibility graph of a collection of pairwise disjoint convex obstacles.

Pseudotriangulations have an interesting connection with arrangements of pseudolines: A pseudotriangle in $\mathbb{R}^2$ has a unique interior tangent parallel to each direction. Dualizing the supporting lines of these tangents of a pseudotriangle we obtain the points of an x-monotone curve. Now consider a collection of pseudotriangles with pairwise disjoint interiors (for example those of a pseudotriangulation). Any two of them have exactly one common interior tangent, so the dual pseudolines form a pseudoline arrangement. Every line arrangement can be obtained with this construction. Pocchiola and Vegter also show that some nonstretchable arrangements can be obtained.

Streinu [Str05] studied minimal pseudotriangulations of point sets in the context of her algorithmic solution of the *Carpenter's Rule problem* previously settled existentially by Connelly, Demaine, and Rote [CDR00]. She obtained the following characterization of minimal pseudotriangulations

THEOREM 5.7.4 [Str05]

The following properties are equivalent for a geometric graph T on a set P of n vertices.

(i) *T is a pseudotriangulation of P with the minimum possible number of edges, i.e., a minimal pseudotriangulation.*

(ii) *Every vertex of T has an angle of size $> \pi$, i.e., T is a* pointed *pseudotriangulation.*

(iii) *T is a pseudotriangulation of P with $2n-3$ edges.*

(iv) *T is noncrossing, pointed, and has $2n-3$ edges.*

There has been a huge amount of research related to pseudotriangulations. Aichholzer et al. proved Conjecture 5.7.7 from the previous edition of this chapter, see [RSS08, Thm. 3.7]. A detailed treatment and an ample collection of references can be found in the survey article of Rote, Santos, and Streinu [RSS08]. We only mention some directions of research on pseudotriangulations:

- Polytopes of pseudotriangulations.
- Combinatorial properties, flips of pseudotriangulations and enumeration.
- Connections with rigidity theory.
- Combinatorial pseudotriangulations and their stretchability.

Pilaud and Pocciola [PP12] extended the correspondence between pseudoline arrangements and pseudotriangulations to some other classes of geometric graphs. The ***k*-kernel** of a Euclidean arrangement of pseudolines is obtained by deleting the first k and the last k levels of the arrangement. If the arrangement is given by a wiring diagram this corresponds to deleting the first k and the last k horizontal lines.

Call an arrangement $\mathcal{B}$ of $n-2k$ pseudolines a ***k*-descendant** of an arrangement $\mathcal{A}$ of n pseudolines if $\mathcal{B}$ can be drawn on the lines of the k-kernel of $\mathcal{A}$ such that at a crossing of the k-kernel the lines of $\mathcal{B}$ either cross or touch each other, Figure 5.7.1 shows an example.

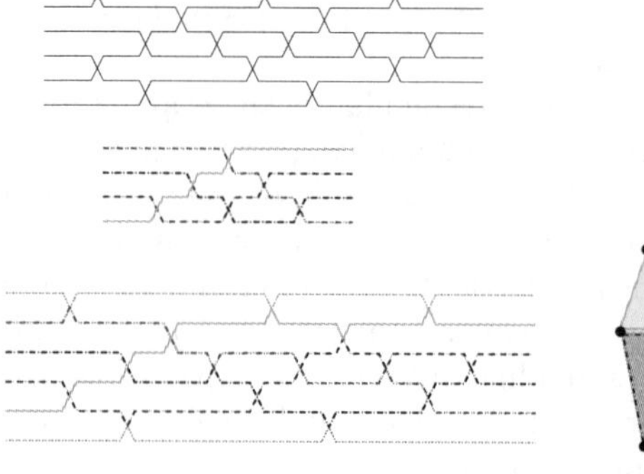

FIGURE 5.7.1
Arrangements $\mathcal{A}$ and $\mathcal{B}$ of 6 and 4 pseudolines, resp., and a drawing of $\mathcal{B}$ on the 1-kernel of $\mathcal{A}$. It corresponds to the pointed pseudotriangulation on the dual point set of $\mathcal{A}$.

THEOREM 5.7.5 [PP12]

(i) *For a point set P and its dual arrangement $\mathcal{A}_P$ there is a bijection between pointed pseudotriangulations of P and 1-descendants of $\mathcal{A}_P$.*

(ii) *For a set C_n in convex position and its dual arrangement $\mathcal{C}_n$ (a cyclic arrangement) there is a bijection between k-triangulations of C_n and k-descendants of $\mathcal{C}_n$.*

Both parts of the theorem can be proven by showing that the objects on both sides of the bijection have corresponding flip-structures. Pilaud and Pocciola also consider the k-descendants of the dual arrangement of nonconvex point sets; their duals are *k-pseudotriangulations.*

Another generalization relates pseudotriangulations of the free space of a set of disjoint convex bodies in the plane with k-descendants of arrangements of *double pseudolines.* This is based on the duality between sets of disjoint convex bodies and arrangements of double pseudolines in the projective plane. A thorough study of this duality can be found in [HP13].

PSEUDOPOLYGONS

A polygon is a cyclic sequence of vertices and noncrossing edges on a configuration of points. Similarly a ***pseudopolygon*** is based on vertices and edges taken from generalized configuration of points (see Section 5.2), O'Rourke and Streinu studied pseudopolygons in the context of visibility problems. They prove

THEOREM 5.7.6 [OS96]

There is a polynomial-time algorithm to decide whether a graph is realizable as the vertex-edge pseudo-visibility graph of a pseudopolygon.

The recognition problem for vertex-edge visibility graphs of polygons is likely to be hard. So this is another instance where relaxing a problem from straight to pseudo helps.

5.8 SOURCES AND RELATED MATERIAL

FURTHER READING

[BLS$^+$99]: A comprehensive account of oriented matroid theory, including a great many references; most references not given explicitly in this chapter can be traced through this book.

[Ede87]: An introduction to computational geometry, focusing on arrangements and their algorithms.

[Fel04] Covers combinatorial aspects of arrangements. Also includes a chapter on pseudotriangulations.

[GP91, GP93]: Two surveys on allowable sequences and order types and their complexity.

[Grü72]: A monograph on planar arrangements and their generalizations, with excellent problems (many still unsolved) and a very complete bibliography up to 1972.

RELATED CHAPTERS

Chapter 1: Finite point configurations
Chapter 4: Helly-type theorems and geometric transversals
Chapter 6: Oriented matroids
Chapter 28: Arrangements
Chapter 37: Computational and quantitative real algebraic geometry

REFERENCES

[AAK01] O. Aichholzer, F. Aurenhammer, and H. Krasser. Enumerating order types for small point sets with applications. *Order* 19:265–281, 2002. See also `www.ist.tugraz.at/staff/aichholzer/research/rp/triangulations/ordertypes`

[ACH+14] O. Aichholzer, J. Cardinal, T. Hackl, F. Hurtado, M. Korman, A. Pilz, R.I. Silveira, R. Uehara, P. Valtr, B. Vogtenhuber, and E. Welzl. Cell-paths in mono- and bichromatic line arrangements in the plane. *Disc. Math. & Theor. Comp. Sci.*, 16:317–332, 2014.

[AK07] O. Aichholzer and H. Krasser. Abstract order type extension and new results on the rectilinear crossing number. *Comput. Geom.* 36:2–15, 2007.

[APP+14] E. Ackerman, J. Pach, R. Pinchasi, R. Radoičić, and G. Tóth. A note on coloring line arrangements. *Electr. J. Comb.*, 21:P2.23, 2014.

[AS05] P.K. Agarwal and M. Sharir. Pseudo-line arrangements: duality, algorithms, and applications. *SIAM J. Comput.*, 34:526–552, 2005.

[AZ99] M. Aigner and G.M. Ziegler. *Proofs from THE BOOK*, 5th edition. Springer, Berlin, 2014.

[BBL08] N. Bartholdi, J. Blanc, and S. Loisel. On simple arrangements of lines and pseudo-lines in $\mathbb{P}^2$ and $\mathbb{R}^2$ with the maximum number of triangles. In J.E. Goodman, J. Pach, and R. Pollack, editors, *Surveys on Discrete and Computational Geometry: Twenty Years Later*, pages 105–116, AMS, Providence, 2008.

[BCC+13] P. Bose, J. Cardinal, S. Collette, F. Hurtado, M. Korman, S. Langerman, and P. Taslakian. Coloring and guarding arrangements. *Disc. Math. Theor. Comp. Sci.*, 15:139–154, 2013.

[BEPY91] M. Bern, D. Eppstein, P. Plassmann, and F. Yao. Horizon theorems for lines and polygons. In J.E. Goodman, R. Pollack, and W. Steiger, editors, *Discrete and Computational Geometry: Papers from the DIMACS Special Year*, pages 45–66, AMS, Providence, 1991.

[Ber08] L.W. Berman. Symmetric simplicial pseudoline arrangements. *Electr. J. Comb.*, 15:R13, 2008.

[Bla11] J. Blanc. The best polynomial bounds for the number of triangles in a simple arrangement of n pseudo-lines. *Geombinatorics*, 21:5–17, 2011.

[BLS+99] A. Björner, M. Las Vergnas, B. Sturmfels, N. White, and G.M. Ziegler. *Oriented Matroids.* 2nd edition, vol. 46 of *Encyclopedia of Mathematics*, Cambridge University Press, 1999.

[Bok08] J. Bokowski. On heuristic methods for finding realizations of surfaces. In Bobenko et al., editors, *Discrete Differential Geometry*, vol. 38 of *Oberwolfach Seminars*, pages 255–260, Birkhäuser, 2008.

[BRS+05] J. Balogh, O. Regev, C. Smyth, W. Steiger, and M. Szegedy. Long monotone paths in line arrangements. *Discrete Comput. Geom.*, 32:167–176, 2005.

[BS89a] J. Bokowski and B. Sturmfels. An infinite family of minor-minimal nonrealizable 3-chirotopes. *Math. Zeitschrift*, 200:583–589, 1989.

[BS89b] J. Bokowski and B. Sturmfels. *Computational Synthetic Geometry.* Vol. 1355 of *Lecture Notes in Math.*, Springer, Berlin, 1989.

[Can69] R.J. Canham. A theorem on arrangements of lines in the plane. *Israel J. Math.*, 7:393–397, 1969.

[CDR00] R. Connelly, E.D. Demaine, and G. Rote. Straightening polygonal arcs and convexifying polygonal cycles. *Proc. 41st IEEE Sympos. Found. Comput. Sci.*, pages 432–442, 2000.

[CE92] B. Chazelle and H. Edelsbrunner. An optimal algorithm for intersecting line segments in the plane. *J. ACM* 39:1–54, 1992.

[CEG+90] K.L. Clarkson, H. Edelsbrunner, L.J. Guibas, M. Sharir, and E. Welzl. Combinatorial complexity bounds for arrangements of curves and spheres. *Discrete Comput. Geom.*, 5:99–160, 1990.

[Cun12] M. Cuntz. Simplicial arrangements with up to 27 lines. *Discrete Comput. Geom.*, 48:682–701, 2012.

[DP98] T.K. Dey and J. Pach. Extremal problems for geometric hypergraphs. *Discrete Comput. Geom.*, 19:473–484, 1998.

[Dum05] A. Dumitrescu. Monotone paths in line arrangements with a small number of directions. *Discrete Comput. Geom.*, 33:687–697, 2005.

[Ede87] H. Edelsbrunner. *Algorithms in Combinatorial Geometry.* Springer, Berlin, 1987.

[EG89] H. Edelsbrunner and L.J. Guibas. Topologically sweeping an arrangement. *J. Comput. System Sci.*, 38:165–194, 1989; Corrigendum: 42:249–251, 1991.

[EP95] P. Erdős and G. Purdy. Extremal problems in combinatorial geometry. In Grötschel et al., editors, *Handbook of Combinatorics*, vol. 1, Elsevier, Amsterdam, 1995.

[Epp14] D. Eppstein. Drawing arrangement graphs in small grids, or how to play planarity. *J. Graph Algorithms Appl.*, 18:211–231, 2014.

[EW86] H. Edelsbrunner and E. Welzl. On the maximal number of edges of many faces in arangements. *J. Combin. Theory Ser. A*, 41:159–166, 1986.

[Fel97] S. Felsner. On the number of arrangements of pseudolines. *Discrete Comput. Geom.*, 18:257–267, 1997.

[Fel04] S. Felsner. *Geometric Graphs and Arrangements.* Vieweg, Wiesbaden, 2004.

[FHNS00] S. Felsner, F. Hurtado, M. Noy, and I. Streinu. Hamiltonicity and colorings of arrangement graphs. *Discrete Appl. Math.*, 154:2470–2483, 2006.

[Fin] L. Finschi. Homepage of Oriented Matroids. `http://www.om.math.ethz.ch`.

[FK99] S. Felsner and K. Kriegel. Triangles in Euclidean arrangements. *Discrete Comput. Geom.*, 22:429–438, 1999.

[FMM13] K. Fukuda, H. Miyata, and S. Moriyama. Complete enumeration of small realizable oriented matroids. *Discrete Comput. Geom.*, 49:359–381, 2013.

[FR98] D. Forge and J.L. Ramírez Alfonsín. Straight line arrangements in the real projective plane. *Discrete Comput. Geom.*, 20:155–161, 1998.

[FR01] D. Forge and J.L. Ramírez Alfonsín. On counting the k-face cells of cyclic arrangements. *European. J. Combin.*, 22:307–312, 2001.

[FV11] S. Felsner and P. Valtr. Coding and counting arrangements of pseudolines. *Discrete Comput. Geom.*, 46:405–416, 2011.

[FW01] S. Felsner and H. Weil. Sweeps, arrangements, and signotopes. *Discrete Appl. Math.*, 109:67–94, 2001.

[FZ01] S. Felsner and G.M. Ziegler. Zonotopes associated with higher Bruhat orders. *Discrete Math.*, 241:301–312, 2001.

[Goo80] J.E. Goodman. Proof of a conjecture of Burr, Grünbaum, and Sloane. *Discrete Math.*, 32:27–35, 1980.

[GP80a] J.E. Goodman and R. Pollack. On the combinatorial classification of nondegenerate configurations in the plane. *J. Combin. Theory Ser. A*, 29:220–235, 1980.

[GP80b] J.E. Goodman and R. Pollack. Proof of Grünbaum's conjecture on the stretchability of certain arrangements of pseudolines. *J. Combin. Theory Ser. A*, 29:385–390, 1980.

[GP81a] J.E. Goodman and R. Pollack. A combinatorial perspective on some problems in geometry. *Congressus Numerantium*, 32:383–394, 1981.

[GP81b] J.E. Goodman and R. Pollack. Three points do not determine a (pseudo-) plane. *J. Combin. Theory Ser. A*, 31:215–218, 1981.

[GP82a] J.E. Goodman and R. Pollack. Helly-type theorems for pseudoline arrangements in P^2. *J. Combin. Theory Ser. A*, 32:1–19, 1982.

[GP82b] J.E. Goodman and R. Pollack. A theorem of ordered duality. *Geometriae Dedicata*, 12:63–74, 1982.

[GP84] J.E. Goodman and R. Pollack. Semispaces of configurations, cell complexes of arrangements. *J. Combin. Theory Ser. A*, 37:257–293, 1984.

[GP85] J.E. Goodman and R. Pollack. A combinatorial version of the isotopy conjecture. *Ann. New York Acad. Sci.*, 440:31–33, 1985.

[GP91] J.E. Goodman and R. Pollack. The complexity of point configurations. *Discrete Appl. Math.*, 31:167–180, 1991.

[GP93] J.E. Goodman and R. Pollack. Allowable sequences and order types in discrete and computational geometry. In J. Pach, editor, *New Trends in Discrete and Computational Geometry*, vol. 10 of *Algorithms Combin.*, pages 103–134, Springer, Berlin, 1993.

[GPS89] J.E. Goodman, R. Pollack, and B. Sturmfels. Coordinate representation of order types requires exponential storage. *Proc. 21th Sympos. Comput. Geom.*, pages 405–410, ACM Press, 1989.

[GPW96] J.E. Goodman, R. Pollack, and R. Wenger. There are uncountably many universal topological planes. *Geom. Dedicata*, 59:157–162, 1996.

[GPWZ94] J.E. Goodman, R. Pollack, R. Wenger, and T. Zamfirescu. Arrangements and topological planes. *Amer. Math. Monthly*, 101:866–878, 1994.

[Grü71] B. Grünbaum. Arrangements of hyperplanes. *Congressus Numerantium*, 3:41–106, 1971.

[Grü72] B. Grünbaum. *Arrangements and Spreads.* Volume 10 of *CBMS Regional Conf. Ser. in Math.* AMS, Providence, 1972.

[Grü09] B. Grünbaum. A catalogue of simplicial arrangements in the real projective plane. *Ars Math. Contemp.*, 2:1–25, 2009.

[Grü13] B. Grünbaum. Simplicial arrangements revisited. *Ars Math. Contemp.*, 6:419–433, 2013.

[Gut16] L. Guth. *Polynomial Methods in Combinatorics.* AMS, Providence, 2016.

[Har85] H. Harborth. Some simple arrangements of pseudolines with a maximum number of triangles. *Ann. New York Acad. Sci.*, 440:31–33, 1985.

[HKM15] U. Hoffmann, L. Kleist, and T. Miltzow. Upper and lower bounds on long dual paths in line arrangements. In *Proc. 40th Sympos. Math. Found. Comp. Sci.*, vol. 9235 of *LNCS*, pages 407–419, Springer, Berlin, 2015.

[HM02] H. Harborth and M. Möller. An arrangement graph without a Hamiltonian cycle decomposition. *Bull. Inst. Combin. Appl.*, 34:7–9, 2002.

[HP13] L. Habert and M. Pocchiola. LR characterization of chirotopes of finite planar families of pairwise disjoint convex bodies. *Discrete Comput. Geom.*, 50:552–648, 2013.

[Jam85] R.E. Jamison. A survey of the slope problem. *Ann. New York Acad. Sci.*, 440: 34–51, 1985.

[Knu92] D.E. Knuth. *Axioms and Hulls*. Vol. 606 of *LNCS*, Springer, Berlin, 1992.

[KSYM11] J. Kawahara, T. Saitoh, R. Yoshinaka, and S. Minato. Counting primitive sorting networks by πDDs. Technical Report, Hokkaido University, TCS-TR-A-11-54, 2011. `http://www-alg.ist.hokudai.ac.jp/~thomas/TCSTR/tcstr_11_54/tcstr_11_54.pdf`

[LLM+07] J. Leaños, M. Lomelí, C. Merino, G. Salazar, and J. Urrutia. Simple Euclidean arrangements with no ($\geq$ 5)-gons. *Discrete Comput. Geom.*, 38:595–603, 2007.

[Lev26] F. Levi. Die Teilung der projektiven Ebene durch Gerade oder Pseudogerade. *Ber. Math.-Phys. Kl. Sächs. Akad. Wiss.*, 78:256–267, 1926.

[Lom90] H. Lombardi. Nullstellensatz réel effectif et variantes. *C. R. Acad. Sci. Paris, Sér. I*, 310:635–640, 1990.

[LPS14] B. Lund, G.B. Purdy, and J.W. Smith. A pseudoline counterexample to the strong dirac conjecture. *Electr. J. Comb.*, 21:P2.31, 2014.

[LRS89] D. Ljubić, J.-P. Roudneff, and B. Sturmfels. Arrangements of lines and pseudolines without adjacent triangles. *J. Combin. Theory Ser. A*, 50:24–32, 1989.

[Mar93] N. Martinov.Classification of arrangements by the number of their cells. *Discrete Comput. Geom.*, 9:39–46, 1993.

[Mat91] J. Matoušek. Lower bounds on the length of monotone paths in arrangements. *Discrete Comput. Geom.*, 6:129–134, 1991.

[Mel41] E. Melchior. Über Vielseite der projektiven Ebene. *Deutsche Math.*, 5:461–475, 1941.

[Mnë85] N.E. Mnëv. On manifolds of combinatorial types of projective configurations and convex polyhedra. *Soviet Math. Dokl.*, 32:335–337, 1985.

[Mnë88] N.E. Mnëv. The universality theorems on the classification problem of configuration varieties and convex polytopes varieties. In O.Y. Viro, editor, *Topology and Geometry—Rohlin Seminar*, vol. 1346 of *Lecture Notes in Math.*, pages 527–544, Springer, Berlin, 1988.

[Nym04] K. Nyman. Linear inequalities for rank 3 geometric lattices. *Discrete Comput. Geom.*, 31:229–242, 2004.

[OS96] J. O'Rourke and I. Streinu. Pseudo-visibility graphs in pseudo-polygons: Part II. Tech. Report 042, Dept. Comput. Sci., Smith College, Northampton, 1996.

[Pin11] R. Pinchasi. The zone theorem revisited. Manuscript, `http://www2.math.technion.ac.il/~room/ps_files/zonespl.pdf`, 2011.

[PP01] J. Pach and R. Pinchasi. On the number of balanced lines. *Discrete Comput. Geom.*, 25:611–628, 2001.

[PP12] V. Pilaud and M. Pocchiola. Multitriangulations, pseudotriangulations and primitive sorting networks. *Discrete Comput. Geom.*, 48:142–191, 2012.

[PS09] J. Pach and M. Sharir. *Combinatorial Geometry and its Algorithmic Applications: The Alcalá Lectures*. vol. 152 of *Mathematical Surveys and Monographs*, AMS, Providence, 2009.

[PV94] M. Pocchiola and G. Vegter. Order types and visibility types of configurations of disjoint convex plane sets. Extended abstract, Tech. Report 94-4, Labo. d'Inf. de l'ENS, Paris, 1994.

[Ric89] J. Richter. Kombinatorische Realisierbarkeitskriterien für orientierte Matroide. *Mitt. Math. Sem. Gießen*, 194:1–112, 1989.

[Ric96a] J. Richter-Gebert. *Realization Spaces of Polytopes*. Vol. 1643 of *Lecture Notes in Math.*, Springer, Berlin, 1996.

[Ric96b] J. Richter-Gebert. Two interesting oriented matroids. *Documenta Math.*, 1:137–148, 1996.

[Rin56] G. Ringel. Teilungen der Ebene durch Geraden oder topologische Geraden. *Math. Z.*, 64:79–102, 1956.

[Rin57] G. Ringel. Über Geraden in allgemeiner Lage. *Elem. Math.*, 12:75–82, 1957.

[Rou87] J.-P. Roudneff. Quadrilaterals and pentagons in arrangements of lines. *Geom. Dedicata*, 23:221–227, 1987.

[Rou88a] J.-P. Roudneff. Arrangements of lines with a minimal number of triangles are simple. *Discrete Comput. Geom.*, 3:97–102, 1988.

[Rou88b] J.-P. Roudneff. Tverberg-type theorems for pseudoconfigurations of points in the plane. *European J. Combin.*, 9:189–198, 1988.

[Rou96] J.-P. Roudneff. The maximum number of triangles in arrangements of (pseudo-) lines. *J. Combin. Theory Ser. B*, 66:44–74, 1996.

[RS88] J.-P. Roudneff and B. Sturmfels. Simplicial cells in arrangements and mutations of oriented matroids. *Geom. Dedicata*, 27:153–170, 1988.

[RSS08] G. Rote, F. Santos, and I. Streinu. Pseudo-triangulations–a survey. In J.E. Goodman J. Pach, and R. Pollack, editors, *Surveys on Discrete and Computational Geometry: Twenty Years Later*, pages 343–410, AMS, Providence, 2008.

[RT03] R. Radoičić and G. Tóth. Monotone paths in line arrangements. *Comput. Geom.*, 24:129–134, 2003.

[SE88] P. Salamon and P. Erdős. The solution to a problem of Grünbaum. *Canad. Math. Bull.*, 31:129–138, 1988.

[SH91] J. Snoeyink and J. Hershberger. Sweeping arrangements of curves. In J.E. Goodman, R. Pollack, and W. Steiger, editors, *Discrete and Computational Geometry: Papers from the DIMACS Special Year*, vol. 6 of *DIMACS Series in Discrete Math. Theor. Comp. Sci.*, pages 309–349, AMS, Providence, 1991.

[Sha79] R.W. Shannon. Simplicial cells in arrangements of hyperplanes. *Geom. Dedicata*, 8:179–187, 1979.

[Sho91] P. Shor. Stretchability of pseudolines is NP-hard. In P. Gritzmann and B. Sturmfels, editors, *Applied Geometry and Discrete Mathematics—The Victor Klee Festschrift*, volume 4 of *DIMACS Series in Discrete Math. Theor. Comp. Sci.*, pages 531–554, AMS, Providence, 1991.

[SS94] W. Steiger and I. Streinu. A pseudo-algorithmic separation of lines from pseudo-lines. *Inf. Process. Lett.*, 53:295–299, 1995.

[SS03] M. Sharir and S. Smorodinsky. Extremal configurations and levels in pseudoline arrangements. In *Proc. 8th Workshop on Algorithms and Data Structures*, vol. 2748 of *LNCS*, pages 127–139, Springer, Berlin, 2003.

[ST83] E. Szemerédi and W.T. Trotter. Extremal problems in discrete geometry. *Combinatorica*, 3:381–392, 1983.

[Sta84] R.P. Stanley. On the number of reduced decompositions of elements of Coxeter groups. *European J. Combin.*, 5:359–372, 1984.

[Str97] I. Streinu. Clusters of stars. *Proc. 13th Sympos. on Comput. Geom.*, pages 439–441, ACM Press, 1997.

[Str05] I. Streinu. Pseudo-triangulations, rigidity and motion planning. *Discrete Comput. Geom.*, 34:587–635, 2005.

[Suv88] P.Y. Suvorov. Isotopic but not rigidly isotopic plane systems of straight lines. In O.Ya. Viro and A.M. Varshik, editors, *Topology and Geometry—Rohlin Seminar*, volume 1346 of *Lecture Notes in Math.*, pages 545–556, Springer, Berlin, 1988.

[Szé97] L.A. Székely. Crossing numbers and hard Erdős problems in discrete geometry. *Combin. Probab. Comput.*, 6:353–358, 1997.

[TT03] H. Tamaki and T. Tokuyama. A characterization of planar graphs by pseudo-line arrangements. *Algorithmica*, 35:269–285, 2003.

[Ung82] P. Ungar. $2N$ noncollinear points determine at least $2N$ directions. *J. Combin. Theory Ser. A*, 33:343–347, 1982.

6 ORIENTED MATROIDS

Jürgen Richter-Gebert and Günter M. Ziegler

INTRODUCTION

The theory of *oriented matroids* provides a broad setting in which to model, describe, and analyze combinatorial properties of geometric configurations. Mathematical objects of study that appear to be disjoint and independent, such as *point and vector configurations*, *arrangements of hyperplanes*, *convex polytopes*, *directed graphs*, and *linear programs* find a common generalization in the language of oriented matroids.

The oriented matroid of a finite set of points P extracts relative position and orientation information from the configuration; for example, it can be given by a list of signs that encodes the orientations of all the bases of P. In the passage from a concrete point configuration to its oriented matroid, metrical information is lost, but many structural properties of P have their counterparts at the—purely combinatorial—level of the oriented matroid. (In computational geometry, the oriented matroid data of an unlabelled point configuration are sometimes called the *order type*.) From the oriented matroid of a configuration of points, one can compute not only that face lattice of the convex hull, but also the set of all its triangulations and subdivisions (cf. Chapter 16).

We first introduce oriented matroids in the context of several models and motivations (Section 6.1). Then we present some equivalent axiomatizations (Section 6.2). Finally, we discuss concepts that play central roles in the theory of oriented matroids (Section 6.3), among them *duality*, *realizability*, the study of *simplicial cells*, and the treatment of *convexity*.

6.1 MODELS AND MOTIVATIONS

This section discusses geometric examples that are usually treated on the level of concrete coordinates, but where an "oriented matroid point of view" gives deeper insight. We also present these examples as standard models that provide intuition for the behavior of general oriented matroids.

6.1.1 ORIENTED BASES OF VECTOR CONFIGURATIONS

GLOSSARY

Vector configuration X: A matrix $X = (x_1, \ldots, x_n) \in (\mathbb{R}^d)^n$, usually assumed to have full rank d.

Matroid of X: The pair $M_X = (E, \mathcal{B}_X)$, where $E := \{1, 2, \ldots, n\}$ and $\mathcal{B}_X$ is the set of all subsets of E that correspond to bases of (the column space of) X.

Matroid: A pair $M = (E, \mathcal{B})$, where E is a finite set, and $\mathcal{B} \subset 2^E$ is a nonempty collection of subsets of E (the ***bases*** of M) that satisfies the ***Steinitz exchange axiom***: For all $B_1, B_2 \in \mathcal{B}$ and $e \in B_1 \backslash B_2$, there exists an $f \in B_2 \backslash B_1$ such that $(B_1 \backslash e) \cup f \in \mathcal{B}$.

Chirotope of X: The map

$$\begin{array}{rrcl} \chi_X: & E^d & \to & \{-1, 0, +1\} \\ & (\lambda_1, \ldots, \lambda_d) & \mapsto & \operatorname{sign}(\det(x_{\lambda_1}, \ldots, x_{\lambda_d})). \end{array}$$

Signs: The elements of the set $\{-, 0, +\}$, used as a shorthand for the corresponding elements of $\{-1, 0, +1\}$.

Ordinary (unoriented) *matroids*, as introduced in 1935 by Whitney (see Kung [Kun86], Oxley [Oxl92]), can be considered as an abstraction of vector configurations in finite dimensional vector spaces over arbitrary fields. All the bases of a matroid M have the same cardinality d, which is called the ***rank*** of the matroid. Equivalently, we can identify M with the characteristic function of the bases $B_M \colon E^d \to \{0, 1\}$, where $B_M(\lambda) = 1$ if and only if $\{\lambda_1, \ldots, \lambda_d\} \in \mathcal{B}$.

One can obtain examples of matroids as follows: Take a finite set of vectors

$$X = \{x_1, x_2, \ldots, x_n\} \subset V$$

of rank d in a d-dimensional vector space V over an arbitrary field K and consider the set of bases of V formed by subsets of the points in X. In other words, the pair

$$M_X = (E, \mathcal{B}_X) = \big(\, \{1, \ldots, n\},\ \{\{\lambda_1, \ldots, \lambda_d\} \mid \det(x_{\lambda_1}, \ldots, x_{\lambda_d}) \neq 0\}\, \big)$$

forms a matroid.

The basic information about the incidence structure of the points in X is contained in the underlying matroid M_X. However, the matroid alone without additional orientation information contains only very restricted information about a geometric configuration. For example, any configuration of n points in the plane in ***general position*** (i.e., no three points on a line) after ***homogenization*** (that is, appending a coordinate 1 to the each point) yields a vector configuration in $\mathbb{R}^3$ such that any three vectors are linearly independent. Thus all such point configurations yield the same matroid $M = U_{3,n}$: Here the matroid retains no information beyond the dimension and size of the configuration, and the fact that it is in general position.

In contrast to matroids, the theory of ***oriented matroids*** considers the structure of dependencies in vector spaces over *ordered* fields. Roughly speaking, an oriented matroid is a matroid where in addition every basis is equipped with an orientation. These oriented bases have to satisfy an oriented version of the Steinitz exchange axiom (to be described later). For the affine setting, oriented matroids not only describe the incidence structure between the points of X and the hyperplanes spanned by points of X (this is the matroid information); they also encode the positions of the points relative to the hyperplanes: "Which points lie on the positive side of a hyperplane, which points lie on the negative side, and which lie on the hyperplane?" If $X \in V^n$ is a configuration of n points in a d-dimensional vector space V over an ordered field K, we can describe the corresponding oriented matroid χ_X by the chirotope, which encodes the orientation of the $(d+1)$-tuples of points in X. The chirotope is very closely related to the oriented matroid of X, but it encodes much more information than the corresponding matroid, including orientation and convexity information about the underlying configuration.

6.1.2 CONFIGURATIONS OF POINTS

GLOSSARY

Affine point configuration: A matrix $X' = (x'_1, \dots, x'_n) \in (\mathbb{R}^{d-1})^n$, usually with the assumption that $x'_1, \dots, x'_n$ affinely span $\mathbb{R}^{d-1}$.

Associated vector configuration: The matrix $X \in (\mathbb{R}^d)^n$ of full rank d obtained from an ordered/labeled point configuration $X' = (x'_1, \dots, x'_n)$ by adding a row of ones. This corresponds to the embedding of the affine space $\mathbb{R}^{d-1}$ into the linear vector space $\mathbb{R}^d$ via $p \longmapsto x = \binom{p}{1}$.

Homogenization: The step from a point configuration to the associated vector configuration.

Oriented matroid of an affine point configuration: The oriented matroid of the associated vector configuration.

Covector of a vector configuration X: Any partition of $X = (x_1, \dots, x_n)$ induced by an oriented linear hyperplane into the points on the hyperplane, on its positive side, and on its negative side. The partition is denoted by a sign vector $C \in \{-, 0, +\}^n$.

Oriented matroid of X: The collection $\mathcal{L}_X \subseteq \{-, 0, +\}^n$ of all covectors of X.

Let $X := (x_1, \dots, x_n) \in (\mathbb{R}^d)^n$ be an $n \times d$ matrix and let $E := \{1, \dots, n\}$. We interpret the columns of X as n vectors in the d-dimensional real vector space $\mathbb{R}^d$. For a linear functional $y^T \in (\mathbb{R}^d)^*$ we set

$$C_X(y) = (\mathrm{sign}(y^T x_1), \dots, \mathrm{sign}(y^T x_n)).$$

Such a sign vector is called a ***covector*** of X. We denote the collection of all covectors of X by

$$\mathcal{L}_X := \{C_X(y) \mid y \in \mathbb{R}^d\}.$$

The pair $\mathcal{M}_X = (E, \mathcal{L}_X)$ is called the ***oriented matroid*** of X. Here each sign vector $C_X(y) \in \mathcal{L}_X$ describes the positions of the vectors $x_1, \dots, x_n$ relative to the linear hyperplane $H_y = \{x \in \mathbb{R}^d \mid y^T x = 0\}$: the sets

$$\begin{aligned} C_X(y)^0 &:= \{e \in E \mid C_X(y)_e = 0\} \\ C_X(y)^+ &:= \{e \in E \mid C_X(y)_e > 0\} \\ C_X(y)^- &:= \{e \in E \mid C_X(y)_e < 0\} \end{aligned}$$

describe how H_y partitions the set of points X. Thus $C_X(y)^0$ contains the points on H_y, while $C_X(y)^+$ and $C_X(y)^-$ contain the points on the positive and on the negative side of H_y, respectively. In particular, if $C_X(y)^- = \emptyset$, then all points not on H_y lie on the positive side of H_y. In other words, in this case H_y determines a face of the positive cone

$$\mathrm{pos}(x_1, \dots, x_n) := \{\lambda_1 x_1 + \lambda_2 x_2 + \dots + \lambda_n x_n \mid 0 \le \lambda_i \in \mathbb{R} \text{ for } 1 \le i \le n\}$$

of all points of X. The face lattice of the cone $\text{pos}(X)$ can be recovered from $\mathcal{L}_X$. It is simply the set $\mathcal{L}_X \cap \{+,0\}^E$, partially ordered by the order induced from the relation "$0 < +$."

If, in the configuration X, we have $x_{i,d} = 1$ for all $1 \le i \le n$, then we can consider X as representing homogeneous coordinates of an *affine* point set X' in $\mathbb{R}^{d-1}$. Here the affine points correspond to the original points x_i after removal of the dth coordinate. The face lattice of the convex polytope $\text{conv}(X') \subset \mathbb{R}^{d-1}$ is then isomorphic to the face lattice of $\text{pos}(X)$. Hence, $\mathcal{M}_X$ can be used to recover the *convex hull* of X'.

Thus oriented matroids are generalizations of point configurations in linear or affine spaces. For general oriented matroids we weaken the assumption that the hyperplanes spanned by points of the configuration are flat to the assumption that they only satisfy certain topological incidence properties. Nonetheless, this kind of picture is sometimes misleading since not all oriented matroids have this type of representation (compare the "Type II representations" of [BLS+93, Sect. 5.3]).

6.1.3 ARRANGEMENTS OF HYPERPLANES AND OF HYPERSPHERES

GLOSSARY

Hyperplane arrangement $\mathcal{H}$: Collection of (oriented) linear hyperplanes in $\mathbb{R}^d$, given by normal vectors $x_1, \dots, x_n$.

Hypersphere arrangement induced by $\mathcal{H}$: Intersection of $\mathcal{H}$ with the unit sphere S^{d-1}.

Covectors of $\mathcal{H}$: Sign vectors of the cells in $\mathcal{H}$; equivalently, $\mathbf{0}$ together with the sign vectors of the cells in $\mathcal{H} \cap S^{d-1}$.

We obtain a different picture if we polarize the situation and consider *hyperplane arrangements* rather than configurations of points. For a real matrix $X := (x_1, \dots, x_n) \in (\mathbb{R}^d)^n$ consider the system of hyperplanes $\mathcal{H}_X := (H_1, \dots, H_n)$ with

$$H_i := \{y \in \mathbb{R}^d \mid y^T x_i = 0\}.$$

Each vector x_i induces an orientation on H_i by defining

$$H_i^+ := \{y \in \mathbb{R}^d \mid y^T x_i > 0\}$$

to be the ***positive side*** of H_i. We define H_i^- analogously to be the ***negative side*** of H_i. To avoid degenerate cases we assume that X contains at least one proper basis (i.e., the matrix X has rank d). The hyperplane arrangement $\mathcal{H}_X$ subdivides $\mathbb{R}^d$ into polyhedral cones. Without loss of information we can intersect with the unit sphere S^{d-1} and consider the sphere system

$$\mathcal{S}_X := \left(H_1 \cap S^{d-1}, \dots, H_n \cap S^{d-1}\right) = \mathcal{H}_X \cap S^{d-1}.$$

Our assumption that X contains at least one proper basis translates to the fact that the intersection of all $H_1 \cap \ldots \cap H_n \cap S^{d-1}$ is empty. $\mathcal{H}_X$ induces a cell decomposition $\Gamma(\mathcal{S}_X)$ on S^{d-1}. Each face of $\Gamma(\mathcal{S}_X)$ corresponds to a sign vector in $\{-,0,+\}^E$ that indicates the position of the cell with respect to the $(d-2)$-spheres

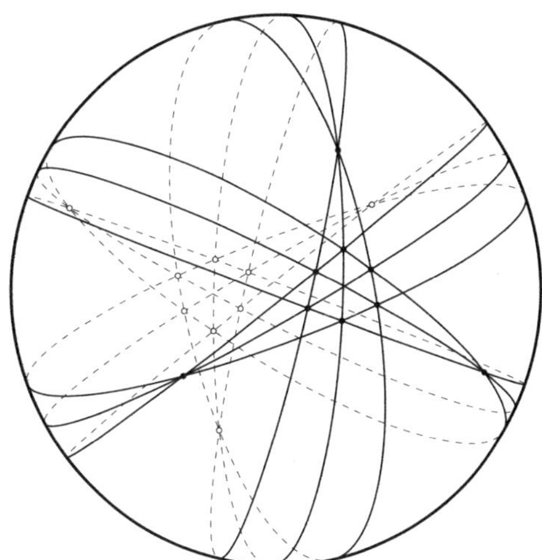

FIGURE 6.1.1
An arrangement of nine great circles on S^2. The arrangement corresponds to a Pappus configuration.

$H_i \cap S^{d-1}$ (and therefore with respect to the hyperplanes H_i) of the arrangement. The list of all these sign vectors is exactly the set $\mathcal{L}_X$ of covectors of $\mathcal{H}_X$.

While the visualization of oriented matroids by sets of points in $\mathbb{R}^n$ does not fully generalize to the case of nonrepresentable oriented matroids, the picture of hyperplane arrangements has a well-defined extension that also covers all the nonrealizable cases. We will see that as a consequence of the topological representation theorem of Folkman and Lawrence (Section 6.2.4) every rank d oriented matroid can be represented as an arrangement of oriented *pseudospheres* (or pseudohyperplanes) embedded in the S^{d-1} (or in $\mathbb{R}^d$, respectively). Arrangements of pseudospheres are systems of topological $(d-2)$-spheres embedded in S^{d-1} that satisfy certain intersection properties that clearly hold in the case of "straight" arrangements.

6.1.4 ARRANGEMENTS OF PSEUDOLINES

GLOSSARY

Pseudoline: Simple closed curve p in the projective plane $\mathbb{R}\mathrm{P}^2$ that is topologically equivalent to a line (i.e., there is a self-homeomorphism of $\mathbb{R}\mathrm{P}^2$ mapping p to a straight line).

Arrangement of pseudolines: Collection of pseudolines $\mathcal{P} := (p_1, \ldots, p_n)$ in the projective plane, any two of them intersecting exactly once.

Simple arrangement: No three pseudolines meet in a common point. (Equivalently, the associated oriented matroid is ***uniform***.)

Equivalent arrangements: Arrangements $\mathcal{P}_1$ and $\mathcal{P}_2$ that generate isomorphic cell decompositions of $\mathbb{R}\mathrm{P}^2$. (In this case there exists a self-homeomorphism of $\mathbb{R}\mathrm{P}^2$ mapping $\mathcal{P}_1$ to $\mathcal{P}_2$.)

Stretchable arrangement of pseudolines: An arrangement that is equivalent to an arrangement of projective lines.

An *arrangement of pseudolines* in the projective plane is a collection of pseudolines such that any two pseudolines intersect in exactly one point, where they cross. (See Grünbaum [Grü72] and Richter [Ric89].) We will always assume that $\mathcal{P}$ is ***essential***, i.e., that the intersection of all the pseudolines p_i is empty.

An arrangement of pseudolines behaves in many respects just like an arrangement of n lines in the projective plane. (In fact, there are only very few combinatorial theorems known that are true for straight arrangements, but not true in general for pseudoarrangements.) Figure 6.1.1 shows a small example of a nonstretchable arrangement of pseudolines. (It is left as a challenging exercise to the reader to prove the nonstretchability.) Up to isomorphism this is the only simple nonstretchable arrangement of 9 pseudolines [Ric89] [Knu92]; every arrangement of 8 (or fewer) pseudolines is stretchable [GP80].

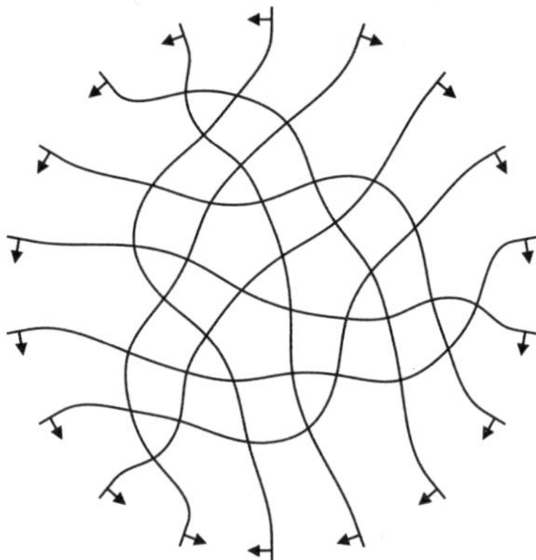

FIGURE 6.1.2
A nonstretchable arrangement of nine pseudolines. It was obtained by Ringel [Rin56] *as a perturbation of the Pappus configuration.*

To associate with a projective arrangement $\mathcal{P}$ an oriented matroid we represent the projective plane (as customary) by the 2-sphere with antipodal points identified. With this, every arrangement of pseudolines gives rise to an arrangement of *great pseudocircles* on S^2. For each great pseudocircle on S^2 we choose a positive side. Each cell induced by $\mathcal{P}$ on S^2 now corresponds to a unique sign vector. The collection of all these sign vectors again forms a set of covectors $\mathcal{L}_\mathcal{P}\backslash\{\mathbf{0}\}$ of an oriented matroid of rank 3. Conversely, as a special case of the topological representation theorem (see Theorem 6.2.4 below), *every* oriented matroid of rank 3 has a representation by an *oriented* pseudoline arrangement.

Thus we can use pseudoline arrangements as a standard picture to represent rank 3 oriented matroids. The easiest picture is obtained when we restrict ourselves to the upper hemisphere of S^2 and assume w.l.o.g. that each pseudoline crosses the equator exactly once, and that the crossings are distinct (i.e., no intersection of the great pseudocircles lies on the equator). Then we can represent this upper hemisphere by an arrangement of mutually crossing, oriented affine pseudolines in the plane $\mathbb{R}^2$. (We did this implicitly while drawing Figure 6.1.2.) For a reasonably elementary proof of the fact that rank 3 oriented matroids are equivalent to arrangements of pseudolines see Bokowski, Mock, and Streinu [BMS01].

By means of this equivalence, all problems concerning pseudoline arrangements can be translated to the language of oriented matroids. For instance, the problem of stretchability is equivalent to the realizability problem for oriented matroids.

6.2 AXIOMS AND REPRESENTATIONS

In this section we define oriented matroids formally. It is one of the main features of oriented matroid theory that the same object can be viewed under quite different aspects. This results in the fact that there are many different equivalent axiomatizations, and it is sometimes very useful to "jump" from one point of view to another. Statements that are difficult to prove in one language may be easy in another. For this reason we present here several different axiomatizations. We also give a (partial) dictionary that indicates how to translate among them. For a complete version of the basic equivalence proofs—which are highly nontrivial—see [BLS+93, Chapters 3 and 5].

We will give axiomatizations of oriented matroids for the following four types of representations:

- collections of covectors,
- collections of cocircuits,
- signed bases, and
- arrangements of pseudospheres.

In the last part of this section these concepts are illustrated by an example.

GLOSSARY

Sign vector: Vector C in $\{-,0,+\}^E$, where E is a finite index set, usually $\{1,\dots,n\}$. For $e \in E$, the e-component of C is denoted by C_e.

Positive, negative, and zero part of C:

$$\begin{aligned} C^+ &:= \{e \in E \mid C_e = +\}, \\ C^- &:= \{e \in E \mid C_e = -\}, \\ C^0 &:= \{e \in E \mid C_e = 0\}. \end{aligned}$$

Support of C: $\underline{C} := \{e \in E \mid C_e \neq 0\}$.

Zero vector: $\mathbf{0} := (0,\dots,0) \in \{-,0,+\}^E$.

Negative of a sign vector: $-C$, defined by $(-C)^+ := C^-$, $(-C)^- := C^+$ and $(-C)^0 = C^0$.

Composition of C and D: $(C \circ D)_e := \begin{cases} C_e & \text{if } C_e \neq 0, \\ D_e & \text{otherwise.} \end{cases}$

Separation set of C and D: $S(C,D) := \{e \in E \mid C_e = -D_e \neq 0\}$.

We partially order the set of sign vectors by "$0 < +$" and "$0 < -$". The partial order on sign vectors, denoted by $C \leq D$, is understood componentwise; equivalently, we have

$$C \leq D \iff \left[\, C^+ \subset D^+ \text{ and } C^- \subset D^- \,\right].$$

For instance, if $C := (+,+,-,0,-,+,0,0)$ and $D := (0,0,-,+,+,-,0,-)$, then we have:

$$C^+ = \{1,2,6\}, \quad C^- = \{3,5\}, \quad C^0 = \{4,7,8\}, \quad \underline{C} = \{1,2,3,5,6\},$$

$$C \circ D = (+,+,-,+,-,+,0,-), \quad C \circ D \geq C, \quad S(C,D) = \{5,6\}.$$

Furthermore, for $x \in \mathbb{R}^n$, we denote by $\sigma(x) \in \{-,0,+\}^E$ the image of x under the componentwise sign function σ that maps $\mathbb{R}^n$ to $\{-,0,+\}^E$.

6.2.1 COVECTOR AXIOMS

Definition: An ***oriented matroid*** given in terms of its covectors is a pair $\mathcal{M} := (E, \mathcal{L})$, where $\mathcal{L} \subseteq \{-,0,+\}^E$ satisfies

(CV0) $\mathbf{0} \in \mathcal{L}$

(CV1) $C \in \mathcal{L} \implies -C \in \mathcal{L}$

(CV2) $C, D \in \mathcal{L} \Longrightarrow C \circ D \in \mathcal{L}$

(CV3) $C, D \in \mathcal{L},\ e \in S(C,D) \Longrightarrow$
there is a $Z \in \mathcal{L}$ with $Z_e = 0$ and with $Z_f = (C \circ D)_f$ for $f \in E \backslash S(C,D)$.

It is not difficult to check that these covector axioms are satisfied by the sign vector system $\mathcal{L}_X$ of the cells in a hyperplane arrangement $\mathcal{H}_X$, as defined in the previous section. The first two axioms are satisfied trivially. For (CV2) assume that x_C and x_D are points in $\mathbb{R}^d$ with $\sigma(x_C^T \cdot X) = C \in \mathcal{L}_X$ and $\sigma(x_D^T \cdot X) = D \in \mathcal{L}_X$. Then (CV2) is implied by the fact that for sufficiently small $\varepsilon > 0$ we have $\sigma((x_C + \varepsilon x_D)^T \cdot X) = C \circ D$. The geometric content of (CV3) is that if $H_e := \{y \in \mathbb{R}^d \mid y^T x_e = 0\}$ is a hyperplane separating x_C and x_D, then there exists a point x_Z on H_e with the property that x_Z is on the same side as x_C and x_D for all hyperplanes not separating x_C and x_D. We can find such a point by intersecting H_e with the line segment that connects x_C and x_D.

As we will see later the partially ordered set $(\mathcal{L}, \leq)$ describes the face lattice of a cell decomposition of the sphere S^{d-1} by pseudohyperspheres. Each sign vector corresponds to a face of the cell decomposition. We define the ***rank*** d of $\mathcal{M} = (E, \mathcal{L})$ to be the (unique) length of the maximal chains in $(\mathcal{L}, \leq)$ minus one. In the case of realizable arrangements $\mathcal{S}_X$ of hyperspheres, the lattice $(\mathcal{L}_X, \leq)$ equals the face lattice of the cell complex of the arrangement (see Section 6.2.4).

6.2.2 COCIRCUITS

The covectors of (inclusion-)minimal support in $\mathcal{L} \backslash \{\mathbf{0}\}$ correspond to the 0-faces (= vertices) of the cell decomposition that we have just described. We call the set $\mathcal{C}^*(\mathcal{M})$ of all such minimal covectors the ***cocircuits*** of $\mathcal{M}$. An oriented matroid can be described by its set of cocircuits, as shown by the following theorem.

THEOREM 6.2.1 *Cocircuit Characterization*

A collection $\mathcal{C}^* \subset \{-,0,+\}^E$ is the set of cocircuits of an oriented matroid $\mathcal{M}$ if and only if it satisfies

(CC0) $\mathbf{0} \notin \mathcal{C}^*$

(CC1) $C \in \mathcal{C}^* \implies -C \in \mathcal{C}^*$

(CC2) *for all $C, D \in \mathcal{C}^*$ we have:* $\underline{C} \subset \underline{D} \Longrightarrow C = D$ *or* $C = -D$

(CC3) $C, D \in \mathcal{C}^*$, $C \neq -D$, *and* $e \in S(C, D) \Longrightarrow$
there is a $Z \in \mathcal{C}^*$ *with* $Z^+ \subset (C^+ \cup D^+) \backslash \{e\}$ *and* $Z^- \subset (C^- \cup D^-) \backslash \{e\}$.

THEOREM 6.2.2 *Covector/Cocircuit Translation*

For every oriented matroid $\mathcal{M}$*, one can uniquely determine the set* $\mathcal{C}^*$ *of cocircuits from the set* $\mathcal{L}$ *of covectors of* $\mathcal{M}$*, and conversely, as follows:*

(i) $\mathcal{C}^*$ *is the set of sign vectors with minimal support in* $\mathcal{L} \backslash \{\mathbf{0}\}$*:*
$\mathcal{C}^* = \{C \in \mathcal{L} \backslash \{\{\mathbf{0}\}\} \mid C' \leq C \Longrightarrow C' \in \{\mathbf{0}, C\}\}$

(ii) $\mathcal{L}$ *is the set of all sign vectors obtained by successive composition of a finite number of cocircuits from* $\mathcal{C}^*$*:*
$\mathcal{L} = \{C_1 \circ \ldots \circ C_k \mid k \geq 0,\ C_1, \ldots, C_k \in \mathcal{C}^*\}$.
(The zero vector is obtained as the composition of an empty set of covectors.)

In part (ii) of this result one may assume additionally that the sign vectors C_i are ***compatible***, that is, they do not have opposite nonzero signs in any component, so that the composition is commutative.

6.2.3 CHIROTOPES

GLOSSARY

Alternating sign map: A map $\chi\colon E^d \longrightarrow \{-, 0, +\}$ such that any transposition of two components changes the sign: $\chi(\tau_{ij}(\lambda)) = -\chi(\lambda)$.

Chirotope: An alternating sign map χ that encodes the basis orientations of an oriented matroid $\mathcal{M}$ of rank d.

We now present an axiom system for *chirotopes*, which characterizes oriented matroids in terms of basis orientations. Here an algebraic connection to determinant identities becomes obvious. Chirotopes are the main tool for translating problems in oriented matroid theory to an algebraic setting [BS89a]. They also form a description of oriented matroids that is very practical for many algorithmic purposes (for instance in computational geometry; see Knuth [Knu92]).

Definition: Let $E := \{1, \ldots, n\}$ and $0 \leq d \leq n$. A ***chirotope of rank d*** is an alternating sign map $\chi\colon E^d \rightarrow \{-, 0, +\}$ that satisfies

(CHI1) the map $|\chi|\colon E^d \rightarrow \{0, 1\}$, $\lambda \mapsto |\chi(\lambda)|$ is a matroid, and

(CHI2) for every $\lambda \in E^{d-2}$ and $a, b, c, d \in E \backslash \lambda$ the set

$$\{\ \chi(\lambda, a, b) \cdot \chi(\lambda, c, d),\ -\chi(\lambda, a, c) \cdot \chi(\lambda, b, d),\ \chi(\lambda, a, d) \cdot \chi(\lambda, b, c)\ \}$$

either contains $\{-1, +1\}$ or equals $\{0\}$.

Where does the motivation of this axiomatization come from? If we again consider a configuration $X := (x_1, \ldots, x_n)$ of vectors in $\mathbb{R}^d$, we can observe the following identity among the $d \times d$ submatrices of X:

$$\begin{aligned} &\det(x_{\lambda_1}, \ldots, x_{\lambda_{d-2}}, x_a, x_b) \cdot \det(x_{\lambda_1}, \ldots, x_{\lambda_{d-2}}, x_c, x_d) \\ -\ &\det(x_{\lambda_1}, \ldots, x_{\lambda_{d-2}}, x_a, x_c) \cdot \det(x_{\lambda_1}, \ldots, x_{\lambda_{d-2}}, x_b, x_d) \\ +\ &\det(x_{\lambda_1}, \ldots, x_{\lambda_{d-2}}, x_a, x_d) \cdot \det(x_{\lambda_1}, \ldots, x_{\lambda_{d-2}}, x_b, x_c) \quad = \quad 0 \end{aligned}$$

for all $\lambda \in E^{d-2}$ and $a, b, c, d \in E\backslash\lambda$. Such a relation is called a ***three-term Grassmann–Plücker identity.*** If we compare this identity to our axiomatization, we see that (CHI2) implies that

$$\begin{array}{rrcl} \chi_X: & E^d & \rightarrow & \{-, 0, +\} \\ & (\lambda_1, \ldots, \lambda_d) & \mapsto & \operatorname{sign}(\det(x_{\lambda_1}, \ldots, x_{\lambda_d})) \end{array}$$

is consistent with these identities. More precisely, if we consider χ_X as defined above for a vector configuration X, the above Grassmann–Plücker identities imply that (CHI2) is satisfied. (CHI1) is also satisfied since for the vectors of X the Steinitz exchange axiom holds. (In fact the exchange axiom is a consequence of higher order Grassmann–Plücker identities.)

Consequently, χ_X is a chirotope for every $X \in (\mathbb{R}^d)^n$. Thus chirotopes can be considered as a combinatorial model of the determinant values on vector configurations. The following is not easy to prove, but essential.

THEOREM 6.2.3 *Chirotope/Cocircuit Translation*

For each chirotope χ of rank d on $E := \{1, \ldots, n\}$ the set

$$\mathcal{C}^*(\chi) = \{(\chi(\lambda, 1), \chi(\lambda, 2), \ldots, \chi(\lambda, n)) \mid \lambda \in E^{d-1}\}$$

forms the set of cocircuits of an oriented matroid. Conversely, for every oriented matroid $\mathcal{M}$ with cocircuits $\mathcal{C}^*$ there exists a unique pair of chirotopes $\{\chi, -\chi\}$ such that $\mathcal{C}^*(\chi) = \mathcal{C}^*(-\chi) = \mathcal{C}^*$.

The retranslation of cocircuits into signs of bases is straightforward but needs extra notation. It is omitted here.

6.2.4 ARRANGEMENTS OF PSEUDOSPHERES

GLOSSARY

A (d-1)-sphere: The standard unit sphere $S^{d-1} := \{x \in \mathbb{R}^d \mid ||x|| = 1\}$, or any homeomorphic image of it.

Pseudosphere: The image $s \subset S^{d-1}$ of the equator $\{x \in S^{d-1} \mid x_d = 0\}$ in the unit sphere under a self-homeomorphism ϕ: $S^{d-1} \rightarrow S^{d-1}$. (This definition describes topologically *tame* embeddings of a $(d-2)$-sphere in S^{d-1}. Pseudospheres behave "nicely" in the sense that they divide S^{d-1} into two open sets, its ***sides***, that are homeomorphic to open $(d-1)$-balls.)

Oriented pseudosphere: A pseudosphere together with a choice of a positive side s^+ and a negative side s^-.

Arrangement of pseudospheres: A set of n pseudospheres in S^{d-1} with the extra condition that any subset of $d+2$ or fewer pseudospheres is ***realizable***: it defines a cell decomposition of S^{d-1} that is isomorphic to a decomposition by an arrangement of $d+2$ linear hyperplanes.

Essential arrangement: An arrangement such that the intersection of all the pseudospheres is empty.

Rank: The codimension in S^{d-1} of the intersection of all the pseudospheres. For an essential arrangement in S^{d-1}, the rank is d.

Topological representation of $\mathcal{M} = (E, \mathcal{L})$: An essential arrangement of oriented pseudospheres such that $\mathcal{L}$ is the collection of sign vectors associated with the cells of the arrangement.

One of the most important interpretations of oriented matroids is given by the topological representation theorem of Folkman and Lawrence [FL78]; see also [BLS+93, Chapters 4 and 5] and [BKMS05]. It states that oriented matroids are in bijection to (combinatorial equivalence classes of) *arrangements of oriented pseudospheres.* Arrangements of pseudospheres are a topological generalization of hyperplane arrangements, in the same way in which arrangements of pseudolines generalize line arrangements. Thus every rank d oriented matroid describes a certain cell decomposition of the $(d-1)$-sphere. Arrangements of pseudospheres are collections of pseudospheres that have intersection properties just like those satisfied by arrangements of proper subspheres.

Definition: A finite collection $\mathcal{P} = (s_1, s_2, \dots, s_n)$ of pseudospheres in S^{d-1} is an ***arrangement of pseudospheres*** if the following conditions hold (we set $E := \{1, \dots, n\}$):

(PS1) For all $A \subset E$ the set $S_A = \bigcap_{e \in A} s_e$ is a topological sphere.

(PS2) If $S_A \not\subset s_e$, for $A \subset E, e \in E$, then $S_A \cap s_e$ is a pseudosphere in S_A with sides $S_A \cap s_e^+$ and $S_A \cap s_e^-$.

Notice that this definition permits two pseudospheres of the arrangement to be identical. An entirely different, but equivalent, definition is given in the glossary.

We see that every essential arrangement of pseudospheres $\mathcal{P}$ partitions the $(d-1)$-sphere into a regular cell complex $\Gamma(\mathcal{P})$. Each cell of $\Gamma(\mathcal{P})$ is uniquely determined by a sign vector in $\{-, 0, +\}^E$ encoding the relative position with respect to each pseudosphere s_i. Conversely, $\Gamma(\mathcal{P})$ characterizes $\mathcal{P}$ up to homeomorphism. An arrangement of pseudospheres $\mathcal{P}$ is ***realizable*** if there exists an arrangement of proper spheres $\mathcal{S}_X$ with $\Gamma(\mathcal{P}) \cong \Gamma(\mathcal{S}_X)$.

The translation of arrangements of pseudospheres to oriented matroids is given by the topological representation theorem of Folkman and Lawrence [FL78], as follows. (For the definition of "loop," see Section 6.3.1.)

THEOREM 6.2.4 *The Topological Representation Theorem (Pseudosphere/ Covector Translation)*

If $\mathcal{P}$ is an essential arrangement of pseudospheres on S^{d-1} then $\Gamma(\mathcal{P}) \cup \{\mathbf{0}\}$ forms the set of covectors of an oriented matroid of rank d. Conversely, for every oriented matroid $(E, \mathcal{L})$ of rank d (without loops) there exists an essential arrangement of pseudospheres $\mathcal{P}$ on S^{d-1} with $\Gamma(\mathcal{P}) = \mathcal{L} \backslash \{\mathbf{0}\}$.

6.2.5 REALIZABILITY AND DUALITY

GLOSSARY

Realizable oriented matroid: Oriented matroid $\mathcal{M}$ such that there is a vector configuration X with $\mathcal{M}_X = \mathcal{M}$.

Realization of $\mathcal{M}$: A vector configuration X with $\mathcal{M}_X = \mathcal{M}$.

Orthogonality: Two sign vectors $C, D \in \{-,0,+\}^E$ are *orthogonal* if the set

$$\{C_e \cdot D_e \mid e \in E\}$$

either equals $\{0\}$ or contains $\{+,-\}$. We then write $C \perp D$.

Vector of $\mathcal{M}$: A sign vector that is orthogonal to all covectors of $\mathcal{M}$; a covector of the dual oriented matroid $\mathcal{M}^*$.

Circuit of $\mathcal{M}$: A vector of minimal nonempty support; a cocircuit of the dual oriented matroid $\mathcal{M}^*$.

Realizability is a crucial (and hard-to-decide) property of oriented matroids that may be discussed in any of the models/axiomatizations that we have introduced: An oriented matroid given by its covectors, cocircuits, chirotope, or by a pseudosphere arrangement is realizable if there is a vector configuration or a hypersphere arrangement which produces these combinatorial data.

There is a natural duality structure relating oriented matroids of rank d on n elements to oriented matroids of rank $n-d$ on n elements. It is an amazing fact that the existence of such a duality relation can be used to give another axiomatization of oriented matroids (see [BLS[+]93, Section 3.4]). Here we restrict ourselves to the definition of the dual of an oriented matroid $\mathcal{M}$.

THEOREM 6.2.5 *Duality*

For every oriented matroid $\mathcal{M} = (E, \mathcal{L})$ of rank d there is a unique oriented matroid $\mathcal{M}^* = (E, \mathcal{L}^*)$ of rank $|E| - d$ given by

$$\mathcal{L}^* \quad = \quad \Big\{D \in \{-,0,+\}^E \mid C \perp D \text{ for every } C \in \mathcal{L}\Big\}.$$

*$\mathcal{M}^*$ is called the **dual** of $\mathcal{M}$. In particular, $(\mathcal{M}^*)^* = \mathcal{M}$.*

In particular, the cocircuits of the dual oriented matroid $\mathcal{M}^*$, which we call the *circuits* of $\mathcal{M}$, also determine $\mathcal{M}$. Hence the collection $\mathcal{C}(\mathcal{M})$ of all circuits of an oriented matroid $\mathcal{M}$, given by

$$\mathcal{C}(\mathcal{M}) \quad := \quad \mathcal{C}^*(\mathcal{M}^*),$$

is characterized by *the same* cocircuit axioms. Analogously, the *vectors* of $\mathcal{M}$ are obtained as the covectors of $\mathcal{M}^*$; they are characterized by the covector axioms.

An oriented matroid $\mathcal{M}$ is realizable if and only if its dual $\mathcal{M}^*$ is realizable. The reason for this is that a matrix $(I_d|A)$ represents $\mathcal{M}$ if and only if $(-A^T|I_{n-d})$ represents $\mathcal{M}^*$. (Here I_d denotes a $d \times d$ identity matrix, $A \in \mathbb{R}^{d\times(n-d)}$, and $A^T \in \mathbb{R}^{(n-d)\times d}$ denotes the transpose of A.)

Thus for a realizable oriented matroid $\mathcal{M}_X$ the vectors represent the linear dependencies among the columns of X, while the circuits represent minimal linear dependencies. Similarly, in the pseudoarrangements picture, circuits correspond to minimal systems of closed hemispheres that cover the whole sphere, while vectors correspond to consistent unions of such covers that never require the use of both hemispheres determined by a pseudosphere. This provides a direct geometric interpretation of circuits and vectors.

6.2.6 AN EXAMPLE

We close this section with an example that demonstrates the different representations of an oriented matroid. Consider the planar point configuration X given in Figure 6.2.1(left).

FIGURE 6.2.1
An example of an oriented matroid on 6 *elements.*

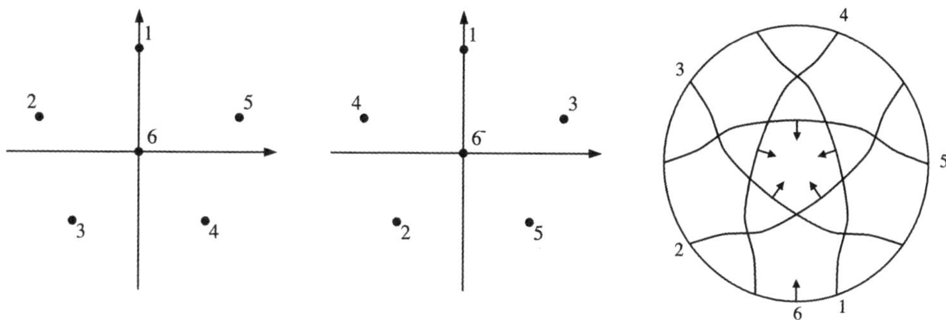

Homogeneous coordinates for X are given by

$$X := \begin{pmatrix} 0 & 3 & 1 \\ -3 & 1 & 1 \\ -2 & -2 & 1 \\ 2 & -2 & 1 \\ 3 & 1 & 1 \\ 0 & 0 & 1 \end{pmatrix}.$$

The chirotope χ_X of $\mathcal{M}$ is given by the orientations:

$$\begin{array}{lllll} \chi(1,2,3) = + & \chi(1,2,4) = + & \chi(1,2,5) = + & \chi(1,2,6) = + & \chi(1,3,4) = + \\ \chi(1,3,5) = + & \chi(1,3,6) = + & \chi(1,4,5) = + & \chi(1,4,6) = - & \chi(1,5,6) = - \\ \chi(2,3,4) = + & \chi(2,3,5) = + & \chi(2,3,6) = + & \chi(2,4,5) = + & \chi(2,4,6) = + \\ \chi(2,5,6) = - & \chi(3,4,5) = + & \chi(3,4,6) = + & \chi(3,5,6) = + & \chi(4,5,6) = + \end{array}$$

Half of the cocircuits of $\mathcal{M}$ are given in the table below (the other half is obtained by negating the data):

$$\begin{array}{lll} (0,0,+,+,+,+) & (0,-,0,+,+,+) & (0,-,-,0,+,-) \\ (0,-,-,-,0,-) & (0,-,-,+,+,0) & (+,0,0,+,+,+) \\ (+,0,-,0,+,+) & (+,0,-,-,0,-) & (+,0,-,-,+,0) \\ (+,+,0,0,+,+) & (+,+,0,-,0,+) & (+,+,0,-,-,0) \\ (+,+,+,0,0,+) & (-,+,+,0,-,0) & (-,-,+,+,0,0) \end{array}$$

Observe that the cocircuits correspond to the point partitions produced by hyperplanes spanned by points. Half of the circuits of $\mathcal{M}$ are given in the next table.

The circuits correspond to sign patterns induced by minimal linear dependencies on the rows of the matrix X. It is easy to check that every pair consisting of a circuit and a cocircuit fulfills the orthogonality condition.

$$
\begin{array}{lll}
(+,-,+,-,0,0) & (+,-,+,0,-,0) & (+,-,+,0,0,-) \\
(+,-,0,+,-,0) & (+,+,0,+,0,-) & (+,-,0,0,-,+) \\
(+,0,-,+,-,0) & (+,0,+,+,0,-) & (+,0,+,0,+,-) \\
(+,0,0,+,-,-) & (0,+,-,+,-,0) & (0,+,-,+,0,-) \\
(0,+,+,0,+,-) & (0,+,0,+,+,-) & (0,0,+,-,+,-)
\end{array}
$$

Figure 6.2.1(right) shows the corresponding arrangement of pseudolines. The circle bounding the configuration represents the projective line at infinity representing line 6.

An affine picture of a realization of the dual oriented matroid is given in Figure 6.2.1(middle). The minus-sign at point 6 indicates that a reorientation at point 6 has taken place. It is easy to check that the circuits and the cocircuits interchange their roles when dualizing the oriented matroid.

6.3 IMPORTANT CONCEPTS

In this section we briefly introduce some very basic concepts in the theory of oriented matroids. The list of topics treated here is tailored toward some areas of oriented matroid theory that are particularly relevant for applications. Thus many other topics of great importance are left out. In particular, see [BLS$^+$93, Section 3.3] for minors of oriented matroids, and [BLS$^+$93, Chapter 7] for basic constructions.

6.3.1 SOME BASIC CONCEPTS

In the following glossary, we list some fundamental concepts of oriented matroid theory. Each of them can be expressed in terms of any one of the representations of oriented matroids that we have introduced (covectors, cocircuits, chirotopes, pseudoarrangements), but for each of these concepts some representations are much more convenient than others. Also, each of these concepts has some interesting properties with respect to the duality operator—which may be more or less obvious, depending on the representation that one uses.

GLOSSARY

Direct sum: An oriented matroid $\mathcal{M} = (E, \mathcal{L})$ has a ***direct sum decomposition***, denoted by $\mathcal{M} = \mathcal{M}(E_1) \oplus \mathcal{M}(E_2)$, if E has a partition into nonempty subsets E_1 and E_2 such that $\mathcal{L} = \mathcal{L}_1 \times \mathcal{L}_2$ for two oriented matroids $\mathcal{M}_1 = (E_1, \mathcal{L}_1)$ and $\mathcal{M}_2 = (E_2, \mathcal{L}_2)$. If $\mathcal{M}$ has no direct sum decomposition, then it is ***irreducible***.

Loops and coloops: A loop of $\mathcal{M} = (E, \mathcal{L})$ is an element $e \in E$ that satisfies $C_e = 0$ for all $C \in \mathcal{L}$. A coloop satisfies $\mathcal{L} \cong \mathcal{L}' \times \{-,0,+\}$, where $\mathcal{L}'$ is obtained by deleting the e-components from the vectors in $\mathcal{L}$. If $\mathcal{M}$ has a direct sum decomposition with $E_2 = \{e\}$, then e is either a loop or a coloop.

Acyclic oriented matroid: Oriented matroid $\mathcal{M} = (E, \mathcal{L})$ for which $(+, \ldots, +)$ is a covector in $\mathcal{L}$; equivalently, the union of the supports of all nonnegative cocircuits is E.

Totally cyclic oriented matroid: An oriented matroid without nonnegative cocircuits; equivalently, $\mathcal{L} \cap \{0, +\}^E = \{\mathbf{0}\}$.

Uniform: An oriented matroid $\mathcal{M}$ of rank d on E is ***uniform*** if all of its cocircuits have support of size $|E| - d + 1$, that is, they have exactly $d - 1$ zero entries. Equivalently, $\mathcal{M}$ is uniform if its chirotope has all values in $\{+, -\}$.

THEOREM 6.3.1 *Duality II*

Let $\mathcal{M}$ be an oriented matroid on the ground set E, and $\mathcal{M}^*$ its dual.

- *$\mathcal{M}$ is acyclic if and only if $\mathcal{M}^*$ is totally cyclic. (However, "most" oriented matroids are neither acyclic nor totally cyclic!)*
- *$e \in E$ is a loop of $\mathcal{M}$ if and only if it is a coloop of $\mathcal{M}^*$.*
- *$\mathcal{M}$ is uniform if and only if $\mathcal{M}^*$ is uniform.*
- *$\mathcal{M}$ is a direct sum $\mathcal{M}(E) = \mathcal{M}(E_1) \oplus \mathcal{M}(E_2)$ if and only if $\mathcal{M}^*$ is a direct sum $\mathcal{M}^*(E) = \mathcal{M}^*(E_1) \oplus \mathcal{M}^*(E_2)$.*

Duality of oriented matroids captures, among other things, the concepts of linear programming duality [BK92] [BLS+93, Chapter 10] and the concept of Gale diagrams for polytopes [Grü67, Section 5.4] [Zie95, Lecture 6]. For the latter, we note here that the vertex set of a d-dimensional convex polytope P with $d+k$ vertices yields a configuration of $d + k$ vectors in $\mathbb{R}^{d+1}$, and thus an oriented matroid of rank $d + 1$ on $d + k$ points. Its dual is a realizable oriented matroid of rank $k - 1$, the ***Gale diagram*** of P. It can be modeled by a signed affine point configuration of dimension $k - 2$, called an ***affine Gale diagram*** of P. Hence, for "small" k, we can represent a (possibly high-dimensional) polytope with "few vertices" by a low-dimensional point configuration. In particular, this is beneficial in the case $k = 4$, where polytopes with "universal" behavior can be analyzed in terms of their 2-dimensional affine Gale diagrams. For further details, see Chapter 15 of this Handbook.

6.3.2 REALIZABILITY AND REALIZATION SPACES

GLOSSARY

Realization space: Let $\chi\colon E^d \to \{-, 0, +\}$ be a chirotope with $\chi(1, \ldots, d) = +$. The *realization space* $\mathcal{R}(\chi)$ is the set of all matrices $X \in \mathbb{R}^{d \cdot n}$ with $\chi_X = \chi$ and $x_i = e_i$ for $i = 1, \ldots, d$, where e_i is the ith unit vector. If $\mathcal{M}$ is the corresponding oriented matroid, we write $\mathcal{R}(\mathcal{M}) = \mathcal{R}(\chi)$.

Rational realization: A realization $X \in \mathbb{Q}^{d \cdot n}$; that is, a point in $\mathcal{R}(\chi) \cap \mathbb{Q}^{d \cdot n}$.

Basic primary semialgebraic set: The (real) solution set of an arbitrary finite system of polynomial equations and strict inequalities with integer coefficients.

Existential Theory of the Reals: The problem of solving arbitrary systems of polynomial equations and inequalities with integer coefficients.

Stable equivalence: A strong type of arithmetic and homotopy equivalence. Two semialgebraic sets are stably equivalent if they can be connected by a sequence of rational coordinate changes, together with certain projections with contractible fibers. (See [RZ95] and [Ric96] for details.) In particular, two stably equivalent semialgebraic sets have the same number of components, they are homotopy equivalent, and either both or neither of them have rational points.

One of the main problems in oriented matroid theory is to design algorithms that find a realization of a given oriented matroid if it exists. However, for oriented matroids with large numbers of points, one cannot be too optimistic, since the realizability problem for oriented matroids is NP-hard. This is one of the consequences of Mnëv's universality theorem below. An upper bound for the worst-case complexity of the realizability problem is given by the following theorem. It follows from general complexity bounds for algorithmic problems about semialgebraic sets by Basu, Pollack, and Roy [BPR00] [BPR03] (see also Chapter 37 of this Handbook).

THEOREM 6.3.2 *Complexity of the Best General Algorithm Known*

The realizability of a rank d oriented matroid on n points can be decided by solving a system of $S = \binom{n}{d}$ real polynomial equations and strict inequalities of degree at most $D = d-1$ in $K = (n-d-1)(d-1)$ variables. Thus, with the algorithms of [BPR00], *the number of bit operations needed to decide realizability is (in the Turing machine model of complexity) bounded by $(S/K)^K \cdot S \cdot D^{O(K)}$ in a situation where d is fixed and n is large.*

THE UNIVERSALITY THEOREM

A basic observation is that all oriented matroids of rank 2 are realizable. In particular, up to change of orientations and permuting the elements in E there is only one uniform oriented matroid of rank 2. The realization space of an oriented matroid of rank 2 is always stably equivalent to $\{0\}$; in particular, if $\mathcal{M}$ is uniform of rank 2 on n elements, then $\mathcal{R}(\mathcal{M})$ is isomorphic to an open subset of $\mathbb{R}^{2n-4}$.

In contrast to the rank 2 case, Mnëv's universality theorem states that for oriented matroids of rank 3, the realization space can be "arbitrarily complicated." Here is what one can observe for oriented matroids on few elements:

- The realization spaces of all realizable uniform oriented matroids of rank 3 and at most 9 elements are contractible (Richter [Ric89]).
- There is a realizable rank 3 oriented matroid on 9 elements that has no realization with rational coordinates (Perles [Grü67, p. 93]).
- There is a realizable rank 3 oriented matroid on 13 elements with disconnected realization space (Tsukamoto [Tsu13]).

The universality theorem is a fundamental statement with various implications for the configuration spaces of various types of combinatorial objects.

THEOREM 6.3.3 *Mnëv's Universality Theorem* [Mnë88]

For every basic primary semialgebraic set V defined over $\mathbb{Z}$ there is a chirotope χ of rank 3 such that V and $\mathcal{R}(\chi)$ are stably equivalent.

Although some of the facts in the following list were proved earlier than Mnëv's universality theorem, they all can be considered as consequences of the construction techniques used by Mnëv.

CONSEQUENCES OF THE UNIVERSALITY THEOREM

1. The full field of algebraic numbers is needed to realize all oriented matroids of rank 3.
2. The realizability problem for oriented matroids is NP-hard (Mnëv [Mnë88], Shor [Sho91]).
3. The realizability problem for oriented matroids is (polynomial-time-)equivalent to the Existential Theory of the Reals (Mnëv [Mnë88]).
4. For every finite simplicial complex Δ, there is an oriented matroid whose realization space is homotopy equivalent to Δ.
5. Realizability of rank 3 oriented matroids cannot be characterized by excluding a finite set of "forbidden minors" (Bokowski and Sturmfels [BS89b]).
6. In order to realize all combinatorial types of integral rank 3 oriented matroids on n elements, even uniform ones, in the integer grid $\{1, 2, \ldots, f(n)\}^3$, the "coordinate size" function $f(n)$ has to grow doubly exponentially in n (Goodman, Pollack, and Sturmfels [GPS90]).
7. The ***isotopy problem*** for oriented matroids (Can one given realization of $\mathcal{M}$ be continuously deformed, through realizations, to another given one?) has a negative solution in general, even for uniform oriented matroids of rank 3 [JMSW89].

6.3.3 TRIANGLES AND SIMPLICIAL CELLS

There is a long tradition of studying *triangles* in arrangements of pseudolines. In his 1926 paper [Lev26], Levi already considered them to be important structures. There are good reasons for this. On the one hand, they form the simplest possible cells of full dimension, and are therefore of basic interest. On the other hand, if the arrangement is simple, triangles locate the regions where a "smallest" local change of the combinatorial type of the arrangement is possible. Such a change can be performed by taking one side of the triangle and "pushing" it over the vertex formed by the other two sides. It was observed by Ringel [Rin56] that any two simple arrangements of pseudolines can be deformed into one another by performing a sequence of such "triangle flips."

Moreover, the realizability of a pseudoline arrangement may depend on the situation at the triangles. For instance, if any one of the triangles in the nonrealizable example of Figure 6.1.2 other than the central one is flipped, we obtain a realizable pseudoline arrangement.

TRIANGLES IN ARRANGEMENTS OF PSEUDOLINES

Let $\mathcal{P}$ be any arrangement of n pseudolines.

1. For any pseudoline ℓ in $\mathcal{P}$ there are at least 3 triangles adjacent to ℓ. Either the $n-1$ pseudolines different from ℓ intersect in one point (i.e., $\mathcal{P}$ is a ***near-pencil***), or there are at least $n-3$ triangles that are not adjacent to ℓ. Thus $\mathcal{P}$ contains at least n triangles (Levi [Lev26]).
2. $\mathcal{P}$ is ***simplicial*** if all its regions are bounded by exactly 3 (pseudo)lines. Except for the near-pencils, there are two infinite classes of simplicial line arrangements and 94 additional "sporadic" simplicial line arrangements (and many more simplicial pseudoarrangements) known: See Grünbaum [Grü71] with the recent updates by Grnbaum [Grü09], and Cuntz [Cun12], who also classified the simplicial pseudoarrangements up to $n = 27$ pseudolines; the first nonstretchable example occurs for $n = 15$.
3. If $\mathcal{P}$ is simple, then it contains at most $\frac{n(n-1)}{3}$ triangles. For infinitely many values of n, there exists a simple arrangement with $\frac{n(n-1)}{3}$ triangles (Harborth and Roudneff, see [Rou96]).
4. Any two simple arrangements $\mathcal{P}_1$ and $\mathcal{P}_2$ can be deformed into one another by a sequence of simplicial flips (Ringel [Rin56]).

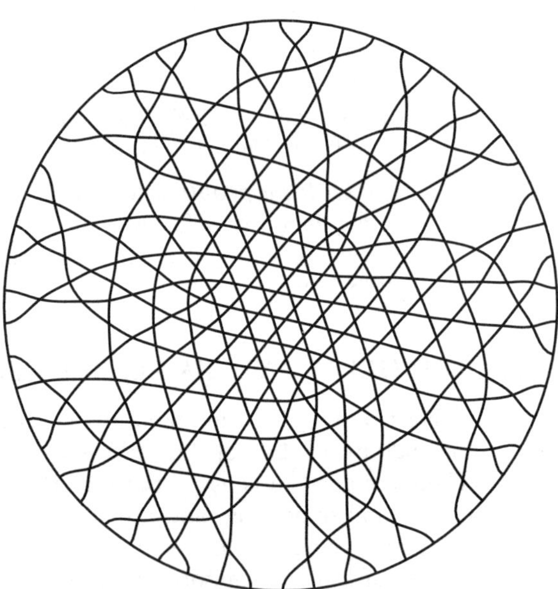

FIGURE 6.3.1
A simple arrangement of 28 pseudolines with a maximal number of 252 triangles.

Every arrangement of pseudospheres in S^{d-1} has a centrally symmetric representation. Thus we can always derive an arrangement of projective pseudohyperplanes (pseudo $(d-2)$-planes in $\mathbb{RP}^{d-1}$) by identifying antipodal points. The proper analogue for the triangles in rank 3 are the $(d-1)$-simplices in projective arrangements of pseudohyperplanes in rank d, i.e., the regions bounded by the minimal number, d, of pseudohyperplanes. We call an arrangement ***simple*** if no more than $d-1$ planes meet in a point.

It was conjectured by Las Vergnas in 1980 [Las80] that (as in the rank 3 case) any two simple arrangements can be transformed into each other by a sequence of flips of simplicial regions. In particular this requires that every simple arrangement contain *at least one* simplicial region (which was also conjectured by Las Vergnas). If we consider the case of realizable arrangements only, it is not difficult to prove that any two members in this subclass can be connected by a sequence of flips of simplicial regions and that each realizable arrangement contains at least one simplicial cell. In fact, Shannon [Sha79] proved that every arrangement (even the nonsimple ones) of n projective hyperplanes in rank d contains at least n simplicial regions. More precisely, for every hyperplane h there are at least d simplices adjacent to h and at least $n-d$ simplices not adjacent to h. The contrast between the Las Vergnas conjecture and the results known for the nonrealizable case is dramatic.

SIMPLICIAL CELLS IN PSEUDOARRANGEMENTS

1. There is an arrangement of 8 pseudoplanes in rank 4 having only 7 simplicial regions (Altshuler and Bokowski [ABS97], Roudneff and Sturmfels [RS88]).
2. Every rank 4 arrangement with $n < 13$ pseudoplanes has at least one simplicial region (Bokowski and Rohlfs [BR01]).
3. For every $k > 2$ there is a rank 4 arrangement of $4k$ pseudoplanes having only $3k+1$ simplicial regions. (This result of Richter-Gebert [Ric93] was improved by Bokowski and Rohlfs [BR01] to arrangements of $7k+c$ pseudoplanes with $5k+c$ simplicial regions, for $k \geq 0$ and $c \geq 4$.)
4. There is a rank 4 arrangement consisting of 20 pseudoplanes for which one plane is not adjacent to any simplicial region (Richter-Gebert [Ric93]; improved to 17 pseudoplanes by Bokowski and Rohlfs [BR01]).

OPEN PROBLEMS

The topic of simplicial cells is interesting and rich in structure even in rank 3. The case of higher dimensions is full of unsolved problems and challenging conjectures. These problems are relevant for various problems of great geometric and topological interest, such as the structure of spaces of triangulations. Three key problems are:

1. Classify simplical arrangements. Is it true, at least, that there are only finitely many types of simplicial arrangements of straight lines outside the three known infinite families?
2. Does every arrangement of pseudohyperplanes contain at least one simplicial region?
3. Is it true that any two simple arrangements (of the same rank and the same number of pseudohyperplanes) can be transformed into one another by a sequence of flips at simplex regions?

6.4 APPLICATIONS IN POLYTOPE THEORY

Oriented matroid theory has become a very important structural tool for the theory of polytopes, but also for the algorithmic treatment and classification of polytopes. In this section we give a brief overview of fundamental concepts and of results.

MATROID POLYTOPES

The convexity properties of a point configuration X are modeled superbly by the oriented matroid $\mathcal{M}_X$. The combinatorial versions of many theorems concerning convexity also hold on the level of general (including nonrealizable) oriented matroids. For instance, there are purely combinatorial versions of Carathéodory's, Radon's, and Helly's theorems [BLS+93, Section 9.2].

In particular, oriented matroid theory provides an entirely combinatorial model of convex polytopes, commonly known as "matroid polytopes," although "oriented matroid polytope" might be more appropriate. (Warning: In Combinatorial Optimization, an entirely different object, the 0/1-polytope spanned by the incidence vectors of the bases of an (unoriented) matroid is also called a "matroid polytope.")

Definition: The ***face lattice*** (sometimes called the ***Las Vergnas face lattice***) of an acyclic oriented matroid $\mathcal{M} = (E, \mathcal{L})$ is the set

$$\mathrm{FL}(\mathcal{M}) := \{C^0 \mid C \in \mathcal{L} \cap \{0,+\}^E\},$$

with the partial order of sign vectors induced from $(\mathcal{L}, \leq)$, which coincides with the partial ordered by inclusion of supports. The elements of $\mathrm{FL}(\mathcal{M})$ are the ***faces*** of $\mathcal{M}$. The acyclic oriented matroid $\mathcal{M} = (E, \mathcal{L})$ is a ***matroid polytope*** if $\{e\}$ is a face for every $e \in E$.

Every polytope gives rise to a matroid polytope: If $P \subset \mathbb{R}^d$ is a d-polytope with n vertices, then the canonical embedding $x \mapsto \binom{x}{1}$ creates a vector configuration X_P of rank $d+1$ from the vertex set of P. The oriented matroid of X_P is a matroid polytope $\mathcal{M}_P$, whose face lattice $\mathrm{FL}(\mathcal{M})$ is canonically isomorphic to the face lattice of P.

Matroid polytopes provide a very precise model of (the combinatorial structure of) convex polytopes. In particular, the topological representation theorem implies that *every* matroid polytope of rank d is the face lattice of a regular piecewise linear (PL) cell decomposition of a $(d-2)$-sphere. Thus matroid polytopes form an excellent combinatorial model for convex polytopes: In fact, much better than the model of PL spheres (which does not have an entirely combinatorial definition).

However, the construction of a polar fails in general for matroid polytopes. The cellular spheres that represent matroid polytopes have dual cell decompositions (because they are piecewise linear), but this dual cell decomposition is not in general a matroid polytope, even in rank 4 (Billera and Munson [BM84]; Bokowski and Schuchert [BS95]). In other words, the order dual of the face lattice of a matroid polytope (as an abstract lattice) is *not in general* the face lattice of a matroid polytope. (Matroid polytopes form an important tool for polytope theory, not only because of the parts of polytope theory that work for them, but also because of those that fail.)

For every matroid polytope one has the dual oriented matroid (which is totally cyclic, hence not a matroid polytope). In particular, the setup for Gale diagrams generalizes to the framework of matroid polytopes; this makes it possible to also include nonpolytopal spheres in a discussion of the realizability properties of polytopes. This amounts to perhaps the most powerful single tool ever developed for polytope theory. It leads to, among other things, the classification of d-dimensional polytopes with at most $d+3$ vertices, the proof that all matroid polytopes of rank $d+1$ with at most $d+3$ vertices are realizable, and the construction of nonrational polytopes as well as of nonpolytopal spheres with $d+4$ vertices.

ALGORITHMIC APPROACHES TO POLYTOPE CLASSIFICATION

For a long time there has been substantial work in classifying the combinatorial types of d-dimensional polytopes with a "small" number n of vertices, for $d \geq 4$.

For $d = 3$ the enumeration problem is reduced to the enumeration of 3-connected planar graphs, by Steinitz's Theorem 15.1.3. For $n \leq d+3$, it may be solved by the "Gale diagram" technique introduced by Perles in the 1960s, which in retrospect may be seen as an incarnation of oriented matroid duality; see Section 15.1.7. For $d = 4$ one can try to do the enumeration via Schlegel diagrams; thus Brückner (1910) attempted to classify the simplicial 4-polytopes with 8 vertices. His work was corrected and completed in the 1960s, see [Grü67].

For the following discussion we restrict our attention to the simplicial case—there are additional technical problems to deal with in the nonsimplicial case, and very little work has been done there as yet (see e.g., Brinkmann & Ziegler [BZ17]).

In the simplicial case, at the core of the enumeration problem lies the following hierarchy:

$$\begin{pmatrix}\text{simplicial}\\ \text{spheres}\end{pmatrix} \supset \begin{pmatrix}\text{uniform}\\ \text{matroid polytopes}\end{pmatrix} \supset \begin{pmatrix}\text{simplicial}\\ \text{convex polytopes}\end{pmatrix}.$$

The classical plan of attack from the 1970s and 1980s, when oriented matroid technology in the current form was not yet available, was to first enumerate all isomorphism types of simplicial spheres with given parameters. Then for each sphere one would try to decide realizability. This has been successfully completed for the classification of all simplicial 3-spheres with 9 vertices (Altshuler, Bokowski, and Steinberg [ABS97]) and of all neighborly 5-spheres with 10 vertices (Bokowski and Shemer [BS87]) into polytopes and nonpolytopes.

In an alternative approach to enumerate all polytopes in a class, largely due to Bokowski and Sturmfels [BS89a], one tries to bypass the first step of enumeration of spheres and enumerates directly all possible oriented matroids/matroid polytopes, and then tries to decide realizability for each single type. Thus one has to effectively deal with three problems:

(1) enumeration of oriented matroids/matroid polytopes,
(2) proving nonrealizability, or
(3) proving realizability.

(1) For the *enumeration problem*, Finschi & Fukuda [FF02] developed an effective approach to generate oriented matroids (including the nonuniform ones!) through single element extensions: Take an oriented matroid and try to add an element. Their algorithms relied on cocircuit graphs. It was later observed that the set of possible single element extensions can be viewed as the set of solutions of a SAT problem. Moreover, SAT-solvers are readily available "off the shelf" and thus much easier to employ than the special purpose software that had previously been developed for such purposes. For example, Miyata & Padrol [MP15] use this as a key step in the enumeration of neighborly oriented matroids. (The use of SAT-solvers in oriented matroid theory had been pioneered by Schewe [Sch10], who first used them for proving that a geometric structure, such as a polytope or a sphere, does not admit a compatible oriented matroid.)

(2) For *proving nonrealizability* of oriented matroids, Biquadratic Final Polynomials (BFPs), introduced by Bokowski and Richter-Gebert [BR90], usually outperform all the other methods. These can be found effectively by linear programming.

(This was up to now used mostly for the uniform case, but may work even more effectively in nonuniform situations; see Brinkmann and Ziegler [BZ17].) Moreover, nonrealizable oriented matroids that do not have a BFP seem to be rare for the parameters of the enumeration problems within reach. (It had been observed by Dress and Sturmfels that every nonrealizable oriented matroid has a final polynomial proof for this fact, but the final polynomials typically are huge and cannot be found efficiently.) Moreover, recent works by Firsching and others have exploited that biquadratic final polynomials typically use only a partial oriented matroid, so BFP proofs can be based on incomplete enumeration trees.

(3) For *proving realizability* of oriented matroids, exact methods—which might solve general semialgebraic problems—are not available or inefficient. Randomized methods can be employed to find realizations, but it seems hard to employ them for nonuniform oriented matroids; see Fukuda, Miyata and Moriyama [FMM13].

Firsching [Fir17] demonstrated recently that current nonlinear optimization software can be used very efficiently to find realizations. Those then (as one is using *numerical* software with rounding errors) have to be checked in exact arithmetic. This turned out to work very well for simplicial polytopes, but also in nonuniform situations, say for inscribed realizations with all vertices on a sphere. If the program does not find a solution or does not terminate (which happens more often), this does not yield a proof, but may be interpreted as suggesting that the oriented matroid might be not realizable.

Another important tool in classification efforts is the use of constructions that preserve realizability, such as stackings, or lexicographic extensions. The latter is in particular useful in the context of neighborly polytopes, see Padrol [Pad13].

Recently there has been a number of complete classification results based on the methods we have mentioned. This includes the classifications of

- simplicial 4-polytopes with 10 vertices [Fir17],
- simplicial 5-polytopes with 9 vertices [FMM13], and
- neighborly 8-polytopes with 12 vertices [MP15].

A comprehensive overview table can be found in [Fir15, p. 17].

Finally, let us note that in the passages from spheres to oriented matroids and to polytopes there are considerable subtleties involved that lead to important structural insights. For a given simplicial sphere, the following may apply:

- There may be *no* matroid polytope that supports it. In this case the sphere is called ***nonmatroidal***. The Barnette sphere [BLS+93, Proposition 9.5.3] is an example.
- There may be *exactly one* matroid polytope. In this (important) case the sphere is called ***rigid***. That is, a matroid polytope $\mathcal{M}$ is rigid if $\mathrm{FL}(\mathcal{M}') = \mathrm{FL}(\mathcal{M})$ already implies $\mathcal{M}' = \mathcal{M}$. For rigid matroid polytopes the face lattice uniquely defines the oriented matroid, and thus every statement about the matroid polytope yields a statement about the sphere. In particular, the matroid polytope and the sphere have the same realization space.

  Rigid matroid polytopes are a priori rare; however, the *Lawrence construction* [BLS+93, Section 9.3] [Zie95, Section 6.6] associates with every oriented matroid $\mathcal{M}$ on n elements in rank d a rigid matroid polytope $\Lambda(\mathcal{M})$ with $2n$ vertices of rank $n+d$. The realizations of $\Lambda(\mathcal{M})$ can be retranslated into realizations of $\mathcal{M}$.

Furthermore, even-dimensional neighborly polytopes are rigid (Shemer [She82], Sturmfels [Stu88]), which is a key property for all approaches to the classification of neighborly polytopes with few vertices, but also for proving the universality for simplicial polytopes (see Theorem 6.3.4 below).

- There may be *many* matroid polytopes.

The situation is similarly complex for the second step, from matroid polytopes to convex polytopes. In fact, for each matroid polytope the following may apply:

- There may be *no* convex polytope—this is the case for a nonrealizable matroid polytope. These exist already with relatively few vertices; namely in rank 5 with 9 vertices [BS95], and in rank 4 with 10 vertices [BLS+93, Proposition 9.4.5].
- There may be essentially *only one*—this is the rare case where the matroid polytope is "projectively unique" (cf. Adiprasito and Ziegler [AZ15]).
- There may be *many* convex polytopes—the space of all polytopes for a given matroid polytope is the realization space of the oriented matroid, and this may be "arbitrarily complicated." This is made precise by Mnëv's universality theorem [Mnë88]. Note that for simplicial polytopes, this has been claimed since the 1980s, but has been proven only recently by Adiprasito and Padrol [AP17]. (One can prove universality for the realization spaces of uniform oriented matroids of rank 3 from the nonuniform case by a scattering technique [BS89a, Thm. 6.2]; for polytopes, this technique does not suffice, as Lawrence extensions (see Chapter 15) destroy simpliciality.)

THEOREM 6.4.1 *The Universality Theorem for Polytopes* [Mnë88] [AP17]
For every [open] basic primary semialgebraic set V defined over $\mathbb{Z}$ there is an integer d and a [simplicial] d-dimensional polytope P on $d+4$ vertices such that V and the realization space of P are stably equivalent.

6.5 SOURCES AND RELATED MATERIAL

FURTHER READING

The basic theory of oriented matroids was introduced in two fundamental papers, Bland and Las Vergnas [BL78] and Folkman and Lawrence [FL78]. We refer to the monograph by Björner, Las Vergnas, Sturmfels, White, and Ziegler [BLS+93] for a broad introduction, and for an extensive development of the theory of oriented matroids. An extensive bibliography is given in Ziegler [Zie96+]. Other introductions and basic sources of information include Bachem and Kern [BK92], Bokowski [Bok93] and [Bok06], Bokowski and Sturmfels [BS89a], and Ziegler [Zie95, Lectures 6 and 7].

RELATED CHAPTERS

Chapter 5: Pseudoline arrangements
Chapter 15: Basic properties of convex polytopes

Chapter 16: Subdivisions and triangulations of polytopes
Chapter 28: Arrangements
Chapter 37: Computational and quantitative real algebraic geometry
Chapter 49: Linear programming
Chapter 60: Geometric applications of the Grassmann–Cayley algebra

REFERENCES

[ABS80] A. Altshuler, J. Bokowski, and L. Steinberg. The classification of simplicial 3-spheres with nine vertices into polytopes and nonpolytopes. *Discrete Math.*, 31:115–124, 1980.

[AP17] K.A. Adiprasito and A. Padrol. The universality theorem for neighborly polytopes. *Combinatorica*, 37:129–136, 2017.

[AZ15] K.A. Adiprasito and G.M. Ziegler. Many projectively unique polytopes, *Invent. Math.*, 119:581–652, 2015.

[BK92] A. Bachem and W. Kern. *Linear Programming Duality: An Introduction to Oriented Matroids.* Universitext, Springer-Verlag, Berlin, 1992.

[BKMS05] J. Bokowski, S. King, S. Mock, and I. Streinu. A topological representation theorem for oriented matroids. *Discrete Comput. Geom.*, 33:645–668, 2005.

[BLS+93] A. Björner, M. Las Vergnas, B. Sturmfels, N. White, and G.M. Ziegler. *Oriented Matroids.* Vol. 46 of *Encyclopedia Math. Appl.*, Cambridge University Press, 1993; 2nd revised edition 1999.

[BL78] R.G. Bland and M. Las Vergnas. Orientability of matroids. *J. Combin. Theory Ser. B*, 24:94–123, 1978.

[BM84] L.J. Billera and B.S. Munson. Polarity and inner products in oriented matroids. *European J. Combin.*, 5:293–308, 1984.

[BMS01] J. Bokowski, S. Mock, and I. Streinu. On the Folkman–Lawrence topological representation theorem for oriented matroids of rank 3, *European J. Combin.*, 22:601–615, 2001.

[Bok93] J. Bokowski. Oriented matroids. In P.M. Gruber and J.M. Wills, editors, *Handbook of Convex Geometry*, pages 555–602, North-Holland, Amsterdam, 1993.

[Bok06] J. Bokowski. *Computational Oriented Matroids: Equivalence Classes of Matroids within a Natural Framework.* Cambridge University Press, 2006.

[BPR96] S. Basu, R. Pollack, and M.-F. Roy. On the combinatorial and algebraic complexity of quantifier elimination. *J. ACM*, 43:1002–1045, 1996.

[BPR03] S. Basu, R. Pollack, and M.-F. Roy. *Algorithms in Real Algebraic Geometry.* Vol. 10 of *Algorithms and Combinatorics*, Springer, Heidelberg, 2003; 2nd revised edition 2006.

[BR90] J. Bokowski and J. Richter. On the finding of final polynomials. *European J. Combin.*, 11:21–43, 1990.

[BR01] J. Bokowski and H. Rohlfs. On a mutation problem of oriented matroids. *European J. Combin.*, 22:617–626, 2001.

[BS87] J. Bokowski and I. Shemer. Neighborly 6-polytopes with 10 vertices. *Israel J. Math.*, 58:103–124, 1987.

[BS89a] J. Bokowski and B. Sturmfels. *Computational Synthetic Geometry.* Vol. 1355 of *Lecture Notes in Math.*, Springer-Verlag, Berlin, 1989.

[BS89b] J. Bokowski and B. Sturmfels. An infinite family of minor-minimal nonrealizable 3-chirotopes. *Math. Z.*, 200:583–589, 1989.

[BS95] J. Bokowski and P. Schuchert. Altshuler's sphere M^9_{963} revisited. *SIAM J. Discrete Math.*, 8:670–677, 1995.

[BZ17] P. Brinkmann and G.M. Ziegler. A flag vector of a 3-sphere that is not the flag vector of a 4-polytope. *Mathematika*, 63:260–271, 2017.

[Cun12] M. Cuntz. Simplicial arrangements with up to 27 lines. *Discrete Comput. Geom.*, 48:682–701, 2012.

[Fir15] M. Firsching. *Optimization Methods in Discrete Geometry*. PhD thesis, FU Berlin, `edocs.fu-berlin.de/diss/receive/FUDISS_thesis_000000101268`, 2015.

[Fir17] M. Firsching. Realizability and inscribability for some simplicial spheres and matroid polytopes. *Math. Program.*, in press, 2017.

[FF02] L. Finschi and K. Fukuda. Generation of oriented matroids—a graph theoretic approach. *Discrete Comput. Geom.*, 27:117–136, 2002.

[FL78] J. Folkman and J. Lawrence. Oriented matroids. *J. Combin. Theory Ser. B*, 25:199–236, 1978.

[FMM13] K. Fukuda, H. Miyata, and S. Moriyama. Complete enumeration of small realizable oriented matroids *Discrete Comput. Geom.*, 49:359–381, 2013.

[GP80] J.E. Goodman and R. Pollack. Proof of Grünbaum's conjecture on the stretchability of certain arrangements of pseudolines. *J. Combin. Theory Ser. A*, 29:385–390, 1980.

[GPS90] J.E. Goodman, R. Pollack, and B. Sturmfels. The intrinsic spread of a configuration in $\mathbb{R}^d$. *J. Amer. Math. Soc.*, 3:639–651, 1990.

[Grü67] B. Grünbaum. *Convex Polytopes*. Interscience, London, 1967; 2nd edition (V. Kaibel, V. Klee, and G. M. Ziegler, eds.), vol 221 of *Graduate Texts in Math.*, Springer-Verlag, New York, 2003.

[Grü71] B. Grünbaum. Arrangements of hyperplanes. In R.C. Mullin et al., eds., *Proc. Second Louisiana Conference on Combinatorics, Graph Theory and Computing*, pages 41–106, Louisiana State University, Baton Rouge, 1971.

[Grü72] B. Grünbaum. *Arrangements and Spreads*. Vol. 10 of *CBMS Regional Conf. Ser. in Math.*, AMS, Providence, 1972.

[Grü09] B. Grünbaum. A catalogue of simplicial arrangements in the real projective plane. *Ars Math. Contemp.* 2:1-25, 2009.

[JMSW89] B. Jaggi, P. Mani-Levitska, B. Sturmfels, and N. White. Constructing uniform oriented matroids without the isotopy property. *Discrete Comput. Geom.*, 4:97–100, 1989.

[Knu92] D.E. Knuth. *Axioms and Hulls*. Vol. 606 of *LNCS*, Springer-Verlag, Berlin, 1992.

[Kun86] J.P.S. Kung. *A Source Book in Matroid Theory*. Birkhäuser, Boston 1986.

[Las80] M. Las Vergnas. Convexity in oriented matroids. *J. Combin. Theory Ser. B*, 29:231–243, 1980.

[Lev26] F. Levi. Die Teilung der projektiven Ebene durch Gerade oder Pseudogerade. *Ber. Math.-Phys. Kl. Sächs. Akad. Wiss.*, 78:256–267, 1926.

[Mnë88] N.E. Mnëv. The universality theorems on the classification problem of configuration varieties and convex polytopes varieties. In O.Ya. Viro, editor, *Topology and Geometry—Rohlin Seminar*, vol. 1346 of *Lecture Notes in Math.*, pages 527–544, Springer-Verlag, Berlin, 1988.

[MP15] H. Miyata and A. Padrol. Enumeration of neighborly polytopes and oriented matroids. *Exp. Math.*, 24:489–505, 2015.

[Oxl92] J. Oxley. *Matroid Theory*. Oxford Univ. Press, 1992; 2nd revised edition 2011.

[Pad13] A. Padrol. Many neighborly polytopes and oriented matroids, *Discrete Comput. Geom.*, 50:865–902, 2013.

[Ric89] J. Richter. Kombinatorische Realisierbarkeitskriterien für orientierte Matroide. *Mitt. Math. Sem. Gießen*, 194:1–112, 1989.

[Ric93] J. Richter-Gebert. Oriented matroids with few mutations. *Discrete Comput. Geom.*, 10:251–269, 1993.

[Ric96] J. Richter-Gebert. *Realization Spaces of Polytopes.* Vol. 1643 of *Lecture Notes in Math.*, Springer-Verlag, Berlin, 1996.

[Rin56] G. Ringel. Teilungen der Ebene durch Geraden oder topologische Geraden. *Math. Z.*, 64:79–102, 1956.

[Rou96] J.-P. Roudneff. The maximum number of triangles in arrangements of pseudolines. *J. Combin. Theory Ser. B*, 66:44–74, 1996.

[RS88] J.-P. Roudneff and B. Sturmfels. Simplicial cells in arrangements and mutations of oriented matroids. *Geom. Dedicata*, 27:153–170, 1988.

[RZ95] J. Richter-Gebert and G.M. Ziegler. Realization spaces of 4-polytopes are universal. *Bull. Amer. Math. Soc.*, 32:403–412, 1995.

[Sch10] L. Schewe. Nonrealizable minimal vertex triangulations of surfaces: Showing nonrealizability using oriented matroids and satisfiability solvers. *Discrete Comput. Geom.*, 43:289–302, 2010.

[Sha79] R.W. Shannon. Simplicial cells in arrangements of hyperplanes. *Geom. Dedicata*, 8:179–187, 1979.

[She82] I. Shemer. Neighborly polytopes. *Israel J. Math.*, 43:291–314, 1982.

[Sho91] P. Shor. Stretchability of pseudolines is NP-hard. In P. Gritzmann and B. Sturmfels, editors, *Applied Geometry and Discrete Mathematics—The Victor Klee Festschrift*, vol 4 of *DIMACS Series Discrete Math. Theor. Comp. Sci.*, pp. 531–554, AMS, Providence, 1991.

[Stu88] B. Sturmfels. Neighborly polytopes and oriented matroids. *European J. Combin.*, 9:537–546, 1988.

[Tsu13] Y. Tsukamoto. New examples of oriented matroids with disconnected realization spaces. *Discrete Comput. Geom.*, 49:287–295, 2013.

[Zie95] G.M. Ziegler. *Lectures on Polytopes.* Vol. 152 of *Graduate Texts in Math.*, Springer, New York, 1995; 7th revised printing 2007.

[Zie96+] G.M. Ziegler. Oriented matroids today: Dynamic survey and updated bibliography. *Electron. J. Combin.*, 3:DS#4, 1996+.

7 LATTICE POINTS AND LATTICE POLYTOPES

Alexander Barvinok

INTRODUCTION

Lattice polytopes arise naturally in algebraic geometry, analysis, combinatorics, computer science, number theory, optimization, probability and representation theory. They possess a rich structure arising from the interaction of algebraic, convex, analytic, and combinatorial properties. In this chapter, we concentrate on the theory of lattice polytopes and only sketch their numerous applications. We briefly discuss their role in optimization and polyhedral combinatorics (Section 7.1). In Section 7.2 we discuss the *decision problem*, the problem of finding whether a given polytope contains a lattice point. In Section 7.3 we address the *counting problem*, the problem of counting all lattice points in a given polytope. The *asymptotic problem* (Section 7.4) explores the behavior of the number of lattice points in a varying polytope (for example, if a dilation is applied to the polytope). Finally, in Section 7.5 we discuss *problems with quantifiers*. These problems are natural generalizations of the decision and counting problems. Whenever appropriate we address algorithmic issues. For general references in the area of computational complexity/algorithms see [AB09]. We summarize the computational complexity status of our problems in Table 7.0.1.

TABLE 7.0.1 Computational complexity of basic problems.

PROBLEM NAME	BOUNDED DIMENSION	UNBOUNDED DIMENSION
Decision problem	polynomial	NP-hard
Counting problem	polynomial	#P-hard
Asymptotic problem	polynomial	#P-hard*
Problems with quantifiers	unknown; polynomial for $\forall\exists$ **	NP-hard

* in bounded codimension, reduces polynomially to volume computation

** with no quantifier alternation, polynomial time

7.1 INTEGRAL POLYTOPES IN POLYHEDRAL COMBINATORICS

We describe some combinatorial and computational properties of integral polytopes. General references are [GLS88], [GW93], [Sch86], [Lag95], [DL97] and [Zie00].

GLOSSARY

$\mathbb{R}^d$: Euclidean d-dimensional space with scalar product $\langle x,y\rangle = x_1y_1 + \ldots + x_dy_d$, where $x = (x_1, \ldots, x_d)$ and $y = (y_1, \ldots, y_d)$.

$\mathbb{Z}^d$: The subset of $\mathbb{R}^d$ consisting of the points with integral coordinates.

Polytope: The convex hull of finitely many points in $\mathbb{R}^d$.

Face of a polytope P: The intersection of P and the boundary hyperplane of a halfspace containing P.

Facet: A face of codimension 1.

Vertex: A face of dimension 0; the set of vertices of P is denoted by Vert P.

$\mathcal{H}$-description of a polytope ($\mathcal{H}$-polytope): A representation of the polytope as the set of solutions of finitely many linear inequalities.

$\mathcal{V}$-description of a polytope ($\mathcal{V}$-polytope): The representation of the polytope by the set of its vertices.

Integral polytope: A polytope with all of its vertices in $\mathbb{Z}^d$.

(0, 1)-polytope: A polytope P such that each coordinate of every vertex of P is either 0 or 1.

An integral polytope $P \subset \mathbb{R}^d$ can be given either by its $\mathcal{H}$-description or by its $\mathcal{V}$-description or (somewhat implicitly) as the convex hull of integral points in some other polytope Q, so $P = \text{conv}\{Q \cap \mathbb{Z}^d\}$. In most cases it is difficult to translate one description into another. The following examples illustrate some typical kinds of behavior.

INTEGRALITY OF $\mathcal{H}$-POLYTOPES

It is an NP-hard problem to decide whether an $\mathcal{H}$-polytope $P \subset \mathbb{R}^d$ is integral. However, if the dimension d is fixed then the straightforward procedure of generating all the vertices of P and checking their integrality has polynomial time complexity. A rare case where an $\mathcal{H}$-polytope P is a priori integral is known under the general name of "total unimodularity." Let A be an $n \times d$ integral matrix such that every minor of A is either 0 or 1 or -1. Such a matrix A is called ***totally unimodular***. If $b \in \mathbb{Z}^n$ is an integral vector then the set of solutions to the system of linear inequalities $Ax \leq b$, when bounded, is an integral polytope in $\mathbb{R}^d$. Examples of totally unimodular matrices include matrices of vertex-edge incidences of oriented graphs and of bipartite graphs. A complete characterization of totally unimodular matrices and a polynomial time algorithm for recognizing a totally unimodular matrix is provided by a theorem of P. Seymour (see [Sch86]). A family of integral polytopes, called ***transportation polytopes***, are much studied in the literature (see [EKK84] and [DLK14]). An example of a transportation polytope is provided by the set of $m \times n$ nonnegative matrices $x = (x_{ij})$ whose row and column sums are given positive integers. Integral points in this polytope are called ***contingency tables***; they play an important role in statistics. A particular transportation polytope, called the ***Birkhoff polytope***, is the set B_n of $n \times n$ nonnegative matrices with all row and column sums equal to 1. Alternatively, it may be described as the convex hull of the $n!$ permutation matrices $\pi(\sigma)_{ij} = \delta_{i\sigma(j)}$ for all permutations σ of the set $\{1, \ldots, n\}$.

The notion of total unimodularity has been generalized in various directions, thus leading to new classes of integral polytopes (see [Cor01] and [Sch03]).

Reflexive polytopes, that is, integral polytopes whose polar dual are also integral polytopes play an important role in mirror symmetry in algebraic geometry [Bat94].

COMBINATORIALLY DEFINED $\mathcal{V}$-POLYTOPES

There are several important situations where the explicit $\mathcal{V}$-description of an integral polytope is too long and a shorter description is desirable although not always available. For example, a $(0,1)$-polytope may be given as the convex hull of the characteristic vectors

$$\chi_S(i) = \begin{cases} 1 & \text{if } i \in S, \\ 0 & \text{otherwise} \end{cases}$$

for some combinatorially interesting family $\mathcal{S}$ of subsets $S \subset \{1,\dots,d\}$ (see [GLS88] for various examples). One of the most famous example is the ***traveling salesman polytope***, the convex hull TSP_n of the $(n-1)!$ permutation matrices $\pi(\sigma)$ where σ is a permutation of the set $\{1,\dots,n\}$ consisting of precisely one cycle (cf. the Birkhoff polytope B_n above). The problem of the $\mathcal{H}$-description of the traveling salesman polytope has attracted a lot of attention (see [GW93], [EKK84] and [Sch03] for some references) because of its relevance to combinatorial optimization. C.H. Papadimitriou proved that it is a co-NP-complete problem to establish whether two given vertices of TSP_n are adjacent, i.e., connected by an edge [Pap78]. L. Billera and A. Sarangarajan proved that every $(0,1)$-polytope can be realized as a face of TSP_n for sufficiently large n [BS96]. Thus the combinatorics of TSP_n contrasts with the combinatorics of the Birkhoff polytope B_n.

Another important polytope arising in this way is the ***cut polytope***, the famous counterexample to the Borsuk conjecture (see [DL97]). It is defined as the convex hull of the set of $n \times n$ matrices x_S, where

$$x_S(i,j) = \begin{cases} 1 & \text{if } |\{i,j\} \cap S| = 1 \text{ and } i \neq j, \\ 0 & \text{otherwise,} \end{cases}$$

where S ranges over all subsets of the set $\{1,\dots,n\}$.

CONVEX HULL OF INTEGRAL POINTS

Let $P \subset \mathbb{R}^d$ be a polytope. Then the convex hull P_I of the set $P \cap \mathbb{Z}^d$, if nonempty, is an integral polytope. Generally, the number of facets or vertices of P_I depends not only on the number of facets or vertices of P but also on the actual numerical size of the description of P (see [CHKM92]). Furthermore, it is an NP-complete problem to check whether a given point belongs to P_I, where P is given by its $\mathcal{H}$-description. If, however, the dimension d is fixed then the complexity of the facial description of the polytope P_I is polynomial in the complexity of the description of P. In particular, the number of vertices of P_I is bounded by a polynomial of degree $d-1$ in the input size of P [CHKM92].

Integrality imposes some restrictions on the combinatorial structure of a polytope. It is known that the combinatorial type of any 2- or 3-dimensional polytope can be realized by an integral polytope. J. Richter-Gebert constructed a 4-dimensional polytope with a nonintegral (and, therefore, nonrational) combinatorial type [Ric96]. Earlier, N. Mnëv had shown that for all sufficiently large d there exist nonrational d-polytopes with $d+4$ vertices [Mnë83]. The number $N_d(V)$ of classes of integral d-polytopes having volume V and nonisomorphic with respect to affine transformations of $\mathbb{R}^d$ preserving the integral lattice $\mathbb{Z}^d$ has logarithmic order

$$c_1(d)V^{\frac{d-1}{d+1}} \leq \log N_d(V) \leq c_2(d)V^{\frac{d-1}{d+1}}$$

for some $c_1(d), c_2(d) > 0$ [BV92].

7.2 DECISION PROBLEM

We consider the following general decision problem: Given a polytope $P \subset \mathbb{R}^d$ and a lattice $\Lambda \subset \mathbb{R}^d$, decide whether $P \cap \Lambda = \emptyset$ and, if the intersection is nonempty, find a point in $P \cap \Lambda$. We describe the main structural and algorithmic results for this problem. General references are [GL87], [GLS88], [GW93], [Sch86], and [Lag95].

GLOSSARY

Lattice: A discrete additive subgroup Λ of $\mathbb{R}^d$, i.e., $x - y \in \Lambda$ for any $x, y \in \Lambda$ and Λ does not contain limit points.

Basis of a lattice: A set of linearly independent vectors $u_1, \ldots, u_k$ such that every vector $y \in \Lambda$ can be (uniquely) represented in the form $y = m_1u_1 + \ldots + m_ku_k$ for some integers $m_1, \ldots, m_k$.

Rank of a lattice: The cardinality of any basis of the lattice. If $\Lambda \subset \mathbb{R}^d$ has rank d, Λ is said to be of ***full rank***.

Determinant of a lattice: For a lattice of rank k the k-volume of the parallelepiped spanned by any basis of the lattice.

Reciprocal lattice: For a full rank lattice $\Lambda \subset \mathbb{R}^d$, the lattice $\Lambda^* = \{x \in \mathbb{R}^d \mid \langle x, y \rangle \in \mathbb{Z} \text{ for all } y \in \Lambda\}$.

Polyhedron: An intersection of finitely many halfspaces in $\mathbb{R}^d$.

Convex body: A compact convex set in $\mathbb{R}^d$ with nonempty interior.

Lattice Polytope: For a given lattice Λ, a polytope with all of its vertices in Λ.

Applying a suitable linear transformation one can reduce the decision problem to the case in which $\Lambda = \mathbb{Z}^k$ and $P \subset \mathbb{R}^k$ is a full-dimensional polytope, $k = \operatorname{rank} \Lambda$.

The decision problem is known to be NP-complete for $\mathcal{H}$-polytopes as well as for $\mathcal{V}$-polytopes, although some special cases admit a polynomial time algorithm. In particular, if one fixes the dimension d then the decision problem becomes polynomially solvable. The main tool is provided by the so-called "flatness results."

FLATNESS THEOREMS

Let $P \subset \mathbb{R}^d$ be a convex body and let $l \in \mathbb{R}^d$ be a nonzero vector. The number

$$\max\{\langle l, x \rangle \mid x \in P\} - \min\{\langle l, x \rangle \mid x \in P\}$$

is called the ***width*** of P with respect to l. For a full rank lattice $\Lambda \subset \mathbb{R}^d$, the minimum width of P with respect to a nonzero vector $l \in \Lambda^*$ is called the ***lattice width*** of P.

The following general result is known under the unifying name of "flatness theorem."

THEOREM 7.2.1

There is a function $f : \mathbb{N} \to \mathbb{R}$ such that for any full rank lattice $\Lambda \subset \mathbb{R}^d$ and any convex body $P \subset \mathbb{R}^d$ with $P \cap \Lambda = \emptyset$, the lattice width of P does not exceed $f(d)$.

There are two types of results relating to the flatness theorem.

First, one may be interested in making $f(d)$ as small as possible. One can observe that $f(d) \geq d$: for some small $\epsilon > 0$, consider $\Lambda = \mathbb{Z}^d$ and the polytope P defined by the inequalities $x_1 + \ldots + x_d \leq d - \epsilon$, $x_i \geq \epsilon$ for $i = 1, \ldots, d$. It is known that one can choose $f(d) = O(d^{3/2})$ and it is conjectured that one can choose $f(d)$ as small as $O(d)$. W. Banaszczyk proved that if P is centrally symmetric, then one can choose $f(d) = O(d \log d)$, which is optimal up to a logarithmic factor. For these and related results, see [BLPS99]. There are results regarding the lattice width of some interesting classes of convex sets. Thus, if $P \subset \mathbb{R}^d$ is an ellipsoid that does not contain lattice points, then the lattice width of P is $O(d)$ [BLPS99]. A lattice polytope with no lattice points other than its vertices is called sometimes ***empty polytope***. J.-M. Kantor [Kan99] showed that the lattice width of a d-dimensional empty simplex can grow linearly in d and then A. Sebő [Seb99] constructed explicit examples of d-dimensional empty simplices of width $d - 2$. If P is a 3-dimensional empty polytope, then the lattice width of P is 1 (see [Sca85] and Section 16.6.1 of this Handbook for more on lattice width).

Second, one may be interested in the best width bound for which the corresponding vector $l \in \Lambda^*$ can be computed in polynomial time. The best bound known is $2^{O(d)}$, where l is polynomially computable even if the dimension d varies; see [GLS88]. For the computational complexity of lattice problems, such as finding the shortest non-zero lattice vector or the nearest lattice vector to a given point, see [MG02].

ALGORITHMS FOR THE DECISION PROBLEMS

Flatness theorems allow one to reduce the dimension in the decision problem: Assuming that $\Lambda = \mathbb{Z}^d$ and that the body P does not contain an integral point, one constructs a vector $l \in \mathbb{Z}^d$ for which P has a small width and reduces the d-dimensional decision problem to a family of $(d-1)$-dimensional decision problems $P_i = \{x \in P \mid \langle l, x \rangle = i\}$, where i ranges between $\min\{\langle l, x \rangle \mid x \in P\}$ and $\max\{\langle l, x \rangle \mid x \in P\}$. This reduction is the main idea of polynomial time algorithms in fixed dimension. The best complexity known for the decision problem in terms of the dimension d is $d^{O(d)}$, see [Dad14] for recent advances.

Constructing l efficiently relies on two major components (see [GLS88]). First, a linear transformation T is computed, such that the image $T(P)$ is "almost round," meaning that $T(P)$ is sandwiched between a pair of concentric balls with the ratio of their radii bounded by some small constant depending only on the dimension d. At this stage, a linear programming algorithm is used. Second, a reasonably short nonzero vector u is constructed in the lattice Λ^* reciprocal to $\Lambda = T(\mathbb{Z}^d)$. A basis reduction algorithm is used at this stage. Then we let $l = (T^*)^{-1}u$.

One can streamline the process by using the generalized lattice reduction [LS92] tailored to a given polytope. A polynomial time algorithm based on counting lattice points in the polytope and not using the flatness argument is sketched in [BP99].

MINKOWSKI'S CONVEX BODY THEOREM

The following classical result, known as "Minkowski's convex body theorem," provides a very useful criterion.

THEOREM 7.2.2

Suppose that $B \subset \mathbb{R}^d$ is a convex body, centrally symmetric about the origin 0, and $\Lambda \subset \mathbb{R}^d$ is a lattice of full rank. If vol $B \geq 2^d \det\Lambda$ *then B contains a nonzero point of Λ.*

For the proof and various generalizations see, for example, [GL87]. An important generalization (Minkowski's Second Theorem) concerns the existence of i linearly independent lattice points in a convex body. Namely, if

$$\lambda_i = \inf\Big\{\lambda > 0 \mid \lambda B \cap \Lambda \text{ contains } i \text{ linearly independent points}\Big\}$$

is the "ith successive minimum," then $\lambda_1 \dots \lambda_d \leq (2^d \det\Lambda)/(\mathrm{vol} B)$.

If B is a symmetric convex body such that vol $B = 2^d \det\Lambda$ but B does not contain a nonzero lattice point in its interior, then B is called ***extremal***. Every extremal body is necessarily a polytope. Moreover, this polytope contains at most $2(2^d - 1)$ facets, and therefore, for every dimension d, there exist only finitely many combinatorially different extremal polytopes. The contracted polytope $P = \{x/2 \mid x \in B\}$ has the property that its lattice translates $P + x$, $x \in \Lambda$, tile the space $\mathbb{R}^d$. Such a tiling polytope is called a ***parallelohedron***. Similarly, for every dimension d there exist only finitely many combinatorially different parallelohedra. Parallelohedra can be characterized intrinsically: a polytope is a parallelohedron if and only if it is centrally symmetric, every facet of it is centrally symmetric, and every class of parallel ridges ($(d-2)$-dimensional faces) consists of four or six ridges. If $q : \mathbb{R}^d \longrightarrow \mathbb{R}$ is a positive definite quadratic form, then the ***Dirichlet-Voronoi cell*** $P_q = \big\{x \mid q(x) \leq q(x - \lambda) \text{ for any } \lambda \in \Lambda\big\}$ is a parallelohedron. The problem of deciding whether a centrally symmetric polyhedron P contains a nonzero point from a given lattice Λ is known to be NP-complete even in the case of the standard cube $P = \{(x_1, \dots, x_d) \mid -1 \leq x_i \leq 1\}$. For fixed dimension d there exists a polynomial time algorithm since the problem obviously reduces to the decision problem (one can add the extra inequality $x_1 + \dots + x_d \geq 1$).

VOLUME BOUNDS

An integral simplex in $\mathbb{R}^d$ containing no integral points other than its vertices has volume $1/2$ if $d = 2$ but already for $d = 3$ can have an arbitrarily large volume (the smallest possible volume of such a simplex is $1/d!$). On the other hand, D. Hensley proved if an integral polytope P contains precisely $k > 0$ integral points then its volume is bounded by a function of k and d. J.C. Lagarias and G.M. Ziegler proved that vol $P \leq k(7(k+1))^{2^{d+1}}$, see [Lag95] and also [Pik01] for some sharpening.

G. Averkov, J. Krümpelmann and B. Nill found the maximum volume of an integral simplex that contains exactly one integer point in its interior [AKN15], thus proving a conjecture of D. Hensley. Namely, let $s_1, \dots, s_d$ be the Sylvester sequence, defined recursively by

$$s_1 = 2 \quad \text{and} \quad s_i = 1 + \prod_{j=1}^{i-1} s_j \quad \text{for} \quad i \geq 2.$$

The volume of an integral simplex in $\mathbb{R}^d$ with precisely one interior point does not exceed $2(s_d - 1)^2/d!$ and the bound is attained for the simplex with vertices at $0, s_1 e_1, \ldots, s_{d-1} e_{d-1}$ and $2(s_d - 1)e_d$, where $e_1, \ldots, e_d$ is the standard basis of $\mathbb{R}^d$.

7.3 COUNTING PROBLEM

We consider the following problem: Given a polytope $P \subset \mathbb{R}^d$, compute exactly or approximately the number of integral points $|P \cap \mathbb{Z}^d|$ in P.

For counting in general convex bodies see [CHKM92]. For applications in the combinatorics of generating functions, see [Sta86]. For applications in representation theory, see [BZ88], [CDW12] and [PP17]. For applications in statistical physics (computing permanents) and statistics (counting contingency tables), see [JS97]. For applications in social sciences, see [GL11]. For general information see surveys [GW93] and [BP99] and books [BR07] and [Bár08].

GLOSSARY

Rational polyhedron: The set

$$P = \{x \in \mathbb{R}^d \mid \langle a_i, x \rangle \leq \beta_i, \ i = 1, \ldots, m\},$$

where $a_i \in \mathbb{Z}^d$ and $\beta_i \in \mathbb{Z}$ for $i = 1, \ldots, m$. Generally, for a ***rational polyhedron with respect to lattice*** $\Lambda \subset \mathbb{R}^d$ of full rank, we have $a_i \in \Lambda^*$ and $\beta_i \in \mathbb{Z}$ for $i = 1, \ldots, m$.

Polyhedral cone: A set $K \subset \mathbb{R}^d$ of the form $K = \{\sum_{i=1}^k \lambda_i u_i \mid \lambda_i \geq 0, \ i = 1, \ldots, k\}$ for some vectors $u_1, \ldots, u_k \in \mathbb{R}^d$. The vectors $u_1, \ldots u_k$ are called ***generators*** of K.

Rational cone: A polyhedral cone having a set of generators belonging to $\mathbb{Z}^d$. Generally, a ***rational cone with respect to lattice*** Λ is a cone generated by vectors from Λ. A rational cone is a rational polyhedron.

Simple cone: A polyhedral cone generated by linearly independent vectors.

Cone of feasible directions at a point: The cone

$$K_v = \{x \mid v + \epsilon x \in P \text{ for all sufficiently small } \epsilon > 0\}$$

for a point v of a polytope P. If v is a vertex, then the cone K_v is generated by the vectors $u_i = v_i - v$, where $[v_i, v]$ is an edge of P.

Unimodular cone: A rational simple cone $K \subset \mathbb{R}^d$ generated by a basis of $\mathbb{Z}^d$. Generally, a ***unimodular cone with respect to lattice*** $\Lambda \subset \mathbb{R}^d$ is a cone generated by a basis of Λ.

Simple polytope: A polytope P such that the cone K_v of feasible directions is simple for every vertex v of P.

Totally unimodular polytope: An integral polytope P such that the cone K_v of feasible directions is unimodular for every vertex v of P.

GENERAL INFORMATION

The counting problem is known to be $\#P$-hard even for an integral $\mathcal{H}$- or $\mathcal{V}$-polytope. However, if the dimension d is fixed, one can solve the counting problem in polynomial time (see [BP99] and [Bár08]).

SOME EXPLICIT FORMULAS IN LOW DIMENSIONS

The classical Pick formula expresses the number of integral points in a convex integral polygon $P \subset \mathbb{R}^2$ in terms of its area and the number of integral points on the boundary ∂P:

$$|P \cap \mathbb{Z}^2| = \text{area}(P) + \frac{1}{2} \cdot |\partial P \cap \mathbb{Z}^2| + 1$$

(see, for example, [Mor93b], [GW93] and [BR07]). This formula almost immediately gives rise to a polynomial time algorithm for counting integral points in integral polygons.

An important explicit formula for the number of integral points in a lattice tetrahedron of a special kind was proven by L. Mordell, see [BR07]. Let a, b, c be pairwise coprime positive integers and $\Delta(a, b, c) \subset \mathbb{R}^3$ be the tetrahedron with vertices $(0,0,0)$, $(a,0,0)$, $(0,b,0)$, and $(0,0,c)$. Then

$$|\Delta(a,b,c) \cap \mathbb{Z}^3| = \frac{abc}{6} + \frac{ab+ac+bc+a+b+c}{4} +$$

$$\frac{1}{12}\left(\frac{ac}{b} + \frac{bc}{a} + \frac{ab}{c} + \frac{1}{abc}\right) - s(bc,a) - s(ac,b) - s(ab,c) + 2. \qquad (7.3.1)$$

Here

$$s(p,q) = \sum_{i=1}^{q} \left(\left(\frac{i}{q}\right)\right)\left(\left(\frac{pi}{q}\right)\right), \qquad \text{where} \qquad ((x)) = x - 0.5(\lfloor x \rfloor + \lceil x \rceil),$$

is the Dedekind sum. A similar formula was found in dimension 4. The reciprocity relation $s(p,q) + s(q,p) = (p/q + q/p + 1/pq - 3)/12$ allows one to compute the Dedekind sum $s(p,q)$ in polynomial time. A version of formula (7.3.1) was used by M. Dyer to construct polynomial time algorithms for the counting problem in dimensions 3 and 4. Formula (7.3.1) was generalized to an arbitrary tetrahedron by J. Pommersheim (see [BP99] and [BR07]).

Computationally efficient formulas for the number of lattice points are known for some particular polytopes, most notably zonotopes. Given integral points $v_1, \ldots, v_n$ in $\mathbb{R}^d$, a ***zonotope*** spanned by $v_1, \ldots, v_n$ is the polytope

$$P = \left\{\lambda_1 v_1 + \ldots + \lambda_n v_n \mid 0 \le \lambda_i \le 1 \quad \text{for} \quad i = 1, \ldots, n\right\}.$$

For each subset $S \subset \{v_1, \ldots, v_n\}$ of linearly independent points, let a_S be the index of the sublattice generated by S in the lattice $\mathbb{Z}^d \cap \text{span}(S)$, where $a_\emptyset = 1$. Then $|P \cap \mathbb{Z}^d| = \sum_S a_S$ (see Chapter 4, Problem 31 of [Sta86]).

EXPONENTIAL SUMS

A powerful tool for solving the counting problem exactly is provided by ***exponential sums***.

Let $P \subset \mathbb{R}^d$ be a polytope and $c \in \mathbb{R}^d$ be a vector. We consider the exponential sum

$$\sum_{x \in P \cap \mathbb{Z}^d} \exp\{\langle c, x\rangle\}.$$

If $c = 0$ we get the number of integral points in P. The reason for introducing the parameter c is that for a "generic" c exponential sums reveal some nontrivial algebraic properties that remain invisible when $c = 0$. To describe these properties we need to consider exponential sums over rational polyhedra and, in particular, over cones.

EXPONENTIAL SUMS OVER RATIONAL POLYHEDRA

Let $K \subset \mathbb{R}^d$ be a rational cone without lines generated by vectors $u_1, \ldots, u_k$ in $\mathbb{Z}^d$. Then the series $\sum\limits_{x \in K \cap \mathbb{Z}^d} \exp\{\langle c, x\rangle\}$ converges for any c such that $\langle c, u_i\rangle < 0$ for all $i = 1, \ldots, k$ and defines a meromorphic function of c, which we denote by $f_K(c)$. In particular, if K is unimodular then

$$f_K(c) = \prod_{i=1}^{k} \frac{1}{1 - \exp\{\langle c, u_i\rangle\}},$$

since the corresponding sum is just the multiple geometric series. Generally speaking, the farther a given cone is from being unimodular, the more complicated the formula for $f_K(c)$ will be.

These results are known in many different forms (see, for example, [Sta86, Section 4.6]). Furthermore, the function $f_K(c)$ can be extended to a finitely additive measure, defined on rational polyhedra in $\mathbb{R}^d$ and taking its values in the space of meromorphic functions in d variables, so that the measure of a rational polyhedron with a line is equal to 0. To state the result precisely, let us associate with every set $A \in \mathbb{R}^d$ its ***indicator function*** $[A] : \mathbb{R}^d \longrightarrow \mathbb{R}$, given by

$$[A](x) = \begin{cases} 1 & \text{if } x \in A, \\ 0 & \text{otherwise.} \end{cases}$$

The following result was proved by A.G. Khovanskii and A. Pukhlikov and, independently, by J. Lawrence, see [Bár08] for an exposition.

THEOREM 7.3.1 *Lawrence-Khovanskii-Pukhlikov Theorem*

There exists a map that associates, to every rational polyhedron $P \subset \mathbb{R}^d$, a meromorphic function $f_P(c)$, $c \in \mathcal{C}^d$, with the following properties:

The correspondence $P \longmapsto f_P$ preserves linear dependencies among indicator functions of rational polyhedra:

$$\sum_{i=1}^{m} \alpha_i [P_i] = 0 \quad \textit{implies} \quad \sum_{i=1}^{m} \alpha_i f_{P_i}(c) = 0$$

for rational polyhedra P_i and integers α_i.

If P does not contain lines, then

$$f_P(c) = \sum_{x \in P \cap \mathbb{Z}^d} \exp\{\langle c, x\rangle\}$$

for all c such that the series converges absolutely.

If P contains a line then $f_P(c) \equiv 0$.

If $P + m$ is a translation of P by an integral vector m then

$$f_{P+m}(c) = \exp\{\langle c, m\rangle\} f_P(c).$$

For example, suppose that $d = 1$ and let us choose $P_+ = [0, +\infty)$, $P_- = (-\infty, 0]$, $P_0 = \{0\}$, and $P = (-\infty, +\infty)$. Then

$$f_{P_+}(c) = \sum_{x=0}^{+\infty} \exp\{cx\} = \frac{1}{1 - \exp\{c\}} \quad \text{and} \quad f_{P_-}(c) = \sum_{x=0}^{-\infty} \exp\{cx\} = \frac{1}{1 - \exp\{-c\}}.$$

Moreover, $f_{P_0} = 1$ and $f_P = 0$ since P contains a line. We see that $[P] = [P_+] + [P_-] - [P_0]$ and that $f_P = f_{P_+} + f_{P_-} - f_{P_0}$.

Let $P \subset \mathbb{R}^d$ be a rational polytope and let $v \in P$ be its vertex. Let us consider the translation $v + K_v$ of the cone K_v of feasible directions at v. The following crucial result was proved by M. Brion, see [BR07] and [Bár08].

THEOREM 7.3.2 *Brion's Theorem*

Let $P \subset \mathbb{R}^d$ be a rational polytope. Then

$$\sum_{x \in P \cap \mathbb{Z}^d} \exp\{\langle c, x\rangle\} = \sum_{v \in \operatorname{Vert} P} f_{v+K_v}(c).$$

If the polytope is integral, we have $f_{v+K_v}(c) = \exp\{\langle c, v\rangle\} f_{K_v}(c)$. We note that if K is a unimodular cone and v is a rational vector then $f_{K+v} = \exp\{\langle c, w\rangle\} f_K(c)$, where $w \in \mathbb{Z}^d$ is a certain "rounding" of v with respect to K. Namely, assume that K is the conic hull of some integral vectors $u_1, \ldots, u_d$ that constitute a basis of $\mathbb{Z}^d$. Let $u_1^*, \ldots, u_d^*$ be the biorthogonal basis such that $\langle u_i^*, u_j\rangle = \delta_{ij}$. Then $w = \sum_{i=1}^d \lceil \langle v, u_i^*\rangle \rceil u_i$.

Essentially, Theorem 7.3.2 can be deduced from Theorem 7.3.1 by noticing that the indicator function of every (rational) polyhedron P can be written as the sum of the indicator functions $[v + K_v]$ modulo indicator functions of (rational) polyhedra with lines, see [BP99] and [Bár08].

Brion's formula allows one to reduce the counting of integral points in polytopes to the counting of points in polyhedral cones, a much easier problem. Below we discuss two instances where the application of exponential sums and Brion's identities leads to an efficient computational solution of the counting problem.

COUNTING IN FIXED DIMENSION

The following result was obtained by A. Barvinok (see [BP99] and [Bár08] for an exposition).

THEOREM 7.3.3

Let us fix the dimension d. Then there exists a polynomial time algorithm that, for any given rational polytope $P \subset \mathbb{R}^d$, computes the number $|P \cap \mathbb{Z}^d|$ of integral points in P.

THE IDEA OF THE ALGORITHM

We assume that the polytope is given by its $\mathcal{V}$-description. Let us choose a "generic" $c \in \mathbb{Q}^d$. We can compute the number $|P \cap \mathbb{Z}^d|$ as the limit of the exponential sum

$$\lim_{t \longrightarrow 0} \sum_{x \in P \cap \mathbb{Z}^d} \exp\{\langle tc, x \rangle\},$$

where t is a real parameter. Using Brion's Theorem 7.3.2, we reduce the problem to the computation of the constant term in the Laurent expansion of the meromorphic function $f_v(t) = f_{v+K_v}(tc)$, where v is a vertex of P and K_v is the cone of feasible directions at v. If K_v is a unimodular cone, we have an explicit formula for $f_{v+K_v}(c)$ (see above) and thus can easily compute the desired term. However, for $d > 1$ the cone K_v does not have to be unimodular. It turns out, nevertheless, that for any given rational cone K one can construct in polynomial time a decomposition

$$[K] = \sum_{i \in I} \epsilon_i [K_i] \quad \text{where} \quad \epsilon_i \in \{-1, 1\},$$

of the "inclusion-exclusion" type, where the cones K_i are unimodular (see below). Thus one can get an explicit expression

$$f_{v+K_v}(c) = \sum_{i \in I} \epsilon_i \cdot f_{v+K_i}(c)$$

and then compute the constant term of the Laurent expansion of $f_v(t)$. The complexity of the algorithm in terms of the dimension d is $d^{O(d)}$. The algorithm have been implemented in packages `LattE` by J.A. De Loera et al. [DLH+04] and `barvinok` by S. Verdoolaege et al. [VSB+07].

COUNTING IN TOTALLY UNIMODULAR POLYTOPES

One can efficiently count the number of integral points in a totally unimodular polytope given by its vertex description even in varying dimension.

THEOREM 7.3.4 [BP99]

There exists an algorithm that, for any d and any given integral vertices $v_1, \ldots, v_m \in \mathbb{Z}^d$ such that the polytope $P = \operatorname{conv}\{v_1, \ldots, v_m\}$ is totally unimodular, computes the number of integral points of P in time linear in the number m of vertices.

Moreover, the same result holds for rational polytopes with unimodular cones of feasible directions at the vertices. The algorithm uses Brion's formulas (Theorem 7.3.2) and the explicit formula above for the exponential sum over a unimodular cone.

EXAMPLE: COUNTING CONTINGENCY TABLES

Suppose A is an $n \times d$ totally unimodular matrix (see Section 7.1). Let us choose $b \in \mathbb{Z}^n$ such that the set P_b of solutions to the system $Ax \leq b$ of linear inequalities is a simple polytope. Then P_b is totally unimodular.

For example, if we know all the vertices of a simple transportation polytope P, we can compute the number of integral points of P in time linear in the number of vertices of P.

One can construct an efficient algorithm for counting integral points in a polytope that is somewhat "close" to totally unimodular and for which the explicit formulas for $f_{K_v}(c)$ are therefore not too long.

One particular application is counting contingency tables (see Section 7.1), see [DLH+04].

CONNECTIONS WITH TORIC VARIETIES

It has been known since the 1970s that the number of integral points in an integral polytope is related to some algebro-geometric invariants of the associated toric variety (see [Oda88]). Naturally, for smooth toric varieties (they correspond to totally unimodular polytopes) computation is much easier. Various formulas for the number of integral points in polytopes were first obtained for totally unimodular polytopes and then, by the use of resolution of singularities, generalized to arbitrary integral polytopes (see, for example, [BP99]). Resolution of singularities of toric varieties reduces to dissection of a polyhedral cone into unimodular cones. However, as one can see, it is impossible to subdivide a rational cone into polynomially (in the input) many unimodular cones even in dimension $d = 2$. For example (see Figure 8.3.1), the plane cone K generated by the points $(1, 0)$ and $(1, n)$ cannot be subdivided into fewer than $2n - 1$ unimodular cones, whereas a polynomial time subdivision would give a polynomial in $\log n$ cones. On the other hand, if we allow a signed linear combination of the inclusion-exclusion type, then one can easily represent this cone as a combination of 3 unimodular cones: $[K] = [K_1] - [K_2] + [K_3]$, where K_1 is generated by the basis $(1, 0)$ and $(0, 1)$, K_2 is generated by $(0, 1)$ and $(1, n)$, and K_3 is generated by $(1, n)$. Moreover, modulo rational cones with lines (cf. Theorem 7.3.1), we need to use only two unimodular cones: $[K] = [K_3] + [K_4]$ modulo rational cones with lines, where K_3 is the cone generated by $(1, n)$ and $(0, -1)$ and K_4 is the cone generated by $(0, 1)$ and $(1, 0)$. Consequently, from Theorem 7.3.1,

$$f_K(c) = \frac{1}{(1 - e^{c_1 + nc_2})(1 - e^{-c_2})} + \frac{1}{(1 - e^{c_1})(1 - e^{c_2})} \quad \text{for} \quad c = (c_1, c_2).$$

As we have mentioned above, once we allow "signed" combinations, any rational polyhedral cone can be decomposed into unimodular cones in polynomial time, provided the dimension is fixed. Moreover, if we allow decompositions modulo rational cones with lines, the algorithm can be sped up further; roughly from $2^{O(d^2)}$ to $2^{O(d \ln d)}$ (see [BP99] and [Bár08]).

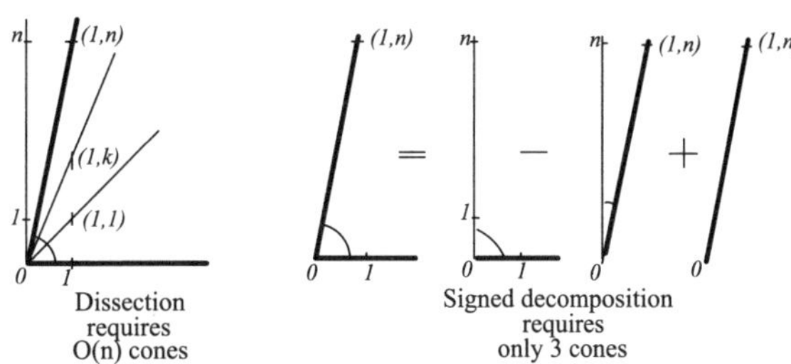

FIGURE 7.3.1
Decomposition of a cone into unimodular cones.

CONNECTIONS WITH VALUATIONS

The number of integral points $\nu(P) = |P \cap \mathbb{Z}^d|$ in an integral polytope $P \subset \mathbb{R}^d$ is a ***valuation***, that is, it preserves linear relations among indicator functions of polytopes; and it is lattice-translation-invariant, i.e., $\nu(P + l) = \nu(P)$ for any $l \in \mathbb{Z}^d$. General properties of valuations and the related notion of the "polytope algebra" have been intensively studied (see, for example, [McM93] and [Mor93a]). Various identities discovered in this area might prove useful in dealing with particular counting problems (see [BP99] and [Bár08]). For example, if the transportation polytope P_b is not simple, one can apply the following recipe. First, triangulating the normal cone at the vertex, we represent it as a combination of unimodular cones (we discard lower-dimensional cones). Then, passing to the dual cones, we get the desired representation of the cone of feasible directions (we discard cones with lines).

Exponential sums can be considered as a way to extend the counting valuation (the number of integer points in a polytope) from rational polytopes to rational polyhedra, possibly unbounded. Similarly, one can extend volume (which is obviously a valuation) from polytopes to possibly unbounded polyhedra by using exponential integrals

$$\int_P \exp\{\langle c, x\rangle\}\ dx$$

and prove the corresponding version of Theorem 7.3.1, see [Bár08]. In [Bar06] a family of intermediate valuations interpolating between discrete (integer points) and continuous (volumes) valuations was introduced: for a given k-dimensional lattice subspace $L \subset \mathbb{R}^d$ (subspace spanned by lattice points), with a polytope $P \subset \mathbb{R}^d$, we associate the sum of k-dimensional volumes of sections $P \cap A$, where A ranges over different lattice translates of L. The corresponding theory of intermediate exponential valuations was developed by V. Baldoni, N. Berline, J.A. De Loera, M. Köppe, M. Vergne, see [BBD$^+$12], [BBK$^+$13] and references therein. These valuations turn out to be useful for efficient asymptotic counting of integer points in polytopes of varying dimension, see Section 7.4 below.

ANALYTIC METHODS

In [BH10], the following approach to approximate counting was suggested. Suppose that the polytope $P \subset \mathbb{R}^d$ is the intersection of an affine subspace of $\mathbb{R}^d$ defined by the system of linear equations $Ax = b$, where A is a $k \times d$ integer matrix of rank $k < d$ and the non-negative orthant $\mathbb{R}^d_+$. Suppose further that the relative interior of P is non-empty. Then the strictly concave function $g : \mathbb{R}^d_+ \longrightarrow \mathbb{R}$,

$$g(x) = \sum_{i=1}^{d} \Big((x_i + 1) \ln (x_i + 1) - x_i \ln x_i \Big) \quad \text{for} \quad x = (x_1, \ldots, x_d)$$

attains its maximum on P at a (necessarily unique) point $z = (z_1, \ldots, z_n)$, which can be easily computed by interior point methods. Suppose that $\xi_1, \ldots, \xi_d$ are independent geometric random variables with expectations $z_1, \ldots, z_d$ respectively. In other words, each random variable ξ_i accepts non-negative integer values and

$$\mathbf{P}\left(\xi_i = k\right) = p_i q_i^k \quad \text{for} \quad p_i = \frac{1}{1 + z_i}, \; q_i = \frac{z_i}{1 + z_i} \quad \text{and} \quad k = 0, 1, 2, \ldots.$$

Let $\xi = (\xi_1, \ldots, \xi_d)$ be the corresponding random vector. It is shown in [BH10] that

$$\mathbf{P}(\xi = m) = e^{-g(z)} \quad \text{for all} \quad m \in P \cap \mathbb{Z}^d$$

and hence one can express the number of integer points in P as

$$\left| P \cap \mathbb{Z}^d \right| = e^{g(z)} \mathbf{P}(A\xi = b). \tag{7.3.2}$$

In fact, the random vector ξ has the largest entropy among all random vectors supported on the set $\mathbb{Z}^d_+$ of non-negative integer vectors and with expectation in the affine subspace $Ax = b$ while $g(x)$ is the entropy of the vector of independent geometric random variables with expectation x. The random k-dimensional vector $A\xi$ is a linear combination of the d columns of matrix A with independent random coefficients and $\mathbf{E}(A\xi) = Az = b$. If $k \ll d$, one can hope that in the spirit of the Local Central Limit Theorem, the distribution of $A\xi$ is close to a multivariate Gaussian distribution, so the right-hand side of (7.3.2) can be estimated via the covariance matrix of $A\xi$. Computing the covariance matrix, we obtain a heuristic formula

$$\left| P \cap \mathbb{Z}^d \right| \approx \frac{e^{g(z)} \left(\det \Lambda\right)}{(2\pi)^{k/2} \sqrt{\det \left(BB^T\right)}}, \tag{7.3.3}$$

where $\Lambda \subset \mathbb{Z}^k$ is the lattice spanned by the columns of matrix A and B is the matrix obtained from A by multiplying the i-th column of A onto $\sqrt{z_i^2 + z_i}$ for $i = 1, \ldots, d$. It is shown in [BH10] that for many classes of polytopes P formula (7.3.3) indeed provides an asymptotically tight approximation for the number of integer points. J.A. De Loera reported encouraging results of some numerical experiments. For example, the exact number of 4×4 non-negative integer matrices with row sums $220, 215, 93$ and 64 and column sums 108, 286, 71 and 127 is $1, 225, 914, 276, 768, 514$, while (7.3.3) gives a 6% error (we approximate the sum of 16 independent 7-dimensional vectors by the Gaussian distribution). The exact number of $3 \times 3 \times 3$ arrays of non-negative integers with sums $[31, 22, 87]$, $[50, 13, 77]$ and $[42, 87, 11]$ along the coordinate hyperplane "slices" is $8, 846, 838, 772, 161, 591$, while (7.3.3) gives a 0.19% error (we approximate the sum of 27 independent 7-dimensional vectors by the Gaussian distribution). It is shown in [Ben14] that (7.3.3) is asymptotically exact for multiway contingency tables (3- and higher-dimensional arrays of non-negative integers with prescribed sums along the coordinate hyperplane slices). In [BH12], it is shown that in the case of classical contingency tables (non-negative integer matrices with prescribed row and column sums) to obtain an asymptotically exact formula, one needs to introduce a correction to (7.3.3) (the so-called Edgeworth correction), which is an explicit, efficiently computable factor based on the third and fourth moments of the random vector $A\xi$.

The number of $m \times n$ non-negative integer matrices with equal row sums s and equal column sums t (and hence the total sum $N = ms = tn$ of entries) was earlier computed by E.R. Canfield and B.D. McKay [CM10]. Asymptotically, this number is

$$\frac{\binom{n+s-1}{s}^m \binom{m+t-1}{t}^n}{\binom{mn+N-1}{N}} \exp\left\{\frac{1}{2}\right\},$$

provided n and m grow roughly proportionately, see [CM10] for details. In [Sha10], non-trivial upper bounds for the number of integer points in polytopes are obtained via anti-concentration inequalities. In particular, by bounding the right-hand side of (7.3.2), the following simple and useful bound is obtained in [Sha10]:

$$\left|P \cap \mathbb{Z}^d\right| \le e^{g(z)} \min_{j_1,\dots,j_k} \prod_{i=1}^{k} \frac{1}{1+z_{j_i}},$$

where the minimum is taken over all collections $j_1, \dots, j_k$ of linearly independent columns of matrix A.

The following simple observation often leads to practically efficient (although theoretically exponential time) algorithms. Suppose that the polytope P is defined as above as the intersection of the non-negative orthant in $\mathbb{R}^d$ with a k-dimensional affine subspace defined by the system $Ax = b$, where $A = (a_{ij})$ is a $k \times d$ integer matrix. Let $z_1, \dots, z_k$ be (complex) variables and let

$$f_A(z_1, \dots, z_k) = \prod_{j=1}^{d} \sum_{m=0}^{+\infty} z_1^{a_{1j}m} z_2^{a_{2j}m} \cdots z_k^{a_{kj}m} = \prod_{j=1}^{d} \frac{1}{1 - z_1^{a_{1j}} z_2^{a_{2j}} \cdots z_k^{a_{kj}}}.$$

Thus $|P \cap \mathbb{Z}^d|$ is equal to the coefficient of $z_1^{b_1} \cdots z_k^{b_k}$ in the expansion of $f_A(z_1, \dots, z_k)$ in a neighborhood of $z_1 = \dots = z_k = 0$. This coefficient may be extracted by numerical differentiation, or by (repeated) application of the residue formula. M. Beck and D. Pixton [BP03] report results on numerical computation for the problem of counting contingency tables using repeated application of the residue formula.

As discussed in [BV97], various identities relating functions f_A mirror corresponding identities among indicator functions of rational polyhedra. In particular, decompositions of f_A into "simple fractions" correspond to decompositions of P into simple cones.

Quite a few useful inequalities for the number of lattice points can be found in [GW93], [Lag95], and [GL87]. Blichfeldt's inequality states that

$$|B \cap \Lambda| \le \frac{d!}{\det\Lambda} \text{vol } B + d,$$

where B is a convex body containing at least $d+1$ affinely independent lattice points. Davenport's inequality implies that

$$|B \cap \mathbb{Z}^d| \le \sum_{i=0}^{d} \binom{d}{i} V_i(B),$$

where the V_i are the intrinsic volumes. A conjectured stronger inequality, $|B \cap \mathbb{Z}^d| \le V_0(K) + \ldots + V_d(K)$, was shown to be false in dimensions $d \ge 207$, although it is correct for $d = 2, 3$. Furthermore, H. Hadwiger proved that

$$|B \cap \mathbb{Z}^d| \ge \sum_{i=0}^{d} (-1)^{d-i} V_i(B),$$

provided $B \subset \mathbb{R}^d$ is a convex body having a nonempty interior (see [Lag95]).

For applications of harmonic analysis, see [BR07].

PROBABILISTIC METHODS

Often, we need the number of integral points only approximately. Probabilistic methods based on Monte-Carlo methods have turned out to be quite successful. The main idea can be described as follows (see [JS97]). Suppose we want to approximate the cardinality of a finite set X (for example, X may be the set of lattice points in a polytope). Suppose, further, that we can present a "filtration" $X_0 \subset X_1 \subset \ldots \subset X_n = X$, where $|X_0| = 1$ (in general, we require $|X_0|$ to be small) and $|X_{i+1}|/|X_i| \le 2$ (in general, we require the ratio $|X_{i+1}|/|X_i|$ to be reasonably small). Finally, suppose that we have an efficient procedure for sampling an element $x \in X_i$ uniformly at random (in practice, we settle for "almost uniform" sampling). Given an $\epsilon > 0$ and a $\delta > 0$, with probability at least $1 - \delta$ one can estimate the ratio $|X_{i+1}|/|X_i|$, within a relative error ϵ/n, by sampling $O(n\epsilon^{-1} \ln \delta^{-1})$ points at random from X_{i+1} and counting how many times the points end up in X_i. Then, by "telescoping," with probability at least $(1 - \delta)^n$, we estimate

$$|X| = |X_n| = \frac{|X_n|}{|X_{n-1}|} \cdots \frac{|X_{i+1}|}{|X_i|} \cdots \frac{|X_2|}{|X_1|}$$

within relative error ϵ.

The bottleneck of the method is the ability to sample a point $x \in X_i$ uniformly at random. To achieve that, a Markov chain on X_i is designed, which converges fast ("mixes rapidly") to the uniform distribution. Usually, there are some natural candidates for such Markov chains and the main difficulty is to establish whether they indeed mix rapidly.

Counting various combinatorial structures can be interpreted as counting vertices in a certain $(0, 1)$-polytope. For example, computing the number of perfect matchings in a given bipartite graph on $n + n$ vertices, or, equivalently, computing the permanent of a given $n \times n$ matrix of 0's and 1's, can be viewed as counting the number of vertices in a particular face of the Birkhoff polytope B_n. M. Jerrum, A. Sinclair, and E. Vigoda [JSV04] have constructed a polynomial-time probabilistic algorithm to approximate the permanent of any given nonnegative matrix. B. Morris and A. Sinclair [MS99] have presented a polynomial-time probabilistic algorithm to compute the number of $(0, 1)$-vectors $(x_1, \ldots, x_n)$ satisfying the inequality $a_1 x_1 + \ldots + a_n x_n \le b_n$, where a_i and b are given positive integers, see also [Dye03] for a deterministic algorithm and [CDR10] for further applications of dynamic programming in derandomization.

R. Kannan and S. Vempala proved [KV99] that if a polytope $P \subset \mathbb{R}^d$ with m facets contains a ball of radius $d\sqrt{\ln m}$ then the number of integer points in P is well approximated by the volume of P and, moreover, sampling a random point from the uniform (or almost uniform) distribution $P \cap \mathbb{Z}^d$ can be achieved by sampling a random point from the uniform (or almost uniform) distribution in P (which is an easier problem, see [JS97] and [Vem10]), rounding the obtained point to an integer point and accepting it if the resulting integer point lies in P. This leads to a polynomial time algorithm for sampling and counting of integer points in such polytopes P.

7.4 ASYMPTOTIC PROBLEMS

If $P \subset \mathbb{R}^d$ is an integral polytope then the number of integral points in the dilated polytope $nP = \{nx \mid x \in P\}$ for a natural number n is a polynomial in n, known as the Ehrhart polynomial. We review several results concerning the Ehrhart polynomial and its generalizations.

GLOSSARY

Cone of feasible directions at a face of a polytope: The cone K_F of feasible directions at any point in the relative interior of the face $F \subset P$.

Tangent cone at a face of a polytope: The translation $x + K_F$ of the cone of feasible directions K_F by any point x in the face $F \subset P$.

Apex of a cone: The largest linear subspace contained in the cone.

Dual cone: The cone $K^* = \big\{x \in \mathbb{R}^d \mid \langle x, y\rangle \leq 0 \text{ for all } y \in K\big\}$, where $K \subset \mathbb{R}^d$ is a given cone.

vol$_k$: The normalized k-volume of a k-dimensional rational polytope $P \subset \mathbb{R}^d$ computed as follows. Let $L \subset \mathbb{R}^d$ be the k-dimensional linear subspace parallel to the affine span of P. Then $\mathrm{vol}_k(P)$ is the Euclidean k-dimensional volume of P in the affine span of P divided by the determinant of the lattice $\Lambda = \mathbb{Z}^d \cap L$.

Lattice subspace: A subspace spanned by lattice points.

EHRHART POLYNOMIALS

The following fundamental result was suggested by Ehrhart.

THEOREM 7.4.1

Let $P \subset \mathbb{R}^d$ be an integral polytope. For a natural number n we denote by $nP = \{nx \mid x \in P\}$ the n-fold dilation of P. Then the number of integral points in nP is a polynomial in n:

$$|nP \cap \mathbb{Z}^d| = E_P(n) \quad \textit{for some polynomial} \quad E_P(x) = \sum_{i=0}^{d} e_i(P) \cdot x^i.$$

Moreover, for positive integers n the value of $(-1)^{\deg E_P} E_P(-n)$ is equal to the number of integral points in the relative interior of the polytope nP (the "reciprocity law").

The polynomial E_P is called the ***Ehrhart polynomial*** and its coefficients $e_i(P)$ are called ***Ehrhart coefficients***. For various proofs of Theorem 7.4.1 see, for example, [Sta83], [Sta86], [BR07] and [Bár08]. The existence of the Ehrhart polynomials and the reciprocity law can be derived from the single fact that the number of integral points in a polytope is a lattice-translation-invariant valuation (see [McM93] and Section 7.3 above).

If P is a rational polytope, one can define $e_k(P) = n^{-k} e_k(P_1)$, where n is a positive integer such that $P_1 = nP$ is an integral polytope. For an integral polytope $P \subset \mathbb{R}^d$, one has $|P \cap \mathbb{Z}^d| = e_0(P) + e_1(P) + \ldots + e_d(P)$. (This formula is no longer true, however, if P is a general rational polytope.) The Ehrhart coefficients constitute a basis of all additive functions (valuations) ν on rational polytopes that are invariant under unimodular transformations (see [McM93] and [GW93]).

GENERAL PROPERTIES

It is known that $e_0(P) = 1$, $e_d(P) = \mathrm{vol}_d(P)$, and $e_{d-1}(P) = \frac{1}{2}\sum_F \mathrm{vol}_{d-1} F$, where the sum is taken over all the facets of P. Thus, computation of the two highest coefficients reduces to computation of the volume. In fact, the computation of any fixed number of the highest Ehrhart coefficients of an $\mathcal{H}$-polytope reduces in polynomial time to the computation of the volumes of faces; see [BP99], [Bár08] and also below.

EXISTENCE OF LOCAL FORMULAS

The Ehrhart coefficients can be decomposed into a sum of "local" summands. The following theorem was proven by P. McMullen (see [McM93], [Mor93a], and [BP99]).

THEOREM 7.4.2

For any natural numbers k and d there exists a real valued function $\mu_{k,d}$, defined on the set of all rational polyhedral cones $K \subset \mathbb{R}^d$, such that for every rational full-dimensional polytope $P \subset \mathbb{R}^d$ we have

$$e_k(P) = \sum_F \mu_{k,d}(K_F) \cdot \mathrm{vol}_k F,$$

where the sum is taken over all k-dimensional faces F of P and K_F is the cone of feasible directions of P at the face F. Moreover, one can choose $\mu_{k,d}$ to be an additive measure on polyhedral cones.

Different explicit and also computationally efficient constructions of $\mu_{d,k}$ were described by R. Morelli [Mor93b], J. Pommersheim and H. Thomas [PT04] and by N. Berline and M. Vergne [BV07].

In general, suppose V is a finite-dimensional real space and let $\Lambda \subset V$ be a lattice that spans V. Then we consider lattice polytopes in V and in every lattice subspace $L \subset V$ we define volume so that $\det(\Lambda \cap L) = 1$. Theorem 7.4.2 holds in this generality, though the function $\mu_{k,d}$ that satisfies the conditions of Theorem 7.4.2 is not unique. To make a canonical choice one has to introduce some additional structure, such as an inner product as in [Mor93b] and [BV07] or fix a flag of subspaces in V as in [PT04]. Essentially, one needs to be able to choose canonically the complement to a given subspace.

For some specific values of k and d convenient choices of $\mu_{k,d}$ has long been known.

EXAMPLE

For a cone $K \subset \mathbb{R}^d$, let $\gamma(K)$ be the spherical measure of K normalized in such a way that $\gamma(\mathbb{R}^d) = 1$. Thus $\gamma(K) = 0.5$ if K is a halfspace. One can choose $\mu_{d,d} = \mu_{d-1,d} = \gamma$ because of the formulas for $e_d(P)$ and $e_{d-1}(P)$ (see above).

On the other hand, one can choose $\mu_{0,d}(K) = \gamma(K^*)$, where K^* is the dual cone, since it is known that $e_0(P) = 1$. We note that if $\mu(K)$ is an additive measure on polyhedral cones then $\nu(K) = \mu(K^*)$ is also an additive measure on polyhedral cones, see, for example, [Bár08]. Moreover, for integral zonotopes (see Section 7.3), one can always choose $\mu_{k,d}(K_F) = \gamma(K_F^*)$ [BP99]. If F is a k-dimensional face of P then K_F^* is a $(d-k)$-dimensional cone and $\gamma(K_F^*)$ is understood as the spherical measure in the span of K_F^*.

BERLINE–VERGNE FORMULAS

We describe an elegant and computationally efficient choice of $\mu_{k,d}$ suggested by N. Berline and M. Vergne in [BV07], see also [Bár08] of an exposition.

Let V be a real vector space endowed with scalar product $\langle \cdot, \cdot \rangle$, let $\Lambda \subset V$ be a lattice that spans V and let $K \subset V$ be a pointed rational polyhedral cone with non-empty interior. Let $d = \dim V$. Our immediate goal is to construct a meromorphic function $\psi(K; c) : V \longrightarrow \mathcal{C}$, which we define recursively for $d = 0, 1, \ldots$. If $\dim V = 0$, we define $\psi(K; c) = 1$. Suppose that $d > 0$ and let $L \subset V$ be a lattice subspace. We introduce the volume form dx_L in L so that $\det(\Lambda \cap L) = 1$. Let V/L be the orthogonal complement of L, let $\Lambda/L \subset V/L$ be the orthogonal projection of Λ onto V/L, which is necessarily a lattice there, and let $K/L \subset V/L$ be the orthogonal projection of K, which is necessarily a rational cone with non-empty interior. We define $\psi(K; c)$ so that the following identity holds:

$$\sum_{m \in K \cap \Lambda} e^{\langle c, m \rangle} = \sum_{L} \psi(K/L; c) \int_{K \cap L} e^{\langle c, x \rangle} dx_L, \tag{7.4.1}$$

where the sum is taken over all subspaces L spanned by the faces of K. We note that unless $L = \{0\}$, we have $\dim V/L < d$ and hence $\psi(K/L; c)$ has been already defined. Since K is a pointed cone, there is a non-empty open set $U \subset V$ for which all the integrals in (7.4.1) converge absolutely, which allows us to define $\psi(K; c)$. Note that the identity (7.4.1) can be understood in terms of exponential valuations (see Section 7.3), in which case one can formally consider the sum over *all* lattice subspaces L: indeed, if $\dim(K \cap L) < \dim L$ then the corresponding integral is 0 and if L intersects the interior of K then K/L contains a line, in which case $\psi(K/L; c) \equiv 0$. Thus only subspaces L spanned by faces of K contribute non-zero terms in (7.4.1). It turns out that $K \longmapsto \psi(K; c)$ extends to a valuation with values in the ring of meromorphic functions, that $\psi(K; c) = 0$ provided K contains a line or has empty interior and that $\psi(K; c)$ is regular at $c = 0$. We then define $\mu_{k,d}(K_F)$ in Theorem 7.4.2 as $\psi(K; 0)$, where $K = K_F/L_F$ and L_F is the apex of K_F.

Using the technique of exponential sums, one can show that $\psi(K;c)$ is computable in polynomial time if $\dim K$ is fixed. Consequently, computation of any fixed number of the highest Ehrhart coefficients reduces in polynomial time to computation of the volumes of faces for an $\mathcal{H}$-polytope (see [BP99] and [Bár08]).

THE h^*-VECTOR

General properties of generating functions (see [Sta86]) imply that for every integral d-dimensional polytope P there exist integers $h_0^*(P), \ldots, h_d^*(P)$ such that

$$\sum_{n=0}^{\infty} E_P(n)x^n = \frac{h_0^*(P) + h_1^*(P)x + \ldots + h_d^*(P)x^d}{(1-x)^{d+1}}.$$

The $(d{+}1)$-vector $h^*(P) = \big(h_0^*(P), \ldots, h_d^*(P)\big)$ is called the ***h^*-vector*** of P. It is clear that $h^*(P)$ is a (vector-valued) valuation on the set of integral polytopes and that $h^*(P)$ is invariant under unimodular transformations of $\mathbb{Z}^d$. Moreover, the functions $h_k^*(P)$ constitute a basis of all valuations on integral polytopes that are invariant under unimodular transformations. Unlike the Ehrhart coefficients $e_k(P)$, the numbers $h_k^*(P)$ are not homogeneous. However, $h_k^*(P)$ are monotone (and, therefore, nonnegative): if $Q \subset P$ are two integral polytopes then $h_k^*(P) \geq h_k^*(Q)$ [Sta93].

The largest k such that $h_k^*(P) \neq 0$ is called the degree of P. Equivalently, k is the smallest non-negative integer such that the dilated polytope $(d-k)P$ has no interior lattice points. C. Haase, B. Nill, and S. Payne proved a decomposition theorem for polytopes of a fixed degree [HNP09], and, as a corollary, established that the volume of P is bounded from above by a function of its degree k and the value of $h_k^*(P)$, independently of the dimension, thus confirming a conjecture of V. Batyrev.

In principle, there is a combinatorial way to calculate $h^*(P)$. Namely, let Δ be a triangulation of P such that every d-dimensional simplex of Δ is integral and has volume $1/d!$ (see Section 7.2). Let $f_k(\Delta)$ be the number of k-dimensional faces of the triangulation Δ. Then

$$h_k^*(P) = \sum_{i=0}^{k} (-1)^{k-i} \binom{d-i}{d-k} f_{i-1}(\Delta),$$

where we let $f_{-1}(\Delta) = 1$. Such a triangulation may not exist for the polytope P but it exists for mP, where m is a sufficiently large integer [KKMS73]. Generally, this triangulation Δ would be too big, but for some special polytopes with nice structure (for example, for the so-called *poset polytopes*) it may provide a reasonable way to compute $h^*(P)$ and hence the Ehrhart polynomial E_P.

Since the number of integral points in a polytope is a valuation, we get the following result proved by P. McMullen (see [McM93]).

THEOREM 7.4.3

Let $P_1, \ldots, P_m$ be integral polytopes in $\mathbb{R}^d$. For an m-tuple of natural numbers $\boldsymbol{n} = (n_1, \ldots, n_m)$, let us define the polytope

$$P(\boldsymbol{n}) = \{n_1x_1 + \ldots + n_mx_m \mid x_1 \in P_1, \ldots x_m \in P_m\}$$

(using "+" for Minkowski addition one can also write $P(\boldsymbol{n}) = n_1P_1+\ldots+n_mP_m$). Then there exists a polynomial $p(x_1,\ldots,x_m)$ of degree at most d such that

$$|P(\boldsymbol{n}) \cap \mathbb{Z}^d| = p(n_1,\ldots,n_m).$$

More generally, the existence of local formulas for the Ehrhart coefficients implies that the number of integral points in an integral polytope $P_h = \{x \in \mathbb{R}^d \mid Ax \leq b+h\}$ is a polynomial in h provided P_h is an integral polytope combinatorially isomorphic to the integral polytope P_0, see, for example, [Bár08]. In other words, if we move the facets of an integral polytope parallel to themselves so that it remains integral and has the same facial structure, then the number of integral points varies polynomially.

INTEGRAL POINTS IN RATIONAL POLYTOPES

If P is a rational (not necessarily integral) polytope then $|nP\cap\mathbb{Z}^d|$ is not a polynomial but a ***quasipolynomial*** (a function of n whose value cycles through the values of a finite list of polynomials). The following result was independently proven by P. McMullen and R. Stanley (see [McM93] and [Sta86]).

THEOREM 7.4.4

Let $P \subset \mathbb{R}^d$ be a rational polytope. For every r, $0 \leq r \leq d$, let ind_r be the smallest natural number k such that all r-dimensional faces of kP are integral polytopes. Then, for every $n \in \mathbb{N}$,

$$|nP \cap \mathbb{Z}^d| = \sum_{r=0}^{d} e_r\big(P, n(\text{mod } \operatorname{ind}_r)\big) \cdot n^r$$

for suitable rational numbers $e_r(P,k)$, $0 \leq k < \operatorname{ind}_r$.

P. McMullen also obtained a generalization of the "reciprocity law" (see [Sta86] and [McM93]).

Let us fix an $n\times d$ integer matrix A such that the set $P_b = \big\{x \mid Ax \leq b\big\}$, $b \in \mathbb{Z}^n$, if nonempty, is a rational polytope. Let $B \subset \mathbb{Z}^n$ be a set of right-hand-side vectors b such that the combinatorial structure of P_b is the same for all $b \in B$. In [BP99] it is shown that as long as the dimension d is fixed, one can find a polynomially computable formula $F(b)$ for the number $|P_b \cap \mathbb{Z}^d|$, where F is a polynomial of degree d in integer parts of linear functions of b. It is based on Brion's Theorem (Theorem 7.3.2) and the "rounding" of rational translations of unimodular cones.

Theorem 7.4.2 and Berline-Vergne formulas extend to rational polytopes. In Theorem 7.4.2, measures $\mu_{k,d}$ depend on the translation class of the tangent cone $x+K_F$ modulo integer translations. Similarly, Berline-Vergne functions $\psi(x+K;c)$ are defined for translations of rational cones and also invariant under lattice translations. Consequently, the computation of $e_r(P,k)$ reduces in polynomial time to the computation of the volume of faces of P as long as the codimension $d-r$ is fixed. A different approach to computing the coefficients $e_r(P;k)$ in fixed codimension is via intermediate valuations, see [Bar06] and [BBD$^+$12].

Interestingly, for a "typical" (and, therefore, nonrational) polytope P the difference $|tP \cap \mathbb{Z}^d| - t^d \operatorname{vol} \mathrm{P}$ has order $O\big((\ln t)^{d-1+\epsilon}\big)$ as $t \longrightarrow +\infty$ [Skr98].

7.5 PROBLEMS WITH QUANTIFIERS

A natural generalization of the decision problem (see Section 7.2) is a problem with quantifiers. We describe some known results and formulate open questions for this class of problems.

FROBENIUS PROBLEM

The most famous problem from this class is the ***Frobenius problem***:

Given k coprime positive integers $a_1, \ldots, a_d$, find the largest integer m that cannot be represented as a non-negative integer combination $a_1 n_1 + \ldots + a_k n_k$.

The problem is known to be NP-hard in general, but a polynomial time algorithm is known for fixed k [Kan92].

PROBLEM WITH QUANTIFIERS

A general ***problem with quantifiers*** can be formulated as follows. Suppose that P is a Boolean combination of convex polyhedra: we start with some polyhedra $P_1, \ldots, P_k \subset \mathbb{R}^d$ given by their facet descriptions and construct P by using the set-theoretical operations of union, intersection, and complement. We want to find out if the formula

$$\exists x_1 \forall x_2 \exists x_3 \ldots \forall x_m : (x_1, \ldots, x_m) \in P \tag{7.5.1}$$

is true. Here x_i is an integral vector from $\mathbb{Z}^{d_i}$, and, naturally, $d_1 + \ldots + d_m = d$, $d_i \geq 0$. The parameters that characterize the size of (8.5.1) can be divided into two classes. The first class consists of the parameters characterizing the *combinatorial size* of the formula. These are the dimension d, the number $m - 1$ of quantifier alternations, the number of linear inequalities and Boolean operations that define the polyhedral set P. The parameters from the other class characterize the *numerical size* of the formula. Those are the bit sizes of the numbers involved in the inequalities that define P.

The following fundamental question remains open.

PROBLEM 7.5.1

Let us fix all the combinatorial parameters of the formula (8.5.1). *Does there exist a polynomial time algorithm that checks whether this formula is true?*

Naturally, "polynomial time" means that the running time of the algorithm is bounded by a polynomial in the numerical size of the formula. The answer to this question is unknown. A polynomial time algorithm is known if the formula contains not more than 1 quantifier alternation, i.e., if $m \leq 2$ ([Kan90]).

Let $P \subset \mathbb{R}^n$ be a rational polytope, let $pr : \mathbb{R}^n \longrightarrow \mathbb{R}^d$ be the projection on the first d coordinates, and let $S = pr\,(P \cap \mathbb{Z}^n)$ be the projection of the set of integer points in P. For any fixed n, A. Barvinok and K. Woods [BW03] constructed a polynomial time algorithm that, given P, computes the exponential sum over S in the form

$$\sum_{m \in S} e^{\langle c,m \rangle} = \sum_{i \in I} \alpha_i \frac{e^{\langle c,a_i \rangle}}{\left(1 - e^{\langle c,b_{i1} \rangle}\right) \cdots \left(1 - e^{\langle c,b_{ik} \rangle}\right)},$$

where $\alpha_i \in \mathbb{Q}$, $a_i \in \mathbb{Z}^d$ and $b_{ij} \in \mathbb{Z}^d \setminus \{0\}$. Besides, k depends only on n and d. Such sets S can be defined by formulas with no quantifier alternations, see also [Woo15] for some related developments. As a corollary, for any fixed d, we obtain a polynomial time algorithm that, for any given coprime positive integers $a_1, \ldots, a_d$, computes the number of non-negative integers m that are not non-negative integer combinations of $a_1, \ldots, a_d$.

7.6 SOURCES AND RELATED MATERIAL

RELATED CHAPTERS

Chapter 3: Tilings
Chapter 15: Basic properties of convex polytopes
Chapter 16: Subdivisions and triangulations of polytopes
Chapter 36: Computational convexity

REFERENCES

[AB09] S. Arora and B. Barak. *Computational Complexity. A Modern Approach.* Cambridge University Press, Cambridge, 2009.

[AKN15] G. Averkov, J. Krümpelmann, and B. Nill. Largest integral simplices with one interior integral point: solution of Hensley's conjecture and related results. *Adv. Math.*, 274:118–166, 2015.

[Bar06] A. Barvinok Computing the Ehrhart quasi-polynomial of a rational simplex. *Math. Comp.*, 75:1449–1466, 2006.

[Bar08] A. Barvinok. *Integer Points in Polyhedra.* Zurich Lectures in Advanced Mathematics, European Mathematical Society (EMS), Zürich, 2008.

[Bat94] V.V. Batyrev. Dual polyhedra and mirror symmetry for Calabi-Yau hypersurfaces in toric varieties. *J. Algebraic Geometry*, 3:493–535, 1994.

[BBD+12] V. Baldoni, N. Berline, J.A. De Loera, M. Köppe, and M. Vergne. Computation of the highest coefficients of weighted Ehrhart quasi-polynomials of rational polyhedra. *Found. Comput. Math.*, 12:435–469, 2012.

[BBK+13] V. Baldoni, N. Berline, M. Köppe, and M. Vergne. Intermediate sums on polyhedra: computation and real Ehrhart theory. *Mathematika*, 59:1–2, 2013.

[Ben14] D. Benson-Putnins. Counting integer points in multi-index transportation polytopes. Preprint, `arXiv:1402.4715`, 2014.

[BH10] A. Barvinok and J.A. Hartigan. Maximum entropy Gaussian approximations for the number of integer points and volumes of polytopes. *Adv. Appl. Math.*, 45:252–289, 2010.

[BH12] A. Barvinok and J.A. Hartigan. An asymptotic formula for the number of non-negative integer matrices with prescribed row and column sums. *Trans. Amer. Math. Soc.*, 364:4323–4368, 2012.

[BLPS99] W. Banaszczyk, A.E. Litvak, A. Pajor, and S.J. Szarek. The flatness theorem for nonsymmetric convex bodies via the local theory of Banach spaces. *Math. Oper. Res.*, 24:728–750, 1999.

[BP99] A. Barvinok and J.E. Pommersheim. An algorithmic theory of lattice points in polyhedra. In *New Perspectives in Algebraic Combinatorics (Berkeley, 1996–97)*, vol. 38 of *MSRI Publications*, pages 91–147, Cambridge Univ. Press, 1999.

[BP03] M. Beck and D. Pixton. The Ehrhart polynomial of the Birkhoff polytope. *Discrete Comput. Geom.*, 30:623–637, 2003.

[BR07] M. Beck and S. Robins. *Computing the Continuous Discretely. Integer-point Enumeration in Polyhedra.* Undergraduate Texts in Math., Springer, New York, 2007.

[BS96] L.J. Billera and A. Sarangarajan. Combinatorics of permutation polytopes. In L.J. Billera, C. Greene, R. Simion, and R. Stanley, editors, *Formal Power Series and Algebraic Combinatorics*, vol. 24 of *DIMACS Series in Discrete Mathematics and Theoretical Computer Science*, pages 1–23, AMS, Providence, 1996.

[BV92] I. Bárány and A.M. Vershik. On the number of convex lattice polytopes. *Geom. Funct. Anal.*, 2:381–393, 1992.

[BV97] M. Brion and M. Vergne. Residue formulae, vector partition functions and lattice points in rational polytopes. *J. AMS*, 10:797–833, 1997.

[BV07] N. Berline and M. Vergne. Local Euler-Maclaurin formula for polytopes. *Mosc. Math. J.*, 7:355–386, 2007.

[BW03] A. Barvinok and K. Woods. Short rational generating functions for lattice point problems. *J. AMS*, 16:957–979, 2003.

[BZ88] A.D. Berenstein and A.V. Zelevinsky. Tensor product multiplicities and convex polytopes in the partition space. *J. Geom. Phys.*, 5:453–472, 1988.

[CDR10] M. Cryan, M. Dyer, and D. Randall. Approximately counting integral flows and cell-bounded contingency tables. *SIAM J. Comput.*, 39:2683–2703, 2010.

[CDW12] M. Christandl, B. Doran, and M. Walter. Computing multiplicities of Lie group representations. In *Proc. 53rd IEEE Sympos. Found. Comp. Sci.*, pages 639–648, 2012.

[CHKM92] W.J. Cook, M. Hartmann, R. Kannan, and C. McDiarmid. On integer points in polyhedra. *Combinatorica*, 12:27–37, 1992.

[CM10] E.R. Canfield and B.D. McKay. Asymptotic enumeration of integer matrices with large equal row and column sums. *Combinatorica*, 30:655–680, 2010.

[Cor01] G. Cornuéjols. *Combinatorial Optimization: Packing and Covering.* Vol. 74 of CBMS-NSF Regional Conference Series in Applied Mathematics, SIAM, Philadelphia, 2001.

[Dad14] D. Dadush. A randomized sieving algorithm for approximate integer programming. *Algorithmica*, 70:208–244, 2014.

[DL97] M.M. Deza and M. Laurent. *Geometry of Cuts and Metrics.* Vol. 15 of *Algorithms Combin.*, Springer-Verlag, Berlin, 1997.

[DLH+04] J.A. De Loera, R. Hemmecke, J. Tauzer, and R. Yoshida. Effective lattice point counting in rational convex polytopes. *J. Symbolic Comput.*, 38:1273–1302, 2004.

[DLK14] J.A. De Loera and E.D. Kim. Combinatorics and geometry of transportation polytopes: an update. In *Discrete Geometry and Algebraic Combinatorics*, vol. 625 of Contemporary Mathematics, pages 37–76, AMS, Providence, 2014.

[Dye03] M. Dyer. Approximate counting by dynamic programming. In *Proc. 35th ACM Sympos. Theory of Comput.*, pages 693–699, 2003.

[EKK84] V.A. Emelichev, M.M. Kovalev, and M.K. Kravtsov. *Polytopes, Graphs and Optimization.* Cambridge University Press, 1984.

[GL87] P.M. Gruber and C.G. Lekkerkerker. *Geometry of Numbers*, 2nd edition. North-Holland, Amsterdam, 1987.

[GL11] W.V. Gehrlein and D. Lepelley. *Voting Paradoxes and Group Coherence: The Condorcet Efficiency of Voting Rules.* Springer-Verlag, Berlin, 2011.

[GLS88] M. Grötschel, L. Lovász, and A. Schrijver. *Geometric Algorithms and Combinatorial Optimization.* Springer-Verlag, Berlin, 1988.

[GW93] P. Gritzmann and J.M. Wills. Lattice points. In P.M. Gruber and J.M. Wills, editors, *Handbook of Convex Geometry*, pages 765–797, Elsevier, Amsterdam, 1993.

[HNP09] C. Haase, B. Nill, and S. Payne. Cayley decompositions of lattice polytopes and upper bounds for h^*-polynomials. *J. Reine Angew. Math.*, 637:207–216, 2009.

[JS97] M. Jerrum and A. Sinclair. The Markov chain Monte Carlo method: an approach to approximate counting and integration. In D.S. Hochbaum, editor, *Approximation Algorithms for NP-Hard Problems*, pages 482–520, PWS, Boston, 1997.

[JSV04] M. Jerrum, A. Sinclair, and E. Vigoda. A polynomial-time approximation algorithm for the permanent of a matrix with non-negative entries. *J. ACM*, 51:671–697, 2004.

[Kan90] R. Kannan. Test sets for integer programs, $\forall\exists$ sentences. In W. Cook and P.D. Seymour, editors, *Polyhedral Combinatorics*, vol. 1 of *DIMACS Series in Discrete Mathematics and Theoretical Computer Science*, pages 39–47, AMS, Providence, 1990.

[Kan92] R. Kannan. Lattice translates of a polytope and the Frobenius problem. *Combinatorica*, 12:161–177, 1992.

[Kan99] J.-M. Kantor. On the width of lattice-free simplices. *Compositio Math.*, 118:235–241, 1999.

[KKMS73] G. Kempf, F.F. Knudsen, D. Mumford, and B. Saint-Donat. *Toroidal Embeddings I.* Vol. 339 of *Lecture Notes in Math.*, Springer-Verlag, Berlin-New York, 1973.

[KV99] R. Kannan and S. Vempala. Sampling lattice points. In *Proc. 29th ACM Sympos. Theory Comput.*, pages 696–700, 1997.

[Lag95] J.C. Lagarias. Point lattices. In R. Graham, M. Grötschel, and L. Lovász, editors, *Handbook of Combinatorics*, pages 919–966, North-Holland, Amsterdam, 1995.

[LS92] L. Lovász and H.E. Scarf. The generalized basis reduction algorithm. *Math. Oper. Res.*, 17:751–764, 1992.

[McM93] P. McMullen. Valuations and dissections. In P.M. Gruber and J.M. Wills, editors, *Handbook of Convex Geometry*, volume B, pages 933–988, North-Holland, Amsterdam, 1993.

[MG02] D. Micciancio and S. Goldwasser. *Complexity of Lattice Problems: A Cryptographic Perspective.* Vol. 671 of Kluwer International Series in Engineering and Computer Science, Kluwer Academic Publishers, Boston, 2002.

[Mnë83] N.E. Mnëv. On the realizability over fields of the combinatorial types of convex polytopes (in Russian). In Differential Geometry, Lie Groups and Mechanics, V. *Zap. Nauchn. Sem. Leningrad. Otdel. Mat. Inst. Steklov. (LOMI)*, 123, pages 203–307, 1983.

[Mor93a] R. Morelli. A theory of polyhedra. *Adv. Math.*, 97:1–73, 1993.

[Mor93b] R. Morelli. Pick's theorem and the Todd class of a toric variety. *Adv. Math.*, 100:183–231, 1993.

[MS99] B. Morris and A. Sinclair. Random walks on truncated cubes and sampling 0-1 knapsack solutions. *SIAM J. Comput.*, 34:195–226, 2004. 1999.

[Oda88] T. Oda. *Convex Bodies and Algebraic Geometry: An Introduction to the Theory of Toric Varieties.* Springer-Verlag, Berlin, 1988.

[Pap78] C.H. Papadimitriou. The adjacency relation on the traveling salesman polytope is NP-complete. *Math. Program.*, 14:312–324, 1978.

[Pik01] O. Pikhurko. Lattice points in lattice polytopes. *Mathematika*, 48:15–24, 2001.

[PP17] I. Pak and G. Panova. On the complexity of computing Kronecker coefficients. *Comput. Complexity*, 26:1–36, 2017.

[PT04] J. Pommersheim and H. Thomas. Cycles representing the Todd class of a toric variety. *J. AMS*, 17:983–994, 2004.

[Ric96] J. Richter-Gebert. *Realization Spaces of Polytopes.* Vol. 1643 of *Lecture Notes in Math.*, Springer-Verlag, Berlin, 1996.

[Sca85] H.E. Scarf. Integral polyhedra in three space. *Math. Oper. Res.*, 10:403–438, 1985.

[Sch86] A. Schrijver. *The Theory of Linear and Integer Programming.* John Wiley & Sons, Chichester, 1986.

[Sch03] A. Schrijver. *Combinatorial Optimization: Polyhedra and Efficiency*, Volume A. Algorithms and Combinatorics 24, Springer-Verlag, Berlin, 2003.

[Seb99] A. Sebő. An introduction to empty lattice simplices. In *Integer Programming and Combinatorial Optimization*, vol. 1610 of *Lecture Notes Comp. Sci.*, pages 400–414, Springer, Berlin, 1999.

[Sha10] A. Shapiro. Bounds on the number of integer points in a polytope via concentration estimates. Preprint, `arXiv:1011.6252`, 2010.

[Skr98] M.M. Skriganov. Ergodic theory on $\mathrm{SL}(n)$, Diophantine approximations and anomalies in the lattice point problem. *Invent. Math.*, 132:1–72, 1998.

[Sta83] R.P. Stanley. *Combinatorics and Commutative Algebra.* Vol. 41 of *Progress in Mathematics*, Birkhäuser, Boston, 1983.

[Sta86] R.P. Stanley. *Enumerative Combinatorics*, Volume 1. Wadsworth and Brooks/Cole, Monterey, 1986.

[Sta93] R.P. Stanley. A monotonicity property of h-vectors and h^*-vectors. *Europ. J. Combin.*, 14:251–258, 1993.

[Vem10] S. Vempala. Recent progress and open problems in algorithmic convex geometry. In *30th International Conference on Foundations of Software Technology and Theoretical Computer Science*, pages 42–64, vol. 8 of LIPIcs, Schloss Dagstuhl, 2010.

[VSB+07] S. Verdoolaege, R. Seghir, K. Beyls, V. Loechner, and M. Bruynooghe. Counting integer points in parametric polytopes using Barvinok's rational functions. *Algorithmica,* 48:37–66, 2007.

[Woo15] K. Woods. Presburger arithmetic, rational generating functions, and quasi-polynomials. *J. Symb. Log.*, 80:233–449, 2015.

[Zie00] G.M. Ziegler. Lectures on 0/1-polytopes. In G. Kalai and G.M. Ziegler, editors, *Polytopes–Combinatorics and Computation (Oberwolfach, 1997)*, pages 1–41, vol. 29 of DMV Seminar, Birkhäuser, Basel, 2000.

8 LOW-DISTORTION EMBEDDINGS OF FINITE METRIC SPACES

Piotr Indyk, Jiří Matoušek, and Anastasios Sidiropoulos

INTRODUCTION

An n-point metric space (X, D) can be represented by an $n \times n$ table specifying the distances. Such tables arise in many diverse areas. For example, consider the following scenario in microbiology: X is a collection of bacterial strains, and for every two strains, one is given their *dissimilarity* (computed, say, by comparing their DNA). It is difficult to see any structure in a large table of numbers, and so we would like to represent a given metric space in a more comprehensible way.

For example, it would be very nice if we could assign to each $x \in X$ a point $f(x)$ in the plane in such a way that $D(x, y)$ equals the Euclidean distance of $f(x)$ and $f(y)$. Such a representation would allow us to see the structure of the metric space: tight clusters, isolated points, and so on. Another advantage would be that the metric would now be represented by only $2n$ real numbers, the coordinates of the n points in the plane, instead of $\binom{n}{2}$ numbers as before. Moreover, many quantities concerning a point set in the plane can be computed by efficient geometric algorithms, which are not available for an arbitrary metric space.

This sounds too good to be generally true: indeed, there are even finite metric spaces that cannot be exactly represented either in the plane or in *any* Euclidean space; for instance, the four vertices of the graph $K_{1,3}$ (a star with 3 leaves) with the shortest-path metric (see Figure 8.0.1). However, it *is* possible to embed the latter metric in a Euclidean space, if we allow the distances to be distorted somewhat. For example, if we place the center of the star at the origin in $\Re^3$ and the leaves at $(1,0,0), (0,1,0), (0,0,1)$, then all distances are preserved *approximately*, up to a factor of $\sqrt{2}$ (Figure 8.0.1b).

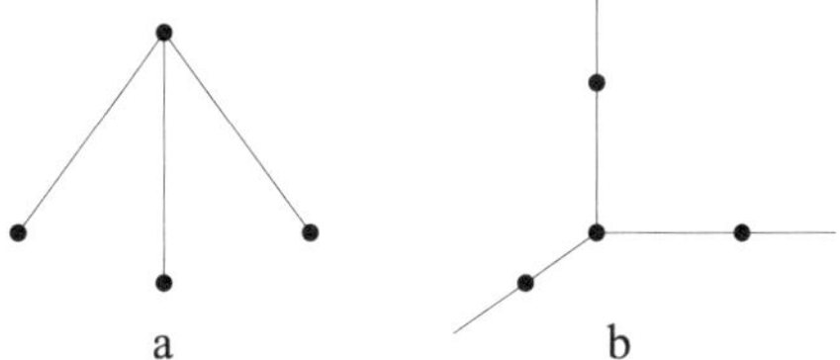

FIGURE 8.0.1
A nonembeddable metric space.

Approximate embeddings have proven extremely helpful for approximate solutions of problems dealing with distances. For many important algorithmic problems, they yield the only known good approximation algorithms.

The normed spaces usually considered for embeddings of finite metrics are the spaces ℓ_p^d, $1 \leq p \leq \infty$, and the cases $p = 1, 2, \infty$ play the most prominent roles.

GLOSSARY

Metric space: A pair (X, D), where X is a set of *points* and $D : X \times X \to [0, \infty)$ is a *distance function* satisfying the following conditions for all $x, y, z \in X$:

(i) $D(x, y) = 0$ if and only if $x = y$,

(ii) $D(x, y) = D(y, x)$ (symmetry), and

(iii) $D(x, y) + D(y, z) \geq D(x, z)$ (triangle inequality).

Separable metric space: A metric space (X, D) containing a countable dense set; that is, a countable set Y such that for every $x \in X$ and every $\varepsilon > 0$ there exists $y \in Y$ with $D(x, y) < \varepsilon$.

Pseudometric: Like metric except that (i) is not required.

Isometry: A mapping $f : X \to X'$, where (X, D) and (X', D') are metric spaces, with $D'(f(x), f(y)) = D(x, y)$ for all x, y.

(Real) normed space: A real vector space Z with a mapping $\|\cdot\|_Z : Z \to [0, \infty]$, the *norm*, satisfying $\|x\|_Z = 0$ iff $x = 0$, $\|\alpha x\|_Z = |\alpha| \cdot \|x\|_Z$ $(\alpha \in \Re)$, and $\|x + y\|_Z \leq \|x\|_Z + \|y\|_Z$. The metric on Z is given by $(x, y) \mapsto \|x - y\|_Z$.

$\boldsymbol{\ell_p^d}$***:*** The space $\Re^d$ with the ℓ_p-norm $\|x\|_p = \left(\sum_{i=1}^d |x_i|^p\right)^{1/p}$, $1 \leq p \leq \infty$ (where $\|x\|_\infty = \max_i |x_i|$).

Finite $\boldsymbol{\ell_p}$ ***metric:*** A finite metric space isometric to a subspace of ℓ_p^d for some d.

$\boldsymbol{\ell_p}$***:*** For a sequence $(x_1, x_2, \ldots)$ of real numbers we set $\|x\|_p = \left(\sum_{i=1}^\infty |x_i|^p\right)^{1/p}$. Then ℓ_p is the space consisting of all x with $\|x\|_p < \infty$, equipped with the norm $\|\cdot\|_p$. It contains every finite ℓ_p metric as a (metric) subspace.

Distortion: A mapping $f : X \to X'$, where (X, D) and (X', D') are metric spaces, is said to have distortion at most c, or to be a *c-embedding*, where $c \geq 1$, if there is an $r \in (0, \infty)$ such that for all $x, y \in X$,

$$r \cdot D(x, y) \leq D'(f(x), f(y)) \leq cr \cdot D(x, y).$$

If X' is a normed space, we usually require $r = \frac{1}{c}$ or $r = 1$.

Order of congruence: A metric space (X, D) has order of congruence at most m if every finite metric space that is not isometrically embeddable in (X, D) has a subspace with at most m points that is not embeddable in (X, D).

8.1 THE SPACES ℓ_p

8.1.1 THE EUCLIDEAN SPACES ℓ_2^d

Among normed spaces, the Euclidean spaces are the most familiar, the most symmetric, the simplest in many respects, and the most restricted. Every finite ℓ_2 metric embeds isometrically in ℓ_p for all p. More generally, we have the following Ramsey-type result on the "universality" of ℓ_2; see, e.g., [MS86]:

THEOREM 8.1.1 Dvoretzky's theorem (a finite quantitative version)

For every d and every $\epsilon > 0$ there exists $n = n(d, \epsilon) \leq 2^{O(d/\epsilon^2)}$ such that ℓ_2^d can be $(1+\epsilon)$-embedded in every n-dimensional normed space.

Isometric embeddability in ℓ_2 has been well understood since the classical works of Menger, von Neumann, Schoenberg, and others (see, e.g., [Sch38]). Here is a brief summary:

THEOREM 8.1.2

(i) *(Compactness) A separable metric space (X, D) is isometrically embeddable in ℓ_2 iff each finite subspace is so embeddable.*

(ii) *(Order of congruence) A finite (or separable) metric space embeds isometrically in ℓ_2^d iff every subspace of at most $d+3$ points so embeds.*

(iii) *For a finite $X = \{x_0, x_1, \ldots, x_n\}$, (X, D) embeds in ℓ_2 iff the $n\times n$ matrix $\left(D(x_0, x_i)^2 + D(x_0, x_j)^2 - D(x_i, x_j)^2\right)_{i,j=1}^{n}$ is positive semidefinite; moreover, its rank is the smallest dimension for such an embedding.*

(iv) *(Schoenberg's criterion) A separable (X, D) isometrically embeds in ℓ_2 iff the matrix $\left(e^{-\lambda D(x_i,x_j)^2}\right)_{i,j=1}^{n}$ is positive semidefinite for all $n \geq 1$, for any points $x_1, x_2, \ldots, x_n \in X$, and for any $\lambda > 0$. (This is expressed by saying that the functions $x \mapsto e^{-\lambda x^2}$, for all $\lambda > 0$, are* positive definite on ℓ_2.*)*

Using similar ideas, the problem of finding the smallest c such that a given finite (X, D) can be c-embedded in ℓ_2 can be formulated as a semidefinite programming problem and thus solved in polynomial time [LLR95]. For embedding into ℓ_2^d no similar result is possible. It has been shown that for any constant $d \geq 1$, approximating even to within a polynomial factor the minimum distortion embedding into ℓ_2^d is NP-hard [MS10]

8.1.2 THE SPACES ℓ_1^d

GLOSSARY

Cut metric: A pseudometric D on a set X such that, for some partition $X = A\dot{\cup}B$, we have $D(x, y) = 0$ if both $x, y \in A$ or both $x, y \in B$, and $D(x, y) = 1$ otherwise.

Hypermetric inequality: A metric space (X, D) satisfies the $(2k+1)$-point hypermetric inequality (also called the $(2k+1)$-gonal inequality) if for every multiset A of k points and every multiset B of $k+1$ points in X, $\sum_{a,a'\in A} D(a, a') + \sum_{b,b'\in B} D(b, b') \leq \sum_{a\in A, b\in B} D(a, b)$. (We get the triangle inequality for $k = 1$.)

Hypermetric space: A space that satisfies the hypermetric inequality for all k.

Cocktail-party graph: The complement of a perfect matching in a complete graph K_{2m}; also called a ***hyperoctahedron graph.***

Half-cube graph: The vertex set consists of all vectors in $\{0, 1\}^n$ with an even number of 0's, and edges connect vectors with Hamming distance 2.

Cartesian product of graphs G and H: The vertex set is $V(G) \times V(H)$, and the edge set is $\{\{(u,v),(u,v')\} \mid u \in V(G), \{v,v'\} \in E(H)\} \cup \{\{(u,v),(u',v)\} \mid \{u,u'\} \in E(G), v \in V(H)\}$. The cubes are Cartesian powers of K_2.

Girth of a graph: The length of the shortest cycle.

The ℓ_1 spaces are important for many reasons, but considerably more complicated than Euclidean spaces; a general reference here is [DL97]. Many important and challenging open problems are related to embeddings in ℓ_1 or in ℓ_1^d.

Unlike the situation in ℓ_2^n, not every n-point ℓ_1-metric lives in ℓ_1^n; dimension of order $\Theta(n^2)$ is sometimes necessary and always sufficient to embed n-point ℓ_1-metrics isometrically (similarly for the other ℓ_p-metrics with $p \neq 2$).

The ℓ_1 metrics on an n-point set X are precisely the elements of the ***cut cone***; that is, linear combinations with nonnegative coefficients of cut metrics on X. Another characterization is this: A metric D on $\{1, 2, \ldots, n\}$ is an ℓ_1 metric iff there exist a measure space (Ω, Σ, μ) and sets $A_1, \ldots, A_n \in \Sigma$ such that $D(i,j) = \mu(A_i \triangle A_j)$.

Every ℓ_1 metric is a hypermetric space (since cut metrics satisfy the hypermetric inequalities), but for 7 or more points, this condition is not sufficient. Hypermetric spaces have an interesting characterization in terms of *Delaunay polytopes* of lattices; see [DL97].

ISOMETRIC EMBEDDABILITY

Deciding isometric embeddability in ℓ_1 is NP-hard. On the other hand, the embeddability of *unweighted* graphs, both in ℓ_1 and in a Hamming cube, has been characterized and can be tested in polynomial time. In particular, we have:

THEOREM 8.1.3

(i) *An unweighted graph G embeds isometrically in some cube $\{0,1\}^m$ with the ℓ_1-metric iff it is bipartite and satisfies the pentagonal inequality.*

(ii) *An unweighted graph G embeds isometrically in ℓ_1 iff it is an isometric subgraph of a Cartesian product of half-cube graphs and cocktail-party graphs.*

A first characterization of cube-embeddable graphs was given by Djokovic [Djo73], and the form in (i) is due to Avis (see [DL97]). Part (ii) is from Shpectorov [Shp93].

ORDER OF CONGRUENCE

The isometric embeddability in ℓ_1^2 is characterized by 6-point subspaces (6 is best possible here), and can thus be tested in polynomial time (Bandelt and Chepoi [BC96]). The proof uses a result of Bandelt and Dress [BD92] of independent interest, about certain canonical decompositions of metric spaces (see also [DL97]).

On the other hand, for no $d \geq 3$ it is known whether the order of congruence of ℓ_1^d is finite; there is a lower bound of d^2 (for odd d) or $d^2 - 1$ (for d even).

8.1.3 THE OTHER p

The spaces ℓ_∞^d are the richest (and thus generally the most difficult to deal with); *every* n-point metric space (X, D) embeds isometrically in ℓ_∞^n. To see this, write $X = \{x_1, x_2, \ldots, x_n\}$ and define $f : X \to \ell_\infty^n$ by $f(x_i)_j = D(x_i, x_j)$.

The other $p \neq 1, 2, \infty$ are encountered less often, but it may be useful to know the cases where all ℓ_p metrics embed with bounded distortion in ℓ_q: This happens iff $p = q$, or $p = 2$, or $q = \infty$, or $1 \leq q \leq p \leq 2$. Isometric embeddings exist in all these cases. Moreover, for $1 \leq q \leq p \leq 2$, the whole of ℓ_p^d can be $(1+\epsilon)$ embedded in ℓ_q^{Cd} with a suitable $C = C(p, q, \epsilon)$ (so the dimension doesn't grow by much); see, e.g., [MS86]. These embeddings are probabilistic. The simplest one is $\ell_2^d \to \ell_1^{Cd}$, given by $x \mapsto Ax$ for a random ± 1 matrix A of size $Cd \times d$ (surprisingly, no good explicit embedding is known even in this case).

8.2 APPROXIMATE EMBEDDINGS OF GENERAL METRICS IN ℓ_p

8.2.1 BOURGAIN'S EMBEDDING IN ℓ_2

The mother of most embeddings mentioned in the next few sections, from both historical and "technological" points of view, is the following theorem.

THEOREM 8.2.1 Bourgain [Bou85]

Any n-point metric space (X, D) can be embedded in ℓ_2 (in fact, in every ℓ_p) with distortion $O(\log n)$.

We describe the embedding, which is constructed probabilistically. We set $m = \lfloor \log_2 n \rfloor$ and $q = \lfloor C \log n \rfloor$ (C a suitable constant) and construct an embedding in ℓ_2^{mq}, with the coordinates indexed by $i = 1, 2, \ldots, m$ and $j = 1, 2, \ldots, q$. For each such i, j, we select a subset $A_{ij} \subseteq X$ by putting each $x \in X$ into A_{ij} with probability 2^{-j}, all the random choices being mutually independent. Then we set $f(x)_{ij} = D(x, A_{ij})$. We thus obtain an embedding in $\ell_2^{O(\log^2 n)}$ (Bourgain's original proof used exponential dimension; the possibility of reducing it was noted later), and it can be shown that the distortion is $O(\log n)$ with high probability.

This yields an $O(n^2 \log n)$ randomized algorithm for computing the desired embedding. The algorithm can be derandomized (preserving the polynomial time and the dimension bound) using the method of conditional probabilities; this result seems to be folklore. Alternatively, it can be derandomized using small sample spaces [LLR95]; this, however, uses dimension $\Theta(n^2)$. Finally, as was remarked above, an embedding of a given space in ℓ_2 with optimal distortion can be computed by semidefinite programming.

The $O(\log n)$ distortion for embedding a general metric in ℓ_2 is tight [LLR95] (and similarly for ℓ_p, $p < \infty$ fixed). Examples of metrics that cannot be embedded any better are the shortest-path metrics of constant-degree expanders. (An n-vertex graph is a ***constant-degree expander*** if all degrees are bounded by some constant r and each subset of k vertices has at least βk outgoing edges, for $1 \leq k \leq \frac{n}{2}$ and for some constant $\beta > 0$ independent of n.)

Another interesting lower bound is due to Linial et al. [LMN02]: The shortest-path metric of *any* k-regular graph ($k \geq 3$) of girth g requires $\Omega(\sqrt{g})$ distortion for embedding in ℓ_2.

8.2.2 NEGATIVE TYPE METRICS

We say that a metric space (X, D) is of ***negative type*** if there exists a mapping $f : X \to L_2$ such that $\|f(x) - f(y)\|_2^2 = D(x, y)$, for all $x, y \in X$.

In [ARV09] the Goemans-Linial semidefinite programming relaxation of the sparsest cut problem was used to obtain a $O(\sqrt{\log n})$-approximation for the uniform case of the problem. Building on this work and a previous bound from [CGR05], it was shown in [ALN08] that any n-point metric space of negative type admits an embedding into ℓ_1 with distortion $O(\sqrt{\log n} \log\log n)$. This implies a $O(\sqrt{\log n} \log\log n)$-approximation for the general sparsest cut problem. In [CKN09] it was shown that there exist n-point metric spaces of negative type that require distortion $(\log n)^{\Omega(1)}$ to be embedded into ℓ_1. This also implies that the integrality of the Goemans-Linial semidefinite programming relaxation is $(\log n)^{\Omega(1)}$. Very recently, that result was improved to $\Omega(\sqrt{\log n})$, matching the upper bound [NY17].

8.2.3 THE DIMENSION OF EMBEDDINGS IN ℓ_∞

If we want to embed all n-point metrics in ℓ_∞^d, there is a tradeoff between the dimension d and the worst-case distortion. The following result was proved in [Mat96] by adapting Bourgain's technique.

THEOREM 8.2.2

For an integer $b > 0$ set $c = 2b-1$. Then any n-point metric space can be embedded in ℓ_∞^d with distortion c, where $d = O(bn^{1/b} \log n)$.

An almost matching lower bound can be proved using graphs without short cycles, an idea also going back to [Bou85]. Let $m(g, n)$ be the maximum possible number of edges of an n-vertex graph of girth $g + 1$. For every fixed $c \geq 1$ and integer $g > c$ there exists an n-point metric space such that any c-embedding in ℓ_∞^d has $d = \Omega(m(g, n)/n)$ [Mat96]. The proof goes by counting: Fix a graph G_0 witnessing $m(g, n)$, and let $\mathcal{G}$ be the set of graphs (considered with the shortest-path metric) that can be obtained from G_0 by deleting some edges. It turns out that if $G, G' \in \mathcal{G}$ are distinct, then they cannot have "essentially the same" c-embeddings in ℓ_∞^d, and there are only "few" essentially different embeddings in ℓ_∞^d if d is small.

It is easy to show that $m(g, n) = O(n^{1+1/\lfloor g/2 \rfloor})$ for all g, and this is conjectured to be the right order of magnitude [Erd64]. This has been verified for $g \leq 7$ and for $g = 10, 11$, while only worse lower bounds are known for the other values of g (with exponent roughly $1 + 4/3g$ for g large). Whenever the conjecture holds for some $g = 2b - 1$, the above theorem is tight up to a logarithmic factor for the corresponding b. Unfortunately, although explicit constructions of graphs of a given girth with many edges are known, the method doesn't provide explicit examples of badly embeddable spaces.

For special classes of metrics improved bounds on the dimension are possible. The shortest path metric of any graph that excludes some fixed minor admits a constant-distortion embedding into $\ell_\infty^{O(\log n)}$ [KLMN05].

DISTANCE ORACLES

An interesting algorithmic result, conceptually resembling the above theorem, was obtained by Thorup and Zwick [TZ01]. They showed that for an integer $b > 0$, every n-point metric space can be stored in a data structure of size $O(n^{1+1/b})$ (with preprocessing time of the same order) so that, within time $O(b)$, the distance between any two points can be approximated within a multiplicative factor of $2b-1$. Mendel and Naor [MN06] have obtained a data structure for the above problem of size $O(n^{1+1/b})$, with query time $O(1)$ that approximates the distance between any two points within a multiplicative factor of $O(b)$.

LOW DIMENSION

The other end of the tradeoff between distortion and dimension d, where d is fixed (and then all ℓ_p-norms on $\Re^d$ are equivalent up to a constant) was investigated in [Mat90]. For all fixed $d \geq 1$, there are n-point metric spaces requiring distortion $\Omega\left(n^{1/\lfloor (d+1)/2 \rfloor}\right)$ for embedding in ℓ_2^d (for $d = 2$, an example is the shortest-path metric of K_5 with every edge subdivided $n/10$ times). On the other hand, every n-point space $O(n)$-embeds in ℓ_2^1 (the real line), and $O(n^{2/d} \log^{3/2} n)$-embeds in ℓ_2^d, $d \geq 3$.

8.2.4 THE JOHNSON-LINDENSTRAUSS LEMMA: FLATTENING IN ℓ_2

The n-point ℓ_2 metric with all distances equal to 1 requires dimension $n-1$ for isometric embedding in ℓ_2. A somewhat surprising and extremely useful result shows that, in particular, this metric can be embedded in dimension only $O(\log n)$ with distortion close to 1.

THEOREM 8.2.3 Johnson and Lindenstrauss [JL84]

For every $\varepsilon > 0$, any n-point ℓ_2 metric can be $(1+\varepsilon)$-embedded in $\ell_2^{O(\log n/\epsilon^2)}$.

There is an almost matching lower bound for the necessary dimension, due to Alon (see [Mat02]): $\Omega(\log n/(\epsilon^2 \log(1/\epsilon)))$. For the special case of linear maps, a matching $\Omega(\min\{n, \log n/\epsilon^2\})$ lower bound has been obtained [LN16].

All known proofs (see, e.g., [Ach01] for references and an insightful discussion) first place the metric under consideration in ℓ_2^n and then map it into ℓ_2^d by a random linear map $A : \ell_2^n \to \ell_2^d$. Here A can be a random orthogonal projection (as in [JL84]). It can also be given by a random $n \times d$ matrix with independent $N(0,1)$ entries [IM98], or even one with independent uniform random ± 1 entries. The proof in the last case, due to [Ach01], is considerably more difficult than the previous ones (which use spherically symmetric distributions), but this version has advantages in applications.

An embedding as in the theorem can be computed deterministically in time $O(n^2 d(\log n + 1/\epsilon)^{O(1)})$ [EIO02] (also see [Siv02]).

The aforementioned embeddings use random dense $n \times d$ matrices A, which means that mapping a point p into Ap takes rectangular (i.e., $\Theta(nd)$) time. In order to reduce the running time several approaches were proposed. The *Fast Johnson-Lindenstrauss transform* approach [AC06, AL13, KW11] uses the product of a diagonal matrix, a Fourier matrix and a projection matrix, which makes it possible to evaluate the matrix-vector product Ap in sub-rectangular time. In particular, the algorithm of [KW11] runs in $O(n \log n)$ time, albeit the reduced dimension d is $O(\log(n) \log^4(d)/\epsilon^2)$.

Another approach to faster dimensionality reduction is *Feature Hashing*, also known as the *Sparse Johnson-Lindenstrauss transform* [WDL+09, DKS10, KN14]. Here, the speedup is achieved by showing that one can make the matrix A sparse without sacrificing the bound on the reduced dimension. In particular [KN14] shows a distribution over $n \times d$ matrices A that have only ϵnd non-zeros and achieve the same bound for d as given in Theorem 8.2.3.

8.2.5 IMPOSSIBILITY OF DIMENSIONALITY REDUCTION IN ℓ_1

It has been shown that no analogue of the Johnson–Lindenstrauss result holds in ℓ_1. The first result of this type was obtained by Brinkman and Charikar [BC05] and was later simplified by Lee and Naor [LN04]. The best known lower bound is due to Andoni et al. [ACNN11] who showed that for any $\epsilon > 0$ and $n > 0$ there exist some n-point subset X of ℓ_1 such that any embedding of X into ℓ_1 with distortion $1+\varepsilon$ requires dimension at least $n^{1-O(1/\log(1/\epsilon))}$.

8.2.6 VOLUME-RESPECTING EMBEDDINGS

Feige [Fei00] introduced the notion of ***volume-respecting*** embeddings in ℓ_2, with impressive algorithmic applications. While the distortion of a mapping depends only on pairs of points, the volume-respecting condition takes into account the behavior of k-tuples. For an arbitrary k-point metric space (S, D), we set $\mathrm{Vol}(S) = \sup_{\text{nonexpanding } f: S \to \ell_2} \mathrm{Evol}(f(S))$, where $\mathrm{Evol}(P)$ is the $(k-1)$-dimensional volume of the convex hull of P (in ℓ_2). Given a nonexpanding $f : X \to \ell_2$ for some metric space (X, D) with $|X| \geq k$, we define the k-distortion of f to be

$$\sup_{S \subseteq X, |S|=k} \left(\frac{\mathrm{Vol}(S)}{\mathrm{Evol}(f(S))} \right)^{1/(k-1)}.$$

If the k-distortion of f is Δ, we call f (k, Δ)-*volume-respecting.*

If $f : X \to \ell_2$ is an embedding scaled so that it is nonexpanding but just so, the 2-distortion coincides with the usual distortion. But note that for $k > 2$, the isometric "straight" embedding of a path in ℓ_2 is not volume-respecting at all. In fact, it is known that for any $k > 2$, *no* $(k, o(\sqrt{\log n}))$-volume-respecting embedding of a *line* exists [DV01].

Extending Bourgain's technique, Feige proved that for every $k > 2$, every n-point metric space has a $(k, O(\log n + \sqrt{k \log n \log k}))$-volume-respecting embedding in ℓ_2. Magen and Zouzias [MZ08] have obtained volume-respecting dimensionality reduction for finite subsets of Euclidean space. They show that any n-point subset of ℓ_2 admits a $O(k, 1+\epsilon)$-volume-respecting embedding into ℓ_2^d, for some $d = O(\max\{k/\epsilon, \epsilon^{-2} \log n\})$.

8.3 APPROXIMATE EMBEDDING OF SPECIAL METRICS IN ℓ_p

GLOSSARY

$\mathcal{G}$-metric: Let $\mathcal{G}$ be a class of graphs and let $G \in \mathcal{G}$. Each positive weight function $w : E(G) \to (0, \infty)$ defines a metric D_w on $V(G)$, namely, the shortest-path metric, where the length of a path is the sum of the weights of its edges. A metric space is a $\mathcal{G}$-metric if it is isometric to a subspace of $(V(G), D_w)$ for some $G \in \mathcal{G}$ and some w.

Tree metric, planar-graph metric: A $\mathcal{G}$-metric for $\mathcal{G}$, the class of all trees or all planar graphs, respectively.

Minor: A graph G is a minor of a graph H if it can be obtained from H by repeated deletions of edges and contractions of edges.

Doubling metric: A metric (X, D) such that for all $r > 0$ any ball of radius r can be covered by a constant number of balls of radius $r/2$.

8.3.1 TREE METRICS, PLANAR-GRAPH METRICS, AND FORBIDDEN MINORS

A major research direction has been improving Bourgain's embedding in ℓ_2 for restricted families of metric spaces.

TREE METRICS

It is easy to show that any tree metric embeds isometrically in ℓ_1. Any n-point tree metric can also be embedded isometrically in $\ell_\infty^{O(\log n)}$ [LLR95]. For ℓ_p embeddings, the situation is rather delicate:

THEOREM 8.3.1

Distortion of order $(\log\log n)^{\min(1/2,1/p)}$ *is sufficient for embedding any n-vertex tree metric in ℓ_p ($p \in (1, \infty)$ fixed)* [Mat99]*, and it is also necessary in the worst case (for the complete binary tree;* [Bou86]*).*

Gupta [Gup00] proved that any n-point tree metric $O(n^{1/(d-1)})$-embeds in ℓ_2^d ($d \geq 1$ fixed), and for $d = 2$ and trees with unit-length edges, Babilon et al. [BMMV02] improved this to $O(\sqrt{n})$. Bădoiu et al. [BCIS06] have shown that any n-point ultrametric $O(n^{1/d})$-embeds in ℓ_2^d.

PLANAR-GRAPH METRICS AND OTHER CLASSES WITH A FORBIDDEN MINOR

The following result was proved by Rao, building on the work of Klein, Plotkin, and Rao.

THEOREM 8.3.2 Rao [Rao99]

Any n-point planar-graph metric can be embedded in ℓ_2 with distortion $O(\sqrt{\log n})$. More generally, let H be an arbitrary fixed graph and let $\mathcal{G}$ be the class of all graphs not containing H as a minor; then any n-point $\mathcal{G}$-metric can be embedded in ℓ_2 with distortion $O(\sqrt{\log n})$.

This bound is tight even for series-parallel graphs (no K_4 minor) [NR02]; the example is obtained by starting with a 4-cycle and repeatedly replacing each edge by two paths of length 2.

A challenging conjecture, one that would have significant algorithmic consequences, states that under the conditions of Rao's theorem, all $\mathcal{G}$-metrics can be

c-embedded in ℓ_1 for some c depending only on $\mathcal{G}$ (but not on the number of points). Apparently, this conjecture was first published in [GNRS04], where it was verified for the forbidden minors K_4 (series-parallel graphs) and $K_{2,3}$ (outerplanar graphs). It has also been verified for graphs that exclude the 4-wheel [CJLV08] and for graphs of bounded pathwidth (equivalently, for graphs that exclude some forest) [LS13]. It has also been shown that the conjecture can be reduced to the conjunction of two apparently simpler problems: the special case of planar graphs (also known as the planar embedding conjecture) and the so-called k-sum conjecture which asserts that constant-distortion embeddability into ℓ_1 is closed under bounded clique-sums [LS09]. The latter conjecture has been verified special case of edge-sums over graphs of bounded size [LP13].

DOUBLING METRICS

It has been shown by Gupta et al. [GKL03] that any n-point doubling metric admits a $O(\sqrt{\log n})$-embedding into ℓ_2 and that there exists a doubling metric that requires distortion $\Omega(\sqrt{\log n})$. Since ℓ_2 embeds into ℓ_1 isometrically, the above upper bound also holds for embedding into ℓ_1. It has been shown that there exists an n-point doubling metric that requires distortion $\Omega(\sqrt{\log n/\log\log n})$ to be embedded into ℓ_1 [LS11].

8.3.2 METRICS DERIVED FROM OTHER METRICS

In this section we focus on metrics derived from other metrics, e.g., by defining a distance between two *sets* or *sequences* of points from the underlying metric.

GLOSSARY

Uniform metric: For any set X, the metric (X, D) is uniform if $D(p,q) = 1$ for all $p \neq q$, $p, q \in X$.

Hausdorff distance: For a metric space (X, D), the Hausdorff metric H on the set 2^X of all subsets of X is given by $H(A,B) = \min(\vec{H}(A,B), \vec{H}(B,A))$, where $\vec{H}(A,B) = \sup_{a\in A} \inf_{b\in B} D(a,b)$.

Earth-mover distance: For a metric space (X, D) and an integer $d \geq 1$, the earth-mover distance of two d-element sets $A, B \subseteq X$ is the minimum weight of a perfect matching between A and B; that is, $\min_{\text{bijective } \pi : A \to B} \sum_{a\in A} D(a, \pi(a))$.

Levenshtein distance (or ***edit distance***): For a metric space $M = (\Sigma, D)$, the distance between two strings $w, w' \in \Sigma^*$ is the minimum cost of a sequence of operations that transforms w into w'. The allowed operations are: character insertion (of cost 1), character deletion (of cost 1), or replacement of a symbol a by another symbol b (of cost $D(a,b)$), where $a, b \in \Sigma$. The total cost of the sequence of operations is the sum of all operation costs.

Fréchet distance: For a metric space $M = (X, D)$, the Fréchet distance (also called the ***dogkeeper's distance***) between two functions $f, g : [0,1] \to X$ is defined as

$$\inf_{\pi : [0,1] \to [0,1]} \sup_{t \in [0,1]} D(f(t), g(\pi(t)))$$

where π is continuous, monotone increasing, and such that $\pi(0) = 0, \pi(1) = 1$.

HAUSDORFF DISTANCE

The Hausdorff distance is often used in computer vision for comparing geometric shapes, represented as sets of points. However, even computing a single distance $H(A, B)$ is a nontrivial task. As noted in [FCI99], for any n-point metric space (X, D), the Hausdorff metric on 2^X can be isometrically embedded in ℓ_∞^n.

The dimension of the host norm can be further reduced if we focus on embedding particular Hausdorff metrics. In particular, let H_M^s be the Hausdorff metric over all s-subsets of M. Farach-Colton and Indyk [FCI99] showed that if $M = (\{1, \ldots, \Delta\}^k, \ell_p)$, then H_M^s can be embedded in $\ell_\infty^{d'}$ with distortion $1 + \epsilon$, where $d' = O(s^2(1/\epsilon)^{O(k)} \log \Delta)$. A different tradeoff was obtained in [BS16], where it is shown that for $M = (\{1, \ldots, \Delta\}^k, \ell_p)$, where k, s are constant, H_M^s embeds into $\ell_\infty^{f(k,s)}$ with $g(k, s)$ distortion for some functions f and g. For a general (finite) metric space $M = (X, D)$ they show that H_M^s can be embedded in $\ell_\infty^{s^{O(1)}|X|^\alpha \log \Delta}$ for any $\alpha > 0$ with constant distortion, where $\Delta = (\min_{p \neq q \in X} D(p, q))/(\max_{p,q \in X} D(p, q))$.

EARTH-MOVER DISTANCE (EMD)

A very interesting relation between embedding EMD in normed spaces and embeddings in probabilistic trees (discussed below in Section 8.4.1) was discovered in [Cha02b]: If a finite metric space can be embedded in a convex combination of dominating trees with distortion c, then the EMD over it can be embedded in ℓ_1 with distortion $O(c)$. Consequently, the EMD over subsets of $(\{1, \ldots, \Delta\}^k, \ell_p)$ can be embedded in ℓ_1 with distortion $O(k \log \Delta)$. Lower bounds have been proven in [KN06, NS07].

LEVENSHTEIN DISTANCE AND ITS VARIANTS

The Levenshtein distance is used in text processing and computational biology. The best algorithm computing the Levenshtein distance of two strings w, w', even approximately, has running time of order $|w| \cdot |w'|$ (for a constant-size Σ). In the simplest (but nevertheless quite common) case of the uniform metric over $\Sigma = \{0, 1\}$, Levenshtein distance over strings of length d admits a $2^{O(\sqrt{\log d \log \log d})}$-embedding into ℓ_1 [OR07]. A lower bound of $\Omega(\log d)$ has also been obtained for this case [KR09, KN06].

If we modify the definition of the distance by permitting the movement of an arbitrarily long contiguous block of characters as a single operation, and if the underlying metric is uniform, then the resulting ***block-edit*** metric can be embedded in ℓ_1 with distortion $O(\log l \cdot \log^* l)$, where l is the length of the embedded strings (see [MS00, CM02] and references therein). The modified metric has applications in computational biology and in string compression. The embedding of a given string can be computed in almost linear time, which yields a very fast approximation algorithm for computing the distance between two strings (the exact distance computation is NP-hard!).

FRÉCHET METRIC

The Fréchet metric is an interesting metric measuring the distances between two *curves*. From the applications perspective, it is interesting to investigate the case where $M = \ell_2^k$ and f, g are continuous, closed polygonal chains, consisting of (say) at most d segments each. Denote the set of such curves by C_d^k. It is not known whether C_d^k, under Fréchet distance, can be embedded in ℓ_∞ with finite dimension (for infinite dimension, an isometric embedding follows from the universality of the ℓ_∞ norm). On the other hand, it is easy to check that for any bounded set $S \subset \ell_\infty^d$, there is an isometry $f : S \to C_{3d}^1$.

8.3.3 OTHER SPECIAL METRICS

GLOSSARY

(1, 2)-B metric: A metric space (X, D) such that for any $x \in X$ the number of points y with $D(x, y) = 1$ is at most B, and all other distances are equal to 2.

Transposition distance: The (unfortunately named) metric D_T on the set of all permutations of $\{1, 2, \ldots, n\}$; $D_T(\pi_1, \pi_2)$ is the minimum number of moves of contiguous subsequences to arbitrary positions needed to transform π_1 into π_2.

BOUNDED DISTANCE METRICS

Trevisan [Tre01] considered approximate embeddings of $(1, 2)$-B metrics in ℓ_p^d (in a sense somewhat different from low-distortion embeddings). Guruswami and Indyk [GI03] proved that any $(1, 2)$-B metric can be isometrically embedded in $\ell_\infty^{O(B \log n)}$.

PERMUTATION METRICS

It was shown in [CMS01] that D_T can be $O(1)$-embedded in ℓ_1; similar results were obtained for other metrics on permutations, including reversal distance and permutation edit distance.

8.4 APPROXIMATE EMBEDDINGS IN RESTRICTED METRICS

GLOSSARY

Dominating metric: Let D, D' be metrics on the same set X. Then D' dominates D if $D(x, y) \geq D'(x, y)$ for all $x, y \in X$.

Convex combination of metrics: Let X be a set, $T_1, T_2, \ldots, T_k$ metrics on it, and $\alpha_1, \ldots, \alpha_k$ nonnegative reals summing to 1. The convex combination of the T_i (with coefficients α_i) is the metric D given by $D(x, y) = \sum_{i=1}^k \alpha_i T_i(x, y)$, $x, y \in X$.

Hierarchically well-separated tree (k-HST): A 1-HST is exactly an *ultrametric*; that is, the shortest-path metric on the leaves of a rooted tree T (with weighted edges) such that all leaves have the same distance from the root. For a k-HST with $k > 1$ we require that, moreover, $\Delta(v) \leq \Delta(u)/k$ whenever v is a child of u in T, where $\Delta(v)$ denotes the diameter of the subtree rooted at v (w.l.o.g. we may assume that each non-leaf has degree at least 2, and so $\Delta(v)$

equals the distance of v to the nearest leaves). *Warning:* This is a newer definition introduced in [BBM06]. Older papers, such as [Bar96, Bar98], used another definition, but the difference is merely technical, and the notion remains essentially the same.

8.4.1 PROBABILISTIC EMBEDDINGS IN TREES

A convex combination $\overline{D} = \sum_{i=1}^{r} \alpha_i T_i$ of some metrics $T_1, \ldots, T_r$ on X can be thought of as a ***probabilistic metric*** (this concept was suggested by Karp). Namely, $\overline{D}(x,y)$ is the expectation of $T_i(x,y)$ for $i \in \{1, 2, \ldots, r\}$ chosen at random according to the distribution given by the α_i. Of particular interest are embeddings in convex combinations of *dominating* metrics. The domination requirement is crucial for many applications. In particular, it enables one to solve many problems over the original metric (X, D) by solving them on a (simple) metric chosen at random from $T_1, \ldots, T_r$ according to the distribution defined by the α_i.

The usefulness of probabilistic metrics comes from the fact that a sum of metrics is much more powerful than each individual metric. For example, it is not difficult to show that there are metrics (e.g., cycles [RR98, Gup01]) that cannot be embedded in tree metrics with $o(n)$ distortion. In contrast, we have the following result:

THEOREM 8.4.1 Fakcharoenphol, Rao, and Talwar [FRT03]

Let (X, D) be any n-point metric space. For every $k > 1$, there exist a natural number r, k-HST metrics $T_1, T_2, \ldots, T_r$ on X, and coefficients $\alpha_1, \ldots, \alpha_r > 0$ summing to 1, such that each T_i dominates D and the (identity) embedding of (X, D) in $(X, \overline{D})$, where $\overline{D} = \sum_{i=1}^{r} \alpha_i T_i$, has distortion $O(k \log n)$.

The first result of this type was obtained by Alon et al [AKPW95]. Their embedding has distortion $2^{O(\sqrt{\log n \log\log n})}$. A few years later Bartal [Bar96] improved the distortion bound considerably, to $O(\log^2 n)$ and later even to $O(\log n \log\log n)$ [Bar98]. Finally, Fakcharoenphol, Rao, and Talwar [FRT03] obtained an embedding with distortion $O(\log n)$. This embeddings uses convex combination of very simple tree metrics (i.e., HST's), which further simplifies the design of algorithms. The $O(\log n)$ distortion is the best possible in general (since any convex combination of tree metrics embeds isometrically in ℓ_1). Embedding the $\sqrt{n} \times \sqrt{n}$ grid into a convex combination of tree metrics requires distortion $\Omega(\log n)$ [AKPW95].

The constructions in [Bar96, Bar98, FRT03] generate trees with Steiner nodes (i.e., nodes that do not belong to X). However, one can get rid of such nodes in *any* tree while increasing the distortion by at most 8 [Gup01]. A lower bound of $8 - o(1)$ on the distortion has also been obtained [CXKR06]. The problem of removing Steiner nodes has also been considered for the case of general graphs. It has been shown that for any graph G with edge weights w and $T \subseteq V(G)$ there exists a some graph G' with edge weights w' and $V(G) = T$ such that G' is isomorphic to a minor of G and for all $u, v \in T$, $D_w(u,v) \le D_{w'}(u,v) \le O(\log^5 |T|) D_w(u,v)$ [KKN15].

An interesting extra feature of the construction of Alon et al. mentioned above is that if the metric D is given as the shortest-path metric of a (weighted) graph G on the vertex set X, then all the T_i are spanning trees of this G. None of the constructions in [Bar96, Bar98, FRT03] share this property. However, more recent work [EEST08] showed an embedding with distortion $O(\log^2 n \log\log n)$, which was later improved to $O(\log n \log\log n (\log\log\log n)^3)$ [ABN08].

The embedding algorithms in Bartal's papers [Bar96, Bar98] are randomized and run in polynomial time. A deterministic algorithm for the same problem was given in [CCG+98]. The latter algorithm constructs a distribution over $O(n \log n)$ trees (the number of trees in Bartal's construction was exponential in n).

Probabilistic embeddings into other classes of graph metrics have also been considered. It has been shown that the metric of any graph of genus $g > 0$ admits an embedding into a convex combination of planar graph metrics with distortion $O(\log(g+1))$ [Sid10]. Similarly, the metric of any graph of bounded genus and with a bounded number of apices admits a $O(1)$-embedding into a convex combination of planar graph metrics [LS09]. In contrast, it has been shown that for any $k \geq 2$ there exist graphs of treewidth $k+1$ such that any embedding into a convex combination of graphs of treewidth k has distortion $\Omega(\log n)$ [CJLV08].

For general graphs the following probabilistic embedding has also been obtained: Let G be a graph with edge weights w and let $T \subseteq V(G)$. Then there exists a $O(\log n \log\log n)$-embedding of (T, D_w) into a convex combination of graph metrics $H_1, \ldots, H_r$ on T where each H_i is minor of G [EGK+10].

8.4.2 RAMSEY-TYPE THEOREMS

Many Ramsey-type questions can be asked in connection with low-distortion embeddings of metric spaces. For example, given classes $\mathcal{X}$ and $\mathcal{Y}$ of finite metric spaces, one can ask whether for every n-point space $Y \in \mathcal{Y}$ there is an m-point $X \in \mathcal{X}$ such that X can be α-embedded in Y, for given n, m, α.

Important results were obtained in [BBM06], and later greatly improved and extended in [BLMN03], for $\mathcal{X}$ the class of all k-HST and $\mathcal{Y}$ the class of all finite metric spaces; they were used for a lower bound in a significant algorithmic problem (metrical task systems). Let us quote some of the numerous results of Bartal et al.:

THEOREM 8.4.2 Bartal, Linial, Mendel, and Naor [BLMN03]

Let $R_{\mathrm{UM}}(n, \alpha)$ denote the largest m such that for every n-point metric space Y there exists an m-point 1-HST (i.e., ultrametric) that α-embeds in Y, and let $R_2(n, \alpha)$ be defined similarly with "ultrametric" replaced with "Euclidean metric."

(i) *There are positive constants C, C_1, c such that for every $\alpha > 2$ and all n,*

$$n^{1-C_1(\log \alpha)/\alpha} \leq R_{\mathrm{UM}}(n, \alpha) \leq R_2(n, \alpha) \leq C n^{1-c/\alpha}.$$

(ii) *(Sharp threshold at distortion 2) For every $\alpha > 2$, there exists $c(\alpha) > 0$ such that $R_2(n, \alpha) \geq R_{\mathrm{UM}}(n, \alpha) \geq n^{c(\alpha)}$ for all n, while for every $\alpha \in (1, 2)$, we have $c'(\alpha) \log n \leq R_{\mathrm{UM}}(n, \alpha) \leq R_2(n, \alpha) \leq 2 \log n + C'(\alpha)$ for all n, with suitable positive $c'(\alpha)$ and $C'(\alpha)$.*

For embedding a k-HST in a given space, one can use the fact that every ultrametric is k-equivalent to a k-HST. For an earlier result similar to the second part of (ii), showing that the largest Euclidean subspace $(1+\varepsilon)$-embeddable in a general n-point metric space has size $\Theta(\log n)$ for all sufficiently small fixed $\varepsilon > 0$, see [BFM86].

TABLE 8.4.1 A summary of approximate embeddings

FROM	TO	DISTORTION	REFERENCE
any	ℓ_p, $1 \leq p < \infty$	$O(\log n)$	[Bou85]
constant-degree expander	ℓ_p, $p < \infty$ fixed	$\Omega(\log n)$	[LLR95]
k-reg. graph, $k \geq 3$, girth g	ℓ_2	$\Omega(\sqrt{g})$	[LMN02]
any	$\ell_\infty^{O(bn^{1/b} \log n)}$	$2b-1$, $b=1,2,\ldots$	[Mat96]
some	$\Omega(n^{1/b})$-dim'l. normed space	$2b-1$, $b=1,2,\ldots$ (Erdős's conj.!)	[Mat96]
any	ℓ_1^1	$\Theta(n)$	[Mat90]
any	ℓ_p^d, d fixed	$O(n^{2/d} \log^{3/2} n)$, $\Omega\left(n^{1/\lfloor (d+1)/2 \rfloor}\right)$	[Mat90]
ℓ_2 metric	$\ell_2^{O(\log n/\epsilon^2)}$	$1+\epsilon$	[JL84]
ℓ_1 metric	$\ell_1^{n^{1-O(1/\log(1/\epsilon))}}$	$1+\epsilon$	[ACNN11]
planar or forbidden minor	ℓ_2	$O(\sqrt{\log n})$	[Rao99]
series-parallel	ℓ_2	$\Omega(\sqrt{\log n})$	[NR02]
planar or forbidden minor	$\ell_\infty^{O(\log n)}$	$O(1)$	[KLMN05]
outerplanar or series-parallel	ℓ_1	$O(1)$	[GNRS04]
tree	ℓ_1	1	(folklore)
tree	$\ell_1^{c(\epsilon) \log n}$	$1+\epsilon$	[LMM13]
bounded pathwidth	ℓ_1	$O(1)$	[LS09]
tree	ℓ_2	$\Theta((\log \log n)^{1/2})$	[Bou86, Mat99]
tree	ℓ_2^d	$O(n^{1/(d-1)})$	[Gup00]
outerplanar, unit edges	ℓ_2^2	$\Theta(n^{1/2})$	[BMMV02, BDHM07]
planar	ℓ_2^2	$\Omega(n^{2/3})$	[BDHM07]
ultrametric	ℓ_2^d	$O(n^{1/d})$	[BCIS06]
doubling metric	ℓ_2	$\Theta(\sqrt{\log n})$	[GKL03]
doubling metric	ℓ_1	$\Omega(\sqrt{\log n / \log \log n})$	[LS11]
Hausdorff metric over (X, D)	$\ell_\infty^{\lvert X \rvert}$	1	[FCI99]
Hausd. over s-subsets of (X, D)	$\ell_\infty^{s^{O(1)} \lvert X \rvert^\alpha \log \Delta}$	$c(\alpha)$	[FCI99]
Hausd. over s-subsets of ℓ_p^k	$\ell_\infty^{s^2 (1/\epsilon)^{O(k)} \log \Delta}$	$1+\epsilon$	[FCI99]
Hausd. over s-subsets of ℓ_p^k	$\ell_\infty^{f(k,s)}$	$g(k, s)$	[BS16]
EMD over (X, D)	ℓ_1	$O(\log \lvert X \rvert)$	[Cha02b]
EMD over $\{0 \ldots n\}^2$	ℓ_1	$\Omega(\sqrt{\log n})$	[NS07]
EMD over $\{0, 1\}^k$	ℓ_1	$\Omega(k)$	[KN06]
Levenshtein metric over $\{0, 1\}^d$	ℓ_1	$2^{O(\sqrt{\log d \log \log d})}$	[OR07]
Levenshtein metric over $\{0, 1\}^d$	ℓ_1	$\Omega(\log d)$	[KR09]
block-edit metric over Σ^d	ℓ_1	$O(\log d \cdot \log^* d)$	[MS00, CM02]
(1,2)-B metric	$\ell_\infty^{O(B \log n)}$	1	[GI03]; for ℓ_p cf. [Tre01]
any	convex comb. of dom. trees (HSTs)	$O(\log n)$	[Bar98, FRT03]
genus-g graph	convex comb. of planar graphs	$O(\log(g+1))$	[Sid10]
any	convex comb. of spanning trees	$O(\log n \log \log n (\log \log \log n)^3)$	[ABN08]

8.4.3 APPROXIMATION BY SPARSE GRAPHS

GLOSSARY

t-Spanner: A subgraph H of a graph G (possibly with weighted edges) is a *t-spanner* of G if $D_H(u,v) \le t \cdot D_G(u,v)$ for every $u, v \in V(G)$.

Sparse spanners are useful as a more economic representation of a given graph (note that if H is a t-spanner of G, then the identity map $V(G) \to V(H)$ is a t-embedding).

THEOREM 8.4.3 Althöfer et al. [ADD+93]

For every integer $t \ge 2$, every n-vertex graph G has a t-spanner with at most $m(t,n)$ edges, where $m(g,n) = O(n^{1+1/\lfloor g/2 \rfloor})$ is the maximum possible number of edges of an n-vertex graph of girth $g+1$.

The proof is extremely simple: Start with empty H, consider the edges of G one by one from the shortest to the longest, and insert each edge into the current H unless it creates a cycle with at most t edges. It is also immediately seen that the bound $m(t,n)$ is the best possible in the worst case.

Rabinovich and Raz [RR98] proved that there are (unweighted) n-vertex graphs G that cannot be t-embedded in graphs (possibly weighted) with fewer than $m(\Omega(t), n)$ edges (for t sufficiently large and n sufficiently large in terms of t). Their main tool is the following lemma, proved by elementary topological considerations: If H is a simple unweighted connected n-vertex graph of girth g and G is a (possibly weighted) graph on at least n vertices with $\chi(G) < \chi(H)$, then H cannot be c-embedded in G for $c < g/4 - 3/2$; here $\chi(G)$ denotes the ***Euler characteristic*** of a graph G, which, for G connected, equals $|E(G)| - |V(G)| + 1$.

Euclidean spanners are spanners of the complete graph on a given point set in ℓ_2^d with edge weights given by the Euclidean distances. Many papers were devoted to computing sparse Euclidean spanners in small (fixed) dimensions. A strong result, subsuming most of the previous work, is due to Arya et al. [ADM+95]: For any fixed $\varepsilon > 0$ and $d > 0$ and for any n-point set in ℓ_2^d, a $(1+\varepsilon)$-spanner of maximum degree $O(1)$ can be computed in $O(n \log n)$ time. Improved spanner constructions have also been obtained in Euclidean space of arbitrary dimension: Any n-point set in ℓ_2 admits a $O(\sqrt{\log n})$-spanner with $O(n \log n \log \log n)$ edges [HPIS13].

8.5 OPEN PROBLEMS AND WORK IN PROGRESS

Since the first edition of this chapter (2002) many of the problems in the area of low-distortion embeddings of metric spaces have been resolved and have given rise to several new algorithmic tools. Many of the most challenging problems remain open. Instead of stating open problems here, we refer to a list compiled by the second author [MN11] which is available on the Web.

8.6 SOURCES AND RELATED MATERIAL

Discrete metric spaces have been studied from many different points of view, and the area is quite wide and diverse. The low-distortion embeddings treated in this chapter constitute only one particular (although very significant) direction. For recent results in some other directions the reader may consult [Cam00, DDL98, DD96], for instance. For more detailed overviews of the topics surveyed here, with many more references, the reader is referred to [Mat02][Chapter 15] (including proofs of basic results) and [Ind01] (with emphasis on algorithmic applications). Approximate embeddings of normed spaces are treated, e.g., in [MS86]. A general reference for isometric embeddings, especially embeddings in ℓ_1, is [DL97].

RELATED CHAPTERS

Chapter 32: Proximity algorithms
Chapter 43: Nearest neighbors in high-dimensional spaces

REFERENCES

[ABN08] I. Abraham, Y. Bartal, and O. Neiman. Nearly tight low stretch spanning trees. In *Proc. 49th IEEE Sympos. Found. Comp. Sci.*, pages 781–790, 2008.

[AC06] N. Ailon and B. Chazelle. Approximate nearest neighbors and the fast Johnson-Lindenstrauss transform. In *Proc. 38th ACM Sympos. Theory Comput.*, pages 557–563, 2006.

[Ach01] D. Achlioptas. Database-friendly random projections. In *Proc. 20th ACM Sympos. Principles of Database Sys.*, pages 274–281, 2001.

[ACNN11] A. Andoni, M.S. Charikar, O. Neiman, and H.L. Nguyen. Near linear lower bound for dimension reduction in ℓ_1. In *Proc. 52nd IEEE Sympos. Found. Comp. Sci.*, pages 315–323, 2011.

[ADD$^+$93] I. Althöfer, G. Das, D.P. Dobkin, D. Joseph, and J. Soares. On sparse spanners of weighted graphs. *Discrete Comput. Geom.*, 9:81–100, 1993.

[ADM$^+$95] S. Arya, G. Das, D.M. Mount, J.S. Salowe, and M. Smid. Euclidean spanners: Short, thin, and lanky. In *Proc. 27th ACM Sympos. Theory Comput.*, pages 489–498, 1995.

[AKPW95] N. Alon, R.M. Karp, D. Peleg, and D. West. A graph-theoretic game and its application to the k-server problem. *SIAM J. Comput.*, 24:78–100, 1995.

[AL13] N. Ailon and E. Liberty. An almost optimal unrestricted fast Johnson-Lindenstrauss transform. *ACM Trans. Algorithms*, 9:21, 2013.

[ALN08] S. Arora, J.R. Lee, and A. Naor. Euclidean distortion and the sparsest cut. *J. Amer. Math. Soc.*, 21:1–21, 2008.

[ARV09] S. Arora, S. Rao, and U. Vazirani. Expander flows, geometric embeddings and graph partitioning. *J. ACM*, 56:5, 2009.

[Bar96] Y. Bartal. Probabilistic approximation of metric spaces and its algorithmic applications. In *Proc. 37th IEEE Sympos. Found. Comp. Sci.*, pages 184–193, 1996.

[Bar98] Y. Bartal. On approximating arbitrary metrics by tree metrics. In *Proc. 30th ACM Sympos. Theory Comput*, pages 161–168, 1998.

[BBM06] Y. Bartal, B. Bollobás, and M. Mendel. Ramsey-type theorems for metric spaces with applications to online problems. *J. Comput. Syst. Sci.*, 72:890–921, 2006.

[BC96] H.-J. Bandelt and V. Chepoi. Embedding metric spaces in the rectilinear plane: a six-point criterion. *Discrete Comput. Geom*, 15:107–117, 1996.

[BC05] B. Brinkman and M. Charikar. On the impossibility of dimension reduction in ℓ_1. *J. ACM*, 52:766–788, 2005.

[BCIS06] M. Bădoiu, J. Chuzhoy, P. Indyk, and A. Sidiropoulos. Embedding ultrametrics into low-dimensional spaces. In *Proc. 22nd Sympos. Comput. Geom.*, pages 187–196, ACM Press, 2006.

[BD92] H.-J. Bandelt and A.W.M. Dress. A canonical decomposition theory for metrics on a finite set. *Adv. Math.*, 92:47–105, 1992.

[BDHM07] M.H. Bateni, E.D. Demaine, M.T. Hajiaghayi, and M. Moharrami. Plane embeddings of planar graph metrics. *Discrete Comput. Geom.*, 38:615–637, 2007.

[BFM86] J. Bourgain, T. Figiel, and V. Milman. On Hilbertian subsets of finite metric spaces. *Israel J. Math.*, 55:147–152, 1986.

[BLMN03] Y. Bartal, N. Linial, M. Mendel, and A. Naor. On metric Ramsey-type phenomena. In *Proc. 35th ACM Sympos. Theory Comput.*, pages 463—472, 2003.

[BMMV02] R. Babilon, J. Matoušek, J. Maxová, and P. Valtr. Low-distortion embeddings of trees. In *Proc. 9th Sympos. Graph Drawing*, vol. 2265 of *LNCS*, pages 343–351, Springer, Berlin, 2002.

[Bou85] J. Bourgain. On Lipschitz embedding of finite metric spaces in Hilbert space. *Israel J. Math.*, 52:46–52, 1985.

[Bou86] J. Bourgain. The metrical interpretation of superreflexivity in Banach spaces. *Israel J. Math.*, 56:222–230, 1986.

[BS16] A. Bačkurs and A. Sidiropoulos. Constant-distortion embeddings of Hausdorff metric into constant-dimensional ℓ_p spaces. In *Proc. 19th Workshop on Approx. Alg. Combin. Opt.*, vol. 60 of *LIPIcs*, article 1, Schloss Dahstuhl, 2016.

[Cam00] P. Cameron, editor. *Discrete Metric Spaces.* Academic Press, London, 2000.

[CCG$^+$98] M. Charikar, C. Chekuri, A. Goel, S. Guha, and S.A. Plotkin. Approximating a finite metric by a small number of tree metrics. In *Proc. 39th IEEE Sympos. Found. Comp. Sci.*, pages 379–388, 1998.

[CGR05] S. Chawla, A. Gupta, and H. Räcke. Embeddings of negative-type metrics and an improved approximation to generalized sparsest cut. In *Proc. 16th ACM-SIAM Sympos. Discrete Algorithms*, pages 102–111, 2005.

[Cha02] M. Charikar. Similarity estimation techniques from rounding. In *Proc. 34th ACM Sympos. Theory Comput.*, pages 380–388, 2002.

[CJLV08] A. Chakrabarti, A. Jaffe, J.R. Lee, and J. Vincent. Embeddings of topological graphs: Lossy invariants, linearization, and 2-sums. In *Proc. 49th IEEE Sympos. Found. Comp. Sci.*, pages 761–770, 2008.

[CKN09] J. Cheeger, B. Kleiner, and A. Naor. A $(\log n)^{\Omega(1)}$ integrality gap for the sparsest cut sdp. In *Proc. 50th IEEE Sympos. Found. Comp. Sci.*, pages 555–564, 2009.

[CM02] G. Cormode and S. Muthukrishnan. The string edit distance matching problem with moves. In *Proc. 13th ACM-SIAM Sympos. Discrete Algorithms*, pages 667–676, 2002.

[CMS01] G. Cormode, M. Muthukrishnan, and C. Sahinalp. Permutation editing and matching via embeddings. In *Proc. 28th Internat. Coll. Automata, Languages and Prog.*, vol. 2076 of *LNCS*, pages 481–492, Springer, Berlin, 2001.

[CXKR06] T.-H.H. Chan, D. Xia, G. Konjevod, and A. Richa. A tight lower bound for the Steiner point removal problem on trees. In *Proc. 9th Workshop on Approx. Alg. Combin. Opt.*, vol. 4110 of *LNCS*, pages 70–81, Springer, Berlin, 2006.

[DD96] W. Deuber and M. Deza, editors. *Discrete Metric Spaces.* Academic Press, London, 1996.

[DDL98] W. Deuber, M. Deza, and B. Leclerc, editors. *Discrete Metric Spaces.* North-Holland, Amsterdam, 1998.

[Djo73] D.Z. Djoković. Distance-preserving subgraphs of hypercubes. *J. Combin. Theory Ser. B*, 14:263–267, 1973.

[DKS10] A. Dasgupta, R. Kumar, and T. Sarlós. A sparse Johnson-Lindenstrauss transform. In *Proc. 42nd ACM Sympos. Theory Computing*, pages 341–350, 2010.

[DL97] M.M. Deza and M. Laurent. *Geometry of Cuts and Metrics.* Vol. 15 of *Algorithms and Combin.*, Springer, Berlin, 1997.

[DV01] J. Dunagan and S. Vempala. On Euclidean embeddings and bandwidth minimization. *Proc. 5th Workshop on Randomization and Approximation*, vol. 2129 of *LNCS*, pages 229–240, Springer, Berlin, 2001.

[EEST08] M. Elkin, Y. Emek, D.A. Spielman, and S.-H. Teng. Lower-stretch spanning trees. *SIAM J. Comput.*, 38:608–628, 2008.

[EGK$^+$10] M. Englert, A. Gupta, R. Krauthgamer, H. Räcke, I. Talgam-Cohen, and K. Talwar. Vertex sparsifiers: New results from old techniques. In *proc. 13th Workshop on Approx. Alg. Combin. Opt.*, vol. 6302 of *LNCS*, pages 152–165, Springer, Berlin. 2010.

[EIO02] L. Engebretsen, P. Indyk, and R. O'Donnell. Derandomized dimensionality reduction with applications. In *Proc. 13th ACM-SIAM Sympos. Discrete Algorithms*, pages 705–712, 2002.

[Erd64] P. Erdős. Extremal problems in graph theory. in *Theory of Graphs and Its Applications (Proc. Sympos. Smolenice, 1963)*, pages 29–36, 1964.

[FCI99] M. Farach-Colton and P. Indyk. Approximate nearest neighbor algorithms for Hausdorff metrics via embeddings. In *Proc. 40th IEEE Sympos. Found. Comp. Sci.*, pages 171–180, 1999.

[Fei00] U. Feige. Approximating the bandwidth via volume respecting embeddings. *J. Comput. Syst. Sci.*, 60:510–539, 2000.

[FRT03] J. Fakcharoenphol, S. Rao, and K. Talwar. A tight bound on approximating arbitrary metrics by tree metrics. In *Proc. 35th ACM Sympos. Theory Comput.*, pages 448–455, 2003.

[GI03] V. Guruswami and P. Indyk. Embeddings and non-approximability of geometric problems. In *Proc. 14th ACM-SIAM Sympos. Discrete Algorithms*, pages 537–538, 2003.

[GKL03] A. Gupta, R. Krauthgamer, and J.R. Lee. Bounded geometries, fractals, and low-distortion embeddings. In *Proc. 44th IEEE Sympos. Found. Comp. Sci.*, pages 534–543, 2003.

[GNRS04] A. Gupta, I. Newman, Y. Rabinovich, and A. Sinclair. Cuts, trees and ℓ_1-embeddings of graphs. *Combinatorica*, 24:233–269, 2004.

[Gup00] A. Gupta. Embedding tree metrics into low dimensional Euclidean spaces. *Discrete Comput. Geom.*, 24:105–116, 2000.

[Gup01] A. Gupta. Steiner nodes in trees don't (really) help. In *Proc. 12th ACM-SIAM Sympos. Discrete Algorithms*, pages 220–227, 2001.

[HPIS13] S. Har-Peled, P. Indyk, and A. Sidiropoulos. Euclidean spanners in high dimensions. In *Proc. 24th ACM-SIAM Sympos. Discrete Algorithms*, pages 804–809, 2013.

[IM98] P. Indyk and R. Motwani. Approximate nearest neighbors: Towards removing the curse of dimensionality. In *Proc. 30th ACM Sympos. Theory Comput.*, pages 604–613, 1998.

[Ind01] P. Indyk. Algorithmic applications of low-distortion embeddings. In *Proc. 42nd IEEE Sympos. Found. Comp. Sci.*, pages 10–33, 2001.

[JL84] W.B. Johnson and J. Lindenstrauss. Extensions of Lipschitz mappings into a Hilbert space. *Contemp. Math.*, 26:189–206, 1984.

[KKN15] L. Kamma, R. Krauthgamer, and H.L. Nguyên. Cutting corners cheaply, or how to remove Steiner points. *SIAM J. Comput.*, 44:975–995, 2015.

[KLMN05] R. Krauthgamer, J.R. Lee, M. Mendel, and A. Naor. Measured descent: A new embedding method for finite metrics. *Geom. Funct. Anal.*, 15:839–858, 2005.

[KN06] S. Khot and A. Naor. Nonembeddability theorems via Fourier analysis. *Math. Ann.*, 334:821–852, 2006.

[KN14] D.M. Kane and J. Nelson. Sparser Johnson-Lindenstrauss transforms. *J. ACM*, 61:4, 2014.

[KR09] R. Krauthgamer and Y. Rabani. Improved lower bounds for embeddings into L_1. *SIAM J. Comput.*, 38:2487–2498, 2009.

[KW11] F. Krahmer and R. Ward. New and improved Johnson-Lindenstrauss embeddings via the restricted isometry property. *SIAM J. Math. Anal.*, 43:1269–1281, 2011.

[LMM13] J.R. Lee, A. de Mesmay, and M. Moharrami. Dimension reduction for finite trees in ℓ_1. *Discrete Comput. Geom.*, 50:977–1032, 2013.

[LLR95] N. Linial, E. London, and Y. Rabinovich. The geometry of graphs and some its algorithmic applications. *Combinatorica*, 15:215–245, 1995.

[LMN02] N. Linial, A. Magen, and N. Naor. Euclidean embeddings of regular graphs—the girth lower bound. *Geom. Funct. Anal.*, 12:380–394, 2002.

[LN04] J.R. Lee and A. Naor. Embedding the diamond graph in Lp and dimension reduction in L_1. *Geom. Funct. Anal.*, 14:745–747, 2004.

[LN16] K.G. Larsen and J. Nelson. The Johnson-Lindenstrauss lemma is optimal for linear dimensionality reduction. In *Proc. 43rd Int. Coll. Automata, Languages, and Progr.*, vol. 55 of *LIPIcs*, article 82, Schloss Daghstuhl, 2016.

[LP13] J.R. Lee and D.E. Poore. On the 2-sum embedding conjecture. In *Proc. 29th Sympos. Comput. Geom.*, pages 197–206, ACM Press, 2013.

[LS09] J.R. Lee and A. Sidiropoulos. On the geometry of graphs with a forbidden minor. In *Proc. 41st ACM Sympos. Theory Comput.*, pages 245–254, 2009.

[LS11] J.R. Lee and A. Sidiropoulos. Near-optimal distortion bounds for embedding doubling spaces into L_1. In *Proc. 43rd ACM Sympos. Theory Comput*, pages 765–772, 2011.

[LS13] J.R. Lee and A. Sidiropoulos. Pathwidth, trees, and random embeddings. *Combinatorica*, 33:349–374, 2013.

[Mat90] J. Matoušek. Bi-Lipschitz embeddings into low-dimensional Euclidean spaces. *Comment. Math. Univ. Carolinae*, 31:589–600, 1990.

[Mat96] J. Matoušek. On the distortion required for embedding finite metric spaces into normed spaces. *Israel J. Math.*, 93:333–344, 1996.

[Mat99] J. Matoušek. On embedding trees into uniformly convex Banach spaces. *Israel J. Math*, 114:221–237, 1999.

[Mat02] J. Matoušek. *Lectures on Discrete Geometry*. Springer, New York, 2002.

[MN06] M. Mendel and A. Naor. Ramsey partitions and proximity data structures. In *Proc. 47th IEEE Sympos. Found. Comp. Sci.*, pages 109–118, 2006.

[MN11] J. Matoušek and A. Naor, editors. Open problems on embedding finite metric spaces. KAM Series (Tech. Report), Dept. Applied Mathematics, Charles University, Prague; revised version at `kam.mff.cuni.cz/~matousek/metrop.ps`, 2011.

[MS86] V.D. Milman and G. Schechtman. *Asymptotic Theory of Finite Dimensional Normed Spaces*. Vol. 1200 of *Lecture Notes in Math.*, Springer-Verlag, Berlin, 1986.

[MS00] S. Muthukrishnan and C. Sahinalp. Approximate nearest neighbors and sequence comparison with block operations. In *Proc. 32nd ACM Sympos. Theory Comput.*, pages 416–424, 2000.

[MS10] J. Matoušek and A. Sidiropoulos. Inapproximability for metric embeddings into $\mathbb{R}^d$. *Trans. Amer. Math. Soc.*, 362:6341–6365, 2010.

[MZ08] A. Magen and A. Zouzias. Near optimal dimensionality reductions that preserve volumes. In *Proc. 11th Workshop Approx. Alg. Combin. Opt.*, vol. 5171 of *LNCS*, pages 523–534, Springer, Berlin, 2008.

[NR02] I. Newman and Y. Rabinovich. A lower bound on the distortion of embedding planar metrics into Euclidean space. *Discrete Comput. Geom.*, 29:77–81, 2002.

[NS07] A. Naor and G. Schechtman. Planar earthmover is not in L_1. *SIAM J. Comput.*, 37:804–826, 2007.

[NY17] A. Naor and R. Young. Vertical perimeter versus horizontal perimeter. Preprint, arXiv:1701.00620, 2017.

[OR07] R. Ostrovsky and Y. Rabani. Low distortion embeddings for edit distance. *J. ACM*, 54:23, 2007.

[Rao99] S. Rao. Small distortion and volume respecting embeddings for planar and Euclidean metrics. In *Proc. 15th Sympos. Comput. Geom.*, pages 300–306, ACM Press, 1999.

[RR98] Y. Rabinovich and R. Raz. Lower bounds on the distortion of embedding finite metric spaces in graphs. *Discrete Comput. Geom.*, 19:79–94, 1998.

[Sch38] I.J. Schoenberg. Metric spaces and positive definite functions. *Trans. Amer. Math. Soc.*, 44:522–53, 1938.

[Shp93] S.V. Shpectorov. On scale embeddings of graphs into hypercubes. *European J. Combin.*, 14:117–130, 1993.

[Sid10] A. Sidiropoulos. Optimal stochastic planarization. In *Proc. 51th IEEE Sympos. Found. Comp. Sci.*, pages 163–170, 2010.

[Siv02] D. Sivakumar. Algorithmic derandomization from complexity theory. In *Proc. 34th ACM Sympos. Theory Comput.*, pages 619–626, 2002.

[Tre01] L. Trevisan. When Hamming meets Euclid: The approximability of geometric TSP and MST. *SIAM J. Comput.*, 30:475–485, 2001.

[TZ01] M. Thorup and U. Zwick. Approximate distance oracles. In *Proc. 33rd ACM Sympos. Theory Comput.*, pages 183–192, 2001.

[WDL+09] K. Weinberger, A. Dasgupta, J. Langford, A. Smola, and J. Attenberg. Feature hashing for large scale multitask learning. In *Proc. 26th ACM Conf. Machine Learning*, pages 1113–1120, 2009.

9 GEOMETRY AND TOPOLOGY OF POLYGONAL LINKAGES

Robert Connelly and Erik D. Demaine

INTRODUCTION

There is a long and involved history of linkages starting at least in the nineteenth century with the advent of complicated and intricate mechanical engineering. Some of these practical problems led to interesting, nontrivial geometric problems, and in modern mathematics there has been significant progress on even very basic questions. Over the years, several different points of view have been taken by various groups of people. We will attempt to survey these different perspectives and results obtained in the exploration of linkages.

9.1 MATHEMATICAL THEORY OF LINKAGES

The underlying principles and definitions are mathematical and in particular geometric. Despite the long history of kinematics, even theoretical kinematics (see, e.g., Bottema and Roth [BR79a]), only since the 1970s does there seem to be any systematic attempt to explore the mathematical and geometric foundations of a theory of linkages.

We begin with some definitions, some of which follow those in rigidity theory described in Chapter 61. The rough, intuitive notions are as follows. A *linkage* is a graph with assigned edge lengths, and we distinguish three special types of linkages: arcs, cycles, and trees. A *configuration* realizes a linkage in Euclidean space, a *reconfiguration* (or *flex*) is a continuum of such configurations, and the *configuration space* embodies all configurations, with paths in the space corresponding to reconfigurations. The configuration space can be considered either allowing or disallowing bars to intersect each other.

GLOSSARY

Bar linkage or ***linkage:*** A graph $G = (V, E)$ and an assignment $\ell : E \to \mathbb{R}^+$ of positive real *lengths* to edges.

Vertex or ***joint:*** A vertex of a linkage.

Bar or ***link:*** An edge e of a linkage, which has a specified fixed length $\ell(e)$.

Polygonal arc: A linkage whose underlying graph is a single path. (Also called an ***open chain*** or a ***ruler***.)

Polygonal cycle: A linkage whose underlying graph is a single cycle. (Also called a ***closed chain*** or a ***polygon***.)

FIGURE 9.1.1
Different types of linkages, according to whether the underlying graph is a path, cycle, or tree, or the graph is arbitrary.

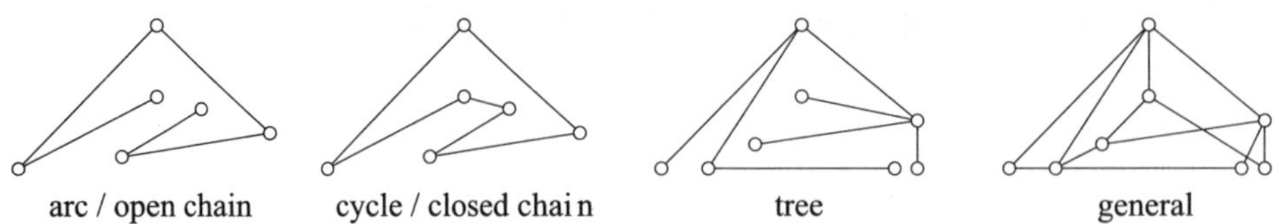

FIGURE 9.1.2
Snapshots of a reconfiguration of a polygonal arc.

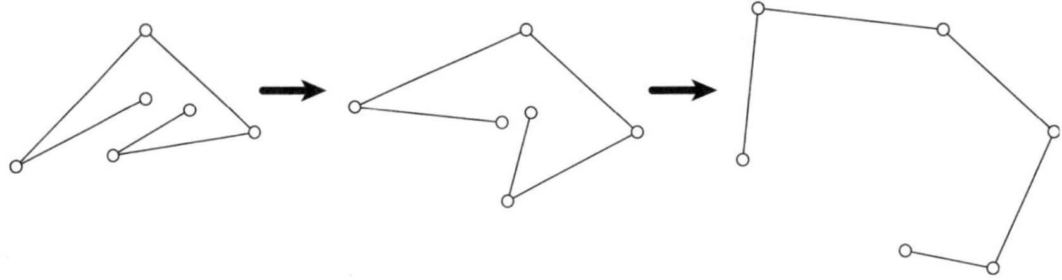

Polygonal tree: A linkage whose underlying graph is a single tree.

Configuration of a linkage in d-space: A mapping $p : V \to \mathbb{R}^d$ specifying a point $p(v) \in \mathbb{R}^d$ for each vertex v of the linkage, such that each bar $\{v, w\} \in E$ has the desired length $\ell(e)$, i.e., $|p(v) - p(w)| = \ell(e)$.

A configuration can be viewed as a point p in $\mathbb{R}^{d|V|}$ by arbitrarily ordering the vertices in V, and assigning the coordinates of the ith vertex $(0 \le i < |V|)$ to coordinates $id+1, id+2, \ldots, id+d$ of p.

Framework or ***bar framework:*** A linkage together with a configuration.

Reconfiguration or ***motion*** or ***flex*** of a linkage: A continuous function $f : [0,1] \to \mathbb{R}^{d|V|}$ specifying a configuration of the linkage for every moment in time between 0 and 1.

Configuration space or ***Moduli space of a linkage:*** The set $\mathcal{M}$ of all configurations (treated as points in $\mathbb{R}^{d|V|}$) of the linkage.

Self-intersecting configuration: A configuration in which two bars intersect but are not incident in the underlying graph of the linkage.

Reconfiguration avoiding self-intersection: A reconfiguration f in which every configuration $f(t)$ does not self-intersect.

Configuration space of a linkage, disallowing self-intersection: The subset $\mathcal{F}$ of the configuration space $\mathcal{M}$ in which every configuration does not self-intersect. (Also called the ***free space*** of the linkage.)

Paths in the configuration space of a linkage capture the key notion of reconfiguration (either allowing or disallowing self-intersection as appropriate). Many important questions about linkages can be most easily phrased in terms of the configuration space. For example, we are often interested in whether the configuration space is connected (every configuration can be reconfigured into every other configuration), or in the topology of the configuration space.

9.2 CONFIGURATION SPACES OF ARCS AND CYCLES WITH POSSIBLE INTERSECTIONS

One fundamental problem is to compute the topology of the configuration space of planar polygonal cycles (polygons), allowing possible self-intersections. There is a long list of results in increasing generality for computing information about the algebraic topological invariants of this configuration space. One approach is Morse Theory, which reveals some of the basic information, in particular, the connectivity and some of the easier invariants such as the Euler characteristic.

CONNECTIVITY

The following is an early result possibly first due to [Hau91], but rediscovered by [Jag92], and then rediscovered again or generalized considerably by many others, in particular, [Kam99, KT99, MS00, KM95, LW95].

THEOREM 9.2.1 *Connectivity for planar polygons* [Hau91]

Let $s_1 \leq s_2 \leq \cdots \leq s_n$ be the cyclic sequence of bar lengths in a polygon, and let $s = s_1 + s_2 + \cdots + s_n$. Then the following occurs:

i) The configuration space is nonempty if and only if $s_n \leq s/2$.

ii) The configuration space, modulo orientation-preserving congruences, is connected if and only if $s_{n-2} + s_{n-1} \leq s/2$. If the space is not connected, there are exactly two connected components, where each configuration in one component is the reflection of a configuration in the other component.

The configuration space is a smooth manifold if and only if there is some configuration p with all its vertices on a line, which in turn is determined by the edge lengths as described above. Also, the configuration space remains congruent no matter how we permute the cyclic sequence of bar lengths. When the linkage is not allowed to self-intersect, it is common to consider the configuration space modulo all congruences of the plane (including reflections); but when self-intersections are allowed, and condition ii) above is satisfied, it is possible to move the linkage from any configuration to its mirror image.

For polygons in dimensions higher than two, the situation is simpler:

THEOREM 9.2.2 *Connectivity for nonplanar polygons* [LW95]

The configuration space of a polygon in d-dimensional space, for $d > 2$, is always connected.

HOMOLOGY, COHOMOLOGY, AND HOMOTOPY

After connectivity, there remains the calculation of the higher homology groups, cohomology groups, and the homotopy type of the configuration space. Here is one special case as an example:

THEOREM 9.2.3 *Configuration space of equilateral polygons* [KT99]

Let $\mathcal{M}$ be the configuration space of a polygon with n equal bar lengths, modulo congruences of the plane. The homology of $\mathcal{M}$ is a torsion-free module given explicitly in [KT99]. When n is odd, $\mathcal{M}$ is a smooth manifold; and when $n = 5$, $\mathcal{M}$ is the compact, orientable two-dimensional manifold of genus 4 (originally shown in [Hav91], as well as in [Jag92]).

See also especially [KM95] for some of the basic techniques. For calculating the configuration space of graphs other than a cycle, see in particular the article [TW84], where a particular linkage, with some pinned vertices, has a configuration space that is an orientable two-dimensional manifold of genus 6.

Another case that has been considered is an equilateral polygon in 3-space with angles between incident edges fixed. This fixed-angle model arises in chemistry [CH88] and in particular in protein folding (see Section 8.6). Alternatively, a fixed angle can be simulated by adding bars between vertices of distance two along the polygon. The configuration space behaves similarly to the planar case:

THEOREM 9.2.4 *Fixed-angle equilateral 3D polygons* [CJ]

Let $\mathcal{M}$ be the configuration space of an equilateral polygon with $n \geq 6$ equal bar lengths and fixed equal angles, modulo congruences of $\mathbb{R}^3$. Suppose further that every turn angle is within an additive ϵ of $2\pi/n$ for ϵ sufficiently small (i.e., configurations are forced nearly planar). Then $\mathcal{M}$ has at most two components. When n is odd, $\mathcal{M}$ is a smooth manifold of dimension $n - 6$. When n is even, $\mathcal{M}$ is singular.

When $n = 6$, the underlying graph is the graph of an octahedron, and there are cases when it is rigid and cases when it is not. This linkage corresponds to cyclohexane in chemistry, and its flexibility was studied by [Bri96] and [Con78].

The restriction of the polygon configurations being almost planar leads to the following problem:

PROBLEM 9.2.5 *General equilateral equi-angular 3D polygons* [Cri92]

How many components does $\mathcal{M}$ have in the theorem above if ϵ is allowed to be large?

9.3 CONFIGURATION SPACES WITHOUT SELF-INTERSECTIONS

When the linkage is not permitted to self-intersect, the main question that has been studied is when it can be locked. Three main classes of linkages have been studied in this context: arcs, cycles, and trees. When the linkage is planar and has cycles, we assume that the clockwise/counterclockwise orientation is given and fixed, for otherwise the linkage is trivially locked: no cycle can be "flipped over" in the plane without self-intersection.

GLOSSARY

Locked linkage: A linkage whose configuration space has multiple connected components when self-intersections are disallowed.

Lockable class of linkages: There is a locked linkage in the class.

Unlockable class of linkages: No linkage in the class is locked.

FIGURE 9.3.1
The problems of arc straightening, cycle convexifying, and tree flattening.

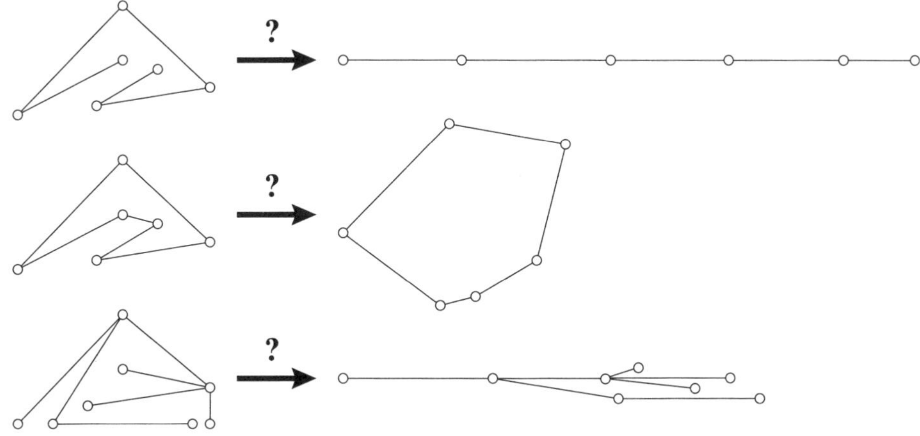

Straightening an arc: A motion bringing a polygonal arc from a given configuration to its ***straight configuration*** in which every joint angle is π.

Convexifying a cycle: A motion bringing a polygonal arc from a given configuration to a ***convex configuration*** in which every joint angle is at most π.

Flattening a tree: A motion bringing a polygonal tree from a given configuration to a ***flat configuration*** in which every joint angle is either 0, π, or 2π, and every bar points "away" from a designated root node.

WHICH LINKAGES ARE LOCKED?

Which of the main classes of linkages can be locked is summarized in Table 9.3.1. In short, the existence of locked arcs and locked unknotted cycles is equivalent to the existence of knots in that dimension: just in 3D. However, this equivalence is by no means obvious, especially in 2D, as evidenced by the existence of locked trees in 2D.

One main approach for determining whether a linkage is locked is to consider the equivalent problem of finding a motion from any configuration to a ***canonical configuration***. Because linkage motions are reversible and concatenable, if every configuration can be canonicalized, then every configuration can be brought to any other configuration, routing through the canonical configuration. Conversely, if

TABLE 9.3.1 Summary of what types of linkages can be locked.

	ARCS AND CYCLES	TREES
2D	Not lockable [CDR03, Str00, CDIO04]	Lockable [BDD+02, CDR02]
3D	Lockable [CJ98, BDD+01, Tou01]	Lockable [arcs are a special case]
4D+	Not lockable [CO01]	Not lockable [CO01]

some configuration cannot be canonicalized, then we know a pair of configurations that cannot reach each other, and therefore the linkage is locked.

This idea leads to the notions of straightening arcs, convexifying cycles, and flattening trees, as defined above. There is only one straight configuration of an arc, but there are multiple convex configurations of cycles and flat configurations of trees; fortunately, it is fairly easy to reconfigure between any pair of convex configurations of a cycle [ADE+01] or between any pair of flat configurations of a tree [BDD+02].

LOCKED LINKAGES

The first results along these lines were negative (see Figure 9.3.2): polygonal arcs in 3D and unknotted polygonal cycles in 3D can be locked [CJ98], and planar polygonal trees can be locked [BDD+02]. Since these results, other examples of unknotted but locked 3D polygonal cycles [BDD+01, Tou01] and locked 2D polygonal trees [CDR02, BCD+09] have been discovered.

More generally, Alt, Knauer, Rote, and Whitesides [AKRW03] constructed a large family of locked 2D trees and 3D arcs in which it is PSPACE-hard to determine whether one configuration can reach another configuration via a continuous motion that avoids self-intersection. Their construction combines several gadgets, many of which resemble the examples in Figure 9.3.2, as well as the "interlocked" linkages of [DLOS03, DLOS02]. However, this work leaves open a closely related problem, deciding whether *every* pair of configurations can reach each other:

PROBLEM 9.3.1 *Complexity of testing whether a linkage is locked* [BDD+01]

What is the complexity of deciding whether a linkage is locked? Particular cases of interest are 3D arcs, 3D cycles restricted to unknotted configurations, and 2D trees.

UNLOCKED LINKAGES

Unlockability was first established in 4D and higher [CO01], where one-dimensional arcs, cycles, and trees have so much freedom that they can never lock. Intuitively, the barriers (self-intersecting configurations) that might prevent e.g., straightening the vertex between the first two bars of an arc have dimension at least 2 lower than the configuration space of that vertex, and hence all barriers can be avoided. Thus, the only problem with straightening an arc vertex-by-vertex is that the configuration that results from straightening one extreme vertex might have self-intersections; in this case, the linkage can be perturbed to remove the problem. Convexifying cycles in 4D and higher is more difficult, but follows a similar idea.

FIGURE 9.3.2
Known examples of locked linkages.

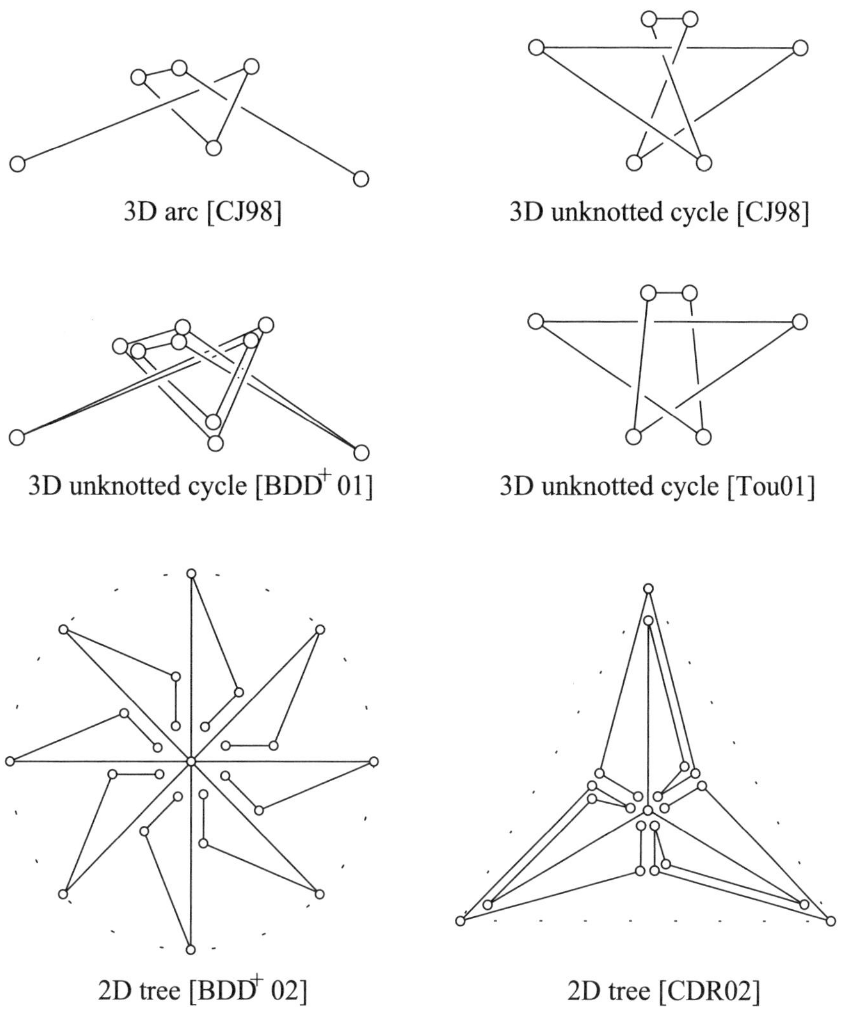

The last cell of Table 9.3.1 to be filled was that 2D arcs and cycles never to lock [CDR03]. Indeed, the following more general theorem holds:

THEOREM 9.3.2 *Straightening 2D arcs and convexifying 2D cycles* [CDR03]
Given a disjoint collection of polygonal arcs and polygonal cycles in the plane, there is a motion that avoids self-intersection and, after finite time, straightens every outermost arc and convexifies every outermost cycle. (An arc or cycle is ***outermost*** *if it is not contained within another cycle.)*

In this theorem, arcs and cycles contained within other cycles may not straighten or convexify—they simply "come along for the ride"—but this is the best we could hope for in general.

There are now three methods for solving this problem. See Figure 9.3.3 for a visual comparison on a simple example. The first method is based on flow through an ordinary differential equation defined implicitly by a convex optimization problem [CDR03]. The second method is more combinatorial and is based on algebraic motions defined by single-degree-of-freedom mechanisms given by pseudotriangulations [Str00]. The third method is based on energy minimization via gradient descent [CDIO04], a technique also adapted to polygon morphing [IOD09].

FIGURE 9.3.3
Convexifying a common polygon via all three convexification methods.

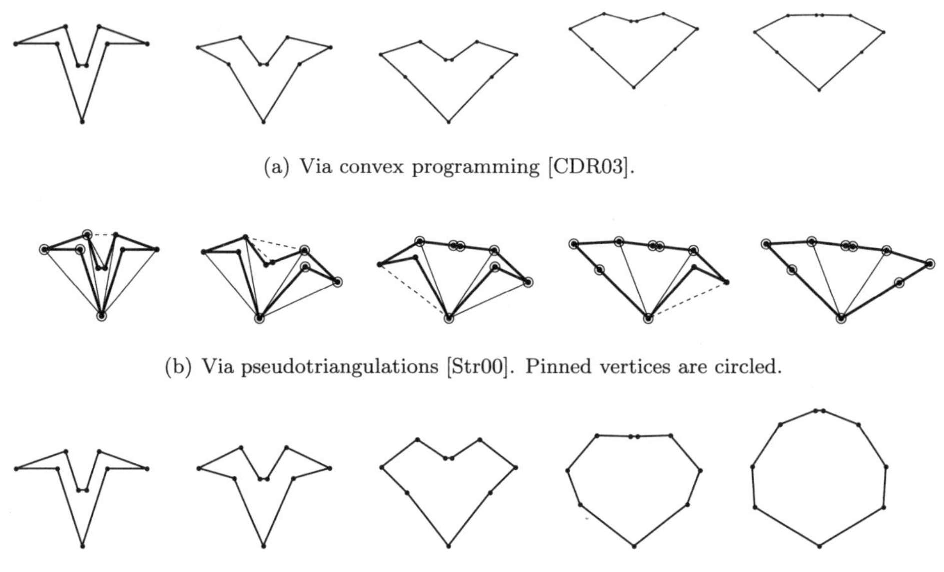

(a) Via convex programming [CDR03].

(b) Via pseudotriangulations [Str00]. Pinned vertices are circled.

(c) Via energy minimization [CDIO04].

The first two motions have the additional property of being ***expansive***—the distance between every pair of vertices never decreases over time—while the third motion only relies on the existence of such a motion. The first and last motions, being flow-based, preserve any initial symmetries of the linkage. Characterizing by continuity, the three motions are respectively piecewise-C^1, piecewise-C^∞, and C^∞. Only the last motion has a corresponding finite-time algorithm to compute a motion that is *piecewise-linear* through configuration space, i.e., the motion can be decomposed into steps where each angle in each step changes at a constant rate. The number of steps has a pseudopolynomial bound, that is, a bound polynomial in the number n of bars and the ratio between the largest and smallest distances among nonincident edges. This algorithm is also easy to implement.

SPECIAL CLASSES OF LINKAGES

In addition to these results for general classes of linkages, various special classes have been shown to have different properties. Polygonal arcs in 3D that lie on

FIGURE 9.3.4
Flipping a polygon until it is convex.

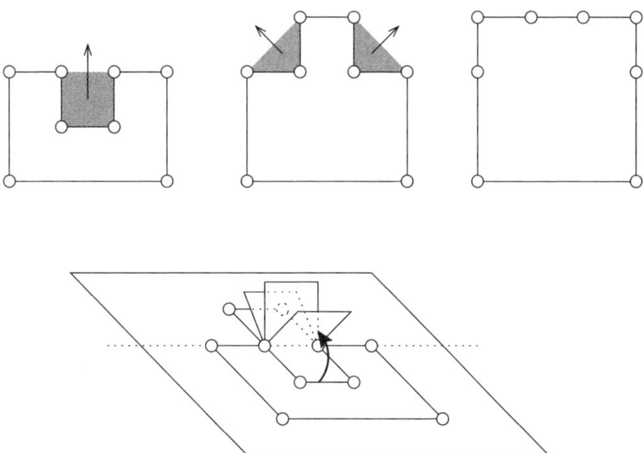

the surface of a convex polyhedron, or having a non-self-intersecting orthogonal projection, are never locked [BDD⁺01]. Polygonal cycles in 3D having a non-self-intersecting orthogonal projection are also never locked [CKM⁺01].

Connelly et al. [CDD⁺10] introduced the idea of *adorned* arc linkages, where each bar has a planar shape (Jordan region) rigidly attached to it. Provided each shape is *slender*—meaning that the distance along the shape boundary to each hinge is monotone—any expansive motion of the underlying arc avoids introducing collisions among the shapes too. This result was later used to prove that hinged dissections can be made continuously foldable without collisions [AAC⁺12].

FLIPS, FLIPTURNS, DEFLATIONS, POPS, POPTURNS

One of the first papers essentially about unlocking linkages is by Erdős [Erd35], who asked whether a particular "flipping" algorithm always convexifies a planar polygon by motions through 3D in a finite number of steps. A ***flip*** rotates by 180° a subchain of the polygon, called a ***pocket***, whose endpoints are consecutive vertices along the convex hull of the polygon. Each such flip never causes the polygon to self-intersect.[1] Nagy [SN39] was the first to claim a proof that a polygon admits only finitely many flips before convexifying. This result was subsequently rediscovered several times; see [Tou05, Grü95]. Unfortunately, a detailed analysis of these proofs [DGOT08] shows that many of the claimed proofs, including Nagy's original, are incorrect or incomplete. Fortunately, multiple correct proofs remain, including the first correct published proof by Reshetnyak [Res57] and a newer, simplified proof [DGOT08]. Thus, pocket flipping is one suitable strategy for convexifying a 2D polygon by motions in 3D.

Joss and Shannon (1973) first proved that the number of flips required to

[1] Erdős [Erd35] originally proposed flipping multiple pockets at once, but such an operation can lead to self-intersection; Nagy [SN39] fixed this problem by proposing flipping only one pocket at once.

convexify a polygon cannot be bounded in terms of the number of vertices, but this work remains unpublished; see [Grü95, Tou05]. However, it may still be possible to bound the number of flips using other metrics:

PROBLEM 9.3.3 *Bounding the number of flips* [M. Overmars, Feb. 1998]

Bound the maximum number of flips a polygon admits in terms of natural measures of geometric closeness such as the sharpest angle, the diameter, and the minimum distance between two nonincident edges.

A related computational problem is to compute the extreme numbers of flips:

PROBLEM 9.3.4 *Maximizing or minimizing flips* [Dem02]

What is the complexity of minimizing or maximizing the length of a convexifying sequence of flips for a given polygon?

Several variations on flips have also been considered. Grünbaum and Zaks [GZ01] generalized Nagy's results to polygons with self-intersections—still they can be convexified by finitely many flips—with further consideration in [DGOT08]. Wegner [Weg93] introduced the notion of ***deflations***, which are the exact reverse of flips, and Fevens et al. [FHM+01] showed that some quadrilaterals admit infinitely many deflations. On the other hand, every pentagon will stop deflating after finitely many carefully chosen deflations, and any infinite deflation sequence of a pentagon results from deflating an induced quadrilateral on four of the vertices [DDF+07].

Flipturns are similar to flips, except that the pocket is temporarily severed from the rest of the linkage and rotated 180° in the plane around the midpoint of the hull edge. Such an operation is not a valid linkage motion, but it has the advantage that the number of flipturns that a polygon admits before convexification is $O(n^2)$ [ACD+02, ABC+00]. This bound is tight up to a constant factor [Bie06], and there is extensive work on finding the precise constants [ACD+02], though some gaps remain to be closed. Also, related to Problem 9.3.4, it is known that maximizing the length of a convexifying flipturn sequence is weakly NP-hard [ACD+02]. Minimizing the number of flipturns leads to the following interesting problem:

PROBLEM 9.3.5 *Number of required flipturns* [Bie06]

Is there a polygon that requires $\Omega(n^2)$ flipturns to convexify, or can all polygons be convexified by $o(n^2)$ carefully chosen flipturns?

The best known lower bound is $\Omega(n)$.

Pops and ***popturns*** are variations on flips and flipturns, respectively, where the "pocket" being reflected or rotated consists of exactly two incident edges of the polygon, not necessarily related to the convex hull. Allowing the polygon to intersect itself, popturns can convexify any polygon [ABB+07], but pops cannot convexify some polygons (even initially non-self-intersecting) [DH10]. Forbidding self-intersection, there is a simple characterization of which polygons can be convexified by popturns [ABB+07].

INTERLOCKED LINKAGES

Combinations of polygonal arcs and cycles in 3D that can or cannot be locked (or more accurately, "interlocked") are studied in [DLOS03, DLOS02]. More precisely, this work studies the shortest (fewest-bar) 3D arcs and cycles that can interlock with each other. For example, three 3-arcs (arcs with three bars each) can interlock, as can a 3-arc and a 4-arc, or a 3-cycle and a 4-arc, or a 3-arc and a 4-cycle. However, two 3-arcs and arbitrarily many 2-arcs never interlock, and nor can a 3-cycle and a 3-arc. Also considered in [DLOS02] is the case that some of the pieces have restricted motion, e.g., all angles being fixed, or just rigid motions being allowed. Glass et al. [GLOZ06] proved that a 2-arc and an 11-arc can interlock.

9.4 UNIVERSALITY RESULTS

TRACING CURVES

The classic motivation of building linkages is to design a planar linkage in which one of the vertices traces a portion of a desired curve given by some polynomial function. In particular, Watt posed the problem of finding a linkage with some vertices pinned so that one vertex would trace out a line (segment). Watt's problem, at first thought to be impossible, was finally solved by Peaucellier in [Pea73], as well as by Lipkin in [Lip71]. See also [Kem77] and [Har74].

Later, Kempe [Kem76] described a linkage that would trace out a portion of any algebraic curve in the plane. However, his description is very brief and it leaves unspecified what portion of the algebraic curve is actually traced out, and whether there are other, possibly unwanted components or pieces of other algebraic curves that can also be traced out. This question also arises for the linkages that trace a line segment.

GLOSSARY

Real algebraic set: A subset of $\mathbb{R}^N$ given by a finite number of polynomial equations with real coefficients.

Real semi-algebraic set: A subset of $\mathbb{R}^N$ given by a finite number of polynomial equations and inequalities with real coefficients.

It is important to realize the distinction between an algebraic set and a semi-algebraic set. For example, a circle (excluding its interior) is an algebraic set, while a (closed) line segment is a semi-algebraic set but not an algebraic set. The linear projection of an algebraic set is always a semi-algebraic set, but it may not be an algebraic set. The configuration space of a linkage is an algebraic set, but the locus of possible positions of one of its vertices is only guaranteed to be a semi-algebraic set, because it represents the projection onto the coordinates corresponding to one of the vertices of the linkage.

ARBITRARY CONFIGURATION SPACES

One of the more precise results related to Kempe's result is the following:

THEOREM 9.4.1 *Creating linkage configuration spaces* [KM95]

Let M be any compact smooth manifold. Then there is a planar linkage whose configuration space is diffeomorphic to a disjoint union of some number of copies of M.

This result was also claimed by Thurston, but there does not seem to be a written proof by him. As a consequence of this result, we obtain the following precise version of what Kempe was trying to claim. This consequence is proved by King [Kin99] using the techniques of Kapovich-Millson [KM02] and Thurston.

THEOREM 9.4.2 *Tracing out an algebraic curve* [Kin99]

Let X be any set in the plane that is the polynomial image of a closed interval. Then there is a linkage in the plane with some pinned vertices such that one of the vertices traces out X exactly.

See [JS99, BM56] for other discussions of how to create linkages to trace out at least a portion of a given algebraic curve. King [Kin98] also generalizes this result to higher dimensions and to the semialgebraic sets arising from projecting the configuration space down to consider some subset of the vertices. See also [KM02] for connections to universality theorems concerning configuration spaces of lines in the plane, for example, as in the work of [Mnë88]. The complexity results of [HJW85] described in Section 8.6 build off a universality construction similar to those mentioned above.

Abbott, Barton, and Demaine [Abb08] characterized the number of bars required to draw a polynomial curve of degree n in d dimensions, as $\Theta(n^d)$. Their constructions also have stronger continuity properties than King's, so that a single motion can construct the entire algebraic set. By giving a polynomial-time construction of a configuration of the linkage, they also prove coNP-hardness of testing rigidity.

Abel et al. [ADD$^+$16] prove that these universality results hold when linkages either do not or are required not to have crossing bars in their configurations. They further prove $\forall\mathbb{R}$-completeness of testing rigidity and global rigidity, and $\exists\mathbb{R}$-completeness of graph realization, even in the noncrossing scenario.

9.5 COMPUTATIONAL COMPLEXITY

There are a variety of algorithmic questions that can be asked about a given linkage. Most of these questions are computationally difficult to answer, either NP-hard or PSPACE-hard. Nonetheless, given the importance of these problems, there is work on developing (exponential-time) algorithms.

GLOSSARY

Ruler folding problem: Given a polygonal arc (i.e., a sequence of bar lengths) and a desired length L, is there a configuration of the arc (ruler) in which the bars lie along a common line segment of length L? If so, find such a configuration. (The problem can also be phrased as reconfiguration, provided the linkage is permitted to self-intersect.)

Reachability problem: Given a configuration of a linkage, a distinguished vertex, and a point in the plane, is it possible to reconfigure the linkage so that the distinguished vertex touches the given point? If so, find such a reconfiguration. In this problem, the linkage has one or more vertices pinned to particular locations in the plane.

Reconfiguration problem: Given two configurations of a linkage, is it possible to reconfigure one into the other? If so, find such a reconfiguration.

Locked decision problem: Given a linkage, is it locked?

HARDNESS RESULTS

One of the simplest complexity results is about the ruler folding problem, obtained via a reduction from set partition:

THEOREM 9.5.1 *Complexity of ruler folding* [HJW85]

The ruler folding problem is NP-complete.

Building on this result, the same authors establish

THEOREM 9.5.2 *Complexity of arc reachability* [HJW85]

The reachability problem is NP-hard for a planar polygonal arc in the presence of four line-segment obstacles and permitting the arc to self-intersect.

For general linkages instead of arcs, stronger complexity results exist:

THEOREM 9.5.3 *Complexity of reachability* [HJW84]

The reachability problem is PSPACE-hard for a planar linkage without obstacles and permitting the linkage to self-intersect.

On the other hand, a similar result holds for a polygonal arc among obstacles:

THEOREM 9.5.4 *Complexity of arc reachability among obstacles* [JP85]

The reachability problem is PSPACE-hard for a planar polygonal arc in the presence of polygonal obstacles and permitting the arc to self-intersect.

Finally, when the linkage is not permitted to self-intersect, and there are no obstacles, hardness is known in cases when the linkage can be locked; see Section 8.6.

THEOREM 9.5.5 *Complexity of non-self-intersecting arc reconfiguration* [AKRW03]

The reconfiguration problem is PSPACE-hard for a 3D polygonal arc or a 2D polygonal tree when the linkage is not permitted to self-intersect.

ALGORITHMS

Algorithms for linkage reconfiguration problems can be obtained from general motion-planning results in Chapter 50 (Section 50.1.1). This connection seems to have first been made explicit in [AKRW03]. To apply the roadmap algorithm of Canny [Can87] (Theorem 50.1.2), we first phrase the algorithmic linkage problems into the motion-planning framework; see also [DO07, ch. 2].

The configuration space of a given linkage is the subset of $\mathbb{R}^{vc}$ in which every point satisfies certain bar-length constraints and, if desired, non-intersection constraints between all pairs of bars. Both types of constraints can be phrased using constant-degree polynomial equations and inequalities, e.g., the former by setting the squared length of each bar to the desired value. (There are also embeddings of the configuration space into Euclidean space with fewer than vc dimensions, dependent on the number of degrees of freedom in the linkage, but the vc-dimensional parameterization is most naturally semi-algebraic.)

Returning to the motion-planning framework, the polynomial equations and inequalities are precisely the obstacle surfaces. The configuration space has dimension $k = vc$, and there are $n \leq b^2$ obstacle surfaces where b is the number of bars, each with degree $d = O(1)$. We can factor out the trivial rigid motions by supposing that one bar of the linkage is pinned, reducing k to $(v-2)c$. Now running the roadmap algorithm produces a representation of the entire configuration space. By path planning within this space, we can solve the reconfiguration problem. By a simple pass through the representation, we can tell whether the space is connected, solving the locked decision problem. By slicing the space with a polynomial specifying that a particular vertex is at a particular point in the plane, we can solve the reachability problem.

Plugging $k \leq vc$, $n \leq b^2$, and $d = O(1)$ into the roadmap algorithm with deterministic running time $O(n^k(\log n)d^{O(k^4)})$ and randomized expected running time $O(n^k(\log n)d^{O(k^2)})$, we obtain

COROLLARY 9.5.6 *Roadmap algorithm applied to linkages* [AKRW03]

The reachability, reconfiguration, and locked decision problems can be solved for an arbitrary linkage with v vertices and b bars in $\mathbb{R}^c$ using $O(b^{2vc}(\log b)2^{O(vc)^4})$ deterministic time or $O(b^{2vc}(\log b)2^{O(vc)^2})$ expected randomized time.

9.6 KINEMATICS

According to Bottema and Roth [BR79b], "kinematics is that branch of mechanics which treats the phenomenon of motion without regard to the cause of the motion. In kinematics there is no reference to mass or force; the concern is only with relative positions and their changes." Kinematics is a subject with a long history and which has had, at various times, notable influence on and has to some extent has been partially identified with such areas as algebraic geometry, differential geometry, mechanics, singularity theory, and Lie theory. It has often been a subject studied from an engineering point of view, and there are many detailed calculations with respect to particular mechanisms of interest. As a representative example, we consider four-bar mechanisms (Figure 9.6.1):

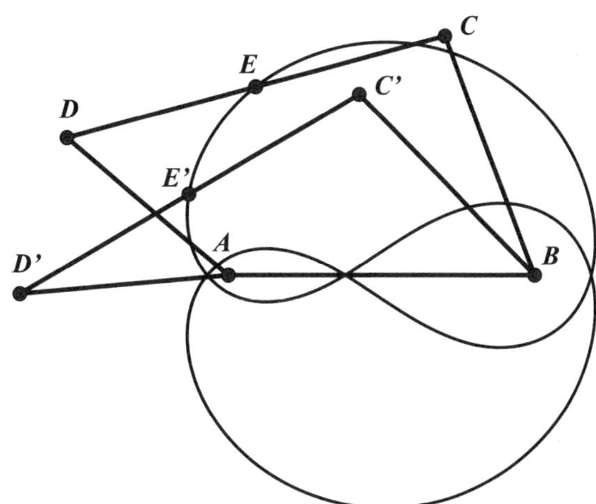

FIGURE 9.6.1
The coupler curve of the midpoint E of the coupler CD as it moves relative to the frame AB in a four-bar mechanism.

GLOSSARY

Mechanism: A linkage with one degree of freedom, modulo global translation and rotation.

Four-bar mechanism: A four-bar polygonal cycle; see Figure 9.6.1 for an example. Sometimes called a ***three-bar mechanism***.

Frame: We generally fix a frame of reference for a mechanism by pinning one bar, fixing its position in the plane. This bar is called the ***frame***. In Figure 9.6.1, bar AB is pinned.

Coupler: A distinguished bar other than the frame. In Figure 9.6.1, we consider the coupler CD.

Coupler motion: The motion of the entire plane induced by the relative motion of the coupler with respect to the frame.

Coupler curve: The path traced during the coupler motion by any point rigidly attached to the coupler (e.g., via two additional bars). Figure 9.6.1 shows the coupler curve of the midpoint E of the coupler bar CD.

FOUR-BAR MECHANISM

Coupler curves can be surprisingly complex. In the generic case, a coupler curve of a four-bar mechanism is an algebraic curve of degree 6. Substantial effort has been put into cataloging the different shapes of coupler curves that can arise from four-bar and other mechanisms. A sample theorem in this context is the following:

THEOREM 9.6.1 *Multiplicity of coupler curves* [Rob75]

Any coupler curve of a four-bar mechanism can be generated by two other four-bar mechanisms.

GLOSSARY

Infinitesimal motion or ***first-order flex:*** The first derivative of a motion at an instantaneous moment in time, assigning a velocity vector to each point involved in the motion. (See Chapter 61 for a more thorough explanation in the context of rigidity.)

Pole or ***instantaneous pole:*** The instantaneous fixed point of a first-order motion of the plane. For a rotation, the pole is the center of rotation. For a translation, the pole is a point at infinity in the projective plane. A combination of rotation and translation can be rewritten as a pure rotation.

Polode: The locus of poles over time during a motion of the plane.

POLES

Some of the central theorems in kinematics treat the instantaneous case. Poles characterize the first-order action of a motion at each moment in time. Together, the polode can be viewed relative to either the fixed plane of the frame (the ***fixed polode***) or the moving plane of the coupler (***moving polode***). Apart from degenerate cases, a planar motion can be described by the moving polode rolling along the fixed polode. A basic theorem in the context of poles is the following:

THEOREM 9.6.2 *Three-Pole Theorem*

For any three motions of the plane, the instantaneous poles of the three mutual relative motions are collinear at any moment in time.

FURTHER READING

For a general introduction and sampling of the field of kinematics, see [Hun78, BR79b, Sta97, McC90, Pot94, McC00]. For relations to singularity theory, see e.g., [GHM97]. For examples, analysis, and synthesis of specific mechanisms such as the four-bar mechanism, see [GN86, Mik01, Sta99, Ale95, BS90, Leb67, Con79, Con78]. For some typical examples from an engineering viewpoint, see e.g., [CP91, Che02, Ler00]. See also Section 60.4 of this Handbook.

9.7 APPLICATIONS

Applications of linkages arise throughout science and engineering. We highlight three modern applications: robotics, manufacturing, and protein folding.

APPLICATIONS IN ENGINEERING

The study of linkages in fact originated in the context of mechanical engineering, e.g., for the purpose of converting circular motion into linear motion. Today, one of the driving applications for linkages is *robotics*, in particular *robotic arms.*

A robotic arm can be modeled as a linkage, typically a polygonal chain. Some robotic arms have hinges that force the bars to remain coplanar, modeled by 2D chains; other arms have universal joints, modeled by 3D chains; other arms pose additional different constraints (such as incident bars being coplanar, yet the whole linkage need not be coplanar), leading to other models of linkage folding. Some planar robotic arms reserve slightly offset planar planes for the bars, modeled by a planar polygonal chain that permits self-intersection. Most other robotic arms are modeled by disallowing self-intersection.

The reachability problem is largely motivated by robotic arms, where the "hand" at one end of the arm must be placed at a particular location, e.g., to pick up an object, but the rest of the configuration is secondary. In other contexts, the entire configuration of the arm is important, and we need to plan a motion to a target configuration, leading to the reconfiguration problem. The locked decision problem is the first question one might ask about the simplicity/complexity of motion planning for a particular type of linkage. However, all of these problems are typically studied in the context of linkages without obstacles, but in robotics there are almost always obstacles. Some obstacles, such as a halfplane representing the floor, can often be avoided; but more generally the problems become much more complicated. See Chapter 51.

Another area with linkage applications is *manufacturing*. Given a straight hydraulic tube or piece of wire, a typical goal is to produce a desired folded configuration. In these contexts, we want to bend the wire as little as possible. In particular, a typical constraint is to bend the wire only monotonically: once it is bent one way, it cannot be bent the other way. This constraint forces straight segments of the target shape to remain straight throughout the motion. Thus, the problem can be modeled as straightening a polygonal chain, either in 2D or 3D depending on the application, with additional constraints. For example, the expansive motions described in Section 8.6 fold all joints monotonically; however, their reliance on bending most joints simultaneously may be undesirable. Arkin et al. [AFMS01] consider the restriction in which only a single joint can be rotated at once, together with additional realistic constraints arising in wire bending.

APPLICATIONS IN BIOLOGY

A crude model of a protein backbone is a polygonal chain in 3D, and a similarly crude model of an entire protein is a polygonal tree in 3D. In both cases, the vertices represent atoms, and the bars represent bonds between atoms (which in reality stay roughly the same length). In proteins, these bar/bond lengths are typically all within a factor of 2 of each other. Two atoms cannot occupy the same space, which can be roughly modeled by disallowing self-intersection. One interesting open problem in this context is the following:

PROBLEM 9.7.1 *Equilateral or near-equilateral locked linkages* [BDD+01]

Is there a locked equilateral arc, cycle, or tree in 3D? More generally, what is the smallest value of $\alpha \geq 1$ for which there is a locked arc/cycle/tree in 3D with all edge lengths between 1 and α?

These crude models may lead to some biological insight, but they do not capture several aspects of real protein folding.

One aspect that can easily be incorporated into linkage folding is that the angles between incident bars is typically fixed. This *fixed-angle constraint* can alternatively be viewed as adding bars between vertices originally at distance two from each another. Soss et al. [Sos01, SEO03, ST00] initiated study of such fixed-angle linkages in computational geometry, in particular establishing NP-hardness of deciding reconfigurability or flattenability. Aloupis et al. [ADD+02, ADM+02] consider when fixed-angle linkages are not locked in the sense that all flat states are reachable from each other by motions avoiding self-intersection. Borcea and Streinu [BS11b, BS11a] gave a polynomial-time algorithm for computing the maximum or minimum attainable distance between the endpoints of a fixed-angle arc; the maximum problem was also solved by Benbernou and O'Rourke [Ben11].

A more challenging aspect of protein folding is the *thermodynamic hypothesis* [Anf72]: that folding is encouraged to follow energy-minimizing pathways. Indeed, the bars are not strictly binding, nor are they completely fixed in length; they are merely encouraged to do so, and sometimes violate these constraints. Unfortunately, these properties are difficult to model, and energy functions defined so far are either incomplete or difficult to manipulate. Also, the implications on linkage-folding problems remain unclear.

One particularly simple energy-based model of protein folding studied in both computer science and biology is the HP (Hydrophilic-Hydrophobic) model; see e.g., [ABD+03, CD93, Dil90, Hay98, ZKL07]. This model is particularly discrete, modeling a protein as an equilateral chain on a lattice, typically square or cubic grid, but possibly also a triangular or tetrahedral lattice. The model captures only hydrophobic bonds and forces, clustering to avoid external water. Finding the optimal folding even in this simple model is NP-complete [BL98, CGP+98], though there are several constant-factor approximation algorithms [HI96, New02, ABD+97] and some practical heuristics [ZKL07]. One interesting open problem is whether designing a protein to fold into particular shape is easier than finding the shape to which a particular protein folds [ABD+03]:

PROBLEM 9.7.2 *HP protein design* [ABD+03]

What is the complexity of deciding whether a given subset of the lattice is an optimal folding of some HP protein, and if so finding such a protein? What if it must be the unique optimal folding of the HP protein?

A result related to the second half of this problem is that arbitrarily long HP proteins with unique optimal foldings exist, at least for open and closed chains in a 2D square grid [ABD+03].

9.8 SOURCES AND RELATED MATERIAL

FURTHER READING

[DO07]: The main book on geometric folding in general, Part I of which focuses on linkage folding. Parts II and III consider folding (reconfiguration) of objects of larger intrinsic dimension, in particular 2 (pieces of paper) and 3 (polyhedra).

[Dem12]: An online class about geometric folding in general, and linkage folding in particular.

[O'R98, Dem00, Dem02]: Older surveys on geometric folding in general.

RELATED CHAPTERS

Chapter 37: Computational and quantitative real algebraic geometry
Chapter 50: Algorithmic motion planning
Chapter 51: Robotics
Chapter 52: Computer graphics
Chapter 53: Modeling motion
Chapter 60: Geometric applications of the Grassmann-Cayley algebra
Chapter 61: Rigidity and scene analysis
Chapter 65: Applications to structural molecular biology

REFERENCES

[AAC+12] T.G. Abbott, Z. Abel, D. Charlton, E.D. Demaine, M.L. Demaine, and S.D. Kominers. Hinged dissections exist. *Discrete Comput. Geom.*, 47:150–186, 2012.

[ABB+07] G. Aloupis, B. Ballinger, P. Bose, M. Damian, E.D. Demaine, M.L. Demaine, R. Flatland, F. Hurtado, S. Langerman, J. O'Rourke, P. Taslakian, and G. Toussaint. Vertex pops and popturns. In *Proc. 19th Canadian Conf. Comput. Geom.*, pages 137–140, 2007.

[Abb08] T.G. Abbott. Generalizations of Kempe's universality theorem. Master's thesis, Massachusetts Institute of Technology, Cambridge, 2008. Joint work with R.W. Barton and E.D. Demaine.

[ABC+00] H.-K. Ahn, P. Bose, J. Czyzowicz, N. Hanusse, E. Kranakis, and P. Morin. Flipping your lid. *Geombinatorics*, 10:57–63, 2000.

[ABD+97] R. Agarwala, S. Batzoglou, V. Dancik, S.E. Decatur, M. Farach, S. Hannenhalli, S. Muthukrishnan, and S. Skiena. Local rules for protein folding on a triangular lattice and generalized hydrophobicity in the HP model. *J. Comput. Biol.*, 4:275–296, 1997.

[ABD+03] O. Aichholzer, D. Bremner, E.D. Demaine, H. Meijer, V. Sacristán, and M. Soss. Long proteins with unique optimal foldings in the H-P model. *Comput. Geom.*, 25:139–159, 2003.

[ACD+02] O. Aichholzer, C. Cortés, E.D. Demaine, V. Dujmović, J. Erickson, H. Meijer, M. Overmars, B. Palop, S. Ramaswami, and G.T. Toussaint. Flipturning polygons. *Discrete Comput. Geom.*, 28:231–253, 2002.

[ADD+02] G. Aloupis, E.D. Demaine, V. Dujmović, J. Erickson, S. Langerman, H. Meijer, I. Streinu, J. O'Rourke, M. Overmars, M. Soss, and G.T. Toussaint. Flat-state connectivity of linkages under dihedral motions. In *Proc. 13th Int. Sympos. Algorithms and Computation*, vol. 2518 of *LNCS*, pages 369–380, Springer, Berlin, 2002.

[ADD+16] Z. Abel, E.D. Demaine, M.L. Demaine, S. Eisenstat, J. Lynch, and T.B. Schardl. Who needs crossings? Hardness of plane graph rigidity. In *Proc. 32nd Sympos. Comput. Geom.*, vol. 51 of *LIPIcs*, article 3, Schloss Dagstuhl, 2016.

[ADE$^+$01] O. Aichholzer, E.D. Demaine, J. Erickson, F. Hurtado, M. Overmars, M.A. Soss, and G.T. Toussaint. Reconfiguring convex polygons. *Comput. Geom.*, 20:85–95, 2001.

[ADM$^+$02] G. Aloupis, E.D. Demaine, H. Meijer, J. O'Rourke, I. Streinu, and G. Toussaint. Flat-state connectedness of fixed-angle chains: Special acute chains. In *Proc. 14th Canadian Conf. Comput. Geom.*, pages 27–30, 2002.

[AFMS01] E.M. Arkin, S.P. Fekete, J.S.B. Mitchell, and S.S. Skiena. On the manufacturability of paperclips and sheet metal structures. In *Abstract 17th European Workshop Comput. Geom.*, pages 187–190, 2001.

[AKRW03] H. Alt, C. Knauer, G. Rote, and S. Whitesides. The complexity of (un)folding. In *Proc. 19th Sympos. Comput. Geom.*, pages 164–170, ACM Press, 2003.

[Ale95] V.A. Aleksandrov. A new example of a bendable polyhedron. *Sibirsk. Mat. Zh.*, 36:1215–1224, 1995.

[Anf72] C. Anfinsen. Studies on the principles that govern the folding of protein chains. In *Les Prix Nobel en 1972*, pages 103–119, Nobel Foundation, 1972.

[BCD$^+$09] B. Ballinger, D. Charlton, E.D. Demaine, M.L. Demaine, J. Iacono, C.-H. Liu, and S.-H. Poon. Minimal locked trees. In *Proc. 11th Algorithms Data Structures Sympos.*, vol. 5664 of *LNCS*, pages 61–73, Springer, Berlin, 2009.

[BDD$^+$01] T. Biedl, E. Demaine, M. Demaine, S. Lazard, A. Lubiw, J. O'Rourke, M. Overmars, S. Robbins, I. Streinu, G. Toussaint, and S. Whitesides. Locked and unlocked polygonal chains in three dimensions. *Discrete Comput. Geom.*, 26:269–281, 2001; full version at arXiv:cs/9910009.

[BDD$^+$02] T. Biedl, E. Demaine, M. Demaine, S. Lazard, A. Lubiw, J. O'Rourke, S. Robbins, I. Streinu, G. Toussaint, and S. Whitesides. A note on reconfiguring tree linkages: Trees can lock. *Discrete Appl. Math.*, 117:293–297, 2002; full version at arXiv:cs/9910024.

[Ben11] N.M. Benbernou. *Geometric Algorithms for Reconfigurable Structures.* PhD thesis, Massachusetts Institute of Technology, Cambridge, 2011.

[Bie06] T. Biedl. Polygons needing many flipturns. *Discrete Comput. Geom.*, 35:131–141, 2006.

[BL98] B. Berger and T. Leighton. Protein folding in the hydrophobic-hydrophilic (*HP*) model is NP-complete. *J. Comput. Biol.*, 5:27–40, 1998.

[BM56] W. Blaschke and H.R. Müller. *Ebene Kinematik.* Verlag von R. Oldenbourg, München, 1956.

[BR79a] O. Bottema and B. Roth. *Theoretical Kinematics.* North-Holland Publishing, Amsterdam, 1979. Reprinted by Dover Publications, 1990.

[BR79b] O. Bottema and B. Roth. *Theoretical Kinematics*, volume 24 of *North-Holland Series in Applied Mathematics and Mechanics.* North-Holland Publishing, Amsterdam, 1979.

[Bri96] R. Bricard. Sur une question de géométrie relative aux polyèdres. *Nouv. Ann. Math.*, 15:331–334, 1896.

[BS90] A.V. Bushmelev and I.K. Sabitov. Configuration spaces of Bricard octahedra. *Ukrain. Geom. Sb.*, 33:36–41, 1990.

[BS11a] C.S. Borcea and I. Streinu. Exact workspace boundary by extremal reaches. In *Proc. 27th Sympos. Comput. Geom.*, pages 481–490, ACM Press, 2011.

[BS11b] C.S. Borcea and I. Streinu. Extremal reaches in polynomial time. In *Proc. 27th Sympos. Comput. Geom.*, pages 472–480, ACM Press, 2011.

[Can87] J.F. Canny. *The Complexity of Robot Motion Planning.* MIT Press, Cambridge, 1987.

[CD93] H.S. Chan and K.A. Dill. The protein folding problem. *Physics Today*, 46:24–32, 1993.

[CDD+10] R. Connelly, E.D. Demaine, M.L. Demaine, S. Fekete, S. Langerman, J.S.B. Mitchell, A. Ribó, and G. Rote. Locked and unlocked chains of planar shapes. *Discrete Comput. Geom.*, 44:439–462, 2010.

[CDIO04] J.H. Cantarella, E.D. Demaine, H.N. Iben, and J.F. O'Brien. An energy-driven approach to linkage unfolding. In *Proc. 20th Sympos. Comput. Geom.*, pages 134–143, ACM Press, 2004.

[CDR02] R. Connelly, E.D. Demaine, and G. Rote. Infinitesimally locked self-touching linkages with applications to locked trees. In J. Calvo, K. Millett, and E. Rawdon, editors, *Physical Knots: Knotting, Linking, and Folding of Geometric Objects in 3-space*, pages 287–311, AMS, Providence, 2002.

[CDR03] R. Connelly, E.D. Demaine, and G. Rote. Straightening polygonal arcs and convexifying polygonal cycles. *Discrete Comput. Geom.*, 30:205–239, 2003.

[CGP+98] P. Crescenzi, D. Goldman, C. Papadimitriou, A. Piccolboni, and M. Yannakakis. On the complexity of protein folding. *J. Comput. Biol.*, 5:423–465, 1998.

[CH88] G.M. Crippen and T.F. Havel. *Distance Geometry and Molecular Conformation*. Vol. 15 of *Chemometrics Series*, Research Studies Press, Chichester, 1988.

[Che02] C.-H. Chen. Kinemato-geometrical methodology for analyzing curvature and torsion of trajectory curve and its applications. *Mech. Mach. Theory*, 37:35–47, 2002.

[CJ] R. Connelly and B. Jaggi. Unpublished.

[CJ98] J. Cantarella and H. Johnston. Nontrivial embeddings of polygonal intervals and unknots in 3-space. *J. Knot Theory Ramifications*, 7:1027–1039, 1998.

[CKM+01] J.A. Calvo, D. Krizanc, P. Morin, M. Soss, and G. Toussaint. Convexifying polygons with simple projections. *Inform. Process. Lett.*, 80:81–86, 2001.

[CO01] R. Cocan and J. O'Rourke. Polygonal chains cannot lock in 4D. *Discrete Comput. Geom.*, 20:105–129, 2001.

[Con78] R. Connelly. The rigidity of suspensions. *J. Differential Geom.*, 13:399–408, 1978.

[Con79] R. Connelly. The rigidity of polyhedral surfaces. *Math. Mag.*, 52:275–283, 1979.

[CP91] C.R. Calladine and S. Pellegrino. First-order infinitesimal mechanisms. *Internat. J. Solids Structures*, 27:505–515, 1991.

[Cri92] G.M. Crippen. Exploring the conformation space of cycloalkanes by linearized embedding. *J. Comput. Chem.*, 13:351–361, 1992.

[DDF+07] E.D. Demaine, M.L. Demaine, T. Fevens, A. Mesa, M. Soss, D.L. Souvaine, P. Taslakian, and G. Toussaint. Deflating the pentagon. In *Proc. Internat. Conf. Computational Geometry and Graph Theory (KyotoCGGT)*, vol. 4535 of *LNCS*, pages 56–67, Springer, Berlin, 2007.

[Dem00] E.D. Demaine. Folding and unfolding linkages, paper, and polyhedra. In *Proc. Japanese Conf. Discrete Comput. Geom.*, vol. 2098 of *LNCS*, pages 113–124, Springer, Berlin, 2000.

[Dem02] E.D. Demaine. *Folding and Unfolding*. PhD thesis, Dept. of Computer Science, U. Waterloo, 2002.

[Dem12] E.D. Demaine. 6.849: Geometric folding algorithms: Linkages, origami, polyhedra. MIT class with online video lectures, 2012. `http://courses.csail.mit.edu/6.849/`.

[DGOT08] E.D. Demaine, B. Gassend, J. O'Rourke, and G.T. Toussaint. All polygons flip finitely... right? In J. Goodman, J. Pach, and R. Pollack, editors, *Surveys on Discrete and Computational Geometry: Twenty Years Later*, vol. 453 of *Contemp. Math.*, pages 231–255, AMS, Providence, 2008.

[DH10] A. Dumitrescu and E. Hilscher. On convexification of polygons by pops. *Discrete Math.*, 310:2542–2545, 2010.

[Dil90] K.A. Dill. Dominant forces in protein folding. *Biochemistry*, 29:7133–7155, 1990.

[DLOS02] E.D. Demaine, S. Langerman, J. O'Rourke, and J. Snoeyink. Interlocked open linkages with few joints. In *Proc. 18th Sympos. Comput. Geom.*, pages 189–198, ACM Press, 2002.

[DLOS03] E.D. Demaine, S. Langerman, J. O'Rourke, and J. Snoeyink. Interlocked open and closed linkages with few joints. *Comput. Geom.*, 26:37–45, 2003.

[DO07] E.D. Demaine and J. O'Rourke. *Geometric Folding Algorithms: Linkages, Origami, Polyhedra.* Cambridge University Press, 2007.

[Erd35] P. Erdős. Problem 3763. *Amer. Math. Monthly*, 42:627, 1935.

[FHM+01] T. Fevens, A. Hernandez, A. Mesa, P. Morin, M. Soss, and G. Toussaint. Simple polygons with an infinite sequence of deflations. *Beitr. Algebra Geom.*, 42:307–311, 2001.

[GHM97] C.G. Gibson, C.A. Hobbs, and W.L. Marar. On versal unfoldings of singularities for general two-dimensional spatial motions. *Acta Appl. Math.*, 47:221–242, 1997.

[GLOZ06] J. Glass, B. Lu, J. O'Rourke, and J.K. Zhong. A 2-chain can interlock with an open 11-chain. *Geombinatorics*, 15:166–176, 2006.

[GN86] C.G. Gibson and P.E. Newstead. On the geometry of the planar 4-bar mechanism. *Acta Appl. Math.*, 7:113–135, 1986.

[Grü95] B. Grünbaum. How to convexify a polygon. *Geombinatorics*, 5:24–30, 1995.

[GZ01] G. Grünbaum and J. Zaks. Convexification of polygons by flips and by flipturns. *Discrete Math.*, 241:333–342, 2001.

[Har74] H. Hart. On certain conversions of motion. *Messenger Math.*, IV:82–88, 1874.

[Hau91] J.-C. Hausmann. Sur la topologie des bras articulés. In *Algebraic topology Poznań 1989*, vol. 1474 of *Lecture Notes in Math.*, pages 146–159, Springer, Berlin, 1991.

[Hav91] T.F. Havel. Some examples of the use of distances as coordinates for Euclidean geometry. *J. Symbolic Comput.*, 11:579–593, 1991.

[Hay98] B. Hayes. Prototeins. *American Scientist*, 86:216–221, 1998.

[HI96] W.E. Hart and S. Istrail. Fast protein folding in the hydrophobic-hydrophilic model within three-eighths of optimal. *J. Comput. Biol.*, 3:53–96, 1996.

[HJW84] J. Hopcroft, D. Joseph, and S. Whitesides. Movement problems for 2-dimensional linkages. *SIAM J. Comput.*, 13:610–629, 1984.

[HJW85] J. Hopcroft, D. Joseph, and S. Whitesides. On the movement of robot arms in 2-dimensional bounded regions. *SIAM J. Comput.*, 14:315–333, 1985.

[Hun78] K.H. Hunt. *Kinematic Geometry of Mechanisms. Oxford Engrg. Sci. Ser.*, Clarendon, Oxford University Press, New York, 1978.

[IOD09] H.N. Iben, J.F. O'Brien, and E.D. Demaine. Refolding planar polygons. *Discrete Comput. Geom.*, 41:444–460, 2009.

[Jag92] B. Jaggi. *Pinktmengen mit vorgeschriebenen Distanzen und ihre Konfigurationsräurne.* Inauguraldissertation, Universität Bern, 1992.

[JP85] D.A. Joseph and W.H. Plantings. On the complexity of reachability and motion planning questions. In *Proc. 1st Sympos. Comput. Geom.*, pages 62–66, ACM Press, 1985.

[JS99] D. Jordan and M. Steiner. Configuration spaces of mechanical linkages. *Discrete Comput. Geom.*, 22:297–315, 1999.

[Kam99] Y. Kamiyama. Topology of equilateral polygon linkages in the Euclidean plane modulo isometry group. *Osaka J. Math.*, 36:731–745, 1999.

[Kem76] A.B. Kempe. On a general method of describing plane curves of the nth degree by linkwork. *Proc. London Math. Soc.*, 7:213–216, 1876.

[Kem77] A.B. Kempe. *How to Draw a Straight Line: A Lecture on Linkages.* Macmillan, London, 1877.

[Kin98] H.C. King. Configuration spaces of linkages in $\mathbb{R}^n$. Preprint, arXiv:math/9811138, 1998.

[Kin99] H.C. King. Planar linkages and algebraic sets. *Turkish J. Math.*, 23:33–56, 1999.

[KM95] M. Kapovich and J. Millson. On the moduli space of polygons in the Euclidean plane. *J. Differential Geom.*, 42:133–164, 1995.

[KM02] M. Kapovich and J.J. Millson. Universality theorems for configuration spaces of planar linkages. *Topology*, 41:1051–1107, 2002.

[KT99] Y. Kamiyama and M. Tezuka. Topology and geometry of equilateral polygon linkages in the Euclidean plane. *Quart. J. Math. Oxford Ser. (2)*, 50:463–470, 1999.

[Leb67] H. Lebesgue. Octaèdres articulés de Bricard. *Enseign. Math. (2)*, 13:175–185, 1967.

[Ler00] J. Lerbet. Some explicit relations in kinematics of mechanisms. *Mech. Res. Comm.*, 27:621–630, 2000.

[Lip71] L. Lipkin. Dispositif articulé pour la transformation rigoureuse du mouvement circulaire en mouvement rectiligne. *Revue Univers. des Mines et de la Métallurgie de Liége*, 30:149–150, 1871.

[LW95] W.J. Lenhart and S.H. Whitesides. Reconfiguring closed polygonal chains in Euclidean d-space. *Discrete Comput. Geom.*, 13:123–140, 1995.

[McC90] J.M. McCarthy. *An Introduction to Theoretical Kinematics.* MIT Press, Cambridge, 1990.

[McC00] J.M. McCarthy. *Geometric Design of Linkages.* Vol. 11 of *Interdisciplinary Applied Mathematics.* Springer-Verlag, New York, 2000.

[Mik01] S.N. Mikhalev. Some necessary metric conditions for the flexibility of suspensions. *Vestnik Moskov. Univ. Ser. I Mat. Mekh.*, 3:15–21, 77, 2001.

[Mnë88] N.E. Mnëv. The universality theorems on the classification problem of configuration varieties and convex polytopes varieties. In *Topology and Geometry—Rohlin Seminar*, vol. 1346 of *Lecture Notes in Math.*, pages 527–543, Springer, Berlin, 1988.

[MS00] O. Mermoud and M. Steiner. Visualisation of configuration spaces of polygonal linkages. *J. Geom. Graph.*, 4:147–157, 2000.

[New02] A. Newman. A new algorithm for protein folding in the HP model. In *Proc. 13th ACM-SIAM Sympos. Discrete Algorithms*, pages 876–884, 2002.

[O'R98] J. O'Rourke. Folding and unfolding in computational geometry. In *Revised Papers from the Japan Conf. Discrete Comput. Geom.*, vol. 1763 of *LNCS*, pages 258–266, Springer, Berlin, 1998.

[Pea73] A. Peaucellier. Note sur une question de géometrie de compas. *Nouv. Ann. de Math.*, 2e serie, XII:71–73, 1873.

[Pot94] H. Pottmann. Kinematische Geometrie. In *Geometrie und ihre Anwendungen*, pages 141–175, Hanser, Munich, 1994.

[Res57] Y.G. Reshetnyak. On a method of transforming a nonconvex polygonal line into a convex one (in Russian). *Uspehi Mat. Nauk*, 12:189–191, 1957.

[Rob75] S. Roberts. On three-bar motion in plane space. *Proc. London Math. Soc.*, 7:14–23, 1875.

[SEO03] M. Soss, J. Erickson, and M. Overmars. Preprocessing chains for fast dihedral rotations is hard or even impossible. *Comput. Geom.*, 26:235–246, 2003.

[SN39] B. de Sz. Nagy. Solution to problem 3763. *Amer. Math. Monthly*, 46:176–177, 1939.

[Sos01] M. Soss. *Geometric and Computational Aspects of Molecular Reconfiguration.* PhD thesis, School of Computer Science, McGill University, 2001.

[ST00] M. Soss and G.T. Toussaint. Geometric and computational aspects of polymer reconfiguration. *J. Math. Chem.*, 27:303–318, 2000.

[Sta97] H. Stachel. Euclidean line geometry and kinematics in the 3-space. In *Proc. 4th International Congress of Geometry (Thessaloniki, 1996)*, pages 380–391, Giachoudis-Giapoulis, Thessaloniki, 1997.

[Sta99] H. Stachel. Higher order flexibility of octahedra. *Period. Math. Hungar.*, 39:225–240, 1999.

[Str00] I. Streinu. A combinatorial approach to planar non-colliding robot arm motion planning. In *Proc. 41st IEEE Sympos. Found. Comp. Sci.*, pages 443–453, 2000.

[Tou05] G. Toussaint. The Erdős-Nagy theorem and its ramifications. *Comput. Geom.*, 31:219–236, 2005.

[Tou01] G. Toussaint. A new class of stuck unknots in Pol_6. *Beitr. Algebra Geom.*, 42:1027–1039, 2001.

[TW84] W. Thurston and J. Weeks. The mathematics of three-dimensional manifolds. *Sci. Amer.*, pages 108–120, 1984.

[Weg93] B. Wegner. Partial inflation of closed polygons in the plane. *Beitr. Algebra Geom.*, 34:77–85, 1993.

[ZKL07] J. Zhang, S.C. Kou, and J.S. Liu. Biopolymer structure simulation and optimization via fragment regrowth Monte Carlo. *J. Chem. Phys.*, 126:225101, 2007.

10 GEOMETRIC GRAPH THEORY

János Pach

INTRODUCTION

In the traditional areas of graph theory (Ramsey theory, extremal graph theory, random graphs, etc.), graphs are regarded as abstract binary relations. The relevant methods are often incapable of providing satisfactory answers to questions arising in geometric applications. Geometric graph theory focuses on combinatorial and geometric properties of graphs drawn in the plane by straight-line edges (or, more generally, by edges represented by simple Jordan arcs). It is a fairly new discipline abounding in open problems, but it has already yielded some striking results that have proved instrumental in the solution of several basic problems in combinatorial and computational geometry (including the k-set problem and metric questions discussed in Sections 1.1 and 1.2, respectively, of this Handbook). This chapter is partitioned into extremal problems (Section 10.1), crossing numbers (Section 10.2), and generalizations (Section 10.3).

10.1 EXTREMAL PROBLEMS

Turán's classical theorem [Tur54] determines the maximum number of edges that an abstract graph with n vertices can have without containing, as a subgraph, a complete graph with k vertices. In the spirit of this result, one can raise the following general question. Given a class $\mathcal{H}$ of so-called *forbidden geometric subgraphs*, what is the maximum number of edges that a geometric graph of n vertices can have without containing a geometric subgraph belonging to $\mathcal{H}$? Similarly, Ramsey's theorem [Ram30] for abstract graphs has some natural analogues for geometric graphs. In this section we will be concerned mainly with problems of these two types.

GLOSSARY

Geometric graph: A graph drawn in the plane by (possibly crossing) straight-line segments; i.e., a pair $(V(G), E(G))$, where $V(G)$ is a set of points ('vertices'), no three of which are collinear, and $E(G)$ is a set of segments ('edges') whose endpoints belong to $V(G)$.

Convex geometric graph: A geometric graph whose vertices are in *convex position*; i.e., they form the vertex set of a convex polygon.

Cyclic chromatic number of a convex geometric graph: The minimum number $\chi_c(G)$ of colors needed to color all vertices of G so that each color class consists of consecutive vertices along the boundary of the convex hull of the vertex set.

Convex matching: A convex geometric graph consisting of disjoint edges, each of which belongs to the boundary of the convex hull of its vertex set.

Parallel matching: A convex geometric graph consisting of disjoint edges, the convex hull of whose vertex set contains only two of the edges on its boundary.

Complete geometric graph: A geometric graph G whose edge set consists of all $\binom{|V(G)|}{2}$ segments between its vertices.

Complete bipartite geometric graph: A geometric graph G with $V(G) = V_1 \cup V_2$, whose edge set consists of all segments between V_1 and V_2.

Geometric subgraph of G***:*** A geometric graph H, for which $V(H) \subseteq V(G)$ and $E(H) \subseteq E(G)$.

Crossing: A common interior point of two edges of a geometric graph.

(k,l)-Grid: $k+l$ vertex-disjoint edges in a geometric graph such that each of the first k edges crosses all of the last l edges. It is called natural if the first k edges are pairwise disjoint segments and the last l edges are pairwise disjoint segments.

Disjoint edges: Edges of a geometric graph that do not cross and do not even share an endpoint.

Parallel edges: Edges of a geometric graph whose supporting lines are parallel or intersect at points not belonging to any of the edges (including their endpoints).

x-Monotone curve: A continuous curve that intersects every vertical line in at most one point.

Outerplanar graph: A (planar) graph that can be drawn in the plane without crossing so that all points representing its vertices lie on the outer face of the resulting subdivision of the plane. A ***maximal outerplanar graph*** is a triangulated cycle.

Hamiltonian path: A path going through all elements of a finite set S. If the elements of S are colored by two colors, and no two adjacent elements of the path have the same color, then it is called an *alternating* path.

Hamiltonian cycle: A cycle going through all elements of a finite set S.

Caterpillar: A tree consisting of a path P and of some extra edges, each of which is adjacent to a vertex of P.

CROSSING-FREE GEOMETRIC GRAPHS

1. Hanani-Tutte theorem: Any graph that can be drawn in the plane so that its edges are represented by simple Jordan arcs such that any two that do not share an endpoint properly cross an even number of times, is planar [Cho34, Tut70]. The analogous result also holds in the projective plane [PSS09].

2. Fáry's theorem: Every planar graph admits a crossing-free straight-line drawing [Fár48, Tut60, Ste22]. Moreover, every 3-connected planar graph and its dual have simultaneous straight-line drawings in the plane such that only dual pairs of edges cross and every such pair is perpendicular [BS93].

3. Koebe's theorem: The vertices of every planar graph can be represented by nonoverlapping disks in the plane such that two of them are tangent to each other if and only if the corresponding two vertices are adjacent [Koe36, Thu78]. This immediately implies Fáry's theorem.

4. Pach-Tóth theorem: Any graph that can be drawn in the plane so that its edges are represented by x-monotone curves with the property that any two of them either share an endpoint or properly cross an even number of times admits a crossing-free straight-line drawing, in which the x-coordinates of the vertices remain the same [PT04]. Fulek *et al.* [FPSŠ13] generalized this result in two directions: it is sufficient to assume that (a) any two edges that do not share an endpoint cross an even number of times, and that (b) the projection of every edge to the x-axis lies between the projections of its endpoints.

5. Grid drawings of planar graphs: Every planar graph of n vertices admits a straight-line drawing such that the vertices are represented by points belonging to an $(n-1) \times (n-1)$ grid [FPP90, Sch90]. Furthermore, such a drawing can be found in $O(n)$ time. For other small-area grid drawings, consult [DF13].

6. Straight-line drawings of outerplanar graphs: For any outerplanar graph H with n vertices and for any set P of n points in the plane in general position, there is a crossing-free geometric graph G with $V(G) = P$, whose underlying graph is isomorphic to H [GMPP91]. For any rooted tree T and for any set P of $|V(T)|$ points in the plane in general position with a specified element $p \in P$, there is a crossing-free straight-line drawing of T such that every vertex of T is represented by an element of P and the root is represented by p [IPTT94]. This theorem generalizes to any *pair* of rooted trees, T_1 and T_2: for any set P of $n = |V(T_1)| + |V(T_2)|$ points in general position in the plane, there is a crossing-free mapping of $T_1 \cup T_2$ that takes the roots to arbitrarily prespecified elements of P. Such a mapping can be found in $O(n^2 \log n)$ time [KK00]. The analogous statement for *triples* of trees is false.

7. Alternating paths: Given n red points and n blue points in general position in the plane, separated by a straight line, they always admit a noncrossing alternating Hamiltonian path [KK03].

TURÁN-TYPE PROBLEMS

By Euler's Polyhedral Formula, if a geometric graph G with $n \geq 3$ vertices has no 2 crossing edges, it cannot have more than $3n - 6$ edges. It was shown in [AAP$^+$97] and [Ack09] that under the weaker condition that no 3 (resp. 4) edges are *pairwise crossing*, the number of edges of G is still $O(n)$. It is not known whether this statement remains true even if we assume only that no 5 edges are pairwise crossing. As for the analogous problem when the forbidden configuration consists of k pairwise *disjoint edges*, the answer is linear for every k [PT94]. In particular, for $k = 2$, the number of edges of G cannot exceed the number of vertices [HP34]. The best lower and upper bounds known for the number of edges of a geometric graph with n vertices, containing no *forbidden* geometric subgraph of a certain type, are summarized in Table 10.1.1. The letter k always stands for a *fixed* positive integer parameter and n tends to infinity. Wherever k does not appear in the asymptotic bounds, it is hidden in the constants involved in the O- and Ω-notations.

Better results are known for *convex* geometric graphs, i.e., when the vertices are in convex position. The relevant bounds are listed in Table 10.1.2. For any convex geometric graph G, let $\chi_c(G)$ denote its *cyclic chromatic number*. Furthermore, let

TABLE 10.1.1 Maximum number of edges of a geometric graph of n vertices containing no forbidden subconfigurations of a certain type.

FORBIDDEN CONFIGURATION	LOWER BOUND	UPPER BOUND	SOURCE
2 crossing edges	$3n-6$	$3n-6$	Euler
3 pairwise crossing edges	$6.5n-O(1)$	$6.5n-O(1)$	[AT07]
4 pairwise crossing edges	$\Omega(n)$	$72(n-2)$	[Ack09]
$k>4$ pairwise crossing edges	$\Omega(n)$	$O(n\log n)$	[Val98]
an edge crossing 2 others	$4n-8$	$4n-8$	[PT97]
an edge crossing 3 others	$5n-12$	$5n-10$	[PT97]
an edge crossing 4 others	$5.5n+\Omega(1)$	$5.5n+O(1)$	[PRTT06]
an edge crossing 5 others	$6n-O(1)$	$6n-12$	[Ack16]
an edge crossing k others	$\Omega(\sqrt{k}n)$	$O(\sqrt{k}n)$	[PT97]
2 crossing edges crossing k others	$\Omega(n)$	$O(n)$	[PRT04]
(k,l)-grid	$\Omega(n)$	$O(n)$	[PPST05]
natural (k,l)-grid	$\Omega(n)$	$O(n\log^2 n)$	[AFP+14]
self-intersecting path of length 3	$\Omega(n\log n)$	$O(n\log n)$	[PPTT02]
self-intersecting path of length 5	$\Omega(n\log\log n)$	$O(n\log n/\log\log n)$	[Tar13], [PPTT02]
self-intersecting cycle of length 4	$\Omega(n^{3/2})$	$O(n^{3/2}\log n)$	[PR04], [MT06]
2 disjoint edges	n	n	[HP34]
noncrossing path of length k	$\Omega(kn)$	$O(k^2n)$	[Tót00]
k pairwise parallel edges	$\Omega(n)$	$O(n)$	[Val98]

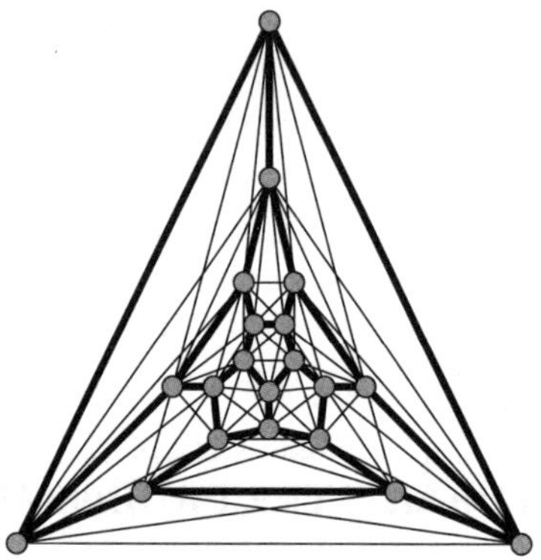

FIGURE 10.1.1
Geometric graph with $n=20$ vertices and $5n-12=88$ edges, none of which crosses 3 others.

$\mathrm{ex}(n,K_k)$ stand for the maximum number of edges of a graph with n vertices that does not have a complete subgraph with k vertices. By Turán's theorem [Tur54] mentioned above, $\mathrm{ex}(n,K_k)=\frac{k-2}{k-1}\binom{n}{2}+O(n)$ is equal to the number of edges of a complete $(k-1)$-partite graph with n vertices whose vertex classes are of size $\lfloor n/(k-1)\rfloor$ or $\lceil n/(k-1)\rceil$. Two disjoint self-intersecting paths of length 3, $xyvz$ and $x'y'v'z'$, in a convex geometric graph are said to be of the *same orientation* if the cyclic order of their vertices is x,v,x',v',y',z',y,z (⋉⋉). They are said to have *opposite orientations* if the cyclic order of their vertices is x,v,v',x',z',y',y,z (*type 1*: ⋊⋉) or v,x,x',v',y',z',z,y (*type 2*: ⋉⋊).

RAMSEY-TYPE PROBLEMS

In classical Ramsey theory, one wants to find large monochromatic subgraphs in a

TABLE 10.1.2 Maximum number of edges of a *convex* geometric graph of n vertices containing no forbidden subconfigurations of a certain type.

FORBIDDEN CONFIGURATION	LOWER BOUND	UPPER BOUND	SOURCE
2 crossing edges	$2n-3$	$2n-3$	Euler
self-intersecting path of length 3	$2n-3$	$2n-3$	Perles
k self-intersecting paths of length 3	$\Omega(n)$	$O(n)$	[BKV03]
2 self-intersecting paths of length 3 with opposite orientations of type 1	$\Omega(n\log n)$	$O(n\log n)$	[BKV03]
2 self-intersecting paths of length 3 with opposite orientations of type 2	$\Omega(n\log n)$	$O(n\log n)$	[BKV03]
2 adjacent edges crossing a 3rd	$\lfloor 5n/2-4\rfloor$	$\lfloor 5n/2-4\rfloor$	Perles-Pinchasi,[BKV03]
k pairwise crossing edges	$2(k-1)n-\binom{2k-1}{2}$	$2(k-1)n-\binom{2k-1}{2}$	[CP92]
noncrossing outerplanar graph of k vertices, having a Hamiltonian cycle	$\mathrm{ex}(n,K_k)$	$\mathrm{ex}(n,K_k)$	Pach [PA95], Perles
convex geometric subgraph G	$\mathrm{ex}(n,K_{\chi_c(G)})$	$\mathrm{ex}(n,K_{\chi_c(G)})+o(n^2)$	[BKV03]
convex matching of k disjoint edges	$\mathrm{ex}(n,K_k)+n-k+1$	$\mathrm{ex}(n,K_k)+n-k+1$	[KP96]
parallel matching of k disjoint edges	$(k-1)n$	$(k-1)n$	[Kup84]
noncrossing caterpillar C of k vertices	$\lfloor (k-2)n/2\rfloor$	$\lfloor (k-2)n/2\rfloor$	Perles [BKV03]

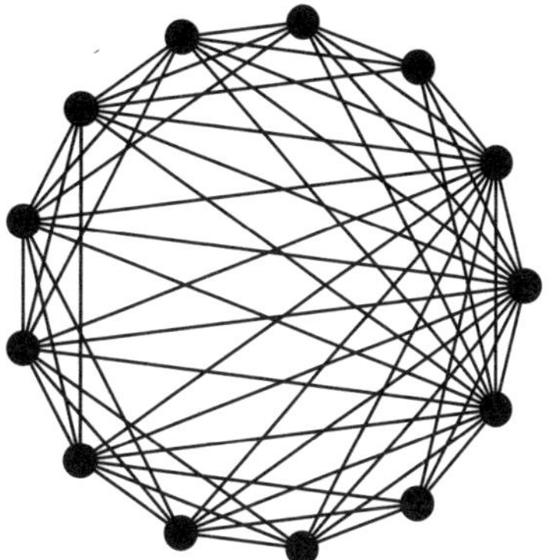

FIGURE 10.1.2
Convex geometric graph with $n=13$ vertices and $6n-\binom{7}{2}=57$ edges, no 4 of which are pairwise crossing [CP92].

complete graph whose edges are colored with several colors [GRS90]. Most questions of this type can be generalized to complete geometric graphs, where the monochromatic subgraphs are required to satisfy certain geometric conditions.

1. Károlyi-Pach-Tóth theorem [KPT97]: If the edges of a finite complete geometric graph are colored by two colors, there exists a noncrossing spanning tree, all of whose edges are of the same color. (This statement was conjectured by Bialostocki and Dierker [BDV04]. The analogous assertion for abstract graph follows from the fact that any graph or its complement is connected.)
2. Geometric Ramsey numbers: Let $\mathcal{G}_1,\dots,\mathcal{G}_k$ be not necessarily different classes of geometric graphs. Let $R(\mathcal{G}_1,\dots,\mathcal{G}_k)$ denote the smallest positive number R with the property that any complete geometric graph of R vertices whose edges are colored with k colors $(1,\dots,k$, say) contains, for some i, an i-colored subgraph belonging to $\mathcal{G}_i$. If $\mathcal{G}_1=\dots=\mathcal{G}_k=\mathcal{G}$, we write $R(\mathcal{G};k)$ instead of $R(\mathcal{G}_1,\dots,\mathcal{G}_k)$. If $k=2$, for the sake of simplicity, let $R(\mathcal{G})$ stand for $R(\mathcal{G};2)$. Some known results on the numbers $R(\mathcal{G}_1,\mathcal{G}_2)$ are listed in Table 10.1.3. In line 3 of the table, we have a better result if we restrict our attention to *convex*

geometric graphs: For any 2-coloring of the edges of a complete convex geometric graph with $2k-1$ vertices, there exists a noncrossing monochromatic path of length $k \geq 2$, and this result cannot be improved. The upper bound $2(k-1)(k-2)+2$ in line 4 is tight for convex geometric graphs [BCK+15]. The bounds in line 4 also hold when $\mathcal{G}_1 = \mathcal{G}_2$ consists of all noncrossing cycles of length k, triangulated from one of their vertices. The geometric Ramsey numbers of convex geometric graphs, when $\mathcal{G}_1 = \mathcal{G}_2$ consists of all isomorphic copies of a given convex geometric graph with at most 4 vertices, can be found in [BH96]. In [CGK+15], polynomial upper bounds are established for the geometric Ramsey numbers of the "ladder graphs" L_k, consisting of two paths of length k with an edge connecting each pair of corresponding vertices.

TABLE 10.1.3 Geometric Ramsey numbers $R(\mathcal{G}_1, \mathcal{G}_2)$ from [KPT97] and [KPTV98].

$\mathcal{G}_1$	$\mathcal{G}_2$	LOWER BOUND	UPPER BOUND
all noncrossing trees of k vertices	all noncrossing trees of k vertices	k	k
k disjoint edges	l disjoint edges	$k+l+\max\{k,l\}-1$	$k+l+\max\{k,l\}-1$
noncrossing paths of length k	noncrossing paths of length k	$\Omega(k)$	$O(k^{3/2})$
noncrossing cycles of length k	noncrossing cycles of length k	$(k-1)^2$	$2(k-1)(k-2)+2$

3. Pairwise disjoint copies: For any positive integer k, let $k\mathcal{G}$ denote the class of all geometric graphs that can be obtained by taking the union of k pairwise disjoint members of $\mathcal{G}$. If k is a power of 2 then

$$R(k\mathcal{G}) \leq (R(\mathcal{G})+1)k-1.$$

In particular, if $\mathcal{G} = \mathcal{T}$ is the class of triangles, we have $R(\mathcal{T}) = 6$. Thus, the above bound yields that

$$R(k\mathcal{T}) \leq 7k-1,$$

provided that k is a power of 2. This result cannot be improved [KPTV98]. Furthermore, for any $k > 0$, we have

$$R(k\mathcal{G}) \leq \left\lceil \frac{3(R(\mathcal{G})+1)}{2} \right\rceil k - \left\lceil \frac{R(\mathcal{G})+1}{2} \right\rceil.$$

For the corresponding quantities for convex geometric graphs, we have

$$R_c(k\mathcal{G}) \leq (R_c(\mathcal{G})+1)k-1.$$

4. Constructive vertex- and edge-Ramsey numbers: Given a class of geometric graphs $\mathcal{G}$, let $R_v(\mathcal{G})$ denote the smallest number R such that *there exists* a (complete) geometric graph of R vertices that, for any 2-coloring of its edges, has a monochromatic subgraph belonging to $\mathcal{G}$. Similarly, let $R_e(\mathcal{G})$ denote the minimum number of *edges* of a geometric graph with this property. $R_v(\mathcal{G})$ and $R_e(\mathcal{G})$ are called the ***vertex-*** and ***edge-Ramsey number*** of $\mathcal{G}$, respectively. Clearly, we have

$$R_v(\mathcal{K}) \leq R(\mathcal{G}), \quad R_e(\mathcal{G}) \leq \binom{R(\mathcal{G})}{2}.$$

(For abstract graphs, similar notions are discussed in [EFRS78, Bec83].) For $\mathcal{P}_k$, the class of noncrossing paths of length k, we have $R_v(\mathcal{P}_k) = O(k^{3/2})$ and $R_e(\mathcal{P}_k) = O(k^2)$.

OPEN PROBLEMS

1. What is the smallest number $u = u(n)$ such that there exists a "universal" set U of u points in the plane with the property that every planar graph of n vertices admits a noncrossing straight-line drawing on a suitable subset of U [FPP90]? It follows from the existence of a small grid drawing (see above) that $u(n) \leq n^2$, and it is shown in [BCD$^+$14] that $u(n) \leq n^2/4$. From below we have only $u(n) > 1.235n$; [Kur04]. Certain subclasses of planar graphs admit universal sets of size $o(n^2)$; [FT15].
2. It was shown by Chalopin and Gonçalvez [CG09] that the vertices of every planar graph G be represented by straight-line segments in the plane so that two segments intersect if and only if the corresponding vertices are adjacent. If the chromatic number of G is 2 or 3, then segments of 2 resp. 3 different directions suffice [FMP94, FM07]. Does there exist a constant c such that the vertices of every planar graph can be represented by segments using at most c different directions?
3. (Erdős, Kaneko-Kano) What is the largest number $A = A(n)$ such that any set of n red and n blue points in the plane admits a noncrossing alternating path of length A? It is known that $A(n) \leq (4/3 + o(1))n$; [KPT07].
4. Is it true that, for any fixed k, the maximum number of edges of a geometric graph with n vertices that does not have k pairwise crossing edges is $O(n)$?
5. (Aronov et al.) Is it true that any complete geometric graph with n vertices has at least $\Omega(n)$ pairwise crossing edges? It was shown in [AEG$^+$94] that one can always find $\sqrt{n/12}$ pairwise crossing edges. On the other hand, any complete geometric graph with n vertices has a noncrossing Hamiltonian path, hence $\lfloor n/2 \rfloor$ pairwise disjoint edges.
6. (Larman-Matoušek-Pach-Töröcsik) What is the smallest positive number $r = r(n)$ such that any family of r closed segments in general position in the plane has n members that are either pairwise disjoint or pairwise crossing? It is known [LMPT94, Kyn12] that $n^{\log 169/\log 8} \approx n^{2.466} \leq r(n) \leq n^5$.

10.2 CROSSING NUMBERS

The investigation of crossing numbers started during WWII with Turán's Brick Factory Problem [Tur77]: how should one redesign the routes of railroad tracks between several kilns and storage places in a brick factory so as to minimize the number of crossings? In the early eighties, it turned out that the chip area required for the realization (VLSI layout) of an electrical circuit is closely related to the crossing number of the underlying graph [Lei83]. This discovery gave an impetus to research in the subject. More recently, it has been realized that general bounds on crossing numbers can be used to solve a large variety of problems in discrete and computational geometry.

GLOSSARY

Drawing of a graph: A representation of the graph in the plane such that its vertices are represented by distinct points and its edges by simple continuous arcs connecting the corresponding point pairs. In a drawing (a) no edge passes through any vertex other than its endpoints, (b) no two edges touch each other (i.e., if two edges have a common interior point, then at this point they properly cross each other), and (c) no three edges cross at the same point.

Crossing: A common interior point of two edges in a graph drawing. Two edges may have several crossings.

Crossing number of a graph: The smallest number of crossings in any drawing of G, denoted by $\text{CR}(G)$. Clearly, $\text{CR}(G) = 0$ if and only if G is planar.

Rectilinear crossing number: The minimum number of crossings in a drawing of G in which every edge is represented by a straight-line segment. It is denoted by $\text{LIN-CR}(G)$.

Pairwise crossing number: The minimum number of crossing pairs of edges over all drawings of G, denoted by $\text{PAIR-CR}(G)$. (Here the edges can be represented by arbitrary continuous curves, so that two edges may cross more than once, but every pair of edges can contribute at most one to $\text{PAIR-CR}(G)$.)

Odd crossing number: The minimum number of those pairs of edges that cross an odd number of times, over all drawings of G. It is denoted by $\text{ODD-CR}(G)$.

Biplanar crossing number: The minimum of $\text{CR}(G_1) + \text{CR}(G_2)$ over all partitions of the graph into two edge-disjoint subgraphs G_1 and G_2.

Bisection width: The minimum number $b(G)$ of edges whose removal splits the graph G into two roughly equal subgraphs. More precisely, $b(G)$ is the minimum number of edges running between V_1 and V_2 over all partitions of the vertex set of G into two disjoint parts $V_1 \cup V_2$ such that $|V_1|, |V_2| \geq |V(G)|/3$.

Cut width: The minimum number $c(G)$ such that there is a drawing of G in which no two vertices have the same x-coordinate and every vertical line crosses at most $c(G)$ edges.

Path width: The minimum number $p(G)$ such that there is a sequence of at most $(p(G)+1)$-element sets $V_1, V_2, \ldots, V_r \subseteq V(G)$ with the property that both endpoints of every edge belong to some V_i and, if a vertex occurs in V_i and V_k $(i < k)$, then it also belongs to every V_j, $i < j < k$.

GENERAL ESTIMATES

Garey and Johnson [GJ83] showed that the determination of the crossing number is an *NP-complete* problem. Analogous results hold for the rectilinear crossing number [Bie91], for the pair crossing number [SSŠ03], and for the odd crossing number [PT00b]. The exact determination of crossing numbers of relatively small graphs of a simple structure (such as complete or complete bipartite graphs) is a hopelessly difficult task, but there are several useful bounds. There is an algorithm

[EGS03] for computing a drawing of a bounded-degree graph with n vertices, for which n plus the number of crossings is $O(\log^3 n)$ times the optimum.

1. For a simple graph G with $n \geq 3$ vertices and e edges, $\text{CR}(G) \geq e - 3n + 6$. From this inequality, a simple probabilistic argument shows that $\text{CR}(G) \geq ce^3/n^2$, for a suitable positive constant c. This important bound, due to Ajtai-Chvátal-Newborn-Szemerédi [ACNS82] and, independently, to Leighton [Lei83], is often referred to as the ***crossing lemma.*** We know that $0.03 \leq c \leq 0.09$ [PT97, PRTT06, Ack16]. The lower bound follows from line 8 in Table 10.1.1. Similar statements hold for $\text{PAIR-CR}(G)$ and $\text{ODD-CR}(G)$ [PT00b].

2. Crossing lemma for multigraphs [Szé97]: Let G be a ***multigraph*** with n vertices and e edges, i.e., the same pair of vertices can be connected by more than one edge. Let m denote the maximum multiplicity of an edge. Then
$$\text{CR}(G) \geq c\frac{e^3}{mn^2} - m^2 n,$$
where c denotes the same constant as in the previous paragraph.

3. Midrange crossing constant: Let $\kappa(n, e)$ denote the minimum crossing number of a graph G with n vertices and at least e edges. That is,
$$\kappa(n,e) = \min_{\substack{n(G)=n \\ e(G)\geq e}} \text{CR}(G).$$
It follows from the crossing lemma that, for $e \geq 4n$, $\kappa(n,e)n^2/e^3$ is bounded from below and from above by two positive constants. Erdős and Guy [EG73] conjectured that if $n \ll e \leq n^2/100$, then $\lim \kappa(n,e)n^2/e^3$ exists. (We use the notation $f(n) \gg g(n)$ to mean that $\lim_{n\to\infty} f(n)/g(n) = \infty$.) This was partially settled in [PST00]: if $n \ll e \ll n^2$, then
$$\lim_{n\to\infty} \kappa(n,e)\frac{n^2}{e^3} = C > 0$$
exists. Moreover, the same result is true with the same constant C, for drawings on every other orientable surface.

4. Graphs with monotone properties: A graph property $\mathcal{P}$ is said to be ***monotone*** if (i) for any graph G satisfying $\mathcal{P}$, every subgraph of G also satisfies $\mathcal{P}$; and (ii) if G_1 and G_2 satisfy $\mathcal{P}$, then their disjoint union also satisfies $\mathcal{P}$. For any monotone property $\mathcal{P}$, let $\text{ex}(n, \mathcal{P})$ denote the maximum number of edges that a graph of n vertices can have if it satisfies $\mathcal{P}$. In the special case when $\mathcal{P}$ is the property that the graph does not contain a subgraph isomorphic to a fixed forbidden subgraph H, we write $\text{ex}(n, H)$ for $\text{ex}(n, \mathcal{P})$.

   Let $\mathcal{P}$ be a monotone graph property with $\text{ex}(n, \mathcal{P}) = O(n^{1+\alpha})$ for some $\alpha > 0$. In [PST00], it was proved that there exist two constants $c, c' > 0$ such that the crossing number of any graph G with property $\mathcal{P}$ that has n vertices and $e \geq cn\log^2 n$ edges satisfies
$$\text{CR}(G) \geq c'\frac{e^{2+1/\alpha}}{n^{1+1/\alpha}}.$$

This bound is asymptotically tight, up to a constant factor. In particular, if $e > 4n$ and G has no cycle of length at most $2r$, then the crossing number of G satisfies

$$\text{CR}(G) \geq c_r \frac{e^{r+2}}{n^{r+1}},$$

where $c_r > 0$ is a suitable constant. For $r = 2, 3$, and 5, these bounds are asymptotically tight, up to a constant factor. If G does not contain a complete bipartite subgraph $K_{r,s}$ with r and s vertices in its classes, $s \geq r$, then we have

$$\text{CR}(G) \geq c_{r,s} \frac{e^{3+1/(r-1)}}{n^{2+1/(r-1)}},$$

where $c_{r,s} > 0$ is a suitable constant. These bounds are tight up to a constant factor if $r = 2, 3$, or if r is arbitrary and $s > (r-1)!$.

5. Crossing number vs. bisection width $b(G)$: For any vertex $v \in V(G)$, let $d(v)$ denote the degree of v in G. It was shown in [PSS96] and [SV94] that

$$\text{CR}(G) + \frac{1}{16} \sum_{v \in V(G)} d^2(v) \geq \frac{1}{40} b^2(G).$$

A similar statement holds with a worse constant for the cut width $c(G)$ of G [DV02]. This, in turn, implies that the same is true for $p(G)$, the path width of G, as we have $p(G) \leq c(G)$ for every G [Kin92].

6. Relations between different crossing numbers: Clearly, we have

$$\text{ODD-CR}(G) \leq \text{PAIR-CR}(G) \leq \text{CR}(G) \leq \text{LIN-CR}(G).$$

It was shown [BD93] that there are graphs with crossing number 4 whose rectilinear crossing numbers are arbitrarily large.

It was established in [PT00b] and [Mat14] that

$$\text{CR}(G) \leq 2\left(\text{ODD-CR}(G)\right)^2,$$

$$\text{CR}(G) \leq O\left(\left(\text{PAIR-CR}(G)\right)^{3/2} \log^2 \text{ODD-CR}(G)\right),$$

respectively. Pelsmajer *et al.* [PSS08] discovered a series of graphs for which $\text{ODD-CR}(G) \neq \text{PAIR-CR}(G)$. In fact, there are graphs satisfying the inequality $\text{PAIR-CR}(G) \geq 1.16\text{ODD-CR}(G)$; see [Tót08]. We cannot rule out the possibility that

$$\text{PAIR-CR}(G) = \text{CR}(G)$$

for every graph G.

7. Crossing numbers of random graphs: Let $G = G(n,p)$ be a ***random graph*** with n vertices, whose edges are chosen independently with probability $p = p(n)$. Let e denote the *expected number* of edges of G, i.e., $e = p \cdot \binom{n}{2}$. It is not hard to see that if $e > 10n$, then almost surely $b(G) \geq e/10$. It therefore follows from the above relation between the crossing number and the bisection width that almost surely we have $\text{LIN-CR}(G) \geq \text{CR}(G) \geq e^2/4000$. Evidently, the order of magnitude of this bound cannot be improved. A similar inequality was proved in [ST02] for the pairwise crossing number, under the stronger condition that $e > n^{1+\epsilon}$ for some $\epsilon > 0$.

8. Biplanar crossing number vs. crossing number: It is known [CSSV08] that the biplanar crossing number of every graph is at most 3/8 times its crossing number. It is conjectured that the statement remains true with 7/24 in place of 3/8.

9. Harary-Kainen-Schwenk conjecture [HKS73]: For every $n \geq m \geq 3$ and cycles C_n and C_m, $\text{CR}(C_n \times C_m)$ is equal to $n(m-2)$. This was proved in [GS04] for every m and for all sufficiently large n. For the crossing number of the skeleton of the n-dimensional hypercube Q_n, we have $1/20 + o(1) \leq \text{CR}(Q_n)/4^n \leq 163/1024$ [FF00, SV93].

OPEN PROBLEMS

1. (Pach-Tóth) Is it true that $\text{PAIR-CR}(G) = \text{CR}(G)$ for every graph G? Does there exist a constant γ such that $\text{CR}(G) \leq \gamma\text{ODD-CR}(G)$?

2. Albertson conjectured that the crossing number of every graph whose chromatic number is at least k is at least as large as the crossing number of K_k. This conjecture was proved for $k \leq 18$, and it is also known that the crossing number of such a graph exceeds ck^4, for a suitable constant c. [ACF09, BT10, Ack16].

3. Zarankiewicz's conjecture [Guy69]: The crossing number of the complete bipartite graph $K_{n,m}$ with n and m vertices in its classes satisfies

$$\text{CR}(K_{n,m}) = \left\lfloor \frac{m}{2} \right\rfloor \cdot \left\lfloor \frac{m-1}{2} \right\rfloor \cdot \left\lfloor \frac{n}{2} \right\rfloor \cdot \left\lfloor \frac{n-1}{2} \right\rfloor.$$

Kleitman [Kle70] verified this conjecture in the special case when $\min\{m, n\} \leq 6$ and Woodall [Woo93] for $m = 7$, $n \leq 10$.

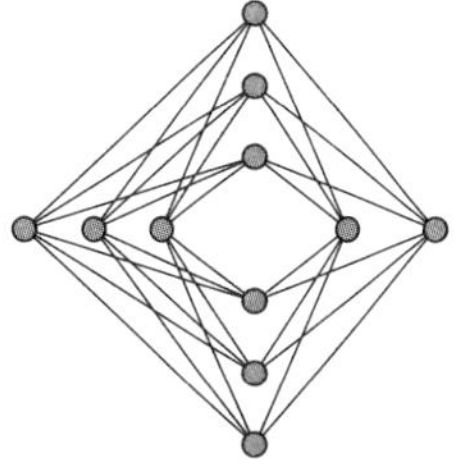

FIGURE 10.2.1
Complete bipartite graph $K_{5,6}$ with 24 crossings.

It is also conjectured that the crossing number of the complete graph K_n satisfies

$$\text{CR}(K_n) = \frac{1}{4} \left\lfloor \frac{n}{2} \right\rfloor \cdot \left\lfloor \frac{n-1}{2} \right\rfloor \cdot \left\lfloor \frac{n-2}{2} \right\rfloor \cdot \left\lfloor \frac{n-3}{2} \right\rfloor.$$

This conjecture is known to be true if we restrict our attention to drawings where every edge is represented by an x-monotone curve [AAF$^+$14, BFK15]. An old construction of Blažek and Koman [BK64] shows that equality can be attained even for drawings of this type.

4. Rectilinear crossing numbers of complete graphs: Determine the value

$$\kappa = \lim_{n\to\infty} \frac{\text{LIN-CR}(K_n)}{\binom{n}{4}}.$$

The best known bounds $0.3799 < \kappa \leq 0.3805$ were found by Ábrego *et al.* [ACF+10, ACF+12]; see also [LVW+04]. All known exact values of LIN-CR(K_n) are listed in Table 10.2.1. For $n < 12$, we have LIN-CR(K_n) = CR(K_n), but LIN-CR(K_{12}) > CR(K_{12}) [ACF+12].

TABLE 10.2.1 Exact values of LIN-CR(K_n).

n	LIN-CR(K_n)	n	LIN-CR(K_n)	n	LIN-CR(K_n)
4	0	13	229	22	2528
5	1	14	324	23	3077
6	3	15	447	24	3699
7	9	16	603	25	4430
8	19	17	798	26	5250
9	36	18	1029	27	6180
10	62	19	1318	30	9726
11	102	20	1657		
12	153	21	2055		

5. Let $G = G(n,p)$ be a *random* graph with n vertices, whose edges are chosen independently with probability $p = p(n)$. Let $e = p \cdot \binom{n}{2}$. Is it true that the pairwise crossing number, the odd crossing number, and the biplanar crossing number are bounded from below by a constant times e^2, provided that $e \gg n$?

10.3 GENERALIZATIONS

The concept of geometric graph can be generalized in two natural directions. Instead of straight-line drawings, we can consider curvilinear drawings. If we put them at the focus of our investigations and we wish to emphasize that they are objects of independent interest rather than planar representations of abstract graphs, we call these drawings *topological graphs.* In this sense, the results in the previous section about crossing numbers belong to the theory of topological graphs. Instead of systems of segments induced by a planar point set, we can also consider systems of simplices in the plane or in higher-dimensional spaces. Such a system is called a *geometric hypergraph.*

GLOSSARY

Topological graph: A graph drawn in the plane so that its vertices are distinct points and its edges are simple continuous arcs connecting the corresponding vertices. In a topological graph (a) no edge passes through any vertex other than its endpoints, (b) any two edges have only a finite number of interior points in common, at which they properly cross each other, and (c) no three edges cross at the same point. (Same as *drawing of a graph.*)

Weakly isomorphic topological graphs: Two topological graphs, G and H, such that there is an incidence-preserving one-to-one correspondence between $\{V(G), E(G)\}$ and $\{V(H), E(H)\}$ in which two edges of G intersect if and only if the corresponding edges of H do.

Thrackle: A topological graph in which any two nonadjacent edges cross precisely once and no two adjacent edges cross.

Generalized thrackle: A topological graph in which any two nonadjacent edges cross an odd number of times and any two adjacent edges cross an even number of times (not counting their common endpoint).

d-Dimensional geometric r-hypergraph H_r^d: A pair (V, E), where V is a set of points in general position in d-space, and E is a set of *closed* $(r-1)$-dimensional simplices induced by some r-tuples of V. The sets V and E are called the ***vertex set*** and ***(hyper)edge set*** of H_r^d, respectively. Clearly, a geometric graph is a 2-dimensional geometric 2-hypergraph.

Forbidden geometric hypergraphs: A class $\mathcal{F}$ of geometric hypergraphs not permitted to be contained in the geometric hypergraphs under consideration. Given a class $\mathcal{F}$ of forbidden geometric hypergraphs, $\mathrm{ex}_r^d(\mathcal{F}, n)$ denotes the maximum number of edges that a d-dimensional geometric r-hypergraph H_r^d of n vertices can have without containing a geometric subhypergraph belonging to $\mathcal{F}$.

Nontrivial intersection: k simplices are said to have a nontrivial intersection if their relative interiors have a point in common.

Crossing of k simplices: A common point of the relative interiors of k simplices, all of whose vertices are *distinct*. The simplices are called ***crossing simplices*** if such a point exists. A set of simplices may be ***pairwise crossing*** but not necessarily crossing. If we want to emphasize that they *all* cross, we say that they cross in the ***strong sense*** or, in brief, that they ***strongly cross***.

TOPOLOGICAL GRAPHS

The fairly extensive literature on topological graphs focuses on very few special questions, and there is no standard terminology. Most of the methods developed for the study of geometric graphs break down for topological graphs, unless we make some further structural assumptions. For example, many arguments go through for x-monotone drawings such that any two edges cross at most once. Sometimes it is sufficient to assume the latter condition.

1. An Erdős-Szekeres type theorem: A classical theorem of Erdős and Szekeres states that every complete geometric graph with n vertices has a complete geometric subgraph, weakly isomorphic to a convex complete graph C_m with $m \geq c \log n$ vertices. For complete *topological* graphs with n vertices, any two of whose edges cross at most once, one can prove the existence of a complete topological subgraph with $m \geq c \log^{1/8} n$ vertices that is weakly isomorphic either to a convex complete graph C_m or to a so-called *twisted* complete graph T_m, as depicted in Figure 10.3.1 [PT03].

2. Every topological complete graph with n vertices, any two of whose edges cross at most once, has a noncrossing subgraph isomorphic to any given tree T with at most $c \log^{1/6} n$ vertices. In particular, it contains a noncrossing path with at least $c \log^{1/6} n$ vertices [PT03].

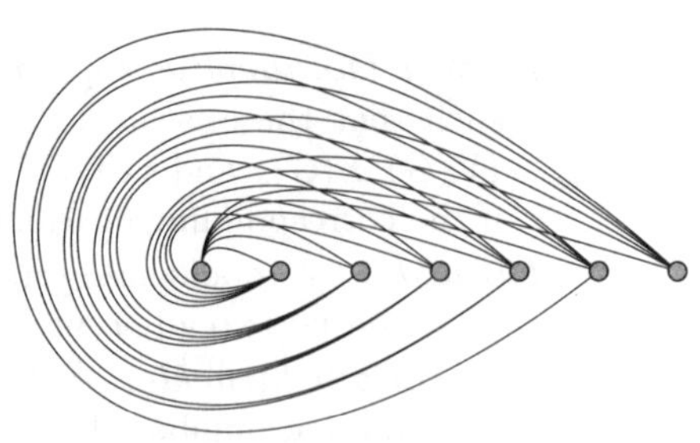

FIGURE 10.3.1
The twisted drawing T_m discovered by Harborth and Mengersen [HM92].

3. Number of topological complete graphs: Let $\overline{\Phi}(n), \Phi(n)$, and $\Phi_d(n)$ denote the number of different (i.e., pairwise weakly nonisomorphic) geometric complete graphs, topological complete graphs, and topological complete graphs in which every pair of edges cross at most d times, resp. We have $\log \overline{\Phi}(n) = \Theta(n \log n)$, $\log \Phi(n) = \Theta(n^4)$, $\Omega(n^2) \leq \log \Phi_1(n) \leq O(n^2 \alpha(n)^{O(1)})$, where $\alpha(n)$ denotes the (extremely slowly growing) inverse of the Ackermann function and $\Omega(n^2 \log n) \leq \log \Phi_d(n) \leq o(n^4)$ for every $d \geq 2$. [PT06, Kyn13].
4. Reducing the number of crossings [SŠ04]: Given an abstract graph $G = (V, E)$ and a set of pairs of edges $P \subseteq \binom{E}{2}$, we say that a topological graph K is a ***weak realization*** of G if no pair of edges not belonging to P cross each other. If G has a weak realization, then it also has a weak realization in which every edge crosses at most $2^{|E|}$ other edges. There is an almost matching lower bound for this quantity [KM91].
5. Every cycle of length different from 4 can be drawn as a thrackle [Woo69]. A bipartite graph can be drawn in the plane as a generalized thrackle if and only if it is planar [LPS97]. Every generalized thrackle with $n > 2$ vertices has at most $2n - 2$ edges, and this bound is sharp [CN00].

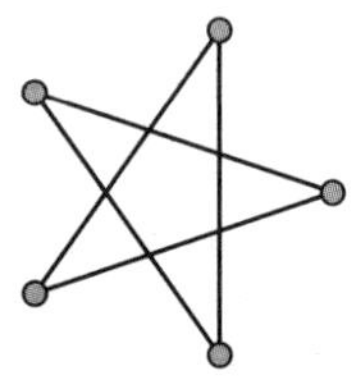

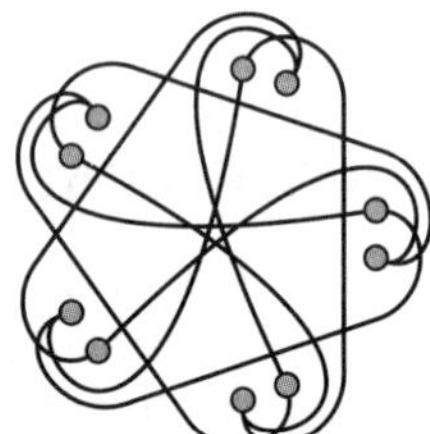

FIGURE 10.3.2
Cycles C_5 and C_{10} drawn as thrackles.

GEOMETRIC HYPERGRAPHS

If we want to generalize the results in the first two sections to higher dimensional geometric hypergraphs, we face some unexpected difficulties. Even if we restrict our attention to systems of triangles induced by 3-dimensional point sets in general position, it is not completely clear how a "crossing" should be defined. If two segments cross, they do not share an endpoint. Should this remain true for triangles? In this subsection, we describe some scattered results in this direction, but it will require further research to identify the key notions and problems.

1. Let $\mathcal{D}_k^r$ denote the class of all geometric r-hypergraphs consisting of k pairwise disjoint edges (closed $(r-1)$-dimensional simplices). Let $\mathcal{I}_k^r$ (respectively, $\mathcal{SI}_k^r$) denote the class of all geometric r-hypergraphs consisting of k simplices, any two of which have a nontrivial intersection (respectively, all of which are

strongly intersecting). Similarly, let $\mathcal{C}_k^r$ (respectively, $\mathcal{SC}_k^r$) denote the class of all geometric r-hypergraphs consisting of k pairwise crossing (respectively, strongly crossing) edges. In Table 10.3.1, we summarize the known estimates on $\mathrm{ex}_r^d(\mathcal{F}, n)$, the maximum number of hyperedges (or, simply, edges) that a d-dimensional geometric r-hypergraph of n vertices can have without containing any forbidden subconfiguration belonging to $\mathcal{F}$. We assume $d \geq 3$. In the first line of the table, the lower bound is conjectured to be tight. The upper bounds in the second line are tight for $d = 2, 3$.

TABLE 10.3.1 Estimates on $\mathrm{ex}_r^d(\mathcal{F}, n)$, the maximum number of edges of a d-dimensional geometric r-hypergraph of n vertices containing no forbidden subconfigurations belonging to $\mathcal{F}$.

r	$\mathcal{F}$	LOWER BOUND	UPPER BOUND	SOURCE
d	$\mathcal{D}_k^d$	$\Omega(n^{d-1})$	$n^{d-(1/k)^{d-1}}$	[AA89]
d	$\mathcal{I}_k^d$ $(k = 2, 3)$	?	$O(n^{d-1})$	[DP98]
d	$\mathcal{I}_k^d$ $(k > 3)$	?	$O(n^{d-1} \log n)$	[Val98]
d	$\mathcal{C}_2^d$	$\Omega(n^{d-1})$	$O(n^{d-1})$	[DP98]
d	$\mathcal{C}_k^d$ $(k > 2)$	?	$O(n^{d-(1/d)^{k-2}})$	[DP98]
$d+1$	$\mathcal{I}_k^{d+1}$	$\Omega(n^{\lceil d/2 \rceil})$	$O(n^{\lceil d/2 \rceil})$	[BF87, DP98]
$d+1$	$\mathcal{SI}_k^{d+1}$	$\Omega(n^{\lceil d/2 \rceil})$	$O(n^{\lceil d/2 \rceil})$	[BF87, DP98]
$d+1$	$\mathcal{C}_2^{d+1}$	$\Omega(n^d)$	$O(n^d)$	[DP98]

2. Akiyama-Alon theorem [AA89]: Let $V = V_1 \cup \ldots \cup V_d$ $(|V_1| = \ldots = |V_d| = n)$ be a dn-element set in general position in d-space, and let E consist of all $(d-1)$-dimensional simplices having exactly one vertex in each V_i. Then E contains n disjoint simplices. This result can be applied to deduce the upper bound in the first line of Table 10.3.1.

3. Assume that, for suitable constants c_1 and $0 \leq \delta \leq 1$, we have $\mathrm{ex}_r^d(\mathcal{SC}_k^r, n) < c_1\binom{n}{r}/n^\delta$ and $e \geq (c_1 + 1)\binom{n}{r}/n^\delta$. Then there exists $c_2 > 0$ such that the minimum number of strongly crossing k-tuples of edges in a d-dimensional r-hypergraph with n vertices and e edges is at least

$$c_2 \binom{n}{kr} e^\gamma / \binom{n}{r}^\gamma,$$

where $\gamma = 1 + (k-1)r/\delta$. This result can be used to deduce the upper bound in line 5 of Table 10.3.1.

4. A Ramsey-type result [DP98]: Let us 2-color all $(d-1)$-dimensional simplices induced by $(d+1)n - 1$ points in general position in $\mathbb{R}^d$. Then one can always find n disjoint simplices of the same color. This result cannot be improved.

5. Convex geometric hypergraphs in the plane [Bra04]: If we choose triangles from points in convex position in the plane, then the concept of isomorphism is much clearer than in the higher-dimensional cases. Thus two triangles without a common vertex can occur in three mutual positions, and we have $\mathrm{ex}(n, \;) = \Theta(n^3)$, $\mathrm{ex}(n, \;) = \Theta(n^2)$, $\mathrm{ex}(n, \;) = \Theta(n^2)$. Similarly, two

triangles with one common vertex can occur again in three positions and we have $\mathrm{ex}(n,) = \Theta(n^3)$, $\mathrm{ex}(n,) = \Theta(n^2)$, $\mathrm{ex}(n,) = \Theta(n^2)$, which is surprising, since the underlying hypergraph has a linear Turán function. Finally, two triangles with two common vertices have two possible positions, and we have $\mathrm{ex}(n,) = \Theta(n^3)$, $\mathrm{ex}(n,) = \Theta(n^2)$. Larger sets of forbidden subconvex geometric hypergraphs occur as the combinatorial core of several combinatorial geometry problems.

OPEN PROBLEMS

1. (Ringel, Harborth) For any k, determine or estimate the smallest integer $n = n(k)$ for which there is a complete topological graph with n vertices, every pair of whose edges intersect at most once (including possibly at their common endpoints), and every edge of which crosses at least k others. It is known that $n(1) = 8$, $7 \le n(2) \le 11$, $7 \le n(3) \le 14$, $7 \le n(4) \le 16$, and $n(k) \le 4k/3 + O(\sqrt{k})$ [HT69]. Does $n(k) = o(k)$ hold?
2. (Harborth) Is it true that in every complete topological graph with n vertices, every pair of whose edges cross at most once (including possibly at their common endpoints), there are at least $2n - 4$ empty triangles [Har98]? (A triangle bounded by all edges connecting three vertices is said to be ***empty***, if there is no point in its interior.) It is known that every complete topological graph with the above property has at least n empty triangles [AHP+15], [Rui15].
3. Conway conjectured that the number of edges of a thrackle cannot exceed its number of vertices. It is known that every thrackle with n vertices has at most $1.4n$ edges [Xu14]. Conway's conjecture is true for thrackles drawn by x-monotone edges [PS11], and for thrackles drawn as outerplanar graphs [CN08].
4. (Kalai) What is the maximum number $\mu(n)$ of hyperedges that a 3-dimensional geometric 3-hypergraph of n vertices can have, if any pair of its hyperedges either are disjoint or share at most one vertex? Is it true that $\mu(n) = o(n^2)$? Károlyi and Solymosi [KS02] showed that $\mu(n) = \Omega(n^{3/2})$.

10.4 SOURCES AND RELATED MATERIAL

SURVEYS

All results not given an explicit reference above may be traced in these surveys.

[PA95]: Monograph devoted to combinatorial geometry. Chapter 14 is dedicated to geometric graphs.

[Pac99, Kár13]: Introduction to geometric graph theory and survey on Ramsey-type problems in geometric graph theory, respectively.

[Pac91, DP98]: The first surveys of results in geometric graph theory and geometric hypergraph theory, respectively.

[PT00a, Pac00, Szé04, SSSV97, Sch14]: Surveys on open problems and on crossing numbers.

[DETT99]: Monograph on graph drawing algorithms.

[BMP05]: Survey of representative results and open problems in discrete geometry, originally started by the Moser brothers.

[Grü72]: Monograph containing many results and conjectures on configurations and arrangements of points and arcs.

RELATED CHAPTERS

Chapter 1: Finite point configurations
Chapter 5: Pseudoline arrangements
Chapter 11: Euclidean Ramsey theory
Chapter 28: Arrangements
Chapter 55: Graph drawing

REFERENCES

[AA89] J. Akiyama and N. Alon. Disjoint simplices and geometric hypergraphs. In G.S. Bloom, R. Graham, and J. Malkevitch. editors, *Combinatorial Mathematics*, vol. 555 of *Ann. New York Acad. Sci.*, pages 1–3, New York Acad. Sci., 1989.

[AAF+14] B.M. Ábrego, O. Aichholzer, S. Fernández-Merchant, P. Ramos, and G. Salazar. Shellable drawings and the cylindrical crossing number of K_n. *Discrete Comput. Geom.*, 52:743–753, 2014.

[AAP+97] P.K. Agarwal, B. Aronov, J. Pach, R. Pollack, and M. Sharir. Quasi-planar graphs have a linear number of edges. *Combinatorica*, 17:1–9, 1997.

[ACF09] M.O. Albertson, D.W. Cranston, and J. Fox. Crossings, colorings, and cliques. *Electron. J. Combin.*, 16:#R45, 2009.

[ACF+10] B.M. Ábrego, M. Cetina, S. Fernández-Merchant, J. Leaños, and G. Salazar. 3-symmetric and 3-decomposable geometric drawings of K_n. *Discrete Appl. Math.*, 158:1240–1258, 2010.

[ACF+12] B.M. Ábrego, M. Cetina, S. Fernández-Merchant, J. Leanos, and G. Salazar. On ($\leq k$)-edges, crossings, and halving lines of geometric drawings of K_n. *Discrete Comput. Geom.*, 48:192–215, 2012.

[Ack09] E. Ackerman. On the maximum number of edges in topological graphs with no four pairwise crossing edges. *Discrete Comput. Geom.*, 41:365–375, 2009.

[Ack16] E. Ackerman. On topological graphs with at most four crossings per edge. Preprint, `arXiv:1509.01932`, 2015.

[ACNS82] M. Ajtai, V. Chvátal, M. Newborn, and E. Szemerédi. Crossing-free subgraphs. *Ann. Discrete Math.*, 12:9–12, 1982.

[AEG+94] B. Aronov, P. Erdős, W. Goddard, D.J. Kleitman, M. Klugerman, J. Pach, and L.J. Schulman. Crossing families. *Combinatorica*, 14:127–134, 1994.

[AFP+14] E. Ackerman, J. Fox, J. Pach, and A. Suk. On grids in topological graphs. *Comput. Geom.*, 47:710–723, 2014.

[AHP+15] O. Aichholzer, T. Hackl, A. Pilz, P. Ramos, V. Sacristán, and B. Vogtenhuber. Empty triangles in good drawings of the complete graph. *Graphs Combin.*, 31:335–345, 2015.

[AT07] E. Ackerman and G. Tardos. On the maximum number of edges in quasi-planar graphs. *J. Combin. Theory Ser. A*, 114:563–571, 2007.

[BCD+14] M.J. Bannister, Z. Cheng, W.E. Devanny, and D. Eppstein. Superpatterns and universal point sets. *J. Graph Algorithms Appl.*, 18:177–209, 2014.

[BCK+15] M. Balko, J. Cibulka, K. Král, and J. Kynć l. Ramsey numbers of ordered graphs. *Electron. Notes Discrete Math.*, 49:419–424, 2015.

[BD93] D. Bienstock and N. Dean. Bounds for rectilinear crossing numbers. *J. Graph Theory*, 17:333–348, 1993.

[BDV04] A. Bialostocki, P. Dierker, and W. Voxman. Either a graph or its complement is connected: a continuing saga. Preprint., Univ. Idaho, Moscow, 2004.

[Bec83] J. Beck. On size Ramsey number of paths, trees, and circuits I. *J. Graph Theory*, 7:115–129, 1983.

[BF87] I. Bárány and Z. Füredi. Empty simplices in Euclidean space. *Canad. Math. Bull.*, 30:436–445, 1987.

[BFK15] M. Balko, R. Fulek, and J. Kynčl. Crossing numbers and combinatorial characterization of monotone drawings of K_n. *Discrete Comput. Geom.*, 53:107–143, 2015.

[BH96] A. Bialostocki and H. Harborth. Ramsey colorings for diagonals of convex polygons. *Abh. Braunschweig. Wiss. Ges.*, 47:159–163, 1996.

[Bie91] D. Bienstock. Some provably hard crossing number problems. *Discrete Comput. Geom.*, 6:443–459, 1991.

[BK64] J. Blažek and M. Koman. A minimal problem concerning complete plane graphs. In: M. Fiedler, editor, *Theory of Graphs and Its Applications*, pages 113–117, Czech Academy of Science, Prague, 1964.

[BKV03] P. Brass, G. Károlyi, and P. Valtr. A Turán-type extremal theory of convex geometric graphs. In B. Aronov, S. Basu, J. Pach, and M. Sharir, editors, *Discrete and Computational Geometry—The Goodman-Pollack Festschrift*, Springer-Verlag, Berlin, 2003.

[BMP05] P. Brass, W.O.J. Moser, and J. Pach. *Research Problems in Discrete Geometry.* Springer, New York, 2005.

[Bra04] P. Brass. Turán-type extremal problems for convex geometric hypergraphs. In J. Pach, editor, *Geometric Graph Theory*, vol. 342 of *Contemp. Math.*, pages 25–33, Amer. Math. Soc., Providence, 2004.

[BS93] G.R. Brightwell and E.R. Scheinerman. Representations of planar graphs. *SIAM J. Discrete Math.*, 6:214–229, 1993.

[BT10] J. Barát and G. Tóth. Towards the Albertson conjecture. *Electron. J. Combin.*, 17:#R73, 2010.

[CG09] J. Chalopin and D. Gonçalvez. Every planar graph is the intersection graph of segments in the plane: extended abstract. *Proc. 41st ACM Sympos. Theory Comput.*, pages 631–638, 2009.

[CGK+15] J. Cibulka, P. Gao, M. Krčál, T. Valla, and P. Valtr. On the geometric Ramsey number of outerplanar graphs. *Discrete Comput. Geom.*, 53:64–79, 2015.

[Cho34] Ch. Chojnacki. Über wesentlich unplättbare Kurven im drei-dimensionalen Raume. *Fund. Math.*, 23:135–142, 1934.

[CN00] G. Cairns and Y. Nikolayevsky. Bounds for generalized thrackles. *Discrete Comput. Geom.*, 23:191–206, 2000.

[CN08] G. Cairns and Y. Nikolayevsky. Outerplanar thrackles. em Graphs Combin., 28:85–96, 2012.

[CP92] V. Capoyleas and J. Pach. A Turán-type theorem on chords of a convex polygon. *J. Combin. Theory Ser. B*, 56:9–15, 1992.

[CSSV08] É. Czabarka, O. Sýkora, L.A. Székely, and I. Vrťo. Biplanar crossing numbers. II. Comparing crossing numbers and biplanar crossing numbers using the probabilistic method. *Random Structures Algorithms*, 33:480–496, 2008.

[DETT99] G. Di Battista, P. Eades, R. Tamassia, and I.G. Tollis. *Graph Drawing: Algorithms for the Visualization of Graphs.* Prentice-Hall, Upper Saddle River, 1999.

[DF13] G. Di Battista and F. Frati. Drawing trees, outerplanar graphs, series-parallel graphs, and planar graphs in small area. In J. Pach, editor, *Geometric Graph Theory*, pages 121-165, Springer, New York, 2013

[DP98] T.K. Dey and J. Pach. Extremal problems for geometric hypergraphs. *Discrete Comput. Geom.*, 19:473–484, 1998.

[DV02] H. Djidjev and I. Vrťo. An improved lower bound on crossing numbers. In *Graph Drawing*, vol. 2265 of *LNCS*, pages 96–101, Springer-Verlag, Berlin, 2002.

[EFRS78] P. Erdős, R.J. Faudree, C.C. Rousseau, and R.H. Schelp. The size Ramsey number. *Period. Math. Hungar.*, 9:145–161, 1978.

[EG73] P. Erdős and R.K. Guy. Crossing number problems. *Amer. Math. Monthly*, 80:52–58, 1973.

[EGS03] G. Even, S. Guha, and B. Schieber. Improved approximations of crossings in graph drawings and VLSI layout areas. *SIAM J. Comput.*, 32:231–252, 2003.

[Fár48] I. Fáry. On straight line representation of planar graphs. *Acta Univ. Szeged. Sect. Sci. Math.*, 11:229–233, 1948.

[FF00] L. Faria and C.M.H. de Figuerado. On Eggleton and Guy conjectured upper bounds for the crossing number of the n-cube. *Math. Slovaca*, 50:271–287, 2000.

[FM07] H. de Fraysseix and P. Ossona de Mendez. Representations by contact and intersection of segments. *Algorithmica*, 47:453–463, 2007.

[FMP94] H. de Fraysseix, P.O. de Mendez, and J. Pach. Representation of planar graphs by segments. In K. Böröczky and G. Fejes Tóth, editors, *Intuitive Geometry*, vol. 63 of *Colloq. Math. Soc. János Bolyai*, pages 109–117, North-Holland, Amsterdam, 1994.

[FPP90] H. de Fraysseix, J. Pach, and R. Pollack. How to draw a planar graph on a grid. *Combinatorica*, 10:41–51, 1990.

[FPSŠ13] R. Fulek, M.J. Pelsmajer, M. Schaefer, and D. Štefankovič. Hanani-Tutte, monotone drawings, and level-planarity. In J. Pach, editor, *Thirty Essays on Geometric Graph Theory*, pages 263–277, Springer, New York, 2013.

[FT15] R. Fulek and C.D. Tóth. Universal point sets for planar three-trees. *J. Discrete Algorithms*, 30:101–112, 2015.

[GJ83] M.R. Garey and D.S. Johnson. Crossing number is NP-complete. *SIAM J. Alg. Discrete Math.*, 4:312–316, 1983.

[GMPP91] P. Gritzmann, B. Mohar, J. Pach, and R. Pollack. Embedding a planar triangulation with vertices at specified points (solution to problem E3341). *Amer. Math. Monthly*, 98:165–166, 1991.

[GRS90] R.L. Graham, B.L. Rothschild, and J.H. Spencer. *Ramsey Theory*, 2nd ed. Wiley, New York, 1990.

[Grü72] B. Grünbaum. *Arrangements and Spreads.* Vol. 10 of *CBMS Regional Conf. Ser. in Math.*, Amer. Math. Soc., Providence, 1972.

[GS04] L.Y. Glebsky and G. Salazar. The crossing number of $C_m \times C_n$ is as conjectured for $n = m(m+1)$. *J. Graph Theory*, 47:53–72, 2004.

[Guy69] R.K. Guy. The decline and fall of Zarankiewicz's theorem. In F. Harary, editor, *Proof Techniques in Graph Theory*, pages 63–69, Academic Press, New York, 1969.

[Har98] H. Harborth. Empty triangles in drawings of the complete graph. *Discrete Math.*, 191:109–111, 1998.

[HKS73] F. Harary, P.C. Kainen, and A.J. Schwenk. Toroidal graphs with arbitrarily high crossing numbers. *Nanta Math.*, 6:58–67, 1973.

[HM92] H. Harborth and I. Mengersen. Drawings of the complete graph with maximum number of crossings. In *Proc. 23rd Southeast. Internat. Conf. Combin. Graph Theory Comput.*, *Congr. Numer.*, 88:225–228, 1992.

[HP34] H. Hopf and E. Pannwitz. Aufg. Nr. 167. *Jahresb.. Deutsch. Math.-Ver.*, 43:114, 1934.

[HT69] H. Harborth and C. Thürmann. Minimum number of edges with at most s crossings in drawings of the complete graph. In *Proc. 25th Southeast. Internat. Conf. Combin. Graph Theory Comput.*, *Congr. Numer.*, 119:79–83, 1969.

[IPTT94] Y. Ikebe, M.A. Perles, A. Tamura, and S. Tokunaga. The rooted tree embedding problem into points in the plane. *Discrete Comput. Geom.*, 11:51–63, 1994.

[Kár13] G. Károlyi. Ramsey-type problems for geometric graphs. In J. Pach, editor, *Thirty Essays on Geometric Graph Theory*, pages 371–382, Springer, New York, 2013.

[Kin92] N. Kinnersley. The vertex separation number of a graph equals its path-width. *Inform. Process. Lett.*, 142:345–350, 1992.

[Kur04] M. Kurowski. A 1.235 lower bound on the number of points needed to draw all n-vertex planar graphs. *Informa. Process. Lett.*, 92:95–98, 2004.

[KK00] A. Kaneko and M. Kano. Straight line embeddings of rooted star forests in the plane. *Discrete Appl. Math.*, 101:167–175, 2000.

[KK03] A. Kaneko and M. Kano. Discrete geometry on red and blue points in the plane—a survey. In B. Aronov, S. Basu, J. Pach, and M. Sharir, editors, *Discrete and Computational Geometry—The Goodman-Pollack Festschrift*, Springer-Verlag, Berlin, 2003.

[Kle70] D.J. Kleitman. The crossing number of $k_{5,n}$. *J. Combin. Theory*, 9:315–323, 1970.

[KM91] J. Kratochvíl and J. Matoušek. String graphs requiring exponential representations. *J. Combin. Theory Ser. B*, 53:1–4, 1991.

[Koe36] P. Koebe. Kontaktprobleme der konformen Abbildung. *Ber. Verh. Sächs. Akad. Wiss. Leipzig Math.-Phys. Klasse*, 88:141–164, 1936.

[KP96] Y.S. Kupitz and M.A. Perles. Extremal theory for convex matchings in convex geometric graphs. *Discrete Comput. Geom.*, 15:195–220, 1996.

[KPT97] G. Károlyi, J. Pach, and G. Tóth. Ramsey-type results for geometric graphs I. *Discrete Comput. Geom.*, 18:247–255, 1997.

[KPTV98] G. Károlyi, J. Pach, G. Tóth, and P. Valtr. Ramsey-type results for geometric graphs II. *Discrete Comput. Geom.*, 20:375–388, 1998.

[KPT07] J. Kynčl, J. Pach, and G. Tóth. Long alternating paths in bicolored point sets. *Discrete Math.*, 308:4315–4321, 2008.

[KS02] G. Károlyi and J. Solymosi. Almost disjoint triangles in 3-space. *Discrete Comput. Geom.*, 28:577–583, 2002.

[Kup84] Y. Kupitz. On pairs of disjoint segments in convex position in the plane. *Ann. Discrete Math.*, 20:203–208, 1984.

[Kyn12] J. Kynčl. Ramsey-type constructions for arrangements of segments. *European J. Combin.*, 33:336–339, 2012.

[Kyn13] J. Kynčl. Improved enumeration of simple topological graphs. *Discrete Comput. Geom.*, 50:727–770, 2013.

[Lei83] T. Leighton. *Complexity Issues in VLSI, Foundations of Computing Series.* MIT Press, Cambridge, 1983.

[LMPT94] D. Larman, J. Matoušek, J. Pach, and J. Törőcsik. A Ramsey-type result for planar convex sets. *Bull. London Math. Soc.*, 26:132–136, 1994.

[LPS97] L. Lovász, J. Pach, and M. Szegedy. On Conway's thrackle conjecture. *Discrete Comput. Geom.*, 18:369–376, 1997.

[LVW$^+$04] L. Lovász, K. Vesztergombi, U. Wagner, and E. Welzl. Convex quadrilaterals and k-sets. In: J. Pach, editor, *Towards a Theory of Geometric Graphs*, vol. 342 of *Contemp. Math.*, pages 139–148, AMS, Providence, 2004.

[Mat14] J. Matoušek. Near-optimal separators in string graphs. *Combin. Probab. Comput.*, 23:135–139, 2014.

[MT06] A. Marcus and G. Tardos. Intersection reverse sequences and geometric applications. *J. Combin. Theory Ser. A*, 113:675–691, 2006.

[PA95] J. Pach and P.K. Agarwal. *Combinatorial Geometry.* Wiley, New York, 1995.

[Pac91] J. Pach. Notes on geometric graph theory. In J. Goodman, R. Pollack, and W. Steiger, editors, *Discrete and Computational Geometry: Papers from the DIMACS Special Year*, pages 273–285, AMS, Providence, 1991.

[Pac99] J. Pach. Geometric graph theory. In J.D. Lamb and D.A. Preece, editors, *Surveys in Combinatorics, 1999*, vol. 267 of *London Math. Soc. Lecture Note Ser.*, pages 167–200, Cambridge University Press, 1999.

[Pac00] J. Pach. Crossing numbers. In J. Akiyama, M. Kano, and M. Urabe, editors, *Discrete and Computational Geometry*, vol. 1763 of *LNCS*, pages 267–273. Springer-Verlag, Berlin, 2000.

[PPST05] J. Pach, R. Pinchasi, M. Sharir, and G. Tóth. Topological graphs with no large grids. *Graphs Combin.*, 21:355–364, 2005.

[PPTT02] J. Pach, R. Pinchasi, G. Tardos, and G. Tóth. Geometric graphs with no self-intersecting path of length three. In M.T. Goodrich and S.G. Kobourov, editors, *Graph Drawing*, vol. 2528 of *LNCS*, pages 295–311, Springer-Verlag, Berlin, 2002.

[PR04] R. Pinchasi and R. Radoičić. Topological graphs with no self-intersecting cycle of length 4. In J. Pach, editor, *Towards a Theory of Geometric Graphs*, vol. 342 of *Contemp. Math.*, pages 233–243, AMS, Providence, 2004.

[PRT04] J. Pach, R. Radoičić, and G. Tóth. A generalization of quasi-planarity. In J. Pach, editor, *Towards a Theory of Geometric Graphs*, vol. 342 of *Contemp. Math.*, pages 177–183, AMS, Providence, 2004.

[PRTT06] J. Pach, R. Radoičić, G. Tardos, and G. Tóth. Improving the crossing lemma by finding more crossings in sparse graphs. *Discrete Comput. Geom.*, 36:527–552, 2006.

[PS11] J. Pach and E. Sterling. Conway's conjecture for monotone thrackles. *Amer. Math. Monthly*, 118:544–548, 2011.

[PSS96] J. Pach, F. Shahrokhi, and M. Szegedy. Applications of crossing number. *Algorithmica*, 16:111–117, 1996.

[PSS08] M.J. Pelsmajer, M. Schaefer, and D. Štefankovič. Odd crossing number and crossing number are not the same. *Discrete Comput. Geom.*, 39:442–454, 2008.

[PSS09] M.J. Pelsmajer, M. Schaefer, and D. Stasi. Strong Hanani-Tutte on the projective plane. *SIAM J. Discrete Math.*, 23:1317–1323, 2009.

[PST00] J. Pach, J. Spencer, and G. Tóth. New bounds on crossing numbers. *Discrete Comput. Geom.*, 24:623–644, 2000.

[PT94] J. Pach and J. Törőcsik. Some geometric applications of Dilworth's theorem. *Discrete Comput. Geom.*, 12:1–7, 1994.

[PT97] J. Pach and G. Tóth. Graphs drawn with few crossings per edge. *Combinatorica*, 17:427–439, 1997.

[PT00a] J. Pach and G. Tóth. Thirteen problems on crossing numbers. *Geombinatorics*, 9:194–207, 2000.

[PT00b] J. Pach and G. Tóth. Which crossing number is it anyway? *J. Combin. Theory Ser. B*, 80:225–246, 2000.

[PT03] J. Pach and G. Tóth. Unavoidable configurations in complete topological graphs. *Discrete Comput. Geom.*, 30:311–320, 2003.

[PT04] J. Pach and G. Tóth. Monotone drawings of planar graphs. *J. Graph Theory*, 46:39–47, 2004. Revised version at `arXiv:1101.0967`.

[PT06] J. Pach and G. Tóth. How many ways can one draw a graph? *Combinatorica*, 26:559–576, 2006.

[Ram30] F. Ramsey. On a problem of formal logic. *Proc. London Math. Soc.*, 30:264–286, 1930.

[Rui15] A.J. Ruiz-Vargas. Empty triangles in complete topological graphs. *Discrete Comput. Geom.*, 53:703–712, 2015.

[Sch90] W. Schnyder. Embedding planar graphs on the grid. In *Proc. 1st ACM-SIAM Sympos. Discrete Algorithms*, pages 138–148, 1990.

[Sch14] M. Schaefer. The graph crossing number and its variants. *Electron. J. Combain.*, Dynamic Survey DS21, 2014.

[SŠ04] M. Schaefer and D. Štefankovič. Decidability of string graphs. *J. Comput. System Sci.*, 68:319–334, 2004.

[SSŠ03] M. Schaefer, E. Sedgwick, and D. Štefankovič. Recognizing string graphs in NP. *J. Comput. System Sci.*, 67:365–380, 2003.

[SSSV97] F. Shahrokhi, O. Sýkora, L.A. Székely, and I. Vrťo. Crossing numbers: bounds and applications. In I. Bárány and K. Böröczky, editors, *Intuitive Geometry*, vol. 6 of *Bolyai Soc. Math. Stud.*, pages 179–206, J. Bolyai Math. Soc., Budapest, 1997.

[ST02] J. Spencer and G. Tóth. Crossing numbers of random graphs. *Random Structures Algorithms*, 21:347–358, 2002.

[Ste22] E. Steinitz. Polyeder und Raumteilungen, part 3AB12. In *Enzykl. Math. Wiss. 3 (Geometrie)*, pages 1–139. 1922.

[SV93] O. Sýkora and I. Vrťo. On the crossing number of the hypercube and the cube connected cycles. *BIT*, 33:232–237, 1993.

[SV94] O. Sýkora and I. Vrťo. On VLSI layouts of the star graph and related networks. *Integration, the VLSI Journal*, 17:83–93, 1994.

[Szé04] L.A. Székely. A successful concept for measuring non-planarity of graphs: the crossing number. *Discrete Math.*, 276:331–352, 2004.

[Szé97] L.A. Székely. Crossing numbers and hard Erdős problems in discrete geometry. *Combin. Probab. Comput.*, 6:353–358, 1997.

[Tar13] G. Tardos. Construction of locally plane graphs with many edges. In . Pach, editor, *Thirty Essays on Geometric Graph Theory*, pages 541–562, Springer, New York, 2013.

[Thu78] W.P. Thurston. *The Geometry and Topology of 3-manifolds.* Princeton Univ. Lect. Notes, 1978.

[Tót00] G. Tóth. Note on geometric graphs. *J. Combin. Theory Ser. A*, 89:126–132, 2000.

[Tót08] G. Tóth. Note on the pair-crossing number and the odd-crossing umber *Discrete Comput. Geom.*, 39:791-799, 2008.

[Tur54] P. Turán. On the theory of graphs. *Colloq. Math.*, 3:19–30, 1954.

[Tur77] P. Turán. A note of welcome. *J. Graph Theory*, 1:7–9, 1977.

[Tut60] W.T. Tutte. Convex representations of graphs. *Proc. London Math. Soc.*, 10:304–320, 1960.

[Tut70] W.T. Tutte. Toward a theory of crossing numbers. *J. Combin. Theory*, 8:45–53, 1970.

[Xu14] Y. Xu. *Generalized Thrackles and Graph Embeddings.* Master Thesis, Simon Fraser University, 2014.

[Val98] P. Valtr. On geometric graphs with no k pairwise parallel edges. *Discrete Comput. Geom.*, 19:461–469, 1998.

[Woo69] D.R. Woodall. Thrackles and deadlock. In D.J.A. Welsh, editor, *Combinatorial Mathematics and Its Applications*, pages 335–348, Academic Press, London, 1969.

[Woo93] D.R. Woodall. Cyclic-order graphs and Zarankiewicz's crossing-number conjecture. *J. Graph Theory*, 17:657–671, 1993.

11 EUCLIDEAN RAMSEY THEORY

R.L. Graham

INTRODUCTION

Ramsey theory typically deals with problems of the following type. We are given a set S, a family $\mathcal{F}$ of subsets of S, and a positive integer r. We would like to decide whether or not for every partition of $S = C_1 \cup \cdots \cup C_r$ into r subsets, it is always true that some C_i contains some $F \in \mathcal{F}$. If so, we abbreviate this by writing $S \stackrel{r}{\longrightarrow} \mathcal{F}$ (and we say S is r-Ramsey). If not, we write $S \stackrel{r}{\not\longrightarrow} \mathcal{F}$. (For a comprehensive treatment of Ramsey theory, see [GRS90].)

In Euclidean Ramsey theory, S is usually taken to be the set of points in some Euclidean space $\mathbb{E}^N$, and the sets in $\mathcal{F}$ are determined by various geometric considerations. The case most studied is the one in which $\mathcal{F} = \mathrm{Cong}(X)$ consists of all *congruent* copies of a fixed finite configuration $X \subset S = \mathbb{E}^N$. In other words, $\mathrm{Cong}(X) = \{gX \mid g \in SO(N)\}$, where $SO(N)$ denotes the special orthogonal group acting on $\mathbb{E}^N$.

Further, we say that X is *Ramsey* if, for all r, $\mathbb{E}^N \stackrel{r}{\longrightarrow} \mathrm{Cong}(X)$ holds provided N is sufficiently large (depending on X and r). This we indicate by writing $\mathbb{E}^N \longrightarrow X$.

Another important case we will discuss (in Section 11.4) is that in which $\mathcal{F} = \mathrm{Hom}(X)$ consists of all *homothetic* copies $aX + \bar{t}$ of X, where a is a positive real and $\bar{t} \in \mathbb{E}^N$. Thus, in this case $\mathcal{F}$ is just the set of all images of X under the group of positive homotheties acting on $\mathbb{E}^N$.

It is easy to see that any Ramsey (or r-Ramsey) set must be finite. A standard compactness argument shows that if $\mathbb{E}^N \stackrel{r}{\longrightarrow} X$ then there is always a *finite* set $Y \subseteq \mathbb{E}^N$ such that $Y \stackrel{r}{\longrightarrow} X$. Also, if X is Ramsey (or r-Ramsey) then so is any homothetic copy $aX + \bar{t}$ of X.

GLOSSARY

$\mathbb{E}^N \stackrel{r}{\longrightarrow}$ **Cong** ***(X)*:** For any partition $\mathbb{E}^N = C_1 \cup \cdots \cup C_r$, some C_i contains a set congruent to X. We say that X is ***r-Ramsey***. When $\mathrm{Cong}(X)$ is understood we will usually write $\mathbb{E}^N \stackrel{r}{\longrightarrow} X$.

$\mathbb{E}^N \longrightarrow$ ***X*:** For every r, $\mathbb{E}^N \stackrel{r}{\longrightarrow} \mathrm{Cong}(X)$ holds, provided N is sufficiently large. We say in this case that X is ***Ramsey***.

11.1 *r*-RAMSEY SETS

In this section we focus on low-dimensional r-Ramsey results. We begin by stating three conjectures.

CONJECTURE 11.1.1

For any nonequilateral triangle T (i.e., the set of 3 vertices of T),

$$\mathbb{E}^2 \xrightarrow{2} T.$$

CONJECTURE 11.1.2 (stronger)

For any partition $\mathbb{E}^2 = C_1 \cup C_2$, every triangle occurs (up to congruence) in C_1, or else the same holds for C_2, with the possible exception of a single equilateral triangle.

The partition $\mathbb{E}^2 = C_1 \cup C_2$ with

$$\begin{aligned} C_1 &= \{(x,y) \mid -\infty < x < \infty, 2m \le y < 2m+1, m = 0, \pm 1, \pm 2, \ldots\} \\ C_2 &= \mathbb{E}^2 \setminus C_1 \end{aligned}$$

into alternating half-open strips of width 1 prevents the equilateral triangle of side $\sqrt{3}$ from occurring in a single C_i. In fact, there are other ways of 2-coloring the plane so as to avoid a monochromatic unit equilateral triangle, such as the so-called "zebra-like" colorings as described in [JKS+09]. It is also shown in [JKS+09] that if the plane is decomposed into the union of an open set and a closed set, then every equilateral triangle occurs at least one of these sets.

CONJECTURE 11.1.3

For any triangle T,

$$\mathbb{E}^2 \overset{3}{\not\rightarrow} T.$$

In the positive direction, we have [EGM+75b]:

THEOREM 11.1.4

(a) $\mathbb{E}^2 \xrightarrow{2} T$ *if T is a triangle satisfying:*

(i) *T has a ratio between two sides equal to $2\sin\theta/2$ with $\theta = 30°, 72°, 90°$, or $120°$*

(ii) *T has a $30°, 90°$, or $150°$ angle* [Sha76]

(iii) *T has angles $(\alpha, 2\alpha, 180° - 3\alpha)$ with $0 < \alpha < 60°$*

(iv) *T has angles $(180° - \alpha, 180° - 2\alpha, 3\alpha - 180°)$ with $60° < \alpha < 90°$*

(v) *T is the degenerate triangle $(a, 2a, 3a)$*

(vi) *T has sides (a, b, c) satisfying*

$$a^6 - 2a^4b^2 + a^2b^4 - 3a^2b^2c^2 + b^2c^2 = 0$$

or

$$a^4c^2 + b^4a^2 + c^4b^2 - 5a^2b^2c^2 = 0$$

(vii) *T has sides (a, b, c) satisfying*

$$c^2 = a^2 + 2b^2 \quad \textit{with} \quad a < 2b \qquad \text{[Sha76]}$$

(viii) *T has sides (a, b, c) satisfying*

$$a^2 + c^2 = 4b^2 \textit{ with } 3b^2 < 2a^2 < 5b^2 \qquad \text{[Sha76]}$$

(ix) *T has sides equal in length to the sides and circumradius of an isosceles triangle*

(b) $\mathbb{E}^3 \xrightarrow{2} T$ *for any nondegenerate triangle* T

(c) $\mathbb{E}^3 \xrightarrow{3} T$ *for any nondegenerate right triangle* T [BT96]

(d) $\mathbb{E}^3 \overset{12}{\not\rightarrow} T$, *a triangle with angles* $(30°, 60°, 90°)$ [Bón93]

(e) $\mathbb{E}^2 \overset{2}{\not\rightarrow} Q^2$ *(4 points forming a square)*

(f) $\mathbb{E}^4 \overset{2}{\not\rightarrow} Q^2$ [Can96a]

(g) $\mathbb{E}^5 \xrightarrow{2} R^2$, *any rectangle* [Tót96]

(h) $\mathbb{E}^n \overset{4}{\not\rightarrow} \circ\overset{1}{—}\circ\overset{1}{—}\circ$ *for any* n *(a degenerate* $(1,1,2)$ *triangle)*

(i) $\mathbb{E}^n \overset{16}{\not\rightarrow} \circ\overset{a}{—}\circ\overset{b}{—}\circ$ *for any* n *(a degenerate* $(a, b, a+b)$ *triangle).*

It is not known whether the 4 in (h) or the 16 in (i) can be replaced by smaller values. Other results of this type can be found in [EGM+73], [EGM+75a], [EGM+75b], [Sha76], and [CFG91].

The 2-point set X_2 consisting of two points a unit distance apart is the simplest set about which such questions can be asked, and has a particularly interesting history (see [Soi91] for details). It is clear that

$$\mathbb{E}^1 \overset{2}{\not\rightarrow} X_2 \quad \text{and} \quad \mathbb{E}^2 \xrightarrow{2} X_2.$$

To see that $\mathbb{E}^2 \xrightarrow{3} X_2$, consider the 7-point Moser graph shown in Figure 11.1.1. All edges have length 1. On the other hand, $\mathbb{E}^2 \overset{7}{\not\rightarrow} X_2$, which can be seen by an appropriate periodic 7-coloring (= partition into 7 parts) of a tiling of $\mathbb{E}^2$ by regular hexagons of diameter 0.9 (see Figure 1.3.1).

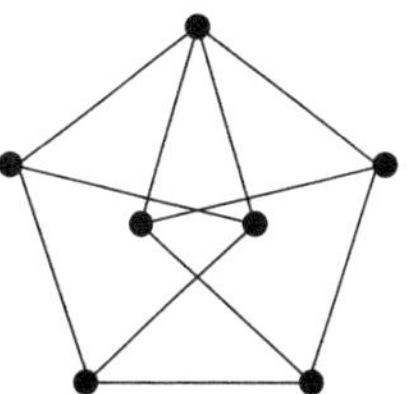

FIGURE 11.1.1
The Moser graph.

Definition: The ***chromatic number*** of $\mathbb{EE}^n$, denoted by $\chi(\mathbb{E}^n)$, is the least m such that $\mathbb{E}^n \overset{m}{\not\rightarrow} X_2$.

By the above remarks,

$$4 \le \chi(\mathbb{E}^2) \le 7.$$

These bounds have remained unchanged for over 50 years.

Some evidence that $\chi(\mathbb{E}^2) \geq 5$ (in the author's opinion) is given by the following result of O'Donnell:

THEOREM 11.1.5 [O'D00a], [O'D00b]

For any $g > 0$, there is 4-chromatic unit distance graph in $\mathbb{E}^2$ with girth greater than g.

Note that the Moser graph has girth 3.

PROBLEM 11.1.6

Determine the exact value of $\chi(\mathbb{E}^2)$.

The best bounds currently known for $\mathbb{E}^n$ are:

$$(1.239 + o(1))^n < \chi(\mathbb{E}^n) < (3 + o(1))^n$$

(see [FW81], [CFG91],[Rai00], [BMP05]).

A "near miss" for showing $\chi(\mathbb{E}^2) < 7$ was found by Soifer [Soi92]. He shows that there exists a partition $\mathbb{E}^2 = C_1 \cup \cdots \cup C_7$ where C_i contains no pair of points at distance 1 for $1 \leq i \leq 6$, while C_7 has no pair at distance $1/\sqrt{5}$.

The best bounds known for $\chi(\mathbb{E}^3)$ are:

$$6 \leq \chi(\mathbb{E}^3) \leq 15.$$

The lower bound is due to Nechushtan [Nec00] and the upper bound is due (independently to Coulson [Col02], and R. Radoičić and G. Tóth [RT02], improving earlier results of Székely and Wormald [SW89] and Bóna/Tóth [BT96]).

See Section 1.3 for more details.

An interesting phenomenon, first pointed out by Székely [Szé84], suggests that the true value of $\chi(\mathbb{E}^2)$ may depend on which axioms for set theory are being used. In ZFC, the standard Zermelo/Fraenkel axioms together with the Axiom of Choice, non-(Lebesgue)-measurable sets exist and can be used to prevent monochromatic configurations from occurring. Indeed, it was shown by Falconer [Fal81] that if the plane is decomposed into four *Lebesgue measurable* sets, then one of the sets must contain a unit distance. In other words, the "measurable" chromatic number of the plane is at least 5. On the other hand, if the Axiom of Choice is replaced by the axiom LM which asserts that every set of reals is Lebesgue measurable, then such constructions are not possible and the chromatic number of the plane may be 4 in these systems. (It is known by a result of Solovay [Sol70] that ZFC and ZF + LM are equally consistent). Further results of this type are given in the papers of Shelah and Soifer [SS03, SS04, Soi05]. Sets for which the chromatic number depends on whether or not the color classes are required to be measurable, are said to have an "ambiguous" chromatic number. In [Pay09], Payne constructs a number of interesting examples of unit-distance graphs in $\mathbb{R}^n$ which have ambiguous chromatic number. This is further evidence that the chromatic number of various configurations in $\mathbb{R}^n$ may depend on the flavor of set theory you prefer!

11.2 RAMSEY SETS

Recall that X is Ramsey (written $\mathbb{E}^N \longrightarrow X$) if, for all r, if $\mathbb{E}^N = C_1 \cup \cdots \cup C_r$ then some C_i must contain a congruent copy of X, provided only that $N \geq N_0(X, r)$.

GLOSSARY

Spherical: X is spherical if it lies on the surface of some sphere.

Rectangular: X is rectangular if it is a subset of the vertices of a rectangular parallelepiped.

Simplex: X is a simplex if it spans $\mathbb{E}^{|X|-1}$.

THEOREM 11.2.1 [EGM+73]

If X and Y are Ramsey then so is $X \times Y$.

Thus, since any 2-point set is Ramsey (for any r, consider the unit simplex S_{2r+1} in $\mathbb{E}^{2r}$ scaled appropriately), then so is any rectangular parallelepiped. This implies:

THEOREM 11.2.2

Any rectangular set is Ramsey.

Frankl and Rödl strengthen this significantly in the following way.

Definition: A set $A \subset \mathbb{E}^n$ is called ***super-Ramsey*** if there exist positive constants c and ϵ and subsets $X = X(N) \subset \mathbb{E}^N$ for every $N \geq N_0(X)$ such that:

(i) $|X| < c^n$;

(ii) $|Y| < |X|/(1+\epsilon)^n$ holds for all subsets $Y \subset X$ containing no congruent copy of A.

THEOREM 11.2.3 [FR90]

(i) *All two-element sets are super-Ramsey.*

(ii) *If A and B are super-Ramsey then so is $A \times B$.*

COROLLARY 11.2.4

If X is rectangular then X is super-Ramsey.

In the other direction we have

THEOREM 11.2.5

Any Ramsey set is spherical.

The simplest nonspherical set is the degenerate $(1,1,2)$ triangle.

Concerning simplices, we have the result of Frankl and Rödl:

THEOREM 11.2.6 [FR90]

Every simplex is Ramsey.

In fact, they show that for any simplex X, there is a constant $c = c(X)$ such that for all r,

$$\mathbb{E}^{c \log r} \xrightarrow{r} X,$$

which follows from their result:

THEOREM 11.2.7

Every simplex is super-Ramsey.

It was an open problem for more than 20 years as to whether the set of vertices of a regular pentagon was Ramsey. This was finally settled by Křiž [Kři91] who proved the following two fundamental results:

THEOREM 11.2.8 [Kři91]

Suppose $X \subseteq \mathbb{E}^N$ has a transitive solvable group of isometries. Then X is Ramsey.

COROLLARY 11.2.9

Any set of vertices of a regular polygon is Ramsey.

THEOREM 11.2.10 [Kři91]

Suppose $X \subseteq \mathbb{E}^N$ has a transitive group of isometries that has a solvable subgroup with at most two orbits. Then X is Ramsey.

COROLLARY 11.2.11

The vertex sets of the Platonic solids are Ramsey.

CONJECTURE 11.2.12

Any 4-point subset of a circle is Ramsey.

Křiž [Kři92] has shown this holds if a pair of opposite sides of the 4-point set are parallel (i.e., form a trapezoid).

Certainly, the outstanding open problem in Euclidean Ramsey theory is to determine the Ramsey sets. The author (bravely?) makes the following:

CONJECTURE 11.2.13 ($1000)

Any spherical set is Ramsey.

If true then this would imply that the Ramsey sets are exactly the spherical sets.

Recently, an alternative conjecture has been suggested by Leader, Russell and Walters [LRW12]. Let us call a finite configuration **C** in Euclidean space *transitive* if it has a transitive group of symmetries. Further, let us say that **C** is *subtransitive* if it is a subset of a transitive configuration.

CONJECTURE 11.2.14 *[LRW12]*

Any Ramsey set is subtransitive.

These authors have also shown [LRW11] that almost all 4-points subsets of a unit circle are *not* subtransitive. Thus, the question as to whether 4-point cyclic subsets are Ramsey sharply separates these two conjectures!

We point out that a result of Spencer [Spe79] shows that any finite configuration C in $\mathbb{E}^n$ is arbitrarily close to a Ramsey set. Let us say that C' is ϵ-close to C if C' can be obtained by moving each point of C by a distance of at most ϵ.

THEOREM 11.2.15 [Spe79]

For every finite configuration $C \subset \mathbb{E}^n$ and every $\epsilon > 0$, there is any ϵ-close configuration C' which is a Ramsey set.

11.3 SPHERE-RAMSEY SETS

Since spherical sets play a special role in Euclidean Ramsey theory, it is natural that the following concept arises.

GLOSSARY

***$S^N(\rho)$*:** A sphere in $\mathbb{E}^N$ with radius ρ.

Sphere-Ramsey: X is sphere-Ramsey if, for all r, there exist $N = N(X,r)$ and $\rho = \rho(X,r)$ such that

$$S^N(\rho) \xrightarrow{r} X.$$

In this case we write $S^N(\rho) \longrightarrow X$.

For a spherical set X, let $\rho(X)$ denote its circumradius, i.e., the radius of the smallest sphere containing X as a subset.

Remark. If X and Y are sphere-Ramsey then so is $X \times Y$.

THEOREM 11.3.1 [Gra83]

If X is rectangular then X is sphere-Ramsey.

In [Gra83], it was conjectured that in fact if X is rectangular and $\rho(X) = 1$ then $S^N(1+\epsilon) \longrightarrow X$ should hold. This was proved by Frankl and Rödl [FR90] in a much stronger "super-Ramsey" form.

Concerning simplices, Matousěk and Rödl proved the following spherical analogue of simplices being Ramsey:

THEOREM 11.3.2 [MR95]

For any simplex X with $\rho(X) = 1$, any r, and any $\epsilon > 0$, there exists $N = N(X, r, \epsilon)$ such that

$$S^N(1+\epsilon) \xrightarrow{r} X.$$

The proof uses an interesting mix of techniques from combinatorics, linear algebra, and Banach space theory.

The following results show that the "blowup factor" of $1+\epsilon$ is really needed.

THEOREM 11.3.3 [Gra83]

Let $X = \{x_1, \ldots, x_m\} \subset \mathbb{E}^N$ such that:

(i) *for some nonempty $I \subseteq \{1, 2, \ldots, m\}$, there exist nonzero a_i, $i \in I$, with*

$$\sum_{i \in I} a_i x_i = 0 \in \mathbb{E}^N,$$

(ii) *for all nonempty $J \subseteq I$,*

$$\sum_{j \in J} a_j \neq 0.$$

Then X is not sphere-Ramsey.

This implies that $X \subset S^N(1)$ is not sphere-Ramsey if the convex hull of X contains the center of $S^N(1)$.

Definition: A simplex $X \subset \mathbb{E}^N$ is called ***exceptional*** if there is a subset $A \subseteq X$, $|A| \geq 2$, such that the affine hull of A translated to the origin has a nontrivial intersection with the linear span of the points of $X \setminus A$ regarded as vectors.

THEOREM 11.3.4 [MR95]

If X is a simplex with $\rho(X) = 1$ and $S^N(1) \longrightarrow X$ then X must be exceptional.

It is not known whether it is true for exceptional X that $S^N(1) \longrightarrow X$. The simplest nontrivial case is for the set of three points $\{a, b, c\}$ lying on some great circle of $S^N(1)$ (with center o) so that the line joining a and b is parallel to the line joining o and c. We close with a fundamental conjecture:

CONJECTURE 11.3.5

If X is Ramsey, then X is sphere-Ramsey.

11.4 EDGE-RAMSEY SETS

In this variant (introduced in [EGM+75b], we color all the line segments $[a, b]$ in $\mathbb{E}^n$ rather than coloring the points. Analogously to our earlier definition, we will say that a configuration E of line segments is ***edge-Ramsey*** if for any r, there is an $N = N(r)$ such any r-coloring of the line segments in $\mathbb{E}^N$ contains a monochromatic congruent copy of E (up to some Euclidean motion). The main results known for edge-Ramsey configurations are the following:

THEOREM 11.4.1 [EGM+75b]

If E is edge-Ramsey then all edges of E must have the same length.

THEOREM 11.4.2 [Gra83]

If E is edge-Ramsey then the endpoints of the edges of E must lie on two spheres.

THEOREM 11.4.3 [Gra83]

If the endpoints of E do not lie on a sphere and the graph formed by E is not bipartite then E is not edge-Ramsey.

It is clear that the edge set of an n-dimensional simplex is edge-Ramsey. Less obvious (but equally true) are the following.

THEOREM 11.4.4 [Can96b]

The edge set of an n-cube is edge-Ramsey.

THEOREM 11.4.5 [Can96b]

The edge set of an n-dimensional cross polytope is edge-Ramsey.

This set, a generalization of the octahedron, has as its edges all $2n(n-1)$ line segments of the form $[(0, 0, ..., \pm1, ..., 0), (0, 0, ..., 0, \pm1, ..., 0)]$ where the two ±1's occur in different positions.

THEOREM 11.4.6 [Can96b]

The edge set of a regular n-gon is not edge-Ramsey if $n = 5$ or $n \geq 7$.

Since regular n-gons are edge-Ramsey for $n = 2$, 3, and 4, the only undecided value is $n = 6$.

PROBLEM 11.4.7 *Is the edge set of a regular hexagon edge-Ramsey?*

The situation is not as simple as one might hope since as pointed out by Cantwell [Can96b]:

(i) If AB is a line segment with C as its midpoint, then the set E_1 consisting of the line segments AC and CB is not edge-Ramsey, even though its graph is bipartite and A, B, C lie on two spheres.

(ii) There exist nonspherical sets that are edge-Ramsey.

PROBLEM 11.4.8 *Characterize edge-Ramsey configurations.*

It is not clear at this point what a reasonable conjecture might be. For more results on these topics, see [Can96b] or [Gra83].

11.5 HOMOTHETIC RAMSEY SETS AND DENSITY THEOREMS

In this section we will survey various results of the type $\mathbb{E}^N \xrightarrow{r} \text{Hom}(X)$, the set of positive homothetic images $aX + \bar{t}$ of a given set X. Thus, we are allowed to dilate and translate X but we cannot rotate it. The classic result of this type is van der Waerden's theorem, which asserts the following:

THEOREM 11.5.1 [Wae27]

If $X = \{1, 2, \ldots, m\}$ then $\mathbb{E} \xrightarrow{r} \text{Hom}(X)$.

(Note that $\text{Hom}(X)$ is just the set of m-term arithmetic progressions.)

By the compactness theorem mentioned in the Introduction there exists, for each m, a minimum value $W(m)$ such that

$$\{1, 2, \ldots, W(m)\} \xrightarrow{2} \text{Hom}(X).$$

The determination or even estimation of $W(m)$ seems to be extremely difficult. The known values are:

m	1	2	3	4	5	6
$W(m)$	1	3	9	35	178	1132

The best general result from below (due to Berlekamp—see [GRS90]) is

$$W(p+1) \geq p \cdot 2^p, \quad p \text{ prime}.$$

The best upper bound known follows from a spectacular result of Gowers [Gow01]:

$$W(m) < 2^{2^{2^{2^{m+9}}}}$$

This settled a long-standing \$1000 conjecture of the author. This result is a corollary of Gowers's new quantitative form of Szemerédi's theorem mentioned in the next section. It improves on the earlier bound of Shelah: [She88]:

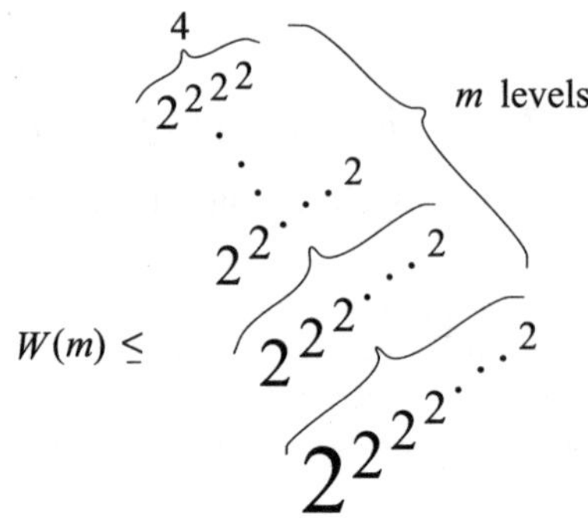

The following conjecture of the author has been open for more than 30 years:

CONJECTURE 11.5.2 (\$1000)

For all m,

$$W(m) \leq 2^{m^2}$$

The generalization to $\mathbb{E}^N$ is due independently to Gallai and Witt (see [GRS90]).

THEOREM 11.5.3

For any finite set $X \subset \mathbb{E}^n$,

$$\mathbb{E}^N \longrightarrow \mathrm{Hom}(X).$$

We remark here that a number of results in (Euclidean) Ramsey theory have stronger so-called *density* versions. As an example, we state the well-known theorem of Szemerédi.

GLOSSARY

$\mathbb{N}$: The set of natural numbers $\{1, 2, 3, \ldots\}$.

$\overline{\boldsymbol{\delta}}(\boldsymbol{A})$: The ***upper density*** of a set $A \subseteq \mathbb{N}$ is defined by:

$$\overline{\delta}(A) = \limsup_{n \longrightarrow \infty} \frac{|A \cap \{1, 2, \ldots, n\}|}{n}.$$

THEOREM 11.5.4 (Szemerédi [Sze75])

If $A \subseteq \mathbb{N}$ has $\overline{\delta}(A) > 0$ then A contains arbitrarily long arithmetic progressions.

That is, $A \cap \mathrm{Hom}\{1, 2, \ldots, m\} \neq \emptyset$ for all m. This clearly implies van der Waerden's theorem since $\mathbb{N} = C_1 \cup \cdots \cup C_r \Rightarrow \max_i \overline{\delta}(C_i) \geq 1/r$.

Furstenberg [Fur77] has given a quite different proof of Szemerédi's theorem, using tools from ergodic theory and topological dynamics. This approach has proved to be very powerful, allowing Furstenberg, Katznelson, and others to prove density versions of the Hales-Jewett theorem (see [FK91]), the Gallai-Witt theorem, and many others. Gowers has proved the following strong quantitative version of Szemerédi's theorem:

THEOREM 11.5.5 [Gow01]

For every $k > 0$, any subset of $1, 2, ..., N$ of size at least $N(\log\log N)^{-c(k)}$ contains a k-term arithmetic progression, where $c(k) = 2^{-2^{k+9}}$.

Recently, the Polymath project, initiated by Gowers, has resulted in several new proofs of the density Hales-Jewett theorem (see [Tao09, Poly12, DKT14]).

There are other ways of expressing the fact that A is relatively dense in $\mathbb{N}$ besides the condition that $\overline{\delta}(A) > 0$. One would expect that these could also be used as a basis for a density version of van der Waerden or Gallai-Witt. Very little is currently known in this direction, however. We conclude this section with several conjectures of this type.

CONJECTURE 11.5.6 (Erdős)

If $A \subseteq \mathbb{N}$ satisfies $\sum_{a \in A} 1/a = \infty$ then A contains arbitrarily long arithmetic progressions.

CONJECTURE 11.5.7 (Graham)

If $A \subseteq \mathbb{N} \times \mathbb{N}$ with $\sum_{(x,y) \in A} 1/(x^2 + y^2) = \infty$ then A contains the 4 vertices of an axis-aligned square.

More generally, I expect that A will always contain a homothetic image of $\{1, 2, \ldots, m\} \times \{1, 2, \ldots, m\}$ for all m. Of course, if we assume A has positive upper density, then this result follows from the density Hales-Jewett theorem [FK91]. A nice combinatorial proof by Solymosi for the square appears in [Sol04].

Finally, we mention a direction in which the group $SO(n)$ is enlarged to allow dilations as well.

Definition: For a set $W \subseteq \mathbb{E}^k$, define the ***upper density*** $\overline{\delta}(W)$ of W by

$$\overline{\delta}(W) := \limsup_{R \longrightarrow \infty} \frac{m(B(o, R) \cap W)}{m(B(o, R))},$$

where $B(o, R)$ denotes the k-ball $\left\{ (x_1, \ldots, x_k) \in \mathbb{E}^k \;\middle|\; \sum_{i=1}^{k} x_i^2 \leq R^2 \right\}$ centered at the origin, and m denotes Lebesgue measure.

THEOREM 11.5.8 (Bourgain [Bou86])

Let $X \subseteq \mathbb{E}^k$ be a simplex. If $W \subseteq \mathbb{E}^k$ with $\overline{\delta}(W) > 0$ then there exists t_0 such that for all $t > t_0$, W contains a congruent copy of tX.

Some restrictions on X are necessary as the following result shows.

THEOREM 11.5.9 (Graham [Gra94])

Let $X \subseteq \mathbb{E}^k$ be nonspherical. Then for any N there exist a set $W \subseteq \mathbb{E}^N$ with $\overline{\delta}(W) > 0$ and a set $T \subseteq \mathbb{R}$ with $\underline{\delta}(T) > 0$ such that W contains no congruent copy of tX for any $t \in T$.

Here $\underline{\delta}$ denotes ***lower density***, defined similarly to $\overline{\delta}$ but with lim inf replacing lim sup.

It is clear that much remains to be done here.

11.6 VARIATIONS

There are quite a few variants of the preceding topics that have received attention in the literature (e.g., see [Sch93]). We mention some of the more interesting ones.

ASYMMETRIC RAMSEY THEOREMS

Typical results of this type assert that for given sets X_1 and X_2 (for example), for every partition of $\mathbb{E}^N = C_1 \cup C_2$, either C_1 contains a congruent copy of X_1, or C_2 contains a congruent copy of X_2. We can denote this by

$$\mathbb{E}^N \xrightarrow{2} (X_1, X_2).$$

Here is a sampling of results of this type (more of which can be found in [EGM$^+$73], [EGM$^+$75a], [EGM$^+$75b]).

(i) $\mathbb{E}^2 \xrightarrow{2} (T_2, T_3)$ where T_i is any subset of $\mathbb{E}^2$ with i points, $i = 2, 3$.

(ii) $\mathbb{E}^2 \xrightarrow{2} (P_2, P_4)$ where P_2 is a set of two points at a distance 1, and P_4 is a set of four collinear points with distance 1 between consecutive points.

(iii) $\mathbb{E}^3 \xrightarrow{2} (T, Q^2)$ where T is an isosceles right triangle and Q^2 is a square.

(iv) $\mathbb{E}^2 \xrightarrow{2} (P_2, T_4)$ where P_2 is as in (ii) and T_4 is any set of four points [Juh79].

(v) There is a set T_8 of 8 points such that

$$\mathbb{E}^2 \overset{2}{\not\to} (P_2, T_8) \quad \text{[CT94].}$$

This strengthens an earlier result of Juhász [Juh79], which proved this for a certain set of 12 points.

POLYCHROMATIC RAMSEY THEOREMS

Here, instead of asking for a copy of the target set X in a single C_i, we require only that it be contained in the union of a small number of C_i, say at most m of the C_i.

Let us indicate this by writing $\mathbb{E}^N \underset{m}{\to} X$.

(i) If $\mathbb{E}^N \underset{m}{\to} X$ then X must be embeddable on the union of m concentric spheres [EGM$^+$73].

(ii) Suppose X_i is finite and $\mathbb{E}^N \underset{m_i}{\longrightarrow} X_i$, $1 \le i \le t$. Then

$$\mathbb{E}^N \underset{m_1 m_2 \cdots m_t}{\xrightarrow{\hspace{2em}}} X_1 \times X_2 \times \cdots \times X_t \qquad \text{[ERS83].}$$

(iii) If X_6 is the 6-point set formed by taking the four vertices of a square together with the midpoints of two adjacent sides then $\mathbb{E}^2 \not\to X_6$ but $\mathbb{E}^2 \underset{2}{\longrightarrow} X_6$.

(iv) If X is the set of vertices of a regular simplex in $\mathbb{E}^N$ together with the trisection points of each of its edges then

$$\mathbb{E}^2 \not\rightarrow X_6 \quad \text{but} \quad \mathbb{E}^2 \underset{3}{\longrightarrow} X_6.$$

It is not known if $\mathbb{E}^2 \underset{2}{\longrightarrow} X_6$. Many other results of this type can be found in [ERS83].

PARTITIONS OF $\mathbb{E}^n$ WITH ARBITRARILY MANY PARTS

Since $\mathbb{E}^2 \overset{7}{\not\rightarrow} P_2$, where P_2 is a set of two points with unit distance, one might ask whether there is any nontrivial result of the type $\mathbb{E}^2 \overset{m}{\longrightarrow} \mathcal{F}$ when m is allowed to go to infinity. Of course, if $\mathcal{F}$ is sufficiently large, then there certainly are. There are some interesting geometric examples for which $\mathcal{F}$ is not too large.

THEOREM 11.6.1 [Gra80a]

For any partition of $\mathbb{E}^n$ into finitely many parts, some part contains, for all $\alpha > 0$ and all sets of lines $L_1, \ldots, L_n$ that span $\mathbb{E}^n$, a simplex having volume α and edges through one vertex parallel to the L_i.

Many other theorems of this type are possible (see [Gra80a]).

PARTITIONS WITH INFINITELY MANY PARTS

Results of this type tend to have a strong set-theoretic flavor. For example: $\mathbb{E}^2 \overset{\aleph_0}{\not\rightarrow} T_3$ where T_3 is an equilateral triangle [Ced69]. In other words, $\mathbb{E}^2$ can be partitioned into countably many parts so that no part contains the vertices of an equilateral triangle. In fact, this was recently strengthened by Schmerl [Sch94b] who showed that for all N,

$$\mathbb{E}^N \overset{\aleph_0}{\not\rightarrow} T_3.$$

In fact, this result holds for *any* fixed triangle T in place of T_3 [Sch94b]. Schmerl also has shown [Sch94a] that there is a partition of $\mathbb{E}^N$ into countably many parts such that no part contains the vertices of *any* isosceles triangle.

Another result of this type is this:

THEOREM 11.6.2 [Kun]

Assuming the Continuum Hypothesis, it is possible to partition $\mathbb{E}^2$ into countably many parts, none of which contains the vertices of a triangle with **rational** *area.*

We also note the interesting result of Erdős and Komjath:

THEOREM 11.6.3 [EK90]

The existence of a partition of $\mathbb{E}^2$ into countably many sets, none of which contains the vertices of a **right** *triangle is equivalent to the Continuum Hypothesis.*

The reader can consult Komjath [Kom97] for more results of this type.

COMPLEXITY ISSUES

S. Burr [Bur82] has shown that the algorithmic question of deciding if a given set $X \subset \mathbb{N} \times \mathbb{N}$ can be partitioned $X = C_1 \cup C_2 \cup C_3$ so that $x, y \in C_i$ implies distance$(x, y) \geq 6$, for $i = 1, 2, 3$, is NP-complete. (Also, he shows that a certain infinite version of this is undecidable.)

Finally, we make a few remarks about the celebrated problem of Esther Klein (who became Mrs. Szekeres), which, in some sense, initiated this whole area (see [Sze73] for a charming history).

THEOREM 11.6.4 [ES35]

There is a minimum function $f : \mathbb{N} \longrightarrow \mathbb{N}$ such that any set of $f(n)$ points in $\mathbb{E}^2$ in general position contains the vertices of a convex n-gon.

This result of Erdős and George Szekeres actually spawned an independent genesis of Ramsey theory. The best bounds currently known for $f(n)$ are:

$$2^{n-2} + 1 \leq f(n) \leq 2^{n+4n^{4/5}}.$$

The lower bound appears in [ES35]. The upper bound is a striking new result of Andrew Suk [Suk17]. It applies for n sufficiently large and is the first significant improvement of the original upper bound $\binom{2n-4}{n-2+1}$ of Erdős and Szekeres.

CONJECTURE 11.6.5

Prove (or disprove) that $f(n) = 2^{n-2} + 1$, $n \geq 3$.

(See Chapter 1 of this Handbook for more details.)

11.7 SOURCES AND RELATED MATERIAL

SURVEYS

The principal surveys for results in Euclidean Ramsey theory are [GRS90], [Gra80b], [Gra85], and [Gra94]. The first of these is a monograph on Ramsey theory in general, with a section devoted to Euclidean Ramsey theory, while the last three are specifically about the topics discussed in the present chapter.

RELATED CHAPTERS

Chapter 1: Finite point configurations
Chapter 13: Geometric discrepancy theory and uniform distribution

REFERENCES

[BMP05] P. Brass, W.O.J. Moser, and J. Pach. *Research Problems in Discrete Geometry.* Springer, New York, 2005

[Bón93] M. Bóna. A Euclidean Ramsey theorem. *Discrete Math.*, 122:349–352, 1993.

[Bou86] J. Bourgain. A Szemerédi type theorem for sets of positive density in $\mathbb{R}^k$. *Israel J. Math.*, 54:307–316, 1986.

[BT96] M. Bóna and G. Tóth. A Ramsey-type problem on right-angled triangles in space. *Discrete Math.*, 150:61–67, 1996

[Bur82] S.A. Burr. An NP-complete problem in Euclidean Ramsey theory. In *Proc. 13th Southeastern Conf. on Combinatorics, Graph Theory and Computing*, vol. 35, pages 131–138, 1982.

[Can96a] K. Cantwell. Finite Euclidean Ramsey theory. *J. Combin. Theory Ser. A*, 73:273–285, 1996.

[Can96b] K. Cantwell. Edge-Ramsey theory. *Discrete Comput. Geom.*, 15:341-352, 1996.

[Ced69] J. Ceder. Finite subsets and countable decompositions of Euclidean spaces. *Rev. Roumaine Math. Pures Appl.*, 14:1247–1251, 1969.

[CFG91] H.T. Croft, K.J. Falconer, and R.K. Guy. *Unsolved Problems in Geometry.* Springer-Verlag, New York, 1991.

[Col02] D. Coulson. A 15-coloring of 3-space omitting distance one. *Discrete Math.*, 256:83–90, 2002.

[CT94] G. Csizmadia and G. Tóth. Note on a Ramsey-type problem in geometry. *J. Combin. Theory Ser. A*, 65:302–306, 1994.

[DKT14] P. Dodos, V. Kamellopoulos, and K. Tyros. A simple proof of the density Hales-Jewett theorem. *Int. Math. Res. Not.*, 12:3340–3352, 2014.

[EGM$^+$73] P. Erdős, R.L. Graham, P. Montgomery, B.L. Rothschild, J. Spencer, and E.G. Straus. Euclidean Ramsey theorems. *J. Combin. Theory Ser. A*, 14:341–63, 1973.

[EGM$^+$75a] P. Erdős, R.L. Graham, P. Montgomery, B.L. Rothschild, J. Spencer, and E.G. Straus. Euclidean Ramsey theorems II. In A. Hajnal, R. Rado, and V. Sós, editors, *Infinite and Finite Sets I*, pages 529–557, North-Holland, Amsterdam, 1975.

[EGM$^+$75b] P. Erdős, R.L. Graham, P. Montgomery, B.L. Rothschild, J. Spencer, and E.G. Straus. Euclidean Ramsey theorems III. In A. Hajnal, R. Rado, and V. Sós, editors, *Infinite and Finite Sets II*, pages 559–583, North-Holland, Amsterdam, 1975.

[EK90] P. Erdős and P. Komjáth. Countable decompositions of $\mathbb{R}^2$ and $\mathbb{R}^3$. *Discrete Comput. Geom.*, 5:325–331, 1990.

[ERS83] P. Erdős, B. Rothschild, and E.G. Straus. Polychromatic Euclidean Ramsey theorems. *J. Geom.*, 20:28–35, 1983.

[ES35] P. Erdős and G. Szekeres. A combinatorial problem in geometry. *Compos. Math.*, 2:463–470, 1935.

[Fal81] K.J. Falconer. The realization of distances in measurable subsets covering RR^n *J. Combin. Theory Ser. A*, 31:184–189, 1981.

[FK91] H. Furstenberg and Y. Katznelson. A density version of the Hales-Jewett theorem. *J. Anal. Math.*, 57:64–119, 1991.

[FR90] P. Frankl and V. Rödl. A partition property of simplices in Euclidean space. *J. Amer. Math. Soc.*, 3:1–7, 1990.

[Fur77] H. Furstenberg. Ergodic behavior of diagonal measures and a theorem of Szemerédi on arithmetic progressions. *J. Anal. Math.*, 31:204–256, 1977.

[FW81] P. Frankl and R.M. Wilson. Intersection theorems with geometric consequences. *Combinatorica*, 1:357–368, 1981.

[Gow01] T. Gowers. A new proof of Szemerédi's theorem. *Geom. Funct. Anal.*, 11:465–588, 2001.

[Gra80a] R.L. Graham. On partitions of $\mathbb{E}^n$. *J. Combin. Theory Ser. A*, 28:89–97, 1980.

[Gra80b] R.L. Graham. Topics in Euclidean Ramsey theory. In J. Nešetřil and V. Rödl, editors, *Mathematics of Ramsey Theory*, Springer-Verlag, Heidelberg, 1980.

[Gra83] R.L. Graham. Euclidean Ramsey theorems on the n-sphere. *J. Graph Theory*, 7:105–114, 1983.

[Gra85] R.L. Graham. Old and new Euclidean Ramsey theorems. In J.E. Goodman, E. Lutwak, J. Malkevitch, and R. Pollack, editors, *Discrete Geometry and Convexity*, vol. 440 of Ann. New York Acad. Sci., pages 20–30, 1985.

[Gra94] R.L. Graham. Recent trends in Euclidean Ramsey theory. *Discrete Math.*, 136:119–127, 1994.

[GRS90] R.L. Graham, B.L. Rothschild, and J. Spencer. *Ramsey Theory*, 2nd edition. Wiley, New York, 1990.

[JKS+09] V. Jelínek, J. Kynčl, R. Stolař, and T. Valla. Monochromatic triangles in two-colored plane. *Combinatorica*, 29:699–718, 2009.

[Juh79] R. Juhász. Ramsey type theorems in the plane. *J. Combin. Theory Ser. A*, 27:152–160, 1979.

[Kom97] P. Komjáth. Set theory: geometric and real. R.L. Graham and J. Nešetřil, editors, *The Mathematics of Paul Erdős, II*, vol. 14 of Algorithms and Combinatorics, pages 461–466, Springer, Berlin, 1997.

[Kři91] I. Kříž. Permutation groups in Euclidean Ramsey theory. *Proc. Amer. Math. Soc.*, 112:899–907, 1991.

[Kři92] I. Kříž. All trapezoids are Ramsey. *Discrete Math.*, 108:59–62, 1992.

[Kun] K. Kunen. Personal communication.

[LRW11] I. Leader, P.A. Russell, and M. Walter. Transitive sets and cyclic quadrilaterals. *J. Combinatorics*, 2:457–462, 2011.

[LRW12] I. Leader, P.A. Russell, and M. Walter. Transitive sets in Euclidean Ramsey theory. *J. Combin. Theory Ser. A* 119:382–396, 2012.

[MR95] J. Matoušek and V. Rödl. On Ramsey sets on spheres. *J. Combin. Theory Ser. A*, 70:30–44, 1995.

[Nec00] O. Nechushtan. A note on the space chromatic number. *Discrete Math.*, 256:499–507, 2002.

[O'D00a] P. O'Donnell. Arbitrary girth, 4-chromatic unit distance graphs in the plane; Part 1: Graph Description. *Geombinatorics*, 9:145–150, 2000.

[O'D00b] P. O'Donnell. Arbitrary girth, 4-chromatic unit distance graphs in the plane; Part 2: Graph embedding. *Geombinatorics*, 9:180–193, 2000.

[Pay09] M.S. Payne. Unit distance graphs with ambiguous chromatic number. *Electr. J. Combin.*, 16:N31, 2009.

[Poly12] D.H.J. Polymath. A new proof of the density Hales-Jewett theorem. *Ann. Math.*, 175:1283–1327, 2012.

[Rai00] A.M. Raigorodskii. On the chromatic number of the space. *Usp. Mat. Nauk*, 55:147–148, 2000.

[RT02] R. Radoičić and G. Tóth. Note on the chromatic number of the space. In B. Aronov et al., editors, *Discrete and Computational Geometry*, vol. 25 of *Algorithms and Combinatorics*, pages 695–698, Springer, Berlin, 2003.

[Sch93] P. Schmitt. Problems in discrete and combinatorial geometry. In P.M. Gruber and J.M. Wills, editors, *Handbook of Convex Geometry*, volume A, North-Holland, Amsterdam, 1993.

[Sch94a] J.H. Schmerl. Personal communication, 1994.

[Sch94b] J.H. Schmerl. Triangle-free partitions of Euclidean space. *Bull. London Math. Soc.*, 26:483–486, 1994.

[Sha76] L. Shader. All right triangles are Ramsey in $\mathbb{E}^2$! *J. Combin. Theory Ser. A*, 20:385–389, 1976.

[She88] S. Shelah. Primitive recursive bounds for van der Waerden numbers. *J. Amer. Math. Soc.*, 1:683–697, 1988.

[Soi91] A. Soifer. Chromatic number of the plane: A historical survey. *Geombinatorics*, 1:13–14, 1991.

[Soi92] A. Soifer. A six-coloring of the plane. *J. Combin. Theory Ser. A*, 61:292–294, 1992.

[Soi05] A. Soifer. Axiom of choice and chromatic number of $\mathbb{R}^n$. *J. Combin. Theory Ser. A*, 110:169–173, 2005.

[Sol04] J. Solymosi. A note on a question of Erdős and Graham. *Combin. Probab. Comput.*, 13:263–267, 2004.

[Sol70] R.M. Solovay. A model of set theory in which every set of reals is Lebesgue measurable. *Ann. of Math.*, 92:1–56, 1970.

[Spe79] J.H. Spencer All finite configurations are almost Ramsey. *J. Combin. Theory Ser. A*, 27:410–403, 1979.

[SS03] S. Shelah and A. Soifer. Axiom of choice and chromatic number of the plane. *J. Combin. Theory Ser. A*, 103:387–391, 2003.

[SS04] S. Shelah and A. Soifer. Axiom of choice and chromatic number: example on the plane. *J. Combin. Theory Ser. A*, 105:359–364, 2004.

[Suk17] A. Suk. On the Erdős-Szekeres convex polygon problem. *J. Amer. Math. Soc.*, 30:1047–1053, 2017.

[Sze73] G. Szekeres. A combinatorial problem in geometry: Reminiscences. In J. Spencer, editor, *Paul Erdős: The Art of Counting, Selected Writings*, pages xix–xxii, MIT Press, Cambridge, 1973.

[Sze75] E. Szemerédi. On sets of integers containing no k elements in arithmetic progression. *Acta Arith.*, 27:199–245, 1975.

[Szé84] L.A. Székely. Measurable chromatic number of geometric graphs and set without some distances in Euclidean space. *Combinatorica*, 4:213–218, 1984.

[SW89] L.A. Székely and N.C. Wormald. Bounds on the measurable chromatic number of $\mathbb{R}^n$. *Discrete Math.*, 75:343–372, 1989.

[Tao09] T. Tao. Polymath1 and three new proofs of the density Hales-Jewett theorem. Available at `http://terrytao.wordpress.com`, 2009.

[Tót96] G. Tóth. A Ramsey-type bound for rectangles. *J. Graph Theory*, 23:53–56, 1996.

[Wae27] B.L. van der Waerden. Beweis einer Baudetschen Vermutung. *Nieuw Arch. Wisk.*, 15:212–216, 1927.

12 DISCRETE ASPECTS OF STOCHASTIC GEOMETRY

Rolf Schneider

INTRODUCTION

Stochastic geometry studies randomly generated geometric objects. The present chapter deals with some discrete aspects of stochastic geometry. We describe work that has been done on finite point sets, their convex hulls, infinite discrete point sets, arrangements of flats, and tessellations of space, under various assumptions of randomness. Typical results concern expectations of geometrically defined random variables, or probabilities of events defined by random geometric configurations. The selection of topics must necessarily be restrictive. We leave out the large number of special elementary geometric probability problems that can be solved explicitly by direct, though possibly intricate, analytic calculations. We pay special attention to either asymptotic results, where the number of points considered tends to infinity, or to inequalities, or to identities where the proofs involve more delicate geometric or combinatorial arguments. The close ties of discrete geometry with convexity are reflected: we consider convex hulls of random points, intersections of random halfspaces, and tessellations of space into convex sets.

There are many topics that one might classify under 'discrete aspects of stochastic geometry', such as optimization problems with random data, the average-case analysis of geometric algorithms, random geometric graphs, random coverings, percolation, shape theory, and several others. All of these have to be excluded here.

12.1 RANDOM POINTS

The setup is a finite number of random points or, alternatively, a point process, in a topological space Σ. Mostly, the space Σ is $\mathbb{R}^d$ $(d \geq 2)$, the d-dimensional Euclidean space, with scalar product $\langle \cdot\,,\cdot \rangle$ and norm $\|\cdot\|$. Later, Σ can also be a space of r-flats in $\mathbb{R}^d$, or a space of convex polytopes. By $B^d := \{x \in \mathbb{R}^d : \|x\| \leq 1\}$ we denote the unit ball of $\mathbb{R}^d$, and by $S^{d-1} := \{x \in \mathbb{R}^d : \|x\| = 1\}$ the unit sphere. The volume of B^d is denoted by κ_d.

GLOSSARY

Random point in Σ: A Borel-measurable mapping from some probability space into Σ.

Distribution of a random point X in Σ: The probability measure μ on Σ such that $\mu(B)$, for a Borel set $B \subset \Sigma$, is the probability that $X \in B$.

i.i.d. random points: Stochastically independent random points (on the same probability space) with the same distribution.

12.1.1 NATURAL DISTRIBUTIONS OF RANDOM POINTS

In geometric problems about i.i.d. random points, a few distributions have been considered as particularly natural.

TABLE 12.1.1 Natural distributions of a random point in $\mathbb{R}^d$.

NAME OF DISTRIBUTION	PROBABILITY DENSITY AT $x \in \mathbb{R}^d$
Uniform in K, μ_K	$\propto$ indicator function of K at x
Standard normal, γ_d	$\propto \exp\left(-\frac{1}{2}\|x\|^2\right)$
Beta type 1	$\propto (1-\|x\|^2)^q \times$ indicator function of B^d at x, $q > -1$
Beta type 2	$\propto \|x\|^{\alpha-1}(1+\|x\|)^{-(\alpha+\beta)}, \alpha, \beta > 0$
Spherically symmetric	function of $\|x\|$

Here $K \subset \mathbb{R}^d$ is a given closed set of positive, finite volume, often a ***convex body*** (a compact, convex set with interior points). In that case, μ_K denotes Lebesgue measure, restricted to K and normalized. Usually, the name of a distribution of a random point is also associated with the random point itself. A ***uniform point*** in K is a random point with distribution μ_K. If F is a smooth compact hypersurface in $\mathbb{R}^d$, a random point is uniform on F if its distribution is proportional to the volume measure on F.

By $\mathbb{P}$ we denote probability and by $\mathbb{E}$ we denote mathematical expectation.

12.1.2 POINT PROCESSES

For randomly generated infinite discrete point sets, suitable models are provided by stochastic point processes. They are considered in spaces more general than the Euclidean, so that the 'points' may be other geometric objects, e.g., hyperplanes.

GLOSSARY

Locally finite: $M \subset \Sigma$ (Σ a locally compact Hausdorff space with countable base) is locally finite if $\operatorname{card}(M \cap B) < \infty$ for every compact set $B \subset \Sigma$.

$\mathcal{M}$: The set of all locally finite subsets of Σ.

M: The smallest σ-algebra on $\mathcal{M}$ for which every function $M \mapsto \operatorname{card}(M \cap B)$, with $B \subset \Sigma$ a Borel set, is measurable.

(Simple) point process X on Σ: A measurable map X from some probability space (Ω, A, P) into the measurable space $(\mathcal{M}, \mathsf{M})$. When convenient, X is interpreted as a measure, identifying $\operatorname{card}(X \cap A)$ with $X(A)$, for Borel sets A.

Distribution of X: The image measure P_X of P under X.

Intensity measure Λ of X: $\Lambda(B) = \mathbb{E}\operatorname{card}(X \cap B)$, for Borel sets $B \subset \Sigma$.

Stationary (or ***homogeneous***): If a translation group operates on Σ (e.g., if Σ is $\mathbb{R}^d$, or the space of r-flats, or the space of convex bodies in $\mathbb{R}^d$), then X is a stationary point process if the distribution P_X is invariant under translations.

The point process X on Σ, with intensity measure Λ (assumed to be finite on compact sets), is a ***Poisson process*** if, for any finitely many pairwise disjoint

Borel sets $B_1, \ldots, B_m$, the random variables card $(X \cap B_1), \ldots,$ card $(X \cap B_m)$ are independent and Poisson distributed, namely

$$\mathbb{P}(card\,(X \cap B) = k) = e^{-\Lambda(B)} \frac{\Lambda(B)^k}{k!}$$

for $k \in \mathbb{N}_0$ and every Borel set B. If $\Sigma = \mathbb{R}^d$ and the process is stationary, then the intensity measure Λ is γ times the Lebesgue measure, and the number γ is called the ***intensity*** of X. Let X be a stationary Poisson process on $\mathbb{R}^d$ and $C \subset \mathbb{R}^d$ a compact set, and let $k \in \mathbb{N}_0$. Under the condition that exactly k points of the process fall into C, these points are stochastically equivalent to k i.i.d. uniform points in C. For an introduction, see also [ScW08, Section 3.2].

A detailed study of geometric properties of stationary Poisson processes in the plane was made by Miles [Mil70].

12.2 CONVEX HULLS OF RANDOM POINTS

We consider convex hulls of finitely many i.i.d. random points in $\mathbb{R}^d$. A random polytope in a convex body K according to the ***uniform model*** (also called *binomial model*) is defined by $K_n := \text{conv}\{X_1, \ldots, X_n\}$, where $X_1, \ldots, X_n$ are i.i.d. uniform random points in K. A random polytope in K according to the ***Poisson model*** is the convex hull, $\widetilde{K}_\lambda$, of the points of a Poisson process with intensity measure $\lambda\mu_K$, for $\lambda > 0$.

A convex body K is of class C_+^k (where $k \geq 2$) if its boundary ∂K is a k-times continuously differentiable, regular hypersurface with positive Gauss–Kronecker curvature; the latter is denoted by κ.

By $\Omega(K)$ we denote the affine surface area of K (see, e.g., [Sch14, Sec. 10.5]).

NOTATION

$X_1, \ldots, X_n$	i.i.d. random points in $\mathbb{R}^d$
μ	the common probability distribution of X_i
φ	a measurable real function defined on polytopes in $\mathbb{R}^d$
$\varphi(\mu, n)$	the random variable $\varphi(\text{conv}\{X_1, \ldots, X_n\})$
$\varphi(K, n)$	$= \varphi(K_n) = \varphi(\mu_K, n)$
f_k	number of k-faces
ψ_j	1 on polytopes with j vertices, 0 otherwise
V_j	jth intrinsic volume (see Chapter 15); in particular:
V_d	d-dimensional volume
S	$= 2V_{d-1}$, surface area; dS element of surface area
$D_j(K, n)$	$= V_j(K) - V_j(K, n)$

12.2.1 SOME GENERAL IDENTITIES

There are a few general identities and recursion formulas for random variables of type $\varphi(K, n)$. Some of them hold for quite general distributions of the random points.

A classical result due to Wendel (1962; reproduced in [ScW08, 8.2.1]) concerns the probability, say $p_{d,n}$, that $0 \notin \text{conv}\{X_1, \ldots, X_n\}$. If the distribution of the i.i.d. random points $X_1, \dots, X_n \in \mathbb{R}^d$ is symmetric with respect to 0 and assigns measure zero to every hyperplane through 0, then

$$p_{d,n} = \frac{1}{2^{n-1}} \sum_{k=0}^{d-1} \binom{n-1}{k}. \tag{12.2.1}$$

This follows from a combinatorial result of Schläfli, on the number of d-dimensional cells in a partition of $\mathbb{R}^d$ by n hyperplanes through 0 in general position. It was proved surprisingly late that the symmetric distributions are extremal: Wagner and Welzl [WaW01] showed that if the distribution of the points is absolutely continuous with respect to Lebesgue measure, then $p_{d,n}$ is at least the right-hand side of (12.2.1).

Some of the expectations of $\varphi(K,n)$ for different functions φ are connected by identities. Two classical results of Efron [Efr65],

$$\mathbb{E}\psi_{d+1}(K,n) = \binom{n}{d+1} \frac{\mathbb{E}V_d(K,d+1)^{n-d-1}}{V_d(K)^{n-d-1}} \tag{12.2.2}$$

and

$$\mathbb{E}f_0(K,n+1) = \frac{n+1}{V_d(K)} \mathbb{E}D_d(K,n), \tag{12.2.3}$$

have found far-reaching generalizations in work of Buchta [Buc05]. He extended (12.2.3) to higher moments of the volume, showing that

$$\frac{\mathbb{E}V_d(K,n)^k}{V_d(K)^k} = \mathbb{E}\prod_{i=1}^{k} \left(1 - \frac{f_0(K,n+k)}{n+i}\right) \tag{12.2.4}$$

for $k \in \mathbb{N}$ (this is now known as the Efron–Buchta identity). As a consequence, the kth moment of $V_d(K,n)/V_d(K)$ can be expressed linearly by the first k moments of $f_0(K,n+k)$. Conversely, the distribution of $f_0(K,n)$ is determined by the moments $\mathbb{E}[V_d(K,j)/V_d(K)]^{n-j}$, $j = d+1, \dots, n$. Further consequences are variance estimates for $D_d(K,n)$ and $f_0(K,n)$ for sufficiently smooth convex bodies K. Relation (12.2.4) holds for more general distributions, if the volume is replaced by the probability content.

Of combinatorial interest is the expectation $\mathbb{E}\psi_i(K,n)$, which is the probability that K_n has exactly i vertices. For this, Buchta [Buc05] proved that

$$\mathbb{E}\psi_i(K,n) = (-1)^i \binom{n}{i} \sum_{j=d+1}^{i} (-1)^j \binom{i}{j} \frac{\mathbb{E}V_d(K,j)^{n-j}}{V_d(K)^{n-j}},$$

which for $i = d+1$ reduces to (12.2.2).

The expected values $\mathbb{E}V_d(\mu,n)$ for different numbers n are connected by a sequence of identities. For an arbitrary probability distribution μ on $\mathbb{R}^d$, Buchta [Buc90] proved the recurrence relations

$$\mathbb{E}V_d(\mu,d+2m) = \frac{1}{2}\sum_{k=1}^{2m-1} (-1)^{k+1} \binom{d+2m}{k} \mathbb{E}V_d(\mu,d+2m-k)$$

and, consequently,

$$\mathbb{E}V_d(\mu,d+2m) = \sum_{k=1}^{m} \left(2^{2k}-1\right) \frac{B_{2k}}{k} \binom{d+2m}{2k-1} \mathbb{E}V_d(\mu,d+2m-2k+1)$$

for $m \in \mathbb{N}$, where the constants B_{2k} are the Bernoulli numbers. Cowan [Cow07] deduced these relations by integrating a new pointwise identity. Of this, he made other applications. For a distribution μ that assigns measure zero to each hyperplane, Cowan [Cow10] proved that

$$\mathbb{E}f_0(\mu,n) = \frac{n}{2} + \frac{1}{2}\sum_{j=1}^{n-1}(-1)^{j-1}\binom{n}{j}\mathbb{E}f_0(\mu,n-j)$$

if $n \geq d+3$ and n is odd, and similar formulas for $\mathbb{E}f_{d-1}(\mu,n)$ and $\mathbb{E}f_{d-2}(\mu,n)$.

For a slightly different model of random polytopes, Beermann and Reitzner [BeR15] were able to go beyond the Efron–Buchta identity, by linking generating functions. Let μ be a probability measure on $\mathbb{R}^d$ which is absolutely continuous with respect to Lebesgue measure. Let η be a Poisson point process on $\mathbb{R}^d$ with intensity measure $t\mu$, for $t > 0$. The convex hull, Π_t, of η is a random polytope. Let $g_{I(\Pi_t)}$ denote the probability-generating function of the random variable $I(\Pi_t) := \eta(\mathbb{R}^d) - f_0(\Pi_t)$ (the number of inner points of Π_t), and let $h_{\mu(\Pi_t)}$ denote the moment-generating function of $\mu(\Pi_t)$ (the μ-content of Π_t). It is proved in [BeR15] that

$$g_{I(\Pi_t)}(z+1) = h_{\mu(\Pi_t)}(tz) \quad \text{for } z \in \mathbb{C}.$$

12.2.2 RANDOM POINTS IN CONVEX BODIES—EXPECTATIONS

We consider the random variables $\varphi(K,n)$, for convex bodies K and basic geometric functionals φ, such as volume and vertex number or, more generally, intrinsic volumes and numbers of k-faces.

A few cases where information on the whole distribution is available, are quoted in [ScW08, Note 2 on p. 312]. This section deals mainly with expectations. The table below lists the rare cases where $\mathbb{E}\,\varphi(K,n)$ is known explicitly.

TABLE 12.2.1 Expected value of $\varphi(K,n)$.

DIMENSION d	CONVEX BODY K	FUNCTIONAL φ	SOURCES
2	polygon	V_2	Buchta [Buc84a]
2	polygon	f_0	Buchta and Reitzner [BuR97a]
2	ellipse	V_2	Buchta [Buc84b]
3	ellipsoid	V_3	Buchta [Buc84b]
≥ 2	ball	S, mean width, f_{d-1}	Buchta and Müller [BuM84]
≥ 2	ball	V_d	Affentranger [Aff88]
3	tetrahedron	V_3	Buchta and Reitzner [BuR01]

The value of $\mathbb{E}V_3(C^3,4)$ for the three-dimensional cube C^3 was determined by Zinani [Zin03].

For information on inequalities for expectations $\mathbb{E}\,\varphi(K,n)$ (some of them classical), we refer to [ScW08, Note 1 on p. 311 and Section 8.6].

We turn to inequalities that exhibit the behavior of $\mathbb{E}\,\varphi(K,n)$ for large n. For a convex body $K \subset \mathbb{R}^d$, Schneider [Sch87] showed the existence of positive constants $a_1(K), a_2(K)$ such that

$$a_1(K)n^{-\frac{2}{d+1}} < \mathbb{E}D_1(K,n) < a_2(K)n^{-\frac{1}{d}} \tag{12.2.5}$$

for $n \in \mathbb{N}$, $n \geq d+1$. Smooth bodies (left) and polytopes (right) show that the orders are best possible.

For general convex bodies K, a powerful method for investigating the random polytopes K_n was invented by Bárány and Larman [BáL88] (for introductions and surveys, see [Bár00], [Bár07], [Bár08]). For K of volume one and for sufficiently small $t > 0$, they introduced the floating body

$$K[t] := \{x \in K \mid V_d(K \cap H) \geq t \text{ for every halfspace } H \text{ with } x \in H\}.$$

Their main result says that K_n and $K[1/n]$ approximate K of the same order and that $K \backslash K_n$ is close to $K \backslash K[1/n]$ in a precise sense. From this, several results on the expectations $\mathbb{E}\,\varphi(K,n)$ for various φ were obtained by Bárány and Larman [BáL88], by Bárány [Bár89], for example

$$c_1(d)(\log n)^{d-1} < \mathbb{E} f_j(K,n) < c_2(d) n^{\frac{d-1}{d+1}} \tag{12.2.6}$$

for $j \in \{0, \ldots, d\}$ with positive constants $c_i(d)$ (the orders are best possible), and by Bárány and Vitale [BáV93].

The inequalities (12.2.5) show that for general K the approximation, measured in terms of $D_1(K, \cdot)$, is not worse than for polytopes and not better than for smooth bodies. For approximation measured by $D_d(K, \cdot)$, the class of polytopes and the class of smooth bodies interchange their roles, since

$$b_1(K) n^{-1} (\log n)^{d-1} < \mathbb{E} D_d(K,n) < b_2(K) n^{-\frac{2}{d+1}},$$

as follows from [BáL88] (or from (12.2.6) for $j = 0$ and (12.2.3)).

Precise asymptotic behavior for some cases of $\mathbb{E}\,\varphi(K,n)$ is known if K is either a polytope or sufficiently smooth. For early results and for work in the plane, we refer to the survey [Sch88, Section 5] and to [BuR97a]. For d-dimensional polytopes P, Bárány and Buchta [BáB93] showed that

$$\mathbb{E} f_0(P,n) = \frac{T(P)}{(d+1)^{d-1}(d-1)!} \log^{d-1} n + O(\log^{d-2} n \log\log n), \tag{12.2.7}$$

where $T(P)$ denotes the number of chains $F_0 \subset F_1 \subset \cdots \subset F_{d-1}$ where F_i is an i-dimensional face of P. They established a corresponding relation for the volume, from which (12.2.7) follows by (12.2.3). This work was the culmination of a series of papers by other authors, among them Affentranger and Wieacker [AfW91], who settled the case of simple polytopes, which is applied in [BáB93]. An extension of (12.2.7) was proved by Reitzner [Rei05a], in the form

$$\mathbb{E} f_k(P,n) = c(d,k,P) \log^{d-1} n + o(\log^{d-1} n),$$

for $k = 0, \ldots, d-1$.

For convex bodies $K \subset \mathbb{R}^d$ of class C_+^2, Reitzner [Rei05a] succeeded with showing that

$$\mathbb{E} f_k(K,n) = c_{d,k} \Omega(K)\, n^{\frac{d-1}{d+1}} + o\left(n^{\frac{d-1}{d+1}}\right)$$

with a constant $c_{d,k}$.

A relation of the form

$$\mathbb{E} D_j(K,n) = c^{(j,d)}(K)\, n^{-\frac{2}{d+1}} + o\left(n^{-\frac{2}{d+1}}\right) \tag{12.2.8}$$

for $j = 1, \ldots, d$ was first proved for convex bodies K of class C^3_+ by Bárány [Bár92] (for $j = 1$, such a result, with explicit $c^{(1,d)}$, was obtained earlier by Schneider and Wieacker [ScWi80]; this has been extended in [BöFRV09] to convex bodies in which a ball rolls freely). Reitzner [Rei04] proved (12.2.8) for bodies of class C^2_+ and showed that the coefficients are given by

$$c^{(j,d)}(K) = c^{(j,d)} \int_{\partial K} \kappa^{\frac{1}{d+1}} H_{d-j}\, dS, \tag{12.2.9}$$

where H_{d-j} denotes the $(d-j)$th normalized elementary symmetric function of the principal curvatures of ∂K and $c^{(j,d)}$ depends only on j and d. Under stronger differentiability assumptions, Reitzner also obtained more precise asymptotic expansions. (Earlier, Gruber [Gru96] had obtained analogous expansions for $\eta(C) - \mathbb{E}\eta(C,n)$, where $\eta(C)$ is the value of the support function of the convex body C at a given vector $u \in S^{d-1}$.) Relation (12.2.8) was extended by Böröczky, Hoffmann and Hug [BöHH08] to convex bodies K satisfying only the assumption that a ball rolls freely in K.

Let $V_d(K) = 1$. The coefficient $c^{(d,d)}(K)$ in (12.2.8) is a (dimension-dependent) constant multiple of the affine surface area $\Omega(K)$. The limit relation

$$\lim_{n\to\infty} n^{\frac{2}{d+1}} \mathbb{E} D_d(K,n) = c^{(d,d)} \int_{\partial K} \kappa^{\frac{1}{d+1}}\, dS$$

was extended by Schütt [Schü94] to arbitrary convex bodies (of volume one), with the Gauss–Kronecker curvature generalized accordingly. A generalization, with a modified proof, is found in Böröczky, Fodor and Hug [BöFH10, Thm. 3.1].

The following concerns the Hausdorff distance δ. For $K \subset \mathbb{R}^d$ of class C^2_+, Bárány [Bár89] showed that there are positive constants $c_1(K), c_2(K)$ such that

$$c_1(K)(n^{-1}\log n)^{\frac{2}{d+1}} < \mathbb{E}\,\delta(K,K_n) < c_2(K)(n^{-1}\log n)^{\frac{2}{d+1}}.$$

OPEN PROBLEMS

PROBLEM 12.2.1 (often posed)

For a convex body $K \subset \mathbb{R}^d$ of given positive volume, is $\mathbb{E}V_d(K, d+1)$ maximal if K is a simplex? This problem is related to some other major open problems in convex geometry, such as the slicing problem. (This problem asks whether there is a constant $c > 0$, independent of the dimension, such that every d-dimensional convex body of volume 1 has a hyperplane section of $(d-1)$-dimensional volume at least c.) See, e.g., Milman and Pajor [MiP89].

PROBLEM 12.2.2 (Vu [Vu06])

For a convex body K in $\mathbb{R}^d$, is $\mathbb{E}f_0(K,n)$ an increasing function of n? For $d = 2$, this was proved in [DeGG$^+$13].

PROBLEM 12.2.3 (M. Meckes)

Is there a universal constant $c > 0$ such that, for convex bodies $K, L \subset \mathbb{R}^d$, the inclusion $K \subset L$ implies $\mathbb{E}V_d(K, d+1) \le c^d\, \mathbb{E}V_d(L, d+1)$? This question is equivalent to the slicing problem; see Rademacher [Rad12]. A main result of [Rad12] is that $K \subset L$ implies $\mathbb{E}V_d(K, d+1) \le \mathbb{E}V_d(L, d+1)$ for $d \le 2$, but not for $d \ge 4$.

12.2.3 VARIANCES, HIGHER MOMENTS, LIMIT THEOREMS

The step from expectations, as considered in the previous section, to higher moments, concentration inequalities and limit theorems requires in general much more sophisticated tools from probability theory. Important progress has been made in recent years, but since it is often more due to modern probabilistic techniques than to refined discrete geometry, the presentation here will be briefer than the topic deserves.

A central limit theorem (CLT) for one of our series of random variables $\varphi(\mu, n)$ is an assertion of the form

$$\lim_{n\to\infty} \left| \mathbb{P}\left(\frac{\varphi(\mu,n) - \mathbb{E}\,\varphi(\mu,n)}{\sqrt{\operatorname{Var}\varphi(\mu,n)}} \le t \right) - \Phi(t) \right| = 0$$

for all $t \in \mathbb{R}$, where $\Phi(t) = (2\pi)^{-1/2} e^{-t^2/2}$ is the distribution function of the standard normal distribution in one dimension. Here, the values of the expectation and the variance Var may, or may not, be known explicitly for each n, or may be replaced by asymptotic values.

First we consider the planar case. The first central limit theorems for random variables $\varphi(K, n)$ were obtained in pioneering work of Groeneboom [Gro88]. For a convex polygon $P \subset \mathbb{R}^2$ with r vertices, he proved that

$$\frac{f_0(P,n) - \frac{2}{3} r \log n}{\sqrt{\frac{10}{27} r \log n}} \xrightarrow{\mathcal{D}} \gamma_1$$

for $n \to \infty$, where $\xrightarrow{\mathcal{D}}$ denotes convergence in distribution and γ_1 is the standard normal distribution in one dimension. Groeneboom also showed a CLT for $f_0(B^2, n)$, where B^2 is the circular disc (an asymptotic variance appearing here was determined by Finch and Hueter [FiH04]).

For a polygon P with r vertices, a result of Cabo and Groeneboom [CaG94], in a version suggested by Buchta [Buc05] (for further clarification, see also Groeneboom [Gro12]), says that

$$\frac{V_2(P)^{-1} D_2(P,n) - \frac{2}{3} r \frac{\log n}{n}}{\sqrt{\frac{28}{27} r \frac{\log n}{n^2}}} \xrightarrow{\mathcal{D}} \gamma_1.$$

A result of Hsing [Hsi94] was made more explicit by Buchta [Buc05] and gives a CLT for $D_2(B^2, n)$.

An unpublished preprint by Nagaev and Khamdamov [NaK91] contains a central limit theorem for the joint distribution of $f_0(P, n)$ and $D_2(P, n)$, for a convex polygon P. This result was re-proved by Groeneboom [Gro12], based on his earlier work [Gro88].

Bräker, Hsing, and Bingham [BrHB98] investigated the asymptotic distribution of the Hausdorff distance between a planar convex body K (either smooth or a polygon) and the convex hull of n i.i.d. uniform points in K. A thorough study of the asymptotic properties of $D_2(\mu, n)$ and $D_1(\mu, n)$ was presented by Bräker and Hsing [BrH98], for rather general distributions μ (including the uniform distribution) concentrated on a convex body K in the plane, where K is either sufficiently smooth and of positive curvature, or a polygon.

For general convex bodies in the plane, central limit theorems for vertex number and area were finally proved by Pardon [Par11], [Par12], first for the Poisson model and then for the uniform model.

Turning to higher dimensions, we remark that Küfer [Küf94] studied the asymptotic behavior of $D_d(B^d, n)$ and showed, in particular, that its variance is at most of order $n^{-\frac{d+3}{d+1}}$, as $n \to \infty$. Schreiber [Schr02] determined the asymptotic orders of the moments of $D_1(B^d, n)$ and proved that $\lim_{n\to\infty} n^{\frac{2}{d+1}} D_1(B^d, n) = a_d$ in probability, with an explicit constant a_d.

Essential progress in higher dimensions began with two papers by Reitzner. For convex bodies $K \in \mathbb{R}^d$ of class C^2_+ he obtained in [Rei03], using the Efron–Stein jackknife inequality, optimal upper bounds for the orders of $\operatorname{Var} V_d(K, n)$ and $\operatorname{Var} f_0(K, n)$. He deduced corresponding strong laws of large numbers, for example,

$$\lim_{n\to\infty} n^{\frac{2}{d+1}} D_d(K, n) = \Gamma_d \Omega(K)$$

with probability one, where Γ_d is an explicit constant. Using a CLT for dependency graphs due to Rinott, Reitzner [Rei05b] showed central limit theorems for V_d and f_0 in the Poisson model (that is, for $\widetilde{K}_n$, the convex hull of the points of a Poisson point process with intensity measure $n\mu_K$). One of Reitzner's results says that, for a convex body K of class C^2_+,

$$\left| \mathbb{P}\left(\frac{V_d(\widetilde{K}_n) - \mathbb{E}V_d(\widetilde{K}_n)}{\sqrt{\operatorname{Var} V_d(\widetilde{K}_n)}} \leq x \right) - \Phi(x) \right| \leq c(K) n^{-\frac{1}{2}+\frac{1}{d+1}} \log^{2+\frac{2}{d+1}} n$$

with a constant $c(K)$. He has a similar result for $f_k(\widetilde{K}_n)$, and also central limit theorems for V_d and f_k in the uniform model, however, with the standard deviation of $V_d(K, n)$ replaced by that of $V_d(\widetilde{K}_n)$.

For the Poisson model in a given polytope, central limit theorems for the volume and the numbers f_k were achieved by Bárány and Reitzner [BáR10a].

Using martingale techniques, Vu [Vu05] obtained concentration inequalities for $V_d(K, n)$ and $f_0(K, n)$. Based on this, he obtained in [Vu06] for K of class C^2_+ tail estimates of type

$$\mathbb{P}\left(|D_d(K, n)| \geq \sqrt{\lambda n^{-\frac{d+3}{d+1}}} \right) \leq 2\exp(-c\lambda) + \exp(-c'n) \tag{12.2.10}$$

for any $0 < \lambda < n^{\alpha}$, where c, c' and α are positive constants. This allowed him to determine the order of magnitude for each moment of $D_d(K, n)$. Using such tail estimates and Reitzner's CLT for the Poisson model, Vu [Vu06] finally proved central limit theorems for $V_d(K, n)$ and $f_k(K, n)$, for K of class C^2_+.

For a convex body K with a freely rolling ball, Böröczky, Fodor, Reitzner, and Vígh [BöFRV09] obtained lower and upper bounds for the variance of $V_1(K, n)$ and proved a strong law of large numbers for $D_1(K, n)$. For K of class C^2_+, Bárány, Fodor, and Vígh [BáFV10] obtained matching lower and upper bounds for the orders of the variances of $V_i(K, n)$ and proved strong laws of large numbers for $D_i(K, n)$. For a polytope P, Bárány and Reitzner [BáR10b] determined asymptotic lower and upper bounds (and lower bounds also for general convex bodies) for the variances of $V_d(P,n)$ and $f_k(P,n)$ and deduced corresponding strong laws of large numbers.

For the uniform model and the Poisson model in the ball B^d, Calka and Schreiber [CaS06] established large deviation results and laws of large numbers for the vertex number f_0. Schreiber and Yukich [ScY08] determined the precise asymptotic behavior of the variance of $f_0(B^d,n)$ and proved a CLT. Calka and Yukich [CaY14] were able, based on prior joint work with Schreiber, to determine the precise asymptotics of the variance of the vertex number in the case of a convex body $K \subset \mathbb{R}^d$ of class C^3_+: for $k \in \{0,\dots,d-1\}$, it is proved in [CaY14] that there exists a constant $F_{k,d} > 0$ such that

$$\lim_{\lambda\to\infty} \lambda^{-\frac{d-1}{d+1}} \operatorname{Var} f_k(\widetilde{K}_\lambda) = F_{k,d}\,\Omega(K)$$

and also that

$$\lim_{n\to\infty} n^{-\frac{d-1}{d+1}} \operatorname{Var} f_k(K,n) = F_{k,d}\,\Omega(K).$$

For the diameter of the random polytope K_n, in case $K = B^d$ (and for more general spherically symmetric distributions), a limit theorem was proved by Mayer and Molchanov [MaM07].

12.2.4 RANDOM POINTS ON CONVEX SURFACES

If μ is a probability distribution on the boundary ∂K of a convex body K and μ has density h with respect to the volume measure of ∂K, we write $\varphi(\partial K,h,n) := \varphi(\mu,n)$ and $D_j(\partial K,h,n) := V_j(K) - V_j(\partial K,h,n)$. Some references concerning $\mathbb{E}\,\varphi(\partial K,h,n)$ are given in [Sch88, p. 224]. Most of them are superseded by an investigation of Reitzner [Rei02a]. For K of class C^2_+ and for continuous $h > 0$ he showed that

$$\mathbb{E}D_j(\partial K,h,n) = b^{(j,d)} \int_{\partial K} h^{-\frac{2}{d-1}} \kappa^{\frac{1}{d-1}} H_{d-j}\, dS \cdot n^{-\frac{2}{d-1}} + o(n^{-\frac{2}{d-1}})$$

as $n \to \infty$ (with H_{d-j} as in (12.2.9)). Under stronger differentiability assumptions on K and h, an asymptotic expansion with more terms was established. Similar results for support functions were obtained earlier by Gruber [Gru96]. The case $j = d$ of Reitzner's result was strengthened by Schütt and Werner [ScWe03], who admitted convex bodies that satisfy bounds on upper and lower curvatures. A still more general result is due to Böröczky, Fodor and Hug [BöFH13], who extended Reitzner's relation to convex bodies K satisfying only the assumption that a ball rolls freely in K.

For a convex body K of class C^2_+, Reitzner [Rei03] obtained an upper variance estimate for $V_d(\partial K,h,n)$ and a law of large numbers for $D_d(\partial K,h,n)$. Under the same assumption, Richardson, Vu, and Wu [RiVW08] established a concentration inequality of type (12.2.10) for $V_d(\partial K,h,n)$, upper estimates for the moments of $V_d(\partial K,h,n)$, and a CLT for the convex hull of the points of a Poisson process on ∂K with intensity measure n times the normalized volume measure on the boundary of K.

The following results concern the Hausdorff metric δ. Let $(X_k)_{k\in\mathbb{N}}$ be an i.i.d. sequence of random points on the boundary ∂K of a convex body K, the distribution of which has a continuous positive density h with respect to the volume measure on ∂K. Let $P_n := \text{conv}\{X_1, \ldots, X_n\}$. For $h = 1$, Dümbgen and Walther [DüW96] showed that $\delta(K, P_n)$ is almost surely of order $O((\log n/n)^{1/(d-1)})$ for general K, and of order $O((\log n/n)^{2/(d-1)})$ under a smoothness assumption. Glasauer and Schneider [GlS96] proved for K of class C^3_+ that

$$\lim_{n\to\infty} \left(\frac{n}{\log n}\right)^{\frac{2}{d-1}} \delta(K, P_n) = \frac{1}{2}\left(\frac{1}{\kappa_{d-1}} \max \frac{\sqrt{\kappa}}{h}\right)^{\frac{2}{d-1}} \quad \text{in probability.}$$

It was conjectured that this holds with almost sure convergence. For $d = 2$, this is true, and similar results hold with the Hausdorff distance replaced by area or perimeter difference; this was proved by Schneider [Sch88].

Bárány, Hug, Reitzner, and Schneider [BáHRS17] investigated spherical convex hulls of i.i.d. uniform points in a closed halfsphere. This model exhibits some new phenomena, since the boundary of the halfsphere is totally geodesic.

12.2.5 GAUSSIAN RANDOM POINTS

Let γ_d denote the standard normal distribution on $\mathbb{R}^d$. For the expectations $\mathbb{E}f_k(\gamma_d, n)$ it is known that

$$\mathbb{E}f_k(\gamma_d, n) \sim \frac{2^d}{\sqrt{d}}\binom{d}{k+1}\beta_{k,d-1}(\pi \log n)^{\frac{d-1}{2}} \tag{12.2.11}$$

as $n \to \infty$, where $\beta_{k,d-1}$ is the interior angle of the regular $(d-1)$-dimensional simplex at one of its k-dimensional faces. This follows from [AfS92], where the Grassmann approach (see Section 12.4.3) was used, together with an equivalence of [BaV94] explained in Section 12.4.3. Relation (12.2.11) was proved by Hug, Munsonius, and Reitzner [HuMR04] in a more direct way, together with similar asymptotic relations for other functionals, such as the k-volume of the k-skeleton. A limit relation for $V_k(\gamma_d, n)$ similar to (12.2.11) was proved by Affentranger [Aff91]. For the random variables $f_k(\gamma_d, n)$ and $V_k(\gamma_d, n)$, Hug and Reitzner [HuR05] obtained variance estimates and deduced laws of large numbers.

For Gaussian polytopes (that is, convex hulls of i.i.d. points with distribution γ_d), Hueter [Hue94], [Hue99] was the first to obtain central limit theorems. In [Hue94], she proved a CLT for $\varphi(\gamma_2, n)$ in the plane, where $\varphi \in \{V_1, V_2, f_0\}$, and in [Hue99] for $f_0(\gamma_d, n)$ (an unjustified criticism in [BáV07] was revoked in [BáV08]). In the plane, Massé [Mas00] derived from [Hue94] that

$$\lim_{n\to\infty} \frac{f_0(\gamma_2, n)}{\sqrt{8\pi \log n}} = 1 \quad \text{in probability.}$$

Bárány and Vu [BáV07] succeeded with proving a CLT for $V_d(\gamma_d, n)$ and for $f_k(\gamma_d, n)$, $k = 0, \ldots, d-1$.

12.2.6 SPHERICALLY SYMMETRIC AND OTHER DISTRIBUTIONS

The following setup has been studied repeatedly. For $0 \le p \le r+1 \le d-1$, one considers $r+1$ independent random points, of which the first p are uniform in the ball B^d and the last $r+1-p$ are uniform on the boundary sphere S^{d-1}. Precise information on the moments and the distribution of the r-dimensional volume of the convex hull is available; see the references in [Sch88, pp. 219, 224] and the work of Affentranger [Aff88].

Among spherically symmetric distributions, the Beta distributions are particularly tractable. For these, again, the r-dimensional volume of the convex hull of $r+1$ i.i.d. random points has frequently been studied. We refer to the references given in [Sch88] and Chu [Chu93]. Affentranger [Aff91] determined the asymptotic behavior, as $n \to \infty$, of the expectation $\mathbb{E}V_j(\mu, n)$, where μ is either the Beta type-1 distribution, the uniform distribution in B^d, or the standard normal distribution in $\mathbb{R}^d$. The asymptotic behavior of $\mathbb{E}f_{d-1}(\mu, n)$ was also found for these cases. Further information is contained in the book of Mathai [Mat99].

For more general spherically symmetric distributions μ, the asymptotic behavior of the random variables $\varphi(\mu, n)$ will essentially depend on the tail behavior of the distribution. Extending work of Carnal [Car70], Dwyer [Dwy91] obtained asymptotic estimates for $\mathbb{E}f_0(\mu, n)$, $\mathbb{E}f_{d-1}(\mu, n)$, $\mathbb{E}V_d(\mu, n)$, and $\mathbb{E}S(\mu, n)$. These investigations were continued, under more general assumptions, by Hashorva [Has11].

Aldous *et al.* [AlFGP91] considered an i.i.d. sequence $(X_k)_{k\in\mathbb{N}}$ in $\mathbb{R}^2$ with a spherically symmetric (or more general) distribution, and under an assumption of slowly varying tail, they determined a limiting distribution for $f_0(\mu, n)$.

Devroye [Dev91] and Massé [Mas99], [Mas00] constructed spherically symmetric distributions μ in the plane for which $f_0(\mu, n)$ has some unexpected properties. For example, Devroye showed that for any monotone sequence $\omega_n \uparrow \infty$ and for every $\epsilon > 0$, there is a radially symmetric distribution μ in the plane for which $\mathbb{E}f_0(\mu, n) \geq n/\omega_n$ infinitely often and $\mathbb{E}f_0(\mu, n) \leq 4 + \epsilon$ infinitely often.

For distributions μ on $\mathbb{R}^d$ with a density with respect to Lebesgue measure, Devroye [Dev91] showed that $\mathbb{E}f_0(\mu, n) = o(n)$ and $\lim_{n\to\infty} n^{-1} f_0(\mu, n) = 0$ almost surely.

12.3 RANDOM GEOMETRIC CONFIGURATIONS

The topic of this section is still finitely many i.i.d. random points, but instead of taking functionals of their convex hull, we consider qualitative properties, in different ways related to convexity.

12.3.1 SYLVESTER'S PROBLEM AND CONVEX POSITION

Sylvester's classical problem asked for $\mathbb{E}\psi_3(K, 4)$ for a convex body $K \subset \mathbb{R}^2$. More generally, one may ask for $\mathbb{E}\psi_{d+1}(K, n)$ for a convex body $K \subset \mathbb{R}^d$ and $n > d+1$, the probability that the convex hull of n i.i.d. uniform points in K is a simplex. (See [ScW08], Section 8.1, for some history, and the notes for Subsection 8.2.3 for more information.)

At the other end, $p(n, K) := \mathbb{E}\psi_n(K, n)$ is of interest, the probability that n i.i.d. uniform points in K are in convex position. Bárány and Füredi [BáF88] determined the limits $\lim_{d\to\infty} p(n(d), B^d)$ for suitable functions $n(d)$. Valtr [Val95] determined $p(n, K)$ if P is a parallelogram, and in [Val96] if P is a triangle. Marckert [Mar17] calculated the values of $p(n, B^2)$ for $n \leq 8$. For convex bodies $K \subset \mathbb{R}^2$ of area one, Bárány [Bár99] obtained the astonishing limit relation

$$\lim_{n\to\infty} n^2 \sqrt[n]{p(n, K)} = \frac{1}{4} e^2 A^3(K),$$

where $A(K)$ is the supremum of the affine perimeters of all convex bodies contained in K. He also established a law of large numbers for convergence to a limit shape. There is a unique convex body $\tilde{K} \subset K$ with affine perimeter $A(K)$. If K_n denotes the convex hull of n i.i.d. uniform points in K and δ is the Hausdorff metric, then Bárány's result says that

$$\lim_{n\to\infty} \mathbb{P}(\delta(K_n, \tilde{K}) > \epsilon \mid f_0(K_n) = n) = 0$$

for every $\epsilon > 0$. For a convex body $K \subset \mathbb{R}^d$ of volume one, Bárány [Bár01] showed that

$$c_1 < n^{\frac{2}{d-1}} \sqrt[n]{p(n,K)} < c_2$$

for $n \geq n_0$, where the constants $n_0, c_1, c_2 > 0$ depend only on d.

12.3.2 FURTHER QUALITATIVE ASPECTS

For a given distribution μ on $\mathbb{R}^2$, let $P_1, \dots, P_j, Q_1, \dots, Q_k$ be i.i.d. points distributed according to μ. Let $p_{jk}(\mu)$ be the probability that the convex hull of $P_1, \dots, P_j$ is disjoint from the convex hull of $Q_1, \dots, Q_k$. Continuing earlier work of several authors, Buchta and Reitzner [BuR97a] investigated $p_{jk}(\mu)$. For the uniform distribution μ in a convex body K, they connected $p_{jk}(\mu)$ to equiaffine inner parallel curves of K, found an explicit representation in the case of polygons, and proved, among other results, that

$$\lim_{n\to\infty} \frac{p_{nn}(\mu)}{n^{3/2}4^{-n}} \geq \frac{8\sqrt{\pi}}{3},$$

with equality if K is centrally symmetric. The investigation was continued by Buchta and Reitzner in [BuR97b].

Of interest for the investigation of random points in planar polygons is the study of ***convex chains***. Let $X_1, \dots, X_n$ be i.i.d. uniform random points in the triangle with vertices $(0,0), (0,1), (1,1)$, let $p_k^{(n)}$ be the probability that precisely k points of $X_1, \dots, X_n$ are vertices of $T_n := \operatorname{conv}\{(0,0), X_1, \dots, X_n, (1,1)\}$, and let N_n be the number of these vertices. Bárány, Rote, Steiger, and Zhang [BáRSZ00] calculated $p_n^{(n)}$, and Buchta [Buc06] found $p_k^{(n)}$ in general. Buchta [Buc12] determined explicitly the first four moments of N_n and proved that

$$\mathbb{E}N_n^m = \left(\frac{2}{3}\log n\right)^m + O\left(\log^{m-1} n\right)$$

as $n \to \infty$, for $m \in \mathbb{N}$. For the same model, Buchta [Buc13] obtained expectation and variance for the missed area.

A random convex n-chain is defined as the boundary of T_n minus the open segment connecting $(0,0)$ and $(1,1)$, conditional on the event that T_n has $n+2$ vertices. Bárány *et al.* [BáRSZ00] established a limit shape of random convex n-chains, as $n \to \infty$, and proved several related limit theorems, among them a CLT. Ambrus and Bárány [AmB09] introduced longest convex chains defined by $X_1, \dots, X_n$ ('long' in terms of number of vertices), and they found the asymptotic behavior of their expected length, as $n \to \infty$, as well as a limit shape.

Related to k-sets (see Chapter 1 of this Handbook) is the following investigation of Bárány and Steiger [BáS94]. If X is a set of n points in general position in $\mathbb{R}^d$, a subset $S \subset X$ of d points is called a k-simplex if X has exactly k points on one side of the affine hull of S. The authors study the expected number of k-simplices for n i.i.d. random points, for different distributions.

An *empty convex polytope* in a finite point set in $\mathbb{R}^d$ is a convex polytope with vertices in the point set, but no point of the set in the interior. For i.i.d. uniform random points $X_1, \dots, X_n$ in a convex body $K \subset \mathbb{R}^d$ (mostly $d = 2$), empty polytopes in $\{X_1, \dots, X_n\}$ were investigated in [BF87], [BaGS13], [BáMR13], [FaHM15].

Various elementary geometric questions can be asked, even about a small number of random points. To give one example: Let $X_1, \dots, X_m$, $m \geq 2$, be independent uniform random points in a convex body $K \subset \mathbb{R}^d$. The probability that the smallest ball containing $X_1, \dots, X_m$ is contained in K is maximal if and only if K is a ball; see [BaS95]. Many examples are treated in the book of Mathai [Mat99].

12.4 RANDOM FLATS AND HALFSPACES

Next to random points, randomly generated r-dimensional flats in $\mathbb{R}^d$ are the objects of study in stochastic geometry that are particularly close to discrete geometry. Like convex hulls of random points, intersections of random halfspaces yield random polytopes in a natural way.

GLOSSARY

$A(d, r)$: Affine Grassmannian of r-flats (r-dimensional affine subspaces) in $\mathbb{R}^d$, with its standard topology.

r-Flat process in $\mathbb{R}^d$: A point process in the space $A(d, r)$.

Isotropic: An r-flat process is isotropic if its distribution is rotation invariant.

12.4.1 POISSON FLAT PROCESSES

A basic model for infinite discrete random arrangements of r-flats in $\mathbb{R}^d$ is provided by a stationary Poisson process in $A(d, r)$. Fundamental work was done by Miles [Mil71] and Matheron [Math75]. We refer to [ScW08, Sec 4.4] for an introduction and references.

A natural geometric question involves intersections. For example, let X be a stationary Poisson hyperplane process in $\mathbb{R}^d$. For $k \in \{0, \dots, d-1\}$, a kth ***intersection density*** χ_k of X can be defined in such a way that, for any Borel set $A \subset \mathbb{R}^d$, the expectation of the total k-dimensional volume inside A of the intersections of any $d-k$ hyperplanes of the process is given by χ_k times the Lebesgue measure of A. Then, given the intensity χ_{d-1}, the maximal kth intersection density χ_k (for $k \in \{0, \dots, d-2\}$) is achieved if the process is isotropic. This result is due to Thomas (1984, see [ScW08, Sec. 4.4]).

For a modified intersection density, reverse inequalities have been proved by Hug and Schneider [HuS11]. For X as above, define Γ_k as the supremum of $\chi_k(\Lambda X)\chi_{d-1}(\Lambda X)^{-k}$ over all nondegenerate linear transformations Λ. For $k \in \{2, \dots, d\}$, the minimum of Γ_k is attained if and only if the hyperplanes of X attain almost surely only d directions.

Intersection densities can also be considered for stationary Poisson r-flats, for example for $d/2 \leq r \leq d-1$ and intersections of any two r-flats. Here nonisotropic extremal cases occur, such as in the case $r = 2$, $d = 4$, solved by Mecke [Mec88]. Various other cases have been treated; see Mecke [Mec91], Keutel [Keu91], and the references given there.

For stationary Poisson r-flat processes with $1 \leq r < d/2$, instead of intersection densities one can introduce a notion of ***proximity***, measuring how close the flats come to each other in the mean. A sharp inequality for this proximity was proved by Schneider [Sch99]. Very thorough investigations of stationary Poisson flat processes, regarding these and many other aspects, including limit theorems, are due to Schulte and Thäle [ScT14] and to Hug, Thäle, and Weil [HTW15].

In Heinrich [Hei09] one finds, together with references to earlier related work, a CLT for the total number of intersection points of a stationary Poisson hyperplane process within expanding convex sampling windows. As an application of very general results on U-statistics for Poisson point processes, such CLTs were obtained by Reitzner and Schulte [ReiS13].

12.4.2 INTERSECTIONS OF RANDOM HALFSPACES

Intersections of random halfspaces appear as solution sets of systems of linear inequalities with random coefficients. Therefore, such random polyhedra play a role in the average case analysis of linear programming algorithms (see the book by Borgwardt [Bor87] and its bibliography).

There are two standard ways to generate a convex polytope, either as the convex hull of finitely many points or as the intersection of finitely many closed halfspaces. We consider here the second way, with random halfspaces.

The following model is, in a heuristic sense, dual to that of the convex hull of i.i.d. random points in a convex body. For a given convex body K let $K_1 := K + B^d$ and let $\mathcal{H}_K$ be the set of hyperplanes meeting K_1 but not the interior of K. Let $\mu^{(K)}$ be the distribution on $\mathcal{H}_K$ arising from the motion invariant measure on the space of hyperplanes by restricting it to $\mathcal{H}_K$ and normalizing. Let $H_1, \ldots, H_n$ be independent random hyperplanes with distribution $\mu^{(K)}$, and let H_i^- be the closed halfspace bounded by H_i and containing K. Choose a polytope Q with $K \subset \operatorname{int} Q$, $Q \subset \operatorname{int} K_1$ and set $K^{(n)} := \bigcap_{i=1}^n H_i^- \cap Q$ (the intersection with Q makes $K^{(n)}$ bounded, and the choice of Q does not affect the asymptotic results). The following contributions concern this model. For K of class C_+^3, Kaltenbach [Kal90] obtained an asymptotic formula (as $n \to \infty$) for $\mathbb{E}V_d(K^{(n)}) - V_d(K)$. Fodor, Hug, and Ziebarth [FHZ16] proved such a limit result for the same functional under the sole assumption that K slides freely in a ball. They also obtained an asymptotic upper estimate for the variance and a corresponding law of large numbers. Böröczky and Schneider [BöS10] considered the mean width for this model and showed the existence of constants c_1, c_2, independent of n, such that

$$c_1 n^{-1} \log^{d-1} n < \mathbb{E}V_1(K^{(n)}) - V_1(K) < c_2 n^{-\frac{2}{d+1}}$$

for all sufficiently large n. For a simplicial polytope P, they also obtained precise asymptotic relations for $\mathbb{E}V_1(P^{(n)}) - V_1(P)$ and for $\mathbb{E}f_0(P^{(n)})$ and $\mathbb{E}f_{d-1}(P^{(n)})$.

For an arbitrary convex body K, Böröczky, Fodor, and Hug [BöFH10] proved that

$$\lim_{n\to\infty} n^{\frac{2}{d+1}}\mathbb{E}\left(V_1(K^{(n)})-V_1(K)\right)=c_d\int_{\partial K}\kappa^{\frac{d}{d+1}}\,dS$$

with an explicit constant c_d, where κ is the generalized Gauss–Kronecker curvature, and a similar relation for $\mathbb{E}f_{d-1}(K^{(n)})$.

A different model for circumscribed random polytopes is obtained in the following way. Let $X_1,\dots,X_n$ be i.i.d. random points on the boundary of a smooth convex body K. Let $K'_{(n)}$ be the intersection of the supporting halfspaces of K at $X_1,\dots,X_n$, and let $K_{(n)}$ be the intersection of $K'_{(n)}$ with some fixed large cube (to make it bounded). For $K=B^d$ and the uniform distribution on ∂K, Buchta [Buc87] found $\mathbb{E}\,f_0(K'_{(n)})$, asymptotically and explicitly. Put $D_{(j)}(K,n):=V_j(K_{(n)})-V_j(K)$. Under the assumption that the distribution of the X_i has a positive density h and that K and h are sufficiently smooth, Böröczky and Reitzner [BöR04] obtained asymptotic expansions, as $n\to\infty$, for $\mathbb{E}D_{(j)}(K,n)$ in the cases $j=d$, $d-1$, and 1.

Let $(X_i)_{i\in\mathbb{N}}$ be a sequence of i.i.d. random points on ∂K, and let $K_{(n)}$ and h be defined as above. If ∂K is of class C^2 and positive curvature and h is positive and continuous, one may ask whether $n^{2/(d-1)}D_{(j)}(K,n)$ converges almost surely to a positive constant, if $n\to\infty$. For $d=2$ and $j=1,2$, this was shown by Schneider [Sch88]. Reitzner [Rei02b] proved such a result for $d\geq 2$ and $j=d$. He deduced that random approximation, in this sense, is very close to best approximation.

Still another model of circumscribed random polytopes starts with a stationary Poisson hyperplane process X in $\mathbb{R}^d$ with a nondegenerate directional distribution. Given a convex body K, let $P_{X,K}$ be the intersection of all closed halfspaces bounded by hyperplanes of X and containing K. Hug and Schneider [HuS14] studied the behavior of the Hausdorff distance $\delta(K,P_{X,K})$ as the intensity of X tends to infinity. This behavior depends heavily on relations between the directional distribution of X and the surface area measure of K.

An interesting class of parametric random polytopes with rotation invariant distribution is obtained from an isotropic Poisson hyperplane process in $\mathbb{R}^n$, which depends on a distance exponent and an intensity, by taking the intersection of the closed halfspaces that are bounded by a hyperplane of the process and contain the origin. Various aspects of these random polytopes were investigated by Hörrmann, Hug, Reitzner, and Thäle [HöHRT15].

12.4.3 PROJECTIONS TO RANDOM FLATS

For combinatorial problems about tuples of random points in $\mathbb{R}^d$, the following approach leads to a natural distribution. Every configuration of $N+1>d$ labeled points in general position in $\mathbb{R}^d$ is affinely equivalent to the orthogonal projection of the set of labeled vertices of a fixed regular simplex $T^N\subset\mathbb{R}^N$ onto a unique d-dimensional linear subspace of $\mathbb{R}^N$. This establishes a one-to-one correspondence between the (orientation-preserving) affine equivalence classes of such configurations and an open dense subset of the Grassmannian $G(N,d)$ of oriented d-dimensional subspaces of $\mathbb{R}^N$. The unique rotation-invariant probability measure on $G(N,d)$ thus leads to a probability distribution on the set of affine equivalence classes of $(N+1)$-tuples of points in general position in $\mathbb{R}^d$. References for this

Grassmann approach, which was proposed by Vershik and by Goodman and Pollack, are given in Affentranger and Schneider [AfS92]. Baryshnikov and Vitale [BaV94] proved that an affine-invariant functional of $(N+1)$-tuples with this distribution is stochastically equivalent to the same functional taken at an i.i.d. $(N+1)$-tuple of standard normal points in $\mathbb{R}^d$.

In particular, every d-polytope with $N+1 \geq d+1$ vertices is affinely equivalent to an orthogonal projection of an N-dimensional regular simplex. Similarly, every centrally symmetric d-polytope with $2N \geq 2d$ vertices is affinely equivalent to an orthogonal projection of an N-dimensional regular cross-polytope. This leads to further geometrically natural models for d-dimensional random polytopes. For this, we take a regular simplex, cube, or cross-polytope in $\mathbb{R}^N$, $N > d$, and project it orthogonally to a uniform random d-subspace of $\mathbb{R}^N$. For more information on this approach, including formulas for the expected k-face numbers of these random polytopes and their asymptotics, and for references, we refer to [ScW08, Section 8.3]. Note 7 of that section contains hints to important applications by Donoho and Tanner. Here it plays a role, for example, to choose d and k in dependence on N so that the random projections are k-neighborly with high probability. See also the later work [DoT10], which describes a particular application to compressed sensing.

12.4.4 RANDOM FLATS THROUGH CONVEX BODIES

The notion of a uniform random point in a convex body K in $\mathbb{R}^d$ is extended by that of a uniform random r-flat through K. A random r-flat in $\mathbb{R}^d$ is a measurable map from some probability space into $A(d,r)$. It is a ***uniform*** (***isotropic uniform***) random r-flat through K if its distribution can be obtained from a translation invariant (resp. rigid-motion invariant) measure on $A(d,r)$, by restricting it to the r-flats meeting K and normalizing to a probability measure. (For details, see [WeW93, Section 2].)

A random r-flat E (uniform or not) through K generates the random section $E \cap K$, which has often been studied, particularly for $r = 1$. References are in [ScWi93, Section 7] and [ScW08, Section 8.6]. Finitely many i.i.d. random flats through K lead to combinatorial questions. Associated random variables, such as the number of intersection points inside K if $d = 2$ and $r = 1$, are hard to attack; for work of Sulanke (1965) and Gates (1984) see [ScWi93].

Of special interest is the case of $n \leq d$ i.i.d. uniform hyperplanes $H_1, \dots, H_n$ through a convex body $K \subset \mathbb{R}^d$. Let $p_n(K,\psi)$ denote the probability that the intersection $H_1 \cap \dots \cap H_n$ also meets K; here ψ (an even probability measure on S^{d-1}) is the directional distribution of the hyperplanes. If ψ is rotation invariant, then $p_n(K,\psi) = n!\kappa_n V_n(K)/(2V_1(K))^n$. For references on this equality and some inequalities for $p_n(K,\psi)$ in general, we refer to [BaS95].

If $N > d$ i.i.d. uniform hyperplanes through K are given, they give rise to a random cell decomposition of $\operatorname{int} K$. For $k \in \{0,\dots,d\}$, the expected number, $\mathbb{E}\nu_k$, of k-dimensional cells of this decomposition was determined in [Sch82],

$$\mathbb{E}\nu_k = \sum_{n=d-k}^{d} \binom{n}{d-k}\binom{N}{n} p_n(K,\psi).$$

12.5 RANDOM MOSAICS

By a ***tessellation*** of $\mathbb{R}^d$, or a ***mosaic*** in $\mathbb{R}^d$, we understand a collection of d-dimensional convex polytopes such that their union is $\mathbb{R}^d$, the intersection of the interiors of any two of the polytopes is empty, and any bounded set meets only finitely many of the polytopes. (For tessellations in general, we refer to [OkBSC00].) Except in Section 12.5.5, all considered mosaics are 'face-to-face,' that is, the intersection of any two of its polytopes is either empty or a face of each of them. A ***random mosaic*** can be modeled by a point process in the space of convex polytopes, such that the properties above are satisfied almost surely. (Alternatively, a random mosaic is often modeled as the random closed set given by the union of its cell boundaries.) For an introduction, we refer to [ScW08, Section 10.1]. Further general references are the basic article by Møller [Møl89] and the surveys given in [MeSSW90, Chapter 3], [WeW93, Section 7], and [ChSKM13, Chapter 9].

NOTATION

X	stationary random mosaic in $\mathbb{R}^d$
$X^{(k)}$	process of its k-dimensional faces, $k = 0, \dots, d$
$d_j^{(k)}$	density of the jth intrinsic volume of the polytopes in $X^{(k)}$
$\gamma^{(k)}$	$= d_0^{(k)}$, k-face intensity of X
$Z^{(k)}$	typical k-face of X
n_{jk}	mean number of k-faces of the typical (j,k)-face star (j-face together with the set of its adjacent k-faces)
Z_0	zero cell (the a.s. unique cell containing 0) of X

Under a natural assumption on the stationary random mosaic X, the notions of 'density' and 'typical' exist with a precise meaning. Here, we can only convey the intuitive idea that one averages over expanding bounded regions of the mosaic and performs a limit procedure. Exact definitions can be found in [ScW08]; we also refer to Chapter 15 of that book for all references missing below.

12.5.1 GENERAL MOSAICS

For arbitrary stationary random mosaics, there are a number of identities relating averages of combinatorial quantities. Basic examples are:

$$\sum_{k=0}^{j}(-1)^k n_{jk} = 1, \qquad \sum_{k=j}^{d}(-1)^{d-k} n_{jk} = 1, \qquad \gamma^{(j)} n_{jk} = \gamma^{(k)} n_{kj},$$

$$\sum_{i=j}^{d}(-1)^i d_j^{(i)} = 0, \qquad \text{and in particular} \qquad \sum_{i=0}^{d}(-1)^i \gamma^{(i)} = 0.$$

If the random mosaic X is normal, meaning that every k-face ($k \in \{0, \dots, d-1\}$) is contained in exactly $d-k+1$ d-polytopes of X, then

$$(1-(-1)^k)\gamma^{(k)} = \sum_{j=0}^{k-1}(-1)^j \binom{d+1-j}{k-j} \gamma^{(j)}$$

for $k = 1, \ldots, d$. Lurking in the background are the polytopal relations of Euler, Dehn–Sommerville, and Gram; for these, see Chapters 15 and 17 of this Handbook. For many special relations between the numbers $\gamma^{(i)}$ and n_{jk} in two and three dimensions, we refer to [ScW08, Thms. 10.1.6, 10.1.7].

12.5.2 VORONOI AND DELAUNAY MOSAICS

A discrete point set in $\mathbb{R}^d$ induces a Voronoi and a Delaunay mosaic (see Chapter 27 for the definitions). Starting from a stationary Poisson point process $\widetilde{X}$ in $\mathbb{R}^d$, one obtains in this way a stationary ***Poisson–Voronoi mosaic*** and ***Poisson–Delaunay mosaic***. Both of these are completely determined by the intensity, $\widetilde{\gamma}$, of the underlying Poisson process $\widetilde{X}$. For a Poisson–Voronoi mosaic and for $k \in \{0, \ldots, d\}$, one has

$$d_k^{(k)} = \frac{2^{d-k+1}\pi^{\frac{d-k}{2}}}{d(d-k+1)!} \frac{\Gamma\left(\frac{d^2-kd+k+1}{2}\right)\Gamma\left(1+\frac{d}{2}\right)^{d-k+\frac{k}{d}}\Gamma\left(d-k+\frac{k}{d}\right)}{\Gamma\left(\frac{d^2-kd+k}{2}\right)\Gamma\left(\frac{d+1}{2}\right)^{d-k}\Gamma\left(\frac{k+1}{2}\right)} \widetilde{\gamma}^{\frac{d-k}{d}}.$$

In particular, the vertex density is given by

$$\gamma^{(0)} = \frac{2^{d+1}\pi^{\frac{d-1}{2}}}{d^2(d+1)} \frac{\Gamma\left(\frac{d^2+1}{2}\right)}{\Gamma\left(\frac{d^2}{2}\right)} \left[\frac{\Gamma\left(1+\frac{d}{2}\right)}{\Gamma\left(\frac{d+1}{2}\right)}\right]^d \widetilde{\gamma}.$$

For many other parameters, their explicit values in terms of $\widetilde{\gamma}$ are known, especially in small dimensions. We refer to [ScW08, Thm. 10.2.5] for a list in two and three dimensions. The lecture notes of Møller [Møl89] are devoted to random Voronoi tessellations.

If a Voronoi mosaic X and a Delaunay mosaic Y are induced by the same stationary Poisson point process, then the k-face intensities $\beta^{(k)}$ of Y and $\gamma^{(k)}$ of X are related by $\beta^{(j)} = \gamma^{(d-j)}$ for $j = 0, \ldots, d$, by duality. For stationary Poisson–Delaunay mosaics in the plane, some explicit parameter values are shown in [ScW08, Thm. 10.2.9].

A generalization of the Voronoi mosaics is provided by Laguerre mosaics, where the generating points are endowed with weights. A systematic study of Laguerre mosaics generated by stationary Poisson point processes is made by Lautensack and Zuyev [LaZ08].

OPEN PROBLEM

PROBLEM 12.5.1 *Find the k-face intensities $\gamma^{(k)}$ of stationary Poisson–Voronoi mosaics explicitly in all dimensions.*

12.5.3 HYPERPLANE TESSELLATIONS

A random mosaic X is called a ***stationary hyperplane tessellation*** if it is induced, in the obvious way, by a stationary hyperplane process (as defined in Section 12.4.1). Such random mosaics have special properties. Under an assumption of general position, they satisfy, for $0 \le j \le k \le d$,

$$d_j^{(k)} = \binom{d-j}{d-k} d_j^{(j)}, \quad \text{in particular} \quad \gamma^{(k)} = \binom{d}{k} \gamma^{(0)},$$

and

$$n_{kj} = 2^{k-j} \binom{k}{j}$$

(see, e.g., [ScW08, Thm. 10.3.1]).

A stationary Poisson hyperplane process $\widehat{X}$, satisfying a suitable assumption of nondegeneracy, induces a stationary random mosaic X, called a ***Poisson hyperplane mosaic***. The process $\widehat{X}$ is determined, up to stochastic equivalence, by its spherical directional distribution $\widehat{\varphi}$, an even probability measure on the unit sphere S^{d-1}, and the intensity $\widehat{\gamma} > 0$. Several parameters of X can be expressed in terms of the ***associated zonoid*** $\Pi_{\widehat{X}}$, which is the convex body defined by its support function,

$$h(\Pi_{\widehat{X}}, u) = \frac{\widehat{\gamma}}{2} \int_{S^{d-1}} |\langle u, v \rangle| \, \widehat{\varphi}(dv), \quad u \in S^{d-1}.$$

Examples are

$$d_j^{(k)} = \binom{d-j}{d-k} \binom{d}{j} V_{d-j}(\Pi_{\widehat{X}}),$$

in particular,

$$\gamma^{(k)} = \binom{d}{k} V_d(\Pi_{\widehat{X}}).$$

In the isotropic case (where the distributions are invariant under rigid motions), the associated zonoid is a ball of radius depending only on $\widehat{\gamma}$ and the dimension. For the preceding relations and for further information, we refer to [ScW08, Sec10.3].

12.5.4 TYPICAL CELLS AND FACES

A stationary random mosaic gives rise to several interesting random polytopes. The almost surely unique cell containing the origin is called the ***zero cell*** (or Crofton cell, in the case of a hyperplane process). The ***typical cell*** is obtained, heuristically, by choosing at random, with equal chances, one of the cells of the mosaic within a large bounded region of the mosaic and translating it so that a suitable center lies at the origin. See also the explanations given in [Cal10, Sec. 5.1.2] and [Cal13, Sec. 6.1.2]. A precise definition avoiding ergodic theory can make use either of the grain distribution of a stationary particle process (see, e.g., [ScW08, Sec. 10.1]) or of Palm distributions (see, e.g., [Sch09]). Similarly the typical k-face can be defined, as well as the weighted typical k-face, where, heuristically, the chance to be chosen is proportional to the k-dimensional volume. The weighted d-cell is, up to translations, stochastically equivalent to the zero cell (e.g., [ScW08, Thm. 10.4.1]). In special cases, information on the distributions is available, so for the typical cell of a stationary Poisson–Delaunay mosaic ([ScW08, Thm. 10.4.4]) or of a stationary Poisson hyperplane mosaic ([ScW08, Thms. 10.4.5, 10.4.6]); less explicitly also for the weighted typical k-face ([Sch09, Thm. 1]).

For the typical k-face $Z^{(k)}$ and the weighted typical k-face $Z_0^{(k)}$ of a stationary hyperplane tessellation, we mention a few aspects of particular geometric interest.

For a mosaic generated by a stationary Poisson hyperplane process $\widehat{X}$ and for $0 \le j \le k \le d$,

$$\mathbb{E}V_j(Z^{(k)}) = \frac{\binom{d-j}{d-k} V_{d-j}(\Pi_{\widehat{X}})}{\binom{d}{k} V_d(\Pi_{\widehat{X}})},$$

as follows from [ScW08, (10.3), (10.43), (10.44)].

For a mosaic generated by a stationary hyperplane process satisfying only some general position and finiteness conditions, one of the results mentioned in Section 12.5.3 says that

$$\mathbb{E}f_j(Z^{(k)}) = 2^{k-j}\binom{k}{j}, \quad 0 \le j \le k \le d,$$

independent of the distribution. Let X be a stationary Poisson hyperplane tessellation. Then the variance of the vertex number of the typical k-face of X satisfies

$$0 \le \operatorname{Var} f_0(Z^{(k)}) \le \left(2^k k! \sum_{j=0}^{k} \frac{\kappa_j^2}{4^j (k-j)!}\right) - 2^{2k}.$$

For $k \ge 2$, equality on the left side holds if and only if X is a parallel mosaic (i.e., the hyperplanes of X belong to d translation classes), and on the right side if (and for $k = d$ only if) X is isotropic with respect to a suitable scalar product on $\mathbb{R}^d$ ([Sch16]). For weighted faces one has, for $k \in \{2, \dots, d\}$,

$$2^k \le \mathbb{E}f_0(Z_0^{(k)}) \le 2^{-k} k! \kappa_k^2,$$

with equality on the left if and only if X is a parallel mosaic and on the right if X is isotropic ([Sch09]).

For the zero cell Z_0 of a stationary, isotropic Poisson hyperplane mosaic, there is a linear relation between $\mathbb{E}f_{d-k}(Z_0)$ and $\mathbb{E}V_k(Z_0)$ (see [Sch09, p. 693]), but more information on these expectations is desirable.

A well-known problem of D.G. Kendall, dating back to the 1940s, asked for the asymptotic shape of the zero cell of a planar isotropic Poisson line mosaic under the condition that is has large area. A solution was found by Kovalenko (1997). We mention here the first general higher-dimensional result of this type. Let $\widehat{X}$ be a stationary Poisson hyperplane process in $\mathbb{R}^d$ with intensity $\widehat{\gamma} > 0$ and spherical directional distribution $\widehat{\varphi}$. Let $B_{\widehat{X}}$ be the Blaschke body of $\widehat{X}$, that is, the unique origin symmetric convex body for which $\widehat{\varphi}$ is the area measure. The deviation of a convex body K from the homothety class of $B_{\widehat{X}}$ is measured by

$$r_B(K) := \inf\{s/r - 1 : rB + z \subset K \subset sB + z,\ z \in \mathbb{R}^d,\ r, s > 0\}.$$

Let Z_0 be the zero cell of the mosaic induced by $\widehat{X}$. Then there is a positive constant c_0 depending only on $B_{\widehat{X}}$ such that for $\varepsilon \in (0,1)$ and $a^{\frac{1}{d}}\widehat{\gamma} \ge \sigma_0 > 0$ one has

$$\mathbb{P}\left(r_B(Z_0) \ge \varepsilon \mid V_d(Z_0) \ge a\right) \le c \exp\left\{-c_0\, \varepsilon^{d+1} a^{\frac{1}{d}} \widehat{\gamma}\right\},$$

where c is a constant depending on $B_{\widehat{X}}, \varepsilon, \sigma_0$. In particular, the conditional probability $\mathbb{P}(r_B(Z_0) \ge \varepsilon \mid V_d(Z_0) \ge a\}$ tends to zero for $a \to \infty$. This was proved by Hug, Reitzner, and Schneider (2004). Many similar results have been obtained, for different size measures, also for Poisson–Voronoi and Poisson–Delaunay tessellations, and for typical and weighted typical faces of lower dimensions. For more information and for references, we refer to [ScW08, pp. 512–514] and to the more recent survey [Sch13, Sections 5–7].

12.5.5 STIT TESSELLATIONS

The two most prominent (because tractable) models of random mosaics, the Poisson–Voronoi mosaic and the Poisson hyperplane tessellation, have in recent years been supplemented by a third one, that of STIT tessellation (the name stands for 'stable under iteration'). A comparison of the three models in two and three dimensions was made by Redenbach and Thäle [ReT13]. We roughly indicate the idea behind the STIT tessellations. Let X and Y be two random mosaics in $\mathbb{R}^n$. For $k \in \mathbb{N}$, let Y_k be a random mosaic stochastically equivalent to Y such that $X, Y_1, Y_2, \dots$ are independent. Choose a (measurable) numeration of the cells of X and replace the kth cell of X by its intersections with the cells of Y_k, for $k \in \mathbb{N}$. This yields the cells of a new random mosaic (which is not face-to-face), denoted by $X \lhd Y$. Define

$$\underbrace{X \lhd \cdots \lhd X}_{m+1} := (\underbrace{X \lhd \cdots \lhd X}_{m}) \lhd X \quad \text{and} \quad I_m(X) := m \cdot (\underbrace{X \lhd \cdots \lhd X}_{m})$$

for $m = 2, 3, \dots$ A stationary random mosaic X is called ***stable with respect to iteration*** (***STIT***) if $I_m(X)$ has the same distribution as X, for $m = 2, 3, \dots$

According to Nagel and Weiss [NaW05], the starting point of their existence proof for stationary STIT tessellations can be described as follows. Let Λ be a nondegenerate, locally finite, translation invariant measure on $A(d, d-1)$, and let $W \subset \mathbb{R}^d$ be a compact set, let $[W] := \{H \in A(d, d-1) : H \cap W \neq \emptyset\}$ and assume that $0 < \Lambda([W]) < \infty$. The window W gets an exponentially distributed 'lifetime', with a parameter depending on $\Lambda([W])$. At the end of this lifetime, W is divided into two cells, by a random hyperplane with distribution $\Lambda(\cdot \cap [W])/\Lambda([W])$. With each of these cells, the splitting process starts again, independent of each other and of the previous splittings. This procedure of repeated cell division (or 'cracking') is stopped at a fixed time $a > 0$, yielding a random tessellation $R(a, W)$ of W (the cells of which are intersections of W with convex polytopes). It is proved in [NaW05] that for any $a > 0$ there exists a random tessellation $Y(a)$ of $\mathbb{R}^d$ such that $Y(a) \cap W$ is equal in distribution to $R(a, W)$, for all windows W satisfying the assumptions. It is also proved that $Y(a)$ is a stationary STIT tessellation, called the crack STIT tessellation determined by Λ and a.

Variants of the construction of STIT tessellations were described by Mecke, Nagel, and Weiss [MeNW08a], [MeNW08b]. In recent years, various aspects of STIT tessellations have been investigated thoroughly. By way of example for the extensive later developments, we refer to [ScT13a], [ScT13b] and the references given there.

12.6 SOURCES AND RELATED MATERIAL

SOURCES FOR STOCHASTIC GEOMETRY IN GENERAL

Matheron [Math75]: A monograph on basic models of stochastic geometry and applications of integral geometry.

Santaló [San76]: The classical work on integral geometry and its applications to geometric probabilities.

Solomon [Sol78]: A selection of topics from geometric probability theory.

Ambartzumian [Amb90]: A special approach to stochastic geometry via factorization of measures, with various applications.

Weil and Wieacker [WeW93]: A comprehensive handbook article on stochastic geometry.

Klain and Rota [KlR97]: An introduction to typical results of integral geometry, their interpretations in terms of geometric probabilities, and counterparts of discrete and combinatorial character.

Mathai [Mat99]: A comprehensive collection of results on geometric probabilities, with extensive references.

Schneider and Weil [ScW08]: An introduction to the mathematical models of stochastic geometry, with emphasis on the application of integral geometry and functionals from convexity.

Chiu, Stoyan, Kendall, and Mecke [ChSKM13]: Third edition of a monograph on theoretical foundations and applications of stochastic geometry.

RELEVANT SURVEYS

Buchta [Buc85]: An early survey on random polytopes.

Schneider [Sch88], Affentranger [Aff92]: Surveys on approximation of convex bodies by random polytopes.

Bauer and Schneider [BaS95]: Information on inequalities and extremum problems for geometric probabilities.

Gruber [Gru97], Schütt [Schü02]: Surveys comparing best and random approximation of convex bodies by polytopes.

Bárány [Bár00], [Bár07], [Bár08]: Surveys containing (among other material) introductions to the floating body and cap covering method and their applications to random polytopes.

Hug [Hug07], Calka [Cal10], [Cal13]: Surveys on random mosaics.

Schneider [Sch03], Reitzner [Rei1], Hug [Hug13]: Surveys on random polytopes.

Schneider [Sch13]: Survey on extremal problems related to random mosaics.

RELATED CHAPTERS

Chapter 1: Finite point configurations
Chapter 2: Packing and covering
Chapter 15: Basic properties of convex polytopes
Chapter 17: Face numbers of polytopes and complexes
Chapter 22: Random simplicial complexes
Chapter 27: Voronoi diagrams and Delaunay triangulations
Chapter 44: Randomization and derandomization

REFERENCES

[Aff88] F. Affentranger. The expected volume of a random polytope in a ball. *J. Microscopy,* 151:277–287, 1988.

[Aff91] F. Affentranger. The convex hull of random points with spherically symmetric distributions. *Rend. Sem. Mat. Univ. Politec. Torino*, 49:359–383, 1991.

[Aff92] F. Affentranger. Aproximación aleatoria de cuerpos convexos. *Publ. Mat.*, 36:85–109, 1992.

[AfS92] F. Affentranger and R. Schneider. Random projections of regular simplices. *Discrete Comput. Geom.*, 7:219–226, 1992.

[AfW91] F. Affentranger and J.A. Wieacker. On the convex hull of uniform random points in a simple d-polytope. *Discrete Comput. Geom.*, 6:291–305, 1991.

[AlFGP91] D.J. Aldous, B. Fristedt, P.S. Griffin, and W.E. Pruitt. The number of extreme points in the convex hull of a random sample. *J. Appl. Probab.*, 28:287–304, 1991.

[Amb90] R.V. Ambartzumian. *Factorization Calculus and Geometric Probability.* Volume 33 of *Encyclopedia Math. Appl.,* Cambridge University Press, 1990.

[AmB09] G. Ambrus and I. Bárány. Longest convex chains. *Random Structures Algorithms*, 35:137–162, 2009.

[BaGS13] J. Balogh, H. González-Aguilar, and G. Salazar. Large convex holes in random point sets. *Comput. Geom.*, 46:725–733, 2013.

[Bár89] I. Bárány. Intrinsic volumes and f-vectors of random polytopes. *Math. Ann.*, 285:671–699, 1989.

[Bár92] I. Bárány. Random polytopes in smooth convex bodies. *Mathematika*, 39:81–92, 1992.

[Bár99] I. Bárány. Sylvester's question: the probability that n points are in convex position. *Ann. Probab.*, 27:2020–2034, 1999.

[Bár00] I. Bárány. The technique of M-regions and cap covering: a survey. *Rend. Circ. Mat. Palermo, Ser. II, Suppl.*, 65:21–38, 2000.

[Bár01] I. Bárány. A note on Sylvester's four-point problem. *Studia Sci. Math. Hungar.*, 38:73–77, 2001.

[Bár07] I. Bárány. Random polytopes, convex bodies, and approximation. In: A.J. Baddeley, I. Bárány, R. Schneider, and W. Weil, *Stochastic Geometry. C.I.M.E. Summer School*, vol. 1892 of *Lecture Notes in Math.*, page 77–118, Springer, Berlin, 2007.

[Bár08] I. Bárány. Random points and lattice points in convex bodies. *Bull. Amer. Math. Soc.*, 45:339–365, 2008.

[BáB93] I. Bárány and C. Buchta. Random polytopes in a convex polytope, independence of shape, and concentration of vertices. *Math. Ann.*, 297:467–497, 1993.

[BáFV10] I. Bárány, F. Fodor, and V. Vígh. Intrinsic volumes of inscribed random polytopes in smooth convex bodies. *Adv. Appl. Prob. (SGSA)*, 42:605–619, 2010.

[BáF87] I. Bárány and Z. Füredi. Empty simplices in Euclidean space. *Canad. Math. Bull.*, 30:436–445, 1987.

[BáF88] I. Bárány and Z. Füredi. On the shape of the convex hull of random points. *Probab. Theory Related Fields*, 77:231–240, 1988.

[BáHRS17] I. Bárány, D. Hug, M. Reitzner, and R. Schneider. Random points in halfspheres. *Random Structures Algorithms*, 50:3–22, 2017.

[BáL88] I. Bárány and D.G. Larman. Convex bodies, economic cap coverings, random polytopes. *Mathematika*, 35:274–291, 1988.

[BáMR13] I. Bárány, J.-F. Marckert, and M. Reitzner. Many empty triangles have a common edge. *Discrete Comput. Geom.*, 50:244–252, 2013.

[BáR10a] I. Bárány and M. Reitzner. Poisson polytopes. *Ann. Probab.*, 38:1507–1531, 2010.

[BáR10b] I. Bárány and M. Reitzner. On the variance of random polytopes. *Adv. Math.*, 225:1986–2001, 2010.

[BáRSZ00] I. Bárány, G. Rote, W. Steiger, and C.-H. Zhang. A central limit theorem for convex chains in the square. *Discrete Comput. Geom.*, 23:35–50, 2000.

[BáS94] I. Bárány and W. Steiger. On the expected number of k-sets. *Discrete Comput. Geom.*, 11:243–263, 1994.

[BáV93] I. Bárány and R.A. Vitale. Random convex hulls: floating bodies and expectations. *J. Approx. Theory*, 75:130–135, 1993.

[BáV07] I. Bárány and V. Vu. Central limit theorems for Gaussian polytopes. *Ann. Probab.*, 35:1593–1621, 2007.

[BáV08] I. Bárány and V. Vu. Correction: "Central limit theorems for Gaussian polytopes." *Ann. Probab.*, 36:1998, 2008.

[BaV94] Y.M. Baryshnikov and R.A. Vitale. Regular simplices and Gaussian samples. *Discrete Comput. Geom.*, 11:141–147, 1994.

[BaS95] C. Bauer and R. Schneider. Extremal problems for geometric probabilities involving convex bodies. *Adv. Appl. Prob.*, 27:20–34, 1995.

[BeR15] M. Beermann and M. Reitzner. Beyond the Efron–Buchta identities: Distributional results for Poisson polytopes. *Discrete Comput. Geom.*, 53:226–244, 2015.

[Bor87] K.H. Borgwardt. *The Simplex Method: A Probabilistic Analysis*. Springer-Verlag, Berlin, 1987.

[BöFH10] K.J. Böröczky, F. Fodor, and D. Hug. The mean width of random polytopes circumscribed around a convex body. *J. London Math. Soc. (2)*, 81:499–523, 2010.

[BöFH13] K.J. Böröczky, F. Fodor, and D. Hug. Intrinsic volumes of random polytopes with vertices on the boundary of a convex body. *Trans. Amer. Math. Soc.*, 365:785–809, 2013.

[BöFRV09] K.J. Böröczky, F. Fodor, M. Reitzner, and V. Vígh. Mean width of inscribed random polytopes in a reasonably smooth convex body. *J. Multivariate Anal.*, 100:2287–2295, 2009.

[BöHH08] K.J. Böröczky, L.M. Hoffmann, and D. Hug. Expectation of intrinsic volumes of random polytopes. *Periodica Math. Hungar.*, 57:143–164, 2008.

[BöR04] K.J. Böröczky and M. Reitzner. Approximation of smooth convex bodies by random circumscribed polytopes. *Ann. Appl. Prob.*, 14:239–273, 2004.

[BöS10] K.J. Böröczky and R. Schneider. The mean width of circumscribed random polytopes. *Canad. Math. Bull.*, 53:614–628, 2010.

[BrH98] H. Bräker and T. Hsing. On the area and perimeter of a random convex hull in a bounded convex set. *Probab. Theory Related Fields*, 111:517–550, 1998.

[BrHB98] H. Bräker, T. Hsing, and N.H. Bingham. On the Hausdorff distance between a convex set and an interior random convex hull. *Adv. Appl. Prob.*, 30:295–316, 1998.

[Buc84a] C. Buchta. Zufallspolygone in konvexen Vielecken. *J. Reine Angew. Math.*, 347:212–220, 1984.

[Buc84b] C. Buchta. Das Volumen von Zufallspolyedern im Ellipsoid. *Anz. Österreich. Akad. Wiss. Math.-Natur. Kl.*, 121:1–4, 1984.

[Buc85] C. Buchta. Zufällige Polyeder—Eine Übersicht. In: E. Hlawka, editor, *Zahlentheoretische Analysis*, vol. 1114 of *Lecture Notes in Math.*, pages 1–13, Springer-Verlag, Berlin, 1985.

[Buc87] C. Buchta. On the number of vertices of random polyhedra with a given number of facets. *SIAM J. Algebraic Discrete Methods*, 8:85–92, 1987.

[Buc90] C. Buchta. Distribution-independent properties of the convex hull of random points. *J. Theoret. Probab.*, 3:387–393, 1990.

[Buc05] C. Buchta. An identity relating moments of functionals of convex hulls. *Discrete Comput. Geom.*, 33:125–142, 2005.

[Buc06] C. Buchta. The exact distribution of the number of vertices of a random convex chain. *Mathematika*, 53:247–254, 2006.

[Buc12] C. Buchta. On the boundary structure of the convex hull of random points. *Adv. Geom.*, 12:179–190, 2012.

[Buc13] C. Buchta. Exact formulae for variances of functionals of convex hulls. *Adv. Appl. Prob. (SGSA)*, 45:917–924, 2013.

[BuM84] C. Buchta and J. Müller. Random polytopes in a ball. *J. Appl. Probab.*, 21:753–762, 1984.

[BuR97a] C. Buchta and M. Reitzner. Equiaffine inner parallel curves of a plane convex body and the convex hulls of randomly chosen points. *Probab. Theory Related Fields*, 108:385–415, 1997.

[BuR97b] C. Buchta and M. Reitzner. On a theorem of G. Herglotz about random polygons. *Rend. Circ. Mat. Palermo, Ser. II, Suppl.*, 50:89–102, 1997.

[BuR01] C. Buchta and M. Reitzner. The convex hull of random points in a tetrahedron: Solution of Blaschke's problem and more general results. *J. Reine Angew. Math.*, 536:1–29, 2001.

[CaG94] A.J. Cabo and P. Groeneboom. Limit theorems for functionals of convex hulls. *Probab. Theory Related Fields*, 100:31–55, 1994.

[Cal10] P. Calka. Tessellations. In: W.S. Kendall and I. Molchanov, editors, *New Perspectives in Stochastic Geometry*, pages 145–169, Oxford University Press, 2010.

[Cal13] P. Calka. Asymptotic methods for random tessellations. In: E. Spodarev, editor, *Stochastic Geometry, Spatial Statistics and Random Fields: Asymptotic Methods*, vol. 2068 of *Lecture Notes in Math.*, pages 183–204, Springer, Berlin, 2013.

[CaS06] P. Calka and T. Schreiber. Large deviation probabilities for the number of vertices of random polytopes in the ball. *Adv. Appl. Prob. (SGSA)*, 38:47–58, 2006.

[CaY14] P. Calka and J.E. Yukich. Variance asymptotics for random polytopes in smooth convex bodies. *Probab. Theory Relat. Fields*, 158:435–463, 2014.

[Car70] H. Carnal. Die konvexe Hülle von n rotations-symmetrisch verteilten Punkten. *Z. Wahrscheinlichkeitsth. Verwandte Geb.*, 15:168–178, 1970.

[ChSKM13] S.N. Chiu, D. Stoyan, W.S. Kendall, and J. Mecke. *Stochastic Geometry and its Applications.* 3rd ed., John Wiley & Sons, Chichester, 2013.

[Chu93] D.P.T. Chu. Random r-content of an r-simplex from beta-type-2 random points. *Canad. J. Statist.*, 21:285–293, 1993.

[Cow07] R. Cowan. Identities linking volumes of convex hulls. *Adv. Appl. Prob. (SGSA)*, 39:630–644, 2007.

[Cow10] R. Cowan. Recurrence relationships for the mean number of faces and vertices for random convex hulls. *Discrete Comput. Geom.*, 43:209–220, 2010.

[DeGG$^+$13] O. Devillers, M. Glisse, X. Goaoc, G. Moroz, and M. Reitzner. The monotonicity of f-vectors of random polytopes. *Electron. Commun. Probab.*, 18:23, 2013.

[Dev91] L. Devroye. On the oscillation of the expected number of extreme points of a random set. *Statist. Probab. Lett.*, 11:281–286, 1991.

[DoT10] D.L. Donoho and J. Tanner. Counting the faces of randomly-projected hypercubes and orthants, with applications. *Discrete Comput. Geom.*, 43:522–541, 2010.

[DüW96] L. Dümbgen and G. Walther. Rates of convergence for random approximations of convex sets. *Adv. Appl. Prob.*, 28:384–393, 1996.

[Dwy91] R.A. Dwyer. Convex hulls of samples from spherically symmetric distributions. *Discrete Appl. Math.*, 31:113–132, 1991.

[Efr65] B. Efron. The convex hull of a random set of points. *Biometrika*, 52:331–343, 1965.

[FaHM15] R. Fabila-Monroy, C. Huemer, and D. Mitsche. Empty non-convex and convex four-gons in random point sets. *Studia Sci. Math. Hungar.*, 52:52–64, 2015.

[FiH04] S. Finch and I. Hueter. Random convex hulls: a variance revisited. *Adv. Appl. Prob. (SGSA)*, 36:981–986, 2004.

[FHZ16] F. Fodor, D. Hug, and I. Ziebarth. The volume of random polytopes circumscribed araound a convex body. *Mathematika*, 62:283–306, 2016.

[GlS96] S. Glasauer and R. Schneider. Asymptotic approximation of smooth convex bodies by polytopes. *Forum Math.*, 8:363–377, 1996.

[Gro88] P. Groeneboom. Limit theorems for convex hulls. *Probab. Theory Related Fields*, 79:327–368, 1988.

[Gro12] P. Groeneboom. Convex hulls of uniform samples from a convex polygon. *Adv. Appl. Prob. (SGSA)*, 44:330–342, 2012.

[Gru96] P.M. Gruber. Expectation of random polytopes. *Manuscripta Math.*, 91:393–419, 1996.

[Gru97] P.M. Gruber. Comparisons of best and random approximation of convex bodies by polytopes. *Rend. Circ. Mat. Palermo, Ser. II, Suppl.*, 50:189–216, 1997.

[Has11] E. Hashorva. Asymptotics of the convex hull of spherically symmetric samples. *Discrete Appl. Math.*, 159:201–211, 2011.

[Hei09] L. Heinrich. Central limit theorems for motion-invariant Poisson hyperplanes in expanding convex bodies. *Rend. Circ. Mat. Palermo, Ser. II, Suppl.*, 81:187–212, 2009.

[HöHRT15] J. Hörrmann, D. Hug, M. Reitzner, and Ch. Thäle. Poisson polyhedra in high dimensions. *Adv. Math.* 281:1–39, 2015.

[Hsi94] T. Hsing. On the asymptotic distribution of the area outside a random convex hull in a disk. *Ann. Appl. Probab.*, 4:478–493, 1994.

[Hue94] I. Hueter. The convex hull of a normal sample. *Adv. Appl. Prob.*, 26:855–875, 1994.

[Hue99] I. Hueter. Limit theorems for the convex hull of random points in higher dimensions. *Trans. Amer. Math. Soc.*, 351:4337–4363, 1999.

[Hug07] D. Hug. Random mosaics. In: A.J. Baddeley, I. Bárány, R. Schneider, and W. Weil, *Stochastic Geometry. C.I.M.E. Summer School*, vol. 1892 of *Lecture Notes in Math.*, pages 247–266, Springer, Berlin, 2007.

[Hug13] D. Hug. Random polytopes. In: E. Spodarev, editor, *Stochastic Geometry, Spatial Statistics and Random Fields: Asymptotic Methods*, vol. 2068 of *Lecture Notes in Math.*, pages 205–238, Springer, Berlin, 2013.

[HuMR04] D. Hug, G.O. Munsonius, and M. Reitzner. Asymptotic mean values of Gaussian polytopes. *Beiträge Algebra Geom.*, 45:531–548, 2004.

[HuR05] D. Hug and M. Reitzner. Gaussian polytopes: variance and limit theorems. *Adv. Appl. Prob. (SGSA)*, 37:297–320, 2005.

[HuS11] D. Hug and R. Schneider. Reverse inequalities for zonoids and their application. *Adv. Math.*, 228:2634–2646, 2011.

[HuS14] D. Hug and R. Schneider. Approximation properties of random polytopes associated with Poisson hyperplane processes. *Adv. Appl. Prob. (SGSA)*, 46:919–936, 2014.

[HTW15] D. Hug, C. Thäle, and W. Weil. Intersection and proximity of processes of flats. *J. Math. Anal. Appl.*, 426:1–42, 2015.

[Kal90] F.J. Kaltenbach. Asymptotisches Verhalten zufälliger konvexer Polyeder. Dissertation, Univ. Freiburg i. Br., 1990.

[Keu91] J. Keutel. Ein Extremalproblem für zufällige Ebenen und für Ebenenprozesse in höherdimensionalen Räumen. Dissertation, Univ. Jena, 1991.

[KlR97] D.A. Klain and G.-C. Rota. *Introduction to Geometric Probability.* Cambridge University Press, 1997.

[Küf94] K.-H. Küfer. On the approximation of a ball by random polytopes. *Adv. Appl. Prob.*, 26:876–892, 1994.

[LaZ08] C. Lautensack and S. Zuyev. Random Laguerre tessellations. *Adv. Appl. Prob.*, 40:630–650, 2008.

[Mar17] J.-F. Marckert. Probability that n random points in a disk are in convex position. *Braz. J. Probab. Stat.*, 31:320–337, 2017.

[Mas99] B. Massé. On the variance of the number of extreme points of a random convex hull. *Statist. Probab. Lett.*, 44:123–130, 1999.

[Mas00] B. Massé. On the LLN for the number of vertices of a random convex hull. *Adv. Appl. Prob.*, 32:675–681, 2000.

[Mat99] A.M. Mathai. *An Introduction to Geometrical Probability: Distributional Aspects with Applications.* Gordon and Breach, Singapore, 1999.

[Math75] G. Matheron. *Random Sets and Integral Geometry.* John Wiley & Sons, Chichester, 1975.

[MaM07] M. Mayer and I. Molchanov. Limit theorems for the diameter of a random sample in the ball. *Extremes*, 10:129–150, 2007.

[Mec88] J. Mecke. An extremal property of random flats. *J. Microscopy*, 151:205–209, 1988.

[Mec91] J. Mecke. On the intersection density of flat processes. *Math. Nachr.*, 151:69–74, 1991.

[MeNW08a] J. Mecke, W. Nagel, and V. Weiss. A global construction of homogeneous random planar tessellations that are stable under iterations. *Stochastics*, 80:51–67, 2008.

[MeNW08b] J. Mecke, W. Nagel, and V. Weiss. The iteration of random tessellations and a construction of a homogeneous process of cell divisions. *Adv. Appl. Prob. (SGSA)*, 40:49–59, 2008.

[MeSSW90] J. Mecke, R. Schneider, D. Stoyan, and W. Weil. *Stochastische Geometrie.* Volume 16 of *DMV Sem.*, Birkhäuser, Basel, 1990.

[Mil70] R.E. Miles. On the homogeneous planar Poisson point process. *Math. Biosci.*, 6:85–127, 1970.

[Mil71] R.E. Miles. Poisson flats in Euclidean spaces. II: Homogeneous Poisson flats and the complementary theorem. *Adv. Appl. Prob.*, 3:1–43, 1971.

[MiP89] V.D. Milman and A. Pajor. Isotropic position and inertia ellipsoids and zonoids of the unit ball of a normed n-dimensional space. In: *Geometric Aspects of Functional Analysis*,pages 64–104, vol. 1376 of *Lecture Notes in Math.*, Springer, Berlin, 1989.

[Møl89] J. Møller. Random tessellations in $\mathbb{R}^d$. *Adv. Appl. Prob.*, 21:37–73, 1989.

[Møl89] J. Møller. *Lectures on Random Voronoi Tessellations.* Lecture Notes in Statistics 87, Springer, New York, 1994.

[NaK91] A.V. Nagaev and I.M. Khamdamov. Limit theorems for functionals of random convex hulls (in Russian). Preprint, Institute of Mathematics, Academy of Sciences of Uzbekistan, Tashkent, 1991.

[NaW05] W. Nagel and V. Weiss. Crack STIT tessellations: characterization of stationary random tessellations stable with respect to iteration. *Adv. Appl. Prob. (SGSA)*, 37:859–883, 2005.

[OkBSC00] A. Okabe, B. Boots, K. Sugihara, and S.N. Chiu. *Spatial Tessellations: Concepts and Applications of Voronoi Diagrams*, 2nd edition. John Wiley & Sons, Chichester, 2000.

[Par11] J. Pardon. Central limit theorems for random polygons in an arbitrary convex set. *Ann. Probab.*, 39:881–903, 2011.

[Par12] J. Pardon. Central limit theorems for uniform model random polygons. *J. Theor. Prob.*, 25:823–833, 2012.

[Rad12] L. Rademacher. On the monotonicity of the expected volume of a random simplex. *Mathematika*, 58:77–91, 2012.

[ReT13] C. Redenbach and Ch. Thäle. Second-order comparison of three fundamental tessellation models. *Statistics*, 47:237–257, 2013.

[Rei02a] M. Reitzner. Random points on the boundary of smooth convex bodies. *Trans. Amer. Math. Soc.*, 354:2243–2278, 2002.

[Rei02b] M. Reitzner. Random polytopes are nearly best approximating. *Rend. Circ. Mat. Palermo, Ser. II, Suppl.*, 70:263–278, 2002.

[Rei03] M. Reitzner. Random polytopes and the Efron-Stein jackknife inequality. *Ann. Probab.*, 31:2136–2166, 2003.

[Rei04] M. Reitzner. Stochastical approximation of smooth convex bodies. *Mathematika*, 51:11–29, 2004.

[Rei05a] M. Reitzner. The combinatorial structure of random polytopes. *Adv. Math.*, 191:178–208, 2005.

[Rei05b] M. Reitzner. Central limit theorems for random polytopes. *Probab. Theory Related Fields*, 133:483–507, 2005.

[Rei1] M. Reitzner. Random polytopes. In: W.S. Kendall and I. Molchanov, editors, *New Perspectives in Stochastic Geometry*, pages 45–76, Oxford University Press, 2010.

[ReiS13] M. Reitzner and M. Schulte. Central limit theorems for U-Statistics of Poisson point processes. *Ann. Probab.*, 41:3879–3909, 2013.

[RiVW08] R.M. Richardson, V. Vu, and L. Wu. An inscribing model for random polytopes. *Discrete Comput. Geom.*, 39:469–499, 2008.

[San76] L.A. Santaló. *Integral Geometry and Geometric Probability*. Volu. 1 of *Encyclopedia of Mathematics*, Addison–Wesley, Reading, 1976.

[Sch82] R. Schneider. Random hyperplanes meeting a convex body. *Z. Wahrsch. Verw. Gebiete*, 61:379–387, 1982.

[Sch87] R. Schneider. Approximation of convex bodies by random polytopes. *Aequationes Math.*, 32:304–310, 1987.

[Sch88] R. Schneider. Random approximation of convex sets. *J. Microscopy*, 151:211–227, 1988.

[Sch99] R. Schneider. A duality for Poisson flats. *Adv. Appl. Prob. (SGSA)*, 331:63–68, 1999.

[Sch08] R. Schneider. Recent results on random polytopes. *Boll. Un. Math. Ital. (9)*, 1:17–39, 2008.

[Sch09] R. Schneider. Weighted faces of Poisson hyperplane tessellations. *Adv. Appl. Prob. (SGSA)*, 41:682–694, 2009.

[Sch13] R. Schneider. Extremal properties of random mosaics. In: I. Bárány, K.J. Böröczky, G. Fejes Tóth, and J. Pach, editors, *Geometry — Intuitive, Discrete, and Convex: A Tribute to László Fejes Tóth*, pages 301–330, vol. 24 of *Bolyai Soc. Math. Studies*, Springer, Berlin, 2013.

[Sch14] R. Schneider. *Convex Bodies: The Brunn-Minkowski Theory*, second expanded edition. Vol. 151 of *Encyclopedia of Mathematics and Its Applications*, Cambridge University Press, 2014.

[Sch16] R. Schneider. Second moments related to Poisson hyperplane tessellations. *J. Math. Anal. Appl.*, 434:1365–1375, 2016.

[ScW08] R. Schneider and W. Weil. *Stochastic and Integral Geometry*. Springer, Berlin, 2008

[ScWi80] R. Schneider and J.A. Wieacker. Random polytopes in a convex body. *Z. Wahrsch. Verw. Gebiete*, 52:69–73, 1980.

[ScWi93] R. Schneider and J.A. Wieacker. Integral geometry. In P.M. Gruber and J.M. Wills, editors, *Handbook of Convex Geometry*, pages 1349–1390, Elsevier, Amsterdam, 1993.

[Schr02] T. Schreiber. Limit theorems for certain functionals of unions of random closed sets. *Teor. Veroyatnost. i Primenen.*, 47:130–142, 2002; *Theory Probab. Appl.*, 47:79–90, 2003.

[ScT13a] T. Schreiber and Ch. Thäle. Geometry of iteration stable tessellations: Connection with Poisson hyperplanes. *Bernoulli*, 19:1637–1654, 2013.

[ScT13b] T. Schreiber and Ch. Thäle. Limit theorems for iteration stable tessellations. *Ann. Probab.*, 41:2261–2278, 2013.

[ScY08] T. Schreiber and J.E. Yukich. Variance asymptotics and central limit theorems for generalized growth processes with applications to convex hulls and maximal points. *Ann. Probab.*, 36:363–396, 2008.

[ScT14] M. Schulte and Ch. Thäle. Distances between Poisson k-flats. *Methodol. Comput. Appl. Probab.*, 16:311–329, 2014.

[Schü94] C. Schütt. Random polytopes and affine surface area. *Math. Nachr.*, 170:227–249, 1994.

[Schü02] C. Schütt. Best and random approximation of convex bodies by polytopes. *Rend. Circ. Mat. Palermo, Ser. II, Suppl.*, (part II) 70:315-334, 2002.

[ScWe03] C. Schütt and E. Werner. Polytopes with vertices chosen randomly from the boundary of a convex body. In: GAFA Seminar Notes, vol. 1807 of *Lecture Notes in Math*, pages 241–422, Springer, Berlin, 2003.

[Sol78] H. Solomon. *Geometric Probability*. SIAM, Philadelphia, 1978.

[Val95] P. Valtr. Probability that n random points are in convex position. In: I. Bárány and J. Pach, editors, *The László Fejes Tóth Festschrift. Discrete Comput. Geom.*, 13:637–643, 1995.

[Val96] P. Valtr. The probability that n random points in a triangle are in convex position. *Combinatorica*, 16:567–573, 1996.

[Vu05] V. Vu. Sharp concentration of random polytopes. *Geom. Funct. Anal.*, 15:1284–1318, 2005.

[Vu06] V. Vu. Central limit theorems for random polytopes in a smooth convex set. *Adv. Math.*, 207:221–243, 2006.

[WaW01] U. Wagner and E. Welzl. A continuous analogue of the upper bound theorem. *Discrete Comput. Geom.*, 26:205–219, 2001.

[WeW93] W. Weil and J.A. Wieacker. Stochastic geometry. In P.M. Gruber and J.M. Wills, editors, *Handbook of Convex Geometry*, pages 1391–1438, Elsevier, Amsterdam, 1993.

[Zin03] A. Zinani. The expected volume of a tetrahedron whose vertices are chosen at random in the interior of a cube. *Monatsh. Math.*, 139:341–348, 2003.

13 GEOMETRIC DISCREPANCY THEORY AND UNIFORM DISTRIBUTION

J. Ralph Alexander, József Beck, and William W.L. Chen

INTRODUCTION

A sequence $s_1, s_2, \ldots$ in $\mathbf{U} = [0,1)$ is said to be *uniformly distributed* if, in the limit, the number of s_j falling in any given subinterval is proportional to its length. Equivalently, $s_1, s_2, \ldots$ is uniformly distributed if the sequence of equiweighted atomic probability measures $\mu_N(s_j) = 1/N$, supported by the initial N-segments $s_1, s_2, \ldots, s_N$, converges weakly to Lebesgue measure on $\mathbf{U}$. This notion immediately generalizes to any topological space with a corresponding probability measure on the Borel sets.

Uniform distribution, as an area of study, originated from the remarkable paper of Weyl [Wey16], in which he established the fundamental result known nowadays as the Weyl criterion (see [Cas57, KN74]). This reduces a problem on uniform distribution to a study of related exponential sums, and provides a deeper understanding of certain aspects of Diophantine approximation, especially basic results such as Kronecker's density theorem. Indeed, careful analysis of the exponential sums that arise often leads to Erdős-Turán-type upper bounds, which in turn lead to quantitative statements concerning uniform distribution.

Today, the concept of uniform distribution has important applications in a number of branches of mathematics such as number theory (especially Diophantine approximation), combinatorics, ergodic theory, discrete geometry, statistics, numerical analysis, etc. In this chapter, we focus on the geometric aspects of the theory.

13.1 UNIFORM DISTRIBUTION OF SEQUENCES

GLOSSARY

Uniformly distributed: Given a sequence $(s_n)_{n\in\mathbb{N}}$, with $s_n \in \mathbf{U} = [0,1)$, let $Z_N([a,b)) = |\{j \le N \mid s_j \in [a,b)\}|$. The sequence is uniformly distributed if, for every $0 \le a < b \le 1$, $\lim_{N\to\infty} N^{-1} Z_N([a,b)) = b - a$.

Fractional part: The fractional part $\{x\}$ of a real number x is $x - \lfloor x \rfloor$.

Kronecker sequence: A sequence of points of the form $(\{N\alpha_1\}, \ldots, \{N\alpha_k\})_{N\in\mathbb{N}}$ in $\mathbf{U}^k$, where $1, \alpha_1, \ldots, \alpha_k \in \mathbb{R}$ are linearly independent over $\mathbb{Q}$.

Discrepancy, or ***irregularity of distribution:*** The discrepancy of a sequence

$(s_n)_{n\in\mathbb{N}}$, with $s_n \in \mathbf{U} = [0,1)$, in a subinterval $[a,b)$ of $\mathbf{U}$, is

$$\Delta_N([a,b)) = |Z_N([a,b)) - N(b-a)|.$$

More generally, the discrepancy of a sequence $(s_n)_{n\in\mathbb{N}}$, with $s_n \in \mathcal{S}$, a topological probability space, in a measurable subset $A \subset \mathcal{S}$, is $\Delta_N(A) = |Z_N(A) - N\mu(A)|$, where $Z_N(A) = |\{j \le N \mid s_j \in A\}|$.

Aligned rectangle, aligned triangle: A rectangle (resp. triangle) in $\mathbb{R}^2$ two sides of which are parallel to the coordinate axes.

Hausdorff dimension: A set S in a metric space has Hausdorff dimension m, where $0 \le m \le +\infty$, if

(i) for any $0 < k < m$, $\mu_k(S) > 0$;

(ii) for any $m < k < +\infty$, $\mu_k(S) < +\infty$.

Here, μ_k is the k-dimensional ***Hausdorff measure***, given by

$$\mu_k(S) = 2^{-k}\kappa_k \liminf_{\epsilon\to 0}\left\{\sum_{i=1}^{\infty}(\text{diam } S_i)^k \;\middle|\; S \subset \bigcup_{i=1}^{\infty} S_i,\ \text{diam } S_i \le \epsilon\right\},$$

where κ_k is the volume of the unit ball in $\mathbb{E}^k$.

Remark. Throughout this chapter, the symbol c will always represent the generic absolute positive constant, depending only on the indicated parameters. The value generally varies from one appearance to the next.

It is not hard to prove that for any irrational number α, the sequence of fractional parts $\{N\alpha\}$ is everywhere dense in $\mathbf{U}$ (here N is the running index). Suppose that the numbers $1, \alpha_1, \ldots, \alpha_k$ are linearly independent over $\mathbb{Q}$. Then Kronecker's theorem states that the k-dimensional Kronecker sequence $(\{N\alpha_1\}, \ldots, \{N\alpha_k\})$ is dense in the unit k-cube $\mathbf{U}^k$. It is a simple consequence of the Weyl criterion that any such Kronecker sequence is uniformly distributed in $\mathbf{U}^k$, a far stronger result than the density theorem. For example, letting $k = 1$, we see that $\{N\sqrt{2}\}$ is uniformly distributed in $\mathbf{U}$.

Weyl's work led naturally to the question: How rapidly can a sequence in $\mathbf{U}$ become uniformly distributed as measured by the discrepancy $\Delta_N([a,b))$ of subintervals? Here, $\Delta_N([a,b)) = |Z_N([a,b)) - N(b-a)|$, where $Z_N([a,b))$ counts those $j \le N$ for which s_j lies in $[a,b)$. Thus we see that Δ_N measures the difference between the actual number of s_j in an interval and the expected number. The sequence is uniformly distributed if and only if $\Delta_N(I) = o(N)$ for all subintervals I. The notion of discrepancy immediately extends to any topological probability space, provided there is at hand a suitable collection of measurable sets $\mathcal{J}$ corresponding to the intervals. If A is in $\mathcal{J}$, set $\Delta_N(A) = |Z_N(A) - N\mu(A)|$.

From the works of Hardy, Littlewood, Ostrowski, and others, it became clear that the smaller the partial quotients in the continued fractions of the irrational number α are, the more uniformly distributed the sequence $\{N\alpha\}$ is. For instance, the partial quotients of quadratic irrationals are characterized by being cyclic, hence bounded. Studying the behavior of $\{N\alpha\}$ for these numbers has proved an excellent indicator of what might be optimal for general sequences in $\mathbf{U}$. Here one has $\Delta_N(I) < c(\alpha)\log N$ for all intervals I and integers $N \ge 2$. Unfortunately, one does not have anything corresponding to continued fractions in higher dimensions, and this has been an obstacle to a similar study of Kronecker sequences (see [Bec94]).

Van der Corput gave an alternative construction of a super uniformly distributed sequence of rationals in $\mathbf{U}$ for which $\Delta_N(I) < c\log N$ for all intervals I and integers $N \geq 2$ (see [KN74, p. 127]). He also asked for the best possible estimate in this direction. In particular, he posed

PROBLEM 13.1.1 *Van der Corput Problem* [Cor35a] [Cor35b]

Can there exist a sequence for which $\Delta_N(I) < c$ for all N and I?

He conjectured, in a slightly different formulation, that such a sequence could not exist. This conjecture was affirmed by van Aardenne-Ehrenfest [A-E45], who later showed that for any sequence in $\mathbf{U}$, $\sup_I \Delta_N(I) > c\log\log N/\log\log\log N$ for infinitely many values of N [A-E49]. Her pioneering work gave the first nontrivial lower bound on the discrepancy of general sequences in $\mathbf{U}$. It is trivial to construct a sequence for which $\sup_I \Delta_N(I) \leq 1$ for infinitely many values of N.

In a classic paper, Roth showed that for any infinite sequence in $\mathbf{U}$, it must be true that $\sup_I \Delta_N(I) > c(\log N)^{1/2}$ for infinitely many N. Finally, in another classic paper, Schmidt used an entirely new method to prove the following result.

THEOREM 13.1.2 *Schmidt* [Sch72b]

The inequality $\sup_I \Delta_N(I) > c\log N$ holds for infinitely many N.

For a more detailed discussion of work arising from the van der Corput conjecture, see [BC87, pp. 3–6].

In light of van der Corput's sequence, as well as $\{N\sqrt{2}\}$, Schmidt's result is best possible. The following problem, which has been described as "excruciatingly difficult," is a major remaining open question from the classical theory.

PROBLEM 13.1.3

Extend Schmidt's result to a best possible estimate of the discrepancy for sequences in $\mathbf{U}^k$ for $k > 1$.

For a given sequence, the results above do not imply the existence of a fixed interval I in $\mathbf{U}$ for which $\sup_N \Delta_N(I) = \infty$. Let I_α denote the interval $[0, \alpha)$, where $0 < \alpha \leq 1$. Schmidt [Sch72a] showed that for any fixed sequence in $\mathbf{U}$ there are only countably many values of α for which $\Delta_N(I_\alpha)$ is bounded. The best result in this direction is due to Halász.

THEOREM 13.1.4 *Halász* [Hal81]

For any fixed sequence in $\mathbf{U}$, let A denote the set of values of α for which $\Delta_N(I_\alpha) = o(\log N)$. Then A has Hausdorff dimension 0.

For a more detailed discussion of work arising from this question, see [BC87, pp. 10–11].

The fundamental works of Roth and Schmidt opened the door to the study of discrepancy in higher dimensions, and there were surprises. In his classic paper, Roth [Rot54] transformed the heart of van der Corput's problem to a question concerning the unit square $\mathbf{U}^2$. In this new formulation, Schmidt's "$\log N$ theorem" implies that if N points are placed in $\mathbf{U}^2$, there is always an aligned rectangle $I = [\gamma_1, \alpha_1) \times [\gamma_2, \alpha_2)$ having discrepancy exceeding $c\log N$. Roth also showed that it was possible to place N points in the square $\mathbf{U}^2$ so that the discrepancy of no aligned rectangle exceeds $c\log N$. One way is to choose $p_j = ((j-1)/N, \{j\sqrt{2}\})$ for $j \leq N$. Thus, the function $c\log N$ describes the *minimax discrepancy* for aligned rectangles. However, Schmidt showed that there is always an aligned right triangle (the part of an aligned rectangle above, or below, a diagonal) with discrepancy exceeding $cN^{1/4-\epsilon}$! Later work has shown that $cN^{1/4}$ exactly describes the minimax discrepancy of aligned right triangles. This paradoxical behavior is not isolated.

Generally, if one studies a collection $\mathcal{J}$ of "nice" sets such as disks, aligned boxes, rotated cubes, etc., in $\mathbf{U}^k$ or some other convex region, it turns out that the minimax discrepancy is either bounded above by $c(\log N)^r$ or bounded below by cN^s, with nothing halfway. In $\mathbf{U}^k$, typically $s = (k-1)/2k$. Thus, there tends to be a logarithmic version of the Vapnik-Chervonenkis principle in operation (see Chapter 40 of this Handbook for a related discussion). Later, we shall see how certain geometric properties place $\mathcal{J}$ in one or the other of these two classes.

13.2 THE GENERAL FREE PLACEMENT PROBLEM FOR *N* POINTS

One can ask for bounds on the discrepancy of N variable points $\mathcal{P} = \{p_1, p_2, \ldots, p_N\}$ that are freely placed in a domain $\mathbf{K}$ in Euclidean t-space $\mathbb{E}^t$. By contrast, when one considers the discrepancy of a *sequence* in $\mathbf{K}$, the initial n-segment of $p_1, \ldots, p_n, \ldots, p_N$ remains fixed for $n \leq N$ as new points appear with increasing N. For a given $\mathbf{K}$, as the unit interval $\mathbf{U}$ demonstrates, estimates for these two problems are quite different as functions of N. The freely placed points in $\mathbf{U}$ need never have discrepancy exceeding 1.

With Roth's reformulation (discussed in Section 13.3), the classical problem is easier to state and, more importantly, it generalizes in a natural manner to a wide class of problems. The bulk of geometric discrepancy problems are now posed as free placement problems. In practically all situations, the domain $\mathbf{K}$ has a very simple description as a cube, disk, sphere, etc., and standard notation is used in the specific situations.

PROBABILITY MEASURES AND DISCREPANCY

In a free placement problem there are two probability measures in play. First, there is the atomic measure μ^+ that assigns weight $1/N$ to each p_j. Second, there is a probability measure μ^- on the Borel sets of $\mathbf{K}$. The measure μ^- is generally the restriction of a natural uniform measure, such as scaled Lebesgue measure. An example would be given by $\mu^- = \sigma/4\pi$ on the unit sphere $\mathbf{S}^2$, where σ is the usual surface measure. It is convenient to define the signed measure $\mu = \mu^+ - \mu^-$ (in the previous section μ^- was denoted by μ). The discrepancy of a Borel set A is, as before, given by $\Delta(A) = |Z(A) - N\mu^-(A)| = N|\mu(A)|$.

The function Δ is always restricted to a very special collection $\mathcal{J}$ of sets, and the challenge lies in obtaining estimates concerning the restricted Δ. It is the central importance of the collection $\mathcal{J}$ that gives the study of discrepancy its distinct character. In a given problem it is sometimes possible to reduce the size of $\mathcal{J}$. Taking the unit interval $\mathbf{U}$ as an example, letting $\mathcal{J}$ be the collection of intervals $[\gamma, \alpha)$ seems to be the obvious choice. But a moment's reflection shows that only intervals of the form $I_\alpha = [0, \alpha)$ need be considered for estimates of discrepancy. At most a factor of 2 is introduced in any estimate of bounds.

NOTIONS OF DISCREPANCY

In most interesting problems $\mathcal{J}$ itself carries a measure ν in the sense of integral geometry, and this adds much more structure. While there is artistic latitude in the choice of ν, more often than not there is a natural measure on $\mathcal{J}$. In the example of $\mathbf{U}$, by identifying $I_\alpha = [0, \alpha)$ with its right endpoint, it is clear that Lebesgue measure on $\mathbf{U}$ is the natural choice for ν.

Given that the measure ν exists, for $0 < W < \infty$ define

$$\|\Delta(\mathcal{P}, \mathcal{J})\|_W = \left(\int_{\mathcal{J}} (\Delta(A))^W d\nu\right)^{1/W} \qquad \text{and} \qquad \|\Delta(\mathcal{P}, \mathcal{J})\|_\infty = \sup_{\mathcal{J}} \Delta(A),$$

and for $0 < W \leq \infty$ define

$$D(\mathbf{K}, \mathcal{J}, W, N) = \inf_{|\mathcal{P}|=N} \{\|\Delta(\mathcal{P}, \mathcal{J})\|_W\}. \tag{13.2.1}$$

The determination of the "minimax" $D(\mathbf{K}, \mathcal{J}, \infty, N)$ is generally the most important as well as the most difficult problem in the study. It should be noted that the function $D(\mathbf{K}, \mathcal{J}, \infty, N)$ is defined even if the measure ν is not. The term $D(\mathbf{K}, \mathcal{J}, 2, N)$ has been shown to be intimately related to problems in numerical integration in some special cases, and is of increasing importance. These various functions $D(\mathbf{K}, \mathcal{J}, W, N)$ measure how well the continuous distribution μ^- can be approximated by N freely placed atoms.

For $W \geq 1$, the inequality

$$\nu(\mathcal{J})^{-1/W} \|\Delta(\mathcal{P}, \mathcal{J})\|_W \leq \|\Delta(\mathcal{P}, \mathcal{J})\|_\infty \tag{13.2.2}$$

provides a general approach for obtaining a lower bound for $D(\mathbf{K}, \mathcal{J}, \infty, N)$. The choice $W = 2$ has been especially fruitful, but good estimates of $D(\mathbf{K}, \mathcal{J}, W, N)$ for any W are of independent interest.

An upper bound on $D(\mathbf{K}, \mathcal{J}, \infty, N)$ generally is obtained by showing the existence of a favorable example. This may be done either by a direct construction, often extremely difficult to verify, or by a probabilistic argument showing such an example does exist without giving it explicitly. These comments would apply as well to upper bounds for any $D(\mathbf{K}, \mathcal{J}, W, N)$.

13.3 ALIGNED RECTANGLES IN THE UNIT SQUARE

The unit square $\mathbf{U}^2 = [0, 1) \times [0, 1)$ is by far the most thoroughly studied 2-dimensional object. The main reason for this is Roth's reformulation of the van der Corput problem. Many of the interesting questions that arose have been answered, and we give a summary of the highlights.

For $\mathbf{U}^2$ one wishes to study the discrepancy of rectangles of the type $I = [\gamma_1, \alpha_1) \times [\gamma_2, \alpha_2)$. It is a trivial observation that only those I for which $\gamma_1 = \gamma_2 = 0$ need be considered, and this restricted family, denoted by $\mathcal{B}^2$, is the choice for $\mathcal{J}$. By considering this smaller collection one introduces at most a factor of 4 on bounds. There is a natural measure ν on $\mathcal{B}^2$, which may be identified with Lebesgue measure on $\mathbf{U}^2$ via the upper right corner points (α_1, α_2). In the same spirit, let $\mathcal{B}^1$ denote the previously introduced collection of intervals $I_\alpha = [0, \alpha)$ in $\mathbf{U}$.

THEOREM 13.3.1 *Roth's Equivalence* [Rot54] [BC87, pp. 6–7]

Let f be a positive increasing function tending to infinity. Then the following two statements are equivalent:

(i) *There is an absolute positive constant c_1 such that for any finite sequence $s_1, s_2, \ldots, s_N$ in $\mathbf{U}$, there always exists a positive integer $n \le N$ such that $\|\Delta(\mathcal{P}_n, \mathcal{B}^1)\|_\infty > c_1 f(N)$. Here, $\mathcal{P}_n$ is the initial n-segment.*

(ii) *There is an absolute positive constant c_2 such that for all positive integers N, $D(\mathbf{U}^2, \mathcal{B}^2, \infty, N) > c_2 f(N)$.*

The equivalence shows that the central question of bounds for the van der Corput problem can be replaced by an elegant problem concerning the free placement of N points in the unit square $\mathbf{U}^2$. The mapping $s_j \to ((j-1)/N, s_j)$ plays a role in the proof of this equivalence. If one takes as $\mathcal{P}_N$ the image in $\mathbf{U}^2$ under the mapping of the initial N-segment of the van der Corput sequence, the following upper bound theorem may be proved.

THEOREM 13.3.2 *Lerch* [BC87, Theorem 4, $K = 2$]

For $N \ge 2$,

$$D(\mathbf{U}^2, \mathcal{B}^2, \infty, N) < c \log N. \tag{13.3.1}$$

The corresponding lower bound is established by the important "$\log N$ theorem" of Schmidt.

THEOREM 13.3.3 *Schmidt* [Sch72b] [BC87, Theorem 3B]

One has

$$D(\mathbf{U}^2, \mathcal{B}^2, \infty, N) > c \log N. \tag{13.3.2}$$

By an explicit lattice construction, Davenport [Dav56] gave the best possible upper bound estimate for $W = 2$. His analysis shows that if the irrational number α has continued fractions with bounded partial quotients, then the $N = 2M$ points in $\mathbf{U}^2$ given by

$$p_j^{\pm} = ((j-1)/M, \{\pm j\alpha\}), \qquad j \le M,$$

can be taken as $\mathcal{P}$ in proving the following theorem. Other proofs have been given by Vilenkin [Vil67], Halton and Zaremba [HZ69], and Roth [Rot76].

THEOREM 13.3.4 *Davenport* [Dav56] [BC87, Theorem 2A]

For $N \ge 2$,

$$D(\mathbf{U}^2, \mathcal{B}^2, 2, N) < c(\log N)^{1/2}. \tag{13.3.3}$$

This complements the following lower bound obtained by Roth in his classic paper.

THEOREM 13.3.5 *Roth* [Rot54] [BC87, Theorem 1A, $K = 2$]

One has

$$D(\mathbf{U}^2, \mathcal{B}^2, 2, N) > c(\log N)^{1/2}. \tag{13.3.4}$$

For $W = 1$, an upper bound $D(\mathbf{U}^2, \mathcal{B}^2, 1, N) < c(\log N)^{1/2}$ follows at once from Davenport's bound (13.3.3) by the monotonicity of $D(\mathbf{U}^2, \mathcal{B}^2, W, N)$ as a function of W. The corresponding lower bound was obtained by Halász.

THEOREM 13.3.6 *Halász* [Hal81] [BC87, Theorem 1C, $K = 2$]

One has

$$D(\mathbf{U}^2, \mathcal{B}^2, 1, N) > c(\log N)^{1/2}. \tag{13.3.5}$$

Halász (see [BC87, Theorem 3C]) deduced that there is always an aligned square of discrepancy larger than $c \log N$. Of course, the square generally will not be a member of the special collection $\mathcal{B}^2$. Ruzsa [Ruz93] has given a clever elementary proof that the existence of such a square follows directly from inequality (13.3.2) above; see also [Mat99].

The ideas developed in the study of discrepancy can be applied to approximations of integrals. We briefly mention two examples, both restricted to 2 dimensions for the sake of simplicity.

A function ψ is termed ***M-simple*** if $\psi(x) = \sum_{j=1}^{M} m_j \chi_{B_j}(x)$, where χ_{B_j} is the characteristic function of the aligned rectangle B_j. In this theorem, the lower bounds are nontrivial because of the logarithmic factors coming from discrepancy theory on $\mathbf{U}^2$.

THEOREM 13.3.7 *Chen* [Che85] [Che87] [BC87, Theorems 5A, 5C]

Let the function f be defined on $\mathbf{U}^2$ by $f(x) = C + \int_{B(x)} g(y)dy$ where C is a constant, g is nonzero on a set of positive measure in $\mathbf{U}^2$, and $B(x_1, x_2) = [0, x_1) \times [0, x_2)$. Then, for any M-simple function ψ,

$$\|f - \psi\|_W > c(f, W)M^{-1}(\log M)^{1/2}, \qquad 1 \le W < \infty;$$
$$\|f - \psi\|_\infty > c(f)M^{-1}\log M.$$

Let $\mathcal{C}$ be the class of all continuous real valued functions on $\mathbf{U}^2$, endowed with the Wiener sheet measure ω. For every function $f \in \mathcal{C}$ and every set $\mathcal{P}$ of N points in $\mathbf{U}^2$, let

$$I(f) = \int_{\mathbf{U}^2} f(x)dx \qquad \text{and} \qquad U(\mathcal{P}, f) = \frac{1}{N}\sum_{p\in\mathcal{P}} f(p).$$

THEOREM 13.3.8 *Woźniakowski* [Woź91]

One has

$$\inf_{|\mathcal{P}|=N} \left(\int_{\mathcal{C}} |U(\mathcal{P}, f) - I(f)|^2 d\omega \right)^{1/2} = \frac{D(\mathbf{U}^2, \mathcal{B}^2, 2, N)}{N}.$$

13.4 ALIGNED BOXES IN A UNIT k-CUBE

The van der Corput problem led to the study of $D(\mathbf{U}^2, \mathcal{B}^2, W, N)$, which in turn led to the study of $D(\mathbf{U}^k, \mathcal{B}^k, W, N)$ for $W > 0$ and general positive integers k. Here, $\mathcal{B}^k$ denotes the collection of boxes $I = [0, \alpha_1) \times \ldots \times [0, \alpha_k)$, and the measure ν is identified with Lebesgue measure on $\mathbf{U}^k$ via the corner points $(\alpha_1, \ldots, \alpha_k)$.

The principle of Roth's equivalence extends so that the discrepancy problem for sequences in $\mathbf{U}^k$ reformulates as a free placement problem in $\mathbf{U}^{k+1}$, so that we discuss only the latter version. Inequalities (13.3.1) – (13.3.5) give the exact order of magnitude of $D(\mathbf{U}^2, \mathcal{B}^2, W, N)$ for the most natural values of W, namely

$1 \le W \le 2$ and $W = \infty$, with the latter being top prize. While much is known, knowledge of $D(\mathbf{U}^k, \mathcal{B}^k, W, N)$ is incomplete, especially for $W = \infty$, while there is ongoing work on the case $W = 1$ which may lead to its complete solution. It should be remarked that if k and N are fixed, then $D(\mathbf{U}^k, \mathcal{B}^k, W, N)$ is a nondecreasing function of the positive real number W.

As was indicated earlier, upper bound methods generally fall into two classes, explicit constructions and probabilistic existence arguments. In practice, careful constructions are made prior to a probabilistic averaging process. Chen's proof of the following upper bound theorem involved extensive combinatorial and number-theoretic constructions as well as probabilistic considerations.

THEOREM 13.4.1 *Chen* [Che80] [BC87, Theorem 2D]
For positive real numbers W, and integers $k \ge 2$ and $N \ge 2$,

$$D(\mathbf{U}^k, \mathcal{B}^k, W, N) < c(W, k)(\log N)^{(k-1)/2}. \tag{13.4.1}$$

A second proof was given by Chen [Che83] (see also [BC87, Section 3.5]), where the idea of digit shifts was first used in the subject. Earlier, Roth [Rot80] (see also [BC87, Theorem 2C]) treated the case $W = 2$. The inequality (13.4.1) highlights one of the truly baffling aspects of the theory, namely the apparent jump discontinuity in the asymptotic behavior of $D(\mathbf{U}^k, \mathcal{B}^k, W, N)$ at $W = \infty$. This discontinuity is most dramatically established for $k = 2$, but is known to occur for any $k \ge 3$ (see (13.4.4) below).

Explicit multidimensional sequences greatly generalizing the van der Corput sequence also have been used to obtain upper bounds for $D(\mathbf{U}^k, \mathcal{B}^k, \infty, N)$. Halton constructed explicit point sets in $\mathbf{U}^k$ in order to prove the next theorem. Faure (see [BC87, Section 3.2]) gave a different proof of the same result.

THEOREM 13.4.2 *Halton* [Hal60] [BC87, Theorem 4]
For integers $k \ge 2$ and $N \ge 2$,

$$D(\mathbf{U}^k, \mathcal{B}^k, \infty, N) < c(k)(\log N)^{k-1}. \tag{13.4.2}$$

In order to prove (13.3.3), Davenport used properties of special lattices in 2 dimensions. However, it took many years before we had success with lattices in higher dimensions. Skriganov has established some most interesting results, which imply the following theorem. Given a region, a lattice is termed ***admissible*** if the region contains no member of the lattice except possibly the origin (see [Cas59]). Examples for the following theorem are given by lattices arising from algebraic integers in totally real algebraic number fields.

THEOREM 13.4.3 *Skriganov* [Skr94]
Suppose Γ is a fixed k-dimensional lattice admissible for the region $|x_1 x_2 \ldots x_k| < 1$.

(i) *Halton's upper bound inequality* (13.4.2) *holds if the N points are obtained by intersecting $\mathbf{U}^k$ with $t\Gamma$, where $t > 0$ is a suitably chosen real scalar.*

(ii) *With the same choice of t as in part* (i)*, there exists $x \in \mathbb{E}^k$ such that Chen's upper bound inequality* (13.4.1) *holds if the N points are obtained by intersecting $\mathbf{U}^k$ with $t\Gamma + x$.*

Later, using p-adic Fourier-Walsh analysis together with ideas originating from coding theory, Chen and Skriganov [CS02] have obtained explicit constructions that give (13.4.1) in the special case $W = 2$, with an explicitly given constant $c(2, k)$. The extension to arbitrary positive real numbers W was given by Skriganov [Skr06]. Alternative proofs of these results, using dyadic Fourier-Walsh analysis, were given by Dick and Pillichshammer [DP14] in the special case $W = 2$, and by Dick [Dic14] for arbitrary positive real numbers W.

Moving to lower bound estimates, the following theorem of Schmidt is complemented by Chen's result (13.4.1). For $W \geq 2$ this lower bound is due to Roth, since D is monotone in W.

THEOREM 13.4.4 *Schmidt* [Sch77a] [BC87, Theorem 1B]

For $W > 1$ and integers $k \geq 2$,

$$D(\mathbf{U}^k, \mathcal{B}^k, W, N) > c(W, k)(\log N)^{(k-1)/2}. \tag{13.4.3}$$

Concerning $W = 1$, there is the result of Halász, which is probably not optimal. It is reasonably conjectured that $(k-1)/2$ is the correct exponent.

THEOREM 13.4.5 *Halász* [Hal81] [BC87, Theorem 1C]

For integers $k \geq 2$,

$$D(\mathbf{U}^k, \mathcal{B}^k, 1, N) > c(k)(\log N)^{1/2}.$$

The next lower bound estimate, although probably not best possible, firmly establishes a discontinuity in asymptotic behavior at $W = \infty$ for all $k \geq 3$.

THEOREM 13.4.6 *Bilyk, Lacey, Vagharshakyan* [BLV08]

For integers $k \geq 3$, there exist constants $\delta_k \in (0, 1/2)$ such that

$$D(\mathbf{U}^k, \mathcal{B}^k, \infty, N) > c(k)(\log N)^{(k-1)/2+\delta_k}. \tag{13.4.4}$$

Earlier, Beck [Bec89] had established a weaker lower bound for the case $k = 3$, of the form

$$D(\mathbf{U}^k, \mathcal{B}^k, \infty, N) > c(3)(\log N)(\log \log N)^{c_3},$$

where c_3 can be taken to be any positive real number less than $1/8$. These bounds represent the first improvements of Roth's lower bound

$$D(\mathbf{U}^k, \mathcal{B}^k, \infty, N) > c(k)(\log N)^{(k-1)/2},$$

established over 60 years ago.

Can the factor $1/2$ be removed from the exponent? This is the "great open problem." Recently, there has been evidence that suggests that perhaps

$$D(\mathbf{U}^k, \mathcal{B}^k, \infty, N) > c(k)(\log N)^{k/2}, \tag{13.4.5}$$

but no more. To discuss this, we need to modify the definition (13.2.1) in light of the idea of digit shifts introduced by Chen [Che83]. Let $\mathcal{S}$ be a finite set of dyadic digit shifts. For every point set $\mathcal{P}$ and every dyadic shift $S \in \mathcal{S}$, let $\mathcal{P}(S)$ denote the image of $\mathcal{P}$ under S. Corresponding to (13.2.1), let

$$E(\mathbf{K}, \mathcal{J}, W, N, \mathcal{S}) = \inf_{|\mathcal{P}|=N} \sup_{S \in \mathcal{S}} \{\|\Delta(\mathcal{P}(S), \mathcal{J})\|_W\}.$$

In Skriganov [Skr16], it is shown that for every N, there exists a finite set $\mathcal{S}$ of dyadic digit shifts, depending only on N and k, such that

$$E(\mathbf{U}^k, \mathcal{B}^k, \infty, N, \mathcal{S}) > c(k)(\log N)^{k/2}, \tag{13.4.6}$$

an estimate consistent with the suggestion (13.4.5). In the same paper, it is also shown that for every fixed $W > 0$ and for every N, there exists a finite set $\mathcal{S}$ of dyadic digit shifts, depending only on N, k and W, such that

$$E(\mathbf{U}^k, \mathcal{B}^k, W, N, \mathcal{S}) > c(W, k)(\log N)^{(k-1)/2}, \tag{13.4.7}$$

somewhat extending (13.4.3).

Beck has also refined Roth's estimate in a geometric direction.

THEOREM 13.4.7 *Beck* [BC87, Theorem 19A]

Let $\mathcal{J}$ be the collection of aligned cubes contained in $\mathbf{U}^k$. Then

$$D(\mathbf{U}^k, \mathcal{J}, \infty, N) > c(k)(\log N)^{(k-1)/2}. \tag{13.4.8}$$

Actually, Beck's method shows $D(\mathbf{U}^k, \mathcal{J}, 2, N) > c(k)(\log N)^{(k-1)/2}$, with respect to a natural measure ν on sets of aligned cubes. This improves Roth's inequality $D(\mathbf{U}^k, \mathcal{B}^k, 2, N) > c(k)(\log N)^{(k-1)/2}$. So far, it has not been possible to extend Ruzsa's ideas to higher dimensions in order to show that the previous theorem follows directly from Roth's estimate. However, Drmota [Drm96] has published a new proof that $D(\mathbf{U}^k, \mathcal{J}, 2, N) > c(k)D(\mathbf{U}^k, \mathcal{B}^k, 2, N)$, and this does imply (13.4.8).

13.5 MOTION-INVARIANT PROBLEMS

In this section and the next three, we discuss collections $\mathcal{J}$ of convex sets having the property that any set in $\mathcal{J}$ may be moved by a direct (orientation preserving) motion of $\mathbb{E}^k$ and yet remain in $\mathcal{J}$. Motion-invariant problems were first extensively studied by Schmidt, and many of his estimates, obtained by a difficult technique using integral equations, were close to best possible. The book [BC87] contains an account of Schmidt's methods. Later, the Fourier transform method of Beck has achieved results that in general surpass those obtained by Schmidt. For a broad class of problems, Beck's Fourier method gives nearly best possible estimates for $D(\mathbf{K}, \mathcal{J}, 2, N)$.

The pleasant surprise is that if $\mathcal{J}$ is motion-invariant, then the bounds on $D(\mathbf{K}, \mathcal{J}, \infty, N)$ turn out to be very close to those for $D(\mathbf{K}, \mathcal{J}, 2, N)$. This is shown by a probabilistic upper bound method, which generally pins $D(\mathbf{K}, \mathcal{J}, \infty, N)$ between bounds differing at most by a factor of $c(k)(\log N)^{1/2}$.

The simplest motion-invariant example is given by letting $\mathcal{J}$ be the collection of all directly congruent copies of a given convex set A. In this situation, $\mathcal{J}$ carries a natural measure ν, which may be identified with Haar measure on the motion group on $\mathbb{E}^k$. A broader choice would be to let $\mathcal{J}$ be all sets in $\mathbb{E}^k$ directly similar to A. Again, there is a natural measure ν on $\mathcal{J}$. However, for the results stated in the next two sections, the various measures ν on the choices for $\mathcal{J}$ will not be discussed in great detail. In most situations, such measures do play an active role in the proofs through inequality (13.2.2) with $W = 2$. A complete exposition of integration in the context of integral geometry, Haar measure, etc., may be found in the book by Santaló [San76].

For any domain $\mathbf{K}$ in $\mathbb{E}^t$ and each collection $\mathcal{J}$, it is helpful to define three auxiliary collections:

Definition:

(i) $\mathcal{J}_{tor}$ consists of those subsets of $\mathbf{K}$ obtained by reducing elements of $\mathcal{J}$ modulo $\mathbb{Z}^k$. To avoid messiness, let us always suppose that $\mathcal{J}$ has been restricted so that this reduction is 1–1 on each member of $\mathcal{J}$. For example, one might consider only those members of $\mathcal{J}$ having diameter less than 1.

(ii) $\mathcal{J}_c$ consists of those subsets of $\mathbf{K}$ that are members of $\mathcal{J}$.

(iii) $\mathcal{J}_i$ consists of those subsets of $\mathbf{K}$ obtained by intersecting $\mathbf{K}$ with members of $\mathcal{J}$.

Note that $\mathcal{J}_c$ and $\mathcal{J}_i$ are well defined for any domain $\mathbf{K}$. However, $\mathcal{J}_{tor}$ essentially applies only to $\mathbf{U}^k$. If viewed as a flat torus, then $\mathbf{U}^k$ is the proper domain for Kronecker sequences and Weyl's exponential sums. There are several general inequalities for discrepancy results involving $\mathcal{J}_{tor}$, $\mathcal{J}_c$, and $\mathcal{J}_i$. For example, we have $D(\mathbf{U}^k, \mathcal{J}_c, \infty, N) \le D(\mathbf{U}^k, \mathcal{J}_{tor}, \infty, N)$ because $\mathcal{J}_c$ is contained in $\mathcal{J}_{tor}$. Also, if the members of $\mathcal{J}$ have diameters less than 1, then we have $D(\mathbf{U}^k, \mathcal{J}_{tor}, \infty, N) \le 2^k D(\mathbf{U}^k, \mathcal{J}_i, \infty, N)$, since any set in $\mathcal{J}_{tor}$ is the union of at most 2^k sets in $\mathcal{J}_i$.

13.6 SIMILAR OBJECTS IN THE UNIT k-CUBE

GLOSSARY

If A is a compact convex set in $\mathbb{E}^k$, let $d(A)$ denote the diameter of A, $r(A)$ denote the radius of the largest k-ball contained in A, and $\sigma(\partial A)$ denote the surface content of ∂A. The collection $\mathcal{J}$ is said to be ***ds-generated*** by A if $\mathcal{J}$ consists of all directly similar images of A having diameters not exceeding $d(A)$.

We state two pivotal theorems of Beck. As usual, if $\mathcal{S}$ is a discrete set, $Z(B)$ denotes the cardinality of $B \cap \mathcal{S}$.

THEOREM 13.6.1 *Beck* [Bec87] [BC87, Theorem 17A]

Let $\mathcal{S}$ be an arbitrary infinite discrete set in $\mathbb{E}^k$, A be a compact convex set with $r(A) \ge 1$, and $\mathcal{J}$ be ds-generated by A. Then there is a set B in $\mathcal{J}$ such that

$$|Z(B) - \operatorname{vol} B| > c(k)(\sigma(\partial A))^{1/2}. \tag{13.6.1}$$

COROLLARY 13.6.2 *Beck* [BC87, Corollary 17B]

Let A be a compact convex body in $\mathbb{E}^k$ with $r(A) \ge N^{-1/k}$, and let $\mathcal{J}$ be ds-generated by A. Then

$$D(\mathbf{U}^k, \mathcal{J}_{tor}, \infty, N) > c(A)N^{(k-1)/2k}. \tag{13.6.2}$$

The deduction of Corollary 13.6.2 from Theorem 13.6.1 involves a simple rescaling argument. Another important aspect of Beck's work is the introduction of upper bound methods based on probabilistic considerations. The following result shows that Theorem 13.6.1 is very nearly best possible.

THEOREM 13.6.3 *Beck* [BC87, Theorem 18A]

Let A be a compact convex body in $\mathbb{E}^k$ with $r(A) \geq 1$, and let $\mathcal{J}$ be ds-generated by A. Then there exists an infinite discrete set $\mathcal{S}_0$ such that for every set B in $\mathcal{J}$,

$$|Z(B) - \operatorname{vol} B| < c(k)(\sigma(\partial A))^{1/2}(\log \sigma(\partial A))^{1/2}. \tag{13.6.3}$$

COROLLARY 13.6.4 *Beck* [BC87, Corollary 18C]

Let A be a compact convex body in $\mathbb{E}^k$, and $\mathcal{J}$ be ds-generated by A. Then

$$D(\mathbf{U}^k, \mathcal{J}_{tor}, \infty, N) < c(A)N^{(k-1)/2k}(\log N)^{1/2}. \tag{13.6.4}$$

Beck (see [BC87, pp. 129–130]) deduced several related corollaries from Theorem 13.6.3. The example sets $\mathcal{P}_N$ for Corollary 13.6.4 can be taken as the initial segments of a certain fixed sequence whose choice definitely depends on A. If $d(A) = \lambda$ and A is either a disk (solid sphere) or a cube, then the right side of (13.6.2) takes the form $c(k)(\lambda^k N)^{(k-1)/2k}$. Montgomery [Mon89] has obtained a similar lower bound for cubes and disks.

The problem of estimating discrepancy for $\mathcal{J}_c$ is even more challenging because of "boundary effects." We state, as an example, a theorem for disks. The right inequality follows from (13.6.4).

THEOREM 13.6.5 *Beck* [Bec87] [BC87, Theorem 16A]

Let $\mathcal{J}$ be ds-generated by a k-disk. Then for every $\epsilon > 0$,

$$c_1(k,\epsilon)N^{(k-1)/2k-\epsilon} < D(\mathbf{U}^k, \mathcal{J}_c, \infty, N) < c_2(k)N^{(k-1)/2k}(\log N)^{1/2}. \tag{13.6.5}$$

Because all the lower bounds above come from $\mathbf{L}^2$ estimates, these various results (13.6.1) – (13.6.5) allow us to make the general statement that for W in the range $2 \leq W \leq \infty$, the magnitude of $D(\mathbf{U}^k, \mathcal{J}, W, N)$ is controlled by $N^{(k-1)/2k}$. Thus there is no extreme discontinuity in asymptotic behavior at $W = \infty$. However, work by Beck and Chen proves that there is a discontinuity at some W satisfying $1 \leq W \leq 2$, and the following results indicate that $W = 1$ is a likely candidate.

THEOREM 13.6.6 *Beck, Chen* [BC93b]

Let $\mathcal{J}$ be ds-generated by a convex polygon A with $d(A) < 1$. Then

$$\begin{aligned} &D(\mathbf{U}^2, \mathcal{J}_{tor}, W, N) < c(A,W)N^{(W-1)/2W}, \qquad 1 < W \leq 2; \\ &D(\mathbf{U}^2, \mathcal{J}_{tor}, 1, N) < c(A)(\log N)^2. \end{aligned} \tag{13.6.6}$$

In fact, Theorem 13.6.6 is motivated by the study of discrepancy with respect to halfplanes, and is established by ideas used to establish Theorem 13.9.8 below. Note the similarities of the inequalities (13.6.6) and (13.9.5). After all, a convex polygon is the intersection of a finite number of halfplanes, and so the proof of Theorem 13.6.6 involves carrying out the idea of the proof of Theorem 13.9.8 a finite number of times.

The next theorem shows that powers of N other than $N^{(k-1)/2k}$ may appear for $2 \leq W \leq \infty$. It deals with what has been termed the ***isotropic discrepancy*** in $\mathbf{U}^k$.

THEOREM 13.6.7 *Schmidt* [Sch75] [BC87, Theorem 15]
Let $\mathcal{J}$ be the collection of all convex sets in $\mathbb{E}^k$. Then

$$D(\mathbf{U}^k, \mathcal{J}_i, \infty, N) > c(k)N^{(k-1)/(k+1)}. \tag{13.6.7}$$

The function $N^{(k-1)/(k+1)}$ dominates $N^{(k-1)/2k}$, so that this largest possible choice for $\mathcal{J}$ does in fact yield a larger discrepancy. Beck has shown by probabilistic techniques that the inequality (13.6.7), excepting a possible logarithmic factor, is best possible for $k = 2$.

The following result shows that for certain rotation-invariant $\mathcal{J}$ the discrepancy of Kronecker sequences (defined in Section 13.1) will not behave as $cN^{(k-1)/2k}$, but as the square of this quantity.

THEOREM 13.6.8 *Larcher* [Lar91]
Let the sequence of point sets $\mathcal{P}_N$ be the initial segments of a Kronecker sequence in $\mathbf{U}^k$, and let $\mathcal{J}$ be ds-generated by a cube of edge length $\lambda < 1$. Then, for each N,

$$\|\Delta(\mathcal{P}_N, \mathcal{J}_i)\|_\infty > c(k)\lambda^{k-1}N^{(k-1)/k}.$$

Furthermore, the exponent $(k-1)/k$ cannot be increased.

13.7 CONGRUENT OBJECTS IN THE UNIT k-CUBE

GLOSSARY

If $\mathcal{J}$ consists of all directly congruent copies of a convex set A, we say that A ***dm-generates*** $\mathcal{J}$. Simple examples are given by the collection of all k-disks of a fixed radius r or by the collection of all k-cubes of a fixed edge length λ.

Given a convex set A, there is some evidence for the conjecture that the discrepancy for the dm-generated collection will be essentially as large as that for the ds-generated collection. However, this is generally very difficult to establish, even in very specific situations. There are the following results in this direction. The upper bound inequalities all come from Corollary 13.6.4 above.

THEOREM 13.7.1 *Beck* [BC87, Theorem 22A]
Let $\mathcal{J}$ be dm-generated by a square of edge length λ. Then

$$c_1(\lambda)N^{1/8} < D(\mathbf{U}^2, \mathcal{J}_{tor}, \infty, N) < c_2(\lambda)N^{1/4}(\log N)^{1/2}.$$

It is felt that $N^{1/4}$ gives the proper lower bound, and for $\mathcal{J}_i$ this is definitely true. The lower bound in the next result follows at once from the work of Alexander [Ale91] described in Section 13.9.

THEOREM 13.7.2 *Alexander, Beck*

Let $\mathcal{J}$ be dm-generated by a k-cube of edge length λ. Then

$$c_1(\lambda, k)N^{(k-1)/2k} < D(\mathbf{U}^k, \mathcal{J}_i, \infty, N) < c_2(\lambda, k)N^{(k-1)/2k}(\log N)^{1/2}.$$

A similar result probably holds for k-disks, but this has been established only for $k = 2$.

THEOREM 13.7.3 *Beck* [BC87, Theorem 22B]

Let $\mathcal{J}$ be dm-generated by a 2-disk of radius r. Then

$$c_1(r)N^{1/4} < D(\mathbf{U}^2, \mathcal{J}_i, \infty, N) < c_2(r)N^{1/4}(\log N)^{1/2}.$$

13.8 WORK OF MONTGOMERY

It should be reported that Montgomery [Mon89] has independently developed a lower bound method which, as does Beck's method, uses techniques from harmonic analysis. Montgomery's method, especially in dimension 2, obtains for a number of special classes $\mathcal{J}$ estimates comparable to those obtained by Beck's method. In particular, Montgomery has considered $\mathcal{J}$ that are ds-generated by a region whose boundary is a piecewise smooth simple closed curve.

13.9 HALFSPACES AND RELATED OBJECTS

GLOSSARY

Segment: Given a compact subset $\mathbf{K}$ and a closed halfspace H in $\mathbb{E}^k$, $K \cap H$ is called a segment of $\mathbf{K}$.

Slab: The region between two parallel hyperplanes.

Spherical slice: The intersection of two open hemispheres on a sphere.

Let H be a closed halfspace in $\mathbb{E}^k$. Then the collection $\mathcal{H}^k$ of all closed halfspaces is dm-generated by H, and if we associate H with the oriented hyperplane ∂H, there is a well-known invariant measure ν on $\mathcal{H}^k$. Further information concerning this and related measures may be found in Chapter 12 of Santaló [San76]. For a compact domain $\mathbf{K}$ in $\mathbb{E}^k$, it is clear that only the collection $\mathcal{H}_i^k$, the *segments* of $\mathbf{K}$, are proper for study, since $\mathcal{H}_c^k$ is empty and $\mathcal{H}_{tor}^k$ is unsuitable.

In this section, it is necessary for the domain $\mathbf{K}$ to be somewhat more general; hence we make only the following broad assumptions:

(i) $\mathbf{K}$ lies on the boundary of a fixed convex set $\mathbf{M}$ in $\mathbb{E}^{k+1}$;

(ii) $\sigma(\mathbf{K}) = 1$, where σ is the usual k-measure on $\partial\mathbf{M}$.

Since $\mathbb{E}^k$ is the boundary of a halfspace in $\mathbb{E}^{k+1}$, any set in $\mathbb{E}^k$ of unit Lebesgue k-measure satisfies these assumptions. The normalization of assumption (ii) is for convenience, and, by rescaling, the inequalities of this section may be applied to any uniform probability measure on a domain $\mathbf{K}$ in $\mathbb{E}^{k+1}$. Such rescaling only affects dimensional constants; for standard domains, such as the unit k-sphere $\mathbf{S}^k$ and the unit k-disk $\mathbf{D}^k$, this will be done without comment.

Although in applications $\mathbf{K}$ will have a simple geometric description, the next theorem treats the general situation and obtains the essentially exact magnitude of $D(\mathbf{K}, \mathcal{H}_i^{k+1}, 2, N)$. If $\mathbf{K}$ lies in $\mathbb{E}^k$, then $\mathcal{H}^{k+1}$ may be replaced by $\mathcal{H}^k$. If ν is properly normalized, this change invokes no rescaling.

THEOREM 13.9.1 *Alexander* [Ale91]

Let $\mathcal{K}$ be the collection of all $\mathbf{K}$ satisfying assumptions (i) *and* (ii) *above. Then*

$$c_1(k)N^{(k-1)/2k} < \inf_{K \in \mathcal{K}} D(\mathbf{K}, \mathcal{H}_i^{k+1}, 2, N) < c_2(\mathbf{M})N^{(k-1)/2k}. \tag{13.9.1}$$

The upper bound of (13.9.1) can be proved by an indirect probabilistic method introduced by Alexander [Ale72] for $\mathbf{K} = \mathbf{S}^2$, but the method of Beck and Chen [BC90] also may be applied for standard choices of $\mathbf{K}$ such as $\mathbf{U}^k$ and $\mathbf{D}^k$. When $\mathbf{M} = \mathbf{K} = \mathbf{S}^k$, the segments are the spherical caps. For this important special case the upper bound is due to Stolarsky [Sto73], while the lower bound is due to Beck [Bec84] (see also [BC87, Theorem 24B]).

Since the ν-measure of the halfspaces that separate $\mathbf{M}$ is less than $c(k)d(\mathbf{M})$, inequality (13.2.2) may be applied to obtain a lower bound for $D(\mathbf{K}, \mathcal{H}_i^{k+1}, \infty, N)$. The upper bound in the following theorem should be taken in the context of actual applications such as $\mathbf{M}$ being a k-sphere $\mathbf{S}^k$, a compact convex body in $\mathbb{E}^k$, or more generally, a compact convex hypersurface in $\mathbb{E}^{k+1}$.

THEOREM 13.9.2 *Alexander, Beck*

Let $\mathcal{K}$ be the collection of $\mathbf{K}$ satisfying assumptions (i) *and* (ii) *above. Furthermore, suppose that $\mathbf{M}$ is of finite diameter. Then*

$$c_3(k)(d(\mathbf{M}))^{-1/2}N^{(k-1)/2k} < \inf_{K \in \mathcal{K}} D(\mathbf{K}, \mathcal{H}_i^{k+1}, \infty, N) < c_4(\mathbf{M})N^{(k-1)/2k}(\log N)^{1/2}. \tag{13.9.2}$$

For $\mathbf{M} = \mathbf{K} = \mathbf{S}^k$, inequalities (13.9.2) are due to Beck, improving a slightly weaker lower bound by Schmidt [Sch69]. Consideration of $\mathbf{K} = \mathbf{U}^2$ makes it obvious that there exists an aligned right triangle with discrepancy at least $cN^{1/4}$, as stated in Section 13.1. For the case $\mathbf{M} = \mathbf{K} = \mathbf{D}^2$, a unit 2-disk (Roth's *disk-segment problem*), Beck [Bec83] (see also [BC87, Theorem 23A]) obtained inequalities (13.9.2), excepting a factor $(\log N)^{-7/2}$ in the lower bound. Later, Alexander [Ale90] improved the lower bound, and Matoušek [Mat95] obtained essentially the same upper bound. Matoušek's work on $\mathbf{D}^2$ makes it seem likely that Beck's factor $(\log N)^{1/2}$ in his general upper bound theorem might be removable in many specific situations, but this is very challenging.

THEOREM 13.9.3 *Alexander, Matoušek*

For Roth's disk-segment problem,

$$c_1 N^{1/4} < D(\mathbf{D}^2, \mathcal{H}_i^2, \infty, N) < c_2 N^{1/4}. \tag{13.9.3}$$

Alexander's lower bound method, by the nature of the convolutions employed, gives information on the discrepancy of slabs. This is especially apparent in the work of Chazelle, Matoušek, and Sharir, who have developed a more direct and geometrically transparent version of Alexander's method. The following theorem on the discrepancy of thin slabs is a corollary to their technique. It is clear that if a slab has discrepancy Δ, then one of the two bounding halfspaces has discrepancy at least $\Delta/2$.

THEOREM 13.9.4 *Chazelle, Matoušek, Sharir* [CMS95]

Let N points lie in the unit cube $\mathbf{U}^k$. Then there exists a slab $\mathbf{T}$ of width $c_1(k)N^{-1/k}$ such that $\Delta(\mathbf{T}) > c_2(k)N^{(k-1)/2k}$.

Alexander [Ale94] has investigated the effect of the dimension k on the discrepancy of halfspaces, and obtained somewhat complicated inequalities that imply the following result.

THEOREM 13.9.5 *Alexander*

For the lower bounds in inequalities (13.9.1) *and* (13.9.2) *above, there is an absolute positive constant c such that one may choose $c_1(k) > ck^{-3/4}$ and $c_3(k) > ck^{-1}$.*

Schmidt [Sch69] studied the discrepancy of spherical slices (the intersection of two open hemispheres) on $\mathbf{S}^k$. Associating a hemisphere with its pole, Schmidt identified ν with the normalized product measure on $\mathbf{S}^k \times \mathbf{S}^k$. Blümlinger [Blü91] demonstrated a surprising relationship between halfspace (spherical cap) and slice discrepancy for $\mathbf{S}^k$. However, his definition for ν in terms of Haar measure on $SO(k+1)$ differed somewhat from Schmidt's.

THEOREM 13.9.6 *Blümlinger*

Let $\mathcal{S}^k$ be the collection of slices of $\mathbf{S}^k$. Then

$$c(k)D(\mathbf{S}^k, \mathcal{H}_i^{k+1}, 2, N) < D(\mathbf{S}^k, \mathcal{S}^k, 2, N). \tag{13.9.4}$$

For the next result, the left inequality follows from inequalities (13.2.2), (13.9.1), and (13.9.4). Blümlinger uses a version of Beck's probabilistic method to establish the right inequality.

THEOREM 13.9.7 *Blümlinger*

For slice discrepancy on $\mathbf{S}^k$,

$$c_1(k)N^{(k-1)/2k} < D(\mathbf{S}^k, \mathcal{S}^k, \infty, N) < c_2(k)N^{(k-1)/2k}(\log N)^{1/2}.$$

Grabner [Gra91] has given an Erdős-Turán type upper bound on spherical cap discrepancy in terms of spherical harmonics. This adds to the considerable body of results extending inequalities for exponential sums to other sets of orthonormal functions, and thereby extends the Weyl theory.

All of the results so far in this section treat $2 \le W \le \infty$. For W in the range $1 \le W < 2$ there is mystery, but we do have the following result, related to inequality (13.6.6), showing that a dramatic change in asymptotic behavior occurs in the range $1 \le W \le 2$. For $\mathbf{U}^2$, Beck and Chen show that regular grid points will work for the upper bound example for $W = 1$, and they are able to modify their method to apply to any bounded convex domain in $\mathbb{E}^2$.

THEOREM 13.9.8 *Beck, Chen* [BC93a]

Let $\mathbf{K}$ *be a bounded convex domain in* $\mathbb{E}^2$. *Then*

$$\begin{aligned} D(\mathbf{K}, \mathcal{H}_i^2, W, N) &< c(\mathbf{K}, W)N^{(W-1)/2W}, \qquad 1 < W \leq 2; \\ D(\mathbf{K}, \mathcal{H}_i^2, 1, N) &< c(\mathbf{K})(\log N)^2. \end{aligned} \tag{13.9.5}$$

13.10 BOUNDARIES OF GENERATORS FOR HOMOTHETICALLY INVARIANT J

We have already noted several factors that play a role in determining whether $D(\mathbf{K}, \mathcal{J}, W, N)$ behaves like N^r as opposed to $(\log N)^s$. Beck's work shows that if $\mathcal{J}$ is dm-generated, $D(\mathbf{K}, \mathcal{J}, \infty, N)$ behaves as N^r. However, the work of Beck and Chen clearly shows that if W is sufficiently small, then even for motion-invariant $\mathcal{J}$, it may be that $D(\mathbf{K}, \mathcal{J}, W, N)$ is bounded above by $(\log N)^s$.

Beck [Bec88] has extensively studied $D(\mathbf{U}^2, \mathcal{J}_{tor}, \infty, N)$ under the assumption that $\mathcal{J}$ is homothetically invariant, and in this section we shall record some of the results obtained.

It turns out that the boundary shape of a generator is the critical element in determining to which, if either, class $\mathcal{J}$ belongs. Remarkably, for the "typical" homothetically invariant class $\mathcal{J}$, $D(\mathbf{U}^2, \mathcal{J}_{tor}, \infty, N)$ oscillates infinitely often to be larger than $N^{1/4-\epsilon}$ and smaller than $(\log N)^{4+\epsilon}$.

GLOSSARY

The convex set A ***h-generates*** $\mathcal{J}$ if $\mathcal{J}$ consists of all homothetic images B of A with $d(B) \leq d(A)$.

Blaschke-Hausdorff metric: The metric on the space CONV(2) of all compact convex sets in $\mathbb{E}^2$ in which the distance between two sets is the minimum distance from any point of one set to the other.

A set is of ***first category*** if it is a countable union of nowhere dense sets.

If one considers the two examples of $\mathcal{J}$ being h-generated by an aligned square and by a disk, previously stated results make it very likely that shape strongly affects discrepancy for homothetically invariant $\mathcal{J}$. The first two theorems quantify this phenomenon for two very standard boundary shapes, first polygons, then smooth closed curves.

THEOREM 13.10.1 *Beck* [Bec88] [BC87, Corollary 20D]

Let $\mathcal{J}$ *be h-generated by a convex polygon* A. *Then, for any* $\epsilon > 0$,

$$D(\mathbf{U}^2, \mathcal{J}_{tor}, \infty, N) = o((\log N)^{4+\epsilon}). \tag{13.10.1}$$

Beck and Chen [BC89] have given a less complicated argument that obtains $o((\log N)^{5+\epsilon})$ on the right side of (13.10.1).

THEOREM 13.10.2 *Beck* [Bec88] [BC87, Corollary 19F]

Let A *be a compact convex set in* $\mathbb{E}^2$ *with a twice continuously differentiable boundary curve having strictly positive curvature. If* A *h-generates* $\mathcal{J}$, *then for* $N \geq 2$,

$$D(\mathbf{U}^2, \mathcal{J}_{tor}, \infty, N) > c(A)N^{1/4}(\log N)^{-1/2}. \tag{13.10.2}$$

For sufficiently smooth positively curved bodies, Drmota [Drm93] has extended (13.10.2) into higher dimensions and also removed the logarithmic factor. Thus, he obtains a lower bound of the form $c(A)N^{(k-1)/2k}$, along with the standard upper bound obtained by Beck's probabilistic method.

Let CONV(2) denote the usual locally compact space of all compact convex sets in $\mathbb{E}^2$ endowed with the Blaschke-Hausdorff metric. There is the following surprising result, which quantifies the oscillatory behavior mentioned above.

THEOREM 13.10.3 *Beck* [BC87, Theorem 21]

Let $\epsilon > 0$ be given. For all A in CONV(2)*, excepting a set of first category, if $\mathcal{J}$ is h-generated by A, then each of the following two inequalities is satisfied infinitely often:*

(i) $D(\mathbf{U}^2, \mathcal{J}_{tor}, \infty, N) < (\log N)^{4+\epsilon}$.

(ii) $D(\mathbf{U}^2, \mathcal{J}_{tor}, \infty, N) > N^{1/4}(\log N)^{-(1+\epsilon)/2}$.

In fact, the final theorem of this section will say more about the rationale of such estimates.

The next theorem gives the best lower bound estimate known if it is assumed only that the generator A has nonempty interior, certainly a minimal hypothesis.

THEOREM 13.10.4 *Beck* [Bec88] [BC87, Corollary 19G]

If $\mathcal{J}$ is h-generated by a compact convex set A having positive area, then

$$D(\mathbf{U}^2, \mathcal{J}_{tor}, \infty, N) > c(A)(\log N)^{1/2}.$$

Possibly the right side should be $c(A)\log N$, which would be best possible as the example of aligned squares demonstrates. Lastly, we discuss the important theorem underlying most of these results about h-generated $\mathcal{J}$. Let A be a member of CONV(2) with nonempty interior, and for each integer $l \geq 3$ let A_l be an inscribed l-gon of maximal area. The Nth ***approximability number*** $\xi_N(A)$ is defined as the smallest integer l such that the area of $A \setminus A_l$ is less than l^2/N.

THEOREM 13.10.5 *Beck* [Bec88] [BC87, Corollary 19H, Theorem 20C]

Let A be a member of CONV(2) *with nonempty interior. Then if $\mathcal{J}$ is h-generated by A, we have*

$$c_1(A)(\xi_N(A))^{1/2}(\log N)^{-1/4} < D(\mathbf{U}^2, \mathcal{J}_{tor}, \infty, N) < c_2(A,\epsilon)\xi_N(A)(\log N)^{4+\epsilon}. \tag{13.10.3}$$

The proof of the preceding fundamental theorem, which is in fact the join of two major theorems, is long, but the import is clear; namely, that for h-generated $\mathcal{J}$, if one understands $\xi_N(A)$, then one essentially understands $D(\mathbf{U}^2, \mathcal{J}_{tor}, \infty, N)$. If $\xi_N(A)$ remains nearly constant for long intervals, then A acts like a polygon and D will drift below $(\log N)^{4+2\epsilon}$. If, at some stage, ∂A behaves as if it consists of circular arcs, then $\xi_N(A)$ will begin to grow as $cN^{1/2}$.

For still more information concerning the material in this section, along with the proofs, see [BC87, Chapter 7]. Károlyi [Kár95a, Kár95b] has extended the idea of approximability number to higher dimensions and obtained upper bounds analogous to those in (13.10.3).

13.11 $D(\mathbf{K}, \mathcal{J}, 2, N)$ IN LIGHT OF DISTANCE GEOMETRY

Although knowledge of $D(\mathbf{K}, \mathcal{J}, \infty, N)$ is our highest aim, in the great majority of problems this is achieved by first obtaining bounds on $D(\mathbf{K}, \mathcal{J}, 2, N)$. In this section, we briefly show how this function fits nicely into the theory of metric spaces of negative type. In our situation, the distance between points will be given by a Crofton formula with respect to the measure ν on $\mathcal{J}$. This approach evolved from a paper written in 1971 by Alexander and Stolarsky investigating extremal problems in distance geometry, and has been developed in a number of subsequent papers by both authors studying special cases. However, we reverse history and leap immediately to a formulation suitable for our present purposes. We avoid mention of certain technical assumptions concerning $\mathcal{J}$ and ν which cause no difficulty in practice.

Assume that $\mathbf{K}$ is a compact convex set in $\mathbb{E}^k$ and that $\mathcal{J} = \mathcal{J}_c$. This latter assumption causes no loss of generality since one can always just redefine $\mathcal{J}$. Let ν, as usual, be a measure on $\mathcal{J}$, with the further assumption that $\nu(\mathcal{J}) < \infty$.

Definition: If p and q are points in $\mathbf{K}$, the set A in $\mathcal{J}$ is said to ***separate*** p and q if A contains exactly one of these two points. The ***distance function*** ρ on $\mathbf{K}$ is defined by the Crofton formula $\rho(p,q) = (1/2)\,\nu\{J \mid J \text{ separates } p \text{ and } q\}$, and if μ is any signed measure on $\mathbf{K}$ having finite positive and negative parts, one defines the functional $\mathbf{I}(\mu)$ by

$$\mathbf{I}(\mu) = \iint \rho(p,q)\,d\mu(p)\,d\mu(q).$$

With these definitions one obtains the following representation for $\mathbf{I}(\mu)$.

THEOREM 13.11.1 *Alexander* [Ale91]

One has

$$\mathbf{I}(\mu) = \int_{\mathcal{J}} \mu(A)\mu(\mathbf{K} \setminus A)\,d\nu(A). \tag{13.11.1}$$

For μ satisfying the condition of total mass zero, $\int_K d\mu = 0$, the integrand in (13.11.1) becomes $-(\mu(A))^2$. The signed measures $\mu = \mu^+ - \mu^-$ that we are considering, with μ^- being a uniform probability measure on $\mathbf{K}$ and μ^+ consisting of N atoms of equal weight $1/N$, certainly have total mass zero. Here one has $\Delta(A) = N\mu(A)$. Hence there is the following corollary.

COROLLARY 13.11.2

For the signed measures μ presently considered, if $\mathcal{P}$ denotes the N points supporting μ^+, then

$$-N^2\mathbf{I}(\mu) = \int_{\mathcal{J}} (\Delta(A))^2\,d\nu(A) = (\|\Delta(\mathcal{P}, \mathcal{J})\|_2)^2. \tag{13.11.2}$$

Thus if one studies the metric ρ, it may be possible to prove that $-\mathbf{I}(\mu) > f(N)$, whence it follows that $(D(\mathbf{K}, \mathcal{J}, 2, N))^2 > N^2 f(N)$. If $\mathcal{J}$ consists of the halfspaces of $\mathbb{E}^k$, then ρ is the Euclidean metric. In this important special case, Alexander [Ale91] was able to make good estimates. Chazelle, Matoušek, and Sharir [CMS95] and A.D. Rogers [Rog94] contributed still more techniques for treating the halfspace problem.

If μ_1 and μ_2 are any two signed measures of total mass 1 on $\mathbf{K}$, then one can define the ***relative discrepancy*** $\Delta(A) = N(\mu_1(A) - \mu_2(A))$. The first equality of (13.11.2) still holds if $\mu = \mu_1 - \mu_2$. A signed measure μ_0 of total mass 1 is termed ***optimal*** if it solves the integral equation $\int_K \rho(x,y)d\mu(y) = \lambda$ for some positive number λ. If an optimal measure μ_0 exists, then $\mathbf{I}(\mu_0) = \lambda$ maximizes $\mathbf{I}$ on the class of all signed Borel measures of total mass 1 on $\mathbf{K}$. In the presence of an optimal measure, one has the following very pretty identity.

THEOREM 13.11.3 *Generalized Stolarsky Identity*

Suppose that the measure μ_0 is optimal on $\mathbf{K}$, and that μ is any signed measure of total mass 1 on $\mathbf{K}$. If Δ is the relative discrepancy with respect to μ_0 and μ, then

$$N^2\mathbf{I}(\mu) + \int_{\mathcal{J}} (\Delta(A))^2 d\nu(A) = N^2\mathbf{I}(\mu_0). \qquad (13.11.3)$$

The first important example of this formula is due to Stolarsky [Sto73] where he treated the sphere $\mathbf{S}^k$, taking as μ the uniform atomic measure supported by N variable points. For $\mathbf{S}^k$ it is clear that the uniform probability measure μ_0 is optimal. His integrals involving the spherical caps are equivalent, up to a scale factor, to integrals with respect to the measure on the halfspaces of $\mathbb{E}^k$ for which ρ is the Euclidean metric. Stolarsky's tying of a geometric extremal problem to Schmidt's work on the discrepancy of spherical caps was a major step forward in the study of discrepancy and of distance geometry.

Very little has been done to investigate the deeper nature of the individual metrics ρ determined by classes $\mathcal{J}$ other than halfspaces. They are all metrics of ***negative type***, which essentially means that $\mathbf{I}(\mu) \leq 0$ if μ has total mass 0. There is a certain amount of general theory, begun by Schoenberg and developed by a number of others, but it does not apply directly to the problem of estimating discrepancy.

13.12 UNIFORM PLACEMENT OF POINTS ON SPHERES

As demonstrated by Stolarsky, formula (13.11.3) shows that if one places N points on $\mathbf{S}^k$ so that the sum of all distances is maximized, then $D(\mathbf{S}^k, \mathcal{H}_i^k, 2, N)$ is achieved by this arrangement. Berman and Hanes [BH77] have given a pretty algorithm that searches for optimal configurations. For $k = 2$, while the exact configurations are not known for $N \geq 5$, this algorithm appears to be successful for $N \leq 50$. For such an N surprisingly few rival configurations will be found. Lubotzky, Phillips, and Sarnak [LPS86] have given an algorithm, based on iterations of a specially chosen element in $SO(3)$, which can be used to place many thousands of reasonably well distributed points on $\mathbf{S}^2$. Difficult analysis shows that these points are well placed, but not optimally placed, relative to $\mathcal{H}_i^2$. On the other hand, it is shown that these points are essentially optimally placed with respect to a nongeometric operator discrepancy. Data concerning applications to numerical integration are also included in the paper. More recently, Rakhmanov, Saff, and Zhou [RSZ94] have studied the problem of placing points uniformly on a sphere relative to optimizing certain functionals, and they state a number of interesting conjectures.

In yet another theoretical direction, the existence of very well distributed point sets on $\mathbf{S}^k$ allows the sphere, after difficult analysis, to be closely approximated by equi-edged zonotopes (sums of line segments). The papers of Wagner [Wag93] and of Bourgain and Lindenstrauss [BL93] treat this problem.

13.13 COMBINATORIAL DISCREPANCY

GLOSSARY

A ***2-coloring*** of $\mathbf{X}$ is a mapping $\chi : \mathbf{X} \to \{-1, 1\}$. For each such χ there is a natural integer-valued set function μ_χ on the finite subsets of $\mathbf{X}$ defined by $\mu_\chi(A) = \sum_{x\in A} \chi(x)$, and if $\mathcal{J}$ is a given family of finite subsets of $\mathbf{X}$ we define

$$D(\mathbf{X}, \mathcal{J}) = \min_{\chi} \max_{A\in\mathcal{J}} |\mu_\chi(A)|.$$

Degree: If $\mathcal{J}$ is a collection of subsets of a finite set X, $\deg \mathcal{J} = \max\{|\mathcal{J}(x)| \mid x \in \mathbf{X}\}$, where $\mathcal{J}(x)$ is the subcollection consisting of those members of $\mathcal{J}$ that contain x.

The collection $\mathcal{J}$ ***shatters*** a set $S \subset X$ if, for any given subset $B \subset S$, there exists A in $\mathcal{J}$ such that $B = A \cap S$. The ***VC-dimension*** of $\mathcal{J}$ is defined by $\dim_{vc} \mathcal{J} = \max\{|S| \mid S \subset \mathbf{X},\ \mathcal{J} \text{ shatters } S\}$. For $m \leq |\mathbf{X}|$, the ***primal shatter function*** $\pi_{\mathcal{J}}$ is defined by

$$\pi_{\mathcal{J}}(m) = \max_{\substack{Y\subset\mathbf{X}\\|Y|\leq m}} |\{Y \cap A \mid A \in \mathcal{J}\}|.$$

The ***dual shatter function*** is defined by $\pi^*_{\mathcal{J}}(m) = \pi_{\mathcal{J}^*}(m)$, where $\mathbf{X}^* = \mathcal{J}$, and $\mathcal{J}^* = \{\mathcal{J}(x) \mid x \in \mathbf{X}\}$.

Techniques in combinatorial discrepancy theory have proved very powerful in this geometric setting. Here one 2-colors a discrete set and studies the discrepancy of a special class $\mathcal{J}$ of subsets as measured by $|\#\text{red} - \#\text{blue}|$. If one 2-colors the first N positive integers, then the beautiful "1/4 theorem" of Roth [Rot64] says that there will always be an arithmetic progression having discrepancy at least $cN^{1/4}$. This result should be compared to van der Waerden's theorem, which says that there is a long monochromatic progression, whose discrepancy obviously will be its length. However, it is known that this length need not be more than $\log N$, and the minimax might be as small as $\log\log\ldots\log N$ (here the number of iterated logarithms may be arbitrarily large). Moreover, general results concerning combinatorial discrepancy, for example, those that use the Vapnik-Chervonenkis dimension, are very useful in computational geometry; cf. Chapter 47.

Combinatorial discrepancy theory involves discrepancy estimates arising from 2-colorings of a set $\mathbf{X}$. Upper bound estimates of combinatorial discrepancy have proved to be very helpful in obtaining upper bound estimates of geometric discrepancy. In this final section we briefly discuss various properties of the collection $\mathcal{J}$ that lead to useful upper bound estimates of combinatorial discrepancy.

The simplest property of the collection $\mathcal{J}$ is its cardinality $|\mathcal{J}|$. Here, Spencer obtained a fine result.

THEOREM 13.13.1 *Spencer* [AS93]
Let $\mathbf{X}$ *be a finite set. If* $|\mathcal{J}| \geq |\mathbf{X}|$, *then*

$$D(\mathbf{X}, \mathcal{J}) \leq c\left(|\mathbf{X}| \log\left(1 + \frac{|\mathcal{J}|}{|\mathbf{X}|}\right)\right)^{1/2}.$$

Applications and extensions of the following theorem may be found in [BC87, Chapter 8].

THEOREM 13.13.2 *Beck, Fiala* [BF81] [BC87, Lemma 8.5.]
Let $\mathbf{X}$ *be a finite set. Then*

$$D(\mathbf{X}, \mathcal{J}) \leq 2 \deg \mathcal{J} - 1.$$

Since $\pi_{\mathcal{J}}(m) = 2^m$ if and only if $\dim_{vc} \mathcal{J} \geq m$, the function $\pi_{\mathcal{J}}$ contains much more information than does VC-dimension alone. If $\dim_{vc} \mathcal{J} = d$, then $\pi_{\mathcal{J}}(m)$ is polynomially bounded by cm^d. However, in many geometric situations this bound on the shatter function can be improved, leading to better discrepancy bounds. Detailed discussions may be found in the papers by Haussler and Welzl [HW87] and by Chazelle and Welzl [CW89].

Dual objects are defined in the usual manner (see Glossary). We state several results.

THEOREM 13.13.3 *Matoušek, Welzl, Wernisch* [MWW93]
Suppose that $(\mathbf{X}, \mathcal{J})$ *is a finite set system with* $|\mathbf{X}| = n$. *If* $\pi_{\mathcal{J}}(m) \leq c_1 m^d$ *for* $m \leq n$, *then*

$$\begin{aligned} D(\mathbf{X}, \mathcal{J}) &\leq c_2 n^{(d-1)/2d} (\log n)^{1+1/2d}, \qquad d > 1, \\ D(\mathbf{X}, \mathcal{J}) &\leq c_3 (\log n)^{5/2}, \qquad d = 1. \end{aligned} \tag{13.13.1}$$

If $\pi^*_{\mathcal{J}}(m) \leq c_4 m^d$ *for* $m \leq |\mathcal{J}|$, *then*

$$\begin{aligned} D(\mathbf{X}, \mathcal{J}) &\leq c_5 n^{(d-1)/2d} \log n, \qquad d > 1, \\ D(\mathbf{X}, \mathcal{J}) &\leq c_6 (\log n)^{3/2}, \qquad d = 1. \end{aligned} \tag{13.13.2}$$

Matoušek [Mat95] has shown that the factor $(\log n)^{1+1/2d}$ may be dropped from inequality (13.13.1) for $d > 1$, and has applied this result to halfspaces with great effect (see inequality (13.9.3)). One part of Matoušek's argument depends on combinatorial results of Haussler [Hau95].

13.14 RECENT NEW DIRECTIONS

We mention three recent developments of great interest.

Beck [Bec14] has identified some super irregularity phenomena in long and narrow hyperbolic regions, where the discrepancy is found to be of comparable size to the expectation.

Khinchin's conjecture on strong uniformity is false, although for a long time many believed it to be true. Motivated by continuous versions of the conjecture, Beck [Bec15] has made interesting and deep studies into super uniform motions.

Matoušek and Nikolov [MN15] has recently obtained a very strong lower bound of $c(\log n)^{d-1}$ for the combinatorial discrepancy of n-point sets in the d-dimensional unit cube with respect to axis-parallel boxes. This is the combinatorial analogue of the problem studied in Theorem 13.4.6.

13.15 SOURCES AND RELATED MATERIAL

FURTHER READING

There are a few principal surveys on discrepancy theory. The first survey [Sch77b] covers the early development. The Cambridge tract [BC87] is a comprehensive account up to the mid-1980s. The account [DT97] contains a comprehensive list of results and references but few detailed proofs. The exquisite account [Mat99] is most suitable for beginners, as the exposition is very down to earth. The recent volume [CST14] is a collection of essays by experts in various areas.

Among related texts, [KN74] deals mostly with uniform distribution, [Cha00] deals with the discrepancy method and is geared towards computer science, while [DP10] concerns mostly numerical integration.

Auxiliary texts relating to this chapter include [Cas57], [Cas59], and [San76] and [AS93].

RELATED CHAPTERS

Chapter 1: Finite point configurations
Chapter 2: Packing and covering
Chapter 11: Euclidean Ramsey theory
Chapter 12: Discrete aspects of stochastic geometry
Chapter 40: Range searching
Chapter 44: Randomization and derandomization
Chapter 47: Epsilon-nets and epsilon-approximations
Chapter 52: Computer graphics

REFERENCES

[A-E45] T. van Aardenne-Ehrenfest. Proof of the impossibility of a just distribution of an infinite sequence of points over an interval. *Nederl. Akad. Wetensch. Proc.*, 48:266–271, 1945 (*Indagationes Math.*, 7:71–76, 1945).

[A-E49] T. van Aardenne-Ehrenfest. On the impossibility of a just distribution. *Nederl. Akad. Wetensch. Proc.*, 52:734–739, 1949 (*Indagationes Math.*, 11:264–269, 1949).

[Ale72] J.R. Alexander. On the sum of distances between n points on a sphere. *Acta Math. Hungar.*, 23:443–448, 1972.

[Ale90] J.R. Alexander. Geometric methods in the study of irregularities of distribution. *Combinatorica*, 10:115–136, 1990.

[Ale91] J.R. Alexander. Principles of a new method in the study of irregularities of distribution. *Invent. Math.*, 103:279–296, 1991.

[Ale94] J.R. Alexander. The effect of dimension on certain geometric problems of irregularities of distribution. *Pacific J. Math.*, 165:1–15, 1994.

[AS93] N. Alon and J. Spencer. *The Probabilistic Method.* Wiley, New York, 1993.

[BC87] J. Beck and W.W.L. Chen. *Irregularities of Distribution.* Vol. 89 of *Cambridge Tracts in Math.*, Cambridge University Press, 1987.

[BC89] J. Beck and W.W.L. Chen. Irregularities of point distribution relative to convex polygons. In G. Halász and V.T. Sós, editors, *Irregularities of Partitions*, vol. 8 of *Algorithms Combin.*, pages 1–22, Springer, Berlin, 1989.

[BC90] J. Beck and W.W.L. Chen. Note on irregularities of distribution II. *Proc. London Math. Soc.*, 61:251–272, 1990.

[BC93a] J. Beck and W.W.L. Chen. Irregularities of point distribution relative to half planes I. *Mathematika*, 40:102–126, 1993.

[BC93b] J. Beck and W.W.L. Chen. Irregularities of point distribution relative to convex polygons II. *Mathematika*, 40:127–136, 1993.

[Bec83] J. Beck. On a problem of K.F. Roth concerning irregularities of point distribution. *Invent. Math.*, 74:477–487, 1983.

[Bec84] J. Beck. Sums of distances between points on a sphere – an application of the theory of irregularities of distribution to discrete geometry. *Mathematika*, 31:33–41, 1984.

[Bec87] J. Beck. Irregularities of distribution I. *Acta Math.*, 159:1–49, 1987.

[Bec88] J. Beck. Irregularities of distribution II. *Proc. London Math. Soc.*, 56:1–50, 1988.

[Bec89] J. Beck. A two-dimensional van Aardenne-Ehrenfest theorem in irregularities of distribution. *Compos. Math.*, 72:269–339, 1989.

[Bec94] J. Beck. Probabilistic diophantine approximation I: Kronecker sequences. *Ann. of Math.*, 140:449–502, 1994.

[Bec14] J. Beck. Superirregularity. In W.W.L. Chen, A. Srivastav and G. Travaglini, editors, *A Panorama of Discrepancy Theory*, vol. 2107 of *Lecture Notes in Math.*, pages 221–316, Springer, Berlin, 2014.

[Bec15] J. Beck. From Khinchin's conjecture on strong uniformity to superuniform motions. *Mathematika*, 61:591–707, 2015.

[BF81] J. Beck and T. Fiala. Integer-making theorems. *Discrete Appl. Math.*, 3:1–8, 1981.

[BH77] J. Berman and K. Hanes. Optimizing the arrangement of points on the unit sphere. *Math. Comp.*, 31:1006–1008, 1977.

[BL93] J. Bourgain and J. Lindenstrauss. Approximating the ball by a Minkowski sum of segments with equal length. *Discrete Comput. Geom.*, 9:131–144, 1993.

[Blü91] M. Blümlinger. Slice discrepancy and irregularities of distribution on spheres. *Mathematika*, 38:105–116, 1991.

[BLV08] D. Bilyk, M.T. Lacey, and A. Vagharshakyan. On the small ball inequality in all dimensions. *J. Funct. Anal.*, 254:2470–2502, 2008.

[Cas57] J.W.S. Cassels. *An Introduction to Diophantine Approximation.* Volume 45 of *Cambridge Tracts in Math.*, Cambridge University Press, 1957.

[Cas59] J.W.S. Cassels. *An Introduction to the Geometry of Numbers.* Vol. 99 of *Grundlehren Math. Wiss.*, Springer-Verlag, Berlin, 1959.

[Cha00] B. Chazelle. *The Discrepancy Method.* Cambridge University Press, 2000.

[Che80] W.W.L. Chen. On irregularities of distribution. *Mathematika*, 27:153–170, 1980.

[Che83] W.W.L. Chen. On irregularities of distribution II. *Quart. J. Math. Oxford*, 34:257–279, 1983.

[Che85] W.W.L. Chen. On irregularities of distribution and approximate evaluation of certain functions. *Quart. J. Math. Oxford*, 36:173–182, 1985.

[Che87] W.W.L. Chen. On irregularities of distribution and approximate evaluation of certain functions II. In A.C. Adolphson, J.B. Conrey, A. Ghosh and R.I. Yager, editors, *Analytic Number Theory and Diophantine Problems*, vol. 70 of *Progress in Mathematics*, pages 75–86. Birkhäuser, Boston, 1987.

[CMS95] B. Chazelle, J. Matoušek, and M. Sharir. An elementary approach to lower bounds in geometric discrepancy. In I. Bárány and J. Pach, editors, *The László Fejes Tóth Festschrift, Discrete Comput. Geom.*, 13:363–381, 1995.

[Cor35a] J.G. van der Corput. Verteilungsfunktionen I. *Proc. Kon. Ned. Akad. v. Wetensch.*, 38:813–821, 1935.

[Cor35b] J.G. van der Corput. Verteilungsfunktionen II. *Proc. Kon. Ned. Akad. v. Wetensch.*, 38:1058–1066, 1935.

[CS02] W.W.L. Chen and M.M. Skriganov. Explicit constrictions in the classical mean squares problem in irregularities of point distribution. *J. Reine Angew. Math.*, 545:67–95, 2002.

[CST14] W.W.L. Chen, A. Srivastav, and G. Travaglini, editors. *A Panorama of Discrepancy Theory*. Vol. 2107 of *Lecture Notes in Math.*, Springer, Berlin, 2014.

[CW89] B. Chazelle and E. Welzl. Quasi-optimal range searching in spaces of finite VC-dimension. *Discrete Comput. Geom.*, 4:467–489, 1989.

[Dav56] H. Davenport. Note on irregularities of distribution. *Mathematika*, 3:131–135, 1956.

[Dic14] J. Dick. Discrepancy bounds for infinite dimensional order two digital sequences over $\mathbb{F}_2$. *J. Number Theory*, 136:204–232, 2014.

[DP10] J. Dick and F. Pillichshammer. *Digital Nets and Sequences: Discrepancy Theory and Quasi-Monte Carlo Integration*. Cambridge University Press, 2010.

[DP14] J. Dick and F. Pillichshammer. Optimal L_2 discrepancy bounds for higher order digital sequences over the finite field $\mathbb{F}_2$. *Acta Arith.*, 162:65–99, 2014.

[Drm93] M. Drmota. Irregularities of distribution and convex sets. *Grazer Math. Ber.*, 318:9–16, 1993.

[Drm96] M. Drmota. Irregularities of distribution with respect to polytopes. *Mathematika*, 43:108–119, 1996.

[DT97] M. Drmota and R.F. Tichy. *Sequences, Discrepancies and Applications*. Vol. 1651 of *Lecture Notes in Math.*, Springer, Berlin, 1997.

[Gra91] P.J. Grabner. Erdős-Turán type discrepancy bounds. *Monatsh. Math.*, 111:127–135, 1991.

[Hal60] J.H. Halton. On the efficiency of certain quasirandom sequences of points in evaluating multidimensional integrals. *Num. Math.*, 2:84–90, 1960.

[Hal81] G. Halász. On Roth's method in the theory of irregularities of point distributions. In H. Halberstam and C. Hooley, editors, *Recent Progress in Analytic Number Theory, Vol. 2*, pages 79–94, Academic Press, London, 1981.

[Hau95] D. Haussler. Sphere packing numbers for subsets of the Boolean n-cube with bounded Vapnik-Chervonenkis dimension. *J. Combin. Theory Ser. A*, 69:217–232, 1995.

[HZ69] J.H. Halton and S.K. Zaremba. The extreme and L^2 discrepancies of some plane sets. *Monats. Math.*, 73:316–328, 1969.

[HW87] D. Haussler and E. Welzl. ϵ-nets and simplex range queries. *Discrete Comput. Geom.*, 2:127–151, 1987.

[Kár95a] G. Károlyi. Geometric discrepancy theorems in higher dimensions. *Studia Sci. Math. Hungar.*, 30:59–94, 1995.

[Kár95b] G. Károlyi. Irregularities of point distributions with respect to homothetic convex bodies. *Monatsh. Math.*, 120:247–279, 1995.

[KN74] L. Kuipers and H. Niederreiter. *Uniform Distribution of Sequences.* Wiley, New York, 1974.

[Lar91] G. Larcher. On the cube discrepancy of Kronecker sequences. *Arch. Math. (Basel)*, 57:362–369, 1991.

[LPS86] A. Lubotzky, R. Phillips, and P. Sarnak. Hecke operators and distributing points on a sphere. *Comm. Pure Appl. Math.*, 39:149–186, 1986.

[Mat95] J. Matoušek. Tight upper bounds for the discrepancy of half-spaces. *Discrete Comput. Geom.*, 13:593–601, 1995.

[Mat99] J. Matoušek. *Geometric Discrepancy: An Illustrated Guide.* Vol. 18 of *Algorithms Combin.*, Springer-Verlag, Berlin, 1999.

[MN15] J. Matoušek and A. Nikolov. Combinatorial discrepancy for boxes via the γ_2 norm. In *Proc. 31st Sympos. Comput. Geom.*, vol. 34 of *LiPICS*, pages 1–15, Schloss Dagstuhl, 2015.

[Mon89] H.L. Montgomery. Irregularities of distribution by means of power sums. In *Congress of Number Theory (Zarautz)*, pages 11–27, Universidad del País Vasco, Bilbao, 1989.

[MWW93] J. Matoušek, E. Welzl, and L. Wernisch. Discrepancy and approximations for bounded VC-dimension. *Combinatorica*, 13:455–467, 1993.

[Rog94] A.D. Rogers. A functional from geometry with applications to discrepancy estimates and the Radon transform. *Trans. Amer. Math. Soc.*, 341:275–313, 1994.

[Rot54] K.F. Roth. On irregularities of distribution. *Mathematika*, 1:73–79, 1954.

[Rot64] K.F. Roth. Remark concerning integer sequences. *Acta Arith.*, 9:257–260, 1964.

[Rot76] K.F. Roth. On irregularities of distribution II. *Comm. Pure Appl. Math.*, 29:749–754, 1976.

[Rot80] K.F. Roth. On irregularities of distribution IV. *Acta Arith.*, 37:67–75, 1980.

[RSZ94] E.A. Rakhmanov, E.B. Saff, and Y.M. Zhou. Minimal discrete energy on the sphere. *Math. Res. Lett.*, 1:647–662, 1994.

[Ruz93] I.Z. Ruzsa. The discrepancy of rectangles and squares. *Grazer Math. Ber.*, 318:135–140, 1993.

[San76] L.A. Santaló. *Integral Geometry and Geometric Probability.* Vol. 1 of *Encyclopedia of Mathematics*, Addison-Wesley, Reading, 1976.

[Sch69] W.M. Schmidt. Irregularities of distribution III. *Pacific J. Math.*, 29:225–234, 1969.

[Sch72a] W.M. Schmidt. Irregularities of distribution VI. *Compos. Math.*, 24:63–74, 1972.

[Sch72b] W.M. Schmidt. Irregularities of distribution VII. *Acta Arith.*, 21:45–50, 1972.

[Sch75] W.M. Schmidt. Irregularities of distribution IX. *Acta Arith.*, 27:385–396, 1975.

[Sch77a] W.M. Schmidt. Irregularities of distribution X. In H. Zassenhaus, editor, *Number Theory and Algebra*, pages 311–329, Academic Press, New York, 1977.

[Sch77b] W.M. Schmidt. *Irregularities of Distribution.* Vol. 56 of *Lecture Notes on Math. and Physics*, Tata, Bombay, 1977.

[Skr94] M.M. Skriganov. Constructions of uniform distributions in terms of geometry of numbers. *St. Petersburg Math. J. (Algebra i. Analiz)*, 6:200–230, 1994.

[Skr06] M.M. Skriganov. Harmonic analysis on totally disconnected groups and irregularities of point distributions. *J. Reine Angew. Math.*, 600:25–49, 2006.

[Skr16] M.M. Skriganov. Dyadic shift randomization in classical discrepancy theory. *Mathematika*, 62:183–209, 2016.

[Sto73] K.B. Stolarsky. Sums of distances between points on a sphere II. *Proc. Amer. Math. Soc.*, 41:575–582, 1973.

[Vil67] I.V. Vilenkin. Plane nets of integration. *USSR Comput. Math. and Math. Phys.*, 7:258–267, 1967.

[Wag93] G. Wagner. On a new method for constructing good point sets on spheres. *Discrete Comput. Geom.*, 9:111–129, 1993.

[Wey16] H. Weyl. Über die Gleichverteilung von Zahlen mod Eins. *Math. Ann.*, 77:313–352, 1916.

[Woź91] H. Woźniakowski. Average case complexity of multivariate integration. *Bull. Amer. Math. Soc.*, 24:185–194, 1991.

14 POLYOMINOES

Gill Barequet, Solomon W. Golomb, and David A. Klarner[1]

INTRODUCTION

A *polyomino* is a finite, connected subgraph of the square-grid graph consisting of infinitely many unit cells matched edge-to-edge, with pairs of adjacent cells forming edges of the graph. Polyominoes have a long history, going back to the start of the 20th century, but they were popularized in the present era initially by Solomon Golomb, then by Martin Gardner in his *Scientific American* columns "Mathematical Games," and finally by many research papers by David Klarner. They now constitute one of the most popular subjects in mathematical recreations, and have found interest among mathematicians, physicists, biologists, and computer scientists as well.

14.1 BASIC CONCEPTS

GLOSSARY

Cell: A unit square in the Cartesian plane with its sides parallel to the coordinate axes and with its center at an integer point (u, v). This cell is denoted $[u, v]$ and identified with the corresponding member of $\mathbb{Z}^2$.

Adjacent cells: Two cells, $[u, v]$ and $[r, s]$, with $|u - r| + |v - s| = 1$.

Square-grid graph: The graph with vertex set $\mathbb{Z}^2$ and an edge for each pair of adjacent cells.

Polyomino: A finite set S of cells such that the induced subgraph of the square-grid graph with vertex set S is connected. A polyomino of size n, that is, with exactly n cells, is called an ***n-omino***. Polyominoes are also known as ***animals*** on the square lattice.

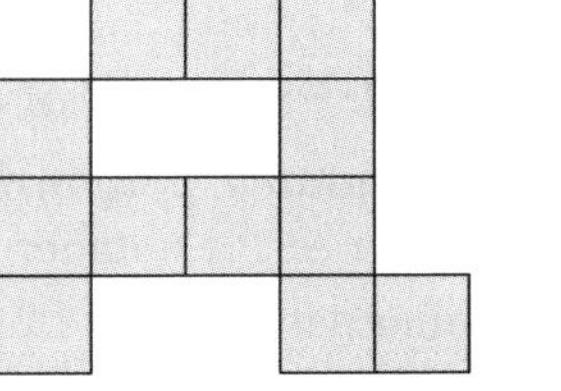
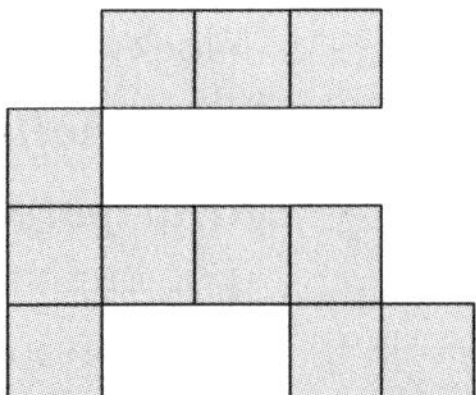

FIGURE 14.1.1
Two sets of cells: the set on the left is a polyomino, the one on the right is not.

[1]This is a revision, by G. Barequet, of the chapter of the same title originally written by the late D.A. Klarner for the first edition, and revised by the late S.W. Golomb for the second edition.

14.2 EQUIVALENCE OF POLYOMINOES

Notions of equivalence for polyominoes are defined in terms of groups of affine maps that act on the set $\mathbb{Z}^2$ of cells in the plane.

GLOSSARY

Translation by (r,s): The mapping from $\mathbb{Z}^2$ to itself that maps $[u,v]$ to $[u+r, v+s]$; it sends any subset $S \subset \mathbb{Z}^2$ to its *translate* $S+(r,s) = \{[u+r, v+s] : [u,v] \in S\}$.

Translation-equivalent: Sets S, S' of cells such that S' is a translate of S.

Fixed polyomino: A translation-equivalence class of polyominoes; $\boldsymbol{t(n)}$ denotes the number of fixed n-ominoes. (($A(n)$ is also widely used in the literature.)

Representatives of the six fixed 3-ominoes are shown in Figure 14.2.1.

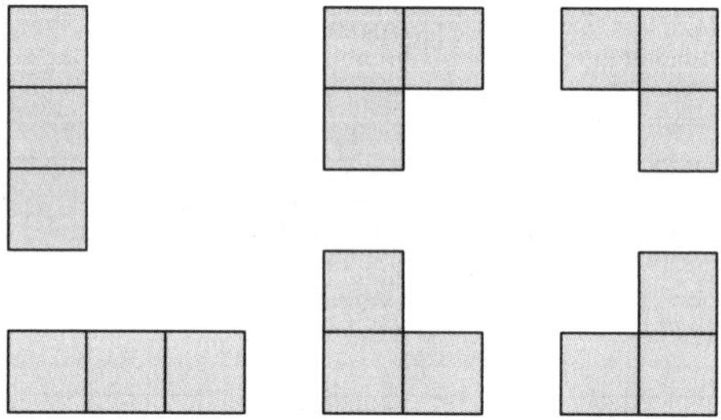

FIGURE 14.2.1
The six fixed 3-ominoes.

The ***lexicographic cell ordering*** $\prec$ on $\mathbb{Z}^2$ is defined by: $[r,s] \prec [u,v]$ if $s < v$, or if $s = v$ and $r < u$.

Standard position: The translate $S-(u,v)$ of S, where $[u,v]$ is the lexicographically minimum cell in S.

A finite set $S \subset \mathbb{Z}^2$ is in standard position if and only if $[0,0] \in S$, $v \geq 0$ for all $[u,v] \in S$, and $u \geq 0$ for all $[u,0] \in S$.

Rotation-translation group: The group $\mathcal{R}$ of mappings of $\mathbb{Z}^2$ to itself of the form $[u,v] \mapsto [u,v]\begin{bmatrix} 0 & -1 \\ 1 & 0 \end{bmatrix}^k + (r,s)$. (The matrix $\begin{bmatrix} 0 & -1 \\ 1 & 0 \end{bmatrix}$, which is denoted by R, maps $[u,v]$ to $[v,-u]$ by right multiplication, hence represents a clockwise rotation of 90°.)

Rotationally equivalent: Sets S, S' of cells with $S' = \rho S$ for some $\rho \in \mathcal{R}$.

Chiral polyomino, or ***handed polyomino:*** A rotational-equivalence class of polyominoes; $\boldsymbol{r(n)}$ denotes the number of chiral n-ominoes.

The top row of 5-ominoes in Figure 14.2.2 consists of the set of cells $F = \{[0,-1], [-1,0], [0,0], [0,1], [1,1]\}$, together with FR, FR^2, and FR^3. All four of these 5-ominoes are rotationally equivalent. The bottom row in Figure 14.2.2 shows these same four 5-ominoes reflected about the x-axis. These four 5-ominoes are rotationally equivalent as well, but none of them is rotationally equivalent to any of the 5-ominoes shown in the top row. Representatives of the seven chiral 4-ominoes are shown in Figure 14.2.3.

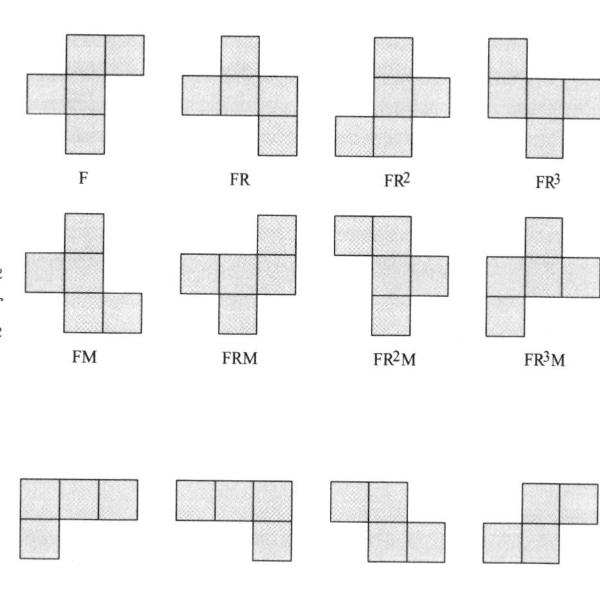

FIGURE 14.2.2
The 5-ominoes in the top row are rotationally equivalent, and so are their reflections in the bottom row, but the two sets are rotationally distinct.

FIGURE 14.2.3
The seven chiral 4-ominoes.

Congruence group: The group $\mathbb{S}$ of motions generated by the matrix $M = \begin{bmatrix} 1 & 0 \\ 0 & -1 \end{bmatrix}$ (reflection in the x-axis) and the rotation-translation group $\mathcal{R}$. (A typical element of $\mathbb{S}$ has the form $[u,v] \mapsto [u,v]R^kM^i + (r,s)$, for some $k = 0, 1, 2$, or 3, some $i = 0$ or 1, and some $r, s \in \mathbb{Z}$.)

Congruent: Sets S and S' of cells such that $S' = \sigma(S)$ for some $\sigma \in \mathbb{S}$.

Free polyomino: A congruence class of polyominoes; $\boldsymbol{s(n)}$ denotes the number of free n-ominoes. The twelve free 5-ominoes are shown in Figure 14.2.4.

THEOREM 14.2.1 *Embedding Theorem*

For each n, let U_n consist of the $n^2 - n + 1$ cells of the form $[u,v]$, where $\begin{cases} 0 \le u \le n, & \text{for } v = 0 \\ |u| + v \le n, & \text{for } v > 0 \end{cases}$. (See Figure 14.2.5 for the case $n = 5$.) *Then, all n-ominoes in standard position are edge-connected subsets of U_n that contain $[0,0]$.*

COROLLARY 14.2.2

The number of fixed n-ominoes is finite for each n.

The same result can be obtained by a simple argument due to Eden [Ede61]: Every polyomino P of size n can be built according to a set of $n-1$ "instructions" taken from a superset of size $3(n-1)$. Starting with a single square, each instruction tells us how to choose a lattice cell c, neighboring a cell already in P, and add c to P.

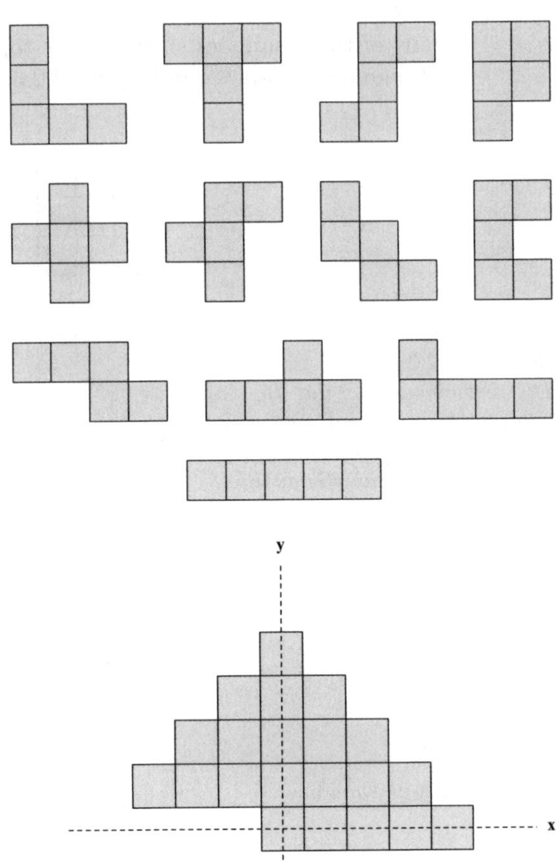

FIGURE 14.2.4
The twelve free 5-ominoes.

FIGURE 14.2.5
A set of n^2-n+1 cells (for $n=5$) that contains every n-omino in standard position.

(Some of these instruction sets are illegal, and some other sets produce the same polyominoes.) Hence, the number of polyominoes of size n is less than $\binom{3(n-1)}{n-1}$.

14.3 HOW MANY n-OMINOES ARE THERE?

Table 14.3.1, calculated by Redelmeier [Red81], indicates the values of $t(n)$, $r(n)$, and $s(n)$ for $n=1,\ldots,24$. Jensen and Guttmann [JG00, Jen01] and Jensen [Jen03] extended the enumeration of polyominoes up to $n=56$. See also sequence A001168 in the OEIS (On-line Encyclopedia of Integer Sequences) [oeis].

It is easy to see that for each n, we have

$$\frac{t(n)}{8} \leq s(n) \leq r(n) \leq t(n).$$

The values of $t(n)$ seem to be growing exponentially, and indeed they have exponential bounds.

THEOREM 14.3.1 [Kla67]

$\lim_{n\to\infty}(t(n))^{1/n} = \lambda$ *exists.*

TABLE 14.3.1 The number of fixed, chiral, and free n-ominoes for $n \leq 24$.

n	$t(n)$	$r(n)$	$s(n)$
1	1	1	1
2	2	1	1
3	6	2	2
4	19	7	5
5	63	18	12
6	216	60	35
7	760	196	108
8	2725	704	369
9	9910	2500	1285
10	36446	9189	4655
11	135268	33896	17073
12	505861	126759	63600
13	1903890	476270	238591
14	7204874	1802312	901971
15	27394666	6849777	3426576
16	104592937	26152418	13079255
17	400795844	100203194	50107909
18	1540820542	385221143	192622052
19	5940738676	1485200848	742624232
20	22964779660	5741256764	2870671950
21	88983512783	22245940545	11123060678
22	345532572678	86383382827	43191857688
23	1344372335524	336093325058	168047007728
24	5239988770268	1309998125640	654999700403

This constant (often referred to in the literature as the "growth constant" of polyominoes) has since then been called "Klarner's constant." Only three decades later, Madras proved the existence of the asymptotic growth ratio.

THEOREM 14.3.2 [Mad99]

$\lim_{n\to\infty} t(n+1)/t(n)$ *exists (and is hence equal to* λ*).*

The currently best known lower [BRS16] and upper [KR73] bounds on the constant λ are 4.0025 and 4.6496, respectively. The proof of the lower bound uses an analysis of polyominoes on *twisted cylinders* with the help of a supercomputer, and the proof of the upper bound uses a composition argument of so-called *twigs*. The currently best (unproved) *estimate* of λ, 4.0625696 ± 0.0000005, is by Jensen [Jen03].

ALGORITHMS

Redelmeier

Considerable effort has been expended to find a formula for the number of fixed n-ominoes (say), with no success. Redelmeier's algorithm, which produced the entries in Table 14.3.1, generates *all* fixed n-ominoes one by one and counts them. The recursive algorithm searches G, the underlying cell-adjacency graph of the square lattice, and counts all connected subgraphs of G (up to a predetermined size) that contain some canonical vertex, say, $(0, 0)$, which is assumed to always correspond to the leftmost cell in the bottom row of the polyomino. (This prevents multiple counting of translations of the same polyomino.) See more details in Section 14.4. Although the running time of the algorithm is necessarily exponential, it takes only $O(n)$ space. Redelmeier's algorithm was extended to other lattices, parallelized, and enhanced further; see [AB09a, AB09b, LM11, Mer90, ML92].

Jensen

A faster transfer-matrix algorithm for counting polyominoes was described by Jensen [Jen03], but its running time is still exponential in the size of counted polyominoes. This algorithm does not produce all polyominoes. Instead, it maintains all possible polyomino *boundaries* (see Figure 14.3.1), which are the possible configurations of

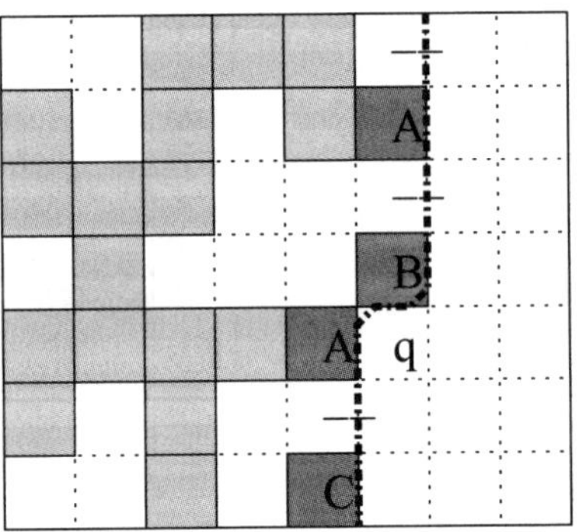

FIGURE 14.3.1
A partially built 21-cell polyomino with boundary `-A-BA-C`.

the right sides of the polyominoes, associated with information about the connectivity between cells of the boundary through polyomino cells found to the left of the boundary. Polyominoes are "built" column by column from left to right, and in every column, cells are considered from top to bottom. The algorithm maintains a database indexed by the boundaries, where for each possible boundary σ, the database keeps only the *counts* of "partial" polyominoes of all possible sizes, which can have σ as their boundary. ("Partial" in the sense that polyominoes may still be invalid in this stage due to having more than one component, but they may become valid (connected) later when more cells are added to the polyomino.) The figure shows a polyomino with the boundary `-A-BA-C`, where letters represent components of the polyominoes corresponding to this boundary. When considering the next cell q (just below the "kink" in the boundary), there are two options: either to add q to the polyomino, or to not add it. Choosing either option will change the boundary from σ to σ' (possibly $\sigma' = \sigma$), and the algorithm will then update the contents of the entry with index σ' in the database. (Note again that the database does not keep the polyominoes, but only counts of polyominoes for each possible boundary.) The algorithm counts all connected polyominoes produced in this process, up to a prescribed size, and for all possible heights (7 in the figure). Analysis of the algorithm [BM07] reveals that the major factor that influences the performance of the algorithm (in terms of both running time and memory consumption) is the number of possible boundaries. It turns out that the number of possible boundaries of length b is proportional to 3^b, up to a small polynomial factor. Due to symmetry, the algorithm needs to consider boundaries of length up to only $\lceil n/2 \rceil$ for counting polyominoes of size n. Hence, the time complexity of the algorithm is roughly $3^{n/2} \approx 1.73^n$, which is significantly less than the total number of polyominoes (about 4.06^n).

UNSOLVED PROBLEMS

PROBLEM 14.3.3

Can $t(n)$ be computed in time polynomial with n?

A related problem concerns the constant λ defined above:

PROBLEM 14.3.4

Is there a polynomial-time algorithm to find, for each n, an approximation λ_n of λ satisfying

$$10^{-n} < |\lambda_n - \lambda| < 10^{-n+1} \text{ ?}$$

PROBLEM 14.3.5

Define some decreasing sequence $\beta = (\beta_1, \beta_2, \dots)$ that tends to λ, and give an algorithm to compute β_n for every n.

Define the two sequences $\tau_1(n) = (t(n))^{1/n}$ and $\tau_2(n) = t(n+1)/t(n)$. A folklore polyomino-concatenation argument shows that for all n we have $(t(n))^2 \le t(2n)$, hence, $(t(n))^{1/n} \le (t(2n))^{1/(2n)}$, which implies that $(t(n))^{1/n} \le \lambda$ for all n, that is, $\tau_1(n)$ approaches λ from below (but is not necessarily monotone). However, it seems, given the first 56 elements of $t(n)$, that both $\tau_1(n)$ and $\tau_2(n)$ are monotone increasing. This gives two more unsolved problems:

PROBLEM 14.3.6

Show that $\tau_1(n) < \tau_1(n+1)$ for all n.

PROBLEM 14.3.7

Show that $\tau_2(n) < \tau_2(n+1)$ for all n.

14.4 GENERATING POLYOMINOES

The algorithm we describe to generate all n-ominoes, which is essentially due to Redelmeier [Red81], also provides a way of encoding n-ominoes. Starting with all n-ominoes in standard position, with each cell and each neighboring cell numbered, it constructs without repetitions all numbered $(n+1)$-ominoes in standard position.

GLOSSARY

Border cell of an n-omino S: A cell $[u, v]$, with $v \geq 0$ or with $v = 0$ and $u \geq 0$, adjacent to some cell of S. The set of all border cells, which is denoted by $B(S)$, can be shown by induction to have no more than $2n$ elements.

The algorithm, illustrated in Figure 14.4.1 for $n = 1$, 2, and 3, begins with cell 1 in position $[0, 0]$, with its border cells marked 2 and 3, and then adds these—one at a time and in this order—each time numbering *new* border cells in their lexicographic order. Whenever a number used for a border cell is not larger than the largest internal number, it is circled, and the corresponding cell is *not* added at the next stage.

FIGURE 14.4.1

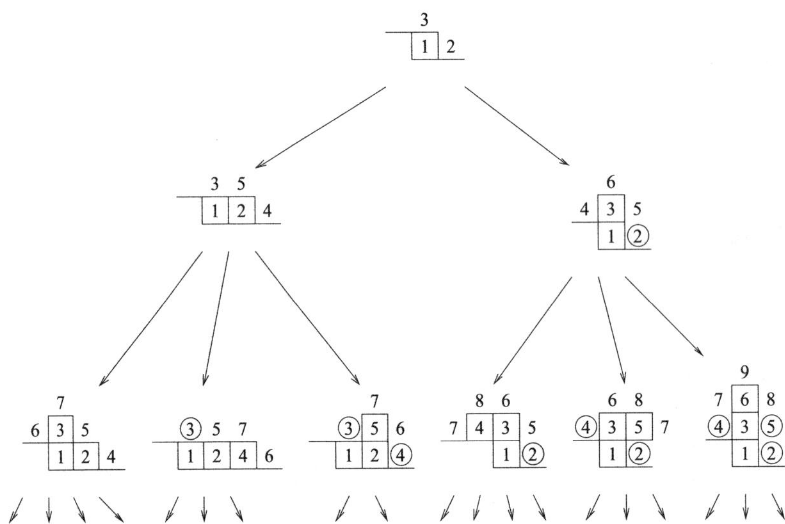

Figure 14.4.2 shows all the 4-ominoes produced in this way, with their border cells marked for the next step of the algorithm.

This process assigns a unique set of positive integers to each n-omino S, also illustrated in Figure 14.4.2. The *set character functions* for these integer sets, in turn, truncated after their final 1's, provide a binary codeword $\chi(S)$ for each n-omino S. For example, the code words for the first three 4-ominoes in Figure 14.4.2 would be 1111, 11101, and 111001.

PROBLEM 14.4.1

Which binary strings arise as codewords for n-ominoes?

The following is easy to see:

THEOREM 14.4.2

$t(n+1) = \sum n + |B(S)| - |\chi(S)|$, *where the sum extends over all n-ominoes S in standard position, and $|\chi(S)|$ is the number of bits in the codeword of S.*

FIGURE 14.4.2

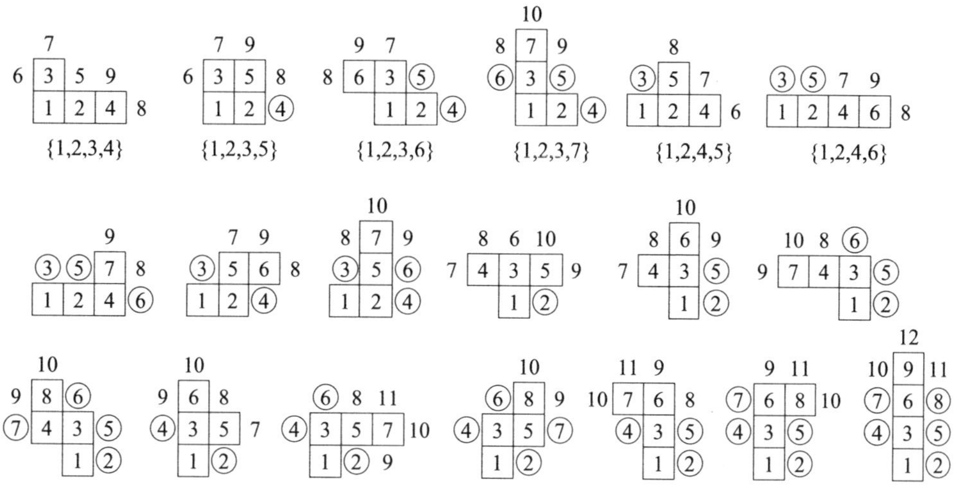

{1,2,3,4} {1,2,3,5} {1,2,3,6} {1,2,3,7} {1,2,4,5} {1,2,4,6}

PROBLEM 14.4.3

Is the generating function $T(z) = \sum_{n=1}^{\infty} t(n)z^n$ a rational function? Is $T(z)$ even algebraic?

14.5 SPECIAL TYPES OF POLYOMINOES

Particular kinds of polyominoes arise in various contexts. We will look at several of the most interesting ones. See more details in a survey by Bousquet-Mélou and Brak [Gut09, §3].

GLOSSARY

A ***composition*** of n with k parts is an ordered k-tuple $(p_1, \ldots, p_k)$ of positive integers with $p_1 + \cdots + p_k = n$.

A ***row-convex*** polyomino: One each of whose horizontal cross-sections is continuous.

A ***column-convex*** polyomino: One each of whose vertical cross-sections is continuous.

A ***convex*** polyomino: A polyomino which is both row-convex and column-convex.

Simply connected polyomino: A polyomino without holes. (Golomb calls these nonholey polyominoes ***profane***.)

A ***width-k*** polyomino: One each of whose vertical cross-sections fits in a $k \times 1$ strip of cells.

A ***directed*** polyomino is defined recursively as follows: Any single cell is a directed polyomino. An $(n+1)$-omino is directed if it can be obtained by adding a new cell immediately above, or to the right of, a cell belonging to some directed n-omino.

COMPOSITIONS AND ROW-CONVEX POLYOMINOES

There is a natural 1-1 correspondence between compositions of n and a certain class of n-ominoes in standard position, as indicated in Figure 14.5.1 for the case $n = 4$.

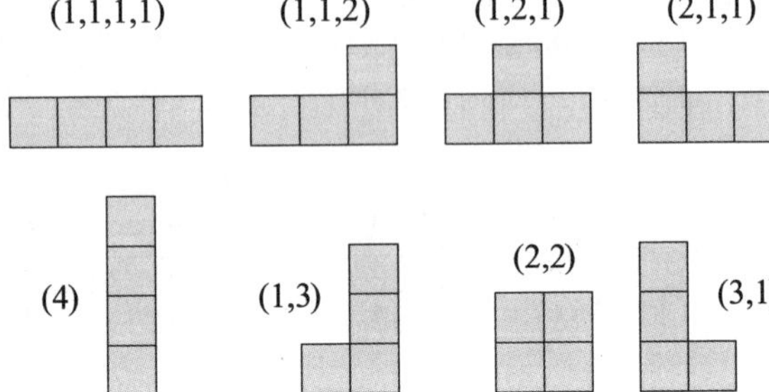

FIGURE 14.5.1
Compositions of 4 corresponding to certain 4-ominoes.

Let us, instead, assign to each composition $(a_1, \ldots, a_k)$ of n an n-omino with a *horizontal* strip of a_i cells in row i. This can be done in many ways, and the results are all the row-convex n-ominoes. Since there are $m + n - 1$ ways to form an $(m+n)$-omino by placing a strip of n cells atop a strip of m cells, it follows that for each composition $(a_1, \ldots, a_k)$ of n into positive parts, there are

$$(a_1 + a_2 - 1)(a_2 - a_3 - 1) \cdots (a_{k-1} - a_k - 1)$$

n-ominoes having a strip of a_i cells in the ith row for each i (see Figure 14.5.2 for an example arising from the composition $6 = 3 + 1 + 2$).

FIGURE 14.5.2
The 6 *row-convex* 6*-ominoes corresponding to the composition* $(3, 1, 2)$ *of* 6.

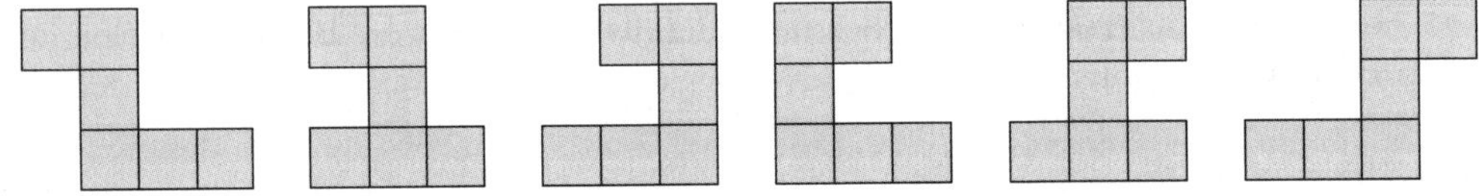

It follows that if $b(n)$ is the number of row-convex n-ominoes, then

$$b(n) = \sum (a_1 + a_2 - 1)(a_2 - a_3 - 1) \cdots (a_{k-1} - a_k - 1),$$

where the sum extends over all compositions $(a_1, \ldots, a_k)$ of n into k parts, for all k. It is known that $b(n)$, and the generating function $B(z) = \sum_{n=1}^{\infty} b(n)z^n$, are given by

THEOREM 14.5.1 [Kla67]

$b(n+3) = 5b(n+2) - 7b(n+1) + 4b(n)$, and $B(z) = \dfrac{z(1-z)^3}{1 - 5z + 7z^2 - 4z^3}$.

COROLLARY 14.5.2

$\lim_{n\to\infty}(b(n))^{1/n} = \beta$, *where* β *is the largest real root of* $z^3 - 5z^2 + 7z - 4 = 0$; $\beta \approx 3.20557$.

CONVEX POLYOMINOES

The existence of a generating function for $c(n)$ with special properties [KR74], enabled Bender to prove the following asymptotic formula:

THEOREM 14.5.3 [Ben74]

$c(n) \sim kg^n$, where $k \approx 2.67564$ and $g \approx 2.30914$.

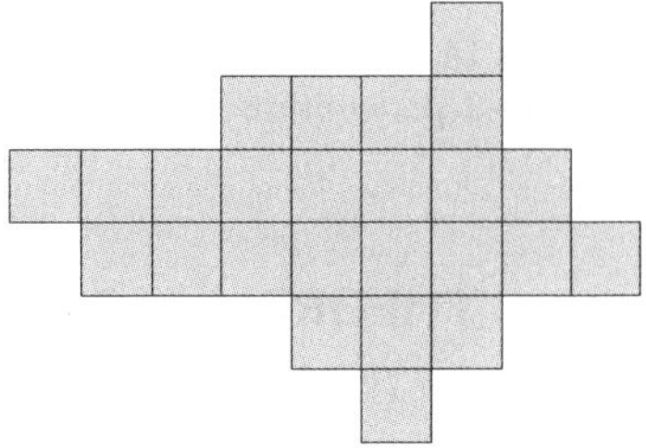

FIGURE 14.5.3
A typical convex polyomino.

The following problem concerns polyominoes radically different from convex ones.

PROBLEM 14.5.4

Find the smallest natural number n_0 such that there exists an n_0-omino with no *row or column consisting of just a single strip of cells.* (An example of a 21-omino with this property is shown in Figure 14.5.4.)

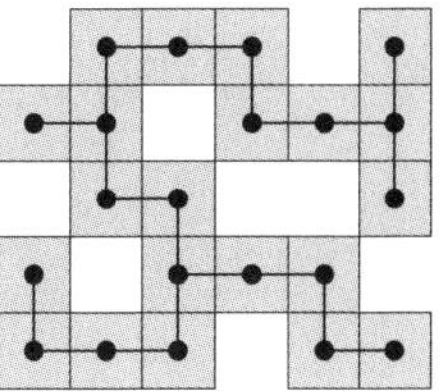

FIGURE 14.5.4
A 21-*omino with no row or column a single strip of cells.*

PROBLEM 14.5.5

How many polyominoes of size $n \geq n_0$ with the above property exist?

SIMPLY-CONNECTED POLYOMINOES

Simply-connected (profane) polyominoes are the interiors of *self-avoiding* polygons. Counts of these polygons (measured by area) are currently known up to $n = 42$ [Gut09, p. 475]. Let $t^*(n)$, $s^*(n)$, and $r^*(n)$ denote the numbers of profane fixed, free, and chiral n-ominoes, respectively. It is easy to see that $(t^*(n))^{1/n}$, $(s^*(n))^{1/n}$, and $(r^*(n))^{1/n}$ all approach the same limit, λ^*, as $n \to \infty$, and that $\lambda^* \leq \lambda$ $(= \lim_{n\to\infty}(t(n))^{1/n}$ as defined in Section 14.3). Van Rensburg and Whittington [RW89, Thm. 5.6] showed that $\lambda^* < \lambda$.

WIDTH-k POLYOMINOES

A typical width-3 polyomino is shown in Figure 14.5.5.

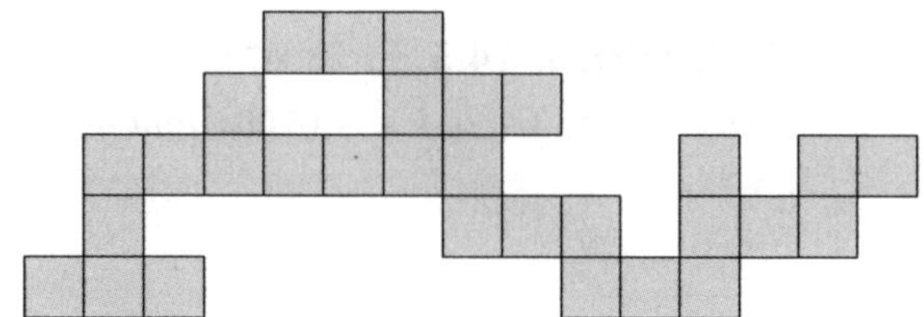

FIGURE 14.5.5
A width-3 polyomino.

THEOREM 14.5.6 [Rea62]

Let $t(n,k)$ be the number of fixed width-k n-ominoes, and $T_k(z) = \sum_{n=1}^{\infty} t(n,k)z^n$. Then $T_k(z) = P_k(z)/Q_k(z)$ for some polynomials $P_k(z), Q_k(z)$ with integer coefficients, no common zeroes, and $Q_k(0) = 1$. Equivalently, the sequence $t(n,k)$, $n = 1, 2, \ldots$, satisfies a linear, homogeneous difference equation with constant coefficients for each fixed k; the order of the equation is roughly 3^k. Furthermore, the sequence $(t(n,k))^{1/n}$ converges to a limit τ_k as $n \to \infty$, and $\lim_{k\to\infty} \tau_k = \lambda$ (see Section 14.3).

For example, for the fixed width-2 n-ominoes (shown in Figure 14.5.6 for small values of n), we have

$$T_2(z) = \frac{z}{1 - 2z - z^2} = z + 2z^2 + 5z^3 + 12z^4 + \ldots,$$

and $t(n+2,2) = 2t(n+1,2) + t(n,2)$ for $n \geq 1$.

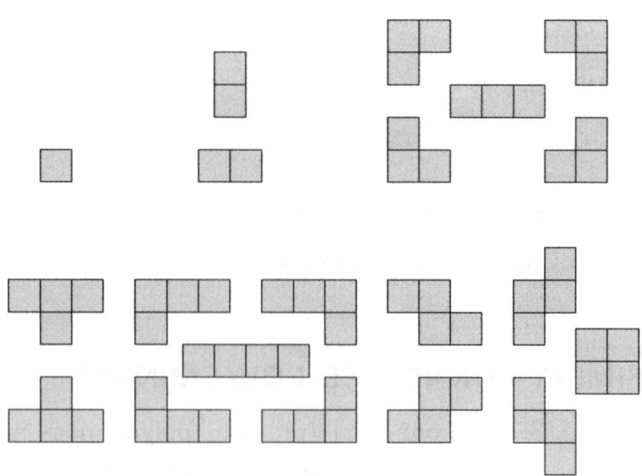

FIGURE 14.5.6
Width-2 n-ominoes for $n = 1, 2, 3, 4$.

DIRECTED POLYOMINOES

A portion of the family tree for directed polyominoes, constructed similarly to the one in Figure 14.4.1, is shown in Figure 14.5.7. As in Section 14.4, codewords can be defined for directed polyominoes, and converted into binary words. Let $\mathcal{V}$ be the language formed by all of these.

FIGURE 14.5.7
A family tree for fixed directed polyominoes.

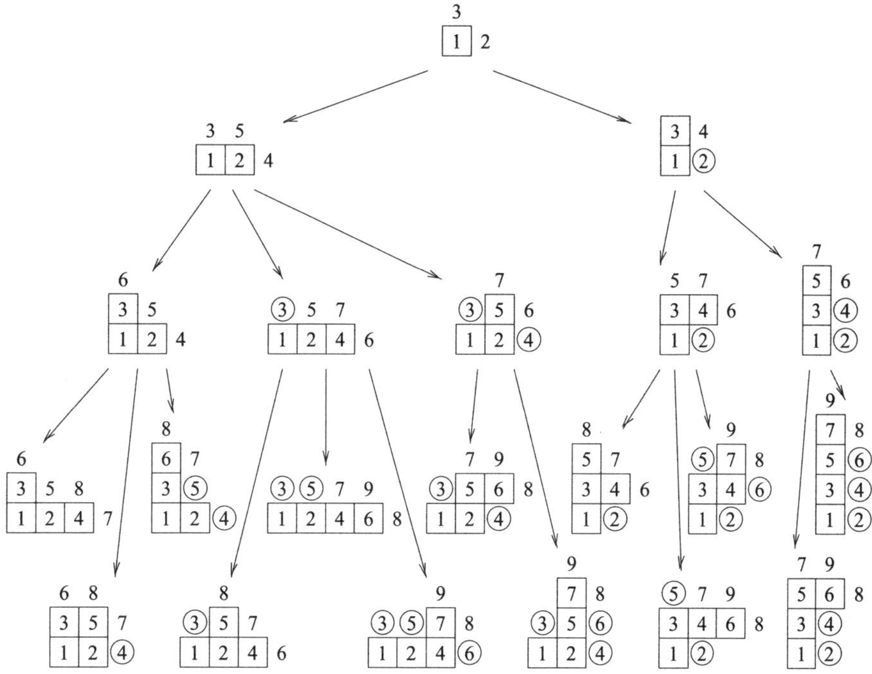

PROBLEM 14.5.7

Characterize the words in $\mathcal{V}$. In particular, is $\mathcal{V}$ an unambiguous context-free language?

THEOREM 14.5.8 [Dha82]; see also [Bou94]

If $d(n)$ is the number of directed n-ominoes in standard position, and $D(z) = \sum d(n)z^n$, then

$$D(z) = \frac{1}{2}\left(\sqrt{\frac{1+z}{1-3z}} - 1\right).$$

COROLLARY 14.5.9

$$d(n) = \sum_{k=0}^{n-1} \binom{k}{\lfloor k/2 \rfloor}\binom{n-1}{k},$$

and $d(n)$ satisfies the recurrence relation

$$d(n) = 3^{n-1} - \sum_{k=1}^{n-1} d(k)d(n-k),$$

which can be represented also as

$$d(n) = (3(n-2)d(n-2) + 2nd(n-1))/n \quad \textit{with} \quad d(1) = 1,\ d(2) = 2.$$

14.6 TILING WITH POLYOMINOES

We consider the special case of the tiling problem (see Chapter 3) in which the space we wish to tile is a set S of cells in the plane and the tiles are polyominoes. Usually S will be a rectangular set.

GLOSSARY

π-type: If S is a finite set of cells, $\mathcal{C}$ a collection of subsets of S, $\pi = (S_1, \ldots, S_k)$ a partition (or cover) of S, and $T \subset S$, the π-type of T is defined as

$$\tau(\pi, T) = (|S_1 \cap T|, \ldots, |S_k \cap T|).$$

Basis: If every rectangle in a set R can be tiled with translates of rectangles belonging to a finite subset $B \subset R$, and if B is minimal with this property, B is called a basis of R.

THEOREM 14.6.1 [Kla70]

Suppose S is a finite set and $\mathcal{C}$ a collection of subsets of S. Then, $\mathcal{C}$ tiles S if and only if, for every partition (or cover) π of S, $\tau(\pi, S)$ is a non-negative integer combination of the types $\tau(\pi, T)$ where T ranges over $\mathcal{C}$.

For example, one can use this to show that a 13×17 rectangular array of squares cannot be tiled with 2×2 and 3×3 squares: Let π be the partition of the 13 array S into "black" and "white" cells shown in Figure 14.6.1, and $\mathcal{C}$ the set of all 2×2 and 3×3 squares in S.

FIGURE 14.6.1
A coloring of the 13×17 *rectangle.*

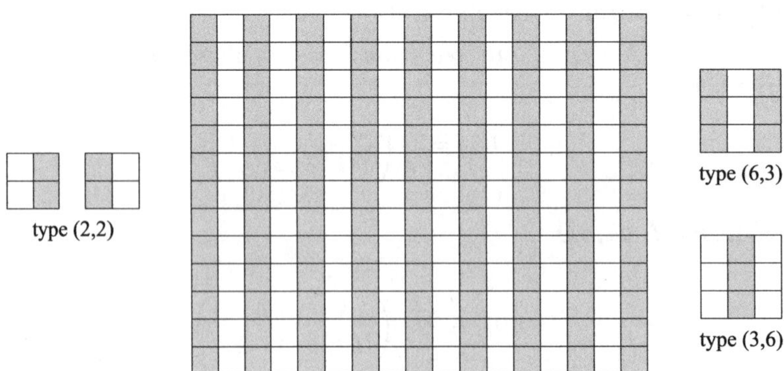

Then, each 2×2 square in $\mathcal{C}$ has type $(2,2)$, while the 3×3 squares have types $(6,3)$ and $(3,6)$. If a tiling were possible, with x 2×2 squares, and with y_1 and y_2 3×3 squares of types $(6,3)$ and $(3,6)$ (respectively), then we would have

$$(9 \cdot 13, 8 \cdot 13) = x(2,2) + y_1(6,3) + y_2(3,6),$$

which gives $13 = 3(y_1 - y_2)$, a contradiction.

THEOREM 14.6.2

Let C be a finite union of translation classes of polyominoes, and let w be a fixed positive integer. Then, one can construct a finite automaton that generates all C-tilings of $w \times n$ rectangles for all possible values of n.

COROLLARY 14.6.3

If w is fixed and C is given, then it is possible to decide whether there exists some n for which C tiles a $w \times n$ rectangle.

For example, if we want to tile a $3 \times n$ rectangle with copies of the L-tetromino shown in Figure 14.6.2 in all eight possible orientations, the automaton of Figure 14.6.2 shows that it is necessary and sufficient for n to be a multiple of 8.

FIGURE 14.6.2
An automaton for tiling a $3 \times n$ rectangle with L-tetrominoes.

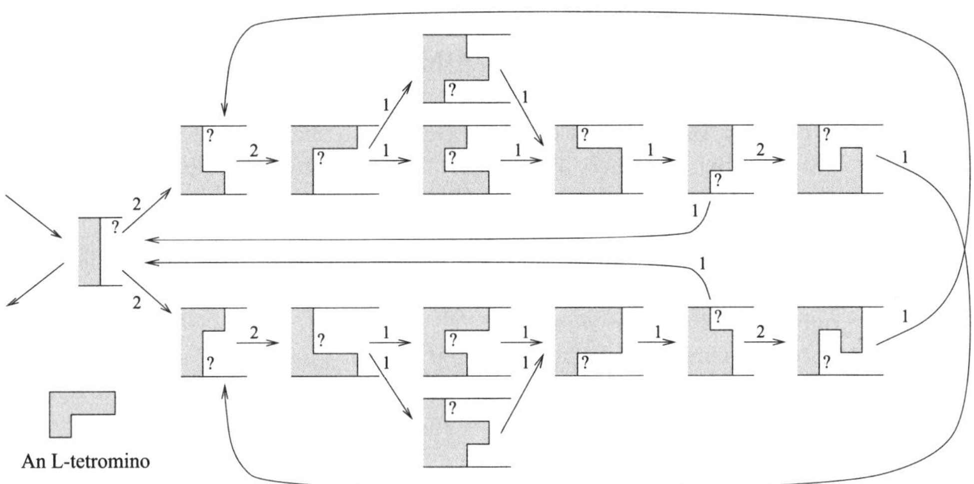

THEOREM 14.6.4 [KG69, BK75]

Let R be an infinite set of oriented rectangles with integer dimensions. Then, R has a finite basis.

(This theorem, which was originally conjectured by F. Göbel, extends to higher dimensions as well [BK75].)

For example, let R be the set of all rectangles that can be tiled with the L-tetromino of Figure 14.6.2, and let $B = \{2 \times 4, 4 \times 2, 3 \times 8, 8 \times 3\} \subset R$. Then, one can show the following three facts:

(a) R is the set of all $a \times b$ rectangles with $a, b > 1$ and $8|ab$;
(b) B is a basis of R;
(c) Each member of B is tilable with the L-tetromino.

PROBLEM 14.6.5

The smallest rectangle that can be tiled with the Y-pentomino (see Figure 14.6.3) *is* 5×10. *Find a basis B for the set R of all rectangles that can be tiled with Y-pentominoes.*

Reid [Rei05] showed that the cardinality of the basis of the Y-pentomino (Problem 14.6.5) is 40. In addition, he proved that there exist polyominoes with arbitrarily large bases.

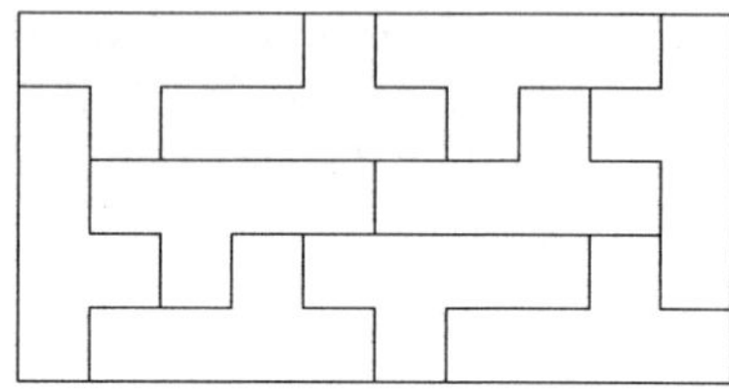

FIGURE 14.6.3
A 5×10 rectangle tiled with Y-pentominoes.

14.7 RECTANGLES OF POLYOMINOES

Here we consider the question of which polyomino shapes have the property that some finite number of copies, allowing all rotations and reflections, can be assembled to form a rectangle. Klarner [Kla69] defined the ***order*** of a polyomino P as the minimum number of congruent copies of P that can be assembled (allowing translation, rotation, and reflection) to form a rectangle. For those polyominoes that will not tile any rectangle, the order is undefined. (A polyomino has order 1 if and only if it is itself a rectangle.)

A polyomino has order 2 if and only if it is "half a rectangle," since two identical copies of it must form a rectangle. This necessarily means that the two copies will be 180° rotations of each other when forming a rectangle. Some examples are shown in Figure 14.7.1.

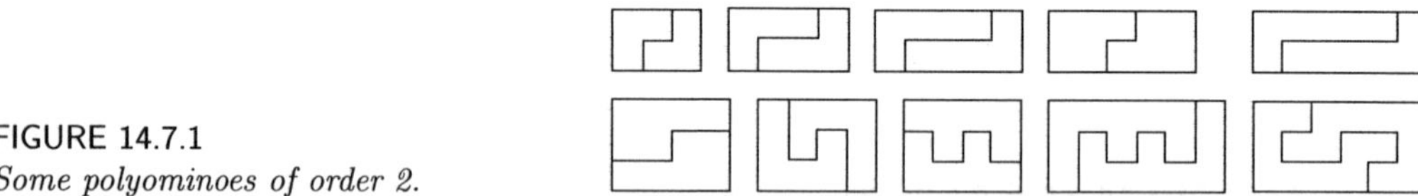

FIGURE 14.7.1
Some polyominoes of order 2.

There are no polyominoes of order 3 [SW92]. In fact, the only way any rectangle can be divided up into three identical copies of a "well-behaved" geometric figure is to partition it into three *rectangles* (see Figure 14.7.2), and by definition a rectangle has order 1.

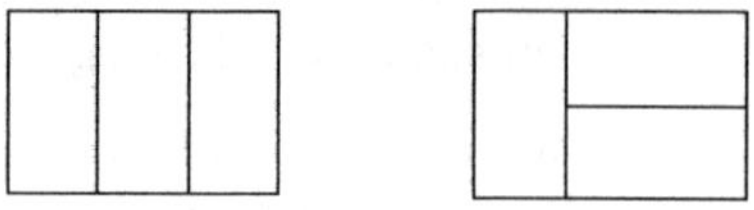

FIGURE 14.7.2
How three identical rectangles can form a rectangle.

There are various ways in which four identical polyominoes can be combined to form a rectangle. One way, illustrated in Figure 14.7.3, is to have four 90° rotations of a single shape forming a square. Another way to combine four identical shapes to form a rectangle uses the fourfold symmetry of the rectangle itself: left-right, up-down, and 180° rotational symmetry. Some examples of this appear in Figure 14.7.4. More complicated order-4 patterns were found by Klarner [Kla69].

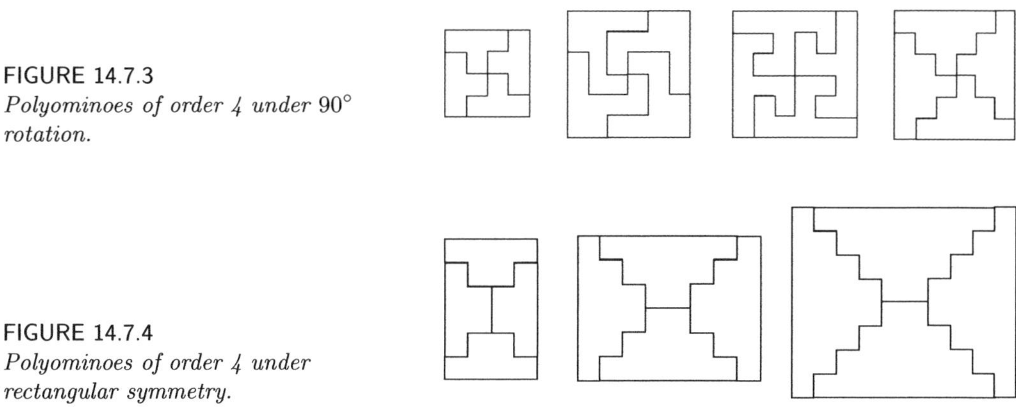

FIGURE 14.7.3
Polyominoes of order 4 under 90° *rotation.*

FIGURE 14.7.4
Polyominoes of order 4 under rectangular symmetry.

Beyond order 4, there is a systematic construction [Gol89] that gives examples of order $4s$ for every positive integer s. Isolated examples of polyominoes with orders of the form $4s+2$ are also known. Figure 14.7.5 shows examples of order 10 [Gol66] and orders 18, 24, and 28 [Kla69]; see also Marshall [Mar97]. More examples are found in the Polyominoes chapter of the second edition of this handbook.

FIGURE 14.7.5
Four "sporadic" polyominoes of orders 10, 18, 24, and 28, respectively.

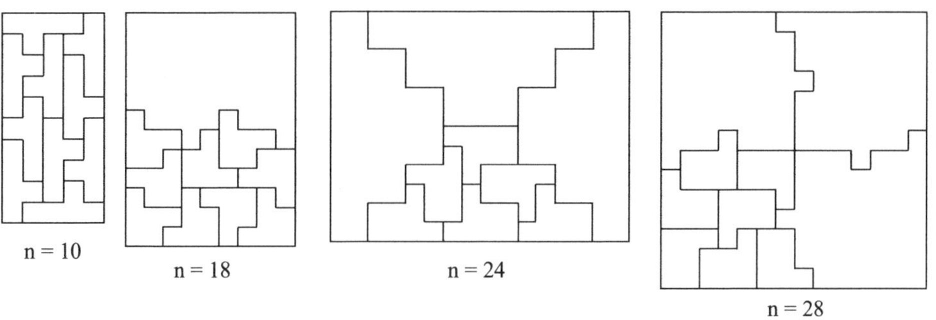

No polyomino whose order is an odd number greater than 1 has ever been found, but the possibility that such polyominoes exist (with orders greater than 3) has never been ruled out. The smallest even order for which no example is known is 6. Figure 14.7.6 shows one way in which six copies of a polyomino can be fitted together to form a rectangle, but the polyomino in question (as shown) actually has order 2.

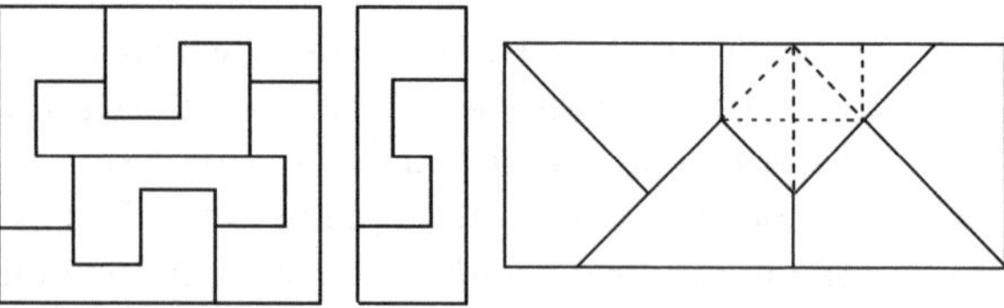

FIGURE 14.7.6
A 12-omino of order 2 that suggests an order-6 tiling, and Michael Reid's order-6 "heptabolo" (a figure made of seven congruent isosceles right triangles). Is there any polyomino *of order 6?*

Yang [Yan14] proved that *rectangular tileability*, the problem of whether or not a given set of polyominoes tiles some (possibly large) rectangle, is undecidable. However, the status of rectangular tileability for a single polyomino is unknown.

PROBLEM 14.7.1

Given a polyomino P, is there a rectangle which P tiles?

14.8 HIGHER DIMENSIONS

GLOSSARY

Polycube: In higher dimensions, the generalization of a polyomino is a d-dimensional *polycube*, which is a connected set of cells in the d-dimensional cubical lattice, where connectivity is through $(d-1)$-dimensional faces of the cubical cells.

Proper polycube: A polycube P is *proper* in d dimensions if it spans d dimensions, that is, the convex hull of the centers of all cells of P is d-dimensional.

Counts of polycubes were given by Lunnon [Lun75], Gaunt et al. [GSR76, Gau80], and Barequet and Aleksandrowicz [AB09a, AB09b]. See sequences A001931 and A151830–35 for counts of polycubes in dimensions 3 through 9, respectively, in the OEIS [oeis]. The most comprehensive counts to date were obtained by a parallel version of Redelmeier's algorithm adapted to dimensions greater than 2.

Similarly to two dimensions, let λ_d denote the growth constant of polycubes in d dimensions. It was proven [BBR10] that $\lambda_d = 2ed - o(d)$, where e is the base of natural logarithms, and evidence shows that $\lambda_d \sim (2d-3)e + O(1/d)$ as $d \to \infty$. Gaunt and Peard [GP00, p. 7521, Eq. (3.8)]) provide the following *semi-rigorously* proved expansion of λ_d in $1/d$:[2]

$$\lambda_d \sim 2ed - 3e - \frac{31e}{48d} - \frac{37e}{16d^2} - \frac{279613e}{46080d^3} - \frac{325183e}{10240d^4} - \frac{54299845e}{2654208d^5} + O\left(\frac{1}{d^6}\right).$$

PROBLEM 14.8.1

Find a general formula for λ_d or a generating function for the sequence of coefficients of the expansion above.

Following Lunnon [Lun75], let $\mathrm{CX}(n,d)$ denote the number of fixed polycubes

[2] The reference above provides a much more general formula. To obtain the expansion for the model of strongly embedded site animals (see below), one needs to substitute in the cited formula $y := 1$, $z := 0$, and $\sigma := 2d-1$, take the exponent of the formula, and expand the result as a power series of $1/d$, i.e., around infinity.

of size n in d dimensions, and let $\mathrm{DX}(n,d)$ denote the number of those of them that are also proper in d dimensions. Lunnon [Lun75] observed that $\mathrm{CX}(n,d) = \sum_{i=0}^{d}\binom{d}{i}\mathrm{DX}(n,i)$. Indeed, every d-dimensional polycube is proper in some $0 \le i \le d$ dimensions (the singleton cube is proper in zero dimensions!), and then these i dimensions can be chosen in $\binom{d}{i}$ ways. However, a polycube of size n cannot obviously be proper in more than $n-1$ dimensions. Hence, this formula can be rewritten as

$$\mathrm{CX}(n,d) = \sum_{i=0}^{\min(n-1,d)} \binom{d}{i}\mathrm{DX}(n,i).$$

Assume now that the value of n is fixed, and let $\mathrm{CX}_n(d)$ denote the number of fixed d-dimensional polycubes of size n, considered as a function of d only. A simple consequence [BBR10] of Lunnon's formula is that $\mathrm{CX}_n(d)$ is a polynomial in d of degree $n-1$. The first few polynomials are $\mathrm{CX}_1(d) = 1$, $\mathrm{CX}_2(d) = d$, $\mathrm{CX}_3(d) = 2d^2 - d$, $\mathrm{CX}_4(d) = \frac{16}{3}d^3 - \frac{15}{2}d^2 + \frac{19}{6}d$, $\mathrm{CX}_5(d) = \frac{50}{3}d^4 - 42d^3 + \frac{239}{6}d^2 - \frac{27}{2}d$, and so on. It was also shown that the leading coefficient of $\mathrm{CX}_n(d)$ is $2^{n-1}n^{n-3}/(n-1)!$.

Interestingly, there is a pattern in the "diagonal formulae" of the form $\mathrm{DX}(n,n-k)$, for small values of $k \ge 1$. Using Cayley trees, it is easy to show that $\mathrm{DX}(n,n-1) = 2^{n-1}n^{n-3}$ (sequence A127670 in the OEIS [oeis]). Barequet et al. [BBR10] proved that $\mathrm{DX}(n,n-2) = 2^{n-3}n^{n-5}(n-2)(2n^2-6n+9)$ (sequence A171860), and Asinowski et al. [ABBR12] proved that $\mathrm{DX}(n,n-3) = 2^{n-6}n^{n-7}(n-3)(12n^5 - 104n^4 + 360n^3 - 679n^2 + 1122n - 1560)/3$ (sequence A191092). Barequet and Shalah [BS17] provided computer-generated proofs for the formulae for $\mathrm{DX}(n,n-4)$ and $\mathrm{DX}(n,n-5)$, as well as a complex recipe for producing these formulae for any value of k. Formulae for $k = 6$ and $k = 7$ were conjectured by Peard and Gaunt [PG95] and by Luther and Mertens [LM11], respectively.

14.9 MISCELLANEOUS

COUNTING BY PERIMETER

Polyominoes are rarely counted by *perimeter* instead of by area. Delest and Viennot [DV84], and Kim [Kim88], proved by completely different methods that the number of convex polyominoes with perimeter $2(m+4)$ (for $m \ge 0$) is

$$(2m+11)4^m - 4(2m+1)\binom{2m}{m}.$$

OTHER MODELS OF ANIMALS

In the literature of statistical physics, polyominoes are referred to as *strongly embedded site animals.* This terminology comes from considering the dual graph of the lattice. When switching to the dual setting, that is, to the cell-adjacency graph, polyomino cells turn into vertices (sites) and adjacencies of cells turn into edges (bonds) of the graph. In the dual setting, connected sets of sites are called *site animals.* Instead of counting animals by the number of their sites, one can count them by the number of their bonds, in which case they are called *bond animals.* The term "strongly embedded" refers to the situation in which if two neighboring sites belong to the animal, then the bond connecting them must also belong to the animal. If this restriction is relaxed, then *weakly-embedded* animals are considered. These extensions have applications in computational chemistry; see, for example, the book by Vanderzande [Van98].

NONCUBICAL LATTICES

In the plane, one can also consider *polyhexes* and *polyiamonds*, which are connected sets of cells in the hexagonal and triangular lattices, respectively. Algorithms and counts for polyhexes and polyiamonds were given by Lunnon [Lun72], Barequet and Aleksandrowicz [AB09a], and Vöge and Guttmann[VG03]. Counts of polyhexes and polyiamonds are currently known up to sizes 46 and 75, respectively [Gut09, pp. 477 and 479]. See also sequences A001207 and A001420, respectively, in the OEIS [oeis].

14.10 SOURCES AND RELATED MATERIAL

FURTHER READING

An excellent introductory survey of the subject, with an abundance of references, is by Golomb [Gol94]. Another notable book on polyominoes is by Martin [Mar91]. A deep and comprehensive collection of essays on the subject is edited by A.J. Guttmann [Gut09]. Finally, there are many articles, puzzles, and problems concerning polyominoes to be found in the magazine *Recreational Mathematics.*

RELATED CHAPTERS

Chapter 3: Tilings

REFERENCES

[AB09a] G. Aleksandrowicz and G. Barequet. Counting d-dimensional polycubes and nonrectangular planar polyominoes. *Internat. J. Comput. Geom. Appl.*, 19:215–229, 2009.

[AB09b] G. Aleksandrowicz and G. Barequet. Counting polycubes without the dimensionality curse. *Discrete Math.*, 309:4576–4583, 2009.

[ABBR12] A. Asinowski, G. Barequet, R. Barequet, and G. Rote. Proper n-cell polycubes in $n-3$ dimensions. *J. Integer Seq.*, 15, article 12.8.4, 2012.

[BBR10] R. Barequet, G. Barequet, and G. Rote. Formulae and growth rates of high-dimensional polycubes. *Combinatorica*, 30:257–275, 2010.

[Ben74] E.A. Bender. Convex n-ominoes. *Discrete Math.*, 8:219–226, 1974.

[BK75] N.G. de Bruijn and D.A. Klarner. A finite basis theorem for packing boxes with bricks. In: *Papers Dedicated to C.J. Bouwkamp*, Philips Research Reports, 30:337–343, 1975.

[BM07] G. Barequet and M. Moffie. On the complexity of Jensen's algorithm for counting fixed polyominoes. *J. Discrete Algorithms*, 5:348–355, 2007.

[Bou94] M. Bousquet-Mélou. Polyominoes and polygons. *Contemp. Math.*, 178:55–70, 1994.

[BRS16] G. Barequet, G. Rote, and M. Shalah. $\lambda > 4$: An improved lower bound on the growth constant of polyominoes. *Comm. ACM*, 59:88–95 2016.

[BS17] G. Barequet and M. Shalah. Counting n-cell polycubes proper in $n-k$ dimensions. *European J. Combin.*, 63:146–163, 2017.

[Dha82] D. Dhar. Equivalence of the two-dimensional directed-site animal problem to Baxter's hard square lattice gas model. *Phys. Rev. Lett.*, 49:959–962, 1982.

[DV84] M. Delest and X. Viennot. Algebraic languages and polyominoes enumeration. *Theoret. Comput. Sci.*, 34:169–206, 1984.

[Ede61] M. Eden. A two-dimensional growth process. In *Proc. 4th Berkeley Sympos. Math. Stat. Prob.*, IV, Berkeley, pages 223–239, 1961.

[Gau80] D.S. Gaunt. The critical dimension for lattice animals. *J. Phys. A: Math. Gen.*, 13:L97–L101, 1980.

[Gol66] S.W. Golomb. Tiling with polyominoes. *J. Combin. Theory*, 1:280–296, 1966.

[Gol89] S.W. Golomb. Polyominoes which tile rectangles. *J. Combin. Theory, Ser. A*, 51:117–124, 1989.

[Gol94] S.W. Golomb. *Polyominoes*, 2nd edition. Princeton University Press, 1994.

[GP00] D.S. Gaunt and P.J. Peard. $1/d$-expansions for the free energy of weakly embedded site animal models of branched polymers. *J. Phys. A: Math. Gen.*, 33:7515–7539, 2000.

[GSR76] D.S. Gaunt, M.F. Sykes, and H. Ruskin. Percolation processes in d-dimensions. *J. Phys. A: Math. Gen.*, 9:1899–1911, 1976.

[Gut09] A.J. Guttmann. *Polygons, Polyominoes, and Polycubes.* Springer, Dodrecht, 2009.

[JG00] I. Jensen and A.J. Guttmann. Statistics of lattice animals (polyominoes) and polygons. *J. Phys. A: Math. Gen.*, 33:L257–L263, 2000.

[Jen01] I. Jensen. Enumerations of lattice animals and trees. *J. Stat. Phys.*, 102:865–881, 2001.

[Jen03] I. Jensen. Counting polyominoes: A parallel implementation for cluster computing. In *Proc. Int. Conf. Comput. Sci.*, III, vol. 2659 of *Lecture Notes Comp. Sci*, pages 203–212, Springer, Berlin, 2003.

[KG69] D.A. Klarner and F. Göbel. Packing boxes with congruent figures. *Indag. Math.*, 31:465–472, 1969.

[Kim88] D. Kim. The number of convex polyominos with given perimeter. *Discrete Math.*, 70:47–51, 1988.

[Kla67] D.A. Klarner. Cell growth problems. *Canad. J. Math.*, 19:851–863, 1967.

[Kla69] D.A. Klarner. Packing a rectangle with congruent N-ominoes. *J. Combin. Theory*, 7:107–115, 1969.

[Kla70] D.A. Klarner. A packing theory. *J. Combin. Theory*, 8:272–278, 1970.

[KR73] D.A. Klarner and R.L. Rivest. A procedure for improving the upper bound for the number of n-ominoes. *Canad. J. Math.*, 25:585–602, 1973.

[KR74] D.A. Klarner and R.L. Rivest. Asymptotic bounds for the number of convex n-ominoes. *Discrete Math.*, 8:31–40, 1974.

[LM11] S. Luther and S. Mertens. Counting lattice animals in high dimensions. *J. Stat. Mech. Theory Exp.*, 9:546–565, 2011.

[Lun72] W.F. Lunnon. Counting hexagonal and triangular polyominoes. In R.C. Read, editor, *Graph Theory and Computing*, pages 87–100, Academic Press, New York, 1972.

[Lun75] W.F. Lunnon. Counting multidimensional polyominoes. *The Computer Journal*, 18:366–367, 1975.

[Mad99] N. Madras. A pattern theorem for lattice clusters. *Ann. Comb.*, 3:357–384, 1999.

[Mar91] G.E. Martin. *Polyominoes. A Guide to Puzzles and Problems in Tiling.* Math. Assoc. Amer., Washington, D.C., 1991.

[Mar97] W.R. Marshall. Packing rectangles with congruent polyominoes. *J. Combin. Theory, Ser. A*, 77:181–192, 1997.

[Mer90] S. Mertens. Lattice animals: A fast enumeration algorithm and new perimeter polynomials. *J. Stat. Phys.*, 58:1095–1108, 1990.

[ML92] S. Mertens and M.E. Lautenbacher. Counting lattice animals: A parallel attack. *J. Stat. Phys.*, 66:669–678, 1992.

[oeis] The On-Line Encyclopedia of Integer Sequences. Published electronically at `http://oeis.org` .

[PG95] P.J. Peard and D.S. Gaunt. $1/d$-expansions for the free energy of lattice animal models of a self-interacting branched polymer. *J. Phys. A: Math. Gen.*, 28:6109–6124, 1995.

[Rea62] R.C. Read. Contributions to the cell growth problem. *Canad. J. Math.*, 14:1–20, 1962.

[Red81] D.H. Redelmeier Counting polyominoes: Yet another attack. *Discrete Math.*, 36:191–203, 1981.

[Rei05] M. Reid. Klarner systems and tiling boxes with polyominoes. *J. Combin. Theory, Ser. A*, 111:89–105, 2005.

[RW89] E.J.J. van Rensburg and S.G. Whittington. Self-avoiding surfaces. *J. Phys. A: Math. Gen.*, 22:4939–4958, 1989.

[SW92] I. Stewart and A. Wormstein. Polyominoes of order 3 do not exist. *J. Combin. Theory, Ser. A*, 61:130–136, 1992.

[Van98] C. Vanderzande. *Lattice Models of Polymers.* Cambridge University Press, 1998.

[VG03] M. Vöge and A.J. Guttmann. On the number of hexagonal polyominoes. *Theoret. Comput. Sci.*, 307:433–453, 2003.

[Yan14] J. Yang. Rectangular tileability and complementary tileability are undecidable. *European J. Combin.*, 41:20–34, 2014.

Part II

POLYTOPES AND POLYHEDRA

15 BASIC PROPERTIES OF CONVEX POLYTOPES

Martin Henk, Jürgen Richter-Gebert, and Günter M. Ziegler

INTRODUCTION

Convex polytopes are fundamental geometric objects that have been investigated since antiquity. The beauty of their theory is nowadays complemented by their importance for many other mathematical subjects, ranging from integration theory, algebraic topology, and algebraic geometry to linear and combinatorial optimization.

In this chapter we try to give a short introduction, provide a sketch of "what polytopes look like" and "how they behave," with many explicit examples, and briefly state some main results (where further details are given in subsequent chapters of this Handbook). We concentrate on two main topics:

- Combinatorial properties: faces (vertices, edges, ..., facets) of polytopes and their relations, with special treatments of the classes of low-dimensional polytopes and of polytopes "with few vertices;"
- Geometric properties: volume and surface area, mixed volumes, and quermassintegrals, including explicit formulas for the cases of the regular simplices, cubes, and cross-polytopes.

We refer to Grünbaum [Grü67] for a comprehensive view of polytope theory, and to Ziegler [Zie95] respectively to Gruber [Gru07] and Schneider [Sch14] for detailed treatments of the combinatorial and of the convex geometric aspects of polytope theory.

15.1 COMBINATORIAL STRUCTURE

GLOSSARY

$\mathcal{V}$-polytope: The convex hull of a finite set $X = \{x^1, \dots, x^n\}$ of points in $\mathbb{R}^d$,

$$P = \operatorname{conv}(X) \;:=\; \Big\{\sum_{i=1}^{n} \lambda_i x^i \mid \lambda_1, \dots, \lambda_n \geq 0,\ \sum_{i=1}^{n} \lambda_i = 1\Big\}.$$

$\mathcal{H}$-polytope: The solution set of a finite system of linear inequalities,

$$P = P(A, b) \;:=\; \big\{x \in \mathbb{R}^d \mid a_i^T x \leq b_i \text{ for } 1 \leq i \leq m\big\},$$

with the extra condition that the set of solutions is bounded, that is, such that there is a constant N such that $||x|| \leq N$ holds for all $x \in P$. Here $A \in \mathbb{R}^{m \times d}$ is a real matrix with rows a_i^T, and $b \in \mathbb{R}^m$ is a real vector with entries b_i.

Polytope: A subset P of some $\mathbb{R}^d$ that can be presented as a $\mathcal{V}$-polytope or (equivalently, by the main theorem below) as an $\mathcal{H}$-polytope.

Affine hull aff(S) ***of a set*** S***:*** The inclusion-minimal affine subspace of $\mathbb{R}^d$ that contains S, which is given by $\{\sum_{j=1}^{p} \lambda_j x^j \mid p > 0,\ x^1, \ldots, x^p \in S, \lambda_1, \ldots, \lambda_p \in \mathbb{R}, \sum_{j=1}^{p} \lambda_j = 1\}$.

Dimension: The dimension of an arbitrary subset S of $\mathbb{R}^d$ is defined as the dimension of its affine hull: $\dim(S) := \dim(\mathrm{aff}(S))$.

d-polytope: A d-dimensional polytope. The prefix "d-" denotes "d-dimensional." A subscript in the name of a polytope usually denotes its dimension. Thus "d-cube C_d" will refer to a d-dimensional incarnation of the cube.

Interior and ***relative interior:*** The interior $\mathrm{int}(P)$ is the set of all points $x \in P$ such that for some $\varepsilon > 0$, the ε-ball $B_\varepsilon(x)$ around x is contained in P.

Similarly, the relative interior $\mathrm{relint}(P)$ is the set of all points $x \in P$ such that for some $\varepsilon > 0$, the intersection $B_\varepsilon(x) \cap \mathrm{aff}(P)$ is contained in P.

Affine equivalence: For polytopes $P \subseteq \mathbb{R}^d$ and $Q \subseteq \mathbb{R}^e$, the existence of an affine map $\pi : \mathbb{R}^d \longrightarrow \mathbb{R}^e$, $x \longmapsto Ax + b$ that maps P bijectively to Q. The affine map π does not need to be injective or surjective. However, it has to restrict to a bijective map $\mathrm{aff}(P) \longrightarrow \mathrm{aff}(Q)$. In particular, if P and Q are affinely equivalent, then they have the same dimension.

THEOREM 15.1.1 *Main Theorem of Polytope Theory* (cf. [Zie95, Sect. 1.1])

The definitions of $\mathcal{V}$-polytopes and of $\mathcal{H}$-polytopes are equivalent. That is, every $\mathcal{V}$-polytope has a description by a finite system of inequalities, and every $\mathcal{H}$-polytope can be obtained as the convex hull of a finite set of points (its vertices*).*

Any $\mathcal{V}$-polytope can be viewed as the image of an $(n-1)$-dimensional simplex under an affine map $\pi : x \mapsto Ax + b$, while any $\mathcal{H}$-polytope is affinely equivalent to an intersection $\mathbb{R}^m_{\geq 0} \cap L$ of the positive orthant in m-space with an affine subspace [Zie95, Lecture 1]. To see the Main Theorem at work, consider the following two statements. The first one is easy to see for $\mathcal{V}$-polytopes, but not for $\mathcal{H}$-polytopes, and for the second statement we have the opposite effect:

1. *Projections:* Every image of a polytope P under an affine map is a polytope.
2. *Intersections:* Every intersection of a polytope with an affine subspace is a polytope.

However, the computational step from one of the main theorem's descriptions of polytopes to the other—a "convex hull computation"—is often far from trivial. Essentially, there are three types of algorithms available: inductive algorithms (inserting vertices, using a so-called beneath-beyond technique), projection algorithms (known as Fourier–Motzkin elimination or double description algorithms), and reverse search methods (as introduced by Avis and Fukuda [AF92]). For explicit computations one can use public domain codes as the software package `polymake` [GJ00] that we use here, or `sage` [SJ05]; see also Chapters 26 and 67.

In each of the following definitions of d-simplices, d-cubes, and d-cross-polytopes we give both a $\mathcal{V}$- and an $\mathcal{H}$-presentation. From this one can see that the $\mathcal{H}$-presentation can have exponential size (number of inequalities) in terms of the size (number of vertices) of the $\mathcal{V}$-presentation (e.g., for the d-cross-polytopes), and vice versa (e.g., for the d-cubes).

Definition: A (regular) d-dimensional ***simplex*** (or ***d-simplex***) in $\mathbb{R}^d$ is given by

$$
\begin{aligned}
T_d &:= \operatorname{conv}\{e^1, e^2, \ldots, e^d, \frac{1-\sqrt{d+1}}{d}(e^1 + \cdots + e^d)\} \\
&= \{x \in \mathbb{R}^d \mid \sum_{i=1}^{d} x_i \leq 1,\ -(1+\sqrt{d+1}+d)x_k + \sum_{i=1}^{d} x_i \leq 1 \text{ for } 1 \leq k \leq d\},
\end{aligned}
$$

where $e^1, \ldots, e^d$ denote the coordinate unit vectors in $\mathbb{R}^d$.

The simplices T_d are ***regular polytopes*** (with a symmetry group that is flag-transitive—see Chapter 18): The parameters have been chosen so that all edges of T_d have length $\sqrt{2}$. Furthermore, the origin $0 \in \mathbb{R}^d$ is in the interior of T_d: This is clear from the $\mathcal{H}$-presentation.

For the combinatorial theory one considers polytopes that differ only by an affine change of coordinates or—more generally—a projective transformation to be equivalent. Combinatorial equivalence is, however, still stronger than projective equivalence. In particular, we refer to any d-polytope that can be presented as the convex hull of d+1 affinely independent points as a ***d-simplex***, since any two such polytopes are equivalent with respect to an affine map. Other standard choices include

$$
\begin{aligned}
\Delta_d &:= \operatorname{conv}\{0, e^1, e^2, \ldots, e^d\} \\
&= \{x \in \mathbb{R}^d \mid \sum_{i=1}^{d} x_i \leq 1,\quad x_k \geq 0 \text{ for } 1 \leq k \leq d\}
\end{aligned}
$$

and the (d–1)-dimensional simplex in $\mathbb{R}^d$ given by

$$
\begin{aligned}
\Delta'_{d-1} &:= \operatorname{conv}\{e^1, e^2, \ldots, e^d\} \\
&= \{x \in \mathbb{R}^d \mid \sum_{i=1}^{d} x_i = 1,\quad x_k \geq 0 \text{ for } 1 \leq k \leq d\}.
\end{aligned}
$$

FIGURE 15.1.1
A 3-simplex, a 3-cube, and a 3-cross-polytope (octahedron).

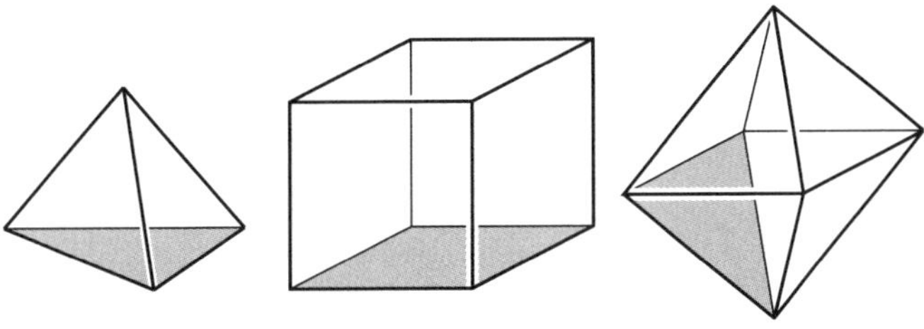

Definition: A ***d-cube*** (a.k.a. the d-dimensional ***hypercube***) is

$$
\begin{aligned}
C_d &:= \operatorname{conv}\{\alpha_1 e^1 + \alpha_2 e^2 + \cdots + \alpha_d e^d \mid \alpha_1, \ldots, \alpha_d \in \{+1, -1\}\} \\
&= \{x \in \mathbb{R}^d \mid -1 \leq x_k \leq 1 \text{ for } 1 \leq k \leq d\}.
\end{aligned}
$$

Again, there are other natural choices, among them the d-dimensional ***unit cube***

$$\begin{aligned}[0,1]^d &= \operatorname{conv}\{\sum_{i\in S} e^i \mid S \subseteq \{1,2,\ldots,d\}\} \\ &= \{x \in \mathbb{R}^d \mid 0 \le x_k \le 1 \text{ for } 1 \le k \le d\}.\end{aligned}$$

A ***d-cross-polytope*** in $\mathbb{R}^d$ (for $d = 3$ known as the ***octahedron***) is given by

$$\begin{aligned}C_d^{\Delta} &:= \operatorname{conv}\{\pm e^1, \pm e^2, \ldots, \pm e^d\} \\ &= \{x \in \mathbb{R}^d \mid \sum_{i=1}^{d} \alpha_i x_i \le 1 \text{ for all } \alpha_1, \ldots, \alpha_d \in \{-1,+1\}\}.\end{aligned}$$

To illustrate concepts and results we will repeatedly use the unnamed polytope with six vertices shown in Figure 15.1.2.

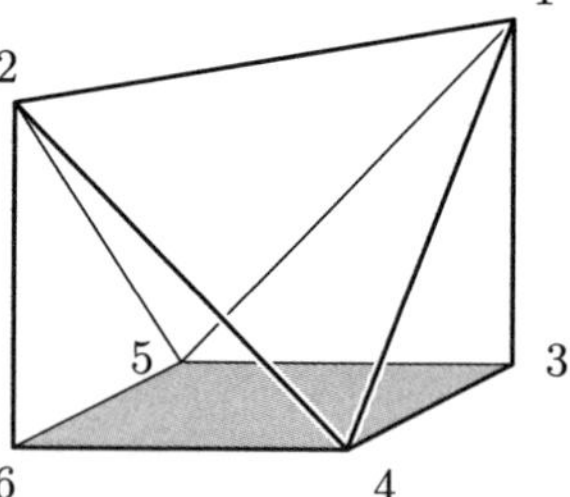

FIGURE 15.1.2
Our unnamed "typical" 3-polytope. It has 6 vertices, 11 edges and 7 facets.

This polytope without a name can be presented as a $\mathcal{V}$-polytope by listing its six vertices. The following coordinates make it into a ***subpolytope*** of the 3-cube C_3: The vertex set consists of all but two vertices of C_3. Our list below (on the left) shows the vertices of our unnamed polytope in a format used as input for the `polymake` program, i.e., the vertices are given in homogeneous coordinates with an additional 1 as first entry. From these data the `polymake` program produces a description (on the right) of the polytope as an $\mathcal{H}$-polytope, i.e., it computes the facet-defining hyperplanes with respect to homogeneous coordinates. For instance, the entries in the last row of the section FACETS describe the halfspace $1\,x_0 - 1\,x_1 + 1\,x_2 - 1\,x_3 \ge 0$, which with $x_0 \equiv 1$ corresponds to the facet-defining inequality $x_1 - x_2 + x_3 \le 1$ of our 3-dimensional unnamed polytope.

```
POINTS                FACETS
1  1  1  1            1  0 -1  0
1 -1 -1  1            1 -1  0  0
1  1  1 -1            1  1  0  0
1  1 -1 -1            1  0  1  0
1 -1  1 -1            1  0  0  1
1 -1 -1 -1            1  1 -1 -1
                      1 -1  1 -1
```

Any unbounded pointed polyhedron (that is, the set of solutions of a system of inequalities, which is not bounded but does have a vertex) is, via a projective transformation, equivalent to a polytope with a distinguished facet; see [Zie95, Sect. 2.9 and p. 75]. In this respect, we do not lose anything on the combinatorial level if we restrict the following discussion to the setting of *full-dimensional* convex polytopes, that is, d-polytopes embedded in $\mathbb{R}^d$.

15.1.1 FACES

GLOSSARY

Support function: Given a polytope $P \subseteq \mathbb{R}^d$, the function

$$h(P,\cdot)\colon \mathbb{R}^d \to \mathbb{R}, \qquad h(P,x) := \sup\{\langle x,y\rangle \mid y \in P\},$$

where $\langle x,y\rangle$ denotes a fixed inner product on $\mathbb{R}^d$. Since P is compact one may replace sup by max.

Supporting hyperplane of P: A hyperplane

$$H(P,v) := \{x \in \mathbb{R}^d \mid \langle x,v\rangle = h(P,v)\},$$

for $v \in \mathbb{R}^d\backslash\{0\}$. Note that $H(P,\mu v) = H(P,v)$ for $\mu \in \mathbb{R}$, $\mu > 0$. For a vector u of the $(d-1)$-dimensional ***unit sphere*** S^{d-1}, $h(P,u)$ is the signed distance of the supporting plane $H(P,u)$ from the origin. For $v = 0$ we set $H(P,0) := \mathbb{R}^d$, which is not a hyperplane.

Face: The intersection of P with a supporting hyperplane $H(P,v)$. If P is full-dimensional, then this is a nontrivial face of P. We call it a ***k-face*** if the dimension of $\mathrm{aff}(P \cap H(P,v))$ is k. Each face is itself a polytope.

The set of all k-faces is denoted by $\mathcal{F}_k(P)$ and its cardinality by $f_k(P)$.

The empty set $\emptyset$ and the polytope P itself are also defined to be faces of P, called the ***trivial faces*** of P, of dimensions -1 and $\dim(P)$, respectively. Thus the ***nontrivial faces*** F of a d-polytope have dimensions $0 \le \dim(F) \le d-1$. All faces other than P are referred to as ***proper faces***.

The faces of dimension 0 and 1 are called ***vertices*** and ***edges***, respectively. The $(d-1)$-faces and $(d-2)$-faces of a d-polytope P are called ***facets*** and ***ridges***, respectively.

f-vector: The vector of face numbers $\boldsymbol{f}(P) = (f_0(P), f_1(P), \ldots, f_{d-1}(P))$ associated with a d-polytope.

Facet-vertex incidence matrix: The matrix $M \in \{0,1\}^{f_{d-1}(P)\times f_0(P)}$ that has an entry $M(F,v) = 1$ if the facet F contains the vertex v, and $M(F,v) = 0$ otherwise.

Graded poset: A partially ordered set $(P,\le)$ with a unique minimal element $\hat{0}$, a unique maximal element $\hat{1}$, and a ***rank function*** $r : P \longrightarrow \mathbb{N}_0$ that satisfies

(1) $r(\hat{0}) = 0$, and $p < p'$ implies $r(p) < r(p')$, and

(2) $p < p'$ and $r(p') - r(p) > 1$ implies that there is a $p'' \in P$ with $p < p'' < p'$.

Lattice L: A partially ordered set $(P,\le)$ in which every pair of elements $p, p' \in P$ has a unique maximal lower bound, called the ***meet*** $p \wedge p'$, and a unique minimal upper bound, called the ***join*** $p \vee p'$.

Atom, coatom: If L is a graded lattice, the minimal elements of $L\backslash\{\hat{0}\}$ (i.e., the elements of rank 1) are the atoms of L. Similarly, the maximal elements of $L\backslash\{\hat{1}\}$ (i.e., the elements of rank $r(\hat{1})-1$) are the coatoms of L. A graded lattice is ***atomic*** if every element is a join of a set of atoms, and it is ***coatomic*** if every element is a meet of a set of coatoms.

Face lattice $L(P)$**:** The set of all faces of P, partially ordered by inclusion.

Combinatorially equivalent: Polytopes whose face lattices are isomorphic as abstract (i.e., unlabeled) partially ordered sets/lattices.

Equivalently, P and P' are combinatorially equivalent if their facet-vertex incidence matrices differ only by column and row permutations.

Combinatorial type: An equivalence class of polytopes under combinatorial equivalence.

THEOREM 15.1.2 *Face Lattices of Polytopes* (cf. [Zie95, Sect. 2.2])

The face lattices of convex polytopes are finite, graded, atomic, and coatomic lattices. The meet operation $G \wedge H$ *is given by intersection, while the join* $G \vee H$ *is the intersection of all facets that contain both* G *and* H. *The rank function on* $L(P)$ *is given by* $r(G) = \dim(G) + 1$.

The minimal nonempty faces of a polytope are its vertices: They correspond to atoms of the lattice $L(P)$. Every face is the join of its vertices, hence $L(P)$ is atomic. Similarly, the maximal proper faces of a polytope are its facets: They correspond to the coatoms of $L(P)$. Every face is the intersection of the facets it is contained in, hence face lattices of polytopes are coatomic.

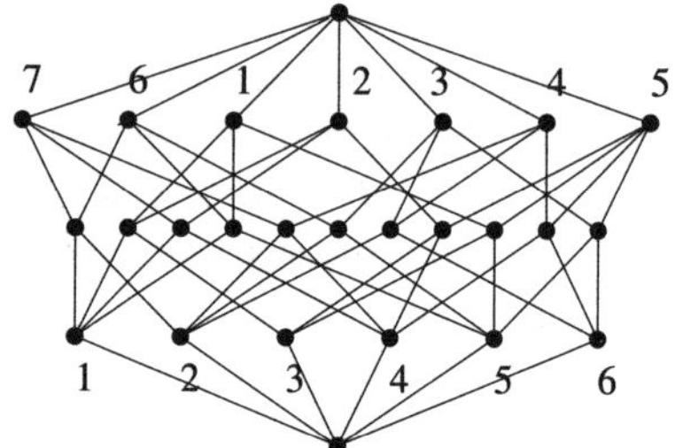

FIGURE 15.1.3
The face lattice of our unnamed 3-polytope. The seven coatoms (facets) and the six atoms (vertices) have been labeled in the order of their appearance in the lists on page 386. *Thus, the downwards-path from the coatom "4" to the atom "6" represents the fact that the fourth facet contains the sixth vertex.*

The face lattice is a complete encoding of the combinatorial structure of a polytope. However, in general the encoding by a facet-vertex incidence matrix is more efficient. The following matrix—also provided by `polymake`—represents our unnamed 3-polytope:

$$M = \begin{array}{c} \\ 1 \\ 2 \\ 3 \\ 4 \\ 5 \\ 6 \\ 7 \end{array} \begin{array}{c} \begin{array}{cccccc} 1 & 2 & 3 & 4 & 5 & 6 \end{array} \\ \begin{pmatrix} 1 & 0 & 1 & 0 & 1 & 0 \\ 1 & 0 & 1 & 1 & 0 & 0 \\ 0 & 1 & 0 & 0 & 1 & 1 \\ 0 & 1 & 0 & 1 & 0 & 1 \\ 0 & 0 & 1 & 1 & 1 & 1 \\ \mathbf{1} & \mathbf{1} & 0 & 0 & 1 & 0 \\ \mathbf{1} & \mathbf{1} & 0 & 1 & 0 & 0 \end{pmatrix} \end{array}$$

How do we decide whether a set of vertices $\{v^1, \ldots, v^k\}$ is (the vertex set of) a face of P? This is the case if and only if no other vertex v^0 is contained in all the facets that contain $\{v^1, \ldots, v^k\}$. This criterion makes it possible, for example, to derive the edges of a polytope P from a facet-vertex matrix. The four 1's printed in boldface in the above matrix M thus certify that the vertices $1, 2$ lie on a face in the unnamed polytope, which is contained in the intersection of the facets 6, 7. By dimension reasons, the face in question is an edge that connects the vertices 1, 2.

For low-dimensional polytopes, the criterion can be simplified: If $d \leq 4$, then two vertices are connected by an edge if and only if there are at least $d-1$ different facets that contain them both. However, the same is no longer true for 5-dimensional polytopes, where vertices may be nonadjacent despite being contained in many common facets. The best way to see this is by using polarity, which is discussed next.

15.1.2 BASIC CONSTRUCTIONS

GLOSSARY

Polarity: If $P \subseteq \mathbb{R}^d$ is a d-polytope with the origin in its interior, then the ***polar*** of P is the d-polytope

$$P^{\Delta} \;:=\; \{y \in \mathbb{R}^d \mid \langle y, x\rangle \leq 1 \text{ for all } x \in P\}.$$

Stacking onto a facet F: A polytope $\mathrm{conv}(P \cup x^F)$, where x^F is a point of the form $y^F - \varepsilon(y^P - y^F)$, where y^P is in the interior of P, y^F is in the relative interior of the facet F, and $\varepsilon > 0$ is small enough.

Note that this definition specifies the combinatorial type of the resulting polytope completely, but not the geometric realization; similarly some of the following constructions are not specified completely.

Vertex figure P/v: If v is a vertex of P, then $P/v := P \cap H$ is a polytope obtained by intersecting P with a hyperplane H that has v on one side and all the other vertices of P on the other side.

Cutting off a vertex: A polytope $P \cap H^-$ obtained by intersecting P with a closed halfspace H^- that does not contain the vertex v, but contains all other vertices of P in its interior.

Quotient of P: A k-polytope obtained from a d-polytope by repeatedly ($d-k$ times) taking a vertex figure.

Simplicial polytope: A polytope all of whose facets (equivalently, proper faces) are simplices. Examples: the d-cross-polytopes.

Simple polytope: A polytope all of whose vertex figures (equivalently, proper quotients) are simplices. Examples: the d-cubes.

Polarity is a fundamental construction in the theory of polytopes. One always has $P^{\Delta\Delta} = P$, under the assumption that P has the origin in its interior. This condition can always be obtained by a change of coordinates. In particular, we speak of (combinatorial) polarity between d-polytopes Q and R that are combinatorially equivalent to P and P^{Δ}, respectively.

Any $\mathcal{V}$-presentation of P yields an $\mathcal{H}$-presentation of P^{Δ}, and vice versa, via

$$P = \mathrm{conv}\{v^1, \ldots, v^n\} \quad\Longleftrightarrow\quad P^{\Delta} = \{x \in \mathbb{R}^d \mid \langle v^i, x\rangle \leq 1 \text{ for } 1 \leq i \leq n\}.$$

There are basic relations between polytopes and polytopal constructions under polarity. For example, the fact that the d-cross-polytopes C_d^{Δ} are the polars of the

d-cubes C_d is built into our notation. More generally, the polars of simple polytopes are simplicial, and vice versa. This can be deduced from the fact that the facets F of a polytope P correspond to the vertex figures P^Δ/v of its polar P^Δ. In fact, F and P^Δ/v are combinatorially polar in this situation. More generally, one has a correspondence between faces and quotients under polarity.

At a combinatorial level, all this can be derived from the fact that the face lattices $L(P)$ and $L(P^\Delta)$ are anti-isomorphic: $L(P^\Delta)$ may be obtained from $L(P)$ by reversing the order relations. Thus, lower intervals in $L(P)$, corresponding to faces of P, translate under polarity into upper intervals of $L(P^\Delta)$, corresponding to quotients of P^Δ.

15.1.3 BASIC CONSTRUCTIONS, II

GLOSSARY

For the following constructions, let
$P \subseteq \mathbb{R}^d$ be a d-dimensional polytope with n vertices and m facets, and
$P' \subseteq \mathbb{R}^{d'}$ a d'-dimensional polytope with n' vertices and m' facets.

Scalar multiple: For $\lambda \in \mathbb{R}$, the polytope λP defined by $\lambda P := \{\lambda x \mid x \in P\}$. Here P and λP are combinatorially (in fact, affinely) equivalent for all $\lambda \neq 0$. In particular, $(-1)P = -P = \{-p \mid p \in P\}$, and $(+1)P = P$.

Minkowski sum: The polytope $P + P' := \{p + p' \mid p \in P,\, p' \in P'\}$.
It is also useful to define the difference as $P - P' = P + (-P')$. The polytopes $P + \lambda P'$ are combinatorially equivalent for all $\lambda > 0$, and similarly for $\lambda < 0$.
If $P' = \{p'\}$ is one single point, then $P - \{p'\}$ is the image of P under the translation that takes p' to the origin.

Product: The $(d+d')$-dimensional polytope $P \times P' := \{(p, p') \in \mathbb{R}^{d+d'} \mid p \in P,\, p' \in P'\}$. $P \times P'$ has $n \cdot n'$ vertices and $m + m'$ facets.

Join: The $(d+d'+1)$-polytope obtained as the convex hull $P * P'$ of $P \cup P'$, after embedding P and P' in a space where their affine hulls are skew. For example, $P * P' := \mathrm{conv}(\{(p, 0, 0) \in \mathbb{R}^{d+d'+1} \mid p \in P\} \cup \{(0, p', 1) \in \mathbb{R}^{d+d'+1} \mid p' \in P'\})$.
$P*P'$ has dimension $d+d'+1$ and $n+n'$ vertices. Its k-faces are the joins of i-faces of P and $(k-i-1)$-faces of P', hence $f_k(P * P') = \sum_{i=-1}^{k} f_i(P) f_{k-i-1}(P')$.

Subdirect sum: The $(d+d')$-dimensional polytope
$P \oplus P' := \mathrm{conv}(\{(p, 0) \in \mathbb{R}^{d+d'} \mid p \in P\} \cup \{(0, p') \in \mathbb{R}^{d+d'} \mid p' \in P'\})$.
Thus the subdirect sum $P \oplus P'$ is a projection of the join $P * P'$. See McMullen [McM76].

Direct sum: If both P and P' have the origin in their interiors—this is the "usual" situation for creating subdirect sums—then $P \oplus P'$ is the direct sum of P and P'. It is $(d + d')$-dimensional, and has $n + n'$ vertices and $m \cdot m'$ facets.

Prism: The product $\mathrm{prism}(P) := P \times I$, where I denotes the real interval $I = [-1, +1] \subseteq \mathbb{R}$. It has dimension $d + 1$, $2n$ vertices and $m + 2$ facets.

Pyramid: The join $\mathrm{pyr}(P) := P * \{0\}$ of P with a point (a 0-dimensional polytope $P' = \{0\} \subseteq \mathbb{R}^0$). It has dimension $d + 1$, $n + 1$ vertices and $m + 1$ facets.

Bipyramid: The subdirect sum $\mathrm{bipyr}(P) := P \oplus I$, where P must have the origin in its interior. It has dimension $d+1$, $n+2$ vertices and $2m$ facets.

One-point suspension, obtained by ***splitting the vertex v:*** The subdirect sum

$$\mathrm{ops}(P, v) := \mathrm{conv}(P \times \{0\} \cup \{v\} \times [-1, 1]),$$

where v is a vertex of P. It has dimension $d+1$, $n+1$ vertices and $2m - m_v$ facets, if v lies in m_v facets of P.

Lawrence extension: If $p \in \mathbb{R}^d$ is a point outside the polytope P, then the subdirect sum $(P - \{p\}) \oplus [1, 2]$ is a *Lawrence extension of P at p.* For $p \in P$ this is just a pyramid.

Wedge over a facet F of P:

$$\mathrm{wedge}(P, F) := P \times \mathbb{R} \cap \{a^T x + |x_{d+1}| \leq b\},$$

where F is a facet of P defined by $a^T x \leq b$. It has dimension $d+1$, $m+1$ facets, and $2n - n_F$ vertices, if F has n_F vertices. More generally, the wedge construction can be performed (defined by the same formula) for a face F. If F is not a facet, then the wedge will have $m+2$ facets.

In contrast to the other constructions in this section, the combinatorial type of the Minkowski sum $P + P'$ is *not* determined by the combinatorial types of its constituents P and P', and the combinatorial type of a Lawrence extension depends on the position of the extension point p with respect to P (see below).

Of course, the many constructions listed in the glossary above are not independent of each other. For instance, some of these constructions are related by polarity: for polytopes P and P' with the origin in their interiors, the product and the direct sum are related by polarity,

$$P \times P' \;=\; (P^{\Delta} \oplus P'^{\Delta})^{\Delta},$$

and this specializes to polarity relations among the pyramid, bipyramid, and prism constructions,

$$\mathrm{pyr}(P) \;=\; (\mathrm{pyr}(P^{\Delta}))^{\Delta} \qquad \text{and} \qquad \mathrm{prism}(P) \;=\; (\mathrm{bipyr}(P^{\Delta}))^{\Delta}.$$

Similarly, "cutting off a vertex" is polar to "stacking onto a facet." The wedge construction is a *subdirect product* in the sense of McMullen [McM76] and the polar dual construction to a subdirect sum.

It is interesting to study—and this has not been done systematically—how the basic polytope operations generate complicated convex polytopes from simpler ones. For example, starting from a one-dimensional polytope $I = C_1 = [-1, +1] \subset \mathbb{R}$, the direct product construction generates the cubes C_d, while direct sums generate the cross-polytopes C_d^{Δ}.

Even more complicated centrally symmetric polytopes, the *Hanner polytopes*, are obtained from copies of the interval I by using products and direct sums. They are interesting since they achieve with equality the conjectured bound that all centrally symmetric d-polytopes have at least 3^d nonempty faces (see Kalai [Kal89] and Sanyal, Werner and Ziegler [SWZ09]).

Every polytope can be viewed as a region of a hyperplane arrangement: For this, take as $\mathcal{A}_P$ the set of all hyperplanes of the form $\mathrm{aff}(F)$, where F is a facet of P.

For additional points, such as the points outside the polytope used for Lawrence extensions, or those used for stackings, it is often enough to know in which region, or in which lower-dimensional region, of the arrangement $\mathcal{A}_P$ they lie.

The combinatorial type of a Lawrence extension depends on the position of p in the arrangement $\mathcal{A}_P$. Thus the Lawrence extensions obtainable from P depend on the realization of P, not only on its combinatorial type.

The Lawrence extension may seem like quite a simple little construction. However, it has the amazing property that it can encode crucial information about the position of a point *outside* a d-polytope into the boundary structure of a $(d+1)$-polytope, and thus is an essential ingredient in some remarkable constructions, such as universality results (see e.g., Ziegler [Zie08], Richter-Gebert [Ric96], and Adiprasito, Padrol and Theran [APT15]), and high-dimensional projectively unique polytopes (Adiprasito and Ziegler [AZ15], Adiprasito and Padrol [AP16]).

15.1.4 MORE EXAMPLES

There are many interesting classes of polytopes arising from diverse areas of mathematics (as well as physics, optimization, crystallography, etc.). Some of these are discussed below. More classes of examples appear in other chapters of this Handbook. For example, regular and semiregular polytopes are discussed in Chapter 18, while polytopes that arise as Voronoi cells of lattices appear in Chapters 3, 7, and 64.

GLOSSARY

Graph of a polytope: The graph $G(P) = (V(P), E(P))$ with vertex set $V(P) = \mathcal{F}_0(P)$ and edge set $E(P) = \{\{v^1, v^2\} \subseteq \binom{V}{2} \mid \text{conv}\{v^1, v^2\} \in \mathcal{F}_1(P)\}$.

Zonotope: Any d-polytope Z that can be represented as the image of an n-dimensional cube C_n $(n \geq d)$ under an affine map; equivalently, any polytope that can be written as a Minkowski sum of n line segments (1-dimensional polytopes). The smallest n such that Z is an image of C_n is the ***number of zones*** of Z.

Moment curve: The curve γ in $\mathbb{R}^d$ defined by $\gamma : \mathbb{R} \longrightarrow \mathbb{R}^d$, $t \longmapsto (t, t^2, \ldots, t^d)^T$.

Cyclic polytope: The convex hull of a finite set of points on a moment curve, or any polytope combinatorially equivalent to it.

k-neighborly polytope: A polytope such that each subset of at most k vertices forms the vertex set of a face. Thus every polytope is 1-neighborly, and a polytope is 2-neighborly if and only if its graph is complete.

Neighborly polytope: A d-dimensional polytope that is $\lfloor d/2 \rfloor$-neighborly.

(0,1)-polytope: A polytope all of whose vertex coordinates are 0 or 1, that is, whose vertex set is a subset of the vertex set $\{0,1\}^d$ of the unit cube.

ZONOTOPES

Zonotopes appear in quite different guises. They can equivalently be defined as the Minkowski sums of finite sets of line segments (1-dimensional polytopes), as the affine projections of d-cubes, or as polytopes all of whose faces (equivalently, all 2-faces) exhibit central symmetry. Thus a 2-dimensional polytope is a zonotope if and only if it is centrally symmetric. By a classical result of McMullen, a d-dimensional polytope is a zonotope if and only if its 2-faces are centrally symmetric.

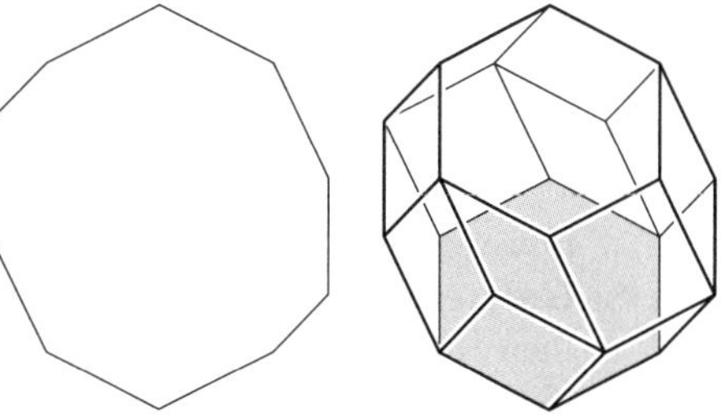

FIGURE 15.1.4
A 2-dimensional and a 3-dimensional zonotope, each with 5 zones, and thus obtainable as affine projections of a 5-dimensional cube. The 2-dimensional one is a projection of the 3-dimensional one. Every projection of a zonotope is a zonotope.

Among the most prominent zonotopes are the permutohedra: The ***permutohedron*** Π_{d-1} is constructed by taking the convex hull of all d-vectors whose coordinates are $\{1, 2, \ldots, d\}$, in any order. The permutohedron Π_{d-1} is a $(d-1)$-dimensional polytope (contained in the hyperplane $\{x \in \mathbb{R}^d \mid \sum_{i=1}^{d} x_i = d(d+1)/2\}$) with $d!$ vertices and $2^d - 2$ facets.

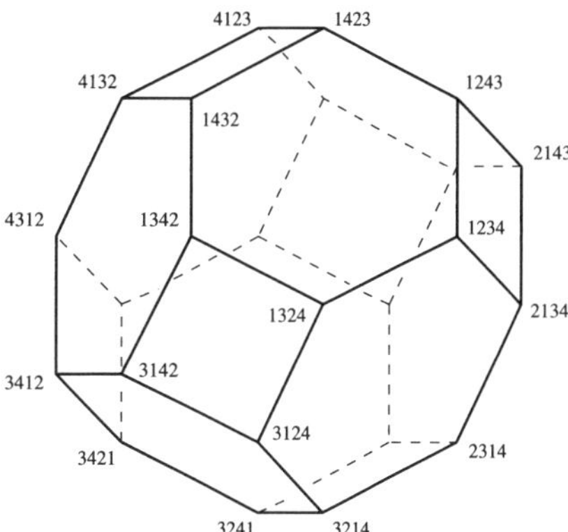

FIGURE 15.1.5
The 3-dimensional permutohedron Π_3. The vertices are labeled by the permutations that, when applied to the coordinate vector in $\mathbb{R}^4$, yield $(1, 2, 3, 4)^T$.
Note that the coordinates of each vertex are given by reading the inverses *(!) of the permutation that is used to label the vertex.*

One unusual feature of permutohedra is that they are simple zonotopes: These are rare in general, and the (unsolved) problem of classifying them is equivalent to the problem of classifying all simplicial arrangements of hyperplanes (see Section 6.3.3).

Zonotopes are important because their theory is equivalent to the theories of vector configurations (realizable oriented matroids) and of hyperplane arrangements. In fact, the system of line segments that generates a zonotope can be considered as a vector configuration, and the hyperplanes that are orthogonal to the line segments provide the associated hyperplane arrangement. We refer to [BLS+93, Section 2.2] and [Zie95, Lecture 7].

Finally, we mention in passing a surprising bijective correspondence between the tilings of a zonotope with smaller zonotopes and oriented matroid liftings (realizable or not) of the oriented matroid of a zonotope. This correspondence is known as the *Bohne–Dress theorem*; we refer to Richter-Gebert and Ziegler [RZ94].

CYCLIC POLYTOPES

Cyclic polytopes can be constructed by taking the convex hull of $n > d$ points on the moment curve in $\mathbb{R}^d$. The "standard construction" is to define a cyclic polytope $C_d(n)$ as the convex hull of n integer points on this curve, such as

$$C_d(n) \;:=\; \mathrm{conv}\{\gamma(1), \gamma(2), \ldots, \gamma(n)\}.$$

However, the combinatorial type of $C_d(n)$ is given by the—entirely combinatorial—***Gale evenness criterion***: If $C_d(n) = \mathrm{conv}\{\gamma(t_1), \ldots, \gamma(t_n)\}$, with $t_1 < \cdots < t_n$, then $\gamma(t_{i_1}), \ldots, \gamma(t_{i_d})$ determine a facet if and only if the number of indices in $\{i_1, ..., i_d\}$ lying between any two indices *not* in that set is even. Thus, the combinatorial type does not depend on the specific choice of points on the moment curve [Zie95, Example 0.6; Theorem 0.7].

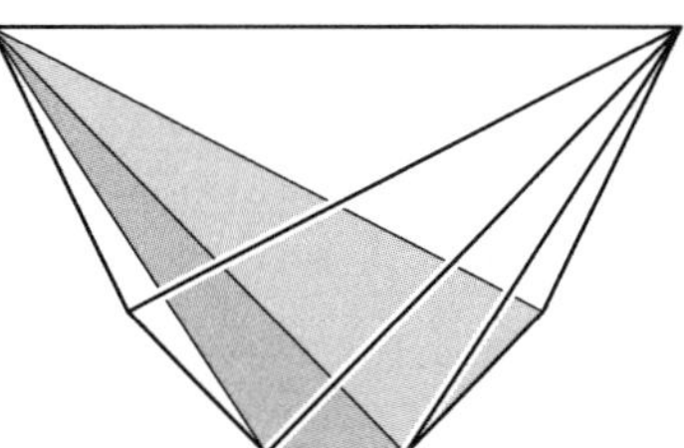

FIGURE 15.1.6
A 3-dimensional cyclic polytope $C_3(6)$ with 6 vertices. (In a projection of γ to the x_1x_2-plane, the curve γ and hence the vertices of $C_3(6)$ lie on the parabola $x_2 = x_1^2$.)

The first property of cyclic polytopes to notice is that they are simplicial. The second, more surprising, property is that they are neighborly. This implies that among all d-polytopes P with n vertices, the cyclic polytopes maximize the number $f_i(P)$ of i-dimensional faces for $i < \lfloor d/2 \rfloor$. The same fact holds for all i: This is part of McMullen's upper bound theorem (see below). In particular, cyclic polytopes have a very large number of facets,

$$f_{d-1}\big(C_d(n)\big) \;=\; \binom{n - \lceil \frac{d}{2} \rceil}{\lfloor \frac{d}{2} \rfloor} \;+\; \binom{n - 1 - \lceil \frac{d-1}{2} \rceil}{\lfloor \frac{d-1}{2} \rfloor}.$$

For example, any cyclic 4-polytope $C_4(n)$ has $n(n-3)/2$ facets. Thus $C_4(8)$ has 8 vertices, any two of them adjacent, and 20 facets. This is more than the 16 facets of the 4-cross-polytope, which also has 8 vertices!

NEIGHBORLY POLYTOPES

Here are a few observations about neighborly polytopes. For more information, see [BLS⁺93, Section 9.4] and the references quoted there.

The first observation is that if a polytope is k-neighborly for some $k > \lfloor d/2 \rfloor$, then it is a simplex. Thus, if one ignores the simplices, then $\lfloor d/2 \rfloor$-neighborly polytopes form the extreme case, which motivates calling them simply "neighborly." However, only in even dimensions $d = 2m$ do the neighborly polytopes have very special structure. For example, one can show that even-dimensional neighborly polytopes are necessarily simplicial, but this is not true in general. For the latter, note that, for example, all 3-dimensional polytopes are neighborly by definition, and that if P is a neighborly polytope of dimension $d = 2m$, then pyr(P) is neighborly of dimension $2m+1$.

All simplicial neighborly d-polytopes with n vertices have the same number of facets (in fact, the same f-vector $(f_0, f_1, \ldots, f_{d-1})$) as $C_d(n)$. They constitute the class of polytopes with the maximal number of i-faces for all i: This is the statement of McMullen's upper bound theorem. We refer to Chapter 17 for a thorough discussion of f-vector theory.

Every even-dimensional neighborly polytope with $n \leq d+3$ vertices is combinatorially equivalent to a cyclic polytope. This covers, for instance, the polar of the product of two triangles, $(\Delta_2 \times \Delta_2)^\Delta$, which is easily seen to be a 4-dimensional neighborly polytope with 6 vertices; see Figure 15.1.9. The first example of an even-dimensional neighborly polytope that is not cyclic appears for $d = 4$ and $n = 8$. It can easily be described in terms of its affine Gale diagram; see below.

Neighborly polytopes may at first glance seem to be very peculiar and rare objects, but there are several indications that they are not quite as unusual as they seem. In fact, the class of neighborly polytopes is believed to be very rich. Thus, Shemer [She82] has shown that for fixed even d the number of nonisomorphic neighborly d-polytopes with n vertices grows superexponentially with n (see also [Pad13]). Also, many of the (0,1)-polytopes studied in combinatorial optimization turn out to be at least 2-neighborly. Both these effects illustrate that "neighborliness" is not an isolated phenomenon.

OPEN PROBLEMS

1. Can every neighborly d-polytope $P \subseteq \mathbb{R}^d$ with n vertices be extended by a new vertex $v \in \mathbb{R}^d$ to a neighborly polytope $P' := \text{conv}(P \cup \{v\})$ with $n+1$ vertices? This has been asked by Shemer [She82, p. 314]. It has been verified only recently for small parameters d and n by Miyata and Padrol [MP15].
2. It is a classic problem of Perles whether every simplicial polytope is a quotient of a neighborly polytope. For polytopes with at most $d+4$ vertices this was confirmed by Kortenkamp [Kor97]. Adiprasito and Padrol [AP16] disproved a related conjecture, that every polytope is a subpolytope of a stacked polytope.
3. Some computer experiments with random polytopes suggest that
   - one obtains a neighborly polytope with high probability (which increases rapidly with the dimension of the space),
   - the most probable combinatorial type is a cyclic polytope,
   - but still this probability of a cyclic polytope tends to zero.

   However, none of this has been proved. See Bokowski and Sturmfels [BS89, p. 101], Bokowski, Richter-Gebert, and Schindler [BRS92], Vershik and Sporychev [VS92], and Donoho and Tanner [DT09].

(0,1)-POLYTOPES

There is a $(0,1)$-polytope (given in terms of a $\mathcal{V}$-presentation) associated with every finite set system $\mathcal{S} \subseteq 2^E$ (where E is a finite set, and 2^E denotes the collection of all of its subsets), via

$$P[\mathcal{S}] := \operatorname{conv}\{\sum_{i \in F} e^i \mid F \in \mathcal{S}\} \subseteq \mathbb{R}^E.$$

The combinatorial optimization contains a multitude of extensive studies on (partial) $\mathcal{H}$-descriptions of special $(0,1)$-polytopes, such as for example

- the ***traveling salesman polytopes*** T^n, where E is the edge set of a complete graph K_n, and $\mathcal{F}$ is the set of all $(n-1)!$ Hamilton cycles (simple circuits through all the vertices) in E (see Grötschel and Padberg [GP85]);
- the ***cut*** and ***equicut polytopes***, where E is the edge set of—for example—a complete graph, and $\mathcal{S}$ represents the family of all cuts, or all equicuts (given by a partition of the vertex set into two blocks of equal size) of the graph (see Deza and Laurent [DL97]).

Besides their importance for combinatorial optimization, there is a great deal of interesting polytope theory associated with such polytopes. It turns out that some of these polytopes are so complicated that a complete $\mathcal{H}$-description or any other "full understanding" will remain out of reach. For example, Billera and Sarangarajan [BS96] showed that *every* $(0,1)$-polytope appears as a face of a traveling salesman polytope. Cut polytopes seem to have particularly many facets, though that has not been proven. Equicut polytopes were used by Kahn and Kalai [KK93] in their striking disproof of Borsuk's conjecture (see also [AZ98]).

Despite the detailed structure theory for the "special" $(0,1)$-polytopes of combinatorial optimization, there is very little known about "general" $(0,1)$-polytopes. For example, what is the "typical," or the maximal, number of facets of a $(0,1)$-polytope? Based on a random construction Bárány and Pór [BP01] proved the existence of d-dimensional $(0,1)$-polytopes with $(c\,d/\log d)^{d/4}$ facets, where c is a universal constant. This lower bound has been improved to $(c\,d/\log d)^{d/2}$ by Gatzouras et al. [GGM05]. The best known upper bounds are of order $(d-2)!$. Another question, which is not only intrinsically interesting but might also provide new clues for basic questions of linear and combinatorial optimization, is: What is the maximal number of faces in a 2-dimensional projection of a $(0,1)$-polytope? For a survey on $(0,1)$-polytopes see [Zie00].

15.1.5 THREE-DIMENSIONAL POLYTOPES AND PLANAR GRAPHS

GLOSSARY

d-connected graph: A connected graph that remains connected if any $d-1$ vertices are deleted.

Drawing of a graph: A representation in the plane where the vertices are represented by distinct points, and simple Jordan arcs (typically: polygonal, or at least piecewise smooth) are drawn between the pairs of adjacent vertices.

Planar graph: A graph that can be drawn in the plane with Jordan arcs that are disjoint except for their endpoints.

Realization space: The set of all coordinatizations of a combinatorial structure, modulo affine coordinate transformations; see Section 6.3.2.

Isotopy property: A combinatorial structure (such as a combinatorial type of polytope) has the isotopy property if any two realizations with the same orientation can be deformed into each other by a continuous deformation that maintains the combinatorial type. Equivalently, the isotopy property holds for a combinatorial structure if and only if its realization space is connected.

THEOREM 15.1.3 *Steinitz's Theorem* [Ste22, Satz 43, S. 77] [SR34]

For every 3-dimensional polytope P, the graph $G(P)$ is a planar, 3-connected graph. Conversely, for every planar 3-connected graph with at least 4 vertices, there is a unique combinatorial type of 3-polytope P with $G(P) \cong G$.

Furthermore, the realization space $\mathcal{R}(P)$ of a combinatorial type of 3-polytope is homeomorphic to $\mathbb{R}^{f_1(P)-6}$, and contains rational points. In particular, 3-polytopes have the isotopy property, and they can be realized with integer vertex coordinates.

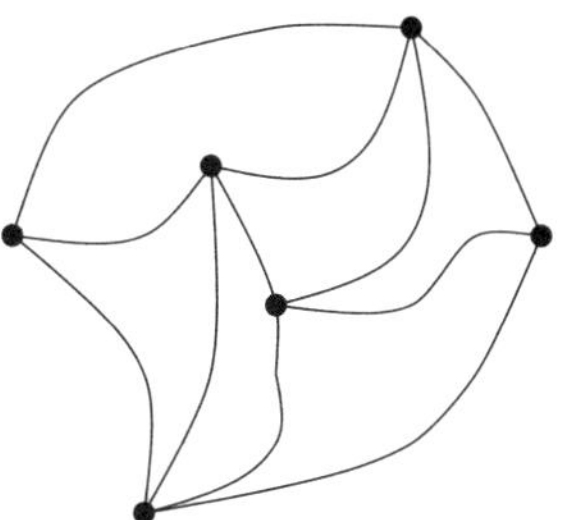

FIGURE 15.1.7
A (planar drawing of a) 3-connected, planar, unnamed graph. The formidable task of any proof of Steinitz's theorem is to construct a 3-polytope with this graph.

There are three essentially different strategies known that yield proofs of Steinitz's theorem. The first approach, due to Steinitz, provides a construction sequence for any type of 3-polytope, starting from a tetrahedron, and using only local operations such as cutting off vertices and polarity. See [Zie95, Lecture 4]. The second "Tutte type" of proof realizes any combinatorial type by a global minimization argument, which as an intermediate step provides a special planar representation of the graph by a framework with a positive self-stress; see Richter-Gebert [Ric96]. The third approach, spear-headed by Thurston, derives Steinitz's Theorem from the Koebe–Andreev–Thurston circle packing theorem. It yields that every 3-polytope has an (essentially unique!) realization with its edges tangent to the unit sphere. This proof can be derived from a variational principle obtained by Bobenko and Springborn [BS04]. For an exposition see Ziegler [Zie07, Lecture 1].

OPEN PROBLEMS

Because of Steinitz's theorem and its extensions and corollaries, the theory of 3-dimensional polytopes is quite complete and satisfactory. Nevertheless, some basic open problems remain.

1. It can be shown that every combinatorial type of 3-polytope with n vertices and a triangular facet can be realized with integer coordinates belonging to $\{1, 2, \dots, 29^n\}^3$ (Ribó Mor, Rote and Schulz [RRS11], improving upon previous bounds by Onn and Sturmfels [OS94] and Richter-Gebert [Ric96, Sect. 13.2]), but it is not clear whether this can be replaced by a polynomial upper bound. No nontrivial lower bounds seem to be available.

2. If P has a nontrivial group G of symmetries, then it also has a symmetric realization (Mani [Man71]). However, it is not clear whether for all 3-polytopes the space of all G-symmetric realizations $\mathcal{R}^{G}(P)$ is still homeomorphic to some $\mathbb{R}^k$. (It does not contain rational points in general, e.g., for the regular icosahedron!)

15.1.6 FOUR-DIMENSIONAL POLYTOPES AND SCHLEGEL DIAGRAMS

GLOSSARY

Subdivision of a polytope P: A collection of polytopes $P_1, \ldots, P_\ell \subseteq \mathbb{R}^d$ such that $P = P_1 \cup \cdots \cup P_\ell$, and for $i \neq j$ we have that $P_i \cap P_j$ is a proper face of P_i and P_j (possibly empty). In this case we write $P = \uplus P_i$.

Triangulation of a polytope: A subdivision into simplices. (See Chapter 16.)

Schlegel diagram: A $(d-1)$-dimensional representation $\mathcal{D}(P,F)$ of a d-dimensional polytope P given by a subdivision of a facet F, obtained as follows: Take a point of view outside of P but very close to a relative interior point of the facet F, and then let $\mathcal{D}(P,F)$ be the decomposition of F given by all the other facets of P, as seen from this point of view in the "window" F.

$(d-1)$-diagram: A subdivision $\mathcal{D}$ of a $(d-1)$-polytope F such that the intersection of any polytope in $\mathcal{D}$ with the boundary of F is a face of F (which may be empty).

Basic primary semialgebraic set defined over $\mathbb{Z}$: The solution set $S \subseteq \mathbb{R}^k$ of a finite set of equations and strict inequalities of the form $f_i(x) = 0$ or $g_j(x) > 0$, where the f_i and g_j are polynomials in k variables with integer coefficients.

Stable equivalence: Equivalence relation between semialgebraic sets generated by rational changes of coordinates and certain types of "stable" projections with contractible fibers. See Richter-Gebert [Ric96, Section 2.5].
In particular, if two sets are stably equivalent, then they have the same homotopy type, and they have the same arithmetic properties with respect to subfields of $\mathbb{R}$; e.g., either both or neither of them contain a rational point.

The situation for 4-polytopes is fundamentally different from that for 3-dimensional polytopes. One reason is that there is no similar reduction of 4-polytope theory to a combinatorial (graph) problem.

The main results about graphs of d-polytopes are that they are d-connected (Balinski [Bal61]), and that each contains a subdivision of the complete graph on $d+1$ vertices, $K_{d+1} = G(T_d)$ (Grünbaum [Grü67, p. 200]). In particular, all graphs of 4-polytopes are 4-connected, and none of them is planar; see also Chapter 19.

Schlegel diagrams provide a reasonably efficient tool for the visualization of 4-polytopes: We have a fighting chance to understand some important properties in terms of the 3-dimensional (!) geometry of Schlegel diagrams.

A $(d-1)$-diagram is a polytopal complex that "looks like" a Schlegel diagram, although there are diagrams (even 2-diagrams) that are not Schlegel diagrams.

The situation is somewhat nicer for *simple* polytopes. The combinatorial structure of a simple polytope is entirely determined by the abstract graph: This is due

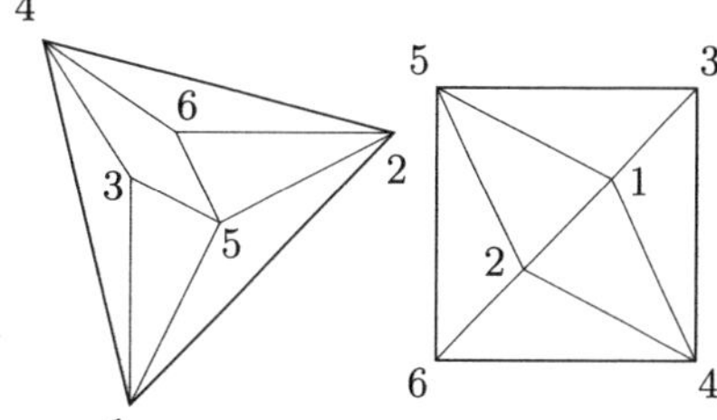

FIGURE 15.1.8
Two Schlegel diagrams of our unnamed 3-polytope, the first based on a triangle facet, the second on the "bottom square."

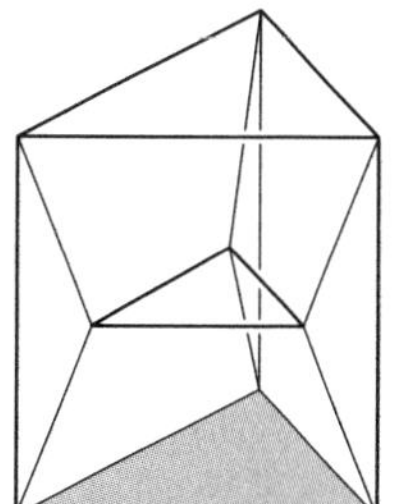

FIGURE 15.1.9
A Schlegel diagram of the product of two triangles. (This is a 4-dimensional polytope with 6 triangular prisms as facets, any two of them adjacent!)

to Blind and Mani-Levitska [BM87], with a simple proof by Kalai [Kal88a] and an efficient (polynomial-time) reconstruction algorithm by Friedman [Fri09]. Moreover, the geometry of higher-dimensional simple d-polytopes can be understood in terms of $(d-1)$-diagrams: For $d \geq 4$ all simple $(d-1)$-diagrams "are Schlegel," that is, they represent genuine d-dimensional polytopes (Whiteley, see Rybnikov [Ryb99]).

The fundamental difference between the theories for polytopes in dimensions 3 and 4 is most apparent in the contrast between Steinitz's theorem and the following result, which states simply that all the "nice" properties of 3-polytopes established in Steinitz's theorem fail dramatically for 4-dimensional polytopes. Indeed, Richter-Gebert showed that the realization spaces of 4-polytopes exhibit the same type of "universality" that was established by Mnëv for the realization spaces of planar point configurations/line arrangements, as well as for d-polytopes with $d+4$ vertices, as discussed in Chapter 6 (see Thm. 6.3.3):

THEOREM 15.1.4 *Universality for 4-Polytopes* [Ric96]

The realization space of a 4-dimensional polytope can be "arbitrarily wild": For every basic primary semialgebraic set S defined over $\mathbb{Z}$ there is a 4-dimensional polytope $P[S]$ whose realization space $\mathcal{R}(P[S])$ is stably equivalent to S.

In particular, this implies the following.

- *The isotopy property fails for 4-dimensional polytopes.*
- *There are nonrational 4-polytopes: combinatorial types that cannot be realized with rational vertex coordinates.*
- *The coordinates needed to represent all combinatorial types of rational 4-polytopes with integer vertices grow doubly exponentially with $f_0(P)$.*

The complete proof of this universality theorem is given in [Ric96]. One key component of the proof corresponds to another failure of a 3-dimensional phenomenon in dimension 4: For any facet (2-face) F of a 3-dimensional polytope P, the shape of F can be arbitrarily prescribed; in other words, the canonical map of realization spaces $\mathcal{R}(P) \longrightarrow \mathcal{R}(F)$ is always surjective. Richter-Gebert shows that a similar statement fails in dimension 4, even if F is a 2-dimensional pentagonal face: See Figure 15.1.10 for the case of a hexagon.

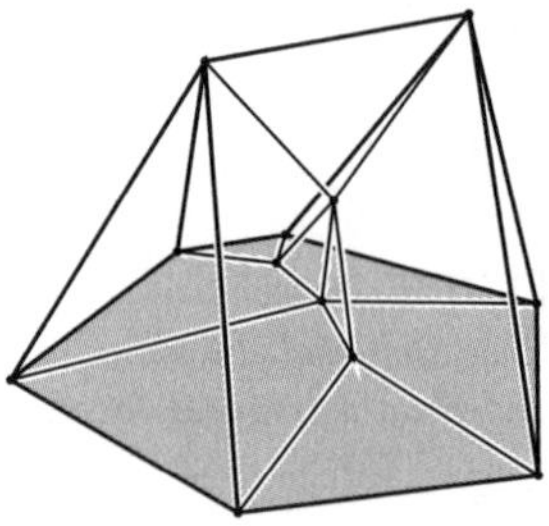

FIGURE 15.1.10
Schlegel diagram of a 4-dimensional polytope with 8 facets and 12 vertices, for which the shape of the base hexagon cannot be prescribed arbitrarily.

A problem that is left open is the structure of the realization spaces of simplicial 4-polytopes. All that is available now is a universality theorem for simplicial polytopes without a dimension bound, which had been claimed by Mnëv and by others since the 1980s, but for which a proof was provided only recently by Adiprasito and Padrol [AP17], and a single example of a simplicial 4-polytope that violates the isotopy property, by Bokowski et al. [BEK84] (see [Bok06, p. 142] [Fir15b, Sect. 1.3.4] for correct coordinates, as a rational inscribed polytope).

15.1.7 POLYTOPES WITH FEW VERTICES AND GALE DIAGRAMS

GLOSSARY

Polytope with few vertices: A polytope that has only a few more vertices than its dimension; usually a d-polytope with at most $d+4$ vertices.

(Affine) Gale diagram: A configuration of n not necessarily pairwise distinct, signed/bicolored ("positive"/"black" and "negative"/"white") points in affine space $\mathbb{R}^{n-d-2}$, which encodes a d-polytope with n vertices uniquely up to projective transformations.

The computation of a Gale diagram involves only simple linear algebra. For this, let $V \in \mathbb{R}^{d\times n}$ be a matrix whose columns consist of coordinates for the vertices of a d-polytope. For simplicity, we assume that P is not a pyramid, and that the vertices $\{v^1,\ldots,v^{d+1}\}$ affinely span $\mathbb{R}^d$. Let $\widetilde{V} \in \mathbb{R}^{(d+1)\times n}$ be obtained from V by adding an extra (terminal) row of ones. The vector configuration given by the columns of $\widetilde{V}$ represents the *oriented matroid* of P; see Chapter 6.

Now perform row operations on the matrix $\widetilde{V}$ to get it into the form $\widetilde{V} \sim (I_{d+1}|A)$, where I_{d+1} denotes a unit matrix, and $A \in \mathbb{R}^{(d+1)\times(n-d-1)}$. (The row operations do not change the oriented matroid.) The columns of the matrix $\widetilde{V}^* := (-A^T|I_{n-d-1}) \in \mathbb{R}^{(n-d-1)\times n}$ then represent the dual oriented matroid. We find a vector $a \in \mathbb{R}^{n-d-1}$ that has nonzero scalar product with all the columns of $\widetilde{V}^*$, divide each column w^* of $\widetilde{V}^*$ by the value $\langle a, w^*\rangle$, and delete from the resulting matrix any row that affinely depends on the others, thus obtaining a matrix $W \in \mathbb{R}^{(n-d-2)\times n}$. The columns of W give a bicolored point configuration in $\mathbb{R}^{n-d-2}$, where *black* points are used for the columns where $\langle a, w^*\rangle > 0$, and *white* points for the others. This bicolored point configuration represents an affine Gale diagram of P.

An affine configuration of bicolored points (consisting of n points that affinely span $\mathbb{R}^e$) represents a polytope (with n vertices, of dimension $n-e-2$) if and only

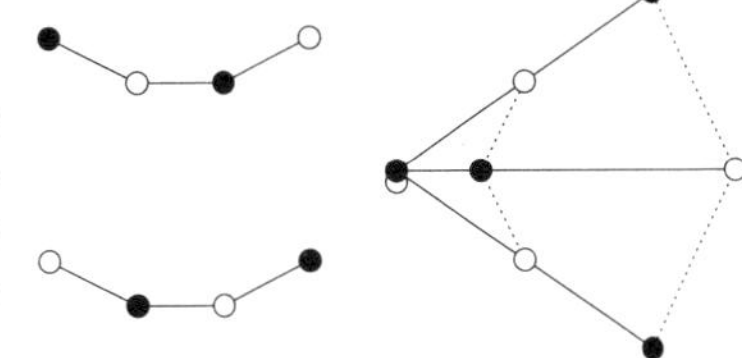

FIGURE 15.1.11
Two affine Gale diagrams of 4-dimensional polytopes: for a noncyclic neighborly polytope with 8 vertices, and for the polar (with 8 vertices) of the polytope with 8 facets from Figure 15.1.10, for which the shape of a hexagonal face cannot be prescribed arbitrarily.

if for any hyperplane spanned by some of the points, and for each side of it, the number of black points on this side, plus the number of white points on the other side, is at least 2.

The final information one needs is how to read off properties of a polytope from its affine Gale diagram: A set of points represents a face if and only if the bicolored points *not* in the set support an affine dependency, with positive coefficients on the black points, and with negative coefficients on the white points. Equivalently, the convex hull of all the black points not in our set, and the convex hull of all the white points not in the set, intersect in their relative interiors.

Affine Gale diagrams have been *very* successfully used to study and classify polytopes with few vertices.

d+1 vertices: The only d-polytopes with d+1 vertices are the d-simplices.

d+2 vertices: This corresponds to the situation of 0-dimensional affine Gale diagrams. There are exactly $\lfloor d^2/4 \rfloor$ combinatorial types of d-polytopes with d+2 vertices: They are of the form $\Delta_{a-1} * (\Delta_b \oplus \Delta_c)$ with $d = a + b + c$, $a \geq 0$, $b \geq c \geq 0$, that is, multiple (a-fold) pyramids over simplicial polytopes that are direct sums of simplices $\Delta_b \oplus \Delta_c$. Among these, the $\lfloor d/2 \rfloor$ types with $a = 0$ are the simplicial ones.

d+3 vertices: All d-polytopes with d+3 vertices are realizable with (small) integral coordinates and satisfy the isotopy property: All this can be easily analyzed in terms of 1-dimensional affine Gale diagrams. In addition, formulas for the numbers of

- polytopes [Fus06, Thm. 1],
- simplicial polytopes [Grü67, Sect. 6.2, Thm. 6.3.2, p. 113 and p. 424],
- neighborly polytopes [McM74], and
- simplicial neighborly polytopes [AM73]

have been produced using Gale diagrams. This leads to subtle enumeration problems, some of them connected to colored necklace counting problems.

d+4 vertices: Here anything can go wrong: The universality theorem for oriented matroids of rank 3 yields a universality theorem for d-polytopes with d+4 vertices. See Section 6.3.4.

We refer to [Zie95, Lecture 6] for a detailed introduction to affine Gale diagrams.

15.2 METRIC PROPERTIES

The combinatorial data of a polytope—vertices, edges, . . . , facets—have their counterparts in genuine geometric data, such as face volumes, surface areas, quermassintegrals, and the like. In this second half of the chapter, we give a brief sketch of some key geometric concepts related to polytopes.

However, the topics of combinatorial and of geometric invariants are not disjoint at all: Much of the beauty of the theory stems from the subtle interplay between the two sides. Thus, the computation of volumes inevitably leads to the construction of triangulations (explicitly or implicitly), mixed volumes lead to mixed subdivisions of Minkowski sums (one "hot topic" for current research in the area), quermassintegrals relate to face enumeration, and so on.

A concrete and striking example of the interplay is related to Kalai's 3^d conjecture on the face numbers of centrally symmetric polytopes $P \subset \mathbb{R}^d$ mentioned before: Using convex geometric methods, Figiel, Lindenstrauss and Milman [FLM77] proved that $\ln f_0(P) \ln f_{d-1}(P) \geq \frac{1}{16}d$.

Furthermore, the study of polytopes yields a powerful approach to the theory of convex bodies: Sometimes one can extend properties of polytopes to arbitrary convex bodies by approximation [Sch14]. However, there are also properties valid for polytopes that fail for convex bodies in general. This bug/feature is designed to keep the game interesting.

15.2.1 VOLUME AND SURFACE AREA

GLOSSARY

Volume of a d-simplex T: $V(T) = \dfrac{1}{d!}\left|\det\begin{pmatrix} v^0 & \cdots & v^d \\ 1 & \cdots & 1 \end{pmatrix}\right|$, where $T = \operatorname{conv}\{v^0,\ldots,v^d\}$ with $v^0,\ldots,v^d \in \mathbb{R}^d$.

Volume of a d-polytope: $V(P) := \sum_{T\in\Delta(P)} V(T)$, where $\Delta(P)$ is any triangulation of P.

k-volume $V^k(P)$ of a k-polytope $P \subseteq \mathbb{R}^d$: The volume of P, computed with respect to the k-dimensional Euclidean measure induced on $\operatorname{aff}(P)$.

Surface area of a d-polytope P: $F(P) := \sum_{T\in\Delta(P),\, F\in\mathcal{F}_{d-1}(P)} V^{d-1}(T\cap F)$, where $\Delta(P)$ is a triangulation of P.

The volume $V(P)$ (i.e., the d-dimensional Lebesgue measure) and the surface area $F(P)$ of a d-polytope $P \subseteq \mathbb{R}^d$ can be derived from any triangulation of P, since volumes of simplices are easy to compute. The crux for this is in the (efficient?) generation of a triangulation, a topic on which Chapters 16 and 29 of this Handbook have more to say.

The following recursive approach only implicitly generates a triangulation, but derives explicit volume formulas. Let $P \subseteq \mathbb{R}^d$ ($P \neq \emptyset$) be a polytope. If $d = 0$ then we set $V(P) = 1$. Otherwise we set $\mathcal{S}_{d-1}(P) := \{u \in S^{d-1} \mid \dim(H(P,u)\cap P) = d-1\}$, and use this to define the volume of P as

$$V(P) \;:=\; \frac{1}{d} \sum_{u\in\mathcal{S}_{d-1}(P)} h(P,u)\cdot V^{d-1}(H(P,u)\cap P).$$

Thus, for any d-polytope the volume is a sum of its facet volumes, each weighted by $1/d$ times its signed distance from the origin. This can be interpreted geometrically as follows: Assume for simplicity that the origin is in the interior of P.

Then the collection $\{\text{conv}(F \cup \{0\}) \mid F \in \mathcal{F}_{d-1}(P)\}$ is a subdivision of P into d-dimensional pyramids, where the base of $\text{conv}(F \cup \{0\})$ has $(d-1)$-dimensional volume $V^{d-1}(F)$—to be computed recursively, the height of the pyramid is $h(P, u^F)$, and thus its volume is $\frac{1}{d}h(P, u^F) \cdot V^{d-1}(F)$; compare to Figure 15.2.1. The formula remains valid even if the origin is outside P or on its boundary.

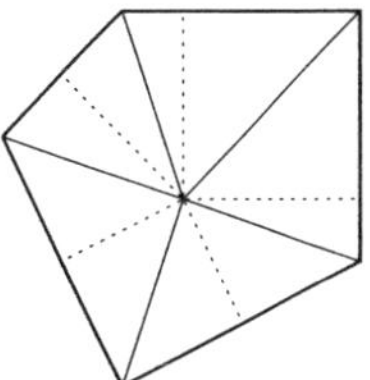

FIGURE 15.2.1
This pentagon, with the origin in its interior, is decomposed into five pyramids (triangles), each with one of the pentagon facets (edges) F_i as its base. For each pyramid, the height, of length $h(P, u^{F_i})$, is drawn as a dotted line.

A classical and beautiful result of Minkowski states that the volumes of the facets $V^{d-1}(F)$ along with their normals (directions) u^F determine a polytope as uniquely as possible:

THEOREM 15.2.1 *Minkowski* (cf. [Sch14, pp. 455])

Let $u^1, \ldots, u^n \in S^{d-1}$ be pairwise distinct unit vectors linearly spanning $\mathbb{R}^d$ and let $f_1, \ldots, f_n > 0$ be positive real numbers. There exists a d-dimensional polytope P with n facets $F_1, \ldots, F_n$ such that $F_i = H(P, u^i) \cap P$ and $V^{d-1}(F_i) = f_i$, $1 \le i \le n$, if and only if

$$\sum_{i=1}^{n} f_i\, u^i = 0.$$

Moreover, P is uniquely determined up to translations.

This statement is a particular instance of the modern and "hot" L_p-Minkowski problem for which we refer to [BLYZ13] and the references within.

Note that $V(P) \ge 0$. This holds with strict inequality if and only if the polytope P has full dimension d. The surface area $F(P)$ can also be expressed as

$$F(P) \;=\; \sum_{u \in \mathcal{S}_{d-1}(P)} V^{d-1}(H(P, u) \cap P).$$

Thus for a d-polytope the surface area is the sum of the $(d-1)$-volumes of its facets. If $\dim(P) = d - 1$, then $F(P)$ is twice the $(d-1)$-volume of P. One has $F(P) = 0$ if and only if $\dim(P) < d - 1$.

TABLE 15.2.1

POLYTOPE	$f_k(\cdot)$	VOLUME	SURFACE AREA
C_d	$2^{d-k}\binom{d}{k}$	2^d	$2d \cdot 2^{d-1}$
C_d^{Δ}	$2^{k+1}\binom{d}{k+1}$	$\frac{2^d}{d!}$	$2^d \frac{\sqrt{d}}{(d-1)!}$
T_d	$\binom{d+1}{k+1}$	$\frac{\sqrt{d+1}}{d!}$	$(d+1) \cdot \frac{\sqrt{d}}{(d-1)!}$

Both the volume and the surface area are continuous with respect to the Hausdorff metric (as defined in Chap. 2). They are monotone and invariant with respect to rigid motions. The volume is homogeneous of degree d, i.e., $V(\mu P) = \mu^d V(P)$ for $\mu \geq 0$, whereas the surface area is homogeneous of degree $d-1$. For further properties of the functionals $V(\cdot)$ and $F(\cdot)$ see [Had57] and [Sch14].

Table 15.2.1 gives the numbers of k-faces, the volume, and the surface area of the d-cube C_d (with edge length 2), of the cross-polytope C_d^{Δ} with edge length $\sqrt{2}$, and of the regular simplex T_d with edge length $\sqrt{2}$.

15.2.2 MIXED VOLUMES

GLOSSARY

Volume polynomial: The volume of the Minkowski sum $\lambda_1 P_1 + \lambda_2 P_2 + \cdots + \lambda_r P_r$, which is a homogeneous polynomial in $\lambda_1, \ldots, \lambda_r \geq 0$. Here the P_i may be convex polytopes of any dimension, or more general (closed, bounded) convex sets.

Mixed volumes: The suitably normalized coefficients of the volume polynomial of $P_1, \ldots, P_r$.

Normal cone: The normal cone $N(F, P)$ of a face F of a polytope P is the set of all vectors $v \in \mathbb{R}^d$ such that the supporting hyperplane $H(P, v)$ contains F, i.e.,

$$N(F, P) = \{v \in \mathbb{R}^d \mid F \subseteq H(P, v) \cap P\}.$$

THEOREM 15.2.2 *Mixed Volumes* (cf. [Sch14, pp. 275])

Let $P_1, \ldots, P_r \subseteq \mathbb{R}^d$ be polytopes, $r \geq 1$, and $\lambda_1, \ldots, \lambda_r \geq 0$. The volume of $\lambda_1 P_1 + \cdots + \lambda_r P_r$ is a homogeneous polynomial in $\lambda_1, \ldots, \lambda_r$ of degree d. Thus it can be written in the form

$$V(\lambda_1 P_1 + \cdots + \lambda_r P_r) = \sum_{(i(1),\ldots,i(d)) \in \{1,2,\ldots,r\}^d} \lambda_{i(1)} \cdots \lambda_{i(d)} \cdot V(P_{i(1)}, \ldots, P_{i(d)}),$$

where the coefficients in this expansion are chosen to be symmetric in their indices. Furthermore, the coefficient $V(P_{i(1)}, \ldots, P_{i(d)})$ depends only on $P_{i(1)}, \ldots, P_{i(d)}$. It is called the ***mixed volume*** *of the polytopes $P_{i(1)}, \ldots, P_{i(d)}$.*

With the abbreviation

$$V(P_1, k_1; \ldots; P_r, k_r) := V(\underbrace{P_1, \ldots, P_1}_{k_1 \text{ times}}, \ldots, \underbrace{P_r, \ldots, P_r}_{k_r \text{ times}}),$$

the polynomial becomes

$$V(\lambda_1 P_1 + \cdots + \lambda_r P_r) = \sum_{\substack{k_1,\ldots,k_r \geq 0 \\ k_1 + \cdots + k_r = d}} \binom{d}{k_1, \ldots, k_r} \lambda_1^{k_1} \cdots \lambda_r^{k_r} V(P_1, k_1; \ldots; P_r, k_r).$$

In particular, the volume of the polytope P_i is given by the mixed volume $V(P_1, 0; \ldots; P_i, d; \ldots; P_r, 0)$. The theorem is also valid for arbitrary convex bodies: This is a good example of a result where the general case can be derived from the polytope case by approximation. For more about the properties of mixed volumes from different points of view see Schneider [Sch14], Sangwine-Yager [San93], and McMullen [McM93].

The definition of the mixed volumes as coefficients of a polynomial is somewhat unsatisfactory. Schneider gave the following explicit rule, which generalizes an earlier result of Betke [Bet92] for the case $r = 2$. It uses information about the normal cones at certain faces. For this, note that $N(F, P)$ is a finitely generated cone, which can be written explicitly as the sum of the orthogonal complement of aff(P) and the positive hull of those unit vectors u that are both parallel to aff(P) and induce supporting hyperplanes $H(P, u)$ that contain a facet of P including F. Thus, for $P \subseteq \mathbb{R}^d$ the dimension of $N(F, P)$ is $d - \dim(F)$.

THEOREM 15.2.3 *Schneider's Summation Formula* [Sch94]

Let $P_1, \ldots, P_r \subseteq \mathbb{R}^d$ be polytopes, $r \geq 2$. Let $x^1, \ldots, x^r \in \mathbb{R}^d$ with $x^1 + \cdots + x^r = 0$, $(x^1, \ldots, x^r) \neq (0, \ldots, 0)$, and

$$\bigcap_{i=1}^{r} \left(\mathrm{relint} N(F_i, P_i) - x^i\right) \;=\; \emptyset$$

whenever F_i is a face of P_i and $\dim(F_1) + \cdots + \dim(F_r) > d$. Then

$$\binom{d}{k_1, \ldots, k_r} V(P_1, k_1; \ldots ; P_r, k_r) = \sum_{(F_1, \ldots, F_r)} V(F_1 + \cdots + F_r),$$

where the summation extends over the r-tuples $(F_1, \ldots, F_r)$ of k_i-faces F_i of P_i with $\dim(F_1 + \cdots + F_r) = d$ and $\bigcap_{i=1}^{r} \left(N(F_i, P_i) - x^i\right) \neq \emptyset$.

The choice of the vectors $x^1, \ldots, x^r$ implies that the selected k_i-faces $F_i \subseteq P_i$ of a summand $F_1 + \cdots + F_r$ are contained in complementary subspaces. Hence one may also write

$$\binom{d}{k_1, \ldots, k_r} V(P_1, k_1; \ldots ; P_r, k_r) = \sum_{(F_1, \ldots, F_r)} [F_1, \ldots, F_r] \cdot V^{k_1}(F_1) \cdots V^{k_r}(F_r),$$

where $[F_1, \ldots, F_r]$ denotes the volume of the parallelepiped that is the sum of unit cubes in the affine hulls of $F_1, \ldots, F_r$.

Finally, we remark that the selected sums of faces in the formula of the theorem form a subdivision of the polytope $P_1 + \cdots + P_r$, i.e.,

$$P_1 + \cdots + P_r \;=\; \biguplus_{(F_1, \ldots, F_r)} (F_1 + \cdots + F_r)\,.$$

See Figure 15.2.2 for an example.

VOLUMES OF ZONOTOPES

If all summands in a Minkowski sum $Z = P_1 + \cdots + P_r$ are line segments, say $P_i = p^i + [0, 1]z^i = \mathrm{conv}\{p^i, p^i + z^i\}$ with $p^i, z^i \in \mathbb{R}^d$ for $1 \leq i \leq r$, then the resulting polytope Z is a zonotope. In this case the summation rule immediately gives $V(P_1, k_1; \ldots ; P_r, k_r) = 0$ if the vectors

$$\underbrace{z^1, \ldots, z^1}_{k_1 \text{ times}}, \;\ldots,\; \underbrace{z^r, \ldots, z^r}_{k_r \text{ times}}$$

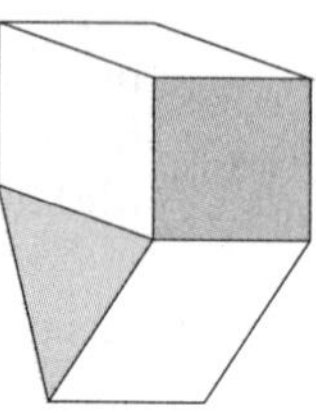

FIGURE 15.2.2
Here the Minkowski sum of a square P_1 and a triangle P_2 is decomposed into translates of P_1 and of P_2 (this corresponds to two summands with $F_1 = P_1$ and $F_2 = P_2$, respectively), together with three "mixed" faces that arise as sums $F_1 + F_2$, where F_1 and F_2 are faces of P_1 and P_2 (corresponding to summands with $\dim(F_1) = \dim(F_2) = 1$*).*

are linearly dependent. This can also be seen directly from dimension considerations. Otherwise, for $k_{i(1)} = k_{i(2)} = \cdots = k_{i(d)} = 1$, say,

$$V(P_1, k_1; \ldots ; P_r, k_r) = \frac{1}{d!} \left|\det\left(z^{i(1)}, z^{i(2)}, \ldots, z^{i(d)}\right)\right|.$$

Therefore, one obtains McMullen's formula for the volume of the zonotope Z (cf. Shephard [She74]):

$$V(Z) = \sum_{1 \le i(1) < i(2) < \cdots < i(d) \le r} \left|\det(z^{i(1)}, \ldots, z^{i(d)})\right|.$$

15.2.3 QUERMASSINTEGRALS AND INTRINSIC VOLUMES

GLOSSARY

***i*th quermassintegral $W_i(P)$:** The mixed volume $V(P, d{-}i; B_d, i)$ of a polytope P and the d-dimensional unit ball B_d.

κ_d: The volume of B_d. Hence $\kappa_0 = 1$, $\kappa_1 = 2$, $\kappa_2 = \pi$, etc.

***i*th intrinsic volume $V_i(P)$:** The $(d{-}i)$th quermassintegral, scaled by the constant $\binom{d}{i}/\kappa_{d-i}$.

Outer parallel body of P at distance λ: The convex body $P + \lambda B_d$ for some $\lambda > 0$.

External angle $\gamma(F,P)$ at a face F of a polytope P: The volume of the intersection $\left(\text{lin}(F - x^F) + N(F,P)\right) \cap B_d$ divided by κ_d, for $x^F \in \text{relint}(F)$, where $\text{lin}(\cdot)$ denotes the linear hull. Thus $\gamma(F,P)$ is the "fraction of $\mathbb{R}^d$ taken up by $\text{lin}(F - x^F) + N(F,P)$." Equivalently, the external angle at a k-face F is the fraction of the spherical volume of S covered by $N(F,P) \cap S$, where S denotes the $(d{-}k{-}1)$-dimensional unit sphere in $\text{lin}(N(F,P))$.

Internal angle $\beta(F,G)$ for faces $F \subseteq G$: The "fraction" of $\text{lin}\{G - x^F\}$ taken up by the cone $\text{pos}\{x - x^F \mid x \in G\}$, for $x^F \in \text{relint}(F)$. A detailed discussion of relations between external and internal angles can be found in McMullen [McM75].

The quermassintegrals are generalizations of both the volume and the surface area of P. In fact, they can also be seen as the continuous convex geometry analogs of face numbers.

For a polytope $P \subseteq \mathbb{R}^d$ and the d-dimensional unit ball B_d, the mixed volume formula, applied to the outer parallel body $P + \lambda B_d$, gives

$$V(P + \lambda B_d) = \sum_{i=0}^{d} \binom{d}{i} \lambda^i W_i(P),$$

with the convention $W_i(P) = V(P, d-i; B_d, i)$. This formula is known as the ***Steiner polynomial***. The mixed volume $W_i(P)$, the ith quermassintegral of P, is an important quantity and of significant geometric interest [Had57, Sch14]. As special cases, $W_0(P) = V(P)$ is the volume, $dW_1(P) = F(P)$ is the surface area, and $W_d(P) = \kappa_d$.

For the geometric interpretation of $W_i(P)$ for polytopes, we use a normalization of the quermassintegrals due to McMullen [McM75]: For $0 \le i \le d$, the ith intrinsic volume of P is defined by

$$V_i(P) := \frac{\binom{d}{i}}{\kappa_{d-i}} W_{d-i}(P).$$

With this notation the Steiner polynomial can be written as

$$V(P + \lambda B_d) = \sum_{i=0}^{d} \lambda^{d-i} \kappa_{d-i} V_i(P).$$

See Figure 15.2.3 for an example. $V_d(P)$ is the volume of P, $V_{d-1}(P)$ is half the surface area, and $V_0(P) = 1$. One advantage of this normalization is that the intrinsic volumes are unchanged if P is embedded in some Euclidean space of different dimension. Thus, for $\dim(P) = k \le d$, $V_k(P)$ is the ordinary k-volume of P with respect to the Euclidean structure induced in $\text{aff}(P)$.

FIGURE 15.2.3

The Minkowski sum of a square P with a ball λB_2 yields the outer parallel body. This outer parallel body can be decomposed into pieces, whose volumes, $V(P)$, $\lambda V_1(P)\kappa_1$, and $\lambda^2\kappa_2$, correspond to the three terms in the Steiner polynomial.

$$V(P + \lambda B_2) = V_2(P) + \lambda V_1(P)\kappa_1 + \lambda^2\kappa_2$$

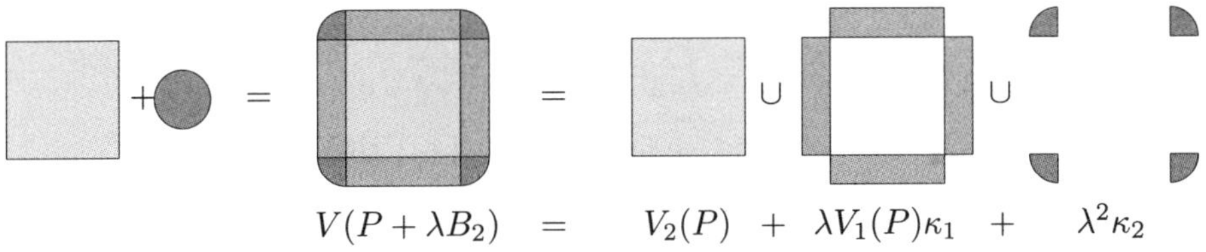

For a $(\dim(P)-2)$-face F, the concept of external angle (see the glossary) reduces to the "usual" concept: then the external angle is given by $\frac{1}{2\pi}\arccos\langle u^{F_1}, u^{F_2}\rangle$ for unit normal vectors $u^{F_1}, u^{F_2} \in S^{d-1}$ to the facets F_1, F_2 with $F_1 \cap F_2 = F$. One has $\gamma(P, P) = 1$ for the polytope itself and $\gamma(F, P) = 1/2$ for each facet F. Using this concept, we get

$$V_k(P) = \sum_{F \in \mathcal{F}_k(P)} \gamma(F, P) \cdot V^k(F).$$

Internal and external angles are also useful tools in order to express combinatorial properties of polytopes (see the application below). One classical example is ***Gram's equation*** [Gra74] [Grü67, Sect. 14.1].

$$\sum_{k=0}^{d-1}(-1)^k \sum_{F\in\mathcal{F}_k(P)} \beta(F,P) = (-1)^{d-1}.$$

This formula is quite similar to the Euler relation for the face numbers of a polytope (see Chapter 17). It was discovered by Shephard and by Welzl that Gram's equation follows directly from Euler's relation applied to a random projection [Grü67, p. 315a].

SOME COMPUTATIONS

In principle, one can use the external angle formula to determine the intrinsic volumes of a given polytope, but in general it is hard to calculate external angles. Indeed, for the computation of spherical volumes there are explicit formulas only in small dimensions.

In what follows, we give formulas for the intrinsic volumes of the polytopes C_d, C_d^Δ, and T_d. For this, we identify the k-faces of C_d with the k-cube C_k and the k-faces of C_d^Δ and of T_d with T_k, for $0 \le k < d$.

The case of the cube C_d is rather trivial. Since $\gamma(C_k, C_d) = 2^{-(d-k)}$ one gets (see Table 15.2.1)

$$V_k(C_d) = 2^k \binom{d}{k}.$$

For the regular simplex T_d we have

$$V_k(T_d) = \binom{d+1}{k+1} \cdot \frac{\sqrt{k+1}}{k!} \cdot \gamma(T_k, T_d).$$

An explicit formula for the external angles of a regular simplex by Ruben [Rub60] [Had79] is:

$$\gamma(T_k, T_d) = \sqrt{\frac{k+1}{\pi}} \int_{-\infty}^{\infty} e^{-(k+1)x^2} \left(\frac{1}{\sqrt{\pi}} \int_{-\infty}^{x} e^{-y^2} dy\right)^{d-k} dx.$$

For the regular cross-polytope we find for $k \le d-1$ that

$$V_k(C_d^\Delta) = 2^{k+1} \binom{d}{k+1} \cdot \frac{\sqrt{k+1}}{k!} \cdot \gamma(T_k, C_d^\Delta).$$

For this, the external angles of C_d^Δ were determined by Betke and Henk [BH93]:

$$\gamma(T_k, C_d^\Delta) = \sqrt{\frac{k+1}{\pi}} \int_{0}^{\infty} e^{-(k+1)x^2} \left(\frac{2}{\sqrt{\pi}} \int_{0}^{x} e^{-y^2} dy\right)^{d-k-1} dx.$$

AN APPLICATION

External angles and internal angles play a crucial role in work by Affentranger and Schneider [AS92] (see also [BV02]), who computed the expected number of

k-faces of the orthogonal projection of a polytope $P \subseteq \mathbb{R}^d$ onto a randomly chosen isotropic subspace of dimension n. Let $E[f_k(P;n)]$ be that number. Then for $0 \le k < n \le d-1$ it was shown that

$$E[f_k(P;n)] \;=\; 2\sum_{m\ge 0}\;\sum_{F\in\mathcal{F}_k(P)}\;\sum_{\substack{G\in\mathcal{F}_{n-1-2m}(P)\\ F\subseteq G}} \beta(F,G)\gamma(G,P),$$

where $\beta(F,G)$ is the internal angle of the face F with respect to a face $G \supseteq F$.

In the sequel we apply the above formula to the polytopes C_d, C_d^Δ, and T_d. For the cubes one has $\beta(C_k, C_l) = (1/2)^{l-k}$, while the number of l-faces of C_d containing any given k-face is equal to $\binom{d-k}{l-k}$. Hence

$$E[f_k(C_d;n)] = 2\binom{d}{k}\sum_{m\ge 0}\binom{d-k}{n-1-k-2m}.$$

In particular, $E[f_k(C_d;d-1)] = (2^{d-k}-2)\binom{d}{k}$.

For the cross-polytope C_d^Δ the number of l-faces that contain a k-face is equal to $2^{l-k}\binom{d-k-1}{l-k}$. Thus

$$E[f_k(C_d^\Delta;n)] = 2\binom{d}{k+1}\sum_{m\ge 0}2^{n-2m}\binom{d-k-1}{n-1-k-2m}\beta(T_k,T_{n-1-2m})\gamma(T_{n-1-2m},C_d^\Delta).$$

In the same way one obtains for T_d

$$E[f_k(T_d;n)] = 2\binom{d+1}{k+1}\sum_{m\ge 0}\binom{d-k}{n-1-k-2m}\beta(T_k,T_{n-1-2m})\gamma(T_{n-1-2m},T_d).$$

For the last two formulas one needs the internal angles $\beta(T_k,T_l)$ of the regular simplex T_d, for $0 \le k \le l \le d$, For this, one has the following complex integral [BH99]:

$$\beta(T_k,T_l) = \frac{(k{+}1{+}l)^{1/2}(k{+}1)^{(l-1)/2}}{\pi^{(l+1)/2}}\int_{-\infty}^{\infty} e^{-w^2}\left(\int_0^{\infty} e^{-(k+1)y^2+2iwy}dy\right)^l dw.$$

Using this formula one can determine the asymptotic behavior of $E[f_k(C_d^\Delta;n)]$ and $E[f_k(T_d;n)]$ as n tends to infinity [BH99].

15.3 SOURCES AND RELATED MATERIAL

FURTHER READING

The classic account of the combinatorial theory of convex polytopes was given by Grünbaum in 1967 [Grü67]. It inspired and guided a great part of the subsequent research in the field. Besides the related chapters of this Handbook, we refer to [Zie95] and the handbook surveys by Klee and Kleinschmidt [KK16] and by Bayer and Lee [BL93] for further reading.

For the geometric theory of convex bodies we refer to the Handbook of Convex Geometry [GW93], to Schneider [Sch14] for an excellent monograph, and as an introduction to modern convex geometry we recommend [Bal97]. For high-dimensional/asymptotic methods and results see also the recent book [AGM15].

As for the algorithmic aspects of computing volumes, etc., we refer to Chapter 36 of this Handbook, on Computational Convexity, and to the additional references given there. For further and recent aspects related to random polytopes we refer to Chapter 12 “Discrete Aspects of Stochastic Geometry” of this Handbook.

RELATED CHAPTERS

Chapter 6: Oriented matroids
Chapter 7: Lattice points and lattice polytopes
Chapter 16: Subdivisions and triangulations of polytopes
Chapter 17: Face numbers of polytopes and complexes
Chapter 18: Symmetry of polytopes and polyhedra
Chapter 19: Polytope skeletons and paths
Chapter 26: Convex hull computations
Chapter 29: Triangulations and mesh generation
Chapter 36: Computational convexity
Chapter 64: Crystals, periodic and aperiodic
Chapter 67: Software

REFERENCES

[AF92] D. Avis and K. Fukuda. A pivoting algorithm for convex hulls and vertex enumeration of arrangements and polyhedra. *Discrete Comput. Geom.*, 8:295–313, 1992.

[AGM15] S. Artstein-Avidan, A. Giannopoulos, and V.D. Milman. *Asymptotic Geometric Analysis. Part I.* vol. 202 of *Math. Surveys and Monogr.*, AMS, Providence, 2015.

[AM73] A. Altshuler and P. McMullen. The number of simplicial neighbourly d-polytopes with $d+3$ vertices. *Mathematika*, 20:263–266, 1973.

[AP16] K.A. Adiprasito and A. Padrol. A universality theorem for projectively unique polytopes and a conjecture of Shephard. *Israel J. Math.*, 211:239–255, 2016.

[AP17] K.A. Adiprasito and A. Padrol. The universality theorem for neighborly polytopes. *Combinatorica*, 37:129–136, 2017.

[APT15] K.A. Adiprasito, A. Padrol, and L. Theran. Universality theorems for inscribed polytopes and Delaunay triangulations. *Discrete Comput. Geom.*, 54:412–431, 2015.

[AS92] F. Affentranger and R. Schneider. Random projections of regular simplices. *Discrete Comput. Geom.*, 7:219–226, 1992.

[AZ15] K.A. Adiprasito and G.M. Ziegler. Many projectively unique polytopes. *Invent. Math.*, 119:581–652, 2015.

[AZ98] M. Aigner and G.M. Ziegler. *Proofs from THE BOOK,* Springer, Berlin, 1998; 5th edition 2014.

[Bal61] M.L. Balinski. On the graph structure of convex polyhedra in n-space. *Pacific J. Math.*, 11:431–434, 1961.

[Bal97] K. Ball. An elementary introduction to modern convex geometry. In S. Levy, editor, *Flavors of Geometry*, vol. 31 of *MSRI Publications*, pages 1–58, Cambridge University Press, 1997.

[BEK84] J. Bokowski, G. Ewald, and P. Kleinschmidt. On combinatorial and affine automorphisms of polytopes. *Israel J. Math.*, 47:123–130, 1984.

[Bet92] U. Betke. Mixed volumes of polytopes. *Arch. Math.*, 58:388–391, 1992.

[BH93] U. Betke and M. Henk. Intrinsic volumes and lattice points of crosspolytopes. *Monatshefte Math.*, 115:27–33, 1993.

[BH99] K.J. Böröczky, Jr. and M. Henk. Random projections of regular polytopes. *Arch. Math.*, 73:465–473, 1999.

[BL93] M.M. Bayer and C.W. Lee. .Combinatorial aspects of convex polytopes. In P.M. Gruber and J.M. Wills, editors, *Handbook of Convex Geometry*, pages 485–534, North-Holland, Amsterdam, 1993.

[BLS+93] A. Björner, M. Las Vergnas, B. Sturmfels, N. White, and G.M. Ziegler. *Oriented Matroids.* Vol. 46 of *Encyclopedia Math. Appl.*, Cambridge University Press, 1993; 2nd revised edition 1999.

[BLYZ13] K.J. Böröczky, E. Lutwak, D. Yang, and G. Zhang. The logarithmic Minkowski problem. *J. Amer. Math. Sco.*, 26:831–852, 2013.

[BM87] R. Blind and P. Mani-Levitska. On puzzles and polytope isomorphisms. *Aequationes Math.*, 34:287–297, 1987.

[Bok06] J. Bokowski. *Computational Oriented Matroids: Equivalence Classes of Matroids within a Natural Framework.* Cambridge University Press, 2006.

[BP01] I. Bárány and A. Pór. $0-1$ polytopes with many facets. *Adv. Math.*, 161:209–228, 2001.

[BRS92] J. Bokowski, J. Richter-Gebert, and W. Schindler. On the distribution of order types. *Comput. Geom.*, 1:127–142, 1992.

[BS04] A.I. Bobenko and B.A. Springborn. Variational principles for circle patterns, and Koebe's theorem. *Trans. Amer. Math. Soc.*, 356:659–689, 2004.

[BS89] J. Bokowski and B. Sturmfels. *Computational Synthetic Geometry.* Vol. 1355 of *Lecture Notes Math.*, Springer, Berlin, 1989.

[BS96] L.J. Billera and A. Sarangarajan. All 0-1 polytopes are traveling salesman polytopes. *Combinatorica*, 16:175–188, 1996.

[BV94] Y.M. Baryshnikov and R.A. Vitale. Regular simplices and Gaussian samples. *Discrete Comput. Geom.*, 11:141–147, 1994.

[DL97] M.M. Deza and M. Laurent. *Geometry of Cuts and Metrics.* Vol. 5 of *Algorithms Combin.*, Springer, Berlin, 1997.

[DT09] D.L. Donoho and J. Tanner. Counting faces of randomly projected polytopes when the projection radically lowers dimension. *J. Amer. Math. Soc.*, 22:1–53, 2009.

[Fir15b] M. Firsching. *Optimization Methods in Discrete Geometry.* PhD thesis, FU Berlin, `edocs.fu-berlin.de/diss/receive/FUDISS_thesis_000000101268`, 2015.

[FLM77] T. Figiel, J. Lindenstraussm and V.D. Milman. The dimension of almost spherical sections of convex bodies. *Acta Math.*, 139:53–94, 1977.

[Fri09] E.J. Friedman. Finding a simple polytope from its graph in polynomial time. *Discrete Comput. Geom.*, 41:249–256, 2009.

[Fus06] É. Fusy. Counting d-polytopes with $d+3$ vertices. *Electron. J. Combin.*, 13:#23, 2006.

[GGM05] D. Gatzouras, A. Giannopoulos, and N. Markoulakis. Lower bound for the maximal number of facets of a $0/1$ polytope. *Discrete Comput. Geom.*, 34:331–349, 2005.

[GJ00] E. Gawrilow and M. Joswig. `polymake`: a framework for analyzing convex polytopes. In G. Kalai and G.M. Ziegler, editors, *Polytopes—Combinatorics and Computation*, vol. 29 of *DMV Seminars*, pages 43–74, Birkhäuser, Basel, 2000. `http://www.polymake.org`

[GP85] M. Grötschel and M. Padberg. Polyhedral theory. In E.L. Lawler et al., editors, *The Traveling Salesman Problem*, pages 251–360, Wiley, Chichester, 1985.

[Gra74] J.P. Gram. Om Rumvinklerne i et Polyeder. *Tidsskr. Math.*, 4:161–163, 1874.

[Gru07] P.M. Gruber. *Convex and Discrete Geometry.* Vol. 336 of *Grundlehren Series*, Springer, Berlin, 2007.

[Grü67] B. Grünbaum. *Convex Polytopes.* Interscience, London 1967. Second edition, V. Kaibel, V. Klee, and G. M. Ziegler, eds., vol. 221 of *Graduate Texts in Math.*, Springer, New York, 2003.

[GW93] P.M. Gruber and J.M. Wills, editors. *Handbook of Convex Geometry.* Volumes A and B, North-Holland, Amsterdam, 1993.

[Had57] H. Hadwiger. *Vorlesungen über Inhalt, Oberfläche und Isoperimetrie.* Springer, Berlin, 1957.

[Had79] H. Hadwiger. Gitterpunktanzahl im Simplex und Wills'sche Vermutung. *Math. Annalen*, 239:271–288, 1979.

[Kal88] G. Kalai. A simple way to tell a simple polytope from its graph. *J. Combin. Theory Ser. A*, 49:381–383, 1988.

[Kal89] G. Kalai. The number of faces of centrally-symmetric polytopes (Research Problem). *Graphs Combin.*, 5:389–391, 1989.

[KK93] J. Kahn and G. Kalai. A counterexample to Borsuk's conjecture. *Bull. Amer. Math. Soc.*, 29:60–62, 1993.

[KK95] V. Klee and P. Kleinschmidt. Polyhedral complexes and their relatives. In R. Graham, M. Grötschel, and L. Lovász, editors, *Handbook of Combinatorics*, pages 875–917, North-Holland, Amsterdam, 1995.

[Kor97] U.H. Kortenkamp. Every simplicial polytope with at most d+4 vertices is a quotient of a neighborly polytope. *Discrete Comput. Geom.*, 18:455-462, 1997.

[Man71] P. Mani. Automorphismen von polyedrischen Graphen. *Math. Ann.* 192:279–303, 1971.

[McM74] P. McMullen. The number of neighbourly d-polytopes with $d+3$ vertices. *Mathematika*, 21:26–31, 1974.

[McM75] P. McMullen. Non-linear angle-sum relations for polyhedral cones and polytopes. *Math. Proc. Comb. Phil. Soc.*, 78:247–261, 1975.

[McM76] P. McMullen. Constructions for projectively unique polytopes, *Discrete Math.*, 14:347–358, 1976.

[McM93] P. McMullen. Valuations and dissections. In P.M. Gruber and J.M. Wills, editors, *Handbook of Convex Geometry*, Volume B, pages 933–988, North-Holland, Amsterdam, 1993.

[MP15] H. Miyata and A. Padrol. Enumeration of neighborly polytopes and oriented matroids. *Exp. Math.*, 24:489–505, 2015.

[OS94] S. Onn and B. Sturmfels. A quantitative Steinitz' theorem. *Beiträge Algebra Geom.*, 35:125–129, 1994.

[Pad13] A. Padrol. Many neighborly polytopes and oriented matroids, *Discrete Comput. Geom.*, 50:865–902, 2013.

[Ric96] J. Richter-Gebert. *Realization Spaces of Polytopes.* Vol. 1643 of *Lecture Notes in Math.*, Springer, Berlin, 1996.

[RRS11] A. Ribó Mor, G. Rote, and A. Schulz. Small grid embeddings of 3-polytopes, *Discrete Comput. Geom.*, 45:65–87, 2011.

[Rub60] H. Ruben. On the geometrical moments of skew-regular simplices in hyperspherical space; with some applications in geometry and mathematical statistics. *Acta. Math.*, 103:1–23, 1960.

[Ryb99] K. Rybnikov. Stresses and liftings of cell-complexes. *Discrete Comput. Geom.*, 21:481–517, 1999.

[RZ94] J. Richter-Gebert and G.M. Ziegler. Zonotopal tilings and the Bohne-Dress theorem. In H. Barcelo and G. Kalai, editors, *Jerusalem Combinatorics '93*, vol. 178 of *Contemp. Math.*, pages 211–232, AMS, Providence, 1994.

[San93] J.R. Sangwine-Yager. Mixed volumes. In P.M. Gruber and J.M. Wills, editors, *Handbook of Convex Geometry*, Volume A, pages 43–71, North-Holland, Amsterdam, 1993.

[Sch14] R. Schneider. *Convex Bodies: The Brunn-Minkowski Theory*, 2nd expanded edition. Vol. 151 of *Encyclopedia Math. Appl.*, Cambridge University Press, 2014.

[Sch94] R. Schneider. Polytopes and the Brunn-Minkowski theory. In T. Bisztriczky et al. editors, *Polytopes: Abstract, Convex and Computational*, vol. 440 of *NATO Adv. Sci. Inst. Ser. C: Math. Phys. Sci.*, pages 273–299, Kluwer, Dordrecht, 1994.

[She82] I. Shemer. Neighborly polytopes. *Israel J. Math.*, 43:291–314, 1982.

[She74] G.C. Shephard. Combinatorial properties of associated zonotopes. *CanadJ. Math.*, 26:302–321, 1974.

[SJ05] W. Stein and D. Joyner. SAGE: system for algebra and geometry experimentation. *SIGSAM Bull.*, 39:61–64, 2005; `http://www.sagemath.org/`

[Ste22] E. Steinitz. Polyeder und Raumeinteilungen. In: F. Meyer and H. Mohrmann, editors, *Encyclopädie der Mathematischen Wissenschaften*, vol. 3, Geometrie, erster Teil, zweite Hälfte, pages 1–139, Teubner, Leipzig, 1922.

[SR34] E. Steinitz and H. Rademacher. *Vorlesungen über die Theorie der Polyeder.* Springer, Berlin, 1934; reprint, Springer, Berlin, 1976.

[SWZ09] R. Sanyal, A. Werner, and G.M. Ziegler. On Kalai's conjectures about centrally symmetric polytopes. *Discrete Comput. Geom.*, 41:183–198, 2009.

[VS92] A.M. Vershik and P.V. Sporyshev. Asymptotic behavior of the number of faces of random polyhedra and the neighborliness problem. *Selecta Math. Soviet.*, 11:181–201, 1992.

[Zie00] G.M. Ziegler. Lectures on 0/1-polytopes. In G. Kalai and G.M. Ziegler, editors, *Polytopes—Combinatorics and Computation*, vol. 29 of *DMV Seminars*, pages 1–41, Birkhäuser, Basel, 2000.

[Zie07] G.M. Ziegler. Convex Polytopes: Extremal constructions and f-vector shapes. In E. Miller, V. Reiner, and B. Sturmfels, editors, *Geometric Combinatorics*, vol. 13 of *IAS/Park City Math. Ser.*, pages 617–691, AMS, 2007.

[Zie08] G.M. Ziegler. Non-rational configurations, polytopes, and surfaces, *Math. Intelligencer*, 30:36–42, 2008.

[Zie95] G.M. Ziegler. *Lectures on Polytopes.* Vol. 152 of *Graduate Texts in Math.*, Springer, New York, 1995; 7th revised printing 2007.

16 SUBDIVISIONS AND TRIANGULATIONS OF POLYTOPES

Carl W. Lee and Francisco Santos

INTRODUCTION

We are interested in the set of all subdivisions or triangulations of a given polytope P and with a fixed finite set V of points that can be used as vertices. V must contain the vertices of P, and it may or may not contain additional points; these additional points are vertices of some, but not all, the subdivisions that we can form. This setting has interest in several contexts:

- In computational geometry there is often a set of *sites* V and one wants to find the triangulation of V that is optimal with respect to certain criteria.
- In algebraic geometry and in integer programming one is interested in triangulations of a lattice polytope P using only lattice points as vertices.
- Subdivisions of some particular polytopes using only vertices of the polytope turn out to be interesting mathematical objects. For example, for a convex n-gon and for the prism over a d-simplex they are isomorphic to the face posets of two remarkable polytopes, the associahedron and the permutahedron.

Our treatment is very combinatorial. In particular, instead of regarding a subdivision as a set of polytopes we regard it as a set of subsets of V, whose convex hulls subdivide P. This may appear to be an unnecessary complication at first, but it has advantages in the long run. It also relates this chapter to Chapter 6 (oriented matroids). For more application-oriented treatments of triangulations see Chapters 27 and 29. A general reference for the topics in this chapter is [DRS10].

16.1 BASIC CONCEPTS

GLOSSARY

Affine span: The affine span of a set $V \subset \mathbb{R}^d$ is the smallest affine space, or flat, containing V. It is denoted by $\mathrm{aff}(V)$.

Convex hull: The convex hull of a set $V \subset \mathbb{R}^d$ is the smallest convex set containing V. It is denoted by $\mathrm{conv}\,(V)$.

Polytope: A polytope P is the convex hull of a finite set V of points. Its ***dimension*** is the dimension of its affine span $\mathrm{aff}(P) = \mathrm{aff}(V)$. A ***face*** of P is the set $P^f := \{x \in P : f(x) \geq f(y)\ \forall y \in P\}$ that maximizes a linear functional f. The empty set and P are considered faces and every face is a polytope, of dimension ranging from -1 (empty set), 0 (vertices), 1 (edges), 2, ..., to $d-1$ (facets), and d (P itself). The set of vertices will be denoted by vert (P). The ***boundary*** of a d-dimensional polytope is the union of all its proper faces. See Chapter 15.

Polytopal complex: A polytopal complex is a finite, nonempty collection $S = \{P_1, \ldots, P_k\}$ of polytopes in $\mathbb{R}^d$ such that every face of each $P_i \in S$ is in S, and such that $P_i \cap P_j$ is always a common face of both (possibly empty). The dimension of S, dim (S), is the largest dimension of P_i. S is ***pure*** if all maximal polytopes in S have the same dimension [Zie95]. The ***k-skeleton*** of P_i is the k-dimensional complex consisting of faces of dimension at most k.

Faces of a set: Let S be a subset of a finite set V of points in $\mathbb{R}^d$. We say S is a face of V if there is a face F of the polytope $P = \operatorname{conv}(V)$ for which $S = V \cap F$. Note that S may include points that are not vertices of F. The dimension of S is the dimension of conv (S), and faces of dimension 0, 1, and dim $(V) - 1$ are referred to as vertices, edges, and facets, respectively, of the set V.

Subdivision: Suppose V is a finite set of points in $\mathbb{R}^d$ such that $P = \operatorname{conv}(V)$ is d-dimensional. A subdivision of V is a finite collection $S = \{S_1, \ldots, S_m\}$ of subsets of V, called ***cells***, such that:

(DP) for each $i \in \{1, \ldots, m\}$, $P_i := \operatorname{conv}(S_i)$ is d-dimensional (a d-polytope);

(UP) P is the union of $P_1, \ldots, P_m$; and

(IP) if $i \neq j$ then $F := S_i \cap S_j$ is a common (possibly empty) proper face of S_i and S_j and $P_i \cap P_j = \operatorname{conv}(F)$.

We will also say that S is a subdivision of the polytope P. The collection of polytopes $P_1, \ldots, P_m$, together with their faces, is a pure polytopal complex.

Trivial subdivision: The trivial subdivision of V is the subdivision $\{V\}$.

Simplex: A d-dimensional simplex is a d-polytope with exactly $d+1$ vertices. Equivalently, it is the convex hull of a set of affinely independent points in $\mathbb{R}^d$. We will also refer to the set of vertices of a d-simplex as a d-simplex.

Triangulation: A subdivision of V is a triangulation if every cell is a simplex.

Faces: The faces of a subdivision $\{S_1, \ldots, S_m\}$ are $S_1, \ldots, S_m$ and all their faces.

EXAMPLES

In Figure 16.1.1, (a) shows a set of six points in $\mathbb{R}^2$. The collection of three polygons in (b) is not a subdivision of that set since not every pair of polygons meets along a common edge or vertex; (c) shows a subdivision that is not a triangulation; and (d) gives a triangulation.

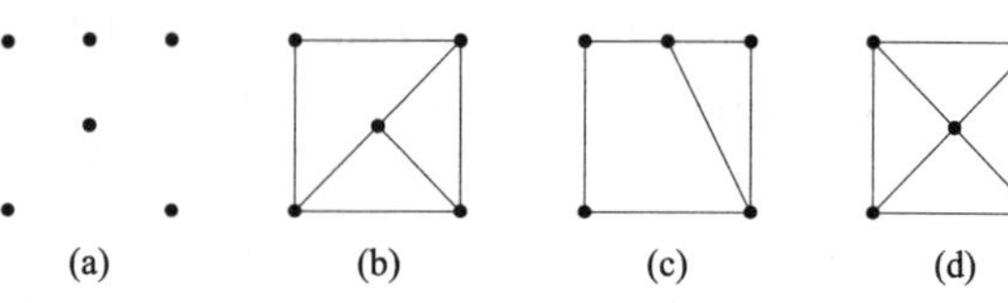

FIGURE 16.1.1
(a) *A set of points.*
(b) *A nonsubdivision.*
(c) *A subdivision.*
(d) *A triangulation.*

16.2 BASIC CONSTRUCTIONS AND PROPERTIES

GLOSSARY

The ***size of a subdivision*** is its number of cells (full dimensional faces). That is, the size of $S = \{S_1, \ldots, S_m\}$ is m.

Diameter of a subdivision: Let $S = \{S_1, \ldots, S_m\}$ be a subdivision and let $P_i = \operatorname{conv}(S_i)$, $1 \le i \le m$. Polytopes $P_i \ne P_j$ are ***adjacent*** if they share a common facet. A sequence $P_{i_0}, \ldots, P_{i_k}$ is a ***path*** if P_{i_j} and $P_{i_{j-1}}$ are adjacent for each j, $1 \le j \le k$. The ***length*** of such a path is k. The ***distance*** between P_i and P_j is the length of the shortest path connecting them. The diameter of S is the maximum distance occurring between pairs of polytopes P_i, P_j.

Refinement of a subdivision: Suppose $S = \{S_1, \ldots, S_l\}$ and $T = \{T_1, \ldots, T_m\}$ are two subdivisions of V. Then T is a refinement of S if for each j, $1 \le j \le m$, there exists i, $1 \le i \le l$, such that $T_j \subseteq S_i$. In this case we will write $T \le S$.

Visible facet: Let $P = \operatorname{conv}(V)$ be a d-polytope in $\mathbb{R}^d$, F a facet of P, and v a point in $\mathbb{R}^d$. We say that F (or that $F \cap V$, as a facet of V) is visible from v if the unique hyperplane containing F has v and the interior of P in opposite sides. If P is a k-polytope in $\mathbb{R}^d$ with $k < d$ and $v \in \operatorname{aff}(P)$, then the above definition is modified in the obvious way, taking $\operatorname{aff}(P)$ as the ambient space.

Placing a vertex: Suppose $S = \{S_1, \ldots, S_m\}$ is a subdivision of V and $v \notin V$. The subdivision T of $V \cup \{v\}$ that results from placing v is obtained as follows:

- If $v \notin \operatorname{aff}(V)$, then $T = \{S_i \cup \{v\} : S_i \in S\}$ (cone over S with apex v).
- If $v \in \operatorname{aff}(V)$, then T equals S together with the faces $F \cup \{v\}$ for each $(d-1)$-face F in S that is contained in a facet of $\operatorname{conv}(V)$ visible from v.

Note that if $v \in \operatorname{conv}(V)$, then $S = T$ (that is, T does not use v).

Pulling a vertex: Suppose $S = \{S_1, \ldots, S_m\}$ is a subdivision of V and $v \in S_1 \cup \cdots \cup S_m$. The result of pulling v is the refinement T of V obtained by modifying each $S_i \in S$ as follows. It was described in [Hud69, Lemma 1.4].

- If $v \notin S_i$, then $S_i \in T$.
- If $v \in S_i$, then for every facet F of S_i not containing v, $F \cup \{v\} \in T$.

Pushing a vertex: Suppose $S = \{S_1, \ldots, S_m\}$ is a subdivision of V (where $\dim(\operatorname{conv}(V)) = d$) and $v \in S_1 \cup \cdots \cup S_m$. The result of pushing v is the refinement T of V obtained by modifying each $S_i \in S$ as follows:

- If $v \notin S_i$, then $S_i \in T$.
- If $v \in S_i$ and $S_i \setminus \{v\}$ is $(d-1)$-dimensional (i.e., $\operatorname{conv}(S_i)$ is a pyramid with apex v), then $S_i \in T$.
- If $v \in S_i$ and $S_i \setminus \{v\}$ is d-dimensional, then $S_i \setminus \{v\} \in T$ and for every facet F of $S_i \setminus \{v\}$ that is visible from v, $F \cup \{v\} \in T$.

Lexicographic subdivisions: A subdivision T of V is lexicographic if it can be obtained from the trivial subdivision by pushing and/or pulling some of the points in V in some order. If only pushings (resp. pullings) are used, we call it a pushing (resp. pulling) subdivision.

16.2.1 LEXICOGRAPHIC SUBDIVISIONS

Refinement of subdivisions of V is a partial order with a unique maximal element, the trivial subdivision, and whose minimal elements are the triangulations of V: Every subdivision that is not a triangulation can be lexicographically refined.

Lexicographic triangulations were introduced by Sturmfels [Stu91] and studied in detail by Lee [Lee91]. The following results show that they are a quite versatile way of constructing subdivisions. For more details see [DRS10, Sect. 4.3]:

1. Placing and pushing are closely related: the triangulation obtained by placing the points of V in a certain order is the same as obtained starting with the trivial subdivision of V and pushing the points of V in the opposite order.
2. Placing all elements of V in any given order produces a triangulation of V. If the order is chosen so that no point is in the convex hull of the previously placed ones (e.g., ordering the points with respect to a generic linear functional) then the triangulation obtained uses all points of V as vertices. This shows that every finite set V is the vertex set of some triangulation of V.
3. Pulling or pushing a vertex v in a lexicographic subdivision in which v had already been pushed or pulled produces no effect. Hence, every lexicographic subdivision can be determined as an ordered subset of V indicating, for each point in it, whether it is to be pulled or pushed.
4. After all but an affinely independent subset of V have been pulled or pushed the lexicographic subdivision is a triangulation. For pullings the converse does not hold, but for pushings it does: For every ordering $\{v_1, \ldots, v_n\}$ of the points in V such that the last $d+1$ are affinely independent, each of the first $n-d-1$ pushings produces a proper refinement. In particular, the poset of subdivisions of V has chains of length at least $n-d-1$, for every V.
5. In a pulling subdivision, all cells contain the first point that is pulled.
6. In a pushing subdivision S, if all points except those of a subset $F \subset V$ are pushed, then F is a face of S.
7. Both operations may produce subdivisions that do not use all the points of V: if a $v \in V$ is pushed before it is a vertex, then it disappears from all cells containing it. The same happens if $v \in V$ is in the relative interior of a face F of a subdivision and another point of F is pulled. In particular, in a pulling triangulation at most one point in the interior of $\operatorname{conv}(V)$ is used.
8. If $\operatorname{card}(V) \leq d+3$, then every triangulation of V is lexicographic [Lee91]:
   - If $\operatorname{card}(V) = d+1$, then V has a unique triangulation, the trivial one.
   - If $\operatorname{card}(V) = d+2$, let $(\lambda_1, \ldots, \lambda_{d+2})$ be the unique (up to rescaling) affine dependence among $V = \{v_1, \ldots, v_{d+2}\}$. (That is, the solution to $\sum_i \lambda_i = 0$ and $\sum_i \lambda_i v_i = 0$.) Then, V has exactly two triangulations

$$T^+ = \{V \setminus \{v_i\} \mid \lambda_i > 0\} \qquad \text{and} \qquad T^- = \{V \setminus \{v_i\} \mid \lambda_i < 0\}.$$

T^+ (resp. T^-) is obtained by pushing any v_i with $\lambda_i > 0$ (resp. $\lambda_i < 0$) or by pulling any v_i with $\lambda_i < 0$ (resp. $\lambda_i > 0$). See Figure 16.3.1.

- If card $(V) = d+3$, then V has at most $d+3$ triangulations, with equality if (but not only if) no $d+1$ points lie in a hyperplane. They can all be obtained by pushing the points in specific orders, but not always by pulling them. See an example in Figure 16.4.1.

The triangulations in Figure 16.3.2, with $d+4$ points, are not lexicographic.

EXAMPLES

FIGURE 16.2.1
(a) *Pulling point 1 already gives a triangulation.*
(b) *Pushing triangulation for the order* 1234567. *Equivalently, placing triangulation for the order* 7654321.
(c) *Pushing at* 1 *then pulling at* 2.

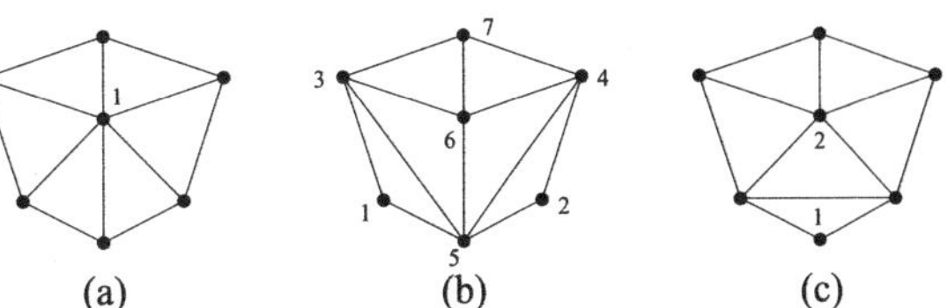

Figure 16.2.1 gives three lexicographic triangulations. (a) is obtained by pulling point 1, but cannot be obtained by pushing alone. (b) is obtained by pushing the points in the indicated order, but cannot be obtained by pulling points alone. (c) is obtained by pushing point 1 and then pulling point 2.

16.2.2 NUMBER, SIZE, AND DIAMETER OF TRIANGULATIONS

Size and diameter of subdivisions are monotone with respect to refinement, so the maximum is always achieved at a triangulation.

Every d-dimensional triangulation with n vertices has size bounded above by $O(n^{\lceil d/2 \rceil})$, achieved for example for (some) triangulations of cyclic polytopes. See Section 16.7. Note that asymptotic bounds in this section consider d fixed.

Every set V has triangulations of diameter at most $2(n-d-1)$, since pushing a point in a subdivision increases its diameter by at most two units [Lee91]. No good upper bound is known for the diameters of all triangulations. In particular, no upper bound for the diameter that is polynomial in both d and n is known (obtaining them is essentially as difficult as solving the polynomial Hirsch conjecture for polytopes), and no construction of triangulations with diameter greater than a small constant times $n-d$ is known. See [San13] and the references therein.

A triangulation of V is completely determined if we know its faces up to dimension $d/2$ [Dey93]. Hence the number of different triangulations V can have is bounded above by $2^{\binom{n}{d/2+1}}$. This bound is not far from the number of triangulations of a cyclic d-polytope with n vertices, which is in $2^{\Omega(n^{\lfloor d/2 \rfloor})}$ [DRS10, Sect. 6.1.6].

16.2.3 TRIANGULATIONS AND ORIENTED MATROIDS

Checking whether a given collection S of subsets of V is a subdivision (or a triangulation) can be done knowing the oriented matroid M of V alone. (We refer to

Chapter 6 or to [BLS+99] for details on oriented matroids). Indeed, properties (1) and (2) of the following theorem are respectively equivalent to (DP) and (IP) in the definition of subdivision. Once (IP) and (DP) hold, (3) is equivalent to (UP).

THEOREM 16.2.1 [DRS10, Theorems 4.1.31 and 4.1.32]
A collection $S = \{S_1, \ldots, S_m\}$ of subsets of $V \subset \mathbb{R}^d$ is a subdivision if and only if:

1. *Every S_i is spanning in (that is, contains a basis of) M.*
2. *For every oriented circuit $C = (C^+, C^-)$ in M with $C^+ \subset S_i$ for some $S_i \in S$, either $C^- \subset S_i$ or C^- is not contained in any $S_j \in S$.*
3. *S is not empty and for every S_i in S and facet F of S_i, either F is contained in a facet of M or there is another $S_j \in M$ having F as a facet and lying in the other side of F. (Facets of V can easily be detected via cocircuits of M.)*

This led Billera and Munson to introduce triangulations of (perhaps not realizable) oriented matroids [BM84], including notions of placing, pulling and pushing for them. See also [BLS+99, Ch. 9] and [San02]. The oriented matroid approach to triangulations is implemented in the software package TOPCOM [PR03, Ram02], which is currently part of the distribution of `polymake` (see Chapter 67).

16.3 REGULAR TRIANGULATIONS AND SUBDIVISIONS

One way to construct a subdivision of a point set $V \subset \mathbb{R}^d$ is to lift it to $\mathbb{R}^{d+1}$ and then look at the projection of the lower facets (facets visible from below) of the lifted point set. This allows any convex hull algorithm in $\mathbb{R}^{d+1}$ (see Chapter 26 of this Handbook) to be used to compute subdivisions in $\mathbb{R}^d$. The subdivisions obtained in this way are called regular, and they have some special properties.

GLOSSARY

Regular subdivision: Let $V = \{v_1, \ldots, v_n\} \subset \mathbb{R}^d$ and let $\alpha = (\alpha_1, \ldots, \alpha_n) \in \mathbb{R}^n$ be any vector. The ***regular subdivision*** of V obtained by the ***lifting vector*** α is defined as follows [GKZ94, Lee91, Zie95, DRS10]:

(i) Let $\tilde{v}_i = (v_i, \alpha_i)$ for each i and compute the facets of $\tilde{V} = \{\tilde{v}_1, \ldots, \tilde{v}_n\}$.

(ii) Project the lower facets of $\tilde{V}$ onto $\mathbb{R}^d$.

Here, a ***lower facet*** of $\tilde{V}$ is a facet that is visible from below. That is, a facet whose outer normal vector has its last coordinate negative. Observe that the "projection" step is combinatorially trivial. For each lower facet $\{\tilde{v}_{i_1}, \ldots, \tilde{v}_{i_k}\}$ of $\tilde{V}$ we simply make $\{v_{i_1}, \ldots, v_{i_k}\}$ a cell in the subdivision.

Combinatorially isomorphic subdivisions: Let V and V' be point sets. A subdivision S of V and a subdivision S' of V' are ***combinatorially isomorphic*** if there is a bijection between V and V' such that for every face F of S the corresponding subset $F' \subseteq V'$ is a face of S', and vice-versa. See Figure 16.3.2.

Shellable: A pure polytopal complex S is ***shellable*** if it is 0-dimensional (i.e., a nonempty finite set of points) or else $\dim(S) = k > 0$ and S has a ***shelling***, i.e., an ordering $P_1, \ldots, P_m$ of its maximal faces such that for $2 \leq j \leq m$ the intersection of P_j with $P_1 \cup \cdots \cup P_{j-1}$ is nonempty and is the beginning segment of a shelling of the $(k-1)$-dimensional boundary complex of P_j [Zie95].

Nonconvex polytope: The region of $\mathbb{R}^d$ enclosed by a $(d-1)$-dimensional pure polytopal complex homeomorphic to a sphere.

16.3.1 PROPERTIES AND EXAMPLES OF REGULAR SUBDIVISIONS

All regular subdivisions are shellable. To see this, consider a ray in the direction $(0, \ldots, 0, -1)$ emitted from a point in the interior of conv $(\tilde{V})$ and in sufficiently general position. The order in which the ray crosses the supporting hyperplanes of the lower facets of $\tilde{V}$ is a shelling order. (This is an example of a *line shelling* of conv $(\tilde{V})$; see [BM71, Zie95]). In contrast, there exist nonshellable subdivisions, starting in dimension 3. The first example was Rudin's nonshellable triangulation of a tetrahedron [Rud58]. For some additional discussion, including a nonshellable triangulation of the 3-cube, see Ziegler [Zie95].

Regular subdivisions include all lexicographic ones. The subdivision obtained by pushing/pulling $v_{i_1}, \ldots, v_{i_k}$ in that order coincides with the regular subdivision constructed by choosing $|\alpha_{i_1}| \gg \cdots \gg |\alpha_{i_k}| \gg 0$, where $\alpha_i > 0$ if v_i is pushed and $\alpha_i < 0$ if v_i is pulled, and choosing $\alpha_i = 0$ if v_i is neither pulled nor pushed. Pulling or pushing points in a regular subdivision produces a regular subdivision.

Non-regular subdivisions exist for all dimensions $d \geq 2$ and number of points $n \geq d+4$ [Lee91] (see Figure 16.3.2(b) for a smallest example). In dimension 2 all subdivisions are combinatorially isomorphic to regular ones as a consequence of Steinitz's Theorem (see [Grü67, Zie95] and Chapter 15 of this Handbook). For $d \geq 3$ and $n \geq 7$ the same is not true [Lee91] (see Figure 16.3.3(b)).

There are, in general, many fewer regular than nonregular triangulations:

- For fixed $n-d$, the number of regular subdivisions of V is bounded above by a polynomial of degree $(n-d-1)^2$ in d [BFS90]. In contrast, the number of non-regular triangulations can grow exponentially, even fixing $n-d=4$ [DHSS96].
- For fixed d, the number of regular triangulations is bounded above by $2^{O(n \log n)}$ while the number of nonregular ones can grow as $2^{\Omega(n^{\lfloor d/2 \rfloor})}$ [DRS10, Sec. 6.1].

A prime example of a regular subdivision is the Delaunay subdivision, obtained with $\alpha_i = \|v_i\|^2$. In fact, regular subdivisions are sometimes called *weighted Delaunay subdivisions*. The regular subdivision obtained with $\alpha_i = -\|v_i\|^2$ is the "farthest site" Delaunay subdivision. See Chapter 27 of this Handbook.

Regularity of a subdivision of V cannot be decided based only on the oriented matroid of V: the two point sets in Figure 16.3.2 have the same oriented matroid, yet the triangulation in (a) is regular and the triangulation in (b) is not.

Checking regularity is equivalent to feasibility of a linear program on n variables (the α_i's) with one constraint for each pair of adjacent cells (local convexity of the lift) [DRS10, Sec. 8.2]. On the other hand, checking whether a triangulation or subdivision is combinatorially isomorphic to a regular one is very hard, as difficult as the existential theory of the reals (determining feasibility of systems of real polynomial inequalities). See comments on the Universality Theorem in Chapters 6 and 15 of this Handbook, and in [Zie95]).

EXAMPLES

Figure 16.3.1 shows the two triangulations (both regular) of the vertices of a 3-dimensional bipyramid over a triangle. In (a) there are two tetrahedra, sharing a common internal triangle; in (b) there are three, sharing a common internal edge.

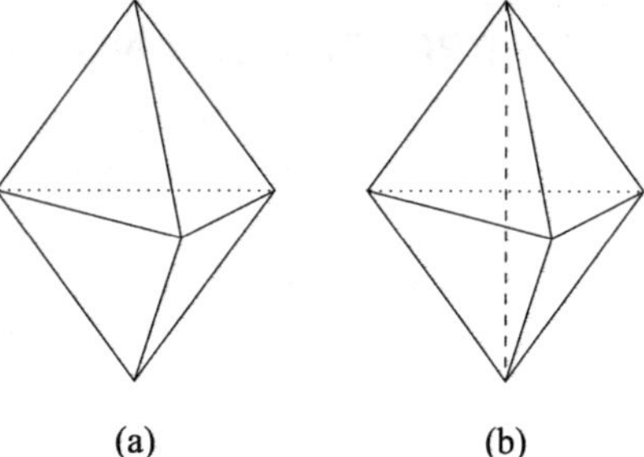

FIGURE 16.3.1
The two triangulations of a set of 5 *points in* $\mathbb{R}^3$.

Figure 16.3.2 shows the same triangulation for two different sets of 6 points in $\mathbb{R}^2$ having the same oriented matroid. Only the first triangulation is regular.

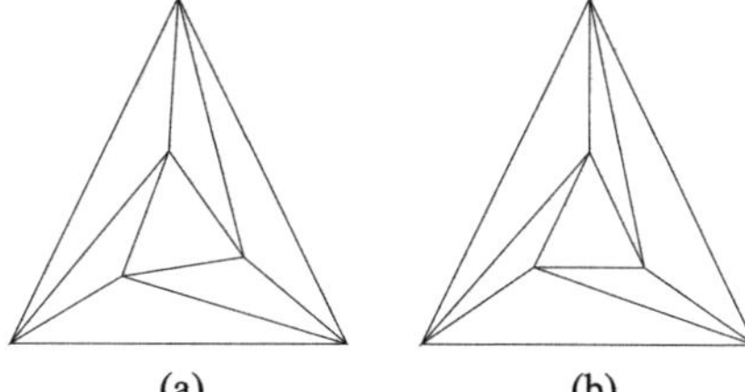

FIGURE 16.3.2
A regular and a nonregular (but combinatorially isomorphic) triangulation.

Figure 16.3.3 shows two 3-polytopes, both with 7 vertices. The "capped triangular prism" in (a) admits two nonregular triangulations: $\{1257, 1457, 1236, 1267, 1345, 1346, 1467\}$ and $\{1245, 1247, 1237, 1367, 1356, 1456, 1467\}$. Both triangulations are combinatorially isomorphic to regular ones. The polytope in (b) is obtained from the capped triangular prism by slightly rotating the top triangle. It has one nonregular triangulation, not combinatorially isomorphic to a regular one: $\{1245, 1247, 1237, 1367, 1356, 1456, 1467, 2457, 2367, 2345\}$. See [Lee91].

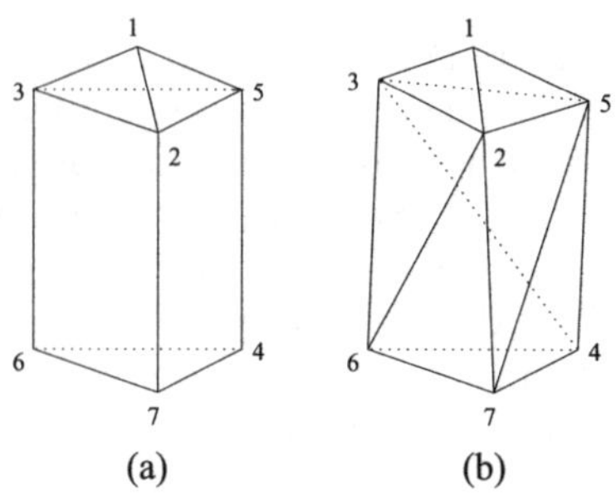

FIGURE 16.3.3
Two polytopes with nonregular triangulations.

16.3.2 TRIANGULATING REGIONS BETWEEN POLYTOPES

Not every nonconvex polytope can be triangulated without additional vertices. One classical example is Schönhardt's 3-polytope [Sch28] [DRS10, Example 3.6.1]: a nonconvex octahedron obtained by slightly rotating with respect to one another the two triangular facets of a triangular prism. However, regular triangulations can sometimes be used to triangulate nonconvex regions:

- Suppose P and Q are two d-polytopes in $\mathbb{R}^d$ with Q contained in P. If we start with the trivial subdivision and push all vertices of P, then we get a subdivision in which the region of P outside Q is triangulated [GP88].
- Now suppose P and Q are two disjoint d-polytopes in $\mathbb{R}^d$ with vertex sets V and W, respectively. One can triangulate the region in $\mathrm{conv}\,(P \cup Q)$ that is exterior to both P and Q by the following procedure [GP88]:
    1. Let H be a hyperplane for which P and Q are contained in opposite open halfspaces. Construct a regular subdivision of $V \cup W$ by setting α_i equal to the distance of v_i to H for each $v_i \in V \cup W$. For example, if $H = \{x \mid a \cdot x = \beta\}$, then α_i can be taken to equal $|a \cdot v_i - \beta|$.
    2. Refine this arbitrarily and ignore the simplices within P or Q.

However, if we are given *three* mutually disjoint polytopes P, Q and R it may be impossible to triangulate, without additional vertices, the region in $\mathrm{conv}\,(P \cup Q \cup R)$ that is exterior to the three. As an example, removing the three tetrahedra $2457, 2367, 2345$ from the convex hull of 234567 in Figure 16.3.3(b) gives Schönhardt's nontriangulable 3-polytope [Sch28].

16.4 SECONDARY AND FIBER POLYTOPES

This section deals with the structure of the collection of all regular subdivisions of a given finite set of points $V = \{v_1, \ldots, v_n\} \subset \mathbb{R}^d$. The main result is that the poset of regular triangulations of V is isomorphic to the face poset of a certain polytope, the secondary polytope of V. This polytope plays an important role in the study of generalized discriminants and determinants [GKZ94] and Gröbner bases [Stu96]. Secondary polytopes are studied in detail in [DRS10, Ch. 5].

16.4.1 SECONDARY POLYTOPES

GLOSSARY

Volume vector: Suppose T is a triangulation of $V = \{v_1, \ldots, v_n\}$. Define the volume vector $z(T) = (z_1, \ldots, z_n) \in \mathbb{R}^n$ by $z_i = \sum_{v_i \in F \in T} \mathrm{vol}\,(F)$, where the sum is taken over all d-simplices F in T having v_i as a vertex. $z(T)$ is sometimes called the ***GKZ-vector*** of T, to honor Gelfand, Kapranov, and Zelevinsky.

Secondary polytope: The secondary polytope $\Sigma(V)$ is the convex hull of the volume vectors of all triangulations of V.

Link: The link of a face F of a triangulation T is $\{G \mid F \cup G \in T, F \cap G = \emptyset\}$.

THEOREM 16.4.1 Gelfand, Kapranov, and Zelevinsky [GKZ94]

1. *$\Sigma(V)$ has dimension $n-d-1$. Its affine span is defined by the $d+1$ equations*

$$\sum_{i=1}^{n} z_i = (d+1)\mathrm{vol}\,(P), \quad \text{and} \quad \sum_{i=1}^{n} z_i v_i = (d+1)\mathrm{vol}\,(P)c, \qquad (16.4.1)$$

where c is the centroid of $P = \mathrm{conv}\,(V)$.

2. *The poset of (nonempty) faces of $\Sigma(V)$ is isomorphic to the poset of all regular subdivisions of V, partially ordered by refinement:*
   - *For a given regular subdivision S of V, the volume vectors of all regular triangulations that refine S are the vertices of a face F_S of $\Sigma(V)$. This face contains also the volume vectors of nonregular triangulations refining S, but these are never vertices of it.*
   - *A lifting vector $(\alpha_1, \ldots, \alpha_n)$ produces S as a regular subdivision if and only if it lies in the relatively open normal cone of F_S in $\Sigma(V)$.*

The secondary polytope $\Sigma(V)$ can also be expressed as a discrete or continuous Minkowski sum of polytopes coming from a representation of V as a projection of the vertices of an $(n-1)$-dimensional simplex. See Section 16.4.3.

The following are consequences of Theorem 16.4.1:

1. The vertices of $\Sigma(V)$ are the volume vectors of the regular triangulations.
2. Two nonregular triangulations can have the same volume vector, but two regular ones, or a regular and a nonregular one, cannot. This implies that the triangulation of Figure 16.3.2(b) is nonregular: Flipping three diagonals in it produces another triangulation with the same volume vector.

Lifting vectors, as used in the definition of regular subdivision, correspond to linear functionals in the ambient space of $\Sigma(V)$: Suppose $S = \{S_1, \ldots, S_m\}$ is a regular subdivision of $V = \{v_1, \ldots, v_n\} \subset \mathbb{R}^d$ determined by lifting numbers $\alpha_1, \ldots, \alpha_n$. Let $f : \mathrm{conv}\,(V) \to \mathbb{R}$ be the piecewise-linear convex function whose graph is given by the lower facets of $Q = \mathrm{conv}\,(\{(v_1, \alpha_1), \ldots, (v_n, \alpha_n)\})$. Define c_j to be the centroid of the polytope $P_j = \mathrm{conv}\,(S_j)$, $1 \le j \le m$. Then the inequality

$$\sum_{i=1}^{n} \alpha_i z_i \ge (d+1) \sum_{j=1}^{m} \mathrm{vol}\,(P_j) f(c_j)$$

is valid on the secondary polytope and holds with equality at the volume vector of a triangulation T if and only if T refines S. This allows for a local monotone algorithm to construct the regular triangulation corresponding to a certain lifting vector α: start with any regular triangulation T of V (for example, a lexicographic one) and do flips in it (see Section 16.4.2) always decreasing $\sum_{i=1}^{n} \alpha_i z_i$ and keeping the regularity property. When such flips no longer exist we have the desired regular triangulation. For Delaunay triangulations this procedure was first described in [ES96].

There are two ubiquitous polytopes that can be constructed as secondary polytopes (see Chapter 15 of this Handbook):

- When V is the set of vertices of a convex n-gon, $\Sigma(V)$ is the ***associahedron*** of dimension $n-3$ [Lee89]. Explicit coordinates and inequalities for $\Sigma(V)$ can be found in [Zie95]. See also Section 16.7.3.
- When V is the vertex set of the Cartesian product of a d-simplex and a segment, $\Sigma(V)$ is a d-dimensional ***permutahedron***, affinely isomorphic to the convex hull of the $(d+1)!$ vectors obtained by permuting the coordinates in $(1, 2, 3, \ldots, d)$. See Section 16.7.1.

16.4.2 THE GRAPH OF TRIANGULATIONS OF V

GLOSSARY

(Oriented) circuits, Radon partitions: A circuit is a set $C = \{c_1, \ldots, c_k\}$ of affinely dependent points such that every proper subset is affinely independent. This implies, in particular, that $k = \dim(C) + 2$ and that there is a unique (up to rescaling) affine dependence $\lambda = (\lambda_1, \ldots, \lambda_k)$ of C. (That is, a solution to $\sum_i \lambda_i = 0$ and $\sum_i \lambda_i c_i = 0$.) Since λ has no zero entries, this produces a natural (and unique) way of partitioning C as the disjoint union of $C^+ = \{c_i \mid \lambda_i > 0\}$ and $C^- = \{c_i \mid \lambda_i < 0\}$ with the property that $\operatorname{conv}(C^+) \cap \operatorname{conv}(C^-) \neq \emptyset$. The pair (C^+, C^-) is the ***oriented circuit*** or ***Radon partition*** of C.

Triangulations of a circuit: A circuit C with Radon partition (C^+, C^-) has exactly two triangulations

$$T^+ = \{C \setminus \{c_i\} \mid c_i \in C^+\} \qquad \text{and} \qquad T^- = \{C \setminus \{c_i\} \mid c_i \in C^-\}.$$

See Figure 16.3.1 for an example.

Adjacent triangulations, bistellar flips, graph of triangulations: Let T be a triangulation of V. Suppose there is a circuit C in V such that T contains one of the two triangulations, say T^+, of C, and suppose further that the links in T of all the cells of T^+ are identical. Then it is possible to construct a new triangulation T' of V by removing T^+ (together with its link) and inserting T^- (with the same link). This operation is called a ***(geometric bistellar) flip***, and T' is said to be adjacent to T. The set of all triangulations of V, under adjacency by flips, forms the graph of triangulations, or flip-graph, of V.

Flips correspond to "next-to-minimal" elements in the refinement poset of subdivisons of V: If T_1 and T_2 are two adjacent triangulations of V, then there is a subdivision S whose only two proper refinements are T_1 and T_2. Conversely, if all proper refinements of a subdivision S are triangulations then S has exactly two such refinements, which are adjacent triangulations [DRS10, Sec. 2.4].

In particular, all edges of the secondary polytope $\Sigma(V)$ correspond to adjacency between regular triangulations. That is, the 1-skeleton of $\Sigma(V)$ is a subgraph of the graph of triangulations of V. But it may not be an induced subgraph: sometimes two regular triangulations T_1 and T_2 are adjacent but the intermediate subdivision S is not regular, hence the flip between T_1 and T_2 does not correspond to an edge of $\Sigma(V)$ [DRS10, Examples 5.3.4 and 5.4.16].

Since the 1-skeleton of every $(n-d-1)$-polytope is $(n-d-1)$-connected and has all vertices of degree at least $n-d-1$, all regular triangulations of V have at least $n-d-1$ flips and the adjacency graph of regular triangulations of V is $(n-d-1)$-connected. For general triangulations the following is known:

- When $|V| \leq \dim(V)+3$ all triangulations are regular [Lee91]. When $|V| = \dim(V)+4$ every triangulation has at least three flips and the graph of all triangulations of V is 3-connected [AS00].
- When $\dim(V) \leq 2$ the graph of triangulations is known to be connected [Law72], and triangulations are known to have at least $n-3$ flips. Whether the flip-graph is $(n-3)$-connected for every V is an open question.
- Flip-deficient triangulations (that is, triangulations with fewer than $n-d-1$ flips) exist starting in dimension three and with $|V| = 8$ [DRS10, Ex. 7.1.1]. Triangulations exist in dimension three with $O(\sqrt{n})$ flips, in dimension four with $O(1)$ flips, and in dimension six without flips [San00].
- Point sets exist with disconnected graphs of triangulations in dimension five and higher [San05b]. In dimension six they can be constructed in general position [San06]. Whether they exist in dimensions three and four is open.

The graph of triangulations of V is known to be connected for the vertex sets of cyclic polytopes [Ram97], of Cartesian products of two simplices if one of them has dimension at most three [San05a, Liu16a] and of regular cubes up to dimension four [Pou13]. It is known to be disconnected for the vertex set of the Cartesian product of a 4-simplex and a k-simplex, for sufficiently large k [Liu16b].

Figure 16.4.1 shows the five regular triangulations of a set of 5 points in $\mathbb{R}^2$, marking which pairs of triangulations are adjacent.

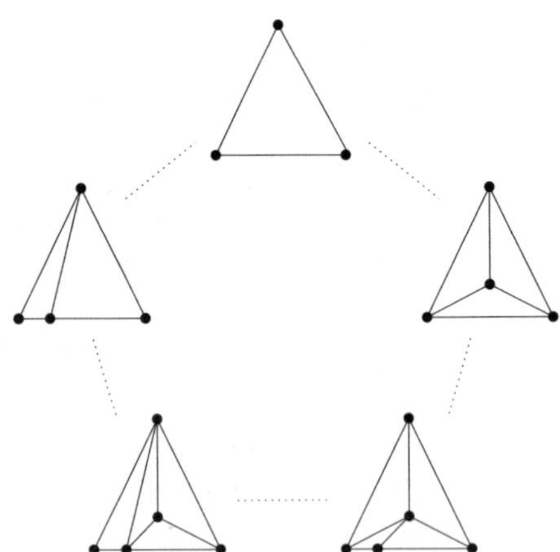

FIGURE 16.4.1
A polygon of regular triangulations.

See [DHSS96] for properties of the polytope that is the convex hull of the $(0,1)$ incidence vectors of all triangulations of V, and for the relationship of it to $\Sigma(V)$. For the vertex-set of a convex n-gon this polytope was first described in [DHH85].

16.4.3 FIBER POLYTOPES

GLOSSARY

Fiber polytope: Let $\pi : P \to Q$ be an affine surjective map (a projection) from a polytope P to a polytope Q. A ***section*** of π is a continuous map $\gamma : Q \to P$ with $\pi(\gamma(x)) = x$ for all $x \in Q$. The fiber polytope of π is defined to be the set of all average values of the sections of π:

$$\Sigma(P,Q) = \left\{ \frac{1}{\mathrm{vol}\,(Q)} \int_Q \gamma(x)dx \mid \gamma \text{ is a section of } \pi \right\}.$$

Equivalently, $\Sigma(P,Q)$ equals the Minkowski average of all fibers of the map π.

π-induced and π-coherent subdivisions: A subdivision S of Q is π-induced if each cell in S equals the image of the vertex set of a face of P. It is π-coherent if π factors as $Q \to P' \to P$ for a polytope P' of dimension $\dim(P) + 1$ and S equals the lower part of the convex hull of P'.

Baues poset: The poset of all nontrivial π-induced subdivisions, under refinement, is called the Baues poset of π.

The fiber polytope $\Sigma(P,Q)$ has dimension $\dim(P) - \dim(Q)$, and its face poset equals the refinement poset of π-coherent subdivisions. Billera, Kapranov and Sturmfels [BKS94] conjectured the Baues poset to be homotopy equivalent to a $(\dim P - \dim Q - 1)$-sphere, and they proved the conjecture for $\dim Q = 1$. Although the conjecture in its full generality was soon disproved [RZ96], the following so-called ***Generalized Baues Problem*** received attention: *When is the Baues poset homotopy equivalent to a sphere of dimension* $\dim P - \dim Q - 1$*?* See [Rei99] for a survey of this problem and [BS92, BKS94, DRS10, HRS00, RGZ94, Zie95] for general information on fiber polytopes. The following three cases are of special interest:

1. When P is a simplex all the subdivisions of $V = \pi(\mathrm{vert}(P))$ are π-induced subdivisions, and π-coherent ones are the regular ones [BFS90, BS92, Zie95]. $\Sigma(P,Q)$ is the secondary polytope of V. Examples in which the Baues poset is disconnected exist [San06].
2. When $\dim(Q) = 1$ the finest π-induced subdivisions are ***monotone paths*** in the 1-skeleton of P with respect to the linear functional that is constant on each fiber of π. $\Sigma(P,Q)$ is the ***monotone path polytope*** of P and the Baues complex is homotopy equivalent to a $(\dim(P) - 2)$-sphere [BKS94].
3. When P is a regular k-cube, Q is a zonotope (a Minkowski sum of segments). π-induced subdivisions are the ***zonotopal tilings*** of Q, the finest ones being ***cubical tilings***. The Bohne-Dress Theorem [RGZ94, HRS00] states that the Baues poset coincides with the *extension space* of the oriented matroid dual to that of Q. Examples of disconnected extension spaces (with $\dim(Q) = 3$) have been recently announced [Liu16c].

The d-dimensional permutahedron stands out as a fiber polytope belonging to the three cases above: it is the monotone path polytope of the $(d+1)$-cube projected to a line via the sum of coordinates, and it is the secondary polytope of the vertex set of $\Delta_d \times I$ for a d-simplex Δ_d and a segment I. This coincidence is a special instance of the ***combinatorial Cayley Trick***: If $\pi_i : P_i \to Q_i$, $i = 1, \ldots, k$ are polytope projections with $Q_i \in \mathbb{R}^d$ for all i then the fiber polytopes of the following two projections, defined from the π_i's in the natural way, coincide [HS95, HRS00]:

$$\begin{aligned}\Pi_C : P_1 * \cdots * P_k &\to \operatorname{conv}(Q_1 \times \{e_1\} \cup \cdots \cup Q_k \times \{e_k\}) \subset \mathbb{R}^{d+k},\\ \Pi_M : P_1 \times \cdots \times P_k &\to Q_1 + \cdots + Q_k \subset \mathbb{R}^d.\end{aligned}$$

Here $Q_1 + \cdots + Q_k$ is a ***Minkowski sum***, $P_1 \times \cdots \times P_k$ is a ***Cartesian product***, $P_1 * \cdots * P_k := \operatorname{conv}((P_1 \times \cdots \times \{0\} \times \{e_1\}) \cup \cdots \cup (\{0\} \times \cdots \times P_k \times \{e_k\})) \subset \mathbb{R}^{d_1 + \cdots + d_k + k}$ is the ***join*** of the polytopes $P_i \subset \mathbb{R}^{d_i}$ and $\operatorname{conv}(Q_1 \times \{e_1\} \cup \cdots \cup Q_k \times \{e_k\})$ is called the ***Cayley polytope*** or ***Cayley sum*** of $Q_1, \ldots, Q_k$.

When all the P_i's are simplices we have that $P_1 * \cdots * P_k$ is a simplex, so that all subdivisions of the Cayley polytope are Π_C-induced, and the Π_M-induced subdivisions of $Q_1 + \cdots + Q_k$ are the ***mixed subdivisions***, of interest in algebraic geometry [HS95]. Hence, the Cayley Trick gives a bijection between all subdivisions of a Cayley polytope and mixed subdivisions of the corresponding Minkowski sum.

If we further assume that all P_i's are segments then Π_M-induced subdivisions are zonotopal tilings $Q_1 + \cdots + Q_k$, and the Cayley polytope of a set of segments is called a ***Lawrence polytope***. (Equivalently, a Lawrence polytope is a polytope with a centrally symmetric Gale diagram. See Chapters 6 and 15 for the definition of Gale diagrams). In particular, the Cayley Trick and the Bohne-Dress Theorem imply the following three posets to be isomorphic, for a set $Q_1, \ldots, Q_k$ of segments:

- The Baues poset of the Lawrence polytope $\operatorname{conv}(Q_1 \times \{e_1\} \cup \cdots \cup Q_k \times \{e_k\})$.
- The poset of zonotopal tilings of the zonotope $Q_1 + \cdots + Q_k$.
- The poset of extensions of the oriented matroid dual to $\{Q_1, \ldots, Q_k\}$.

16.5 FACE VECTORS OF SUBDIVISIONS

In this section we examine some properties of the numbers of faces of different dimensions of a triangulation or subdivision. More information on f-vectors, g-vectors, and h-vectors can be found in Chapter 17 of this Handbook. But note that the symbol d is here shifted by one unit with respect to the conventions there, since there the h- and g-vector are usually meant for the boundary of a d-polytope.

GLOSSARY

Boundary and interior: Every face of dimension $d-1$ of a pure d-dimensional complex $S \subset \mathbb{R}^d$ is contained in exactly one or two cells. Those contained in one cell, together with their faces, form the ***boundary*** ∂S of S, which is a pure polytopal $(d-1)$-complex. The faces of S that are not in the boundary form the ***interior*** of S, which is not a polytopal complex. The boundary complex of a subdivision S of V equals $\{F \mid \text{is a face of } S \text{ and } F \subseteq G \text{ for some facet } G \text{ of } V\}$.

f-vector: Let $f_j(S)$ denote the number of j-dimensional faces of S, $-1 \le j \le d$. Note that $f_{-1}(S) = 1$ since the empty set is the unique face of S of dimension -1. The f-vector of S is $f(S) = (f_0(S), \ldots, f_d(S))$. In an analogous way we define $f(\partial S)$ and $f(\operatorname{int} S)$. Note that $f_{-1}(\partial S) = 1$ and $f_{-1}(\operatorname{int} S) = 0$.

Simplicial polytope: A polytope all of whose faces are simplices.

(Geometric) simplicial complex: A polytopal complex all of whose faces are simplices.

h-vector and g-vector: For a d-dimensional simplicial complex S we define the h-vector $h(S) = (h_0(S), \ldots, h_{d+1}(S))$ with generating function $h(S,x) = \sum_{i=0}^{d+1} h_i x^{d+1-i}$ as

$$\sum_{i=0}^{d+1} h_i x^{d+1-i} = \sum_{i=0}^{d+1} f_{i-1}(S)(x-1)^{d+1-i}.$$

We define the g-vector $g(S) = (g_0(S), \ldots, g_{\lfloor (d+1)/2 \rfloor}(S))$ as

$$g_i(S) = h_i(S) - h_{i-1}(S), \quad 1 \le i \le \lfloor (d+1)/2 \rfloor.$$

Take $h_i(S) = 0$ if $i < 0$ or $i > d+1$, and $g_i(S) = 0$ if $i < 0$ or $i > \lfloor (d+1)/2 \rfloor$.

16.5.1 h-VECTORS and g-VECTORS

The f-vector and the h-vector of a simplicial complex carry the same information on S, since the definition of the h-vector can be inverted to give

$$\sum_{i=0}^{d+1} f_{i-1} x^{d+1-i} = \sum_{i=0}^{d+1} h_i(S)(x+1)^{d+1-i}.$$

But the h-vector more directly captures topological properties of S. For example:

$$(-1)^d h_{d+1} = -f_{-1} + f_0 - f_1 + \cdots + (-1)^{d-1} f_{d-1} + (-1)^d f_d = \chi(S) - 1,$$

where $\chi(S)$ is the ***Euler characteristic*** of S. In particular, $h_{d+1} = 0$ if S is a ball and $h_{d+1} = 1$ if S is a sphere, of whatever dimension.

For every triangulation T of a point configuration the following hold:

1. The sum $\sum_{i=0}^{d+1} h_i(T)$ of the components of the h-vector equals $f_d(T)$.
2. $h_0(T) = f_{-1}(T) = 1$ and $h_i(T) \ge 0$ for all i [Sta96].
3. The h-vector of ∂T is symmetric; i.e., $h_i(\partial T) = h_{d-i}(\partial T)$, $0 \le i \le d$. These are the Dehn-Sommerville equations; see [MS71, Sta96, Zie95] and Chapter 17 of this Handbook. The case $h_d(\partial T) = h_0(\partial T) = 1$ is Euler's formula.
4. The h-vectors of T, ∂T, and $\operatorname{int} T$ are related in the following ways [MW71]:

$$h_i(T) - h_{d+1-i}(T) = h_i(\partial T) - h_{i-1}(\partial T) = g_i(\partial T), \; 0 \le i \le d+1.$$

$$h_i(T) = h_{d+1-i}(\operatorname{int} T), \; 0 \le i \le d+1.$$

   In particular, the h-vectors and the f-vectors of ∂T and $\operatorname{int} T$ are completely determined by the h-vector (and hence the f-vector) of T.

5. Assume further that T is shellable and that $P_1, \ldots, P_m$ is a shelling order of the d-dimensional simplices in T. In particular, each P_j meets $\bigcup_{i=1}^{j-1} P_i$ in some positive number s_j of facets of P_j, $2 \le j \le m$. Define also $s_1 = 0$. Then $h_i(T)$ equals $\operatorname{card}\{j \mid s_j = i\}$, $0 \le i \le d+1$ [McM70, MS71, Sta96].

6. Assume further that T is regular. Then, for every integer $0 \leq k \leq d+2$, the vector $(h_0(T)-h_{d+k+1}(T), h_1(T)-h_{d+k}(T), h_2(T)-h_{d+k-1}(T), \ldots, h_{\lfloor(d+k+1)/2\rfloor}(T)-h_{\lfloor(d+k+2)/2\rfloor}(T))$ is an M-sequence [BL81]. (See Chapter 17 of this Handbook for the definition of M-sequence.)

Properties (1) to (5) above hold for any *simplicial ball* (simplicial complex that is topologically a d-ball). Property (6) follows from the *g-theorem*, and it would hold for all simplicial balls if the *g-conjecture* holds for simplicial sphere (see Chapter 17 for details). In the other direction, Billera and Lee [BL81] conjectured the conditions in part (6) to be also *sufficient* for a vector to be the h-vector of a regular triangulation. In dimensions up to four the conditions indeed characterize h-vectors of balls [LS11, Kol11], but in dimensions five and higher Kolins [Kol11] has shown that some vectors satisfying property (6) are not the h-vectors of any ball, let alone regular triangulation.

TABLE 16.5.1 h- and g-vectors of polytopal complexes.

S	h-vector	g-vector
$\{\emptyset\}$	(1)	(1)
Set of n points	$(1, n-1)$	(1)
Line segment	$(1, 0, 0)$	$(1, -1)$
Boundary of convex n-gon	$(1, n-2, 1)$	$(1, n-3)$
Trivial subdivision of convex n-gon	$(1, n-3, 0, 0)$	$(1, n-4)$
Boundary of tetrahedron	$(1, 1, 1, 1)$	$(1, 0)$
Trivial subdivision of tetrahedron	$(1, 0, 0, 0, 0)$	$(1, -1, 0)$
Boundary of cube	$(1, 5, 5, 1)$	$(1, 4)$
Trivial subdivision of cube	$(1, 4, 0, 0, 0)$	$(1, 3, -4)$
Triangulation of cube into 6 tetrahedra (See Figure 16.7.3(a))	$(1, 4, 1, 0, 0)$	$(1, 3, -3)$
Boundary of triangular prism	$(1, 3, 3, 1)$	$(1, 2)$
Trivial subdivision of triangular prism	$(1, 2, 0, 0, 0)$	$(1, 1, -2)$
Triangulation of triangular prism into 3 tetrahedra (See Figure 16.7.1)	$(1, 2, 0, 0, 0)$	$(1, 1, -2)$

The definitions of h- and g-vectors can be extended to arbitrary polytopal complexes in the following recursive way:

1. $g_0(S) = h_0(S)$.
2. $g_i(S) = h_i(S) - h_{i-1}(S)$, $1 \leq i \leq \lfloor(d+1)/2\rfloor$.
3. $g(\emptyset, x) = h(\emptyset, x) = 1$. (Here $\emptyset$ denotes the empty polytopal complex, not to be confused with $\{\emptyset\}$, the polytopal complex consisting of a the empty set.)
4. $h(S, x) = \sum\limits_{G \text{ face of } S} g(\partial G, x)(x-1)^{d-\dim(G)}$.

This restricts to the previous definition since $g(\partial\Delta, x) = 1$ for every simplex Δ.

For example, the h-vector of the trivial subdivision of a point set V equals:

$$h_i(\{V\}) = \begin{cases} g_i(\partial(\{V\})), & 1 \leq i \leq \lfloor d/2 \rfloor, \\ 0, & \lfloor d/2 \rfloor < i \leq d. \end{cases}$$

where $\partial(\{V\})$ denotes the complex of proper faces of V [Bay93]. For any subdivision S of V one has: $h_i(S) \geq h_i(P)$ and $h_i(\partial S) \geq h_i(\partial P)$ for all i. In particular, $f_d(S) \geq h_{\lfloor d/2 \rfloor}(\partial S) \geq h_{\lfloor d/2 \rfloor}(\partial P)$ [Bay93, Sta92, Kar04].

Table 16.5.1 lists the h-vectors and g-vectors of some polytopal complexes.

16.5.2 STACKED AND EQUIDECOMPOSABLE POLYTOPES

GLOSSARY

In the following definitions V is the vertex set of a convex d-polytope P.

Shallow triangulation: A triangulation T of V is called ***shallow*** if every face F of T is contained in a face of P of dimension at most $2\dim(F)$.

Weakly neighborly: A polytope P is ***weakly neighborly*** if every set of $k+1$ vertices is contained in a face of dimension at most $2k$ for all $k \leq d/2$.

Equidecomposable: If all triangulations of V have the same f-vector, then P is ***equidecomposable***.

Stacked, k-stacked: If P is simplicial and it has a triangulation in which there are no interior faces of dimension smaller than $d-k$, then P is ***k-stacked***. A 1-stacked polytope is simply called ***stacked***.

Shallow triangulations were introduced in order to understand the case of equality in the last result mentioned in Section 16.5.1: a shallow triangulation T has $h(T) = h(P)$ and $h(\partial T) = h(\partial P)$. See [Bay93, BL93]. Stackedness and neighborliness are somehow opposite properties for a polytope: if P is stacked then its f-vector is as small as can be (for a given dimension and number of vertices) and if it is neighborly (and simplicial, see Chapter 15) then it is as big as can be. Both properties have implications for triangulations of P, in particular for shallow ones.

A polytope P is weakly neighborly if and only if all its triangulations are shallow [Bay93]. In this case P is equidecomposable. Equidecomposability admits the following characterization via circuits.

THEOREM 16.5.1 [DRS10, Section 8.5.3]

The following are equivalent for a point configuration V:

1. *All triangulations of V have the same f-vector (V is equidecomposable).*
2. *All triangulations of V have the same number of d-simplices.*
3. *Every circuit (C^+, C^-) of V is balanced:* $\text{card}(C^+) = \text{card}(C^-)$.

The following are examples of weakly neighborly, hence equidecomposable, polytopes [Bay93]: Cartesian products of two simplices of any dimensions, Lawrence polytopes (see Section 16.4 for the definition), pyramids over weakly neighborly polytopes, and subpolytopes of weakly neighborly polytopes. The only simplicial weakly neighborly polytopes are simplices and even-dimensional neighborly polytopes (those for which every $d/2$ vertices form a face of the polytope; see Chapter 15). The only weakly neighborly 3-polytopes are pyramids and the triangular prism. Regular octahedra are equidecomposable, but not weakly neighborly.

Assume now that P is simplicial. In this case having a shallow triangulation is equivalent to being $\lfloor d/2 \rfloor$-stacked [Bay93]. McMullen [McM04] calls a triangulation of P *small-face-free*, abbreviated *sff*, if it has no interior faces of dimension less than $d/2$. That is, P has an sff triangulation if and only if it is $(\lceil d/2 \rceil - 1)$-stacked. Observe that shallow and sff are the same if d is odd, but they differ by one unit in the dimension of the allowed interior faces if d is even. The sff-triangulation of P is unique, in case it exists [McM04]. Its existence and the minimum size of its interior feces are related to the g-vector of ∂P:

THEOREM 16.5.2 Generalized lower bound theorem [MW71, Sta80, MN13]

For any simplicial d-polytope P and any $k \in \{2, \ldots, \lfloor d/2 \rfloor\}$ *one has* $g_k(\partial P) \geq 0$, *with equality if and only if P is* $(k-1)$*-stacked.*

The inequality was proved by Stanley [Sta80] and the 'if' part of the equality was already established in [MW71]. The 'only if' part was recently proved by Murai and Nevo. The case $k = 2$ is the lower bound theorem, proved by Barnette [Bar73].

16.6 TRIANGULATIONS OF LATTICE POLYTOPES

A ***lattice polytope*** or an *integral polytope* is a polytope with vertices in $\mathbb{Z}^d$ (or, more generally, in a lattice $\Lambda \subset \mathbb{R}^d$). Lattice polytopes and their triangulations have interest in algebraic geometry and in integer optimization. See [BR07, CLO11, Stu96, BG09, HPPS14].

GLOSSARY

Normalized volume: The ***normalized volume*** of a lattice polytope $P \subset \mathbb{R}^d$ is its Euclidean volume multiplied by $d!$. It is always an integer. All references to volume in this section are meant normalized.

Empty simplex: A lattice simplex with no lattice points apart from its vertices.

Unimodular simplex: A lattice simplex $\Delta \subset \mathbb{R}^d$ whose vertices are an affine lattice basis of $\mathbb{Z}^d \cap \text{aff}(\Delta)$. If $\dim(\Delta) = d$ this is equivalent to $\text{vol}\,(\Delta) = 1$.

Unimodular triangulation: A triangulation into unimodular simplices.

Flag triangulation: A triangulation (or a more general simplicial complex) in which all *minimal non-faces* (i.e., minimal subsets of vertices that are not faces) have at most size two. A flag triangulation is the clique complex of its 1-skeleton.

Width: The width of a lattice polytope P with respect to an integer linear functional $f : \mathbb{Z}^d \to \mathbb{Z}$ equals $\max_{p \in P} f(p) - \min_{p \in P} f(p)$. The width of P itself is the minimum width taken over all non-zero integer linear functionals.

16.6.1 EMPTY SIMPLICES AND UNIMODULAR TRIANGULATIONS

Every lattice polytope P can be triangulated into empty simplices, via any triangulation of $A := P \cap \mathbb{Z}^d$ that uses all points (e.g., placing them with respect to a suitable ordering). One central question on lattice polytopes is whether they have unimodular triangulations, and how to construct them.

In dimension two, every lattice polygon has unimodular triangulations since every empty triangle is unimodular (by Pick's Theorem, see [BR07]). The set of unimodular triangulations of a lattice polygon is known to be connected under bistellar flips, and the number of them is at most 2^{3i+b-3}, where i and b are the numbers of lattice points in the interior and the boundary of P, respectively [Anc03].

In dimension three and higher there are empty non-unimodular simplices, which implies that not every polytope has unimodular triangulations. Empty 3-simplices are well understood, but higher dimensional ones are not:

1. Every 3-dimensional empty simplex is equivalent (modulo an affine integral automorphism $\mathbb{Z}^d \to \mathbb{Z}^d$) to the following $\Delta_{p,q}$, for some $0 \le p < q$ with $\gcd(p,q) = 1$ [Whi64]:

$$\Delta_{p,q} := \operatorname{conv}\{(0,0,0),(1,0,0),(0,0,1),(p,q,1)\}.$$

   In particular, they all have width one. Moreover, $\Delta_{p,q} \cong \Delta_{p',q'}$ if and only if $q' = q$ and $p' = \pm p^{\pm 1} \pmod q$. Observe that $\operatorname{vol}(\Delta_{p,q}) = q$.

2. All but finitely many empty 4-simplices have width one or two [BHHS16]. The exceptions have been computed up to volume 1000. The maximum volume among them is 179 and the maximum width is 4, achieved at a unique empty 4-simplex [HZ00]. It is conjectured that no larger exceptions exist.

3. The quotient group of $\mathbb{Z}^d$ by the sublattice generated by vertices of an empty d-simplex is cyclic if $d \le 4$ but not always so if $d \ge 5$ [BBBK11].

4. The maximum width of empty d-simplices lies between $2\lfloor d/2 \rfloor - 1$ and $O(d \log d)$ [Seb99, BLPS99] (see also Chapter 7).

The most general result about existence of unimodular triangulations is:

THEOREM 16.6.1 *(Knudson-Mumford-Waterman [Knu73])*

For each lattice polytope P there is an integer $k \in \mathbb{N}$ such that the dilation kP has a unimodular triangulation.

The original proof of this theorem does not lead to a bound on k. That the following k is valid for lattice d-polytopes of volume v was proved in [HPPS14]:

$$k = (d+1)!^{v!(d+1)^{(d+1)^2 v}}.$$

It is easy to show that the set $\{k \in \mathbb{N} \mid kP \text{ has some unimodular triangulation}\}$ is closed under taking multiples, for every P [DRS10, Thm. 9.3.17]. But it is unknown whether it contains all sufficiently large choices of k. It is also unknown whether there is a global value k_d that works for all polytopes of fixed dimension $d \ge 4$. These two open questions have a positive answer in dimension three, where every $k \in \mathbb{N} \setminus \{1,2,3,5\}$ works for every lattice 3-polytope [KS03, SZ13], or if the requirement is relaxed to kP having unimodular ***covers*** (sets of unimodular simplices contained in P and covering it), which is weaker than unimodular triangulations: there is a $k_d \in O(d^6)$ such that kP has unimodular covers for every $k \ge k_d$ and every d-polytope P [BG99, Theorem 3.23].

There is some literature on the existence of unimodular triangulations for particular lattice polytopes. See [HPPS14] for a recent survey. A notable open question is the following (see definition of smooth polytope in Section 16.6.2):

QUESTION 16.6.2

Does every smooth polytope have a (regular) unimodular triangulation?

16.6.2 RELATION TO TORIC VARIETIES AND GRÖBNER BASES

Let $\mathbb{K}[x_1, \ldots, x_n]$ be the polynomial ring over an algebraically closed field. For a nonnegative integer vector $u \in \mathbb{N}^n$ we denote as x^u the monomial $\prod_i x_i^{u_i}$ and to each $u \in \mathbb{Z}^n$ we associate the binomial $x^{u_+} - x^{u_-}$, where $u = u_+ - u_-$ is the minimal decomposition of u with $u_+, u_- \in \mathbb{N}^n$. (This minimal decomposition is the unique one in which u_+ and u_- have disjoint supports.) General references for the topics in this section are [CLO11, Stu96]. See also Chapter 7 in this Handbook.

GLOSSARY

Let $A \in \mathbb{Z}^{k \times n}$ be an integer matrix.

Toric ideal: The toric ideal of A, denoted I_A, is the ideal generated by the binomials $\{x^{u_+} - x^{u_-} : Au = 0\}$. The variety cut out by I_A is the (perhaps not normal) ***affine toric variety*** of A, denoted X_A.

Smooth polytope: A lattice polytope P is smooth if it is simple and the primitive normals to the facets at each vertex form a lattice basis. Equivalently, if each vertex v of P together with the first lattice point along each edge incident to v forms a unimodular simplex.

Normal polytope: P is ***normal*** or ***integrally closed*** if every integer point in $k \cdot \text{conv}\,(V)$ can be written as the sum of k (perhaps repeated) points of V.

To emphasize the relations to triangulations of point sets, we assume that the columns of A are $\{(a_1, 1), \ldots, (a_n, 1)\}$ for a point set $V = \{a_1, \ldots, a_n\} \in \mathbb{Z}^d$. Then all binomials defining I_A are homogeneous, so besides the affine toric variety X_A we have a projective variety Y_A. We also assume that V generates $\mathbb{Z}^d$ as an affine lattice (it is not contained in a proper sublattice). In these conditions:

1. The normalizations $\widetilde{X}_A$ and $\widetilde{Y}_A$ of X_A and Y_A are the toric varieties associated in the standard way to $\mathbb{R}_{\geq 0}(A)^\vee$ and to the normal fan of $\text{conv}\,(V)$ [Stu96, Cor. 13.6]. Here $\mathbb{R}_{\geq 0}(A)$ is the cone generated by the columns of A and $\mathbb{R}_{\geq 0}(A)^\vee$ is its polar cone. $\widetilde{Y}_A$ is smooth if and only if $\text{conv}\,(V)$ is smooth.
2. X_A is normal if and only if the semigroup $\mathbb{Z}_{\geq 0}(A)$ is normal (that is, $\mathbb{R}_{\geq 0}A \cap \mathbb{Z}A = \mathbb{Z}_{\geq 0}A$). Equivalently, if $V = \text{conv}\,(V) \cap \mathbb{Z}^d$ and $\text{conv}\,(V)$ is normal. In this case Y_A is called ***projectively normal.***
3. Y_A is normal if the same happens for sufficiently large k [Stu96, Thm. 13.11].

If V has a unimodular triangulation then the condition in (2) holds. Hence, a positive answer to Question 16.6.2 is weaker than a positive answer to the following:

QUESTION 16.6.3 *(Oda's question)*

Is every smooth projective toric variety projectively normal? That is, is every smooth lattice polytope normal?

Reduced Gröbner bases of I_A are related to regular triangulations of V as follows: Let $\alpha = (\alpha_1, \ldots, \alpha_n) \in \mathbb{R}^n$ be a generic weight vector. We can use α to define a regular triangulation T_α of V, and also to define a monomial order in $K[x_1, \ldots, x_n]$, which in turn defines a monomial initial ideal $\text{in}_\alpha(I_A)$ (and a Gröbner basis) of I_A. Then:

THEOREM 16.6.4 Sturmfels [Stu91], see also [Stu96, Chapter 8]

For every subset $S \subset V$ we have that S is not a face in T_α if and only if S is the support of a monomial in $\text{in}_\alpha(I_A)$. Said in a more algebraic language: the radical of $\text{in}_\alpha(I_A)$ equals the Stanley-Reisner ideal of T_α.

Moreover, if T_α is unimodular, then $\text{in}_\alpha(I_A)$ is square-free (it equals its own radical). In particular, the maximum degree of a generator in $\text{in}_\alpha(I_A)$ (and in the associated reduced Gröbner basis) equals the maximum size of a set S that is not a face in T_α but such that every proper subset of S is a face.

For example, if V has a regular, unimodular and flag triangulation, then I_A has a Gröbner basis consisting of binomials of degree two. For this reason regular flag unimodular triangulations of lattice polytopes are called ***quadratic***.

Observe that this theorem induces a surjective map from the monomial initial ideals of I_A to the regular triangulations of V. In case α is not generic then T_α may be a subdivision instead of a triangulation, and $\text{in}_\alpha(I_A)$ may not be a monomial ideal, but the above map extends to this case. That is to say, *the Gröbner fan* of I_A refines the secondary fan of V [Stu91].

In [Stu96, Ch. 10], Sturmfels further extends the correspondence in Theorem 16.6.4 to a map from *all* the subdivisions of V to the set of A-graded ideals, which generalize initial ideals of I_A (see the definition in [Stu96]). This map is no longer surjective, but its image contains all regular subdivisions and all unimodular triangulations. The set of all A-graded ideals has a natural algebraic structure called the ***toric Hilbert scheme*** of A and, using a notion of "flip" between radical monomial A-graded ideals, Maclagan and Thomas showed:

THEOREM 16.6.5 Maclagan and Thomas [MT02]

If two unimodular triangulations lie in different connected components of the graph of triangulations of V then the corresponding monomial radical A-graded ideals lie in different connected components of the toric Hilbert scheme of A.

Point configurations with unimodular triangulations not connected by a sequence of flips are known to exist [San05b]. Very recently Gaku Liu [Liu16b] has announced that this happens also for the Cartesian product $\Delta_4 \times \Delta_N$ of a 4-simplex and an N-simplex, which is a smooth polytope whose associated toric variety is the Cartesian product $\mathbb{P}^4 \times \mathbb{P}^N$ of two projective spaces. That is to say, the toric Hilbert scheme of $\mathbb{P}^4 \times \mathbb{P}^N$ is disconnected, for sufficiently large N.

Another relation between triangulations and toric varieties comes from looking at the affine toric variety U_A associated to the cone $\mathbb{R}_{\geq 0}(A)$ (recall X_A was associated to the dual cone $\mathbb{R}_{\geq 0}(A)^\vee$). This can be defined for every integer matrix A for which $\mathbb{R}_{\geq 0}(A)$ is a *pointed cone*, but the case where the columns of A are of the form $(a_i, 1)$ for a point configuration $V = \{a_1, \ldots, a_n\}$ gives us the extra property that U_A is *Gorenstein.* Since this construction depends only on conv (V) and not V itself, we now assume V to be the set of all lattice points in conv (V).

Then, every polyhedral subdivision T of V, considered as a polyhedral fan covering $\mathbb{R}_{\geq 0}(A)$, induces an affine toric variety U_T and a toric morphism $U_T \to U_A$. If T is a triangulation then U_T only has quotient singularities, which are *terminal* if T uses all lattice points in V. If T is a unimodular triangulation then U_T is smooth; that is, it is a resolution of the singularity of U_A at the origin. These resolutions are called *crepant* and they do not always exist (since not every lattice polytope has unimodular triangulations). See [DHZ98, DHZ01] for more details on this.

16.6.3 RELATION TO COUNTING LATTICE POINTS

For an integral polytope P and a nonnegative integer n, let $i(P,n)$ be the number of integer points in nP. Equivalently, it is the number of points $x \in P$ for which nx has integer coordinates. Ehrhart (1962) showed that $i(P,n)$ is a polynomial in n of degree d. This implies that the generating function $J(P,t) = \sum_{n=0}^{\infty} i(P,n)t^n$ can be rewritten as a rational function with denominator $(1-t)^{d+1}$ and numerator of degree at most d. That is:

$$J(P,t) = \frac{\sum_{j=0}^{d} h_j^* t^j}{(1-t)^{d+1}},$$

for a certain rational vector $h^*(P) = (h_0^*, \ldots, h_d^*)$. $i(P,n)$ and $J(P,t)$ are called, respectively, the ***Ehrhart polynomial*** and ***Ehrhart series*** of P.

Stanley [Sta96] proved that $h^*(P)$ is a nonnegative integer vector and that:

THEOREM 16.6.6 Stanley [Sta96]

For every triangulation T of P one has $h(T) \leq h^*(P)$ coordinate-wise, with equality if and only if T is unimodular. In particular, for a unimodular triangulation T:

$$i(P,n) = \sum_{i=0}^{d} \binom{n-1}{i} f_i(T).$$

See [Sta96, BR07] and Chapter 7 of this Handbook for more details on Ehrhart polynomials. As an example, if P is the standard unit 3-cube, then its unimodular triangulations have $h(T) = (1,4,1,0,0)$. Thus $J(P,t) = (1+4t+t^2)/(1-t)^4 = (1+4t+t^2)(1+4t+10t^2+20t^3+35t^4+\cdots) = 1+8t+27t^2+64t^3+125t^4+\cdots$, which agrees with $i(P,n) = (n+1)^3$. On the other hand, $i(P^o,n) = (n-1)^3 = -i(P,-n)$.

16.7 TRIANGULATIONS OF PARTICULAR POLYTOPES

16.7.1 PRODUCT OF TWO SIMPLICES

Consider the $(k+l)$-polytope $\Delta_k \times \Delta_l$, the product of a k-dimensional simplex Δ_k and an l-dimensional simplex Δ_l. We look at triangulations of its vertex set $V_{k,l}$.

Triangulations of the product of two simplices have interest from several perspectives: They can be used as building blocks to triangulate more complicated polytopes [Hai91, OS03, San00]. In toric geometry they correspond to Gröbner bases of the toric ideal of the product of two projective spaces [Stu96]. Via the

Cayley Trick, they correspond to mixed subdivisions of a dilated simplex [HRS00, San05a]. This connection also relates them to tropical geometry, where they correspond to tropical hyperplane arrangements, tropical convexity, and tropical oriented matroids [AD09, Hor16]. In optimization, their regular triangulations are closely related to dual transportation polytopes. See also [BCS88, DeL96, GKZ94] or [DRS10, Sect. 6.2].

If Δ_k and Δ_l are taken unimodular then $\Delta_k \times \Delta_l$ is ***totally unimodular***; that is, all maximal simplices with vertices in $V_{k,l}$ are unimodular. This implies that $\Delta_k \times \Delta_l$ is equidecomposable. In fact, for every triangulation T of $\Delta_k \times \Delta_l$, we have $f_{k+l}(T) = (k+l)!/(k!l!)$, and $h_i(T) = \binom{k}{i}\binom{l}{i}$ for $0 \le i \le k+l$ (with $h_i(T)$ taken to be zero if $i > \min\{k, l\}$) [BCS88]. For small values of k and/or l the following is known. Most of it is proved via the Cayley Trick mentioned in Section 16.4:

1. All triangulations of $\Delta_k \times \Delta_1$ are affinely equivalent. Hence, they are all lexicographic. There are $k!$ of them and the secondary polytope is (affinely equivalent to) the k-dimensional ***permutahedron***, the convex hull of the points obtained permuting the coordinates of $(1, 2, \ldots, k+1)$. See Chapter 15.
2. All triangulations of $\Delta_3 \times \Delta_2$ and $\Delta_4 \times \Delta_2$ are regular. But all $\Delta_k \times \Delta_l$ with $\min\{k, l\} \ge 3$ or $k - 3 \ge l = 2$, have nonregular triangulations [DeL96].
3. The number of triangulations of $\Delta_k \times \Delta_2$ grows as $2^{\Theta(k^2)}$ [San05a].
4. The flip-graphs of $\Delta_k \times \Delta_2$ and of $\Delta_k \times \Delta_3$ are connected [San05a, Liu16a], but that of $\Delta_k \times \Delta_4$ is not, for large k [Liu16b].

The ***staircase triangulation*** of $\Delta_k \times \Delta_l$ is easy to describe explicitly [BCS88, GKZ94, San05a]: By ordering the vertices of Δ_k and (independently) of Δ_l we have a natural bijection between $V_{k,l} = \{(v_i, w_j) : i = 0, \ldots, k, j = 0, \ldots, l\}$ and the integer points in $[0, k] \times [0, l]$. Then, the vertices in each of the $\binom{k+l}{k}$ monotone paths from $(0, 0)$ to (k, l) form a full-dimensional simplex in $\Delta_k \times \Delta_l$, and these simplices form a triangulation. The same triangulation is obtained starting with the trivial subdivision of P and pulling the vertices in any order compatible with the product order. That is, if $i \le i'$ and $j \le j'$, then (v_i, w_j) is pulled before $(v_{i'}, w_{j'})$. Figure 16.7.1 shows the staircase triangulation for $\Delta_2 \times \Delta_1$, a triangular prism. Its three tetrahedra are $\{00, 10, 20, 21\}$, $\{00, 10, 11, 21\}$ and $\{00, 01, 11, 21\}$, where ij is an abbreviation for (v_i, w_j).

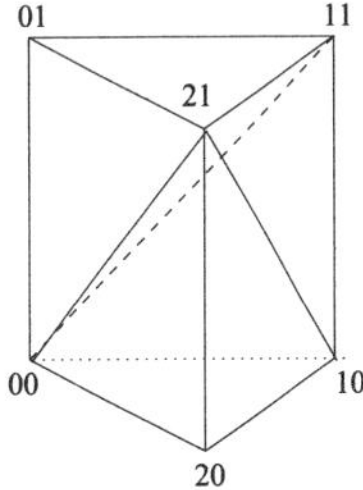

FIGURE 16.7.1
A triangulation of $\Delta_2 \times \Delta_1$.

The staircase triangulation generalizes to a product $\Delta_{l_1} \times \cdots \times \Delta_{l_n}$ of n simplices: consider the natural bijection between vertices of $\Delta_{l_1} \times \cdots \times \Delta_{l_n}$ and integer points in $[0, l_1] \times \cdots \times [0, l_n]$ and take as simplices the monotone paths from $(0, \ldots, 0)$

to $(l_1, \ldots, l_n)$. Staircase triangulations are the natural way to refine a Cartesian product of simplicial complexes to become a triangulation, by triangulating each individual product of simplices [ES52].

16.7.2 d-CUBES

The unit d-dimensional ***cube*** I^d is the d-fold product of the unit interval $I = [0, 1]$ with itself. Here we consider triangulations of it using only its set V of vertices. Up to $d = 4$ they have been completely enumerated: The 3-cube has precisely 74 triangulations, all regular, falling into 6 classes modulo affine symmetries of the cube. Figure 16.7.2 shows the unique (modulo symmetry) triangulation of size 5. The 4-cube has 92 487 256 triangulations in total (247 451 symmetry classes) [PR03, Pou13] of which 87 959 448 are regular (235 277 symmetry classes) [HSYY08]. Nonregular triangulations of it were first described in [DeL96]. Still, its graph of triangulations is connected [Pou13].

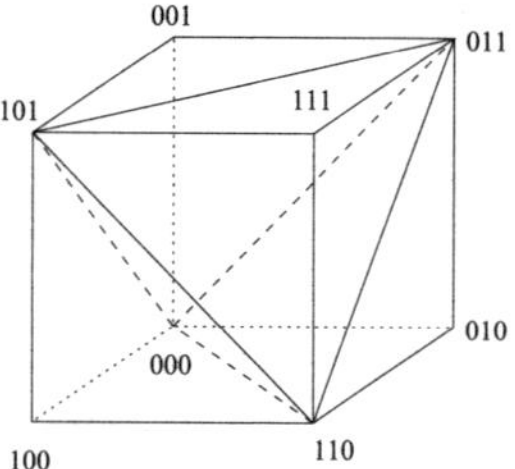

FIGURE 16.7.2
A minimum size triangulation of the 3-cube.

The maximum size of a triangulation of I^d is $d!$, achieved by unimodular triangulations. These include all pulling triangulations. Every unimodular triangulation S of I^d has $h_d(T) = h_{d+1}(T) = 0$ and $h_i(T) = A(d, i)$, $0 \leq i \leq d-1$, where $A(d, i)$ is the Eulerian number (the number of permutations of $\{1, \ldots, d\}$ having exactly i descents). Finding small triangulations of the d-cube is interesting in finite element methods [Tod76]. The minimum possible size is called the ***simplexity*** of the d-cube, which we denote $\varphi(d)$. The following summarizes what we know about it:

- Exact values are known up to dimension 7 [Hug93, HA96]. See Table 16.7.1.

TABLE 16.7.1 Minimal triangulations of d-cubes.

d	1	2	3	4	5	6	7
$\varphi(d)$	1	2	5	16	67	308	1493

- Up to $d = 5$ a minimum triangulation can be obtained by Sallee's ***corner slicing*** idea [Hai91, Sal82]: if the 2^{d-1} vertices of I^d with an odd number of nonzero coordinates are sliced and the rest of I^d is triangulated arbitrarily, a triangulation T of I^d arises in which all cells except the first 2^{d-1} have volume at least $2/d!$. Hence, T has size at most $(d! + 2^{d-1})/2$. This equals $\varphi(d)$ up to $d = 4$ and is off by one for $d = 5$ (where a corner-slicing triangulation of size $\varphi(5) = 67$ still exists). For $d = 6, 7$ the minimum corner-slicing triangulations have sizes 324 and 1820 [Hug93, HA96], much greater than $\varphi(d)$.

- The Hadamard bound for matrices implies that no simplex contained in the cube has volume greater than $(d+1)^{(d+1)/2}/(2^d d!)$ [Hai91]. Hence,

$$\varphi(d) \geq 2^d d!(d+1)^{-(d+1)/2}.$$

A better bound of

$$\varphi(d) \geq \frac{1}{2}\sqrt{6}^d d!(d+1)^{-(d+1)/2}$$

is obtained with the same argument with respect to hyperbolic volume [Smi00].

- For a triangulation S of size $|S|$, let

$$\rho(S) := (|S|/d!)^{1/d}.$$

This parameter is called the ***efficiency*** of S [Tod76]. It is at most one for every S, with equality if and only if S is unimodular. If I^d has a triangulation S of a certain efficiency then any triangulation of I^{kd} obtained by pulling refinement of the k-fold Cartesian product of S with itself has exactly the same efficiency [Hai91]. This shows that $\lim_{d\to\infty}(\varphi(d)/d!)^{1/d}$ exists, and that it is less or equal than the efficiency of any triangulation of any I^d. The best known upper bound for this limit is [OS03]

$$\lim_{d\to\infty} (\varphi(d)/d!)^{1/d} \leq 0.816,$$

but no strictly positive lower bound is known. Observe, for example, that the Hadamard bound only says $(\varphi(d)/d!)^{1/d} \gtrsim \frac{2}{\sqrt{d+1}}$. The improvement in [Smi00] merely changes the constant 2 in the numerator to a $\sqrt{6}$.

SOME SPECIFIC TRIANGULATIONS OF I^d

Standard or ***staircase*** triangulation: Consider the $d!$ monotone paths from $(0, \ldots, 0)$ to $(1, \ldots, 1)$ obtained by changing one coordinate from 0 to 1 at a time. The vertices in each such path form a unimodular simplex, that we call the monotone-path simplex corresponding to that permutation. The $d!$ simplices obtained in this way form a triangulation of I^d, which is nothing but the staircase triangulation of I^d regarded as the product of d segments. It is also known as Kuhn's triangulation [Tod76] and it admits the following alternative descriptions:

- It is the subdivision obtained slicing I^d by all hyperplanes of the form $x_i = x_j$.
- It is the pulling triangulation for any ordering of vertices with the following property: for every face F of I^d, the first vertex of F to be pulled is either the vertex with minimum or maximum support.
- It is the regular triangulation for the height function $f(v) = -(\sum v_i)^2$.
- It is the flag triangulation containing the edge uv for two vertices u and v if and only if $u - v$ is nonnegative (or nonpositive).

- It is a special case of the triangulation of an order polytope by linear extensions: the order polytope $P(\mathcal{O})$ of a poset $\mathcal{O}$ on d elements $\{a_1, \dots, a_d\}$ is the subpolytope of I^d cut by the inequalities $x_i \le x_j$ for every $a_i < a_j$. (This is the whole of I^d when $\mathcal{O}$ is an antichain.) Linear extensions of $\mathcal{O}$ are in bijection to permutations whose associated monotone-path simplex are contained in $P(\mathcal{O})$, and these simplices triangulate $P(\mathcal{O})$.

Alcoved triangulation: If I^d is sliced by all hyperplanes of the form $x_i + \cdots + x_j = m$ for $1 \le i < j \le d$ and $m \in \mathbb{N}$ another regular, unimodular triangulation of I^d is obtained. It was first described by Stanley, who showed a piecewise linear map from I^d to itself sending it to the standard triangulation. It was then studied in detail in [LP07]. It is the flag triangulation whose edges are the pairs uv such that the nonzero coordinates in $u - v$ alternate between $+1$ and -1.

Sallee's middle cut triangulation: Assume $d \ge 2$. Slice the cube into two polytopes by the hyperplane $x_1 + \cdots + x_d = \lfloor d/2 \rfloor$. Refine this subdivision to a triangulation by pulling the vertices in the following order: pull $(v_1, \dots, v_d)$ in step $1 + \sum_{i=0}^{d-1} v_{i+1} 2^i$. This triangulation has size $O(d!/d^2)$ [Sal84].

EXAMPLES

Figure 16.7.3 shows two triangulations of the 3-cube: (a) the one resulting from pulling the vertices in order of increasing distance to the origin, which equals the standard triangulation. And (b), one resulting from pushing the following vertices in order: 000, 100, 101, 001.

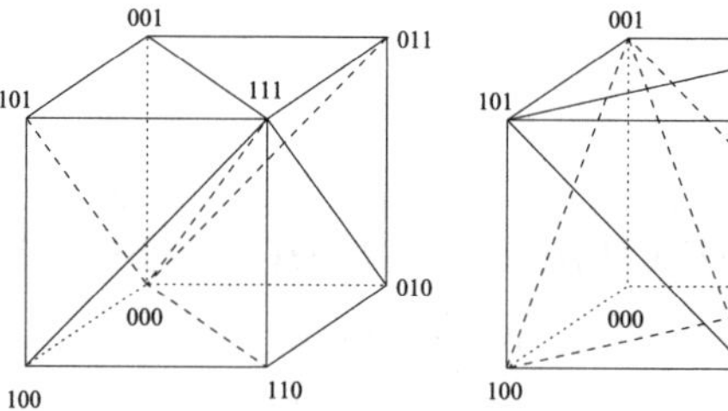

FIGURE 16.7.3
(a) *A pulling triangulation of the 3-cube.*
(b) *A pushing triangulation of the 3-cube.*

16.7.3 CONVEX n-GONS

Let V_n be the set of vertices of a convex n-gon. A subdivision of V_n is determined by a collection of mutually noncrossing internal diagonals, and vice versa. That is, the Baues poset of V_n equals the poset of non-crossing sets of diagonals of the n-gon (with respect to *reverse inclusion*). All subdivisions are regular (in fact, they are all pushing), so this is also the face poset of the secondary polytope of V_n, the ***associahedron*** of dimension $n-3$. See [Lee89, GKZ94, Zie95, DRS10].

The number of triangulations of the n-gon is the Catalan number

$$C_{n-2} = \frac{1}{n-1}\binom{2n-4}{n-2} \in \Theta\left(\frac{4^n}{n^{3/2}}\right).$$

This counts many other combinatorial structures, including the ways to parenthesize a string of $n-1$ symbols, Dyck paths from $(0,0)$ to $(n-2, n-2)$, and rooted binary trees with $n-2$ nodes. More generally, there are $\frac{1}{n-1}\binom{n-3}{j}\binom{n+j-1}{j+1}$ subdivisions of a convex n-gon having exactly j diagonals, $0 \leq j \leq n-3$. This equals the number of $(n-3-j)$-faces of the associahedron.

Two triangulations are adjacent if they share all but one diagonal. In particular, the graph of triangulations is $(n-3)$-regular. Its diameter equals $2n-10$ for all $n > 12$ [STT88, Pou14]. The flip-distance between two triangulations equals the *rotation distance* between the corresponding binary trees [STT88].

The associahedron is an ubiquitous polytope. It was first described by Tamari (1951) and arose in works of Stasheff and Milnor in the 1960's. The first explicit constructions of it as a polytope (and not only as a cell complex) are by Haiman (1984, unpublished) and Lee [Lee89]. Besides being a secondary polytope, it is a *generalized permutahedron*, and dual to a *cluster complex* of type A. The latter means that it can be realized with facet normals equal to the nonnegative roots of type A in a way that captures the combinatorics of the corresponding cluster algebra. See [CSZ15] and the references therein for more details.

16.7.4 CYCLIC POLYTOPES

Cyclic polytopes are neighborly polytopes with very nice combinatorial properties. Being neighborly means their f-vectors and h-vectors are as big as can be, which reflects in their triangulations having number, size and flip-graph diameter also (asymptotically) as big as can be.

GLOSSARY

Cyclic polytope: The standard ***cyclic polytope*** $C_d(n)$ of dimension d with n vertices is the convex hull of the points $c(1), \ldots, c(n)$ on the d-dimensional ***moment curve***, the parametrized curve $c : \mathbb{R} \to \mathbb{R}^d$ defined as

$$c(t) = (t, t^2, \ldots, t^d).$$

The convex hull of any n points in the curve has the same combinatorics (both as a polytope and as an oriented matroid) as $C_d(n)$. Depending on the context, one calls cyclic polytope the convex hull of any n points in this curve, any polytope with the same oriented matroid, or any polytope with the same face lattice.

Cyclic polytopes are simplicial and are the prime examples of ***neighborly polytopes***: d-polytopes in which every $\lfloor d/2 \rfloor$ vertices form a face. For even d this implies they are weakly neighborly in the sense of Section 16.5 (hence equidecomposable).

Consider the natural projection $\pi : C_{d+1}(n) \to C_d(n)$ that forgets the last coordinate. Every triangulation T of $C_d(n)$ is a section of this projection, and every flip to another triangulation T' gives a section that is pointwise above or below T. This allows us to speak of upward and downward flips, and to give a structure of poset to the set of triangulations of $C_d(n)$: T is below T' in the poset if there is a sequence of upward flips from T to T'. This poset is called the ***first Stasheff-Tamari order***, denoted $T <_1 T'$. The ***second Stasheff-Tamari order*** is given by $T <_2 T'$ if the section produced by T is pointwise below that of T'. Clearly, $T <_1 T'$ implies $T <_2 T'$, but the converse is not known in general.

The first Stasheff-Tamari order has a unique minimum and a unique maximum triangulation, which we denote $T_{\hat{0}}$ and $T_{\hat{1}}$: The lower and upper envelope of $C_{d+1}(n)$, which coincide with the pushing and pulling triangulations of $C_d(n)$ with respect to the natural ordering on the vertices. Other properties of this poset are as follows (see [Ram97] or [DRS10, Ch. 6]):

- Every triangulation of $C_d(n)$ lies in some monotone path from the $T_{\hat{0}}$ to $T_{\hat{1}}$.
- Every maximal chain of flips from $T_{\hat{0}}$ to $T_{\hat{1}}$ corresponds to a triangulation of $C_{d+1}(n)$, so that the length of the chain equals the size of the triangulation.
- In odd dimension upward flips decrease size by exactly one, so that $T_{\hat{0}}$ and $T_{\hat{1}}$ are maximum and minimum in size, respectively.

These properties imply the following [DRS10, Corollary 6.1.20]:

1. The minimum and maximum sizes of triangulations of $C_d(n)$ equal the numbers of upper and lower facets of $C_{d+1}(n)$, which are:
$$\binom{n-\lfloor (d+1)/2\rfloor -1}{\lceil (d+1)/2\rceil -1} \quad \text{and} \quad \binom{n-\lceil (d+1)/2\rceil}{\lfloor (d+1)/2\rfloor}.$$
2. If d is odd, the diameter of the flip graph of $C_d(n)$ equals $\binom{n-(d+1)/2-1}{(d+1)/2}$ (the size of every triangulation of $C_{d+1}(n)$). Indeed, this is the length of every monotone chain from $T_{\hat{0}}$ to $T_{\hat{1}}$, and for every two triangulations T and T' there is a cycle in the flip-graph going through them and of twice that length.
3. If d is even, every triangulation is at flip-distance at most $\binom{n-d/2-2}{d/2}$ from $T_{\hat{1}}$, with equality for $T_{\hat{0}}$. Hence, the graph of flips of $C_d(n)$ has diameter between $\binom{n-d/2-2}{d/2}$ and $2\binom{n-d/2-2}{d/2}$.

Exact formulas exist for the number of triangulations of $C_{n-4}(n)$ (the first non-trivial case of few more vertices than the dimension). For even n, $C_{n-4}(n)$ has exactly $(n+4)2^{(n-4)/2}-n$ triangulations. Of these, at most $6\binom{n/2}{4}+3\binom{n/2}{3}+4\binom{n/2}{2}-n/2+2$ are regular, and this number is exact for sufficiently generic coordinatizations of the oriented matroid of $C_d(n)$ [AS02]. Similar but different formulas exist when n is odd. For arbitrary number of vertices the following is proved in [DRS10]:

THEOREM 16.7.1 [DRS10, Thm. 6.1.22]

The cyclic polytope $C_d(n)$ has at least $\Omega(2^{n^{\lfloor d/2\rfloor}})$ triangulations, for d fixed.

16.8 SOURCES AND RELATED MATERIAL

FURTHER READING

Chapter 29 discusses triangulations of more general (e.g., nonconvex) objects. Chapter 27 provides details on Delaunay triangulations and Voronoi diagrams. We refer also to Chapter 15, on basic properties of convex polytopes. For the topics of Sections 16.5 and 16.6; see also Chapters 17 and 7, respectively.

A recent book covering the topics in this chapter is [DRS10]. A section on triangulations and subdivisions of convex polytopes can be found in the survey article [BL93]. The book [Zie95] and the article [Lee91] contain information on regular subdivisions and triangulations; for their important role in generalized discriminants and determinants see the book [GKZ94], and for their significance in computational algebra see the book [Stu96]. Additional references can be found in the above-mentioned sources, as well as the citations given in this chapter.

RELATED CHAPTERS

Chapter 3: Tilings
Chapter 6: Oriented matroids
Chapter 7: Lattice points and lattice polytopes
Chapter 15: Basic properties of convex polytopes
Chapter 17: Face numbers of polytopes and complexes
Chapter 20: Polyhedral maps
Chapter 26: Convex hull computations
Chapter 27: Voronoi diagrams and Delaunay triangulations
Chapter 29: Triangulations and mesh generation
Chapter 36: Computational convexity

REFERENCES

[AD09] F. Ardila and M. Develin. Tropical hyperplane arrangements and oriented matroids. *Math. Z.*, 262:795–816, 2009.

[Anc03] E.E. Anclin. An upper bound for the number of planar lattice triangulations. *J. Combin. Theory Ser. A*, 103:383–386, 2003.

[AS00] M. Azoala and F. Santos. The graph of triangulations of a point configuration with $d+4$ vertices is 3-connected. *Discrete Comput. Geom.*, 23:489–536, 2000.

[AS02] M. Azaola and F. Santos. The number of triangulations of the cyclic polytope $C(n, n-4)$. *Discrete Comput. Geom.*, 27:29–48, 2002.

[Bar73] D.W. Barnette. A proof of the lower bound conjecture for convex polytopes. *Pacific J. Math.*, 46:349–354, 1973.

[Bay93] M.M. Bayer. Equidecomposable and weakly neighborly polytopes. *Israel J. Math.*, 81:301–320, 1993.

[BBBK11] M. Barile, D. Bernardi, A. Borisov, and J.-M. Kantor. On empty lattice simplices in dimension 4. *Proc. Amer. Math. Soc.*, 139:4247–4253, 2011.

[BCS88] L.J. Billera, R. Cushman, and J.A. Sanders. The Stanley decomposition of the harmonic oscillator. *Nederl. Akad. Wetensch. Indag. Math.*, 50:375–393, 1988.

[BFS90] L.J. Billera, P. Filliman, B. Sturmfels. Constructions and complexity of secondary polytopes. *Adv. Math.*, 83:155–179, 1990.

[BG09] W. Bruns and J. Gubeladze. *Polytopes, Rings, and K-Theory.* Monographs in Mathematics, Springer, New York, 2009.

[BG99] W. Bruns and J. Gubeladze. Normality and covering properties of affine semigroups. *J. Reine Angew. Math.*, 510:161–178, 1999.

[BHHS16] M. Blanco, C. Haase, J. Hofmann, and F. Santos. The finiteness threshold width of lattice polytopes. Preprint, arXiv:1607.00798, 2016.

[BKS94] L.J. Billera, M.M. Kapranov, and B. Sturmfels. Cellular strings on polytopes. *Proc. Amer. Math. Soc.*, 122:549–555, 1994.

[BL81] L.J. Billera and C.W. Lee. The numbers of faces of polytope pairs and unbounded polyhedra. *European J. Combin.*, 2:307–322, 1981.

[BL93] M.M. Bayer and C.W. Lee. Combinatorial aspects of convex polytopes. In P.M. Gruber and J.M. Wills, editors, *Handbook of Convex Geometry*, pages 485–534, Elsevier, Amsterdam, 1993.

[BLPS99] W. Banaszczyk, A.E. Litvak, A. Pajor, and S.J. Szarek. The flatness theorem for nonsymmetric convex bodies via the local theory of Banach spaces. *Math. Oper. Res.*, 24:728–750, 1999.

[BLS+99] A. Björner, M. Las Vergnas, B. Sturmfels, N. White, and G.M. Ziegler. *Oriented Matroids*, 2nd edition. Vol. 46 of *Encyclopedia of Mathematics and Its Applications*, Cambridge University Press, 1999.

[BM71] H. Bruggesser and P. Mani. Shellable decompositions of cells and spheres. *Math. Scand.*, 29:197–205, 1971.

[BM84] L.J. Billera and B.S. Munson. Triangulations of oriented matroids and convex polytopes. *SIAM J. Algebraic Discrete Methods*, 5:515–525, 1984.

[BR07] M. Beck and S. Robins. *Computing the Continuous Discretely: Integer-point Enumeration in Polyhedra.* Undergraduate Texts in Mathematics, Springer-Verlag, Berlin, 2007.

[BS92] L.J. Billera and B. Sturmfels. Fiber polytopes. *Ann. of Math. (2)*, 135:527–549, 1992.

[CLO11] D.A. Cox, J.B. Little, and H.K. Schenck. *Toric Varieties.* AMS, Providence, 2011.

[CSZ15] C. Ceballos, F. Santos, and G.M. Ziegler. Many non-equivalent realizations of the associahedron. *Combinatorica*, 35:513–551, 2015.

[DeL96] J.A. De Loera. Nonregular triangulations of products of simplices. *Discrete Comput. Geom.*, 15:253–264, 1996.

[Dey93] T.K. Dey. On counting triangulations in d dimensions. *Comput. Geom.*, 3:315–325, 1993.

[DHH85] G.B. Dantzig, A.J. Hoffman, and T.C. Hu. Triangulations (tilings) and certain block triangular matrices. *Math. Program.*, 31:1–14, 1985.

[DHSS96] J.A. De Loera, S. Hoşten, F. Santos, and B. Sturmfels. The polytope of all triangulations of a point configuration. *Doc. Math.*, 1:103–119, 1996.

[DHZ01] D.I. Dais, C. Haase, and G.M. Ziegler. All toric local complete intersection singularities admit projective crepant resolutions. *Tohoku Math. J. (2)*, 53:95–107, 2001.

[DHZ98] D.I. Dais, M. Henk, and G.M. Ziegler. All abelian quotient c.i.-singularities admit projective crepant resolutions in all dimensions. *Adv. in Math.*, 139:194–239, 1998.

[DRS10] J.A. De Loera, J. Rambau, and F. Santos. *Triangulations: Structures for Algorithms and Applications.* Vol. 25 of *Algorithms and Computation in Mathematics*, Springer-Verlag, Berlin, 2010.

[ES52] S. Eilenberg and N.E. Steenrod. *Foundations of Algebraic Topology.* Princeton University Press, 1952.

[ES96] H. Edelsbrunner and N.R. Shah. Incremental topological flipping works for regular triangulations. *Algorithmica*, 15:223–241, 1996.

[GKZ94] I.M. Gelfand, M.M. Kapranov, and A.V. Zelevinsky. *Discriminants, Resultants and Multidimensional Determinants.* Birkhäuser, Boston, 1994.

[GP88] J.E. Goodman and J. Pach. Cell decomposition of polytopes by bending. *Israel J. Math.*, 64:129–138, 1988.

[Grü67] B. Grünbaum. *Convex Polytopes.* Wiley, London, 1967.

[HA96] R.B. Hughes and M.R. Anderson. Simplexity of the cube. *Discrete Math.*, 158:99–150, 1996.

[Hai91] M. Haiman. A simple and relatively efficient triangulation of the n-cube. *Discrete Comput. Geom.*, 6:287–289, 1991.

[Hor16] S. Horn. A Topological Representation Theorem for tropical oriented matroids, *J. Combin. Theory. Ser. A*, 142:77–112, 2016.

[HPPS14] C. Haase, A. Paffenholz, L.C. Piechnik, and F. Santos. Existence of unimodular triangulations — positive results. Preprint, `arXiv:1405.1687`, 2014.

[HRS00] B. Huber, J. Rambau, and F. Santos. The Cayley trick, lifting subdivisions and the Bohne-Dress theorem on zonotopal tilings. *J. Eur. Math. Soc. (JEMS)*, 2:179–198, 2000.

[HS95] B. Huber and B. Sturmfels. A polyhedral method for solving sparse polynomial systems. *Math. Comp.*, 64:1541–1555, 1995.

[HSYY08] P. Huggins, B. Sturmfels, J. Yu, and D.S. Yuster. The hyperdeterminant and triangulations of the 4-cube. *Math. Comp.*, 77:1653–1679, 2008.

[Hud69] J.F.P. Hudson. *Piecewise Linear Topology.* Benjamin, New York, 1969.

[Hug93] R.B. Hughes. Minimum-cardinality triangulations of the d-cube for $d = 5$ and $d = 6$. *Discrete Math.*, 118:75–118, 1993.

[HZ00] C. Haase and G.M. Ziegler, On the maximal width of empty lattice simplices. *Eur. J. Comb.*, 21:111–119, 2000.

[Kar04] K. Karu. Hard Lefschetz theorem for nonrational polytopes. *Invent. Math.*, 157:419–447, 2004.

[Knu73] F.F. Knudsen. Construction of nice polyhedral subdivisions. In G.R. Kempf, F.F. Knudsen, D. Mumford, and B. Saint-Donat, editors, *Toroidal Embeddings I*, vol. 339 of *Lecture Notes in Math.*, pages 109–164. Springer-Verlag, Berlin, 1973.

[Kol11] S. Kolins. f-vectors of triangulated balls. *Discrete Comput. Geom.*, 46:427–446, 2011.

[KS03] J.-M. Kantor and K.S. Sarkaria. On primitive subdivisions of an elementary tetrahedron. *Pacific J. Math.*, 211:123–155, 2003.

[Law72] C.L. Lawson. Transforming triangulations. *Discrete Math.*, 3:365–372, 1972.

[Lee89] C.W. Lee. The associahedron and triangulations of the n-gon. *European J. Combin.*, 10:551–560, 1989.

[Lee91] C.W. Lee. Regular triangulations of convex polytopes. In P. Gritzmann and B. Sturmfels, editors, *Applied Geometry and Discrete Mathematics: The Victor Klee Festschrift*, vol. 4 of *DIMACS Series in Discrete Math. Theor. Comput. Sci.*, pages 443–456, AMS, Providence, 1991.

[Liu16a] G. Liu. Flip-connectivity of triangulations of the product of a tetrahedron and simplex. Preprint, `arXiv:1601.06031`, 2016.

[Liu16b] G. Liu. A zonotope and a product of two simplices with disconnected flip graphs. Preprint, `arXiv:1605.02366`, 2016.

[Liu16c] G. Liu. A counterexample to the extension space conjecture for realizable oriented matroids. Preprint, `arXiv:1606.05033`, 2016.

[LP07] T. Lam and A. Postnikov. Alcoved polytopes I. *Discrete Comput. Geom.*, 38:453–478, 2007.

[LS11] C.W. Lee and L. Schmidt. On the numbers of faces of low-dimensional regular triangulations and shellable balls. *Rocky Mountain J. Math.*, 41:1939–1961, 2011.

[McM04] P. McMullen. Triangulations of simplicial polytopes. *Beiträge Algebra Geom.*, 45:37–46, 2004.

[McM70] P. McMullen. The maximum numbers of faces of a convex polytope. *Mathematika*, 17:179–184, 1970.

[MN13] S. Murai and E. Nevo. On the generalized lower bound conjecture for polytopes and spheres. *Acta Math.*, 210:185–202, 2013.

[MS71] P. McMullen and G.C. Shephard. *Convex Polytopes and the Upper Bound Conjecture.* Vol. 3 of *London Math. Soc. Lecture Note Ser.*, Cambridge Univ. Press, 1971.

[MT02] D. Maclagan and R.R. Thomas. Combinatorics of the toric Hilbert scheme. *Discrete Comput. Geom.*, 27:249–272, 2002.

[MW71] P. McMullen and D.W. Walkup. A generalized lower-bound conjecture for simplicial polytopes. *Mathematika*, 18:264–273, 1971.

[OS03] D. Orden and F. Santos. Asymptotically efficient triangulations of the d-cube. *Discrete Comput. Geom.*, 30:509–528, 2003.

[Pou13] L. Pournin. The flip-graph of the 4-dimensional cube is connected. *Discrete Comput. Geom.*, 49:511–530, 2013.

[Pou14] L. Pournin. The diameter of associahedra, *Adv. Math.*, 259:13–42, 2014.

[PR03] J. Pfeifle and J. Rambau. Computing triangulations using oriented matroids. In M. Joswig and N. Takayama, editors, *Algebra, Geometry, and Software Systems*, pages 49–75, Springer-Verlag, Berlin, 2003.

[Ram02] J. Rambau. TOPCOM: Triangulations of point configurations and oriented matroids. In A.M. Cohen, X.-S. Gao, and N. Takayama, editors, *Mathematical Software*, pages 330–340, World Scientific, Singapore, 2002. Package available at `http://www.rambau.wm.uni-bayreuth.de/TOPCOM/`.

[Ram97] J. Rambau. Triangulations of cyclic polytopes and higher Bruhat orders. *Mathematika*, 44:162–194, 1997.

[Rei99] V. Reiner. The generalized Baues problem. In L.J. Billera, A. Björner, C. Greene, R. Simion, and R.P. Stanley editors, *New Perspectives in Algebraic Combinatorics*, vol. 38 of *MSRI Publ.*, pages 293–336, Cambridge Univ. Press, 1999.

[RGZ94] J. Richter-Gebert and G.M. Ziegler. Zonotopal tilings and the Bohne-Dress theorem. In H. Barcelo and G. Kalai, editors, *Jerusalem Combinatorics, '93*, pages 211–232, AMS, Providence, 1994.

[Rud58] M.E. Rudin. An unshellable triangulation of a tetrahedron. *Bull. Amer. Math. Soc.*, 64:90–91, 1958.

[RZ96] J. Rambau and G.M. Ziegler. Projections of polytopes and the generalized Baues conjecture. *Discrete Comput. Geom.*, 16:215–237, 1996.

[Sal82] J.F. Sallee. A triangulation of the n-cube. *Discrete Math.*, 40:81–86, 1982.

[Sal84] J.F. Sallee. The middle-cut triangulations of the n-cube. *SIAM J. Algebraic Discrete Methods*, 5:407–419, 1984.

[San00] F. Santos. A point set whose space of triangulations is disconnected. *J. Amer. Math. Soc.*, 13:611–637, 2000.

[San02] F. Santos. Triangulations of oriented matroids. *Mem. Amer. Math. Soc.*, 156:1–80, 2002.

[San05a] F. Santos. The Cayley trick and triangulations of products of simplices. In A. Barvinok, M. Beck, C. Haase, B. Reznick, and V. Welker, editors, *Integer Points in Polyhedra — Geometry, Number Theory, Algebra, Optimization*, pages 151–177, AMS, Providence, 2005.

[San05b] F. Santos. Non-connected toric Hilbert schemes. *Math. Ann.*, 332:645–665, 2005.

[San06] F. Santos. Geometric bistellar flips: the setting, the context and a construction. In *International Congress of Mathematicians, Vol. III*, pages 931–962, Eur. Math. Soc., Zürich, 2006.

[San13] F. Santos, Recent progress on the combinatorial diameter of polytopes and simplicial complexes. *TOP*, 21:426–460, 2013.

[Sch28] E. Schönhardt. Über die Zerlegung von Dreieckspolyedern in Tetraeder. *Math. Ann.*, 98:309–312, 1928.

[Seb99] A. Sebő, An introduction to empty lattice simplices. In *Integer Programming and Combinatorial Optimization*, vol. 1610 of *Lecture Notes Comp. Sci.*, pages 400–414, Springer, Berlin, 1999.

[Smi00] W.D. Smith, A lower bound for the simplexity of the N-cube via hyperbolic volumes. *European J. Combin.*, 21:131–137, 2000.

[Sta80] R.P. Stanley. The number of faces of a simplicial convex polytope. *Adv. Math.*, 35:236–238, 1980.

[Sta92] R.P. Stanley. Subdivisions and local h-vectors. *J. Amer. Math. Soc.*, 5:805–851, 1992.

[Sta96] R.P. Stanley. *Combinatorics and Commutative Algebra*, 2nd edition. Birkhäuser, Boston, 1996.

[STT88] D.D. Sleator, R.E. Tarjan, and W.P. Thurston. Rotation distance, triangulations, and hyperbolic geometry. *J. Amer. Math. Soc.*, 1:647–681, 1988.

[Stu91] B. Sturmfels. Gröbner bases of toric varieties. *Tohoku Math. J.*, 43:249–261, 1991.

[Stu96] B. Sturmfels. *Gröbner Bases and Convex Polytopes*. Vol. 8 of *Univ. Lecture Ser.*, Amer. Math. Soc., Providence, 1996.

[SZ13] F. Santos and G.M. Ziegler. Unimodular triangulations of dilated 3-polytopes. *Trans. Moscow Math. Soc.*, 74:293–311, 2013.

[Tod76] M.J. Todd. *The Computation of Fixed Points and Applications*. Vol. 124 of *Lecture Notes Econom. Math. Syst.*, Springer-Verlag, Berlin, 1976.

[Whi64] G.K. White. Lattice tetrahedra. *Canadian J. Math.*, 16:389–396, 1964.

[Zie95] G.M. Ziegler. *Lectures on Polytopes*. Vol. 152 of *Graduate Texts in Math.*, Springer-Verlag, Berlin, 1995.

17 FACE NUMBERS OF POLYTOPES AND COMPLEXES

Louis J. Billera and Anders Björner

INTRODUCTION

Geometric objects are often put together from simple pieces according to certain combinatorial rules. As such, they can be described as *complexes* with their constituent *cells*, which are usually polytopes and often simplices. Many constraints of a combinatorial and topological nature govern the incidence structure of cell complexes and are therefore relevant in the analysis of geometric objects. Since these incidence structures are in most cases too complicated to be well understood, it is worthwhile to focus on simpler invariants that still say something nontrivial about their combinatorial structure. The invariants to be discussed in this chapter are the *f-vectors* $f = (f_0, f_1, \dots)$, where f_i is the number of i-dimensional cells in the complex.

The theory of f-vectors can be discussed at two levels: (1) the numerical relations satisfied by the f_i numbers, and (2) the algebraic, combinatorial, and topological facts and constructions that give rise to and explain these relations. This chapter will summarize the main facts in the numerology of f-vectors (i.e., at level 1), with emphasis on cases of geometric interest.

The chapter is organized as follows. To begin with, we treat simplicial complexes, first the general case (Section 17.1), then complexes with various Betti number constraints (Section 17.2), and finally triangulations of spheres, polytope boundaries, and manifolds (Section 17.3). Then we move on to nonsimplicial complexes, discussing first the general case (Section 17.4) and then polytopes and spheres (Section 17.5).

17.1 SIMPLICIAL COMPLEXES

GLOSSARY

The convex hull of any set of $j+1$ affinely independent points in $\mathbb{R}^n$ is called a ***j-simplex***. See Chapter 15 for more about this definition, and for the notions of *faces* and *vertices* of a simplex.

A ***geometric simplicial complex*** Γ is a finite nonempty family of simplices in $\mathbb{R}^n$ such that (i) $\sigma \in \Gamma$ implies that $\tau \in \Gamma$ for every face τ of σ, and (ii) if $\sigma, \tau \in \Gamma$ and $\sigma \cap \tau \neq \emptyset$ then $\sigma \cap \tau$ is a face of both σ and τ.

An ***abstract simplicial complex*** Δ is a finite nonempty family of subsets of some ground set V (the ***vertex set***) such that if $A \in \Delta$ and $B \subseteq A$ then

$B \in \Delta$. (Note that always $\emptyset \in \Delta$.) The elements $A \in \Delta$ are called ***faces***. Define the ***dimension*** of a face A and of Δ itself by $\dim A = |A| - 1$; $\dim \Delta = \max_{A \in \Delta} \dim A$. By a ***d-complex*** we mean a d-dimensional complex.

With every geometric simplicial complex Γ we associate an abstract simplicial complex by taking the family of vertex sets of its simplices. Conversely, every d-dimensional abstract simplicial complex Δ can be realized in $\mathbb{R}^n$ for $n \geq 2d+1$ (and sometimes less) by some geometric simplicial complex. The latter is unique up to homeomorphism, so it is correct to think of the realization map as a one-to-one correspondence between abstract and geometric simplicial complexes. We will therefore drop the adjectives "abstract" and "geometric" and speak only of a ***simplicial complex***.

For a simplicial complex Δ, let $\Delta^i = \{i\text{-dimensional faces}\}$ and let $f_i = |\Delta^i|$. The integer sequence $f(\Delta) = (f_0, f_1, \dots)$ is called the ***f-vector*** of Δ. (The entry $f_{-1} = 1$ is usually suppressed.) The subcomplex $\Delta^{\leq i} = \bigcup_{j \leq i} \Delta^j$ is called the ***i-skeleton*** of Δ.

A simplicial complex Δ is called ***pure*** if all maximal faces are of equal dimension. It is called ***r-colorable*** if there exists a partition of the vertex set $V = V_1 \cup \cdots \cup V_r$ such that $|A \cap V_i| \leq 1$ for all $A \in \Delta$ and $1 \leq i \leq r$. Equivalently, Δ is r-colorable if and only if its 1-skeleton $\Delta^{\leq 1}$ is r-colorable in the standard sense of graph theory. An $(r-1)$-complex that is both pure and r-colorable is sometimes called ***balanced***.

The ***clique complex*** of a graph is the collection of vertex sets of all its cliques (complete induced subgraphs). These are also known as ***flag complexes***.

For integers $k, n \geq 1$, there is a unique way of writing

$$n = \binom{a_k}{k} + \binom{a_{k-1}}{k-1} + \cdots + \binom{a_i}{i}$$

so that $a_k > a_{k-1} > \cdots > a_i \geq i \geq 1$. Then define

$$\partial_k(n) = \binom{a_k}{k-1} + \binom{a_{k-1}}{k-2} + \cdots + \binom{a_i}{i-1},$$

and

$$\partial^k(n) = \binom{a_k - 1}{k-1} + \binom{a_{k-1} - 1}{k-2} + \cdots + \binom{a_i - 1}{i-1}.$$

Also let $\partial_k(0) = \partial^k(0) = 0$.

Let $\mathbb{N}^\infty$ denote the set of sequences $(n_0, n_1, \dots)$ of nonnegative integers, and $\mathbb{N}^{(\infty)}$ the subset of sequences such that $n_k = 0$ for all sufficiently large k. We call $n \in \mathbb{N}^{(\infty)}$ a ***K-sequence*** if

$$\partial_{k+1}(n_k) \leq n_{k-1} \qquad \text{for all } k \geq 1.$$

We call $n \in \mathbb{N}^\infty$ an ***M-sequence*** if

$$n_0 = 1 \text{ and } \partial^k(n_k) \leq n_{k-1} \qquad \text{for all } k \geq 2.$$

THE KRUSKAL-KATONA THEOREM AND SOME RELATIVES

The following basic result characterizes the f-vectors of simplicial complexes.

THEOREM 17.1.1 *Kruskal-Katona Theorem*

For $f = (f_0, f_1, \dots) \in \mathbb{N}^{(\infty)}$ the following are equivalent:

(i) *f is the f-vector of a simplicial complex;*

(ii) *f is a K-sequence.*

Theorem 17.1.1 has a generalization to colored complexes, whose statement will require some additional definitions. Fix an integer $r > 0$. Then define $\binom{n}{k}_r$ as follows: partition $\{1, \dots, n\}$ into r subsets $V_1, \dots, V_r$ as evenly as possible (so every subset V_i will have $\lfloor \frac{n}{r} \rfloor$ or $\lfloor \frac{n}{r} \rfloor + 1$ elements), and let $\binom{n}{k}_r$ be the number of k-subsets $F \subseteq \{1, \dots, n\}$ such that $|F \cap V_i| \le 1$ for $1 \le i \le r$. For $k \le r$ every positive integer n can be uniquely written

$$n = \binom{a_k}{k}_r + \binom{a_{k-1}}{k-1}_{r-1} + \cdots + \binom{a_i}{i}_{r-k+i},$$

where $\frac{a_j}{a_{j-1}} > \frac{r-k+j}{r-k+j-1}$ for $j = k, k-1, \dots, i+1$, and $a_i \ge i \ge 1$. Then define

$$\partial_k^{(r)}(n) = \binom{a_k}{k-1}_r + \binom{a_{k-1}}{k-2}_{r-1} + \cdots + \binom{a_i}{i-1}_{r-k+i},$$

and let $\partial_k^{(r)}(0) = 0$.

THEOREM 17.1.2

For $f = (f_0, \dots, f_{d-1})$, $d \le r$, the following are equivalent:

(i) *f is the f-vector of an r-colorable simplicial complex;*

(ii) *$\partial_{k+1}^{(r)}(f_k) \le f_{k-1}$, for all $1 \le k \le d-1$.*

Note that for r sufficiently large Theorem 17.1.2 specializes to Theorem 17.1.1.

THEOREM 17.1.3

The f-vector of any $(r-1)$-dimensional clique complex is the f-vector of some r-colorable complex.

MULTICOMPLEXES AND MACAULAY'S THEOREM

A ***multicomplex*** $\mathcal{M}$ is a nonempty collection of monomials in finitely many variables such that if m is in $\mathcal{M}$ then so is every divisor of m. Let $f_i(\mathcal{M})$ be the number of degree i monomials in $\mathcal{M}$; $f(\mathcal{M}) = (f_0, f_1, \dots)$ is called the ***f-vector*** of $\mathcal{M}$.

THEOREM 17.1.4 *Macaulay's Theorem*

For $f \in \mathbb{N}^\infty$ the following are equivalent:

(i) *f is the f-vector of a multicomplex;*

(ii) f *is an* M*-sequence;*

(iii) $f_i = \dim_{\mathbf{k}} R_i$, $i \geq 0$, *for some finitely generated commutative graded* $\mathbf{k}$*-algebra* $R = \oplus_{i \geq 0} R_i$ *such that* $R_0 \cong \mathbf{k}$ *(a field) and* R_1 *generates* R*.*

A simplicial complex can be viewed as a multicomplex of squarefree monomials. Hence, a K-sequence is (except for a shift in the indexing) an M-sequence: If $(f_0, \ldots, f_{d-1})$ is a K-sequence then $(1, f_0, \ldots, f_{d-1})$ is an M-sequence. For this reason (and others, see, e.g., Theorem 17.2.3), properties of M-sequences are of interest also if one cares mainly about the special case of simplicial complexes.

A multicomplex is ***pure*** if all its maximal (under divisibility) monomials have the same degree.

THEOREM 17.1.5

Let $(f_0, \ldots, f_r)$ *be the* f*-vector of a pure multicomplex,* $f_r \neq 0$*. Then* $f_i \leq f_j$ *for all* $i < j \leq r - i$*.*

COMMENTS

Simplicial complexes (abstract and geometric) are treated in most books on algebraic topology; see, e.g., [Mun84, Spa66]. The Kruskal-Katona theorem (independently discovered by M.-P. Schützenberger, J.B. Kruskal, G.O.H. Katona, L.H. Harper, and B. Lindström during the years 1959-1966) is discussed in many places and several proofs have appeared; see, e.g., [And87, Zie95].

A Kruskal-Katona type theorem for simplicial complexes with vertex-transitive symmetry group appears in [FK96].

Theorem 17.1.2 is from [FFK88]. (*Remark:* The definition of the $\partial_k^{(r)}(\cdot)$ operator is incorrectly stated in [FFK88], in particular the uniqueness claim in [FFK88, Lemma 1.1] is incorrect. The version stated here was suggested to us by J. Eckhoff.)

Theorem 17.1.3 was conjectured by J. Eckhoff and G. Kalai and proved by Frohmader [Fro08].

For Macaulay's theorem we refer to [And87, Sta96]. There is a common generalization of Macaulay's theorem and the Kruskal-Katona theorem due to Clements and Lindström; see [And87]. Theorem 17.1.5 is from [Hib89].

17.2 BETTI NUMBER CONSTRAINTS

GLOSSARY

The ***Euler characteristic*** $\chi(\Delta)$ of a simplicial complex Δ with f-vector $(f_0, \ldots, f_{d-1})$ is $\chi(\Delta) = \sum_{i=0}^{d-1} (-1)^i f_i$.

The ***h-vector*** $(h_0, \ldots, h_d)$ of a $(d-1)$-dimensional simplicial complex is defined by

$$\sum_{i=0}^{d} h_i x^{d-i} = \sum_{i=0}^{d} f_{i-1} (x-1)^{d-i}.$$

The corresponding ***g-vector*** $(g_0, \ldots, g_{\lfloor d/2 \rfloor})$ is defined by $g_0 = 1$ and $g_i = h_i - h_{i-1}$, for $i \geq 1$.

The ***Betti number*** $\beta_i(\Delta)$ is the dimension (as a $\mathbb{Q}$-vector space) of the ith reduced simplicial homology group $\widetilde{H}_i(\Delta, \mathbb{Q})$; see any textbook on algebraic topology (e.g., [Mun84]) for the definition. We call $(\beta_0, \dots, \beta_{\dim \Delta})$ the ***Betti sequence*** of Δ.

The ***link*** $\ell k_\Delta(F)$ of a face F is the subcomplex of Δ defined by $\ell k_\Delta(F) = \{A \in \Delta \mid A \cap F = \emptyset, A \cup F \in \Delta\}$. Note that $\ell k_\Delta(\emptyset) = \Delta$.

A simplicial complex Δ is ***acyclic*** if $\beta_i(\Delta) = 0$ for all i.

A simplicial complex Δ is ***Cohen-Macaulay*** if $\beta_i(\ell k_\Delta(F)) = 0$ for all $F \in \Delta$ and all $i < \dim \ell k_\Delta(F)$.

A simplicial complex Δ is ***m-Leray*** if $\beta_i(\ell k_\Delta(F)) = 0$ for all $F \in \Delta$ and all $i \geq m$.

FIXED BETTI NUMBERS

A simplicial complex is ***connected*** if its 1-skeleton is connected in the sense of graph theory. This is equivalent to demanding $\beta_0 = 0$.

THEOREM 17.2.1

For $f \in \mathbb{N}^{(\infty)}$ the following are equivalent:

(i) *f is the f-vector of a connected simplicial complex;*

(ii) *f is a K-sequence and $\partial^3(f_2) \leq f_1 - f_0 + 1$.*

The most basic relationship between f-vectors and Betti numbers is the ***Euler-Poincaré formula***:

$$\chi(\Delta) = f_0 - f_1 + f_2 - \cdots = 1 + \beta_0 - \beta_1 + \beta_2 - \dots$$

This is in fact the only linear one in the following complete set of relations.

THEOREM 17.2.2

For $f = (f_0, f_1, \dots) \in \mathbb{N}^{(\infty)}$ and $\beta = (\beta_0, \beta_1, \dots) \in \mathbb{N}^{(\infty)}$ the following are equivalent:

(i) *f is the f-vector of some simplicial complex with Betti sequence β;*

(ii) *if $\chi_{k-1} = \sum_{j \geq k} (-1)^{j-k}(f_j - \beta_j)$, $k \geq 0$, then $\chi_{-1} = 1$ and $\partial_{k+1}(\chi_k + \beta_k) \leq \chi_{k-1}$ for all $k \geq 1$.*

By putting $\beta_i = 0$ for all i one gets as a special case a characterization of the f-vectors of acyclic simplicial complexes, viz., $\sum_{i \geq 0} f_{i-1} x^i = (1 + x) \sum_{i \geq 0} f'_{i-1} x^i$, where $(f'_0, f'_1, \dots)$ is a K-sequence.

COHEN-MACAULAY COMPLEXES

Examples of Cohen-Macaulay complexes are triangulations of manifolds whose Betti numbers vanish below the top dimension, in particular triangulations of spheres and balls. Other examples are matroid complexes (the independent sets of a matroid), Tits buildings, and the order complexes (simplicial complex of totally ordered subsets) of several classes of posets, e.g., semimodular lattices (including distributive and geometric lattices). Shellable complexes (see Chapters 16 and 19) are Cohen-Macaulay. Cohen-Macaulay complexes are always pure.

The definition of h-vector given in the glossary shows that the h-vector and the f-vector of a complex mutually determine each other via the formulas:

$$h_i = \sum_{j=0}^{i} (-1)^{i-j} \binom{d-j}{i-j} f_{j-1}, \qquad f_{i-1} = \sum_{j=0}^{i} \binom{d-j}{i-j} h_j,$$

for $0 \le i \le d$. Hence, we may state f-vector results in terms of h-vectors whenever convenient.

THEOREM 17.2.3

For $h = (h_0, \ldots, h_d) \in \mathbb{Z}^{d+1}$ the following are equivalent:

(i) *h is the h-vector of a $(d-1)$-dimensional Cohen-Macaulay complex;*

(ii) *h is the h-vector of a $(d-1)$-dimensional shellable complex;*

(iii) *h is an M-sequence.*

Since there are a total of $\binom{n+k-1}{k}$ monomials of degree k in n variables, and by Theorems 17.1.5 and 17.2.3 the h-vector of a $(d-1)$-dimensional Cohen-Macaulay complex counts certain monomials in $h_1 = f_0 - d$ variables, we derive the inequalities

$$0 \le h_i \le \binom{f_0 - d + i - 1}{i}$$

for the h-vectors of Cohen-Macaulay complexes. The lower bound can be improved for complexes with fixed-point-free involutive symmetry.

THEOREM 17.2.4

Let $h = (h_0, \ldots, h_d)$ be the h-vector of a Cohen-Macaulay complex admitting an automorphism α of order 2, such that $\alpha(F) \neq F$ for all $F \in \Delta \setminus \{\emptyset\}$. Then

$$h_i \ge \binom{d}{i} \qquad \text{for } 0 \le i \le d.$$

Consequently, $f_{d-1} = h_0 + \cdots + h_d \ge 2^d$.

Another condition on a Cohen-Macaulay complex that forces stricter conditions on its h-vector is being r-colorable.

THEOREM 17.2.5

For $h = (h_0, \ldots, h_d) \in \mathbb{Z}^{d+1}$ the following are equivalent:

(i) *h is the h-vector of a $(d-1)$-dimensional and d-colorable Cohen-Macaulay complex;*

(ii) *$(h_1, \ldots h_d)$ is the f-vector of a d-colorable simplicial complex.*

Hence in this case the h-vector is not only an M-sequence, but the special kind of K-sequence characterized in Theorem 17.1.2.

LERAY COMPLEXES

Examples of Leray complexes arise as follows. Let $\mathcal{K} = \{K_1, \ldots, K_t\}$ be a family of convex sets in $\mathbb{R}^m$, and let $\Delta(\mathcal{K}) = \{A \subseteq \{1, \ldots, t\} \mid \bigcap_{i \in A} K_i \neq \emptyset\}$. Then the simplicial complex $\Delta(\mathcal{K})$ is m-Leray.

Fix $m \geq 0$, and let $f = (f_0, \ldots, f_{d-1})$ be the f-vector of a simplicial complex Δ. Define

$$h_k^* = \begin{cases} f_k & \text{for } 0 \leq k \leq m-1 \\ \sum\limits_{j \geq 0} (-1)^j \binom{k+j-m}{j} f_{k+j} & \text{for } k \geq m. \end{cases}$$

The sequence $h^* = (h_0^*, \ldots, h_{d-1}^*)$ is the ***h^*-vector*** of Δ. The two vectors f and h^* mutually determine each other.

THEOREM 17.2.6

For $h^* = (h_0^*, h_1^*, \ldots) \in \mathbb{Z}^{(\infty)}$ the following are equivalent:

(i) *h^* is the h^*-vector of an m-Leray complex;*
(ii) *h^* is the h^*-vector of $\Delta(\mathcal{K})$ for some family $\mathcal{K}$ of convex sets in $\mathbb{R}^m$;*
(iii)
$$\begin{cases} h_k^* \geq 0 & \textit{for } k \geq 0 \\ \partial_{k+1}(h_k^*) \leq h_{k-1}^* & \textit{for } 1 \leq k \leq m-1 \\ \partial_m(h_k^*) \leq h_{k-1}^* - h_k^* & \textit{for } k \geq m. \end{cases}$$

COMMENTS

The Euler-Poincaré formula (due to Poincaré, 1899) is proved in most books on algebraic topology.

A good general source on Cohen-Macaulay complexes is [Sta96]; it contains Theorems 17.2.3, 17.2.4, and 17.2.5, as well as references to the original sources. Theorem 17.2.2 is from [BK88]. A common generalization of Theorems 17.1.1, 17.2.2, and 17.2.3 is given in [Bjö96]. Theorem 17.2.1 is a special case. There are several additional results about h-vectors of Cohen-Macaulay complexes. For instance, for complexes with nontrivial automorphism groups, see [Sta96, Section III.8]; for matroid complexes, see [Sta96, Section III.3]; and for Cohen-Macaulay complexes that are r-colorable for $r < d$, see the references mentioned in [Sta96, Section III.4].

Cohen-Macaulay complexes are pure. However, there is an extension of their theory to a class of nonpure complexes, the so-called *sequentially Cohen-Macaulay* complexes, introduced in [Sta96]. In [ABG17], to which we refer for definitions and references, a numerical characterization is given of the so-called h-triangles (doubly indexed h-numbers) of sequentially Cohen-Macaulay simplicial complexes. This result characterizes the array of numbers of faces of various dimensions and codimensions in such a complex, generalizing Theorem 17.2.3 to the nonpure case.

Cohen-Macaulay complexes are closely related to certain commutative rings [Sta96], and via this connection such complexes have also been of use in the theory of splines; see [Sta96, Section III.5] and also Chapter 56 of this Handbook.

Theorem 17.2.6 was conjectured by Eckhoff and proved by Kalai [Kal84, Kal86].

17.3 SIMPLICIAL POLYTOPES, SPHERES, AND MANIFOLDS

GLOSSARY

A ***triangulated d-ball*** is a simplicial complex Δ whose realization $\|\Delta\|$ is homeomorphic to the ball $\{x \in \mathbb{R}^d \mid x_1^2 + \cdots + x_d^2 \leq 1\}$. A ***triangulated (d−1)-sphere*** is a simplicial complex whose realization is homeomorphic to the sphere $\{x \in \mathbb{R}^d \mid x_1^2 + \cdots + x_d^2 = 1\}$. Equivalently, it is the boundary of a triangulated d-ball. Examples of triangulated $(d-1)$-spheres are given by the boundary complexes of simplicial d-polytopes.

A ***pseudomanifold*** is a pure simplicial complex Δ such that

(i) each face of codimension 1 is contained in precisely two maximal faces; and

(ii) the dual graph (whose vertices are the maximal faces of Δ and whose edges are the faces of codimension 1) is connected.

An ***Eulerian pseudomanifold*** is a pseudomanifold Δ such that Δ itself and the link of each face have the Euler characteristic of a sphere of the corresponding dimension.

A pure $(d-1)$-dimensional simplicial complex Δ is a ***homology manifold*** if it is connected and the link of each nonempty face has the Betti numbers of a sphere of the same dimension. It is a ***homology sphere*** if, in addition, Δ itself has the Betti numbers of a $(d-1)$-sphere. Examples of homology manifolds are given by triangulations of compact connected topological manifolds, i.e., spaces that are locally Euclidean.

The ***cyclic d-polytope with n vertices*** $C_d(n)$ is the convex hull of any n points on the moment curve in $\mathbb{R}^d$. (See Section 15.1.4.)

The following implications hold among these various classes, all of them strict:

polytope boundary $\Rightarrow$ *sphere* $\Rightarrow$ *homology sphere* $\Rightarrow$
Eulerian pseudomanifold $\Rightarrow$ *pseudomanifold*

homology sphere $\Rightarrow$ *homology manifold* $\Rightarrow$ *pseudomanifold*

homology sphere $\Rightarrow$ *Cohen-Macaulay complex*

PSEUDOMANIFOLDS

The following results give the basic lower and upper bounds on f-vectors of pseudomanifolds.

THEOREM 17.3.1 *Lower Bound Theorem*

For a $(d-1)$-dimensional pseudomanifold Δ with n vertices,

$$f_k(\Delta) \geq \begin{cases} \binom{d}{k}n - \binom{d+1}{k+1}k & \text{for } 1 \leq k \leq d-2 \\ (d-1)n - (d-2)(d+1) & \text{for } k = d-1. \end{cases}$$

THEOREM 17.3.2 *Upper Bound Theorem*

Let Δ be a $(d-1)$-dimensional homology manifold with n vertices, such that either

(i) *d is even, or*

(ii) *$d = 2k+1$ is odd, and either $\chi(\Delta) = 2$ or $\beta_k \leq 2\beta_{k-1} + 2\sum_{i=0}^{k-3} \beta_i$.*

Then $f_k(\Delta) \leq f_k(C_d(n))$ for $1 \leq k \leq d-1$.

This upper bound theorem applies when the homology manifold is Eulerian (irrespective of dimension); in particular, it applies to all simplicial polytopes and spheres. By the geometric operation of "pulling vertices" (Section 16.2), one can extend this to all convex polytopes.

THEOREM 17.3.3

If P is any *convex d-polytope with n vertices, then $f(P) \leq f(C_d(n))$.*

The given lower and upper bounds are best possible within the class of simplicial polytope boundaries. The lower bound is attained by the class of stacked polytopes (Sections 16.4.2 and 19.2). To make the upper bound numerically explicit, we give the formula for the f-vector of a cyclic polytope.

THEOREM 17.3.4

For $d \geq 2$ and $0 \leq k \leq d-1$, the number of k-faces of the cyclic polytope $C_d(n)$ with n vertices is

$$f_k(C_d(n)) \quad = \quad \frac{n-\delta(n-k-2)}{n-k-1} \sum_{j=0}^{\lfloor d/2 \rfloor} \binom{n-1-j}{k+1-j} \binom{n-k-1}{2j-k-1+\delta},$$

where $\delta = d - 2\lfloor d/2 \rfloor$.

In particular,

$$f_{d-1}(C_d(n)) \quad = \quad \binom{n - \lceil \frac{d}{2} \rceil}{\lfloor \frac{d}{2} \rfloor} \quad + \quad \binom{n - \lceil \frac{d+1}{2} \rceil}{\lfloor \frac{d-1}{2} \rfloor},$$

which shows that for fixed d the number of facets is $O(n^{\lfloor d/2 \rfloor})$.

POLYTOPES AND SPHERES

For boundaries of simplicial d-polytopes and, more generally, for Eulerian pseudomanifolds, we have the following basic relations.

THEOREM 17.3.5 *Dehn-Sommerville Equations*

For d-dimensional Eulerian pseudomanifolds,

$$h_i = h_{d-i} \qquad \textit{for all } 0 \leq i \leq d.$$

These equations give a complete description of the linear span of all f-vectors of d-polytopes (equivalently, $(d-1)$-spheres). (The affine span is defined by including the relation $h_0 = 1$.)

One consequence of the Dehn-Sommerville equations is the following relation between the h-vector of a triangulated ball K and the g-vector of its boundary ∂K.

THEOREM 17.3.6

For a triangulated d-ball K and its boundary $(d-1)$-sphere ∂K,

$$g_i(\partial K) = h_i(K) - h_{d+1-i}(K) \qquad \text{for } i \geq 1.$$

A complete characterization of the f-vectors of simplicial (and, by duality, simple) convex polytopes is given in terms of the h-vector and g-vector.

THEOREM 17.3.7 *g-Theorem*

A nonnegative integer vector $h = (h_0, \ldots, h_d)$ is the h-vector of a simplicial convex d-polytope if and only if

(i) $h_i = h_{d-i}$, *and*

(ii) $(g_0, \ldots, g_{\lfloor d/2 \rfloor})$ *is an M-sequence.*

One consequence of (ii) is that $g_i \geq 0$, which was known as the *generalized lower bound conjecture.* The case of equality in this conjecture has only recently been settled.

THEOREM 17.3.8

A simplicial d-polytope P satisfies $g_k(P) = h_k(P) - h_{k-1}(P) = 0$ for some k, $1 \leq k \leq \lfloor \frac{d}{2} \rfloor$, if and only if P is k-stacked, i.e., *P can be triangulated so that every face of dimension $d-k$ or less is on the boundary of P.*

For centrally symmetric polytopes, we get a better lower bound.

THEOREM 17.3.9

For centrally symmetric simplicial d-polytopes,

$$g_i = h_i - h_{i-1} \geq \binom{d}{i} - \binom{d}{i-1} \qquad \text{for } i \leq \lfloor d/2 \rfloor.$$

The following arithmetic property of the numbers of k-faces of all simplicial d-polytopes is a consequence of the g-theorem.

THEOREM 17.3.10

Given $0 \leq k < d$ there exist positive integers $G(k,d)$ and $N(k,d)$ such that

(i) $G(k,d)$ divides $f_k(P)$ for every simplicial d-polytope P, and

(ii) if $G(k,d)$ divides n and $n > N(k,d)$, then $n = f_k(P)$ for some simplicial d-polytope P.

MANIFOLDS

For face numbers of triangulations of a $(d-1)$-dimensional manifold X we have the following generalization of the Dehn-Sommerville equations.

THEOREM 17.3.11

If $X = |\Delta|$ is a $(d-1)$-dimensional manifold, then

$$h_{d-i} - h_i = (-1)^i \binom{d}{i} (\chi(\Delta) - \chi(S^{d-1})).$$

If Δ is a $(d-1)$-dimensional simplicial complex, then define

$$h'_i = h_i + \binom{d}{i} \left(\beta_{i-2} - \beta_{i-3} + \cdots \pm \beta_0\right),$$

where the $\beta_j = \beta_j(\Delta)$ are the reduced homology Betti numbers of Δ.

THEOREM 17.3.12

If $X = |\Delta|$ is a $(d-1)$-dimensional manifold, then

$$h'_i \geq \binom{d}{i} \beta_{i-1}.$$

COMMENTS

The Lower Bound Theorem 17.3.1 is due to Kalai and Gromov in the generality given here; see [Kal87] including the note added in proof. The $k = d-1$ case had earlier been done by Klee and the case of polytope boundaries by Barnette. See [Kal87] for a discussion of the history of this result.

The Upper Bound Theorem 17.3.2 is due to Novik [Nov98]. See also [NS12]. The case of polytopes (Theorem 17.3.3) was first proved by McMullen (see [MS71]), and extended to spheres by Stanley (see [Sta96]). The computation of the f-vector of the cyclic polytope can be found in [Grü67, Sections 4.7.3 and 9.6.1] or [MS71]. More results on comparing face numbers of a simplicial polytope to those of the cyclic polytope can be found in [Bjö07].

The Dehn-Sommerville equations for polytopes are classical; proofs can be found in [Grü67, Sta86, Zie95]. The extension to Eulerian pseudomanifolds is due to Klee [Kle64]; an equivariant version appears in [Bar92]. The D-S equations imply an upper bound on the average number of j-faces contained in a k-face of a simple polytope (roughly, the number of j-faces of a k-dimensional cube) due to Nikulin. This has been useful in the theory of hyperbolic reflection groups. See [Nik87, Theorem C] for references and ramifications; see also Theorem 17.5.17, which is a similar result for arrangements and zonotopes.

The g-theorem was conjectured by McMullen and proved by Billera, Lee, and Stanley [BL81a, Sta80]. Another proof of the necessity of these conditions was given by McMullen [McM93]. More recently, a self-contained, elementary proof of necessity was given by Fleming and Karu [FK08]. It is not known whether the second condition of Theorem 17.3.7 holds for general triangulated spheres. The g-theorem has a convenient reformulation as a one-to-one correspondence (via matrix multiplication) between f-vectors of simplicial polytopes and M-sequences, see [Bjö87, Zie95]. Theorem 17.3.8, the equality case of the generalized lower bound theorem, was conjectured by McMullen and Walkup in 1971 and recently proved by Murai and Nevo [MN13]. Theorem 17.3.9 was proved by Stanley [Sta87a]; for another proof see [Nov99]. Theorem 17.3.10 is from Björner and Linusson [BL99], where also an explicit expression for the modulus $G(k,d)$ is given.

Theorems 17.3.11 and 17.3.12 are due to Klee [Kle64] and Novik and Swartz [NS09], respectively. For simplicity, they are here not stated in their maximal generality. That $h'_i \geq 0$ for manifolds was originally shown by Schenzel in 1981.

The question of characterizing f-vectors for compact manifolds more general than spheres is at the present far beyond our reach. However, much interesting work has been done on the more restrictive question of minimizing the number of vertices of triangulations for given manifolds, see e.g., [Küh90, Küh95, BL00, Lut05]. This is of interest for efficient presentations of manifolds to computers. For information about face numbers of manifolds, see [Swa09].

The study of f-vectors of unbounded polyhedra can be approached by studying the f-vectors of ***polytope pairs*** (P, F), where P is a polytope and F is a maximal face of P. See [BL81b, BaL93] for a summary of such results.

17.4 CELL COMPLEXES

GLOSSARY

Convex polytopes and *faces* of such are defined in Chapter 15.

A ***polyhedral complex*** Γ is a finite collection of convex polytopes in $\mathbb{R}^n$ such that (i) if $\pi \in \Gamma$ and σ is a face of π, then $\sigma \in \Gamma$; and (ii) if $\pi, \sigma \in \Gamma$ and $\pi \cap \sigma \neq \emptyset$, then $\pi \cap \sigma$ is a face of both. The ***space*** of Γ is $\|\Gamma\| = \bigcup_{\pi \in \Gamma} \pi$, a subspace of $\mathbb{R}^n$. Examples of polyhedral complexes are given by ***boundary complexes*** ∂P of convex polytopes P (i.e., the collection of all proper faces). A geometric simplicial complex (defined in Section 17.1) is a polyhedral complex all of whose cells are simplices. A ***cubical complex*** is a polyhedral complex all of whose cells are (combinatorially isomorphic to) cubes.

A ***regular cell complex*** Γ is a family of closed balls (homeomorphs of $\{x \in \mathbb{R}^j \mid |x| \leq 1\}$) in a Hausdorff space $\|\Gamma\|$ such that (i) the interiors of the balls partition $\|\Gamma\|$ and (ii) the boundary of each ball in Γ is a union of other balls in Γ. The members of Γ are called (closed) ***cells*** or ***faces***. The ***dimension*** of a cell is its topological dimension and $\dim \Gamma = \max_{\sigma \in \Gamma} \dim \sigma$.

A ***Gorenstein* complex*** is a regular cell complex whose poset of faces has an order complex that is a homology sphere. These include all triangulations of spheres.

A regular cell complex has the ***intersection property*** if, whenever the intersection of two cells is nonempty, then this intersection is also a cell in the complex. Polyhedral complexes are examples of regular cell complexes with the intersection property. Regular cell complexes with the intersection property can be reconstructed up to homeomorphism from the corresponding "abstract" complex consisting of the family of vertex sets of its cells.

For a regular cell complex Γ, let f_i be the number of i-dimensional cells, and let $\beta_i = \dim_{\mathbb{Q}} \widetilde{H}_i(\|\Gamma\|, \mathbb{Q})$. The latter denotes i-dimensional reduced singular homology with rational coefficients of the space $\|\Gamma\|$; see [Mun84, Spa66] for explanations of this concept. Then we have the ***f-vector*** $f = (f_0, f_1, \ldots)$ and the ***Betti sequence*** $\beta = (\beta_0, \beta_1, \ldots)$ of Γ. These definitions generalize those previously given in the simplicial case.

BASIC f-VECTOR RELATIONS

Among the classes of complexes

- simplicial complexes
- polyhedral complexes
- regular cell complexes with the intersection property
- regular cell complexes

each is a proper subclass of its successor. Thus one may wonder how many of the relations for f-vectors of simplicial complexes given in Sections 17.1–17.3 can be extended to these broader classes of complexes. Also, what new phenomena (not visible in the simplicial case) arise? Some answers are given in this section and the following one, but current knowledge is quite fragmentary. We begin here with the most general relations.

THEOREM 17.4.1

$(f_0, \ldots, f_d)$ is the f-vector of a d-dimensional regular cell complex if and only if $f_d \geq 1$ and $f_i \geq 2$ for all $0 \leq i < d$.

THEOREM 17.4.2

f is the f-vector of a regular cell complex with the intersection property if and only if f is a K-sequence.

Let $\beta = (\beta_0, \beta_1, \ldots) \in \mathbb{N}^{(\infty)}$ be fixed, and for every sequence $f = (f_0, f_1, \ldots)$ let

$$\chi_{k-1} = \sum_{j \geq k} (-1)^{j-k} (f_j - \beta_j) \qquad \text{for } k \geq 0.$$

THEOREM 17.4.3

$(f_0, \ldots, f_d)$ is the f-vector of a d-dimensional regular cell complex with Betti sequence β if and only if $\chi_{-1} = 1$ and $\chi_k \geq 1$ for $0 \leq k < d$.

THEOREM 17.4.4

For $f \in \mathbb{N}^{(\infty)}$ the following are equivalent:

(i) *f is the f-vector of a regular cell complex with the intersection property and with Betti sequence β;*

(ii) *$\chi_{-1} = 1$ and $\partial_{k+1}(\chi_k + \beta_k) \leq \chi_{k-1}$ for all $k \geq 1$.*

These results show that the f-vectors of regular cell complexes (with or without Betti number constraints) are considerably more general than the f-vectors of simplicial complexes, but that the two classes of f-vectors agree in the presence of the intersection property.

COMMENTS

Regular cell complexes are known as ***regular CW complexes*** in the topological literature [LW69]. The nonregular CW complexes offer an even more general class of cell complexes [LW69, Mun84, Spa66], but there is very little one can say about

f-vectors in that generality. See [BLS$^+$93, Section 4.7] for a detailed discussion of regular cell complexes from a combinatorial point of view.

For the results of this section see [BK88, BK91, BK89]. A characterization of f-vectors of (cubical) subcomplexes of a cube can be found in [Lin71], and of regular cell decompositions of spheres in [Bay88].

17.5 GENERAL POLYTOPES AND SPHERES

GLOSSARY

A ***flag*** of faces in a (polyhedral) $(d-1)$-complex Δ is a chain $F_1 \subsetneq F_2 \subsetneq \cdots \subsetneq F_k$ of faces F_i in Δ. It is an ***S-flag*** if

$$S = \{\dim F_1, \ldots, \dim F_k\} \subseteq \{0, 1, \ldots, d-1\}.$$

Let $f_S = f_S(\Delta)$ denote the number of S-flags in Δ. The function $S \mapsto f_S$, $S \subseteq \{0, 1, \ldots, d-1\}$, is called the ***flag f-vector*** of Δ. If

$$h_S = \sum_{T \subseteq S} (-1)^{|S|-|T|} f_T,$$

then the function $S \mapsto h_S$, $S \subseteq \{0, 1, \ldots, d-1\}$, is called the ***flag h-vector***.

For $S \subseteq \{0, \ldots, d-1\}$ and noncommuting symbols $\boldsymbol{a}$ and $\boldsymbol{b}$, let $u_S = u_0 u_1 \cdots u_{d-1}$ be the $\boldsymbol{ab}$-word defined by $u_i = \boldsymbol{a}$ if $i \notin S$ and $u_i = \boldsymbol{b}$ otherwise. When Δ is spherical (or, more generally, Eulerian), then the $\boldsymbol{ab}$-polynomial $\sum h_S u_S$ is also a polynomial in $\mathbf{c} = \boldsymbol{a} + \boldsymbol{b}$ and $\mathbf{d} = \boldsymbol{ab} + \boldsymbol{ba}$. (Note that the degree of $\mathbf{c}$ is 1 and the degree of $\boldsymbol{d}$ is 2.) The resulting $\boldsymbol{cd}$-polynomial

$$\sum h_S u_S = \sum \phi_w w,$$

where the right-hand sum is over all $\boldsymbol{cd}$-words w of degree d, is called the ***cd-index*** $\Phi(\Delta)$ of Δ. For 2-, 3-, and 4-polytopes, the $\boldsymbol{cd}$-index is $\mathbf{c}^2 + (f_0 - 2)\boldsymbol{d}$, $\mathbf{c}^3 + (f_0 - 2)\boldsymbol{dc} + (f_2 - 2)\boldsymbol{cd}$, and $\mathbf{c}^4 + (f_0 - 2)\boldsymbol{dc}^2 + (f_1 - f_0)\boldsymbol{cdc} + (f_3 - 2)\boldsymbol{c}^2\boldsymbol{d} + (f_{02} - 2f_2 - 2f_0 + 4)\boldsymbol{d}^2$, respectively.

For *any* convex d-polytope P, we define the ***toric h-vector*** and ***toric g-vector*** recursively by $h(P, x) = \sum_{i=0}^{d} h_i x^{d-i}$ and $g(P, x) = \sum_{i=0}^{\lfloor d/2 \rfloor} g_i x^i$, where $g_i = h_i - h_{i-1}$ and the following relations hold:

(i) $g(\emptyset, x) = h(\emptyset, x) = 1$; and

(ii) $h(P, x) = \sum_{G \text{ face of } P,\, G \neq P} g(G, x)(x-1)^{d-1-\dim G}$.

(Compare to Section 16.4.1, where this toric h-vector is defined for any polyhedral complex. In the notation given there, we have defined h and g for the complex ∂P.) When P is simplicial, this definition coincides with that of the usual h-vector, as defined in Section 17.2. For 2-, 3-, and 4-polytopes, the g-polynomial is $1+(f_0-3)x$, $1+(f_0-4)x$, and $1+(f_0-5)x+(10-3f_0-3f_3+f_{03})x^2$, respectively.

A ***rational polytope*** is one whose vertices all have rational coordinates. Equivalently, all maximal faces are determined by linear forms with rational coefficients.

A ***cubical polytope*** is one that has a cubical boundary complex. For any cubical $(d-1)$-complex with f-vector $(f_0, \ldots, f_{d-1})$, define the ***cubical h-vector*** $h^c = (h_0^c, \ldots, h_d^c)$ by

$$h_i^c = (-1)^i 2^{d-1} + \sum_{j=1}^{i} (-1)^{i-j} 2^{j-1} f_{j-1} \sum_{k=0}^{i-j} \binom{d-j}{k} \qquad \text{for } i = 0, \dots, d.$$

The ***cubical g-vector*** $g^c = (g_0^c, \dots, g_{\lfloor d/2 \rfloor}^c)$ is defined by $g_0^c = h_0^c = 2^{d-1}$ and $g_i^c = h_i^c - h_{i-1}^c$ for $i \geq 1$.

An ***Eulerian polyhedral complex*** is one whose first barycentric subdivision is an Eulerian pseudomanifold. Examples are boundary complexes of polytopes and ***spherical*** polyhedral complexes, i.e., those whose underlying space is homeomorphic to a sphere.

A (central) ***hyperplane arrangement*** is a collection $\mathcal{H}$ of n linear hyperplanes in $\mathbb{R}^d$, given by normal vectors $x_1, \dots, x_n$ (see Section 6.1.3). The arrangement is ***essential*** if the normals x_i span $\mathbb{R}^d$. The associated ***zonotope*** is the Minkowski sum of the n line segments $[-x_i, x_i]$, i.e., $Z = \{\sum \lambda_i x_i \mid -1 \leq \lambda_i \leq 1\}$ (see Section 15.1.4).

LINEAR RELATIONS

We give the linear equalities on the invariants defined above that are known to hold for all boundary complexes of polytopes and, more generally, for all Eulerian polyhedral complexes.

THEOREM 17.5.1

For $(d-1)$-dimensional Eulerian polyhedral complexes, the following relations always hold for the flag h, the toric h, and the flag f:

(i) $h_S = h_{\{0,\dots,d-1\} \setminus S}$ *for all* $S \subseteq \{0, \dots, d-1\}$;

(ii) $h_i = h_{d-i}$ *for* $0 \leq i \leq d$; *and*

(iii) $\sum_{j=i+1}^{k-1} (-1)^{j-i-1} f_{S \cup \{j\}} = (1 - (-1)^{k-i-1}) f_S$ *whenever* $i, k \in S \cup \{-1, d\}$ *with* $i \leq k-2$ *and* $S \cap \{i+1, \dots, k-1\} = \emptyset$.

It is known that the relations in Theorem 17.5.1(iii), the ***generalized Dehn-Sommerville equations***, completely describe the linear span of all flag f-vectors of Eulerian complexes, and so they imply those in (i). Since the toric h is known to be a linear function of the flag f, they imply those in (ii) as well. The linear span of flag f-vectors has dimension e_d, where e_d is the dth Fibonacci number (defined by the recurrence $e_d = e_{d-1} + e_{d-2}$, $e_0 = e_1 = 1$). There are e_d ***cd***-words of degree d. Furthermore, the coefficients ϕ_w of the ***cd***-index, considered as linear expressions in the f_S, form a linear basis for the span of flag f-vectors of d-polytopes. The affine span of all flag f-vectors is defined by including the relation $f_\emptyset = 1$.

For cubical polytopes and spheres, the cubical h-vector satisfies the analogue of the Dehn-Sommerville equations.

THEOREM 17.5.2

For cubical d-polytopes and cubical $(d-1)$-spheres,

$$h_i^c = h_{d-i}^c \qquad \textit{for all } 0 \leq i \leq d.$$

These give all linear relations satisfied by f-vectors of cubical polytopes and spheres. The cubical h-vector satisfies, as well, the equations of Theorem 17.3.6, linking the h of a cubical ball to the g of its boundary sphere.

LINEAR INEQUALITIES

Some linear inequalities that hold for flag f-vectors of all polytope boundaries are given in this section. The list is not thought to be complete, although there are no conjectures for what the complete set might be.

For a Cohen-Macaulay polyhedral complex, i.e., one whose first barycentric subdivision is a Cohen-Macaulay simplicial complex, the flag h is always nonnegative.

THEOREM 17.5.3

For a Cohen-Macaulay polyhedral $(d-1)$*-complex* Γ*, we have* $h_S(\Gamma) \geq 0$ *for all* $S \subseteq \{0, \ldots, d-1\}$.

For general convex polytopes, we also have nonnegativity of the ***cd***-index. In fact, the ***cd***-index of any d-polytope is minimized termwise by the ***cd***-index of the d-simplex $\Delta^{(d)}$.

THEOREM 17.5.4

(i) *If* P *is a convex* d*-polytope (or, more generally a Gorenstein* complex), then*

$$\phi_w(P) \geq 0$$

for all **cd***-words* w *of degree* d.

(ii) *If* P *is a convex* d*-polytope (or, more generally a Gorenstein* complex whose face poset is a lattice), then*

$$\phi_w(P) \geq \phi_w(\Delta^{(d)})$$

for all **cd***-words* w *of degree* d.

Note that Theorem 17.5.4(i) gives the most general possible linear inequalities for flag f-vectors of spherical regular cell complexes (i.e., regular cell complexes homeomorphic to the sphere).

There are also relations between the ***cd***-coefficients ϕ_w for any polytope.

THEOREM 17.5.5

For any d*-polytope* P

$$\phi_{\mathbf{u}\mathbf{d}\mathbf{v}}(P) \geq \phi_{\mathbf{u}\mathbf{c}^2\mathbf{v}}(P),$$

for any **cd***-words* u *and* v *with* $\deg u + \deg v = d - 2$.

For all convex polytopes, it is known, further, that the toric h is unimodal.

THEOREM 17.5.6

For a convex d*-polytope,* $g_i \geq 0$ *for* $i \leq \lfloor d/2 \rfloor$.

Related to this is the following *nonlinear* inequality holding between the g-vectors of a polytope P and any of its faces F. We denote by P/F the ***link*** of F in P, i.e., the polytope whose lattice of faces is (isomorphic to) the interval $[F, P]$ in the face lattice of P.

THEOREM 17.5.7

For a polytope P and any face F, we have the polynomial inequality

$$g(P,t) - g(F,t)g(P/F,t) \geq 0,$$

i.e., all coefficients of this polynomial are nonnegative.

We have a similar relation between the ***cd***-index of a polytope and that of any face.

THEOREM 17.5.8

For a polytope P and any face F, we have the polynomial inequalities

$$\Phi(P) \geq \begin{cases} \mathbf{c} \cdot \Phi(F) \cdot \Phi(P/F) \\ \Phi(F) \cdot \mathbf{c} \cdot \Phi(P/F) \\ \Phi(F) \cdot \Phi(P/F) \cdot \mathbf{c} \end{cases}$$

where $\Phi(P)$, $\Phi(F)$, and $\Phi(P/F)$ are the **cd***-indices of P, F, and P/F, respectively.*

As with f-vectors of polytopes, their flag f-vectors, flag h-vectors and ***cd***-indices satisfy the upper bound theorem.

THEOREM 17.5.9

If P is a d-dimensional polytope with n vertices, then for any S,

$$\begin{aligned} f_S(P) &\leq f_S(C_d(n)), \\ h_S(P) &\leq h_S(C_d(n)), \end{aligned}$$

and termwise as polynomials

$$\Phi(P) \leq \Phi(C_d(n)),$$

where $C_d(n)$ is the cyclic d-polytope with n vertices.

There are the following relations between invariants of a polytope P and its dual polytope P^*. If $w = w_1, \ldots, w_n$ is a ***cd***-word, then $w^* := w_n, \ldots, w_1$, the reverse word.

THEOREM 17.5.10

For a d-polytope P,

(i) $\phi_w(P^*) = \phi_{w^*}(P)$,

(ii) $g_k(P) = 0$ *if and only if* $g_k(P^*) = 0$ *and*

(iii) $\sum_{\emptyset \subseteq F \subseteq P} (-1)^{\dim F} g(F^*,t)\, g(P/F,t) = 0.$

For a $(2k)$-polytope P,

(iv) $g_k(P) = g_k(P^*)$.

Finally, we have the following lower bounds for the number of vertices of polytopes with no triangular faces (this includes the class of cubical polytopes), and for the combined numbers of vertices and facets of centrally symmetric polytopes.

THEOREM 17.5.11

A d-polytope with no triangular 2-face has at least 2^d vertices.

THEOREM 17.5.12

There exists a constant $c > 0$ such that

$$\log f_0 \cdot \log f_{d-1} > cd,$$

for any centrally symmetric d-polytope.

HYPERPLANE ARRANGEMENTS AND ZONOTOPES

An essential hyperplane arrangement $\mathcal{H}$ defines a decomposition of $\mathbb{R}^d$ into polyhedral cones (as in Section 6.1.3). This decomposition $\Gamma_{\mathcal{H}}$, a regular cell complex if intersected with the unit sphere, has a flag f-vector dual to that of its associated zonotope Z, in the sense that $f_S(\Gamma_{\mathcal{H}}) = f_{d-S}(Z)$, where $S = \{i_1, \ldots, i_k\} \subseteq \{1, \ldots, d\}$ and $d - S = \{d - i_k, \ldots, d - i_1\}$.

THEOREM 17.5.13

The flag f-vector of an arrangement (or zonotope) depends only on the matroid (linear dependency structure) of the underlying point configuration $\{x_1, \ldots, x_n\}$.

Although a fairly special subclass of polytopes, the zonotopes nonetheless are varied enough to carry all the linear information carried by flag numbers of general polytopes.

THEOREM 17.5.14

The flag f-vectors of zonotopes (and thus of hyperplane arrangements) satisfy the generalized Dehn-Sommerville equations, and there are no other linear relations not implied by these.

When it comes to linear *inequalities*, however, a difference between zonotopes and general polytopes emerges. As with general convex polytopes, we have nonnegativity of the ***cd***-index for zonotopes. However, the ***cd***-index of any d-zonotope is minimized termwise by the ***cd***-index of the d-cube $C^{(d)}$.

THEOREM 17.5.15

*For a convex d-zonotope Z, $\phi_w(Z) \geq \phi_w(C^{(d)}) \geq 0$ for all **cd**-words w of degree d. Further, if the word w has k **d**'s, then 2^k divides $\phi_w(Z)$.*

There is also a strengthening of Theorem 17.5.5 for zonotopes.

THEOREM 17.5.16

For any d-zonotope Z

$$\phi_{\mathbf{u}\mathbf{d}\mathbf{v}}(Z) - \phi_{\mathbf{u}\mathbf{c}^2\mathbf{v}}(Z) \geq \phi_{\mathbf{u}\mathbf{d}\mathbf{v}}(C^{(d)}) - \phi_{\mathbf{u}\mathbf{c}^2\mathbf{v}}(C^{(d)})$$

*for any **cd**-words u and v with* $\deg u + \deg v = d - 2$.

The following result has the most direct interpretation when it is stated for arrangements, where it bounds the average number of $\{i_1, \ldots, i_k\}$-flags in an i_k-face by the number of $\{i_1-1, \ldots, i_k-1\}$-flags in an (i_k-1)-cube.

THEOREM 17.5.17

For a hyperplane arrangement $\mathcal{H}$ in $\mathbb{R}^d$ and $S = \{i_1, \ldots, i_k\} \subseteq \{1, \ldots, d\}$ with $k \geq 2$,

$$\frac{f_S(\Gamma_{\mathcal{H}})}{f_{i_k}(\Gamma_{\mathcal{H}})} < \binom{i_k - 1}{i_1 - 1, i_2 - i_1, \ldots, i_k - i_{k-1}} 2^{i_k - i_1}.$$

There is a straightforward reformulation of Theorem 17.5.17 for zonotopes that is easily seen not to be valid for all polytopes.

GENERAL 3- AND 4-POLYTOPES

We describe here the situation for flag f-vectors of 3- and 4-polytopes. The equations in Theorem 17.5.1(iii) reduce consideration to (f_0, f_2) when $d = 3$ and to (f_0, f_1, f_2, f_{02}) when $d = 4$.

THEOREM 17.5.18

For 3-polytopes, the following is known about the vector (f_0, f_2).

(i) *An integer vector (f_0, f_2) is the f-vector of a 3-polytope if and only if $f_0 \leq 2f_2 - 4$ and $f_2 \leq 2f_0 - 4$.*

(ii) *An integer vector (f_0, f_2) is the f-vector of a cubical 3-polytope if and only if $f_2 = f_0 - 2$, $f_0 \geq 8$, and $f_0 \neq 9$.*

(iii) *If $(f_0, f_2) = (f_0(Z), f_2(Z))$ for a 3-zonotope Z, then f_0 and f_1 are both even integers, $f_0 \leq 2f_2 - 4$, and $f_2 \leq f_0 - 2$.*

For 4-polytopes, much less is known.

THEOREM 17.5.19

Flag f-vectors (f_0, f_1, f_2, f_{02}) of 4-polytopes satisfy the following inequalities.

(i) $f_{02} \geq 3f_2$

(ii) $f_{02} \geq 3f_1$

(iii) $f_{02} + f_1 + 10 \geq 3f_2 + 4f_0$

(iv) $6f_1 \geq 6f_0 + f_{02}$

(v) $f_0 \geq 5$

(vi) $f_0 + f_2 \geq f_1 + 5$

(vii) $2(f_{02} - 3f_2) \leq \binom{f_0}{2}$

(viii) $2(f_{02} - 3f_1) \leq \binom{f_2 - f_1 + f_0}{2}$

(ix) $f_{02} - 4f_2 + 3f_1 - 2f_0 \leq \binom{f_0}{2}$

(x) $f_{02} + f_2 - 2f_1 - 2f_0 \leq \binom{f_2 - f_1 + f_0}{2}$.

It is not known, for example, whether (i)–(vi) give all linear inequalities holding for flag f-vectors of 4-polytopes.

COMMENTS

It is thought that the best route to an eventual characterization of f-vectors of general polytopes lies in an understanding of their flag f-vectors. The latter inherit many of the algebraic properties of f-vectors of simplicial polytopes that led to their characterization, while having a rich theory of their own.

The relations in Theorem 17.5.1 hold more generally for the case of enumeration of chains in Eulerian posets; see the article by Stanley in [BMSW94]. The relations in Theorem 17.5.1(iii) are proved in [BaB85]. An expression for the toric h in terms of the flag f can be found in the article by Bayer in [BMSW94]. The article by Kalai in the same volume contains an extensive discussion of g-vectors for both simplicial and general polytopes. The existence of the ***cd***-index was established in [BaK91]. Expressions for the (toric) g and h-vectors in terms of the flag h-vector or the ***cd***-index can be found in [BaE00]. The form of the cubical Dehn-Sommerville equations given in Theorem 17.5.2 appeared in [Adi96].

Theorem 17.5.3 can be found in [Sta96, Theorem III.4.4] (where h_S is denoted $\beta(S)$). The nonnegativity of the ***cd***-index for polytopes in Theorem 17.5.4(i) was proved as well for certain shellable spheres by Stanley (see [Sta96, Section III.4]); it was extended to all Gorenstein* complexes (and so all spherical complexes) by Karu [Kar06]. Theorem 17.5.4(ii), that the ***cd***-index is minimized over polytopes by simplices, is shown in [BE00]; the proof for Gorenstein* lattices is in [EK07]. Theorem 17.5.5 is proved in [Ehr05a], where one can find a list of the currently best known inequalities for ***cd***-coefficients of polytopes of low dimensions (through $d = 8$). Theorem 17.5.6 was proved by Stanley [Sta87b] for rational polytopes, and extended to all polytopes by Karu [Kar04]. Relationships between these classes of inequalities and those that can be derived from them are discussed in [Ste04]. Nonnegativity of *certain* ***cd***-coefficients for odd-dimensional simplicial manifolds is shown in [Nov00].

The problem of determining all linear inequalities for flag f-vectors has been considered for classes of partially ordered sets more general than the face posets of polytopes and spheres. In [BH00a], the (Catalan many) extreme rays are determined for the closed convex cone determined by flag f-vectors of all graded posets (posets with a rank function and having minimum and maximum elements). A nice description of the finite minimum set of inequalities is lacking, however. In [BaH01], a partial family of extreme rays is determined for the subcone determined by all Eulerian posets. See [BH00b] for more such results.

There is a notion of convolution product of flag f numbers, originally due to Kalai [Kal88], that can be used to produce new linear inequalities from given ones; see, for example, [BaL93, Section 3.10]. The algebraic properties of this product have been developed in [BiL00]; this has led to a deeper understanding of the combinatorial and algebraic properties of the ***cd***-index via duality of Hopf algebras (see [BHW02]). For a discussion of these developments, including an extension to enumeration in Bruhat intervals in Coxeter groups, see [Bil10] and [BiB11].

Theorem 17.5.7, due to Braden and MacPherson [BM99] (for rational polytopes and [Kar04] in general), gives a connection between the g-vector of a polytope P and that of one of its faces. The analogous Theorem 17.5.8 for ***cd***-indices can be found in [BE00] (as can the upper bound theorem, Theorem 17.5.9). These are examples of "monotonicity theorems" related to face numbers. For similar theorems relating h-vectors of subcomplexes and subdivisions of a simplicial complex Δ, see Sections III.9–10 of [Sta96] and the references given there.

Theorem 17.5.10(i) and (iv) can be found in [BaK91]. Theorem 17.5.10(ii) is proved in [Bra06, Theorem 4.5], where it is attributed to Kalai (unpublished). Theorem 17.5.10(iii) is due to Stanley [Sta92, Proposition 8.1]; this form of it is [Bra06, (20)].

Theorems 17.5.11 and 17.5.12 are due to Blind and Blind [BB90] and Figiel, Lindenstrauss, and Milman [FLM77], respectively.

For the fact that the flag f-vector of a zonotope or arrangement (or, more generally, of an oriented matroid) depends only on the underlying matroid, see [BLS+93, Cor. 4.6.3]. For expressions giving the ***cd***-index of a zonotope in terms of the flag h-vector of its underlying geometric lattice, see [BER97, Corollary 3.2] and [BHW02, Proposition 3.5]. That the only linear relations satisfied by zonotopes are the generalized Dehn-Sommerville equations of Theorem 17.5.1(iii), as well as the divisibility property in Theorem 17.5.15, is proved in [BER98]. The bounds on the ***cd***-indices of zonotopes in Theorem 17.5.15 are proved in [BER97]; the bounds in Theorem 17.5.16 can be found in [Ehr05b]. Theorem 17.5.17 is due to Varchenko for the case $k = 2$ (see [BLS+93, Proposition 4.6.9]) and to Liu. The stronger version given here is due to Stenson; in fact, [Ste05, Theorem 9] gives a stronger inequality (see also [Ste01]).

Theorem 17.5.18(i) can be found in [Grü67, Section 10.3]; 17.5.18(ii) appears in dual form (for 4-valent 3-polytopes) in [Bar83]; 17.5.18(iii) can be derived using the methods of [Grü67, Section 18.2] (see also [BER98]). Theorem 17.5.19 can be found in [Bay87]; see also [HZ00]. An interesting general discussion of f-vectors of 4-polytopes (ordinary and flag) and an up-to-date survey of this topic is given by Ziegler [Zie02]. In particular, a good case is made there that the situation for f-vectors of 4-polytopes is much more complicated than that for polytopes in dimension 3. One reason for this is that *neighborly cubical d*-polytopes begin to exist for $d = 4$: for any $n \geq d \geq 2r + 2$, there is a cubical convex d-polytope whose r-skeleton is combinatorially equivalent to that of the n-dimensional cube [JZ00] (see also [BBC97], where spheres having this property are constructed). In particular, for any $n \geq 4$, there is a cubical 4-polytope with the graph of the n-cube. These polytopes show that the ratio f_3/f_0 is not bounded over cubical 4-polytopes.

17.6 OPEN PROBLEMS

PROBLEM 17.6.1

Characterize the f-vectors of triangulations of the $(d-1)$-sphere. (It has been conjectured that the conditions of the g-theorem provide the answer. See [Swa14] for recent results and an overview of the g-conjecture.)

PROBLEM 17.6.2

Characterize the f-vectors of triangulations of the d-ball. (See [Kol11a, Kol11b, Kol11c, Mur13a, Mur13b] for recent results on this and related questions.)

PROBLEM 17.6.3

Characterize the f-vectors of triangulations of the d-torus. (It is known that $f(\text{2-torus}) = \{(n, 3n, 2n) \mid n \geq 7\}$, but the question is open for $d \geq 3$.)

PROBLEM 17.6.4

Characterize the f-vectors of d-polytopes. (The answer is known for $d \leq 3$, cf. Theorem 17.5.18(i), but for $d \geq 4$ there is not even a conjectured answer.)

PROBLEM 17.6.5 *I. Bárány*

Does there exist a constant $c_d > 0$ such that $f_i \geq c_d \cdot \min\{f_0, f_{d-1}\}$ for all d-polytopes and all i? Will $c_d = 1$ do?

PROBLEM 17.6.6

Characterize the f-vectors of centrally symmetric d-polytopes. (The question is open in the simplicial as well as in the general case. For results and conjectures on upper bounds, see [BN08, BLN13].)

PROBLEM 17.6.7 *Conjecture of G. Kalai*

The total number of faces (counting P but not $\emptyset$) of a centrally symmetric convex d-polytope P is at least 3^d. (Verified in the simplicial case as a consequence of Theorem 17.3.8.)

PROBLEM 17.6.8

Characterize the f-vectors of clique complexes.

PROBLEM 17.6.9 *Conjecture of Charney and Davis* [Sta96, p. 100]

Let $(g_0, \ldots, g_k)$ be the g-vector of a clique complex homeomorphic to the sphere S^{2k-1}. Then $g_k - g_{k-1} + \cdots + (-1)^k g_0 \geq 0$. (The case $k = 2$ was proved by Davis and Okun [DO01].)

PROBLEM 17.6.10 *Conjecture of Ehrenborg*

For d-polytopes P (and more generally for simplicial $(d-1)$-spheres) the **cd***-index satisfies*

$$\phi_{\mathbf{udv}}(P) - \phi_{\mathbf{uc^2v}}(P) \geq \phi_{\mathbf{udv}}(\Delta^{(d)}) - \phi_{\mathbf{uc^2v}}(\Delta^{(d)}),$$

where $\deg u + \deg v = d - 2$, *and $\Delta^{(d)}$ is the d-simplex.* (This is a special case of [Ehr05a, Conj. 6.1].)

PROBLEM 17.6.11 *Adin* [Adi96]

The "generalized lower bound conjecture" for cubical d-polytopes and $(d-1)$-spheres: $g_i^c \geq 0$ for $i \leq \lfloor d/2 \rfloor$. (This has been shown to be the best possible set of linear inequalities for cubical $(d-1)$-spheres [BBC97]. The case $i = 1$ is implied by Theorem 17.5.11.) *More generally, characterize the f-vectors of cubical polytopes.*

PROBLEM 17.6.12

Characterize the flag f-vectors of polytopes and of zonotopes. In particular, determine a complete set of linear inequalities holding for flag f-vectors of polytopes and of zonotopes.

PROBLEM 17.6.13

Characterize (toric) h-vectors of general polytopes.

PROBLEM 17.6.14

Characterize flag f-vectors of colored complexes (here f_S is the number of simplices with color set S) (see [Cho15, Fro12a, Fro12b, Wal07])*; of* pure *colored complexes; of graded posets* (all linear inequalities are known here [BH00a])*; of Eulerian posets* (see [BaH01])*; of Eulerian lattices.*

17.7 SOURCES AND RELATED MATERIAL

FURTHER READING

Surveys of f-vector theory are given in [Bil16, BaL93, Bjö87, BK89, KK95, Sta85]. Books treating f-vectors (among other things) include [And87, BMSW94, Grü67, MS71, Sta96, Zie95].

RELATED CHAPTERS

Chapter 6: Oriented matroids
Chapter 15: Basic properties of convex polytopes
Chapter 16: Subdivisions and triangulations of polytopes
Chapter 56: Splines and geometric modeling

REFERENCES

[Adi96] R.M. Adin. A new cubical h-vector. *Discrete Math.*, 157:3–14, 1996.

[ABG17] K.A. Adiprasito, A. Björner, and A. Goodarzi. Face numbers of sequentially Cohen-Macaulay complexes and Betti numbers of componentwise linear ideals. *J. Eur. Math. Soc.*, to appear, 2017.

[And87] I. Anderson. *Combinatorics of Finite Sets.* Clarendon Press, Oxford, 1987.

[BBC97] E.K. Babson, L.J. Billera, and C.S. Chan. Neighborly cubical spheres and a cubical lower bound conjecture. *Israel J. Math.*, 102:297–315, 1997.

[Bar83] D.W. Barnette. *Map Coloring, Polyhedra, and the Four Color Theorem.* Number 8 of *Dolciani Mathematical Expositions*, MAA, Washington, 1983.

[Bar92] A.I. Barvinok. On equivariant generalization of Dehn-Sommerville equations. *European J. Combin.*, 13:419–428, 1992.

[BN08] A.I. Barvinok and I. Novik. A centrally symmetric version of the cyclic polytope. *Discrete Comput. Geom.*, 39:76–99, 2008.

[BLN13] A.I. Barvinok, S.J. Lee, and I. Novik. Centrally symmetric polytopes with many faces. *Israel J. Math.*, 195:457–472, 2013.

[Bay87] M.M. Bayer. The extended f-vectors of 4-polytopes. *J. Combin. Theory. Ser. A*, 44:141–151, 1987.

[Bay88] M.M. Bayer. Barycentric subdivisions. *Pacific J. Math.*, 135:1–16, 1988.

[BaB85] M.M. Bayer and L.J. Billera. Generalized Dehn-Sommerville relations for polytopes, spheres and Eulerian partially ordered sets. *Invent. Math.*, 79:143–157, 1985.

[BaE00] M.M. Bayer and R. Ehrenborg. The toric h-vector of partially ordered sets. *Trans. Amer. Math. Soc.*, 352:4515–4531, 2000.

[BaH01] M.M. Bayer and G. Hetyei. Flag vectors of Eulerian partially ordered sets. *European J. Combin.*, 22:5–26, 2001.

[BaK91] M.M. Bayer and A. Klapper. A new index for polytopes. *Discrete Comput. Geom.*, 6:33–47, 1991.

[BaL93] M.M. Bayer and C.W. Lee. Combinatorial aspects of convex polytopes. In P.M. Gruber and J.M. Wills, editors, *Handbook of Convex Geometry*, pages 485–534, North-Holland, Amsterdam, 1993.

[Bil10] L.J. Billera. Flag enumeration in polytopes, Eulerian partially ordered sets and Coxeter groups. In *Proc. Internat. Cong. Math. (Hyderabad, 2010)*, Volume IV, Invited Lectures, pages 2389–2415, Hindustan Book Agency, New Delhi, 2010.

[Bil16] L.J. Billera. *"Even more intriguing, if rather less plausible..."*; Face numbers of convex polytopes. In P. Hersh, T. Lam, P. Pylyavskyy and V. Reiner, editors, *The Mathematical Legacy of Richard P. Stanley,* AMS, Providence, 2016.

[BiB11] L.J. Billera and F. Brenti. Quasisymmetric functions and Kazhdan-Lusztig polynomials, *Israel J. Math*, 184:317–348, 2011.

[BE00] L.J. Billera and R. Ehrenborg. Monotonicity of the $\boldsymbol{cd}$-index for polytopes. *Math. Z.*, 233:421–441, 2000.

[BER97] L.J. Billera, R. Ehrenborg, and M. Readdy. The $\boldsymbol{c}$-$2\boldsymbol{d}$-index of oriented matroids. *J. Combin. Theory Ser. A*, 80:79–105, 1997.

[BER98] L.J. Billera, R. Ehrenborg, and M. Readdy. The $\boldsymbol{cd}$-index of zonotopes and arrangements. In B. Sagan and R. Stanley, editors, *Mathematical Essays in Honor of Gian-Carlo Rota*, Birkhäuser, Boston, 1998.

[BH00a] L.J. Billera and G. Hetyei. Linear inequalities for flags in graded posets. *J. Combin. Theory Ser. A*, 89:77–104, 2000.

[BH00b] L.J. Billera and G. Hetyei. Decompositions of partially ordered sets. *Order*, 17:141–166, 2000.

[BHW02] L.J. Billera, S.K. Hsiao, and S. van Willigenburg. Peak quasisymmetric functions and Eulerian enumeration. *Adv. Math.*, 176:248–276, 2003.

[BL81a] L.J. Billera and C.W. Lee. A proof of the sufficiency of McMullen's conditions for f-vectors of simplicial polytopes. *J. Combin. Theory Ser. A*, 31:237–255, 1981.

[BL81b] L.J. Billera and C.W. Lee. The numbers of faces of polytope pairs and unbounded polyhedra. *European J. Combin.*, 2:307–322, 1981.

[BiL00] L.J. Billera and N. Liu. Noncommutative enumeration in graded posets. *J. Algebraic Combin.*, 12:7–24, 2000.

[BMSW94] T. Bisztriczky, P. McMullen, R. Schneider, and A. Ivić Weiss, editors. *Polytopes: Abstract, Convex, and Computational.* Volume 440 of *NATO Adv. Sci. Inst. Ser. C: Math. Phys. Sci.* Kluwer, Dordrecht, 1994.

[Bjö87] A. Björner. Face numbers of complexes and polytopes. In *Proc. Internat. Cong. Math. (Berkeley, 1986)*, pages 1408–1418, AMS, Providence, 1987.

[Bjö96] A. Björner. Nonpure shellability, f-vectors, subspace arrangements, and complexity. In L.J. Billera, C. Greene, R. Simion, and R. Stanley, editors, *Formal Power Series and Algebraic Combinatorics, DIMACS Ser. in Discrete Math. and Theor. Comput. Sci.*, pages 25–53. AMS, Providence, 1996.

[BK88] A. Björner and G. Kalai. An extended Euler-Poincaré theorem. *Acta Math.*, 161:279–303, 1988.

[BK89] A. Björner and G. Kalai. On f-vectors and homology. In G. Bloom, R.L. Graham, and J. Malkevitch, editors, *Combinatorial Mathematics: Proc. 3rd Internat. Conf.,*

New York, 1985, vol. 555 of *Ann. New York Acad. Sci.*, pages 63–80. New York Acad. Sci., 1989.

[BK91] A. Björner and G. Kalai. Extended Euler-Poincaré relations for cell complexes. In P. Gritzmann and B. Sturmfels, editors, *Applied Geometry and Discrete Mathematics—The Victor Klee Festschrift*, pages 81–89, vol. 4 of *DIMACS Series in Discrete Math. and Theor. Comput. Sci.*, AMS, Providence, 1991.

[BLS+93] A. Björner, M. Las Vergnas, B. Sturmfels, N. White, and G.M. Ziegler. *Oriented Matroids.* Volume 46 of *Encyclopedia Math. Appl.*, Cambridge University Press, 1993. Second edition, 1999.

[BL99] A. Björner and S. Linusson. The number of k-faces of a simple d-polytope. *Discrete Comput. Geom.*, 21:1–16, 1999.

[BL00] A. Björner and F.H. Lutz. Simplicial manifolds, bistellar flips and a 16-vertex triangulation of the Poincaré homology 3-sphere. *Experiment. Math.*, 9:275–289, 2000.

[Bjö07] A. Björner. A comparison theorem for f-vectors of simplicial polytopes. *Pure Appl. Math. Quarterly*, 3:347–356, 2007.

[BB90] G. Blind and R. Blind. Convex polytopes without triangular faces. *Israel J. Math.*, 71:129–134, 1990.

[BM99] T.C. Braden and R. MacPherson. Intersection homology of toric varieties and a conjecture of Kalai. *Comment. Math. Helv.*, 74:442–455, 1999.

[Bra06] T.C. Braden. Remarks on the combinatorial intersection cohomology of fans. *Pure Appl. Math. Q.*, 2:1149–1186, 2006.

[Cho15] K.F.E. Chong. *Face Vectors and Hilbert Functions.* Ph.D. Thesis, Cornell Univ., Ithaca, 2015.

[DO01] M.W. Davis and B. Okun. Vanishing theorems and conjectures for the ℓ^2-homology of right-angled Coxeter groups. *Geom. Topol.*, 5:7–74, 2001.

[Ehr05a] R. Ehrenborg. Lifting inequalities for polytopes. *Adv. Math.*, 193:205–222, 2005.

[Ehr05b] R. Ehrenborg. Inequalities for zonotopes. *Combinatorial and Computational Geometry*, pages 277–286, vol. 52 of MSRI Publications, Cambridge University Press, 2005.

[EK07] R. Ehrenborg and K. Karu. Decomposition theorem for the **cd**-index of Gorenstein* posets. *J. Algebraic Combin.*, 26:225–251, 2007.

[FLM77] T. Figiel, J. Lindenstrauss, and V.D. Milman. The dimension of almost spherical sections of convex bodies. *Acta Math.*, 139:53–94, 1977.

[FK08] B. Fleming and K. Karu.Hard Lefschetz theorem for simple polytopes. *J. Algebraic Combin.*, 32:227–239, 2010.

[FFK88] P. Frankl, Z. Füredi, and G. Kalai. Shadows of colored complexes. *Math. Scand.*, 63:169–178, 1988.

[FK96] E. Friedgut and G. Kalai. Every monotone graph property has a sharp threshold. *Proc. AMS*, 124:2993–3002, 1996.

[Fro08] A. Frohmader. Face vectors of flag complexes. *Israel J. Math.*, 164:153–164, 2008.

[Fro12a] A. Frohmader. Flag f-vectors of three-colored complexes. *Electron. J. Combin.*, 19:article 13, 2012.

[Fro12b] A. Frohmader. Flag f-vectors of colored complexes. *J. Combin. Theory Ser. A*, 119:937–941, 2012.

[Grü67] B. Grünbaum. *Convex Polytopes.* Interscience, London, 1967. Revised edition (V. Kaibel, V. Klee, and G.M. Ziegler, editors), vol. 221 of *Grad. Texts in Math.*, Springer, New York, 2003.

[Hib89] T. Hibi. What can be said about pure O-sequences? *J. Combin. Theory Ser. A*, 50:319–322, 1989.

[HZ00] A. Höppner and G.M. Ziegler. A census of flag-vectors of 4-polytopes. In G. Kalai and G.M. Ziegler, editors, *Polytopes—Combinatorics and Computation*, vol. 29 of DMV Sem., pages 105–110, Birkhäuser-Verlag, Basel, 2000.

[JZ00] M. Joswig and G.M. Ziegler. Neighborly Cubical Polytopes. *Discrete Comput. Geom.*, 24:325–344, 2000.

[Kal84] G. Kalai. A characterization of f-vectors of families of convex sets in $\mathbb{R}^d$. Part I: Necessity of Eckhoff's conditions. *Israel J. Math.*, 48:175–195, 1984.

[Kal86] G. Kalai. A characterization of f-vectors of families of convex sets in $\mathbb{R}^d$. Part II: Sufficiency of Eckhoff's conditions. *J. Combin. Theory Ser. A*, 41:167–188, 1986.

[Kal87] G. Kalai. Rigidity and the lower bound theorem I. *Invent. Math.*, 88:125–151, 1987.

[Kal88] G. Kalai. A new basis of polytopes. *J. Combin. Theory Ser. A*, 49:191–208, 1988.

[Kar04] K. Karu. Hard Lefschetz theorem for nonrational polytopes. *Invent. Math.*, 157:419–447, 2004.

[Kar06] K. Karu. The **cd**-index of fans and posets. *Compos. Math.*, 142:701–718, 2006.

[Kle64] V. Klee. A combinatorial analogue of Poincaré's duality theorem. *Canad. J. Math.*, 16:517–531, 1964.

[KK95] V. Klee and P. Kleinschmidt. Convex polytopes and related complexes. In R.L. Graham, M. Grötschel, and L. Lovász, editors, *Handbook of Combinatorics*, pages 875–917, North-Holland, Amsterdam, 1995.

[Kol11a] S.R. Kolins. *Face Vectors of Subdivisions of Balls.* Ph.D. Thesis, Cornell Univ., Ithaca, 2011.

[Kol11b] S.R. Kolins. f-vectors of simplicial posets that are balls. *J. Algebraic Combin.* 34:587–605, 2011.

[Kol11c] S.R. Kolins. f-vectors of triangulated balls. *Discrete Comput. Geom.*, 46:427–446, 2011.

[Küh90] W. Kühnel. Triangulations of manifolds with few vertices. In F. Tricerri, editor, *Advances in Differential Geometry and Topology*, pages 59–114, World Scientific, Singapore, 1990.

[Küh95] W. Kühnel. *Tight Polyhedral Submanifolds and Tight Triangulations.* Vol. 1612 of *Lecture Notes in Math.*, Springer-Verlag, Berlin, 1995.

[Lin71] B. Lindström. The optimal number of faces in cubical complexes. *Ark. Mat.*, 8:245–257, 1971.

[LW69] A.T. Lundell and S. Weingram. *The Topology of CW Complexes.* Van Nostrand, New York, 1969.

[Lut05] F.H. Lutz. Triangulated manifolds with few vertices: Combinatorial manifolds. Preprint, `arXiv:0506372`, 2005.

[McM93] P. McMullen. On simple polytopes. *Invent. Math.*, 113:419–444, 1993.

[MS71] P. McMullen and G.C. Shephard. *Convex Polytopes and the Upper Bound Conjecture.* Volume 3 of *London Math. Soc. Lecture Note Ser.*, Cambridge University Press, 1971.

[Mun84] J.R. Munkres. *Elements of Algebraic Topology.* Addison-Wesley, Reading, 1984.

[Mur13a] S. Murai. Face vectors of simplicial cell decompositions of manifolds. *Israel J. Math.*, 195:187–213, 2013.

[Mur13b] S. Murai. h-vectors of simplicial cell balls. *Trans. Amer. Math. Soc.* 365:1533–1550, 2013.

[MN13] S. Murai and E. Nevo. On the generalized lower bound conjecture for polytopes and spheres. *Acta Math.*, 210:185–202, 2013.

[Nik87] V.V. Nikulin. Discrete reflection groups in Lobachevsky spaces and algebraic surfaces. In *Proc. Internat. Cong. Math. (Berkeley, 1986)*, pages 654–671, AMS, Providence, 1987.

[Nov98] I. Novik. Upper bound theorems for homology manifolds. *Israel J. Math.*, 108:45–82, 1998.

[Nov99] I. Novik. The lower bound theorem for centrally symmetric simple polytopes. *Mathematika*, 46:231–240, 1999.

[Nov00] I. Novik. Lower bounds for the ***cd***-index of odd-dimensional simplicial manifolds. *European J. Combin.*, 21:533–541, 2000.

[NS09] I. Novik and E. Swartz. Socles of Buchsbaum modules, complexes and posets. *Adv. Math.*, 222:2059–2084, 2009.

[NS12] I. Novik and E. Swartz. Face numbers of pseudomanifolds with isolated singularities. *Math. Scand.*, 110:198–222, 2012.

[Spa66] E.H. Spanier. *Algebraic Topology.* McGraw-Hill, New York, 1966.

[Sta80] R.P. Stanley. The number of faces of simplicial convex polytopes. *Adv. Math.*, 35:236–238, 1980.

[Sta85] R.P. Stanley. The number of faces of simplicial polytopes and spheres. In J.E. Goodman, E. Lutwak, J. Malkevitch, and R. Pollack, editors, *Discrete Geometry and Convexity*, vol. 440 of *Ann. New York Acad. Sci.*, pages 212–223, 1985.

[Sta86] R.P. Stanley. *Enumerative Combinatorics*, Vol. I. Wadsworth, Monterey, 1986. Second printing by Cambridge Univ. Press, 1997.

[Sta87a] R.P. Stanley. On the number of faces of centrally-symmetric simplicial polytopes. *Graphs Combin.*, 3:55–66, 1987.

[Sta87b] R.P. Stanley. Generalized h-vectors, intersection cohomology of toric varieties, and related results. In M. Nagata and H. Matsumura, editors, *Commutative Algebra and Combinatorics*, vol. 11 of *Adv. Stud. Pure Math.*, pages 187–213, Kinokuniya, Tokyo and North-Holland, Amsterdam, 1987.

[Sta92] R.P. Stanley. Subdivisions and local h-vectors. *J. Amer. Math. Soc.*, 5:805-851, 1992.

[Sta96] R.P. Stanley. *Combinatorics and Commutative Algebra*, 2nd Ed. Vol. 41 of *Progr. Math.*, Birkhäuser, Boston, 1996.

[Ste01] C. Stenson. *Linear Inequalities for Flag f-vectors of Polytopes.* Ph.D. Thesis, Cornell Univ., Ithaca, 2001.

[Ste04] C. Stenson. Relationships among flag f-vector inequalities for polytopes. *Discrete Comput. Geom.*, 31:257–273, 2004.

[Ste05] C. Stenson. Tight inequalities for polytopes. *Discrete Comput. Geom.*, 34:507–521, 2005.

[Swa09] E. Swartz. Face enumeration—from spheres to manifolds. *J. Eur. Math. Soc.*, 11:449–485, 2009.

[Swa14] E. Swartz. Thirty-five years and counting. Preprint, `arXiv:1411.0987`, 2014.

[Wal07] S.A. Walker. Multicover inequalities on colored complexes. *Combinatorica*, 27:489–501, 2007.

[Zie95] G.M. Ziegler. *Lectures on Polytopes.* Vol. 152 of *Graduate Texts in Math.*, Springer-Verlag, New York, 1995; 7th revised printing 2007.

[Zie02] G.M. Ziegler. Face numbers of 4-polytopes and 3-spheres. In *Proc. Internat. Cong. Math. (Beijing, 2002)*, pages 625–634, Higher Ed. Press, Beijing, 2002.

18 SYMMETRY OF POLYTOPES AND POLYHEDRA

Egon Schulte

INTRODUCTION

Symmetry of geometric figures is among the most frequently recurring themes in science. The present chapter discusses symmetry of discrete geometric structures, namely of polytopes, polyhedra, and related polytope-like figures. These structures have an outstanding history of study unmatched by almost any other geometric object. The most prominent symmetric figures, the regular solids, occur from very early times and are attributed to Plato (427-347 B.C.E.). Since then, many changes in point of view have occurred about these figures and their symmetry. With the arrival of group theory in the 19th century, many of the early approaches were consolidated and the foundations were laid for a more rigorous development of the theory. In this vein, Schläfli (1814-1895) extended the concept of regular polytopes and tessellations to higher dimensional spaces and explored their symmetry groups as reflection groups.

Today we owe much of our present understanding of symmetry in geometric figures (in a broad sense) to the influential work of Coxeter, which provided a unified approach to regularity of figures based on a powerful interplay of geometry and algebra [Cox73]. Coxeter's work also greatly influenced modern developments in this area, which received a further impetus from work by Grünbaum and Danzer [Grü77a, DS82]. In the past three to four decades, the study of regular figures has been extended in several directions that are all centered around an abstract combinatorial polytope theory and a combinatorial notion of regularity [MS02].

History teaches us that the subject has shown an enormous potential for revival. One explanation for this phenomenon is the appearance of polyhedral structures in many contexts that have little apparent relation to regularity such as their occurrence in nature as crystals [Fej64, Sen95, SF88, Wel77].

18.1 REGULAR CONVEX POLYTOPES AND REGULAR TESSELLATIONS IN $\mathbb{E}^d$

Perhaps the most important (but certainly the most investigated) symmetric polytopes are the regular convex polytopes in Euclidean spaces. See [Grü67] and [Zie95] for general properties of convex polytopes, or Chapter 15 in this Handbook. The most comprehensive text on regular convex polytopes and regular tessellations is [Cox73]; many combinatorial aspects are also discussed in [MS02].

GLOSSARY

Convex d-polytope: The intersection P of finitely many closed halfspaces in a Euclidean space, which is bounded and d-dimensional.

Face: The empty set and P itself are ***improper faces*** of dimensions -1 and d, respectively. A ***proper face*** F of P is the (nonempty) intersection of P with a supporting hyperplane of P. (Recall that a hyperplane H ***supports*** P at F if $P \cap H = F$ and P lies in one of the closed halfspaces bounded by H.)

Vertex, edge, i-face, facet: Face of P of dimension 0, 1, i, or $d-1$, respectively.

Vertex figure: A vertex figure of P at a vertex x is the intersection of P with a hyperplane H that strictly separates x from the other vertices of P. (If P is regular, one often takes H to be the hyperplane passing through the midpoints of the edges that contain x.)

Face lattice of a polytope: The set $\mathcal{F}(P)$ of all (proper and improper) faces of P, ordered by inclusion. As a partially ordered set, this is a ranked lattice. Also, $\mathcal{F}(P) \setminus \{P\}$ is called the ***boundary complex*** of P.

Flag: A maximal totally ordered subset of $\mathcal{F}(P)$. Two flags of P are ***adjacent*** if they differ in one proper face.

Isomorphism of polytopes: A bijection $\varphi : \mathcal{F}(P) \mapsto \mathcal{F}(Q)$ between the face lattices of two polytopes P and Q such that φ preserves incidence in both directions; that is, $F \subseteq G$ in $\mathcal{F}(P)$ if and only if $F\varphi \subseteq G\varphi$ in $\mathcal{F}(Q)$. If such an isomorphism exists, P and Q are ***isomorphic***.

Dual of a polytope: A convex d-polytope Q is the dual of P if there is a ***duality*** $\varphi : \mathcal{F}(P) \mapsto \mathcal{F}(Q)$; that is, a bijection reversing incidences in both directions, meaning that $F \subseteq G$ in $\mathcal{F}(P)$ if and only if $F\varphi \supseteq G\varphi$ in $\mathcal{F}(Q)$. A polytope has many duals but any two are isomorphic, justifying speaking of "the dual." (If P is regular, one often takes Q to be the convex hull of the facet centers of P, or a rescaled copy of this polytope.)

Self-dual polytope: A polytope that is isomorphic to its dual.

Symmetry: A Euclidean isometry of the ambient space (affine hull of P) that maps P to itself.

Symmetry group of a polytope: The group $G(P)$ of all symmetries of P.

Regular polytope: A polytope whose symmetry group $G(P)$ is transitive on the flags.

Schläfli symbol: A symbol $\{p_1, \ldots, p_{d-1}\}$ that encodes the local structure of a regular polytope. For each $i = 1, \ldots, d-1$, if F is any $(i{+}1)$-face of P, then p_i is the number of i-faces of F that contain a given $(i{-}2)$-face of F.

Tessellation: A family T of convex d-polytopes in Euclidean d-space $\mathbb{E}^d$, called the ***tiles*** of T, such that the union of all tiles of T is $\mathbb{E}^d$, and any two distinct tiles do not have interior points in common. All tessellations are assumed to be ***locally finite***, meaning that each point of $\mathbb{E}^d$ has a neighborhood meeting only finitely many tiles, and ***face-to-face***, meaning that the intersection of any two tiles is a face of each (possibly the empty face); see Chapter 3.

Face lattice of a tessellation: A ***proper face*** of T is a nonempty face of a tile of T. ***Improper faces*** of T are the empty set and the whole space $\mathbb{E}^d$. The set $\mathcal{F}(T)$ of all (proper and improper) faces is a ranked lattice called the face lattice of T. Concepts such as isomorphism and duality carry over from polytopes.

Symmetry group of a tessellation: The group $G(T)$ of all symmetries of T; that is, of all isometries of the ambient (spherical, Euclidean, or hyperbolic) space that preserve T. Concepts such as regularity and Schläfli symbol carry over from polytopes.

Apeirogon: A tessellation of the real line with closed intervals of the same length. This can also be regarded as an infinite polygon whose edges are given by the intervals.

ENUMERATION AND CONSTRUCTION

The convex regular polytopes P in $\mathbb{E}^d$ are known for each d. If $d = 1$, P is a line segment and $|G(P)| = 2$. In all other cases, up to similarity, P can be uniquely described by its Schläfli symbol $\{p_1, \ldots, p_{d-1}\}$. For convenience one writes $P = \{p_1, \ldots, p_{d-1}\}$. If $d = 2$, P is a convex regular p-gon for some $p \geq 3$, and $P = \{p\}$; also, $G(P) = D_p$, the dihedral group of order $2p$.

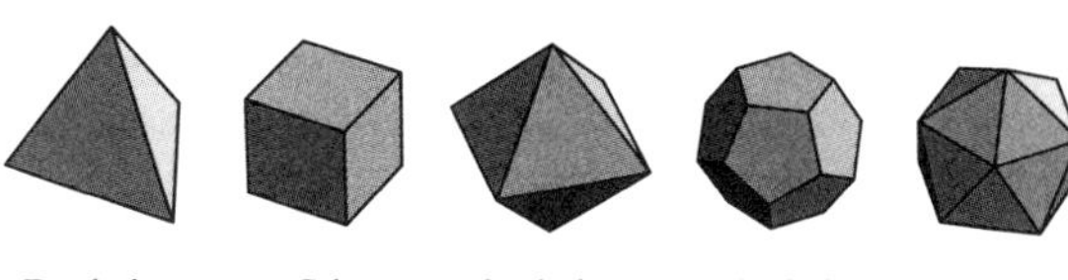

FIGURE 18.1.1
The five Platonic solids.

The regular polytopes P with $d \geq 3$ are summarized in Table 18.1.1, which also includes the numbers f_0 and f_{d-1} of vertices and facets, the order of $G(P)$, and the diagram notation (Section 18.6) for the group (following [Hum90]). Here and below, p^n will be used to denote a string of n consecutive p's. For $d = 3$ the list consists of the five Platonic solids (Figure 18.1.1). The regular d-simplex, d-cube, and d-cross-polytope occur in each dimension d. (These are line segments if $d = 1$, and triangles or squares if $d = 2$.) The dimensions 3 and 4 are exceptional in that there are 2 respectively 3 more regular polytopes. If $d \geq 3$, the facets and vertex figures of $\{p_1, \ldots, p_{d-1}\}$ are the regular $(d-1)$-polytopes $\{p_1, \ldots, p_{d-2}\}$ and $\{p_2, \ldots, p_{d-1}\}$, respectively, whose Schläfli symbols, when superposed, give the original Schläfli symbol. The dual of $\{p_1, \ldots, p_{d-1}\}$ is $\{p_{d-1}, \ldots, p_1\}$. Self-duality occurs only for $\{3^{d-1}\}$, $\{p\}$, and $\{3, 4, 3\}$. Except for $\{3^{d-1}\}$ and $\{p\}$ with p odd, all regular polytopes are centrally symmetric.

TABLE 18.1.1 The convex regular polytopes in $\mathbb{E}^d$ $(d \geq 3)$.

DIMENSION	NAME	SCHLÄFLI SYMBOL	f_0	f_{d-1}	$\|G(P)\|$	DIAGRAM
$d \geq 3$	d-simplex	$\{3^{d-1}\}$	d+1	d+1	(d+1)!	A_d
	d-cross-polytope	$\{3^{d-2}, 4\}$	$2d$	2^d	$2^d d!$	B_d (or C_d)
	d-cube	$\{4, 3^{d-2}\}$	2^d	$2d$	$2^d d!$	B_d (or C_d)
$d = 3$	icosahedron	$\{3, 5\}$	12	20	120	H_3
	dodecahedron	$\{5, 3\}$	20	12	120	H_3
$d = 4$	24-cell	$\{3, 4, 3\}$	24	24	1152	F_4
	600-cell	$\{3, 3, 5\}$	120	600	14400	H_4
	120-cell	$\{5, 3, 3\}$	600	120	14400	H_4

The regular tessellations T in $\mathbb{E}^d$ are also known. When $d = 1$, T is an apeirogon and $G(T)$ is the infinite dihedral group. For $d \geq 2$ see the list in Table 18.1.2. The first $d-1$ entries in $\{p_1, \ldots, p_d\}$ give the Schläfli symbol for the (regular) tiles of T, the last $d-1$ that for the (regular) vertex figures. (The vertex figure at a vertex of a regular tessellation is the convex hull of the midpoints of the edges emanating from that vertex.) The cubical tessellation occurs for each d, while for $d = 2$ and $d = 4$ there is a dual pair of exceptional tessellations.

TABLE 18.1.2 The regular tessellations in $\mathbb{E}^d$ $(d \geq 2)$.

DIMENSION	SCHLÄFLI SYMBOL	TILES	VERTEX-FIGURES
$d \geq 2$	$\{4, 3^{d-2}, 4\}$	d-cubes	d-cross-polytopes
$d = 2$	$\{3, 6\}$	triangles	hexagons
	$\{6, 3\}$	hexagons	triangles
$d = 4$	$\{3, 3, 4, 3\}$	4-cross-polytopes	24-cells
	$\{3, 4, 3, 3\}$	24-cells	4-cross-polytopes

As vertices of the plane polygon $\{p\}$ we can take the points corresponding to the pth roots of unity. The d-simplex can be defined as the convex hull of the $d+1$ points in $\mathbb{E}^{d+1}$ corresponding to the permutations of $(1, 0, \ldots, 0)$. As vertices of the d-cross-polytope in $\mathbb{E}^d$ choose the $2d$ permutations of $(\pm 1, 0, \ldots, 0)$, and for the d-cube take the 2^d points $(\pm 1, \ldots, \pm 1)$. The midpoints of the edges of a 4-cross-polytope are the 24 vertices of a regular 24-cell given by the permutations of $(\pm 1, \pm 1, 0, 0)$. The coordinates for the remaining regular polytopes are more complicated [Cox73, pp. 52,157].

For the cubical tessellation $\{4, 3^{d-2}, 4\}$ take the vertex set to be $\mathbb{Z}^d$ (giving the square tessellation if $d = 2$). For the triangle tessellation $\{3, 6\}$ choose as vertices the integral linear combinations of two unit vectors inclined at $\pi/3$. Locating the face centers gives the vertices of the hexagonal tessellation $\{6, 3\}$. For $\{3, 3, 4, 3\}$ in $\mathbb{E}^4$ take one set of alternating vertices of the cubical tessellation; for example, the integral points with an even coordinate sum. Its dual $\{3, 4, 3, 3\}$ (with 24-cells as tiles) has the vertices at the centers of the tiles of $\{3, 3, 4, 3\}$.

The concept of a regular tessellation extends to other spaces including spherical space (Euclidean unit sphere) and hyperbolic space. Each regular tessellation on the d-sphere is obtained from a convex regular $(d+1)$-polytope by radial projection from its center onto its circumsphere. In the hyperbolic plane there exists a regular tessellations $\{p, q\}$ for each pair p, q with $\frac{1}{p} + \frac{1}{q} < \frac{1}{2}$. There are four regular tessellations in hyperbolic 3-space, and five in hyperbolic 4-space; there are none in hyperbolic d-space with $d \geq 5$ (see [Cox68b] and [MS02, Ch. 6J]). All these tessellations are locally finite and have tiles that are topological balls. For the more general notion of a hyperbolic regular honeycomb see [Cox68b].

The regular polytopes and tessellations have been with us since before recorded history, and a strong strain of mathematics since classical times has centered on them. The classical theory intersects with diverse mathematical areas such as Lie algebras and Lie groups, Tits buildings [Tit74], finite group theory and incidence geometries [Bue95, BC13], combinatorial group theory [CM80, Mag07], geometric and algebraic combinatorics, graphs and combinatorial designs [BCN89], singularity theory, and Riemann surfaces.

SYMMETRY GROUPS

For a convex regular d-polytope P in $\mathbb{E}^d$, pick a fixed (***base***) flag Φ and consider the maximal simplex C (***chamber***) in the barycentric subdivision (***chamber complex***) of P whose vertices are the centers of the nonempty faces in Φ. Then C is a fundamental region for $G(P)$ in P and $G(P)$ is generated by the reflections $R_0, \ldots, R_{d-1}$ in the walls of C that contain the center of P, where R_i is the reflection in the wall opposite to the vertex of C corresponding to the i-face in Φ. If $P = \{p_1, \ldots, p_{d-1}\}$, then

$$\begin{cases} R_i^2 = (R_j R_k)^2 = 1 & (0 \le i,j,k \le d-1,\ |j-k| \ge 2) \\ (R_{i-1}R_i)^{p_i} = 1 & (1 \le i \le d-1) \end{cases}$$

is a presentation for $G(P)$ in terms of these generators. In particular, $G(P)$ is a finite (spherical) Coxeter group with string diagram

$$\bullet \overset{}{\underset{p_1}{\text{———}}} \bullet \underset{p_2}{\text{———}} \bullet \cdots\cdots \bullet \underset{p_{d-2}}{\text{———}} \bullet \underset{p_{d-1}}{\text{———}} \bullet$$

(see Section 18.6).

If T is a regular tessellation of $\mathbb{E}^d$, pick Φ and C as before. Now $G(T)$ is generated by the $d+1$ reflections in all walls of C giving $R_0, \ldots, R_d$ (as above). The presentation for $G(T)$ carries over, but now $G(T)$ is an infinite (Euclidean) Coxeter group.

18.2 REGULAR STAR-POLYTOPES

The regular star-polyhedra and star-polytopes are obtained by allowing the faces or vertex figures to be *starry* (star-like). This leads to very beautiful figures that are closely related to the regular convex polytopes. See Coxeter [Cox73] for a comprehensive account; see also McMullen and Schulte [MS02]. In defining star-polytopes, we shall combine the approach of [Cox73] and McMullen [McM68] and introduce them via the associated starry polytope-configuration.

GLOSSARY

d-polytope-configuration: A finite family Π of affine subspaces, called ***elements***, of Euclidean d-space $\mathbb{E}^d$, ordered by inclusion, such that the following conditions are satisfied. The family Π contains the empty set $\emptyset$ and the entire space $\mathbb{E}^d$ as (*improper*) elements. The dimensions of the other (*proper*) elements takes the values $0, 1, \ldots, d-1$, and the affine hull of their union is $\mathbb{E}^d$. As a partially ordered set, Π is a ranked lattice. For $F, G \in \Pi$ with $F \subsetneq G$ call $G/F := \{H \in \Pi | F \subseteq H \subseteq G\}$ the ***subconfiguration*** of Π defined by F and G; this has itself the structure of a $(\dim(G) - \dim(F) - 1)$-polytope-configuration. As further conditions, each G/F contains at least 2 proper elements if $\dim(G) - \dim(F) = 2$, and as a partially ordered set, each G/F (including Π itself) is connected if $\dim(G) - \dim(F) \ge 3$. (See the definition of an abstract polytope in Section 18.8.) It can be proved that in $\mathbb{E}^d$ every Π satisfies the stronger condition that each G/F contains exactly 2 proper elements if $\dim(G) - \dim(F) = 2$.

Regular polytope-configuration: A polytope-configuration Π whose symmetry group $G(\Pi)$ is flag-transitive. (A flag is a maximal totally ordered subset of Π.)

Regular star-polygon: For positive integers n and k with $(n,k) = 1$ and $1 < k < \frac{n}{2}$, up to similarity the regular star-polygon $\{\frac{n}{k}\}$ is the connected plane polygon whose consecutive vertices are $(\cos(\frac{2\pi kj}{n}), \sin(\frac{2\pi kj}{n}))$ for $j = 0, 1, \ldots, n-1$. If $k = 1$, the same plane polygon bounds a (nonstarry) convex n-gon with Schläfli symbol $\{n\}$ $(= \{\frac{n}{1}\})$. With each regular (convex or star-) polygon $\{\frac{n}{k}\}$ is associated a regular 2-polytope-configuration obtained by replacing each edge by its affine hull.

Star-polytope-configuration: A d-polytope-configuration Π is ***nonstarry*** if it is the family of affine hulls of the faces of a convex d-polytope. It is ***starry***, or a ***star-polytope-configuration***, if it is not nonstarry. For instance, among the 2-polytope-configurations that are associated with a regular (convex or star-) polygon $\{\frac{n}{k}\}$ for a given n, the one with $k = 1$ is nonstarry and those for $k > 1$ are starry. In the first case the corresponding n-gon is convex, and in the second case it is genuinely star-like. In general, the starry polytope configurations are those that belong to genuinely star-like polytopes (that is, star-polytopes).

Regular star-polytope: If $d = 2$, a regular star-polytope is a regular star-polygon. Defined inductively, if $d \geq 3$, a regular d-star-polytope P is a finite family of regular convex $(d-1)$-polytopes or regular $(d-1)$-star-polytopes such that the family consisting of their affine hulls as well as the affine hulls of their "faces" is a regular d-star-polytope-configuration $\Pi = \Pi(P)$. Here, the faces of the polytopes can be defined in such a way that they correspond to the elements in the associated polytope-configuration. The symmetry groups of P and Π are the same.

ENUMERATION AND CONSTRUCTION

Regular star-polytopes P can only exist for $d = 2$, 3, or 4. Like the regular convex polytopes they are uniquely determined by the Schläfli symbol $\{p_1, \ldots, p_{d-1}\}$, but now at least one entry is not integral. Again the symbols for the facets and vertex figures, when superposed, give the original symbol. If $d = 2$, then $P = \{\frac{n}{k}\}$ for some k with $(n,k) = 1$ and $1 < k < \frac{n}{2}$, and $G(P) = D_n$. For $d = 3$ and 4 the star-polytopes are listed in Table 18.2.1 together with the numbers f_0 and f_{d-1} of vertices and facets, respectively.

Great icosahedron

Great stellated dodecahedron

Great dodecahedron

Small stellated dodecahedron

FIGURE 18.2.1
The four Kepler-Poinsot polyhedra.

Every regular d-star-polytope has the same vertices and symmetry group as a regular convex d-polytope. The four regular star-polyhedra (3-star-polytopes) are also known as the ***Kepler-Poinsot polyhedra*** (Figure 18.2.1). They can

TABLE 18.2.1 The regular star-polytopes in $\mathbb{E}^d$ ($d \geq 3$).

DIMENSION	SCHLÄFLI SYMBOL	f_0	f_{d-1}
$d=3$	$\{3, \frac{5}{2}\}$	12	20
	$\{\frac{5}{2}, 3\}$	20	12
	$\{5, \frac{5}{2}\}$	12	12
	$\{\frac{5}{2}, 5\}$	12	12
$d=4$	$\{3, 3, \frac{5}{2}\}$	120	600
	$\{\frac{5}{2}, 3, 3\}$	600	120
	$\{3, 5, \frac{5}{2}\}$	120	120
	$\{\frac{5}{2}, 5, 3\}$	120	120
	$\{3, \frac{5}{2}, 5\}$	120	120
	$\{5, \frac{5}{2}, 3\}$	120	120
	$\{5, 3, \frac{5}{2}\}$	120	120
	$\{\frac{5}{2}, 3, 5\}$	120	120
	$\{5, \frac{5}{2}, 5\}$	120	120
	$\{\frac{5}{2}, 5, \frac{5}{2}\}$	120	120

be constructed from the icosahedron $\{3,5\}$ or dodecahedon $\{5,3\}$ by two kinds of operations, ***stellation*** or ***faceting*** [Cox73]. Loosely speaking, the former operation extends the faces of a polyhedron symmetrically until they again form a polyhedron, while in the latter operation the vertices of a polyhedron are redistributed in classes that are then the vertex sets for the faces of a new polyhedron. Regarded as regular maps on surfaces (Section 18.3), the polyhedra $\{3, \frac{5}{2}\}$ (***great icosahedron***) and $\{\frac{5}{2}, 3\}$ (***great stellated dodecahedron***) are of genus 0, while $\{5, \frac{5}{2}\}$ (***great dodecahedron***) and $\{\frac{5}{2}, 5\}$ (***small stellated dodecahedron***) are of genus 4.

The ten regular star-polytopes in $\mathbb{E}^4$ all have the same vertices and symmetry groups as the 600-cell $\{3,3,5\}$ or 120-cell $\{5,3,3\}$ and can be derived from these by 4-dimensional stellation or faceting operations [Cox73, McM68]. See also [Cox93] for their names, which describe the various relationships among the polytopes. For presentations of their symmetry groups that reflect the finer combinatorial structure of the star-polytopes, see also [MS02].

The dual of $\{p_1, \ldots, p_{d-1}\}$ (which is obtained by dualizing the associated star-polytope-configuration using reciprocation with respect to a sphere) is $\{p_{d-1}, \ldots, p_1\}$. Regarded as abstract polytopes (Section 18.8)), the star-polytopes $\{p_1, \ldots, p_{d-1}\}$ and $\{q_1, \ldots, q_{d-1}\}$ are isomorphic if and only if the symbol $\{q_1, \ldots, q_{d-1}\}$ is obtained from $\{p_1, \ldots, p_{d-1}\}$ by replacing each entry 5 by $\frac{5}{2}$ and each $\frac{5}{2}$ by 5.

18.3 REGULAR SKEW POLYHEDRA

The traditional regular skew polyhedra are finite or infinite polyhedra with convex faces whose vertex figures are skew (antiprismatic) polygons. The standard reference is Coxeter [Cox68a]. Topologically, these polyhedra are regular maps on surfaces. For general properties of regular maps see Coxeter and Moser [CM80], McMullen and Schulte [MS02], or Chapter 20 of this Handbook.

GLOSSARY

(Right) prism, antiprism (with regular bases): A convex 3-polytope whose vertices are contained in two parallel planes and whose set of 2-faces consists of the two ***bases*** (contained in the parallel planes) and the 2-faces in the ***mantle*** that connects the bases. The bases are congruent regular polygons. For a (right) prism, each base is a translate of the other by a vector perpendicular to its affine hull, and the mantle 2-faces are rectangles. For a (right) antiprism, each base is a translate of a congruent reciprocal (dual) of the other by a vector perpendicular to its affine hull, and the mantle 2-faces are isosceles triangles. (The prism or antiprism is ***semiregular*** if its mantle 2-faces are squares or equilateral triangles, respectively; see Section 18.5.)

Map on a surface: A decomposition (tessellation) P of a closed surface S into nonoverlapping simply connected regions, the ***2-faces*** of P, by arcs, the ***edges*** of P, joining pairs of points, the ***vertices*** of P, such that two conditions are satisfied. First, each edge belongs to exactly two 2-faces. Second, if two distinct edges intersect, they meet in one vertex or in two vertices.

Regular map: A map P on S whose combinatorial automorphism group $\Gamma(P)$ is transitive on the flags (incident triples consisting of a vertex, an edge, and a 2-face).

Polyhedron: A map P on a closed surface S embedded (without self-intersections) into a Euclidean space, such that two conditions are satisfied. Each 2-face of P is a convex plane polygon, and any two adjacent 2-faces do not lie in the same plane. See also the more general definition in Section 18.4 below.

Skew polyhedron: A polyhedron P such that for at least one vertex x, the vertex figure of P at x is not a plane polygon; the ***vertex figure*** at x is the polygon whose vertices are the vertices of P adjacent to x and whose edges join consecutive vertices as one goes around x in P.

Regular polyhedron: A polyhedron P whose symmetry group $G(P)$ is flag-transitive. (For a regular skew polyhedron P in $\mathbb{E}^3$ or $\mathbb{E}^4$, each vertex figure must be a 3-dimensional antiprismatic polygon, meaning that it comprises the edges of an antiprism that are not edges of a base. See also Section 18.4.)

ENUMERATION

In $\mathbb{E}^3$ all, and in $\mathbb{E}^4$ all finite, regular skew polyhedra are known [Cox68a]. In these cases the (orientable) polyhedron P is completely determined by the extended Schläfli symbol $\{p, q|r\}$, where the 2-faces of P are convex p-gons such that q meet at each vertex, and r is the number of edges in each edge path of P that leaves, at each vertex, exactly two 2-faces of P on the right. The group $G(P)$ is isomorphic to $\Gamma(P)$ and has the presentation

$$R_0^2 = R_1^2 = R_2^2 = (R_0R_1)^p = (R_1R_2)^q = (R_0R_2)^2 = (R_0R_1R_2R_1)^r = 1$$

(but here not all generators R_i are hyperplane reflections). The polyhedra $\{p, q|r\}$ and $\{q, p|r\}$ are duals, and the vertices of one can be obtained as the centers of the 2-faces of the other.

In $\mathbb{E}^3$ there are just three regular skew polyhedra: $\{4,6|4\}$, $\{6,4|4\}$, and $\{6,6|3\}$. These are the (infinite) ***Petrie-Coxeter polyhedra***. For example, $\{4,6|4\}$ consists of half the square faces of the cubical tessellation $\{4,3,4\}$ in $\mathbb{E}^3$.

TABLE 18.3.1 The finite regular skew polyhedra in $\mathbb{E}^4$.

SCHLÄFLI SYMBOL	f_0	f_2	GROUP ORDER	GENUS
$\{4,4\|r\}$	r^2	r^2	$8r^2$	1
$\{4,6\|3\}$	20	30	240	6
$\{6,4\|3\}$	30	20	240	6
$\{4,8\|3\}$	144	288	2304	73
$\{8,4\|3\}$	288	144	2304	73

The finite regular skew polyhedra in $\mathbb{E}^4$ (or equivalently, in spherical 3-space) are listed in Table 18.3.1. There is an infinite sequence of toroidal polyhedra as well as two pairs of duals related to the (self-dual) 4-simplex $\{3,3,3\}$ and 24-cell $\{3,4,3\}$. For drawings of projections of these polyhedra into 3-space see [BW88, SWi91]; Figure 18.3.1 represents $\{4,8|3\}$.

FIGURE 18.3.1
A projection of $\{4,8|3\}$ *into* $\mathbb{R}^3$.

These projections are examples of ***combinatorially regular polyhedra*** in ordinary 3-space; see [BW93] and Chapter 20 of this Handbook. For regular polyhedra in $\mathbb{E}^4$ with planar, but not necessarily convex, 2-faces, see also [ABM00, Bra00]. For regular skew polyhedra in hyperbolic 3-space, see [Gar67, WL84].

18.4 THE GRÜNBAUM-DRESS POLYHEDRA

A new impetus to the study of regular figures came from Grünbaum [Grü77b], who generalized the regular skew polyhedra by allowing skew polygons as faces as well as vertex figures. This restored the symmetry in the definition of polyhedra. For the classification of these "new" regular polyhedra in $\mathbb{E}^3$, see [Grü77b], [Dre85], and [MS02]. The proper setting for this subject is, strictly speaking, in the context of realizations of abstract regular polytopes (see Section 18.8).

GLOSSARY

Polygon: A figure P in Euclidean space $\mathbb{E}^d$ consisting of a (finite or infinite) sequence of distinct points, called the ***vertices*** of P, joined in successive pairs, and closed cyclically if finite, by line segments, called the ***edges*** of P, such that each compact set in $\mathbb{E}^d$ meets only finitely many edges.

Zigzag polygon: A (zigzag-shaped) infinite plane polygon P whose vertices alternately lie on two parallel lines and whose edges are all of the same length.

Antiprismatic polygon: A closed polygon P in 3-space whose vertices are alternately vertices of each of the two (regular convex) bases of a (right) antiprism Q (Section 18.3), such that the orthogonal projection of P onto the plane of a base gives a regular star-polygon (Section 18.2). This star-polygon (and thus P) has twice as many vertices as each base, and is a convex polygon if and only if the edges of P are just those edges of Q that are not edges of a base.

Prismatic polygon: A closed polygon P in 3-space whose vertices are alternately vertices of each of the two (regular convex) bases of a (right) prism Q (Section 18.3), such that the orthogonal projection of P onto the plane of a base traverses twice a regular star-polygon in that plane (Section 18.2). Each base of Q (and thus the star-polygon) is assumed to have an odd number of vertices. The star-polygon is a convex polygon if and only if each edge of P is a diagonal in a rectangular 2-face in the mantle of Q.

Helical polygon: An infinite polygon in 3-space whose vertices lie on a helix given parametrically by $(a\cos\beta t, a\sin\beta t, bt)$, where $a, b \neq 0$ and $0 < \beta < \pi$, and are obtained as t ranges over the integers. Successive integers correspond to successive vertices.

Polyhedron: A (finite or infinite) family P of polygons in $\mathbb{E}^d$, called the ***2-faces*** of P, such that three conditions are satisfied. First, each edge of a 2-face is an edge of exactly one other 2-face. Second, for any two edges F and F' of (2-faces of) P there exist chains $F = G_0, G_1, \ldots, G_n = F'$ of edges and $H_1, \ldots, H_n$ of 2-faces such that each H_i is incident with G_{i-1} and G_i. Third, each compact set in $\mathbb{E}^d$ meets only finitely many 2-faces.

Regular: A polygon or polyhedron P is regular if its symmetry group $G(P)$ is transitive on the flags.

Chiral: A polyhedron P is chiral if its symmetry group $G(P)$ has two orbits on the flags such that any two adjacent flags are in distinct orbits (see also Section 18.8). Here, two flags are ***adjacent*** if they differ in precisely one element: a vertex, an edge, or a face.

Petrie polygon of a polyhedron: A polygonal path along the edges of a regular polyhedron P such that any two successive edges, but no three, are edges of a 2-face of P.

Petrie dual: The family of all Petrie polygons of a regular polyhedron P. This is itself a regular polyhedron, and its Petrie dual is P itself.

Polygonal complex: A triple $K = (V, E, F)$ consisting of a set V of points in $\mathbb{E}^d$, called **vertices**, a set E of line segments, called **edges**, and a set F of polygons, called **faces**, such that four conditions are satisfied. First, the graph (V, E) is connected. Second, the vertex-figure of K at each vertex of K is connected; here the vertex-figure of K at a vertex v is the graph, possibly with multiple edges, whose vertices are the vertices of K adjacent to v and whose edges are the line segments (u, w), where (u, v) and (v, w) are edges of a common face of K. (There may be more than one such face in K, in which case the edge (u, w) of the vertex-figure at v has multiplicity given by the number of such faces.) Third, each edge of K is contained in at least two faces of K. Fourth, each compact set of $\mathbb{E}^3$ meets only finitely many faces of K. A polygonal complex K is **regular** if its symmetry group $G(K)$ is transitive on the flags.

ENUMERATION

For a systematic discussion of regular polygons in arbitrary Euclidean spaces see [Cox93]. In light of the geometric classification scheme for the new regular polyhedra in $\mathbb{E}^3$ proposed in [Grü77b], it is useful to classify the regular polygons in $\mathbb{E}^3$ into seven groups: convex polygons, plane star-polygons (Section 18.2, apeirogons (Section 18.1), zigzag polygons, antiprismatic polygons, prismatic polygons, and helical polygons. These correspond to the four kinds of isometries in $\mathbb{E}^3$: rotation, rotatory reflection (a reflection followed by a rotation in the reflection plane), glide reflection, and twist.

The 2-faces and vertex figures of a regular polyhedron P in $\mathbb{E}^3$ are regular polygons of the above kind. (The vertex figure at a vertex x is the polygon whose vertices are the vertices of P adjacent to x and whose edges join two such vertices y and z if and only if $\{y, x\}$ and $\{x, z\}$ are edges of a common 2-face in P. For a regular P, this is a single polygon.) It is convenient to group the regular polyhedra in $\mathbb{E}^3$ into 8 classes. The first four are the traditional regular polyhedra: the five Platonic solids; the three planar tessellations; the four regular star-polyhedra (Kepler-Poinsot polyhedra); and the three infinite regular skew polyhedra (Petrie-Coxeter polyhedra). The four other classes and their polyhedra can be described as follows: the class of nine finite polyhedra with finite skew (antiprismatic) polygons as faces; the class of infinite polyhedra with finite skew (prismatic or antiprismatic) polygons as faces, which includes three infinite families as well as three individual polyhedra; the class of polyhedra with zigzag polygons as faces, which contains six infinite families; and the class of polyhedra with helical polygons as faces, which has three infinite families and six individual polyhedra.

Alternatively, these forty-eight polyhedra can be described as follows [MS02]. There are eighteen finite regular polyhedra, namely the nine classical finite regular polyhedra (Platonic solids and Kepler-Poinsot polyhedra), and their Petrie duals. The regular tessellations of the plane, and their Petrie duals (with zigzag 2-faces), are the six planar polyhedra in the list. From those, twelve further polyhedra are obtained as blends (in the sense of Section 18.8) with a line segment or an apeirogon (Section 18.1). The six blends with a line segment have finite skew, or (infinite planar) zigzag, 2-faces with alternate vertices on a pair of parallel planes; the six blends with an apeirogon have helical polygons or zigzag polygons as 2-faces. Finally, there are twelve further polyhedra that are not blends; they fall into a single family and are related to the cubical tessellation of $\mathbb{E}^3$. Each polyhedron can be described by a generalized Schläfli symbol, which encodes the geometric structure of the polygonal faces and vertex figures, tells whether or not the polyhedron is a blend, and signifies a presentation of the symmetry group. For more details see [MS02] (or [Grü77b, Dre85, Joh91]).

The Grünbaum-Dress polyhedra belong to the more general class of regular polygonal complexes in $\mathbb{E}^3$, which were completely classified in [PS10, PS13]. Polygonal complexes are polyhedra-like "skeletal" structures in $\mathbb{E}^3$, in which an edge can be surrounded by any finite number of faces, but at least two, unlike in a polyhedron where this number is exactly two. In addition to polyhedra there are 25 regular polygonal complexes in $\mathbb{E}^3$, all periodic and with crystallographic symmetry groups.

Chiral polyhedra are nearly regular polyhedra. Chirality does not occur in traditional polyhedra but it is striking that it does in skeletal structures. See [Sch04, Sch05] for the full classification of chiral polyhedra in $\mathbb{E}^3$. There are six very large families of chiral polyhedra: three with periodic polyhedra with finite skew faces and vertex-figures, and three with periodic polyhedra with helical faces and planar vertex-figures. Each chiral polyhedron with helical faces is combinatorially isomorphic to a regular polyhedron [PW10].

18.5 SEMIREGULAR AND UNIFORM CONVEX POLYTOPES

The very stringent requirements in the definition of regularity of polytopes can be relaxed in many different ways, yielding a great variety of weaker regularity notions. We shall only consider polytopes and polyhedra that are convex. See Johnson [Joh91] for a detailed discussion, or Martini [Mar94] for a survey.

GLOSSARY

Semiregular: A convex d-polytope P is semiregular if its facets are regular and its symmetry group $G(P)$ is transitive on the vertices of P.

Uniform: A convex polygon is uniform if it is regular. Recursively, if $d \geq 3$, a convex d-polytope P is uniform if its facets are uniform and its symmetry group $G(P)$ is transitive on the vertices of P.

Regular-faced: P is regular-faced if all its facets (and lower-dimensional faces) are regular.

ENUMERATION

Each regular (convex) polytope is semiregular, and each semiregular polytope is uniform. Also, by definition each uniform 3-polytope is semiregular. For $d = 3$ the family of semiregular (uniform) convex polyhedra consists of the Platonic solids, two infinite classes of prisms and antiprisms, as well as the thirteen polyhedra known as Archimedean solids [Fej64]. The seven semiregular polyhedra whose symmetry group is edge-transitive are also called the ***quasiregular*** polyhedra.

Besides the regular polytopes, there are only seven semiregular polytopes in higher dimensions: three for $d = 4$, and one for each of $d = 5, 6, 7, 8$ (for a short proof, see [BB91]). However, there are many more uniform polytopes but a complete list is known only for $d = 4$ [Joh91]. In addition to the regular 4-polytopes and the prisms over uniform 3-polytopes there are exactly 40 uniform 4-polytopes.

For $d = 3$ all, for $d = 4$ all save one, and for $d \geq 5$ many, uniform polytopes can be obtained by a method called ***Wythoff's construction***. This method proceeds from a finite Euclidean reflection group W in $\mathbb{E}^d$, or the even (rotation) subgroup W^+ of W, and constructs the polytopes as the convex hull of the orbit under W or W^+ of a point, the initial vertex, in the fundamental region of the group, which is a d-simplex (chamber) or the union of two adjacent d-simplices in the corresponding chamber complex of W, respectively; see Sections 18.1 and 18.6.

The regular-faced polytopes have also been described for each dimension. In general, such a polytope can have different kinds of facets (and vertex figures). For $d = 3$ the complete list contains exactly 92 regular-faced convex polyhedra and includes all semiregular polyhedra. For each $d \geq 5$, there are only two regular-faced d-polytopes that are not semiregular. Except when $d = 4$, each regular-faced d-polytope has a nontrivial symmetry group.

There are many further generalizations of the notion of regularity [Mar94]. However, in most cases complete lists of the corresponding polytopes are either not known or available only for $d = 3$. The variants that have been considered include: ***isogonal*** polytopes (requiring vertex-transitivity of $G(P)$), or ***isohedral*** polytopes, the reciprocals of the isogonal polytopes, with a facet-transitive group $G(P)$; more generally, ***k-face-transitive*** polytopes (requiring transitivity of $G(P)$ on the k-faces), for a single value or several values of k; ***congruent-faceted***, or ***monohedral***, polytopes (requiring congruence of the facets); and ***equifaceted*** polytopes (requiring combinatorial isomorphism of the facets). Similar problems have also been considered for nonconvex polytopes or polyhedra, and for tilings [GS87].

18.6 REFLECTION GROUPS

Symmetry properties of geometric figures are closely tied to the algebraic structure of their symmetry groups, which often are subgroups of finite or infinite reflection groups. A classical reference for reflection groups is Coxeter [Cox73]. A more recent text is Humphreys [Hum90].

GLOSSARY

Reflection group: A group generated by (hyperplane) reflections in a finite-dimensional space V. In the present context the space is a real or complex vector space (or affine space). A ***reflection*** is a linear (or affine) transformation whose eigenvalues, save one, are all equal to 1, while the remaining eigenvalue is a primitive kth root of unity for some $k \geq 2$; in the real case the eigenvalue is -1. If the space is equipped with further structure, the reflections are assumed to preserve it. For example, if V is real Euclidean, the reflections are Euclidean reflections.

Coxeter group: A group W, finite or infinite, that is generated by finitely many generators $\sigma_1, \ldots, \sigma_n$ and has a presentation of the form $(\sigma_i\sigma_j)^{m_{ij}} = 1$ $(i, j = 1, \ldots, n)$, where the m_{ij} are positive integers or ∞ such that $m_{ii} = 1$ and $m_{ij} =$

$m_{ji} \geq 2$ $(i \neq j)$. The matrix $(m_{ij})_{ij}$ is the ***Coxeter matrix*** of W.

Coxeter diagram: A labeled graph $\mathcal{D}$ that represents a Coxeter group W as follows. The nodes of $\mathcal{D}$ represent the generators σ_i of W. The ith and jth node are joined by a (single) branch if and only if $m_{ij} > 2$. In this case, the branch is labeled m_{ij} if $m_{ij} \neq 3$ (and remains unlabeled if $m_{ij} = 3$).

Irreducible Coxeter group: A Coxeter group W whose Coxeter diagram is connected. (Each Coxeter group W is the direct product of irreducible Coxeter groups, with each factor corresponding to a connected component of the diagram of W.)

Root system: A finite set $\mathcal{R}$ of nonzero vectors, the ***roots***, in $\mathbb{E}^d$ satisfying the following conditions. $\mathcal{R}$ spans $\mathbb{E}^d$, and $\mathcal{R} \cap \mathbb{R}e = \{\pm e\}$ for each $e \in \mathcal{R}$. For each $e \in \mathcal{R}$, the Euclidean reflection S_e in the linear hyperplane orthogonal to e maps $\mathcal{R}$ onto itself. Moreover, the numbers $2(e,e')/(e',e')$, with $e, e' \in \mathcal{R}$, are integers (***Cartan integers***); here $(\ ,\ )$ denotes the standard inner product on $\mathbb{E}^d$. (These conditions define ***crystallographic*** root systems. Sometimes the integrality condition is omitted to give a more general notion of root system.) The group W generated by the reflections S_e $(e \in \mathcal{R})$ is a finite Coxeter group, called the ***Weyl group*** of $\mathcal{R}$.

GENERAL PROPERTIES

Every Coxeter group $W = \langle \sigma_1, \ldots, \sigma_n \rangle$ admits a faithful representation as a reflection group in the real vector space $\mathbb{R}^n$. This is obtained as follows. If W has Coxeter matrix $M = (m_{ij})_{ij}$ and $e_1, \ldots, e_n$ is the standard basis of $\mathbb{R}^n$, define the symmetric bilinear form $\langle\ ,\ \rangle_M$ by

$$\langle e_i, e_j \rangle_M := -\cos(\pi/m_{ij}) \quad (i,j = 1, \ldots, n),$$

with appropriate interpretation if $m_{ij} = \infty$. For $i = 1, \ldots, n$ the linear transformation $S_i : \mathbb{R}^n \mapsto \mathbb{R}^n$ given by

$$xS_i := x - 2\langle e_i, x \rangle_M \, e_i \qquad (x \in \mathbb{R}^n)$$

is the orthogonal reflection in the hyperplane orthogonal to e_i. Let $O(M)$ denote the orthogonal group corresponding to $\langle\ ,\ \rangle_M$. Then $\sigma_i \mapsto S_i$ $(i = 1, \ldots, n)$ defines a faithful representation $\rho : W \mapsto GL(\mathbb{R}^n)$, called the ***canonical representation***, such that $W\rho$ is a subgroup of $O(M)$.

The group W is finite if and only if the associated form $\langle\ ,\ \rangle_M$ is positive definite; in this case, $\langle\ ,\ \rangle_M$ determines a Euclidean geometry on $\mathbb{R}^n$. In other words, each finite Coxeter group is a finite Euclidean reflection group. Conversely, every finite Euclidean reflection group is a Coxeter group. The finite Coxeter groups have been completely classified by Coxeter and are usually listed in terms of their Coxeter diagrams.

The finite irreducible Coxeter groups with string diagrams are precisely the symmetry groups of the convex regular polytopes, with a pair of dual polytopes corresponding to a pair of groups that are related by reversing the order of the generators. See Section 18.1 for an explanation about how the generators act on the polytopes. Table 18.1.1 also lists the names for the corresponding Coxeter diagrams.

For $p_1, \ldots, p_{n-1} \geq 2$ write $[p_1, \ldots, p_{n-1}]$ for the Coxeter group with string diagram $\bullet \frac{}{p_1} \bullet \frac{}{p_2} \bullet \cdots\cdots \bullet \frac{}{p_{n-2}} \bullet \frac{}{p_{n-1}} \bullet$. Then $[p_1, \ldots, p_{n-1}]$ is the automorphism group of the universal abstract regular n-polytope $\{p_1, \ldots, p_{n-1}\}$; see Section 18.8 and [MS02]. The regular honeycombs $\{p_1, \ldots, p_{n-1}\}$ on the sphere (convex regular polytopes) or in Euclidean or hyperbolic space are particular instances of universal polytopes. The spherical honeycombs are exactly the finite universal regular polytopes (with $p_i > 2$ for all i). The Euclidean honeycombs arise exactly when $p_i > 2$ for all i and the bilinear form $\langle\, ,\, \rangle_M$ for $[p_1, \ldots, p_{n-1}]$ is positive semidefinite (but not positive definite). Similarly, the hyperbolic honeycombs correspond exactly to the groups $[p_1, \ldots, p_{n-1}]$ that are Coxeter groups of "hyperbolic type" [MS02].

There are exactly two sources of finite Coxeter groups, to some extent overlapping: the symmetry groups of convex regular polytopes, and the Weyl groups of (crystallographic) root systems, which are important in Lie Theory. Every root system $\mathcal{R}$ has a set of ***simple roots***; this is a subset $\mathcal{S}$ of $\mathcal{R}$, which is a basis of $\mathbb{E}^d$ such that every $e \in \mathcal{R}$ is a linear combination of vectors in $\mathcal{S}$ with integer coefficients that are all nonnegative or all nonpositive. The distinguished generators of the Weyl group W are given by the reflections S_e in the linear hyperplane orthogonal to e ($e \in \mathcal{S}$), for some set $\mathcal{S}$ of simple roots of $\mathcal{R}$. The irreducible Weyl groups in $\mathbb{E}^2$ are the symmetry groups of the triangle, square, or hexagon. The diagrams A_d, B_d, C_d, and F_4 of Table 18.1.1 all correspond to irreducible Weyl groups and root systems (with B_d and C_d corresponding to a pair of dual root systems), but H_3 and H_4 do not (they correspond to a noncrystallographic root system [CMP98]). There is one additional series of irreducible Weyl groups in $\mathbb{E}^d$ with $d \geq 4$ (a certain subgoup of index 2 in B_d), whose diagram is denoted by D_d. The remaining irreducible Weyl groups occur in dimensions 6, 7 and 8, with diagrams E_6, E_7, and E_8, respectively.

Each Weyl group W stabilizes the lattice spanned by a set $\mathcal{S}$ of simple roots, the ***root lattice*** of $\mathcal{R}$. These lattices have many remarkable geometric properties and also occur in the context of sphere packings (see Conway and Sloane [CS02]). The irreducible Coxeter groups W of Euclidean type, or, equivalently, the infinite discrete irreducible Euclidean reflection groups, are intimately related to Weyl groups; they are also called ***affine Weyl groups***.

The complexifications of the reflection hyperplanes for a finite Coxeter group give an example of a complex ***hyperplane arrangement*** (see [BLS+93], [OT92], and Chapter 6). The topology of the set-theoretic complement of these ***Coxeter arrangements*** in complex space has been extensively studied.

For hyperbolic reflection groups, see Vinberg [Vin85]. In hyperbolic space, a discrete irreducible reflection group need not have a fundamental region that is a simplex.

18.7 COMPLEX REGULAR POLYTOPES

Complex regular polytopes are subspace configurations in unitary complex space that share many properties with regular polytopes in real spaces. For a detailed account see Coxeter [Cox93]. The subject originated with Shephard [She52].

GLOSSARY

Complex d-polytope: A d-polytope-configuration as defined in Section 18.2, but now the elements, or ***faces***, are subspaces in unitary complex d-space $\mathbb{C}^d$. However, unlike in real space, the subconfigurations G/F with $\dim(G)-\dim(F)=2$ can contain more than 2 proper elements. A ***complex polygon*** is a complex 2-polytope.

Regular complex polytope: A complex polytope P whose (unitary) symmetry group $G(P)$ is transitive on the flags (the maximal sets of mutually incident faces).

ENUMERATION AND GROUPS

The regular complex d-polytopes P are completely known for each d. Every d-polytope can be uniquely described by a ***generalized Schläfli symbol***

$$p_0\{q_1\}p_1\{q_2\}p_2\dots p_{d-2}\{q_{d-1}\}p_{d-1},$$

which we explain below. For $d=1$, the regular polytopes are precisely the point sets on the complex line, which in corresponding real 2-space are the vertex sets of regular convex polygons; the Schläfli symbol is simply p if the real polygon is a p-gon. When $d\geq 2$ the entry p_i in the above Schläfli symbol is the Schläfli symbol for the complex 1-polytope that occurs as the 1-dimensional subconfiguration G/F of P, where F is an $(i-1)$-face and G an $(i+1)$-face of P such that $F\subseteq G$. As is further explained below, the p_i i-faces in this subconfiguration are cyclically permuted by a hyperplane reflection that leaves the whole polytope invariant. Note that, unlike in real Euclidean space, a hyperplane reflection in unitary complex space need not have period 2 but instead can have any finite period greater than 1. The meaning of the entries q_i is also explained below.

The regular complex polytopes P with $d\geq 2$ are summarized in Table 18.7.1, which includes the numbers f_0 and f_{d-1} of vertices and facets ($(d-1)$-faces) and the group order. Listed are only the nonreal polytopes as well as only one polytope from each pair of duals. A complex polytope is ***real*** if, up to an affine transformation of $\mathbb{C}^d$, all its faces are subspaces that can be described by linear equations over the reals. In particular, $p_0\{q_1\}p_1\dots p_{d-2}\{q_{d-1}\}p_{d-1}$ is real if and only if $p_i=2$ for each i; in this case, $\{q_1,\dots,q_{d-1}\}$ is the Schläfli symbol for the related regular polytope in real space. As in real space, each polytope $p_0\{q_1\}p_1\dots p_{d-2}\{q_{d-1}\}p_{d-1}$ has a dual (reciprocal) and its Schläfli symbol is $p_{d-1}\{q_{d-1}\}p_{d-2}\dots p_1\{q_1\}p_0$; the symmetry groups are the same and the numbers of vertices and facets are interchanged. The polytope $p\{4\}2\{3\}2\dots 2\{3\}2$ is the ***generalized complex d-cube***, and its dual $2\{3\}2\dots 2\{3\}2\{4\}p$ the ***generalized complex d-cross-polytope***; if $p=2$, these are the real d-cubes and d-cross-polytopes, respectively.

The symmetry group $G(P)$ of a complex regular d-polytope P is a finite unitary reflection group in $\mathbb{C}^d$; if $P=p_0\{q_1\}p_1\dots p_{d-2}\{q_{d-1}\}p_{d-1}$, then the notation for the group $G(P)$ is $p_0[q_1]p_1\dots p_{d-2}[q_{d-1}]p_{d-1}$. If $\Phi=\{\emptyset=F_{-1},F_0,\dots,F_{d-1},F_d=\mathbb{C}^d\}$ is a flag of P, then for each $i=0,1,\dots,d-1$ there is a unitary reflection R_i that fixes F_j for $j\neq i$ and cyclically permutes the p_i i-faces in the subconfiguration F_{i+1}/F_{i-1} of P. These generators R_i can be chosen in such a way that in terms of

TABLE 18.7.1 The nonreal complex regular polytopes (up to duality).

DIMENSION	POLYTOPE	f_0	f_{d-1}	$\lvert G(P)\rvert$
$d \geq 1$	$p\{4\}2\{3\}2\ldots 2\{3\}2$	p^d	pd	$p^d d!$
$d = 2$	3{3}3	8	8	24
	3{6}2	24	16	48
	3{4}3	24	24	72
	4{3}4	24	24	96
	3{8}2	72	48	144
	4{6}2	96	48	192
	4{4}3	96	72	288
	3{5}3	120	120	360
	5{3}5	120	120	600
	3{10}2	360	240	720
	5{6}2	600	240	1200
	5{4}3	600	360	1800
$d = 3$	3{3}3{3}3	27	27	648
	3{3}3{4}2	72	54	1296
$d = 4$	3{3}3{3}3{3}3	240	240	155 520

$R_0, \ldots, R_{d-1}$, the group $G(P)$ has a presentation of the form

$$\begin{cases} R_i^{p_i} = 1 & (0 \leq i \leq d-1), \\ R_i R_j = R_j R_i & (0 \leq i < j-1 \leq d-2), \\ R_i R_{i+1} R_i R_{i+1} R_i \ldots = R_{i+1} R_i R_{i+1} R_i R_{i+1} \ldots \\ \text{with } q_{i+1} \text{ generators on each side } (0 \leq i \leq d-2). \end{cases}$$

This explains the entries q_i in the Schläfli symbol. Conversely, any d unitary reflections that satisfy the first two sets of relations, and generate a finite group, determine a regular complex polytope obtained by a complex analogue of Wythoff's construction (see Section 18.5). If P is real, then $G(P)$ is conjugate, in the general linear group of $\mathbb{C}^d$, to a finite (real) Coxeter group (see Section 18.6). Complex regular polytopes are only one source for finite unitary reflection groups; there are also others [Cox93, ST54, MS02].

See Cuypers [Cuy95] for the classification of quaternionic regular polytopes (polytope-configurations in quaternionic space).

18.8 ABSTRACT REGULAR POLYTOPES

Abstract regular polytopes are combinatorial structures that generalize the familiar regular polytopes. The terminology adopted is patterned after the classical theory. Many symmetric figures discussed in earlier sections could be treated (and their structure clarified) in this more general framework. Much of the research in this area is quite recent. For a comprehensive account see McMullen and Schulte [MS02].

GLOSSARY

Abstract d-polytope: A partially ordered set P, with elements called ***faces***, that satisfies the following conditions. P is equipped with a ***rank function*** with range $\{-1, 0, \dots, d\}$, which associates with a face F its ***rank***, denoted by rank F. If rank $F = j$, F is a ***j-face***, or a ***vertex***, an ***edge***, or a ***facet*** if $j = 0, 1$, or $d-1$, respectively. P has a unique minimal element F_{-1} of rank -1 and a unique maximal element F_d of rank d. These two elements are the ***improper*** faces; the others are ***proper***. The ***flags*** (maximal totally ordered subsets) of P all contain exactly $d+2$ faces (including F_{-1} and F_d). If $F < G$ in P, then $G/F := \{H \in P | F \leq H \leq G\}$ is said to be a ***section*** of P. All sections of P (including P itself) are ***connected***, meaning that, given two proper faces H, H' of a section G/F, there is a sequence $H = H_0, H_1, \dots, H_k = H'$ of proper faces of G/F (for some k) such that H_{i-1} and H_i are incident for each $i = 1, \dots, k$. (That is, P is ***strongly connected***.) Finally, if $F < G$ with $0 \leq \text{rank } F + 1 = j = \text{rank } G - 1 \leq d-1$, there are exactly two j-faces H such that $F < H < G$. (In some sense this last condition says that P is topologically real. Note that the condition is violated for nonreal complex polytopes.)

Faces and co-faces: We can safely identify a face F of P with the section $F/F_{-1} = \{H \in P | H \leq F\}$. The section $F_d/F = \{H \in P | F \leq H\}$ is the ***co-face*** of P, or the ***vertex figure*** if F is a vertex.

Regular polytope: An abstract polytope P whose ***automorphism group*** $\Gamma(P)$ (the group of order-preserving permutations of $\mathcal{P}$) is transitive on the flags. (Then $\Gamma(P)$ must be simply flag-transitive.)

C-group: A group Γ generated by involutions $\sigma_1, \dots, \sigma_m$ (that is, a quotient of a Coxeter group) such that the ***intersection property*** holds:

$$\langle \sigma_i | i \in I \rangle \cap \langle \sigma_i | i \in J \rangle = \langle \sigma_i | i \in I \cap J \rangle \text{ for all } I, J \subset \{1, \dots, m\}.$$

The letter "C" stands for "Coxeter." (Coxeter groups are C-groups, but C-groups need not be Coxeter groups.)

String C-group: A C-group $\Gamma = \langle \sigma_1, \dots, \sigma_m \rangle$ such that $(\sigma_i \sigma_j)^2 = 1$ if $1 \leq i < j-1 \leq m-1$. (Then Γ is a quotient of a Coxeter group with a string Coxeter diagram.)

Realization: For an abstract regular d-polytope P with vertex-set $\mathcal{F}_0$, a surjection $\beta : \mathcal{F}_0 \mapsto V$ onto a set V of points in a Euclidean space, such that each automorphism of P induces an isometric permutation of V. Then V is the ***vertex set*** of the realization β.

Chiral polytope: An abstract polytope P whose automorphism group $\Gamma(P)$ has exactly two orbits on the flags, with adjacent flags in different orbits [SWe91]. Here two flags are ***adjacent*** if they differ in exactly one face. Chiral polytopes are an important class of nearly regular polytopes. They are examples of abstract ***two-orbit polytopes***, with just two orbits on flags under the automorphism group [Hub10].

GENERAL PROPERTIES

Abstract 2-polytopes are isomorphic to ordinary n-gons or apeirogons (Section 18.2). All abstract regular polyhedra (3-polytopes) with finite faces and vertex-figures are regular maps on surfaces (Section 18.3); and conversely, most regular maps on surfaces are abstract regular polyhedra with finite faces and vertex-figures. Accordingly, a finite abstract 4-polytope P has facets and vertex figures that are isomorphic to maps on surfaces.

The automorphism group $\Gamma(P)$ of every abstract regular d-polytope P is a string C-group. Fix a flag $\Phi := \{F_{-1}, F_0, \ldots, F_d\}$, the ***base flag*** of P. Then $\Gamma(P)$ is generated by ***distinguished generators*** $\rho_0, \ldots, \rho_{d-1}$ (*with respect to* Φ), where ρ_i is the unique automorphism that keeps all but the i-face of Φ fixed. These generators satisfy relations

$$(\rho_i \rho_j)^{p_{ij}} = 1 \quad (i, j = 0, \ldots, d-1),$$

with $p_{ii} = 1$, $p_{ij} = p_{ji} \geq 2$ $(i \neq j)$, and $p_{ij} = 2$ if $|i-j| \geq 2$; in particular, $\Gamma(P)$ is a string C-group with generators $\rho_0, \ldots, \rho_{d-1}$. The numbers $p_i := p_{i-1,i}$ determine the (*Schläfli*) ***type*** $\{p_1, \ldots, p_{d-1}\}$ of P. The group $\Gamma(P)$ is a (usually proper) quotient of the Coxeter group $[p_1, \ldots, p_{d-1}]$ (Section 18.6).

Conversely, if Γ is a string C-group with generators $\rho_0, \ldots, \rho_{d-1}$, then it is the group of an abstract regular d-polytope P, and $\rho_0, \ldots, \rho_{d-1}$ are the distinguished generators with respect to some base flag of P. The i-faces of P are the right cosets of the subgroup $\Gamma_i := \langle \rho_k | k \neq i \rangle$ of Γ, and in P, $\Gamma_i \varphi \leq \Gamma_j \psi$ if and only if $i \leq j$ and $\Gamma_i \varphi \cap \Gamma_j \psi \neq \emptyset$. For any $p_1, \ldots, p_{d-1} \geq 2$, $[p_1, \ldots, p_{d-1}]$ is a string C-group and the corresponding d-polytope is the ***universal*** regular d-polytope $\{p_1, \ldots, p_{d-1}\}$; every other regular d-polytope of the same type $\{p_1, \ldots, p_{d-1}\}$ is derived from it by making identifications. Examples are the regular spherical, Euclidean, and hyperbolic honeycombs. The one-to-one correspondence between string C-groups and the groups of regular polytopes sets up a powerful dialogue between groups on one hand and polytopes on the other.

For abstract polyhedra (or regular maps) P of type $\{p, q\}$ the group $\Gamma(P)$ is a quotient of the triangle group $[p, q]$ and the above relations for the distinguished generators ρ_0, ρ_1, ρ_2 take the form

$$\rho_0^2 = \rho_1^2 = \rho_2^2 = (\rho_0\rho_1)^p = (\rho_1\rho_2)^q = (\rho_0\rho_2)^2 = 1.$$

There is a wealth of knowledge about regular maps on surfaces in the literature (see Coxeter and Moser [CM80], and Conder, Jones, Širáň and Tucker [CJST]).

A similar dialogue between polytopes and groups also exists for chiral polytopes (see Schulte and Weiss [SWe91, SW94]). If P is an abstract chiral polytope and $\Phi := \{F_{-1}, F_0, \ldots, F_d\}$ is its base flag, then $\Gamma(P)$ is generated by automorphisms $\sigma_1, \ldots, \sigma_{d-1}$, where σ_i fixes all the faces in $\Phi \setminus \{F_{i-1}, F_i\}$ and cyclically permutes consecutive i-faces of P in the (polygonal) section F_{i+1}/F_{i-2} of rank 2. The orientation of each σ_i can be chosen in such a way that the resulting ***distinguished generators*** $\sigma_1, \ldots, \sigma_{d-1}$ of $\Gamma(P)$ satisfy relations

$$\sigma_i^{p_i} = (\sigma_j \sigma_{j+1} \ldots \sigma_k)^2 = 1 \quad (i, j, k = 1, \ldots, d-1 \text{ and } j < k),$$

with p_i determined by the type $\{p_1, \ldots, p_{d-1}\}$ of P. Moreover, a certain intersection property (resembling that for C-groups) holds for $\Gamma(P)$. Conversely, if Γ is a group generated by $\sigma_1, \ldots, \sigma_{d-1}$, and if these generators satisfy the above relations and the intersection property, then Γ is the group of an abstract chiral polytope, or the rotation subgroup of index 2 in the group of an abstract regular polytope. Each isomorphism type of chiral polytope occurs combinatorially in two ***enantiomorphic*** (mirror image) forms; these correspond to two sets of generators σ_i of the group determined by a pair of adjacent base flags.

Following the publication of [MS02], which focused on abstract regular polytopes, there has been a lot of research on other kinds of highly symmetric abstract polytopes including chiral polytopes. Chiral polytopes were known to exist in small ranks [SW94] (see also [CJST] for chiral maps), but the existence in all ranks $d \geq 3$ was only recently established in [Pel10] (see also [CHP08] for ranks 5 and 6).

There is an invariant for chiral (or more general) polytopes, called the chirality group, which in some sense measures the degree of mirror asymmetry (irreflexibility) of the polytope (see [BJS11, Cun14]). For a regular polytope the chirality group would be trivial.

There are several computer based atlases that enumerate all small regular or chiral abstract polytopes of certain kinds (for example, see [Con, Har06, LV06, HHL12]).

Abstract polytopes are closely related to buildings and diagram geometries [Bue95, Tit74]. They are essentially the "thin diagram geometries with a string diagram." The universal regular polytopes $\{p_1, \ldots, p_{d-1}\}$ correspond to "thin buildings." Over the past decade there has been significant progress on polytope interpretations of finite simple groups and closely related groups (see [CLM14, FL11]).

CLASSIFICATION BY TOPOLOGICAL TYPE

Abstract polytopes are not a priori embedded into an ambient space. The traditional enumeration of regular polytopes is therefore replaced by a classification by global or local topological type. At the group level this translates into the enumeration of finite string C-groups with certain kinds of presentations.

The traditional theory of polytopes deals with spherical or locally spherical structures. An abstract polytope P is said to be ***(globally) spherical*** if P is isomorphic to the face lattice of a convex polytope. An abstract polytope P is ***locally spherical*** if all facets and all vertex-figures of P are spherical.

Every ***locally spherical*** abstract regular polytope P of rank $d+1$ is a quotient of a regular tessellation $\{p_1, \ldots, p_d\}$ in spherical, Euclidean, or hyperbolic d-space; in other words, P is a regular tessellation on the corresponding spherical, Euclidean, or hyperbolic space form. In this context, the classical convex regular polytopes are precisely the abstract regular polytopes that are locally spherical and globally spherical. The ***projective regular polytopes*** are the regular tessellations in real projective d-space, and are obtained as quotients of the centrally symmetric convex regular polytopes under the central inversion.

Much work has also been done in the toroidal and locally toroidal case [MS02]. A ***regular toroid*** of rank $d+1$ is the quotient of a regular tessellation $\{p_1, \ldots, p_d\}$ in Euclidean d-space by a lattice that is invariant under all symmetries of the vertex figure of $\{p_1, \ldots, p_d\}$; in other words, a regular toroid of rank $d+1$ is a regular tessellation on the d-torus. If $d = 2$, these are the reflexible regular torus maps of [CM80]. For $d \geq 3$ there are three infinite sequences of ***cubical toroids*** of type $\{4, 3^{d-2}, 4\}$, and for $d = 4$ there are two infinite sequences of ***exceptional toroids*** for each of the types $\{3, 3, 4, 3\}$ and $\{3, 4, 3, 3\}$. Their groups are known in terms of generators and relations.

For $d \geq 2$, the d-torus is the only d-dimensional compact Euclidean space form that can admit a regular or chiral tessellation. Further, chirality can only occur if $d = 2$ (yielding the irreflexible torus maps of [CM80]). Little is known about regular tessellations on hyperbolic space forms (again, see [CM80] and [MS02]).

A main thrust in the theory of abstract regular polytopes is that of the amalgamation of polytopes of lower rank [MS02, Ch. 4]. Let P_1 and P_2 be two regular d-polytopes. An ***amalgamation*** of P_1 and P_2 is a regular $(d+1)$-polytope P with facets isomorphic to P_1 and vertex figures isomorphic to P_2. Let $\langle P_1, P_2 \rangle$ denote the class of all amalgamations of P_1 and P_2. Each nonempty class $\langle P_1, P_2 \rangle$ contains a ***universal polytope*** denoted by $\{P_1, P_2\}$, which "covers" all other polytopes in its class [Sch88]. For example, if P_1 is the 3-cube $\{4, 3\}$ and P_2 is the tetrahedron $\{3, 3\}$, then the universal polytope $\{P_1, P_2\}$ is the 4-cube $\{4, 3, 3\}$, and thus $\{\{4, 3\}, \{3, 3\}\} = \{4, 3, 3\}$; on the other hand, the hemi-4-cube (obtained by identifying opposite faces of the 4-cube) is a nonuniversal polytope in the class $\langle \{4, 3\}, \{3, 3\} \rangle$. The automorphism group of the universal polytope $\{P_1, P_2\}$ is a certain quotient of an amalgamated product of the automorphism groups of P_1 and P_2 [Sch88, MS02].

If we prescribe two topological types for the facets and respectively vertex-figures of polytopes, then the classification of the regular polytopes with these data as local topological types amounts to the enumeration and description of the universal polytopes $\{P_1, P_2\}$ where P_1 and P_2 respectively are of the prescribed topological types [MS02]. The main interest typically lies in classifying all finite universal polytopes $\{P_1, P_2\}$.

A polytope Q in $\langle P_1, P_2 \rangle$ is ***locally toroidal*** if P_1 and P_2 are convex regular polytopes (spheres) or regular toroids, with at least one of the latter kind. For example, if P_1 is the torus map $\{4, 4\}_{(s,0)}$, obtained from an s by s chessboard by identifying opposite edges of the board, and P_2 is the 3-cube $\{4, 3\}$, then the universal locally toroidal 4-polytope $\{\{4, 4\}_{(s,0)}, \{4, 3\}\}$ exists for all $s \geq 2$, but is finite only for $s = 2$ or $s = 3$. The polytope $\{\{4, 4\}_{(3,0)}, \{4, 3\}\}$ can be realized topologically by a decomposition of the 3-sphere into 20 solid tori [CS77, Grü77a].

Locally toroidal regular polytopes can only exist in ranks 4, 5, and 6 [MS02]. The enumeration is complete for rank 5, and nearly complete for rank 4. In rank 6, a list of finite polytopes is known that is conjectured to be complete. The enumeration in rank 4 involves analysis of the Schläfli types $\{4, 4, r\}$ with $r = 3, 4$, $\{6, 3, r\}$ with $r = 3, 4, 5, 6$, and $\{3, 6, 3\}$, and their duals. Here, complete lists of finite universal regular polytopes are known for each type except $\{4, 4, 4\}$ and $\{3, 6, 3\}$; the type $\{4, 4, 4\}$ is almost completely settled, and for $\{3, 6, 3\}$ partial results are known. In rank 5, only the types $\{3, 4, 3, 4\}$ and its dual occur, and these have been settled. In rank 6, there are the types $\{3, 3, 3, 4, 3\}$, $\{3, 3, 4, 3, 3\}$, and $\{3, 4, 3, 3, 4\}$, and their duals. At the group level the classification of toroidal and locally toroidal polytopes amounts to the classification of certain C-groups that are defined in terms of generators and relations. These groups are quotients of Euclidean or hyperbolic Coxeter groups and are obtained from those by either one or two extra defining relations.

Every finite, abstract n-polytope is covered by a finite, abstract regular n-polytope [MS14]. This underlines the significance of abstract regular polytopes as umbrella structures for arbitrary abstract polytopes, including convex polytopes. The article [MPW14] studies coverings of polytopes and their connections with monodromy groups and mixing of polytopes.

The concept of a universal polytope also carries over to chiral polytopes [SW94]. Very little is known about the corresponding classification by topological type.

REALIZATIONS

A good number of the geometric figures discussed in the earlier sections could be described in the general context of realizations of abstract regular polytopes. For an account of realizations see [MS02] or McMullen [McM94, McM].

Let $\beta : \mathcal{F}_0 \mapsto V$ be a realization of a regular d-polytope P, and let $\mathcal{F}_j$ denote the set of j-faces of P ($j = -1, 0, \dots, d$). With $\beta_0 := \beta$, $V_0 := V$, then for $j = 1, \dots, d$, the surjection β recursively induces a surjection $\beta_j : \mathcal{F}_j \mapsto V_j$, with $V_j \subset 2^{V_{j-1}}$, given by

$$F\beta_j := \{G\beta_{j-1} | G \in \mathcal{F}_{j-1}, G \leq F\}$$

for each $F \in \mathcal{F}_j$. It is convenient to identify β and $\{\beta_j\}_{j=0}^{d}$ and also call the latter a realization of $\mathcal{P}$. The realization is ***faithful*** if each β_j is a bijection; otherwise, it is ***degenerate***. Its ***dimension*** is the dimension of the affine hull of V. Each realization corresponds to a (not necessarily faithful) representation of the automorphism group $\Gamma(P)$ of P as a group of Euclidean isometries.

The traditional approach in the study of regular figures starts from a Euclidean (or other) space and describes all figures of a specified kind that are regular according to some geometric definition of regularity. For example, the Grünbaum-Dress polyhedra of Section 18.4 are the realizations in $\mathbb{E}^3$ of abstract regular 3-polytopes P that are both discrete and faithful; their symmetry group is flag-transitive and is isomorphic to the automorphism group $\Gamma(P)$.

Another approach proceeds from a given abstract regular polytope P and describes all the realizations of P. For a finite P, each realization β is uniquely determined by its ***diagonal vector*** Δ, whose components are the squared lengths of the diagonals (pairs of vertices) in the diagonal classes of P modulo $\Gamma(P)$. Each orthogonal representation of $\Gamma(P)$ yields one or more (possibly degenerate) realizations of P. Then taking a sum of two representations of $\Gamma(P)$ is equivalent to an operation for the related realizations called a ***blend***, which in turn amounts to adding the corresponding diagonal vectors. If we identify the realizations with their diagonal vectors, then the space of all realizations of P becomes a closed convex cone $C(P)$, the ***realization cone of*** P, whose finer structure is given by the irreducible representations of $\Gamma(P)$. The extreme rays of $C(P)$ correspond to the ***pure*** (unblended) realizations, which are given by the irreducible representations of $\Gamma(P)$. Each realization of P is a blend of pure realizations.

For instance, a regular n-gon P has $\lfloor \frac{1}{2}n \rfloor$ diagonal classes, and for each $k = 1, \dots, \lfloor \frac{1}{2}n \rfloor$, there is a planar regular star-polygon $\{\frac{n}{k}\}$ if $(n, k) = 1$ (Section 18.2), or a "degenerate star-polygon $\{\frac{n}{k}\}$" if $(n, k) > 1$; the latter is a degenerate realization of P, which reduces to a line segment if $n = 2k$. For the regular icosahedron P there are 3 pure realizations. Apart from the usual icosahedron $\{3, 5\}$ itself, there is another 3-dimensional pure realization, namely the great icosahedron $\{3, \frac{5}{2}\}$ (Section 18.2). The final pure realization is induced by its covering of $\{3, 5\}/2$, the ***hemi-icosahedron*** (obtained from P by identifying antipodal vertices), all of whose diagonals are edges; thus its vertices must be those of a 5-simplex. The regular d-simplex has (up to similarity) a unique realization. The regular d-cross-polytope and d-cube have 2 and d pure realizations, respectively. For the realizations of other polytopes see [BS00, MS02, MW99, MW00].

18.9 SOURCES AND RELATED MATERIAL

SURVEYS

[Ban95]: A popular book on the geometry and visualization of polyhedral and nonpolyhedral figures with symmetries in higher dimensions.

[BLS$^+$93]: A monograph on oriented matroids and their applications.

[BW93]: A survey on polyhedral manifolds and their embeddings in real space.

[BCN89]: A monograph on distance-regular graphs and their symmetry properties.

[Bue95]: A Handbook of Incidence Geometry, with articles on buildings and diagram geometries.

[BC13]: A monograph on diagram geometries.

[CJST]: A monograph on regular maps on surfaces.

[CS02]: A monograph on sphere packings and related topics.

[Cox70]: A short text on certain chiral tessellations of 3-dimensional manifolds.

[Cox73]: A monograph on the traditional regular polytopes, regular tessellations, and reflection groups.

[Cox93]: A monograph on complex regular polytopes and complex reflection groups.

[CM80]: A monograph on discrete groups and their presentations.

[DGS81]: A collection of papers on various aspects of symmetry, contributed in honor of H.S.M. Coxeter's 70th birthday.

[DV64]: A monograph on geometric aspects of the quaternions with applications to symmetry.

[Fej64]: A monograph on regular figures, mainly in 3 dimensions.

[Grü67]: A monograph on convex polytopes.

[GS87]: A monograph on plane tilings and patterns.

[Hum90]: A monograph on Coxeter groups and reflection groups.

[Joh91]: A monograph on uniform polytopes and semiregular figures.

[Mag07]: A book on discrete groups of Möbius transformations and non-Euclidean tessellations.

[Mar94]: A survey on symmetric convex polytopes and a hierarchical classification by symmetry.

[Mon87]: A book on the topology of the three-manifolds of classical plane tessellations.

[McM94]: A survey on abstract regular polytopes with emphasis on geometric realizations.

[McM]: A monograph on geometric regular polytopes and realizations of abstract regular polytopes.

[MS02]: A monograph on abstract regular polytopes and their groups.

[OT92]: A monograph on hyperplane arrangements.

[Rob84]: A text about symmetry classes of convex polytopes.

[Sen95]: An introduction to the geometry of mathematical quasicrystals and related tilings.

[SF88]: A text on interdisciplinary aspects of polyhedra and their symmetries.

[SMT+95]: A collection of twenty-six papers by H.S.M. Coxeter.

[Tit74]: A text on buildings and their classification.

[Wel77]: A monograph on three-dimensional polyhedral geometry and its applications in crystallography.

[Zie95]: A graduate textbook on convex polytopes.

RELATED CHAPTERS

Chapter 3: Tilings
Chapter 6: Oriented matroids
Chapter 15: Basic properties of convex polytopes
Chapter 20: Polyhedral maps
Chapter 64: Crystals, periodic and aperiodic

REFERENCES

[ABM00] J.L. Arocha, J. Bracho, and L. Montejano. Regular projective polyhedra with planar faces, Part I. *Aequationes Math.*, 59:55–73, 2000.

[Ban95] T.F. Banchoff. *Beyond the Third Dimension*. Freeman, New York, 1996.

[BB91] G. Blind and R. Blind. The semiregular polytopes. *Comment. Math. Helv.*, 66:150–154, 1991.

[BC13] F. Buekenhout and A.M. Cohen. *Diagram Geometry*. Springer-Verlag, Berlin, 2013.

[BCN89] A.E. Brouwer, A.M. Cohen, and A. Neumaier. *Distance-Regular Graphs*. Springer-Verlag, Berlin, 1989.

[BJS11] A. Breda D'Azevedo, G.A. Jones, and E. Schulte. Constructions of chiral polytopes of small rank. *Can. J. Math.*, 63:1254–1283, 2011.

[BLS+93] A. Björner, M. Las Vergnas, B. Sturmfels, N. White and G.M. Ziegler. *Oriented Matroids*. Cambridge University Press, 1993.

[Bra00] J. Bracho. Regular projective polyhedra with planar faces, Part II. *Aequationes Math.*, 59:160–176, 2000.

[BS00] H. Burgiel and D. Stanton. Realizations of regular abstract polyhedra of types $\{3,6\}$ and $\{6,3\}$. *Discrete Comput. Geom.*, 24:241–255, 2000.

[Bue95] F. Buekenhout, editor. *Handbook of Incidence Geometry*. Elsevier, Amsterdam, 1995.

[BW88] J. Bokowski and J.M. Wills. Regular polyhedra with hidden symmetries. *Math. Intelligencer*, 10:27–32, 1988.

[BW93] U. Brehm and J.M. Wills. Polyhedral manifolds. In P.M. Gruber and J.M. Wills, editors, *Handbook of Convex Geometry*, pages 535–554, Elsevier, Amsterdam, 1993.

[CHP08] M. Conder, I. Hubard, and T. Pisanski. Constructions for chiral polytopes. *J. London Math. Soc. (2)*, 77:115-129, 2008.

[CJST] M.D.E. Conder, G. Jones, J. Širáň, and T.W. Tucker. *Regular Maps*. In preparation.

[CLM14] T. Connor, D. Leemans, and M. Mixer. Abstract regular polytopes for the O'Nan group. *Int. J. Alg Comput.*, 24:59–68, 2014.

[CM80] H.S.M. Coxeter and W.O.J. Moser. *Generators and Relations for Discrete Groups*, 4th edition. Springer-Verlag, Berlin, 1980.

[CMP98] L. Chen, R.V. Moody, and J. Patera. Non-crystallographic root systems. In J. Patera, editor, *Quasicrystals and Discrete Geometry*, pages 135–178, AMS, Providence, 1998.

[Con] M. Conder. Lists of regular maps, hypermaps and polytopes, trivalent symmetric graphs, and surface actions. `www.math.auckland.ac.nz/~conder`.

[Cox68a] H.S.M. Coxeter. Regular skew polyhedra in 3 and 4 dimensions and their topological analogues. In *Twelve Geometric Essays*, pages 75–105, Southern Illinois University Press, Carbondale, 1968.

[Cox68b] H.S.M. Coxeter. Regular honeycombs in hyperbolic space. In *Twelve Geometric Essays*, pages 199–214, Southern Illinois University Press, Carbondale, 1968.

[Cox70] H.S.M. Coxeter. *Twisted Honeycombs*. Regional Conference Series in Mathematics, vol. 4, AMS, Providence, 1970.

[Cox73] H.S.M. Coxeter. *Regular Polytopes*, 3rd edition. Dover, New York, 1973.

[Cox93] H.S.M. Coxeter. *Regular Complex Polytopes*, 2nd edition. Cambridge University Press, 1993.

[CS77] H.S.M. Coxeter and G.C. Shephard. Regular 3-complexes with toroidal cells. *J. Combin. Theory Ser. B*, 22:131–138, 1977.

[CS88] J.H. Conway and N.J.A. Sloane. *Sphere Packings, Lattices and Groups*. Springer-Verlag, New York, 1988.

[Cun14] G. Cunningham. Variance groups and the structure of mixed polytopes. In R. Connelly, A.I. Weiss, and W. Whiteley, editors, *Rigidity and Symmetry*, pages 97–116, vol. 70 of *Fields Inst. Comm.*, Springer, New York, 2014.

[Cuy95] H. Cuypers. Regular quaternionic polytopes. *Linear Algebra Appl.*, 226/228:311–329, 1995.

[DGS81] C. Davis, B. Grünbaum, and F.A. Sherk. *The Geometric Vein: The Coxeter Festschrift*. Springer-Verlag, New York, 1981.

[Dre85] A.W.M. Dress. A combinatorial theory of Grünbaum's new regular polyhedra. Part II: Complete enumeration. *Aequationes Math.*, 29:222–243, 1985.

[DS82] L. Danzer and E. Schulte. Reguläre Inzidenzkomplexe, I. *Geom. Dedicata*, 13:295–308, 1982.

[DV64] P. Du Val. *Homographies, Quaternions and Rotations*. Oxford University Press, 1964.

[Fej64] L. Fejes Tóth. *Regular Figures*. Macmillan, New York, 1964.

[FL11] M.E. Fernandez and D. Leemans. Polytopes of high rank for the symmetric groups. *Adv. Math.*, 228:3207–3222, 2011.

[Gar67] C.W.L. Garner. Regular skew polyhedra in hyperbolic three-space. *J. Canad. Math. Soc.*, 19:1179–1186, 1967.

[Grü67] B. Grünbaum. *Convex Polytopes*. Interscience, London, 1967. Revised edition (V. Kaibel, V. Klee, and G.M. Ziegler, editors), vol. 221 of *Grad. Texts in Math.*, Springer, New York, 2003.

[Grü77a] B. Grünbaum. Regularity of graphs, complexes and designs. In *Problèmes combinatoires et théorie des graphes*, pages 191–197, vol. 260 of *Colloq. Int. CNRS*, Orsay, 1977.

[Grü77b] B. Grünbaum. Regular polyhedra—old and new. *Aequationes Math.*, 16:1–20, 1977.

[GS87] B. Grünbaum and G.C. Shephard. *Tilings and Patterns*. Freeman, New York, 1987.

[Har06] M.I. Hartley. An atlas of small regular abstract polytopes. *Period. Math. Hungar.*, 53:149–156, 2006. `www.abstract-polytopes.com/atlas`.

[HHL12] M.I. Hartley, I. Hubard, and D. Leemans. Two atlases of abstract chiral polytopes for small groups. *Ars Math. Contemp.*, 5:371–382, 2012. `www.abstract-polytopes.com/chiral`, `www.math.auckland.ac.nz/~dleemans/CHIRAL/`.

[Hub10] I. Hubard. Two-orbit polyhedra from groups. *European J. Combin.*, 31:943–960, 2010.

[Hum90] J.E. Humphreys. *Reflection Groups and Coxeter Groups*. Cambridge University Press, 1990.

[Joh91] N.W. Johnson. *Uniform Polytopes*. Manuscript, 1991.

[LV06] D. Leemans and L. Vauthier. An atlas of abstract regular polytopes for small groups. *Aequationes Math.*, 72:313–320, 2006. `www.math.auckland.ac.nz/~dleemans/abstracts/atlaspoly.html`.

[Mag74] W. Magnus. *Noneuclidean Tessellations and Their Groups*. Academic Press, New York, 1974.

[Mar94] H. Martini. A hierarchical classification of Euclidean polytopes with regularity properties. In T. Bisztriczky, P. McMullen, R. Schneider, and A.I. Weiss, editors, *Polytopes: Abstract, Convex and Computational*, vol. 440 of NATO Adv. Sci. Inst. Ser. C: Math. Phys. Sci., pages 71–96, Kluwer, Dordrecht, 1994.

[McM] P. McMullen. *Geometric Regular Polytopes*. In preparation.

[McM68] P. McMullen. Regular star-polytopes, and a theorem of Hess. *Proc. London Math. Soc. (3)*, 18:577–596, 1968.

[McM94] P. McMullen. Modern developments in regular polytopes. In T. Bisztriczky, P. McMullen, R. Schneider, and A.I. Weiss, editors, *Polytopes: Abstract, Convex and Computational*, vol. 440 of NATO Adv. Sci. Inst. Ser. C: Math. Phys. Sci., pages 97–124, Kluwer, Dordrecht, 1994.

[Mon87] J.M. Montesinos. *Classical Tessellations and Three-Manifolds*. Springer-Verlag, New York, 1987.

[MPW14] B. Monson, D. Pellicer, and G. Williams. Mixing and monodromy of abstract polytopes. *Trans. Amer. Math. Soc.*, 366:2651–2681, 2014.

[MS02] P. McMullen and E. Schulte. *Abstract Regular Polytopes*. Vol. 92 of *Encyclopedia of Mathematics and its Applications*. Cambridge University Press, 2002.

[MS14] B. Monson and E. Schulte. Finite polytopes have finite regular covers. *J. Algebraic Combin.*, 40:75–82, 2014.

[MW00] B.R. Monson and A.I. Weiss. Realizations of regular toroidal maps of type $\{4,4\}$. *Discrete Comput. Geom.*, 24:453–465, 2000.

[MW99] B.R. Monson and A.I. Weiss. Realizations of regular toroidal maps. *Canad. J. Math.*, 51:1240–1257, 1999.

[OT92] P. Orlik and H. Terao. *Arrangements of Hyperplanes*. Springer-Verlag, New York, 1992.

[Pel10] D. Pellicer. A construction of higher rank chiral polytopes. *Discrete Math.*, 310:1222–1237, 2010.

[PS10] D. Pellicer and E. Schulte. Regular polygonal complexes in space, I. *Trans. Amer. Math. Soc.*, 362:6679–6714, 2010.

[PS13] D. Pellicer and E. Schulte. Regular polygonal complexes in space, II. *Trans. Amer. Math. Soc*, 365:2031–2061, 2013.

[PW10] D. Pellicer and A.I. Weiss. Combinatorial structure of Schulte's chiral polyhedra. *Discrete Comput. Geom.*, 44:167–194, 2010.

[Rob84] S.A. Robertson. *Polytopes and Symmetry*. Vol. 90 of *London Math. Soc. Lecture Notes Ser.*, Cambridge University Press, 1984.

[Sch04] E. Schulte. Chiral polyhedra in ordinary space, I. *Discrete Comput. Geom.*, 32:55–99, 2004.

[Sch05] E. Schulte. Chiral polyhedra in ordinary space, II. *Discrete Comput. Geom.*, 34:181–229, 2005.

[Sch88] E. Schulte. Amalgamation of regular incidence-polytopes. *Proc. London Math. Soc. (3)*, 56:303–328, 1988.

[SW94] E. Schulte and A.I. Weiss. Chirality and projective linear groups. *Discrete Math.*, 131:221–261, 1994.

[SWe91] E. Schulte and A.I. Weiss. Chiral polytopes. In P. Gritzmann and B. Sturmfels, editors, *Applied Geometry and Discrete Mathematics: The Victor Klee Festschrift*, vol. 4 of *DIMACS Ser. Discrete Math. Theor. Comp. Sci.*, pages 493–516, AMS, Providence, 1991.

[SWi91] E. Schulte and J.M. Wills. Combinatorially regular polyhedra in three-space. In K.H. Hofmann and R. Wille, editors, *Symmetry of Discrete Mathematical Structures and Their Symmetry Groups*, pages 49–88, Heldermann Verlag, Berlin, 1991.

[Sen95] M. Senechal. *Quasicrystals and Geometry*. Cambridge University Press, 1995.

[SF88] M. Senechal and G. Fleck. *Shaping Space*. Birkhäuser, Boston, 1988.

[She52] G.C. Shephard. Regular complex polytopes. *Proc. London Math. Soc. (3)*, 2:82–97, 1952.

[SMT+95] F.A. Sherk, P. McMullen, A.C. Thompson, and A.I. Weiss, editors. *Kaleidoscopes: Selected Writings of H.S.M. Coxeter*. Wiley, New York, 1995.

[ST54] G.C. Shephard and J.A. Todd. Finite unitary reflection groups. *Canad. J. Math.*, 6:274–304, 1954.

[Tit74] J. Tits. *Buildings of Spherical Type and Finite BN-Pairs*. Springer-Verlag, New York, 1974.

[Vin85] E.B. Vinberg. Hyperbolic reflection groups. *Uspekhi Mat. Nauk*, 40:29–66, 1985. (*Russian Math. Surveys*, 40:31–75, 1985.)

[Wel77] A.F. Wells. *Three-dimensional Nets and Polyhedra*. Wiley, New York, 1977.

[WL84] A.I. Weiss and Z. Lučić. Regular polyhedra in hyperbolic three-space. *Mitteilungen Math. Sem. Univ. Gissen* 165:237-252, 1984.

[Zie95] G.M. Ziegler. *Lectures on Polytopes*. Springer-Verlag, New York, 1995.

19 POLYTOPE SKELETONS AND PATHS

Gil Kalai

INTRODUCTION

The k-dimensional ***skeleton*** of a d-polytope P is the set of all faces of the polytope of dimension at most k. The 1-skeleton of P is called the ***graph*** of P and denoted by $G(P)$. $G(P)$ can be regarded as an abstract graph whose vertices are the vertices of P, with two vertices adjacent if they form the endpoints of an edge of P.

In this chapter, we will describe results and problems concerning graphs and skeletons of polytopes. In Section 19.1 we briefly describe the situation for 3-polytopes. In Section 19.2 we consider general properties of polytopal graphs—subgraphs and induced subgraphs, connectivity and separation, expansion, and other properties. In Section 19.3 we discuss problems related to diameters of polytopal graphs in connection with the simplex algorithm and the Hirsch conjecture. The short Section 19.4 is devoted to polytopal digraphs. Section 19.5 is devoted to skeletons of polytopes, connectivity, collapsibility and shellability, empty faces and polytopes with "few vertices," and the reconstruction of polytopes from their low-dimensional skeletons; we also consider what can be said about the collections of all k-faces of a d-polytope, first for $k = d-1$ and then when k is fixed and d is large compared to k. Section 19.6 describes some results on counting high dimensional polytopes and spheres.

19.1 THREE-DIMENSIONAL POLYTOPES

GLOSSARY

Convex polytopes and their *faces* (and, in particular their *vertices*, *edges*, and *facets*) are defined in Chapter 15 of this Handbook.

A graph is ***d-polytopal*** if it is the graph of some d-polytope.

The following standard graph-theoretic concepts are used: *subgraphs*, *induced subgraphs*, the *complete graph* K_n on n vertices, *cycles*, *trees*, a *spanning tree* of a graph, *degree* of a vertex in a graph, *planar* graphs, d-*connected* graphs, *coloring* of a graph, *subdivision* of a graph, and *Hamiltonian* graphs.

We briefly discuss results on 3-polytopes. Some of the following theorems are the starting points of much research, sometimes of an entire theory. Only in a few cases are there high-dimensional analogues, and this remains an interesting goal for further research.

THEOREM 19.1.1 *Whitney* [Whi32]

Let G be the graph of a 3-polytope P. Then the graphs of faces of P are precisely the induced cycles in G that do not separate G.

THEOREM 19.1.2 *Steinitz* [Ste22]

A graph G is a graph of a 3-polytope if and only if G is planar and 3-connected.

Steinitz's theorem is the first of several theorems that describe the tame behavior of 3-polytopes. These theorems fail already in dimension four; see Chapter 15.

The theory of planar graphs is a wide and rich theory. Let us quote here the fundamental theorem of Kuratowski.

THEOREM 19.1.3 *Kuratowski* [Kur22, Tho81]

A graph G is planar if and only if G does not contain a subdivision of K_5 or $K_{3,3}$.

THEOREM 19.1.4 *Lipton and Tarjan* [LT79], *strengthened by Miller* [Mil86]

The graph of every 3-polytope with n vertices can be separated by at most $2\sqrt{2n}$ vertices, forming a circuit in the graph, into connected components of size at most $2n/3$.

It is worth mentioning that the Koebe circle packing theorem gives a new approach to both the Steinitz and Lipton-Tarjan theorems; see [Zie95, PA95].

Euler's formula $V - E + F = 2$ has many applications concerning graphs of 3-polytopes; in higher dimensions, our knowledge of face numbers of polytopes (see Chapter 17) applies to the study of their graphs and skeletons. Simple applications of Euler's theorem are:

THEOREM 19.1.5

Every 3-polytopal graph has a vertex of degree at most 5. (Equivalently, every 3-polytope has a face with at most five sides.)

Zonotopes are centrally symmetric and so are all their faces and therefore, a 3-zonotope must have a face with four sides. This fact is equivalent to the Gallai-Sylvester theorem (Chapter 1).

THEOREM 19.1.6

Every 3-polytope has either a vertex of degree 3 or a triangular face.

A deeper application of Euler's theorem is:

THEOREM 19.1.7 *Kotzig* [Kot55]

Every 3-polytope has two adjacent vertices the sum of whose degrees is at most 13.

For a simple 3-polytope P, let $p_k = p_k(P)$ be the number of k-sized faces of P.

THEOREM 19.1.8 *Eberhard* [Ebe91]

For every finite sequence (p_k) of nonnegative integers with $\sum_{k\geq 3}(6-k)p_k = 12$, there exists a simple 3-polytope P with $p_k(P) = p_k$ for every $k \neq 6$.

Eberhard's theorem is the starting point of a large number of results and problems, see, e.g., [Juc76, Jen93, GZ74]. While no high-dimensional direct analogues are known or even conjectured, the results and problems on facet-forming polytopes and nonfacets mentioned below seem related.

THEOREM 19.1.9 *Motzkin* [Mot64]

The graph of a simple 3-polytope whose facets have 0 (mod 3) vertices has, all together, an even number of edges.

THEOREM 19.1.10 *Barnette* [Bar66]

Every 3-polytopal graph contains a spanning tree of maximal degree 3.

We will now describe some results and a conjecture on colorability and Hamiltonian circuits.

THEOREM 19.1.11 *Four Color Theorem: Appel-Haken* [AH76, AH89, RSST97]

The graph of every 3-polytope is 4-colorable.

THEOREM 19.1.12 *Tutte* [Tut56]

4-connected planar graphs are Hamiltonian.

Tait conjectured in 1880, and Tutte disproved in 1946, that the graph of every simple 3-polytope is Hamiltonian. This started a rich theory of cubic planar graphs without large paths.

CONJECTURE 19.1.13 *Barnette*

Every graph of a simple 3-polytope whose facets have an even number of vertices is Hamiltonian.

Finally, there are several exact and asymptotic formulas for the numbers of distinct graphs of 3-polytopes. A remarkable enumeration theory was developed by Tutte and was further developed by several authors. We will quote one result.

THEOREM 19.1.14 *Tutte* [Tut62]

The number of rooted simplicial 3-polytopes with v vertices is

$$\frac{2(4v-11)!}{(3v-7)!(v-2)!}.$$

Tutte's theory also provides efficient algorithms to generate random planar graphs of various types.

PROBLEM 19.1.15

What does a random 3-polytopal graph look like?

Motivation to study this problem (and high-dimensional extensions) comes also from physics (specifically, "quantum gravity"). See [ADJ97, Ang02, CS02, GN06, Noy14]. One surprising property of random planar maps of various kinds is that the expected number of vertices of distance at most r from a given vertex behaves like r^4. (Compared to r^2 for the planar grid.)

19.2 GRAPHS OF d-POLYTOPES—GENERALITIES

GLOSSARY

For a graph G, TG denotes any ***subdivision*** of G, i.e., any graph obtained from G by replacing the edges of G by paths with disjoint interiors.

A d-polytope P is ***simplicial*** if all its proper faces are simplices. P is ***simple*** if every vertex belongs to d edges or, equivalently, if the polar of P is simplicial. P is ***cubical*** if all its proper faces are cubes.

A simplicial polytope P is ***stacked*** if it is obtained by the repeated operation of gluing a simplex along a facet.

For the definition of the *cyclic polytope* $C(d, n)$, see Chapter 15.

For two graphs G and H (considered as having disjoint sets V and V' of vertices), $G + H$ denotes the graph on $V \cup V'$ that contains all edges of G and H together with all edges of the form $\{v, v'\}$ for $v \in V$ and $v' \in V'$.

A graph G is ***d-connected*** if G remains connected after the deletion of any set of at most $d - 1$ vertices.

An ***empty simplex*** of a polytope P is a set S of vertices such that S does not form a face but every proper subset of S forms a face.

A graph G whose vertices are embedded in $\mathbb{R}^d$ is ***rigid*** if every small perturbation of the vertices of G that does not change the distance of adjacent vertices in G is induced by an affine rigid motion of $\mathbb{R}^d$. G is ***generically d-rigid*** if it is rigid with respect to "almost all" embeddings of its vertices into $\mathbb{R}^d$. (Generic rigidity is thus a graph theoretic property, but no description of it in pure combinatorial terms is known for $d > 2$; cf. Chapter 61.)

A set A of vertices of a graph G is ***totally separated*** by a set B of vertices, if A and B are disjoint and every path between two distinct vertices in A meets B.

A graph G is an ***ϵ-expander*** if, for every set A of at most half the vertices of G, there are at least $\epsilon \cdot |A|$ vertices not in A that are adjacent to vertices in A.

Neighborly polytopes and $(0, 1)$*-polytopes* are defined in Chapter 15.

The *polar dual* P^Δ of a polytope P is defined in Chapter 15.

SUBGRAPHS AND INDUCED SUBGRAPHS

THEOREM 19.2.1 *Grünbaum* [Grü65]

Every d-polytopal graph contains a TK_{d+1}.

For various extensions of this result see [GLM81].

THEOREM 19.2.2 *Kalai* [Kal87]

The graph of a simplicial d-polytope P contains a TK_{d+2} *if and only if P is not stacked.*

One important difference between the situation for $d = 3$ and for $d > 3$ is that K_n, for every $n > 4$, is the graph of a 4-dimensional polytope (e.g., a cyclic polytope). Simple manipulations on the cyclic 4-polytope with n vertices show:

PROPOSITION 19.2.3 *Perles* (unpublished)

(i) *Every graph G is a spanning subgraph of the graph of a 4-polytope.*

(ii) *For every graph G,* $G + K_n$ *is a d-polytopal graph for some n and some d.*

(Here the graph join $G + H$ of two graphs G and H is obtained from putting G and H on disjoint sets of vertices and adding all edges between them.) This proposition extends easily to higher-dimensional skeletons in place of graphs. It is not known what the minimal dimension is for which $G+K_n$ is d-polytopal, nor even whether $G + K_n$ (for some $n = n(G)$) can be realized in some bounded dimension uniformly for all graphs G.

CONNECTIVITY AND SEPARATION

THEOREM 19.2.4 *Balinski* [Bal61]

The graph of a d-polytope is d-connected.

A set S of d vertices that separates P must form an empty simplex; in this case, P can be obtained by gluing two polytopes along a simplex facet of each.

A graph G is k-linked if for every two disjoint sequences $(v_1, v_2, \dots, v_k)$ and $(w_1, w_2, \dots w_k)$ of vertices of G, there are e vertex-disjoint paths connecting v_i to w_i, $i = 1, 2, \dots, k$.

THEOREM 19.2.5 *Larman and Mani* [LM70]

A graph of a d-polytope is $\lfloor (d+1)/3 \rfloor$*-linked*

Let $k(d)$ be the smallest integer to that every graph of a d-polytopes is $k(d)$-linked. Larman conjectured that $k(d) \geq \lfloor d/2 \rfloor$. However, Gallivan [Gal85]. showed that $k(d) \leq [(2d+3)/5]$.

PROBLEM 19.2.6 *Understand the behavior of* $k(d)$.

THEOREM 19.2.7 *Cauchy, Dehn, Aleksandrov, Whiteley, ...*

(i) *Cauchy's theorem: If P is a simplicial d-polytope, $d \geq 3$, then $G(P)$ (with its embedding in $\mathbb{R}^d$) is rigid.*

(ii) *Whiteley's theorem* [Whi84]*: For a general d-polytope P, let G' be a graph (embedded in $\mathbb{R}^d$) obtained from $G(P)$ by triangulating the 2-faces of P without introducing new vertices. Then G' is rigid.*

COROLLARY 19.2.8

For a simplicial d-polytope P, $G(P)$ is generically d-rigid. For a general d-polytope P and a graph G' (considered as an abstract graph) as in the previous theorem, G' is generically d-rigid.

The main combinatorial application of the above theorem is the Lower Bound Theorem (see Chapter 17) and its extension to general polytopes.

Note that Corollary 19.2.8 can be regarded also as a strong form of Balinski's theorem. It is well known and easy to prove that a generic d-rigid graph is d-connected. Therefore, for simplicial (or even 2-simplicial) polytopes, Corollary 19.2.8 implies directly that $G(P)$ is d-connected.

For general polytopes we can derive Balinski's theorem as follows. Suppose to the contrary that the graph G of a general d-polytope P is not d-connected and therefore its vertices can be separated into two parts (say, red vertices and blue vertices) by deleting a set A of $d-1$ vertices. It is easy to see that every 2-face of P

can be triangulated without introducing a blue-red edge. Therefore, the resulting triangulation is not $(d-1)$-connected and hence it is not generically d-rigid. This contradicts the assertion of Corollary 19.2.8.

Let $\mu(n,d) = f_{d-1}(C(d,n))$ be the number of facets of a cyclic d-polytope with n vertices, which, by the Upper Bound Theorem, is the maximal number of facets possible for a d-polytope with n vertices.

THEOREM 19.2.9 *Klee* [Kle64]

The number of vertices of a d-polytope that can be totally separated by n vertices is at most $\mu(n,d)$.

Klee also showed by considering cyclic polytopes with simplices stacked to each of their facets that this bound is sharp. It follows that there are graphs of simplicial d-polytopes that are not graphs of $(d-1)$-polytopes. (After realizing that the complete graphs are 4-polytopal one's naive thought might be that every d-polytopal graph is 4-polytopal.)

EXPANSION

Expansion properties for the graph of the d-dimensional cube are known and important in various areas of combinatorics. By direct combinatorial methods, one can obtain expansion properties of duals to cyclic polytopes. There are a few positive results and several interesting conjectures on expansion properties of graphs of large families of polytopes.

THEOREM 19.2.10 *Kalai* [Kal91]

Graphs of duals to neighborly d-polytopes with n facets are ϵ-expanders for $\epsilon = O(n^{-4})$.

This result implies that the diameter of graphs of duals to neighborly d-polytopes with n facets is $O(d \cdot n^4 \cdot \log n)$.

CONJECTURE 19.2.11 *Mihail and Vazirani* [FM92, Kai01]

Graphs of $(0,1)$-polytopes P have the following expansion property: For every set A of at most half the vertices of P, the number of edges joining vertices in A to vertices not in A is at least $|A|$.

Dual graphs to cyclic $2k$-polytopes with n vertices for n large look somewhat like graphs of grids in $\mathbb{Z}^k$ and, in particular, have no separators of size $o(n^{1-1/k})$. It was conjectured that graphs of polytopes cannot have very good expansion properties, namely that the graph of every simple d-polytope with n vertices can be separated into two parts, each having at least $n/3$ vertices, by removing $O(n^{1-1/(d-1)})$ vertices. However,

THEOREM 19.2.12 *Loiskekoski and Ziegler* [LZ15]

There are simple 4-dimensional polytopes with n vertices such that all separators of the graph have size at least $\Omega(n/\log^{3/2} n)$.

It is still an open problem if there are examples of simple d-polytopes with n vertices $n \to \infty$ whose graphs have no separators of size $o(n)$, or even of simple d-polytopes with n vertices, whose graphs are expanders.

OTHER PROPERTIES

CONJECTURE 19.2.13 *Barnette*

Every graph of a simple d-polytope, $d \geq 4$, is Hamiltonian.

THEOREM 19.2.14

For a simple d-polytope P, $G(P)$ is 2-colorable if and only if $G(P^\Delta)$ is d-colorable.

This theorem was proved in an equivalent form for $d = 4$ by Goodman and Onishi [GO78]. (For $d = 3$ it is a classical theorem by Ore.) For the general case, see Joswig [Jos02]. This theorem is related to seeking two-dimensional analogues of Hamiltonian cycles in skeletons of polytopes and manifolds; see [Sch94].

19.3 DIAMETERS OF POLYTOPAL GRAPHS

GLOSSARY

A ***d-polyhedron*** is the intersection of a finite number of halfspaces in $\mathbb{R}^d$.

$\Delta(d, n)$ denotes the maximal diameter of the graphs of d-dimensional polyhedra P with n facets. (Here again vertices correspond to 0-faces and edges correspond to 1-faces.)

$\Delta_b(d, n)$ denotes the maximal diameter of the graphs of d-polytopes with n vertices.

Given a d-polyhedron P and a linear functional ϕ on $\mathbb{R}^d$, we denote by $G^{\rightarrow}(P)$ the directed graph obtained from $G(P)$ by directing an edge $\{v, u\}$ from v to u if $\phi(v) \leq \phi(u)$. $v \in P$ is a ***top vertex*** if ϕ attains its maximum value in P on v.

Let $H(d, n)$ be the maximum over all d-polyhedra with n facets and all linear functionals on $\mathbb{R}^d$ of the maximum length of a minimal monotone path from any vertex to a top vertex.

Let $M(d, n)$ be the maximal number of vertices in a monotone path over all d-polyhedra with n facets and all linear functionals on $\mathbb{R}^d$.

For the notions of *simplicial complex*, *polyhedral complex*, *pure simplicial complex*, *flag complex* and the *boundary complex* of a polytope, see Chapter 17.

Given a pure $(d-1)$-dimensional simplicial (or polyhedral) complex K, the ***dual graph*** $G^\Delta(K)$ of K is the graph whose vertices are the facets ($(d-1)$-faces) of K, with two facets F, F' adjacent if $\dim(F \cap F') = d - 2$.

A pure simplicial complex K is ***vertex-decomposable*** if there is a vertex v of K such that $\mathrm{lk}(v) = \{S \setminus \{v\} \mid S \in K, v \in S\}$ and $\mathrm{ast}(v) = \{S \mid S \in K, v \notin S\}$ are both vertex-decomposable. (The complex $K = \{\emptyset\}$ consisting of the empty face alone is vertex-decomposable.)

LOWER BOUNDS

It is a long-outstanding open problem to determine the behavior of the function $\Delta(d,n)$ and $\Delta_b(d,n)$. See [San13b] for a recent surveys on this problem. In 1957, Hirsch conjectured that $\Delta(d,n) \leq n-d$. Klee and Walkup [KW67] showed that the Hirsch conjecture is false for unbounded polyhedra.

THEOREM 19.3.1 *Klee and Walkup*

$$\Delta(d,n) \geq n-d+\min\{\lfloor d/4 \rfloor, \lfloor (n-d)/4 \rfloor\}.$$

The Hirsch conjecture for bounded polyhedra remained open until 2010. The special case asserting that $\Delta_b(d,2d)=d$ is called the ***d*-step conjecture**, and it was shown by Klee and Walkup to imply that $\Delta_b(d,n) \leq n-d$. Another equivalent formulation is that between any pair of vertices v and w of a polytope P there is a ***nonrevisiting path***, i.e., a path $v=v_1,v_2,...,v_m=w$ such that for every facet F of P, if $v_i,v_j \in F$ for $i<j$ then $v_k \in F$ for every k with $i \leq k \leq j$.

THEOREM 19.3.2 *Holt-Klee* [HK98a, HK98b, HK98c], *Fritzsche-Holt* [FH99]

For $n > d \geq 8$

$$\Delta_b(d,n) \geq n-d.$$

THEOREM 19.3.3 *Santos* [San13a], Matschke, Santos, and Weibel [MSW15]

For $d \geq 20$

$$\Delta_b(d,2d) > d.$$

Santos' argument uses a beautiful extension of the nonrevisiting conjecture (in its dual form) for certain *nonsimplicial* polytopes. A counterexample for this conjecture is found (already in dimension five) and it leads to a counterexample of the d-step conjecture. Matschke, Santos, and Weibel [MSW15], improved the construction and it now applies for $d \geq 20$.

UPPER BOUNDS

The known upper bounds for $\Delta(d,n)$ when d is fixed and n is large.

THEOREM 19.3.4 *Larman* [Lar70]

$$\Delta(d,n) \leq n \cdot 2^{d-3}.$$

And a slight improvement by Barnette

THEOREM 19.3.5 *Barnette* [Bar74]

$$\Delta(d,n) \leq \frac{2}{3} \cdot (n-d+5/2) \cdot 2^{d-3}.$$

When n is not very large w.r.t. d the best known upper bounds are quasipolynomials.

THEOREM 19.3.6 *Kalai and Kleitman* [KK92]

$$\Delta(d,n) \leq n \cdot \binom{\log n + d}{d} \leq n^{\log d + 1}.$$

Recently, a small improvement was made by Todd [Tod14] who proved that $\Delta(d,n) \leq (n-d)^{\log d}$.

The major open problem in this area is:

PROBLEM 19.3.7

Is there a polynomial upper bound for $\Delta(d,n)$? Is there a linear upper bound for $\Delta(d,n)$?

Even upper bounds of the form $\exp c \log d \log(n-d)$ and $n2^{cn}$ for some $c < 1$ would be a major progress.

CONJECTURE 19.3.8 *Hähnle, [Häh10, San13b]*

$$\Delta(d,n) \leq d(n-1)$$

Many of the upper bounds for $\Delta(d,n)$ applies to much more general combinatorial objects [Kal92, EHRR10]. Hähnle's conjecture applies to an abstract (much more general) setting of the problem formulated for certain families of monomials, and Hähnle's proposed bound is tight in this greater generality.

SPECIAL CLASSES OF POLYHDRA

Some special classes of polytopes are known to satisfy the Hirsch bound or to have upper bounds for their diameters that are polynomial in d and n.

THEOREM 19.3.9 *Provan and Billera* [PB80]

Let G be the dual graph that corresponds to a vertex-decomposable $(d-1)$-dimensional simplicial complex with n vertices. Then the diameter of G is at most $n-d$.

There are simplicial polytopes whose boundary complexes are not vertex-decomposable. Lockeberg found such an example in 1977 and examples that violate much weaker forms of vertex decomposability were found in [DK12, HPS14].

THEOREM 19.3.10 *Naddef* [Nad89]

The graph of every $(0,1)$ d-polytope has diameter at most d.

Balinski [Bal84] proved the Hirsch bound for dual transportation polytopes, Dyer and Frieze [DF94] showed a polynomial upper bound for unimodular polyhedra; for a recent improved bounds see [BDE$^+$14]. Kalai [Kal92] proved that if the ratio between the number of facets and the dimension is bounded above for the polytope and all its faces then the diameter is bounded above by a polynomial in the dimension, Kleinschmidt and Onn [KO92] proved extensions of Naddef's results to integral polytopes, and Deza, Manoussakis and Onn [DMO17] conjectured a sharp version of their result attained by Minkowski sum of primitive lattice vectors. Deza and Onn [DO95] found upper bounds for the diameter in terms of lattice points in the polytope. A recent result based on the basic fact that locally convex sets of small intrinsic diameter in CAT(1) spaces are convex is:

THEOREM 19.3.11 *Adiprasito and Benedetti* [AB14]

Flag simplicial spheres satisfy the Hirsch conjecture.

LINEAR PROGRAMMING AND ROUTING

The value of $\Delta(d,n)$ is a lower bound for the number of iterations needed in the worst case for Dantzig's simplex algorithm for linear programming with any pivot rule. However, it is still an open problem to find pivot rules where each pivot step can be computed with a polynomial number of arithmetic operations in d and n such that the number of pivot steps needed comes close to the upper bounds for $\Delta(d,n)$ given above. See Chapter 49.

We note that a continuous analog of the Hirsch conjecture proposed and studied by Deza, Terlaky, and Zinchenko [DTZ09] was disproved by Allamigeon, Benchimol, Gaubert, and Joswig [ABGJ17]. The counterexample is a polytope obtained via tropical geometry.

The problem of routing in graphs of polytopes, i.e., finding a path between two vertices, is an interesting computational problem.

PROBLEM 19.3.12

Find an efficient routing algorithm for convex polytopes.

Using linear programming it is possible to find a path in a polytope P between two vertices that obeys the upper bounds given above such that the number of calls to the linear programming subroutine is roughly the number of edges of the path. Finding a routing algorithm for polytopes with a "small" number of arithmetic operations as a function of d and n is an interesting challenge. The subexponential simplex-type algorithms (see Chapter 49) yield subexponential routing algorithms, but improvement for routing beyond what is known for linear programming may be possible.

The upper bounds for $\Delta(d,n)$ mentioned above apply even to $H(d,n)$. Klee and Minty considered a certain geometric realization of the d-cube to show that

THEOREM 19.3.13 *Klee and Minty* [KM72]

$M(d,2d) \geq 2^d$.

Far-reaching extensions of the Klee-Minty construction were found by Amenta and Ziegler [AZ99]. It is not known for $d > 3$ and $n \geq d+3$ what the precise upper bound for $M(d,n)$ is and whether it coincides with the maximum number of vertices of a d-polytope with n facets given by the upper bound theorem (Chapter 17).

19.4 POLYTOPAL DIGRAPHS

Given a d-polytope P and a linear objective function ϕ not constant on edges, direct every edge of $G(P)$ towards the vertex with the higher value of the objective function. A directed graph obtained in this way is called a ***polytopal digraph***.

The following basic result is fundamental for the simplex algorithm and also has many applications for the combinatorial theory of polytopes.

THEOREM 19.4.1 *Folklore* (see, e.g., [Wil88])

A polytopal digraph has one sink (and one source). Moreover, every induced subgraph on the vertices of any face of the polytope has one sink (and one source).

An acyclic orientation of $G(P)$ with the property that every face has a unique sink is called an ***abstract objective function.*** Joswig, Kaibel, and Körner [JKK02] showed that an acyclic orientation for which every 2-dimensional face has a unique sink is already an abstract objective function.

The h-vector of a simplicial polytope P has a simple and important interpretation in terms of the directed graph that corresponds to the polar of P. The number $h_k(P)$ is the number of vertices v of P^Δ of outdegree k. (Recall that every vertex in a simple polytope has exactly d neighboring vertices.) Switching from ϕ to $-\phi$, one gets the Dehn-Sommerville relations $h_k = h_{d-k}$ (including the Euler relation for $k = 0$); see Chapter 17.

Studying polytopal digraphs and digraphs obtained by abstract objective functions is very interesting in three- and higher dimensions. Digraphs whose underlying simple graphs are that of the cube are of special interest and important in the theory of linear programming. The work of Friedmann, Hansen, and Zwick [FHZ11] who found very general constructions based on certain stochastic games is of particular importance. (See Chapter 49.) In three dimensions, polytopal digraphs admit a simple characterization:

THEOREM 19.4.2 *Mihalisin and Klee* [MK00]

Suppose that K is an orientation of a 3-polytopal graph G. Then the digraph K is 3-polytopal if and only if it is acyclic, has a unique source and a unique sink, and admits three independent monotone paths from the source to the sink.

Mihalisin and Klee write in their article "we hope that the present article will open the door to a broader study of polytopal digraphs."

19.5 SKELETONS OF POLYTOPES

GLOSSARY

A pure polyhedral complex K is ***strongly connected*** if its dual graph is connected.

A ***shelling order*** of the facets of a polyhedral $(d-1)$-dimensional sphere is an ordering of the set of facets $F_1, F_2, \dots, F_n$ such that the simplicial complex K_i spanned by $F_1 \cup F_2 \cup \cdots \cup F_i$ is a simplicial ball for every $i < n$. A polyhedral complex is ***shellable*** if there exists a shelling order of its facets.

A simplicial polytope is ***extendably shellable*** if any way to start a shelling can be continued to a shelling.

An ***elementary collapse*** on a simplicial complex is the deletion of two faces F and G so that F is maximal and G is a codimension-1 face of F that is not included in any other maximal face. A polyhedral complex is ***collapsible*** if it can be reduced to the void complex by repeated applications of elementary collapses.

A d-dimensional polytope P is ***facet-forming*** if there is a $(d+1)$-dimensional polytope Q such that all facets of Q are combinatorially isomorphic to P. If no such Q exists, P is called a ***nonfacet***.

A ***rational polytope*** is a polytope whose vertices have rational coordinates. (Not every polytope is combinatorially isomorphic to a rational polytope; see Chapter 15.)

A d-polytope P is ***k-simplicial*** if all its faces of dimension at most k are simplices. P is ***k-simple*** if its polar dual P^Δ is k-simplicial.

Zonotopes are defined in Chapters 15 and 17.

Let K be a polyhedral complex. An ***empty simplex*** S of K is a minimal nonface of K, i.e., a subset S of the vertices of K with S itself not in K, but every proper subset of S in K.

Let K be a polyhedral complex and let U be a subset of its vertices. The ***induced subcomplex*** of K on U, denoted by $K[U]$, is the set of all faces in K whose vertices belong to U. An ***empty face*** of K is an induced polyhedral subcomplex of K that is homeomorphic to a polyhedral sphere. An empty 2-dimensional face is called an ***empty polygon***. An ***empty pyramid*** of K is an induced subcomplex of K that consists of all the proper faces of a pyramid over a face of K.

CONNECTIVITY AND SUBCOMPLEXES

THEOREM 19.5.1 *Grünbaum* [Grü65]

The i-skeleton of every d-polytope contains a subdivision of $\mathrm{skel}_i(\Delta^d)$*, the i-skeleton of a d-simplex.*

THEOREM 19.5.2 *Folklore*

(i) *For $i > 0$,* $\mathrm{skel}_i(P)$ *is strongly connected.*

(ii) *For every face F, let $U_i(F)$ be the set of all i-faces of P containing F. Then if $i > \dim F$, $U_i(F)$ is strongly connected.*

Part (ii) follows at once from the fact that the faces of P containing F correspond to faces of the quotient polytope P/F. However, properties (i) and (ii) together are surprisingly strong, and all the known upper bounds for diameters of graphs of polytopes rely only on properties (i) and (ii) for the dual polytope.

THEOREM 19.5.3 *van Kampen and Flores* [Kam32, Flo32, Wu65]

For $i \geq \lfloor d/2 \rfloor$, $\mathrm{skel}_i(\Delta^{d+1})$ *is not embeddable in S^{d-1} (and hence not in the boundary complex of any d-polytope).*

(This extends the fact that K_5 is not planar.)

A beautiful extension of Balinski's theorem was offered by Lockenberg:

CONJECTURE 19.5.4 *Lockeberg*

For every partition of $d = d_1 + d_2 + \cdots + d_k$ and two vertices v and w of P, there are k disjoint paths between v and w such that the ith path is a path of d_i-faces in which any two consecutive faces have $(d_i - 1)$-dimensional intersection.

(Here by "disjoint" we do not refer to the common first vertex v and last vertex w.)

SHELLABILITY AND COLLAPSIBILITY

THEOREM 19.5.5 *Bruggesser and Mani* [BM71]

Boundary complexes of polytopes are shellable.

The proof of Bruggesser and Mani is based on starting with a point near the center of a facet and moving from this point to infinity, and back from the other direction, keeping track of the order in which facets are seen. This proves a stronger form of shellability, in which each K_i is the complex spanned by all the facets that can be seen from a particular point in $\mathbb{R}^d$. It follows from shellability that

THEOREM 19.5.6

Polytopes are collapsible.

On the other hand,

THEOREM 19.5.7 *Ziegler* [Zie98]

There are d-polytopes, $d \geq 4$, whose boundary complexes are not extendably shellable.

THEOREM 19.5.8

There are triangulations of the $(d-1)$-sphere that are not shellable.

Lickorish [Lic91] produced explicit examples of nonshellable triangulations of S^3. His result was that a triangulation containing a sufficiently complicated knotted triangle was not shellable. Hachimori and Ziegler [HZ00] produced simple examples and showed that a triangulation containing any knotted triangle is not "constructible," constructibility being a strictly weaker notion than shellability. For more on shellability, see [DK78, Bjö92].

FACET-FORMING POLYTOPES AND "SMALL" LOW-DIMENSIONAL FACES

THEOREM 19.5.9 *Perles and Shephard* [PS67]

Let P be a d-polytope such that the maximum number of k-faces of P on any $(d-2)$-sphere in the skeleton of P is at most $(d-1-k)/(d+1-k)f_k(P)$. Then P is a nonfacet.

An example of a nonfacet that is simple was found by Barnette [Bar69]. Some of the proofs of Perles and Shephard use metric properties of polytopes, and for a few of the results alternative proofs using shellability were found by Barnette [Bar80].

THEOREM 19.5.10 *Schulte* [Sch85]

The cuboctahedron and the icosidodecahedron are nonfacets.

PROBLEM 19.5.11

Is the icosahedron facet-forming?

For all other regular polytopes the situation is known. The simplices and cubes in any dimension and the 3-dimensional octahedron are facet-forming. All other regular polytopes with the exception of the icosahedron are known to be nonfacets.

It is very interesting to see what can be said about metric properties of facets (or of low-dimensional faces) of a convex polytope.

THEOREM 19.5.12 *Bárány* (unpublished)

There is an $\epsilon > 0$ such that every d-polytope, $d > 2$, has a facet F for which no balls B_1 of radius R and B_2 of radius $(1+\epsilon)R$ satisfy $B_1 \subset F \subset B_2$.

The stronger statement where balls are replaced by ellipses is open.

Next, we try to understand if it is possible for all the k-faces of a d-polytope to be isomorphic to a given polytope P. The following conjecture asserts that if d is large with respect to k, this can happen only if P is either a simplex or a cube.

CONJECTURE 19.5.13 *Kalai* [Kal90]

For every k there is a $d(k)$ such that every d-polytope with $d > d(k)$ has a k-face that is either a simplex or combinatorially isomorphic to a k-dimensional cube.

Julian Pfeifle showed on the basis of the Wythoff construction (see Chapter 18), that $d(k) > (2k-1)(k-1)$, for $k \geq 3$.

For simple polytopes, it follows from the next theorem that if $d > ck^2$ then every d-polytope has a k-face F such that $f_r(F) \leq f_r(C_k)$. (Here, C_k denotes the k-dimensional cube.)

THEOREM 19.5.14 *Nikulin* [Nik86]

The average number of r-dimensional faces of a k-dimensional face of a simple d-dimensional polytope is at most

$$\binom{d-r}{d-k} \cdot \left(\left(\binom{\lfloor d/2 \rfloor}{r} + \binom{\lfloor (d+1)/2 \rfloor}{r} \right) \Big/ \left(\binom{\lfloor d/2 \rfloor}{k} + \binom{\lfloor (d+1)/2 \rfloor}{k} \right) \right).$$

Nikulin's theorem appeared in his study of reflection groups in hyperbolic spaces. The existence of reflection groups of certain types implies some combinatorial conditions on their fundamental regions (which are polytopes), and Vinberg [Vin85], Nikulin [Nik86], Khovanski [Kho86], and others showed that in high dimensions these combinatorial conditions lead to a contradiction. There are still many open problems in this direction: in particular, to narrow the gap between the dimensions above for which those reflection groups cannot exist and the dimensions for which such groups can be constructed.

THEOREM 19.5.15 *Kalai* [Kal90]

Every d-polytope for $d \geq 5$ has a 2-face with at most 4 vertices.

THEOREM 19.5.16 *Meisinger, Kleinschmidt, and Kalai* [MKK00]

Every (rational) d-polytope for $d \geq 9$ has a 3-face with at most 77 2-faces.

The last two theorems and the next one are proved using the linear inequalities for flag numbers that are known via intersection homology of toric varieties (toric h-vectors); see Chapter 17. Those inequalities were known to hold only for rational polytopes but the work of Karu [Kar04] extended them for non-rational polytopes as well.

The 120-cell shows that there are 4-polytopes all whose 2-faces are pentagons.

CONJECTURE 19.5.17 *Pack*

Every simple 4-polytope without a 2-face with at most 4 vertices has at least 600 vertices.

This conjecture may apply also to general polytopes. It is related in spirit to Theorem 17.5.10 by Blind and Blind identifying the cube as the "smallest" polytope with no triangular 2-faces. The conjecture may also apply to duals of arbitrary triangulations of S^3. However, it does not apply to (duals of) triangulated homology 3-sphere! Lutz, Sulanke, and Sullivan constructed a triangulation of Poincarés dodecahedral sphere that has only 18 vertices and 100 facets and has the property that every edge belongs to at least five simplices.

Here are two closely related questions about 4-polytopes:

PROBLEM 19.5.18 *Ziegler*

What is the maximal number of 2-faces which are not 4-gons for a 4-polytope with n facets?

PROBLEM 19.5.19 *Nevo, Santos, and Wilson [NSW16]*

What is the maximal number of facets which are not simplices for a 4-polytope with n vertices?

Nevo, Santos and Wilson [NSW16] gave lower bounds of $\Omega(n^{3/2})$ for both these questions.

Another possible extension of Theorem 19.5.15 is

CONJECTURE 19.5.20 *For every $\epsilon > 0$, every d-polytope for $d \geq d(\epsilon)$ has an edge such that more than a fraction $(1-\epsilon)$ of 2-faces containing it has at most 4 vertices.*

This conjecture is motivated by recent results (related to the Gallai-Sylvester Theorem) [DSW14] showing it to hold for zonotopes with $f(\epsilon) = 12/\epsilon$.

QUOTIENTS AND DUALITY

We talked about k-faces of d-polytopes, and one can also study, in a similar fashion, quotients of polytopes.

CONJECTURE 19.5.21 *Perles*

For every k there is a $d'(k)$ such that every d-polytope with $d > d'(k)$ has a k-dimensional quotient that is a simplex.

As was mentioned in the first section, $d'(2) = 3$. The 24-cell, which is a regular 4-polytope all of whose faces are octahedra, shows that $d'(3) > 4$.

THEOREM 19.5.22 *Meisinger, Kleinschmidt, and Kalai* [MKK00]

Every d-polytope with $d \geq 9$ has a 3-dimensional quotient that is a simplex.

Of course, Conjecture 19.5.13 implies Conjecture 19.5.21. Another stronger form of Conjecture 19.5.21, raised in [MKK00], is whether, for every k, a high-enough dimensional polytope must contain a face that is a k-simplex, or its polar dual must contain such a face. Adiprasito [Adi11] showed that for the analogous question for polytopal spheres, the answer is negative.

PROBLEM 19.5.23

For which values of k and r are there d-polytopes other than the d-simplex that are both k-simplicial and r-simple?

It is known that this can happen only when $k+r \leq d$. There are infinite families of $(d-2)$-simplicial and 2-simple polytopes, and some examples of $(d-3)$-simplicial and 3-simple d-polytopes.

Concerning this problem Peter McMullen recently noted that the polytopes r_{st}, discussed in Coxeter's classic book on regular polytopes [Cox63] in Sections 11.8 and 11.x, are $(r+2)$-simplicial and $(d-r-2)$-simple, where $d = r+s+t+1$. These so-called ***Gosset-Elte polytopes*** arise by the Wythoff construction from the finite reflection groups (see Chapter 18 of this Handbook); we obtain a finite polytope whenever the reflection group generated by the Coxeter diagram with r, s, t nodes on the three arms is finite, that is, when

$$1/(r+1) + 1/(s+1) + 1/(t+1) > 1.$$

The largest exceptional example, 2_{41}, is related to the Weyl group E_8. The Gosset-Elte polytope 2_{41} is a 4-simple 4-simplicial 8-polytope with 2160 vertices. Are there 5-simplicial 5-simple 10-polytopes?

PROBLEM 19.5.24

For which values of k and d are there self-dual k-simplicial d-polytopes other than the d-simplex?

THEOREM 19.5.25

For $d > 2$, there is no cubical d-polytope P whose dual is also cubical.

I am not aware of a reference for this result but it can easily be proved by exhibiting a covering map from the standard cubical complex realizing $\mathbb{R}^{d-1}$ into the boundary complex of P.

We have considered the problem of finding very special polytopes as "subobjects" (faces, quotients) of arbitrary polytopes. What about realizing arbitrary polytopes as "subobjects" of very special polytopes? There was an old conjecture that every polytope can be realized as a subpolytope (namely the convex hull of a subset of the vertices) of a stacked polytope. However, this conjecture was refuted by Adiprasito and Padrol [AP14]. Perles and Sturmfels asked whether every simplicial d-polytope can be realized as the quotient of some neighborly even-dimensional polytope. (Recall that a $2m$-polytope is ***neighborly*** if every m vertices are the vertices of an $(m-1)$-dimensional face.) Kortenkamp [Kor97] proved that this is the case for d-polytopes with at most $d+4$ vertices. For general polytopes, "neighborly polytopes" should be replaced here by "weakly neighborly" polytopes, introduced by Bayer [Bay93], which are defined by the property that every set of k vertices is contained in a face of dimension at most $2k-1$. The only theorem of this flavor I am aware of is by Billera and Sarangarajan [BS96], who proved that every $(0,1)$-polytope is a face of a traveling salesman polytope.

RECONSTRUCTION

THEOREM 19.5.26 *An extension of Whitney's theorem* [Grü67]

d-polytopes are determined by their $(d-2)$-skeletons.

THEOREM 19.5.27 *Perles* (unpublished, 1973)

Simplicial d-polytopes are determined by their $\lfloor d/2 \rfloor$*-skeletons.*

This follows from the following theorem (here, $\mathrm{ast}(F,P)$ is the complex formed by the faces of P that are disjoint to all vertices in F).

THEOREM 19.5.28 *Perles* (1973)

Let P be a simplicial d-polytope.

(i) *If F is a k-face of P, then* $\mathrm{skel}_{d-k-2}(\mathrm{ast}(F,P))$ *is contractible in* $\mathrm{skel}_{d-k-1}(\mathrm{ast}(F,P))$.

(ii) *If F is an empty k-simplex, then* $\mathrm{ast}(F,P)$ *is homotopically equivalent to* S^{d-k}*; hence,* $\mathrm{skel}_{d-k-2}(\mathrm{ast}(F,P))$ *is not contractible in* $\mathrm{skel}_{d-k-1}(\mathrm{ast}(F,P))$.

An extension of Perles's theorem for manifolds with vanishing middle homology was proved by Dancis [Dan84].

THEOREM 19.5.29 *Blind and Mani-Levitska* [BM87]

Simple polytopes are determined by their graphs.

Blind and Mani-Levitska described their theorem in a dual form and considered $(d-1)$-dimensional "puzzles" whose pieces are simplices and we wish to reconstruct the puzzle based on the "local" information of which two simplices share a facet. Joswig extended their result to more general puzzles where the pieces are general $(d-1)$-dimensional polytopes, and the way in which every two pieces sharing a facet are connected is also prescribed. A simple proof is given in [Kal88a]. This proof also shows that k-dimensional skeletons of simplicial polytopes are also determined by their "puzzle." When this is combined with Perles's theorem it follows that:

THEOREM 19.5.30 *Kalai and Perles*

Simplicial d-polytopes are determined by the incidence relations between i- and $(i+1)$*-faces for every* $i > \lfloor d/2 \rfloor$.

CONJECTURE 19.5.31 *Haase and Ziegler*

Let G be the graph of a simple 4-polytope. Let H be an induced, nonseparating, 3-regular, 3-connected planar subgraph of G. Then H is the graph of a facet of P.

Haase and Ziegler [HZ02] showed that this is not the case if H is not planar. Their proof touches on the issue of embedding knots in the skeletons of 4-polytopes.

PROBLEM 19.5.32

Are simplicial spheres determined by the incidence relations between their facets and subfacets?

THEOREM 19.5.33 *Björner, Edelman, and Ziegler* [BEZ90]

Zonotopes are determined by their graphs.

THEOREM 19.5.34 *Babson, Finschi, and Fukuda* [BFF01]

Duals of cubical zonotopes are determined by their graphs.

In all instances of the above theorems except the single case of the theorem of Blind and Mani-Levitska, the proofs give reconstruction algorithms that are polynomial in the data. It was an open question if a polynomial algorithm exists to determine a simple polytope from its graph. A polynomial "certificate" for reconstruction was recently found by Joswig, Kaibel, and Körner [JKK02]. Finally, Friedman [Fri09] proved

THEOREM 19.5.35 *Friedman* [Fri09]

There is a polynomial type algorithm to determine a simple d-polytope (up to combinatorial isomorphism) from its graph.

An interesting problem was whether there is an e-dimensional polytope other than the d-cube with the same graph as the d-cube.

THEOREM 19.5.36 *Joswig and Ziegler* [JZ00]

For every $d \geq e \geq 4$ there is an e-dimensional cubical polytope with 2^d vertices whose $(\lfloor e/2 \rfloor - 1)$-skeleton is combinatorially isomorphic to the $(\lfloor e/2 \rfloor - 1)$-skeleton of a d-dimensional cube.

Earlier, Babson, Billera, and Chan [BBC97] found such a construction for cubical spheres.

Another issue of reconstruction for polytopes that was studied extensively is the following: In which cases does the combinatorial structure of a polytope determine its geometric structure (up to projective transformations)? Such polytopes are called ***projectively unique***. McMullen [McM76] constructed projectively unique d-polytopes with $3^{d/3}$ vertices. The major unsolved problem whether there are only finitely many projectively unique polytopes in each dimension was settled in

THEOREM 19.5.37 *Adiprasito and Ziegler* [AZ15]

There is an infinite family of projectively unique 69-dimensional polytopes.

The key for this result was the construction of an infinite family of 4-polytopes whose realization spaces have bounded (at most 96) dimension.

A related question is to understand to what extent we can reconstruct the internal structure of a polytope, namely the combinatorial types of subpolytopes (or, equivalently, the oriented matroid described by the vertices) from its combinatorial structure.

THEOREM 19.5.38 *Shemer [She82]*

Neighborly even-dimensional polytopes determine their internal structure.

EMPTY FACES AND POLYTOPES WITH FEW VERTICES

THEOREM 19.5.39 *Perles* (unpublished, 1970)

Let $f(d, k, b)$ be the number of combinatorial types of k-skeletons of d-polytopes with $d + b + 1$ vertices. Then, for fixed b and k, $f(d, k, b)$ is bounded.

This follows from

THEOREM 19.5.40 *Perles* (unpublished, 1970)

The number of empty i-pyramids for d-polytopes with $d + b$ vertices is bounded by a function of i and b.

For another proof of this theorem see [Kal94].

Here is a beautiful recent subsequent result:

THEOREM 19.5.41 *Padrol* [Pad16]

The number of d-polytopes with $d + b$ vertices and $d + c$ facets is bounded by a function of b and c.

EMPTY FACE NUMBERS AND RELATED BETTI NUMBERS

For a d-polytope P, let $e_i(P)$ denote the number of empty i-simplices of P.

PROBLEM 19.5.42

Characterize the sequence of numbers $(e_1(P), e_2(P), \ldots, e_d(P))$ arising from simplicial d-polytopes and from general d-polytopes.

The following theorem, which was motivated by commutative-algebraic concerns, confirmed a conjecture by Kleinschmidt, Kalai, and Lee [Kal94].

THEOREM 19.5.43 *Migliore and Nagel* [MN03, Nag08]

For all simplicial d-polytopes with prescribed h-vector $h = (h_0, h_1, \ldots, h_d)$, the number of i-dimensional empty simplices is maximized by the Billera-Lee polytopes $P_{BL}(h)$.

$P_{BL}(h)$ is the polytope constructed by Billera and Lee [BL81] (see Chapter 17) in their proof of the sufficiency part of the g-theorem. Migliore and Nagel proved that for a prescribed f-vector, the Billera-Lee polytopes maximize even more general parameters that arise in commutative algebra: the sum of the ith Betti numbers of induced subcomplexes on j vertices for every i and j. (These sums correspond to Betti numbers of the resolution of the Stanley-Reisner ring associated with the polytope.) The case $j = i + 2$ reduces to counting empty faces. It is quite possible that the theorem of Migliore and Nagel extends to general simplicial spheres with prescribed h-vector and to general polytopes with prescribed (toric) h-vector. (However, it is not yet known in these cases that the h-vectors are always those of Billera-Lee polytopes; see Chapter 17.)

Recently, valuable connections between these Betti numbers, discrete Morse theory, face numbers, high-notions of chordality, and metrical properties of polytopes are emerging [ANS16, Bag16].

STANLEY-REISNER RINGS

An algebraic object associated to simplicial polytopes, triangulated spheres, and general simplicial complexes is the Stanley-Reisner ring which gives crucial information on face numbers (discussed in Chapter 17). The Stanley-Reisner ring also has applications to the study of skeletons of polytopes discussed in this chapter.

For every simplicial d-polytope P with n vertices one can associate an ideal $I(P)$ of monomials in $n - d$ variables with $h_i(P)$ degree-i monomials for $i = 1, 2, \ldots, d$. (Moreover, $I(P)$ is *shifted* [Kal02, Kal94]). This construction is based on deep properties of the Stanley-Rieser ring associated to simplicial d polytopes which conjecturally extends to arbitrary triangulations of S^{d-1}. The shifted ideal $I(P)$

carries important structural properties of P and its skeletons. For example, there are connections with rigidity [Lee94, Kal94]. Moving from an arbitrary shifted order ideal of monomials to a triangulation is described in [Kal88b] and is the basis for Theorem 19.6.5. There are extensions and analogues of the Stanley-Reisner rings for nonsimplicial polytopes, for restricted classes of polytopes and for other cellular objects.

19.6 COUNTING POLYTOPES, SPHERES AND THEIR SKELETA

THEOREM 19.6.1 *There are only a "few" simplicial d-polytopes with n facets Durhuus and Jonsson: Benedetti and Ziegler* [DJ95, BZ11]

The number of distinct (isomorphism types) simplicial d-polytopes with n facets is at most C_d^n*, where* C_d *is a constant depending on d.*

CONJECTURE 19.6.2 *There are only a "few" triangulations of* S^d *with n facets*

The number of distinct (isomorphism types) triangulations of S^{d-1} *with n facets is at most* C_d^n*, where* C_d *is a constant depending on d.*

See Gromov's paper [Gro00] for some discussion.

CONJECTURE 19.6.3 *There are only a "few" graphs of polytopes*

The number of distinct (isomorphism types) of graphs of d-polytopes with n vertices is at most C_d^n*, where* C_d *is a constant depending on d.*

Both conjectures are open already for $d = 4$. As far as we know, it is even possible that the same constant applies for all dimensions.

We can also ask for the number of simplicial d-polytopes and triangulation of spheres with n vertices. Let $s(d, n)$ denotes the number of triangulations of S^{d-1} with n vertices and let $p_s(d, n)$ be the number of combinatorial types of simplicial d-polytopes with n vertices, and let $p(d, n)$ be the number of combinatorial types of (general) d-polytopes with n vertices.

THEOREM 19.6.4 *Goodman and Pollack; [GP86] Alon [Alo86]*

For some absolute constant C, $\log p_s(d, n) \leq \log p(d, n) \leq C d^2 n \log n$.

THEOREM 19.6.5 *Kalai [Kal88b]*

For a fixed d, $\log s(d, n) = \Omega(n^{[d/2]})$

THEOREM 19.6.6 *Nevo, Santos, and Wilson [NSW16]*

Let k be fixed. Then $\log s(2k + 1, n) = \Omega(n^{k+1})$.

The case of triangulations of 3-spheres is of particular interest. The quadratic lower bound for $\log s(3, n)$ improves an earlier construction from [PZ04]. Stanley's upper bound theorem for triangulated spheres (see Chapter 17) implies that $\log s(d, n) = \Omega(n^{[(k+1)/2] \log n})$.

We already mentioned that there is a rich enumerative theory for 3-polytope. Perles found formulas for the number of d-polytopes, and simplicial d-polytopes with $d+3$ vertices. His work used Gale's diagram and is presented in Grünbaum's classical book on polytopes [Grü67]. There is also a formula for the number of stacked d-polytopes with n labelled vertices [BP71].

19.7 CONCLUDING REMARKS AND EXTENSIONS TO MORE GENERAL OBJECTS

The reader who compares this chapter with other chapters on convex polytopes may notice the sporadic nature of the results and problems described here. Indeed, it seems that our main limits in understanding the combinatorial structure of polytopes still lie in our ability to raise the right questions. Another feature that comes to mind (and is not unique to this area) is the lack of examples, methods of constructing them, and means of classifying them.

We have considered mainly properties of general polytopes and of simple or simplicial polytopes. There are many classes of polytopes that are either of intrinsic interest from the combinatorial theory of polytopes, or that arise in various other fields, for which the problems described in this chapter are interesting.

Most of the results of this chapter extend to much more general objects than convex polytopes. Finding combinatorial settings for which these results hold is an interesting and fruitful area. On the other hand, the results described here are not sufficient to distinguish polytopes from larger classes of polyhedral spheres, and finding delicate combinatorial properties that distinguish polytopes is an important area of research. Few of the results on skeletons of polytopes extend to skeletons of other convex bodies [LR70, LR71, GL81], and relating the combinatorial theory of polytopes with other aspects of convexity is a great challenge.

19.8 SOURCES AND RELATED MATERIAL

FURTHER READING

Grünbaum [Grü75] is a survey on polytopal graphs and many results and further references can be found there. More material on the topic of this chapter and further relevant references can also be found in [Grü67, Zie95, BMSW94, KK95, BL93, DRS10]. Several chapters of [BMSW94] are relevant to the topic of this chapter: The authors chapter [Kal94] expands on various topics discussed here and in Chapter 17. Martini's chapter on the regularity properties of polytopes contains further references on facet-forming polytopes and nonfacets. The original papers on facet-forming polytopes and nonfacets contain many more results, and describe relations to questions on tiling spaces with polyhedra.

RELATED CHAPTERS

Chapter 15: Basic properties of convex polytopes
Chapter 17: Face numbers of polytopes and complexes

Chapter 17 discusses f-vectors of polytopes (and more general cellular structures), and related parameters such as h-vectors and g-vectors for simplicial polytopes, and flag-f-vectors, toric h-vectors, toric g-vectors and the cd-index of general polytopes. There are many relations between these parameters and the combinatorial study of graphs and skeleta of polytopes.

Chapter 49: Linear programming
Chapters 7, 16, 18, 20, and 61 are also related to some parts of this chapter.

REFERENCES

[AB14] K. Adiprasito and B. Benedetti. The Hirsch conjecture holds for normal flag complexes. *Math. Oper. Res.*, 39:1340–1348, 2014.

[ABGJ17] X. Allamigeon, P. Benchimol, S. Gaubert, and M. Joswig. Log-barrier interior point methods are not strongly polynomial. Preprint, `arXiv:1708.01544`, 2017.

[Adi11] K.A. Adiprasito. A note on the simplex cosimplex problem. Unpublished manuscript, 2011.

[ADJ97] J. Ambjørn, B. Durhuus, and T. Jonsson. *Quantum Geometry: A Statistical Field Theory Approach.* Cambridge University Press, 1997.

[AH76] K. Appel and W. Haken. Every planar map is four colorable. *Bull. Amer. Math. Soc.*, 82:711–712, 1976.

[AH89] K. Appel and W. Haken. *Every Planar Map Is Four Colorable.* Vol. 98 of *Contemp. Math.*, AMS, Providence, 1989.

[Alo86] N. Alon. The number of polytopes, configurations and real matroids. *Mathematika*, 33:62–71, 1986.

[Ang02] O. Angel. Growth and percolation on the uniform infinite planar triangulation. *Geom. Funct. Anal.*, 13:935–974, 2003.

[ANS16] K. Adiprasito, E. Nevo, and J.A. Samper. A geometric lower bound theorem. *Geom. Funct. Anal.*, 26:359–378, 2016.

[AP14] K.A. Adiprasito and A. Padrol. A universality theorem for projectively unique polytopes and a conjecture of Shephard. *Israel J. Math.*, 211:239–255, 2014.

[AZ99] N. Amenta and G.M. Ziegler. Deformed products and maximal shadows. In B. Chazelle, J.E. Goodman, and R.Pollack, editors, *Advances in Discrete and Computational Geometry*, vol. 223 of *Contemp. Math.*, pages 57–90, AMS, Providence, 1999.

[AZ15] K.A. Adiprasito and G.M. Ziegler. Many projectively unique polytopes. *Invent. Math.*, 199:581–652, 2015.

[Bag16] B. Bagchi. The μ-vector, Morse inequalities and a generalized lower bound theorem for locally tame combinatorial manifolds. *European. J. Combin.*, 51:69-83, 2016.

[Bal61] M.L. Balinski. On the graph structure of convex polyhedra in n-space. *Pacific J. Math.*, 11:431–434, 1961.

[Bal84] M.L. Balinski. The Hirsch conjecture for dual transportation polyhedra. *Math. Oper. Res.*, 9:629–633, 1984.

[Bar66] D.W. Barnette. Trees in polyhedral graphs. *Canad. J. Math.*, 18:731–736, 1966.

[Bar69] D.W. Barnette. A simple 4-dimensional nonfacet. *Israel J. Math.*, 7:16–20, 1969.

[Bar74] D.W. Barnette. An upper bound for the diameter of a polytope. *Discrete Math.*, 10:9–13, 1974.

[Bar80] D.W. Barnette. Nonfacets for shellable spheres. *Israel J. Math.*, 35:286–288, 1980.

[Bay93] M.M. Bayer. Equidecomposable and weakly neighborly polytopes. *Israel J. Math.*, 81:301–320, 1993.

[BBC97] E.K. Babson, L.J. Billera, and C.S. Chan. Neighborly cubical spheres and a cubical lower bound conjecture. *Israel J. Math.*, 102:297–315, 1997.

[BDE$^+$14] N. Bonifas, M. Di Summa, F. Eisenbrand, N. Hähnle, and M. Niemeier. On subdeterminants and the diameter of polyhedra. *Discrete Comput. Geom.*, 52:102–115, 2014.

[BEZ90] A. Björner, P.H. Edelman, and G.M. Ziegler. Hyperplane arrangements with a lattice of regions. *Discrete Comput. Geom.*, 5:263–288, 1990.

[BFF01] E.K. Babson, L. Finschi, and K. Fukuda. Cocircuit graphs and efficient orientation reconstruction in oriented matroids. *European J. Combin.*, 22:587–600, 2001.

[Bjö92] A. Björner. Homology and shellability of matroids and geometric lattices. In N. White, editor, *Matroid Applications*, vol. 40 of *Encyclopedia of Mathematics*, pages 226–283, Cambridge Univ. Press, 1992.

[BL81] L.J. Billera and C.W. Lee. A proof of the sufficiency of McMullen's conditions for f-vectors of simplicial convex polytopes. *J. Combin. Theory Ser. A*, 31:237–255, 1981.

[BL93] M.M. Bayer and C.W. Lee. Combinatorial aspects of convex polytopes. In P.M. Gruber and J.M. Wills, editors, *Handbook of Convex Geometry*, pages 485–534, North-Holland, Amsterdam, 1993.

[BM71] H. Bruggesser and P. Mani. Shellable decompositions of cells and spheres. *Math. Scand.*, 29:197–205, 1971.

[BM87] R. Blind and P. Mani-Levitska. On puzzles and polytope isomorphisms. *Aequationes Math.*, 34:287–297, 1987.

[BMSW94] T. Bisztriczky, P. McMullen, R. Schneider, and A.I. Weiss, editors. *Polytopes: Abstract, Convex and Computational.* Vol. 440 of *NATO Adv. Sci. Inst. Ser. C: Math. Phys. Sci.*, Kluwer, Dordrecht, 1994.

[BP71] L.W. Beineke and R.E. Pippert. The number of labeled dissections of a k-ball. *Math. Ann.*, 191:87–98, 1971.

[BS96] L.J. Billera and A. Sarangarajan. All 0-1 polytopes are traveling salesman polytopes. *Combinatorica*, 16:175–188, 1996.

[BZ11] B. Benedetti and G.M. Ziegler. On locally constructible spheres and balls. *Acta Mathematica*, 206:205–243, 2011.

[Cox63] H.S.M. Coxeter. *Regular Polytopes.* Macmillan, New York, second edition, 1963. Corrected reprint, Dover, New York, 1973.

[CS02] P. Chassaing and G. Schaeffer. Random planar lattices and integrated superBrownian excursion. *Probability Theory and Related Fields*, 128:161–212, 2004.

[Dan84] J. Dancis. Triangulated n-manifolds are determined by their $[n/2]+1$-skeletons. *Topology Appl.*, 18:17–26, 1984.

[DF94] M. Dyer and A. Frieze. Random walks, totally unimodular matrices, and a randomised dual simplex algorithm. *Math. Programming* 64:1–16, 1994.

[DK78] G. Danaraj and V. Klee. Which spheres are shellable? In B. Alspach, P. Hell, and D.J. Miller, editors, *Algorithmic Aspects of Combinatoric (Vancouver Island BC, 1976)*, vol. 2 of *Ann. Discrete Math.*, pages 33–52, 1978.

[DJ95] B. Durhuus and T. Jonsson. Remarks on the entropy of 3-manifolds. *Nuclear Physics B*, 445:182–192, 1995.

[DK12] J. De Loera and S. Klee. Transportation problems and simplicial polytopes that are not weakly vertex-decomposable. *Math. Oper. Res.*, 37:670–674, 2012.

[DMO17] A. Deza, G. Manoussakis, and S. Onn. Primitive zonotopes. *Discrete Comput. Geom.*, in print, 2017.

[DRS10] J.A. De Loera, J. Rambau, F. Santos, *Triangulations: Structures for Algorithms and Applications*, vol. 25 of *Algorithms and Computation in Mathematics*, Springer-Verlag, Berlin, 2010.

[DO95] M.M. Deza and S. Onn. Lattice-free polytopes and their diameter. *Discrete Comput. Geom.*, 13:59–75, 1995.

[DSW14] Z. Dvir, S. Saraf, and A. Wigderson. Improved rank bounds for design matrices and a new proof of Kelly's theorem. *Forum Math. Sigma*, 2:e4, 2014

[DTZ09] A. Deza, T. Terlaky, and Y. Zinchenko. A continuous d-step conjecture for polytopes. *Discrete Comput. Geom.* 41:318–327, 2009.

[Ebe91] V. Eberhard. *Zur Morphologie der Polyeder.* Teubner, Leipzig, 1891.

[EHRR10] F. Eisenbrand, N. Hähnle, A. Razborov, and T. Rothvoss. Diameter of polyhedra: limits of abstraction. *Math. Oper. Res.*, 35:786–794, 2010.

[FH99] K. Fritzsche and F.B. Holt. More polytopes meeting the conjectured Hirsch bound. *Discrete Math.*, 205:77–84, 1999.

[FHZ11] O. Friedmann, T.D. Hansen, and U. Zwick. Subexponential lower bounds for randomized pivoting rules for the simplex algorithm. In *Proc. 43rd ACM Sympos. Theory Comput.*, pages 283–292, 2011.

[Flo32] A. Flores. Über n-dimensionale Komplexe die im R_{2n+1} absolut selbstverschlungen sind. *Ergeb. Math. Kolloq.*, 6:4–7, 1932/1934.

[FM92] T. Feder and M. Mihail. Balanced matroids. In *Proc. 24th ACM Sympos. Theory Comput.*, pages 26–38, 1992.

[Fri09] E.J. Friedman. Finding a simple polytope from its graph in polynomial time. *Discrete Comput. Geom.* 41:249–256, 2009.

[Gal85] S. Gallivan. Disjoint edge paths between given vertices of a convex polytope. *J. Combin. Theory Ser. A*, 39:112–115, 1985.

[GL81] S. Gallivan and D.G. Larman. Further results on increasing paths in the one-skeleton of a convex body. *Geom. Dedicata*, 11:19–29, 1981.

[GLM81] S. Gallivan, E.R. Lockeberg, and P. McMullen. Complete subgraphs of the graphs of convex polytopes. *Discrete Math.*, 34:25–29, 1981.

[GN06] O. Giménez and M. Noy, Asymptotic enumeration and limit laws of planar graphs. *J. Amer. Math. Soc.*, 22:309–329, 2009.

[GO78] J.E. Goodman and H. Onishi. Even triangulations of S^3 and the coloring of graphs. *Trans. Amer. Math. Soc.*, 246:501–510, 1978.

[GP86] J.E. Goodman and R. Pollack. Upper bounds for configurations and polytopes in R^d. *Discrete Comput. Geom.*, 1:219–227, 1986.

[Gro00] M. Gromov. Spaces and questions. In N. Alon, J. Bourgain, A. Connes, M. Gromov and V. Milman, editors, *Visions in Mathematics: GAFA 2000 Special Volume, Part I*, pages 118–161, Birkhäuser, Basel, 2000.

[Grü65] B. Grünbaum. On the facial structure of convex polytopes. *Bull. Amer. Math. Soc.*, 71:559–560, 1965.

[Grü67] B. Grünbaum. *Convex Polytopes.* Interscience, London, 1967. Revised edition (V. Kaibel, V. Klee, and G.M. Ziegler, editors), Springer-Verlag, New York, 2003.

[Grü75] B. Grünbaum. Polytopal graphs. In D.R. Fulkerson, editor, *Studies in Graph Theory*, pages 201–224, Math. Assoc. Amer., Washington, 1975.

[GZ74] B. Grünbaum and J. Zaks. The existence of certain planar maps. *Discrete Math.*, 10:93–115, 1974.

[Häh10] N. Hähnle. Post in G. Kalai, *Polymath 3: Polynomial Hirsch Conjecture,* September 30, 2010. `http://gilkalai.wordpress.com/2010/09/29/polymath-3-polynomialhirsch-conjecture`.

[HK98a] F. Holt and V. Klee. Counterexamples to the strong d-step conjecture for $d \geq 5$. *Discrete Comput. Geom.*, 19:33–46, 1998.

[HK98b] F. Holt and V. Klee. Many polytopes meeting the conjectured Hirsch bound. *Discrete Comput. Geom.*, 20:1–17, 1998.

[HK98c] F. Holt and V. Klee. A proof of the strict monotone 4-step conjecture. In B. Chazelle, J.E. Goodman, and R. Pollack, editors, *Advances in Discrete and Computational Geometry*, vol. 223 of *Contemp. Math.*, pages 201–216, AMS, Providence, 1998.

[HPS14] N. Hähnle, V. Pilaud, and S. Klee. Obstructions to weak decomposability for simplicial polytopes. *Proc. Amer. Math. Soc.*, 142:3249–3257, 2014.

[HZ00] M. Hachimori and G.M. Ziegler. Decompositions of simplicial balls and spheres with knots consisting of few edges. *Math. Z.*, 235:159–171, 2000.

[HZ02] C. Haase and G.M. Ziegler. Examples and counterexamples for the Perles conjecture. *Discrete Comput. Geom.*, 28:29–44, 2002.

[Jen93] S. Jendrol'. On face vectors and vertex vectors of convex polyhedra. *Discrete Math.*, 118:119–144, 1993.

[JKK02] M. Joswig, V. Kaibel, and F. Körner. On the k-systems of a simple polytope. *Israel J. Math.*, 129:109–118, 2002.

[Jos02] M. Joswig. Projectivities in simplicial complexes and colorings of simple polytopes. *Math. Z.*, 240:243–259, 2002.

[Juc76] E. Jucovič. On face-vectors and vertex-vectors of cell-decompositions of orientable 2-manifolds. *Math. Nachr.*, 73:285–295, 1976.

[JZ00] M. Joswig and G.M. Ziegler. Neighborly cubical polytopes. *Discrete Comput. Geom.*, 24:325–344, 2000.

[Kai01] V. Kaibel. On the expansion of graphs of 0/1-polytopes. Tech. Rep., TU Berlin, 2001; see also `arXiv:math.CO/0112146`.

[Kal87] G. Kalai. Rigidity and the lower bound theorem I. *Invent. Math.*, 88:125-151, 1987.

[Kal88a] G. Kalai. A simple way to tell a simple polytope from its graph. *J. Combin. Theory Ser. A*, 49:381–383, 1988.

[Kal88b] G. Kalai. Many triangulated spheres. *Discrete Comput. Geom.*, 3:1–14, 1988.

[Kal90] G. Kalai. On low-dimensional faces that high-dimensional polytopes must have. *Combinatorica*, 10:271–280, 1990.

[Kal91] G. Kalai. The diameter of graphs of convex polytopes and f-vector theory. In P. Gritzmann and B. Sturmfels, editors, *Applied Geometry and Discrete Mathematics—the Victor Klee Festschrift*, vol. 4 of *DIMACS Ser. Discrete Math. Theoret. Comput. Sci.*, pages 387–411, AMS, Providence, 1991.

[Kal92] G. Kalai. Upper bounds for the diameter and height of graphs of convex polyhedra. *Discrete Comput. Geom.*, 8:363–372, 1992.

[Kal94] G. Kalai. Some aspects in the combinatorial theory of convex polytopes. In [BMSW94], pages 205–230, 1994.

[Kal02] G. Kalai. Algebraic shifting, computational commutative algebra and combinatorics advanced studies. In *Advanced Studies in Pure Mathematics*, vol. 33, pages 121–165, Math. Soc. Japan, Tokyo, 2002.

[Kam32] R.E. van Kampen. Komplexe in euklidischen Räumen. *Abh. Math. Sem. Hamburg*, 9:72–78, 1932. Berichtigung dazu, *ibid.*, 152–153.

[Kar04] K. Karu. Hard Lefschetz theorem for nonrational polytopes. *Invent. Math.*, 157:419–447, 2004.

[Kho86] A.G. Khovanskii. Hyperplane sections of polyhedra, toric varieties and discrete groups in Lobachevskii space. *Funktsional. Anal. i Prilozhen.*, 20:50–61,96, 1986.

[KK92] G. Kalai and D.J. Kleitman. A quasi-polynomial bound for the diameter of graphs of polyhedra. *Bull. Amer. Math. Soc.*, 26:315–316, 1992.

[KK95] V. Klee and P. Kleinschmidt. Polyhedral complexes and their relatives. In R. Graham, M. Grötschel, and L. Lovász, editors, *Handbook of Combinatorics*, pages 875–917. North-Holland, Amsterdam, 1995.

[Kle64] V. Klee. A property of d-polyhedral graphs. *J. Math. Mech.*, 13:1039–1042, 1964.

[KM72] V. Klee and G.J. Minty. How good is the simplex algorithm? In O. Shisha, editor, *Inequalitites, III*, pages 159–175, Academic Press, New York, 1972.

[KO92] P. Kleinschmidt and S. Onn. On the diameter of convex polytopes. *Discrete Math.*, 102:75–77, 1992.

[Kor97] U.H. Kortenkamp. Every simplicial polytope with at most $d+4$ vertices is a quotient of a neighborly polytope. *Discrete Comput. Geom.*, 18:455–462, 1977.

[Kot55] A. Kotzig. Contribution to the theory of Eulerian polyhedra. *Mat.-Fyz. Časopis. Slovensk. Akad. Vied*, 5:101–113, 1955.

[Kur22] C. Kuratowski, Sur l'operation A de l'analysis situs. *Fund. Math.* 3:182–199, 1922.

[KW67] V. Klee and D.W. Walkup. The d-step conjecture for polyhedra of dimension $d < 6$. *Acta Math.*, 117:53–78, 1967.

[Lar70] D.G. Larman. Paths on polytopes. *Proc. London Math. Soc.*, 20:161–178, 1970.

[Lee94] C.W. Lee. Generalized stresses and motions. In *Polytopes: Abstract, Convex and Computational*, vol. 440 of *NATO ASI Series*, pages 249–271, Springer, Dordrecht, 1994.

[Lic91] W.B.R. Lickorish. Unshellable triangulations of spheres. *European J. Combin.*, 12:527–530, 1991.

[LM70] D.G. Larman and P. Mani. On the existence of certain configurations within graphs and the 1-skeletons of polytopes. *Proc. London Math. Soc.*, 20:144–160, 1970.

[LR70] D.G. Larman and C.A. Rogers. Paths in the one-skeleton of a convex body. *Mathematika*, 17:293–314, 1970.

[LR71] D.G. Larman and C.A. Rogers. Increasing paths on the one-skeleton of a convex body and the directions of line segments on the boundary of a convex body. *Proc. London Math. Soc.*, 23:683–698, 1971.

[LT79] R.J. Lipton and R.E. Tarjan. A separator theorem for planar graphs. *SIAM J. Applied Math.*, 36:177–189, 1979.

[LZ15] L. Loiskekoski and G.M. Ziegler. Simple polytopes without small separators. Preprint, `arXiv:1510.00511`, 2015; *Israel J. Math.*, to appear.

[McM76] P. McMullen. Constructions for projectively unique polytopes. *Discrete Math.*, 14:347–358, 1976.

[Mil86] G.L. Miller. Finding small simple cycle separators for 2-connected planar graphs. *J. Comput. System Sci.*, 32:265–279, 1986.

[MK00] J. Mihalisin and V. Klee. Convex and linear orientations of polytopal graphs. *Discrete Comput. Geom.*, 24:421–436, 2000.

[MKK00] G. Meisinger, P. Kleinschmidt, and G. Kalai. Three theorems, with computer-aided proofs, on three-dimensional faces and quotients of polytopes. *Discrete Comput. Geom.*, 24:413–420, 2000.

[MN03] J. Migliore and U. Nagel. Reduced arithmetically Gorenstein schemes and simplicial polytopes with maximal Betti numbers. *Adv. Math.*, 180:1–63, 2003.

[Mot64] T.S. Motzkin. The evenness of the number of edges of a convex polyhedron. *Proc. Nat. Acad. Sci. U.S.A.*, 52:44–45, 1964.

[MSW15] B. Matschke, F. Santos, and C. Weibel. The width of 5-dimensional prismatoids. *Proc. London Math. Soc.*, 110:647–672, 2015.

[Nad89] D. Naddef. The Hirsch conjecture is true for $(0,1)$-polytopes. *Math. Programming*, 45:109–110, 1989.

[Nag08] U. Nagel. Empty simplices of polytopes and graded Betti numbers. *Discrete Comput. Geom.*, 39:389–410, 2008.

[Nik86] V.V. Nikulin. Discrete reflection groups in Lobachevsky spaces and algebraic surfaces. In *Proc. Internat. Cong. Math.*, vol. 1, pages 654-671, Berkeley, 1986.

[Noy14] M. Noy. Random planar graphs and beyond. In *Proc. International Congress Math.*, vol. 4, pages 407–430, Kyung Moon, Seoul, 2014.

[NSW16] E. Nevo, F. Santos, and S. Wilson. Many triangulated odd-spheres. *Math. Ann.*, 364:737–762, 2016.

[PA95] J. Pach and P.K. Agarwal. *Combinatorial Geometry.* Wiley-Interscience, New York, 1995.

[Pad16] A. Padrol. Polytopes with few vertices and few facets. *J. Combin. Theory Ser. A*, 142:177–180, 2016.

[PB80] J.S. Provan and L.J. Billera. Decompositions of simplicial complexes related to diameters of convex polyhedra. *Math. Oper. Res.*, 5:576–594, 1980.

[PS67] M.A. Perles and G.C. Shephard. Facets and nonfacets of convex polytopes. *Acta Math.*, 119:113–145, 1967.

[PZ04] J. Pfeifle and G.M. Ziegler. Many triangulated 3-spheres, *Math. Ann.*, 330:829–837, 2004.

[RSST97] N. Robertson, D. Sanders, P. Seymour, and R. Thomas. The four-colour theorem. *J. Combin. Theory Ser. B*, 70:2–44, 1997.

[San13a] F. Santos. A counter-example to the Hirsch Conjecture. *Ann. Math.*, 176:383–412, 2012.

[San13b] F. Santos. Recent progress on the combinatorial diameter of polytopes and simplicial complexes. *TOP*, 21:426–460, 2013.

[Sch85] E. Schulte. The existence of nontiles and nonfacets in three dimensions. *J. Combin. Theory Ser. A*, 38:75–81, 1985.

[Sch94] C. Schulz. Polyhedral manifolds on polytopes. In M.I. Stoka, editor, *First International Conference on Stochastic Geometry, Convex Bodies and Empirical Measures (Palermo, 1993). Rend. Circ. Mat. Palermo (2) Suppl.*, 35:291–298, 1994.

[She82] I. Shemer. Neighborly polytopes. *Israel J. Math.*, 43:291–314, 1982.

[Ste22] E. Steinitz. Polyeder und Raumeinteilungen. In W.F. Meyer and H. Mohrmann, editors, *Encyklopädie der mathematischen Wissenschaften, Dritter Band: Geometrie, III.1.2., Heft 9, Kapitel III A B 12*, pages 1–139. Teubner, Leipzig, 1922.

[Tho81] C. Thomassen. Kuratowski's theorem. *J. Graph Theory*, 5:225–241, 1981.

[Tod14] M.J. Todd. An improved Kalai-Kleitman bound for the diameter of a polyhedron. *SIAM J. Discrete Math.*, 28:1944–1947, 2014.

[Tut56] W.T. Tutte. A theorem on planar graphs. *Trans. Amer. Math. Soc.*, 82:99–116, 1956.

[Tut62] W.T. Tutte. A census of planar triangulations. *Canad. J. Math.*, 14:21–38, 1962.

[Vin85] E.B. Vinberg. Hyperbolic groups of reflections (Russian). *Uspekhi Mat. Nauk*, 40:29–66,255, 1985.

[Whi32] H. Whitney. Non-separable and planar graphs. *Trans. Amer. Math. Soc.*, 34:339–362, 1932.

[Whi84] W. Whiteley. Infinitesimally rigid polyhedra. I. Statics of frameworks. *Trans. Amer. Math. Soc.*, 285:431–465, 1984.

[Wil88] K. Williamson Hoke. Completely unimodal numberings of a simple polytope. *Discrete Appl. Math.*, 20:69–81, 1988.

[Wu65] W.-T. Wu. *A Theory of Imbedding, Immersion, and Isotopy of Polytopes in a Euclidean Space.* Science Press, Beijing, 1965.

[Zie95] G.M. Ziegler. *Lectures on Polytopes.* Vol. 152 of *Graduate Texts in Math.*, Springer-Verlag, New York, 1995.

[Zie98] G.M. Ziegler. Shelling polyhedral 3-balls and 4-polytopes. *Discrete Comput. Geom.*, 19:159–174, 1998.

20 POLYHEDRAL MAPS

Ulrich Brehm and Egon Schulte

INTRODUCTION

Historically, polyhedral maps on surfaces made their first appearance as convex polyhedra. The famous Kepler-Poinsot (star) polyhedra marked the first occurrence of maps on orientable surfaces of higher genus (namely 4), and started the branch of topology dealing with regular maps. Further impetus to the subject came from the theory of automorphic functions and from the Four-Color-Problem (Coxeter and Moser [CM80], Barnette [Bar83]).

A more systematic investigation of general polyhedral maps and nonconvex polyhedra began only around 1970, and was inspired by Grünbaum's book "Convex Polytopes" [Grü67]. Since then, the subject has grown into an active field of research on the interfaces of convex and discrete geometry, graph theory, and combinatorial topology. The underlying topology is mainly elementary, and many basic concepts and constructions are inspired by convex polytope theory.

20.1 POLYHEDRA

Tessellations on surfaces are natural objects of study in topology that generalize convex polyhedra and plane tessellations. For general properties of convex polyhedra, polytopes, and tessellations, see Grünbaum [Grü67], Coxeter [Cox73], Grünbaum and Shephard [GS87], and Ziegler [Zie95], or Chapters 3, 15, 16, 17, 18, and 19 of this Handbook. For a survey on polyhedral manifolds see Brehm and Wills [BW93], which also has an extensive list of references. The long list of definitions that follows places polyhedral maps in the general context of topological and geometric complexes. For an account of 2- and 3-dimensional geometric topology, see Moise [Moi77].

GLOSSARY

Polyhedral complex: A finite set Γ of convex polytopes, the ***faces*** of Γ, in real n-space $\mathbb{R}^n$, such that two conditions are satisfied. First, if $Q \in \Gamma$ and F is a face of Q, then $F \in \Gamma$. Second, if $Q_1, Q_2 \in \Gamma$, then $Q_1 \cap Q_2$ is a face of Q_1 and Q_2 (possibly the empty face $\emptyset$). The subset $||\Gamma|| := \bigcup_{Q\in\Gamma} Q$ of $\mathbb{R}^n$, equipped with the induced topology, is called the ***underlying space*** of Γ. The ***dimension*** $d := \dim \Gamma$ of Γ is the maximum of the dimensions (of the affine hulls) of the elements in Γ. We also call Γ a ***polyhedral d-complex.*** A face of Γ of dimension 0, 1, or i is a ***vertex***, an ***edge***, or an ***i-face*** of Γ. A face that is maximal (with respect to inclusion) is called a ***facet*** of Γ. (In our applications, the facets are just the d-faces of Γ.)

Face poset: The set $P(\Gamma)$ of all faces of Γ, partially ordered by inclusion. As a partially ordered set, $P(\Gamma) \cup \{||\Gamma||\}$ is a ranked lattice.

(Geometric) simplicial complex: A polyhedral complex Γ all of whose nonempty faces are simplices. An ***abstract simplicial complex*** Δ is a family of subsets of a finite set V, the ***vertex set*** of Δ, such that $\{x\} \in \Delta$ for all $x \in V$, and such that $F \subseteq G \in \Delta$ implies $F \in \Delta$. Each abstract simplicial complex Δ is isomorphic (as a poset ordered by inclusion) to the face poset of a geometric simplicial complex Γ. Once such an isomorphism is fixed, we set $||\Delta|| := ||\Gamma||$, and the terminology introduced for Γ carries over to Δ. (One often omits the qualifications "geometric" or "abstract.")

Link: The link of a vertex x in a simplicial complex Γ is the subcomplex consisting of the faces that do not contain x of all the faces of Γ containing x.

Polyhedron: A subset P of $\mathbb{R}^n$ such that $P = ||\Gamma||$ for some polyhedral complex Γ. In general, given P, there is no canonical way to associate with it the complex Γ. However, once Γ is specified, the terminology for Γ regarding $P(\Gamma)$ is also carried over to P. (For other meanings of the term "polyhedron" see also Chapter 18.)

Subdivision: If Γ_1 and Γ_2 are polyhedral complexes, Γ_1 is a subdivision of Γ_2 if $||\Gamma_1|| = ||\Gamma_2||$ and each face of Γ_1 is a subset of a face of Γ_2. If Γ_1 is a simplicial complex, this is a ***simplicial subdivision***.

Combinatorial d-manifold: For $d = 1$, this is a simplicial 1-complex Δ such that $||\Delta||$ is a 1-sphere. Inductively, if $d \geq 2$, it is a simplicial d-complex Δ such that $||\Delta||$ is a topological d-manifold (without boundary) and each vertex link is a ***combinatorial*** $(\boldsymbol{d}-\mathbf{1})$***-sphere*** (that is, a combinatorial $(d-1)$-manifold whose underlying space is a $(d-1)$-sphere).

Polyhedral d-manifold: A polyhedral d-complex Γ admitting a simplicial subdivision that is a combinatorial d-manifold. If $d = 2$, this is simply a polyhedral 2-complex Γ for which $||\Gamma||$ is a compact 2-manifold (without boundary).

Triangulation: A triangulation (simplicial decomposition) of a topological space X is a simplicial complex Γ such that X and $||\Gamma||$ are homeomorphic.

Ball complex: A finite family $\mathcal{C}$ of topological balls (homeomorphic images of Euclidean unit balls) in a Hausdorff space, the ***underlying space*** $||\mathcal{C}||$ of $\mathcal{C}$, whose relative interiors partition $||\mathcal{C}||$ in such a way that the boundary of each ball in $\mathcal{C}$ is the union of other balls in $\mathcal{C}$. The ***dimension*** of $\mathcal{C}$ is the maximum of the dimensions of the balls in $\mathcal{C}$.

Embedding: For a ball complex $\mathcal{C}$, a continuous mapping $\gamma : ||\mathcal{C}|| \mapsto \mathbb{R}^n$ that is a homeomorphism of $||\mathcal{C}||$ onto its image. $\mathcal{C}$ is said to be ***embedded*** in $\mathbb{R}^n$.

Polyhedral embedding: For a ball complex $\mathcal{C}$, an embedding γ that maps each ball in $\mathcal{C}$ onto a convex polytope.

Immersion: For a ball complex $\mathcal{C}$, a continuous mapping $\gamma : ||\mathcal{C}|| \mapsto \mathbb{R}^n$ that is locally injective (hence the image may have self-intersections). $\mathcal{C}$ is said to be ***immersed*** in $\mathbb{R}^n$.

Polyhedral immersion: For a ball complex $\mathcal{C}$, an immersion γ that maps each ball in $\mathcal{C}$ onto a convex polytope.

Map on a surface: An embedded finite graph M (without loops or multiple edges) on a compact 2-manifold (surface) S such that two conditions are satisfied: The closures of the connected components of $S \setminus M$, the ***faces*** of M, are closed

2-cells (closed topological disks), and each vertex of M has valency at least 3. (Note that some authors use a broader definition of maps; e.g., see [CM80].)

Polyhedral map: A map M on S such that the intersection of any two distinct faces is either empty, a common vertex, or a common edge.

Figure 20.1.1 shows a polyhedral map on a surface of genus 3, known as ***Dyck's regular map***. We will further discuss this map in Sections 20.4 and 20.5.

Type: A map M on S is of type $\{p, q\}$ if all its faces are topological p-gons such that q meet at each vertex. The symbol $\{p, q\}$ is the ***Schläfli symbol*** for M.

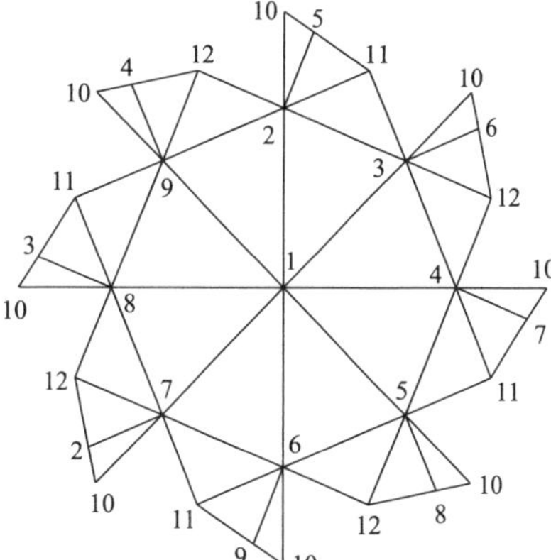

FIGURE 20.1.1
Dyck's regular map, of type $\{3, 8\}$.
Vertices with the same label are identified.

BASIC RESULTS

Simplicial complexes are important in topology, geometry, and combinatorics. Each abstract simplicial d-complex Δ with n vertices is isomorphic to the face poset of a geometric simplicial complex Γ in $\mathbb{R}^{2d+1}$ that is obtained as the image under a projection (Schlegel diagram—see Chapter 15) of a simplicial d-subcomplex in the boundary complex of the cyclic convex $(2d+2)$-polytope $C(n, 2d+2)$ with n vertices; see [Grü67] or Chapter 15 of this Handbook.

Let $\mathcal{C}$ be a ball complex and $P(\mathcal{C})$ the associated poset (i.e., $\mathcal{C}$ ordered by inclusion). Let $\Delta(P(\mathcal{C}))$ denote the ***order complex*** of $P(\mathcal{C})$; that is, the simplicial complex whose vertex set is $\mathcal{C}$ and whose k-faces are the k-chains $x_0 < x_1 < \ldots < x_k$ in $P(\mathcal{C})$. Then $||\mathcal{C}||$ and $||\Delta(P(\mathcal{C}))||$ are homeomorphic. This means that the poset $P(\mathcal{C})$ already carries complete topological information about $||\mathcal{C}||$. See [Bjö95] or [BW93], as well as Chapter 17, for further information.

Each polyhedral d-complex is a d-dimensional ball complex. The set $\mathcal{C}$ of vertices, edges, and faces of a map M on a 2-manifold S is a 2-dimensional ball complex. In particular, a map M is a polyhedral map if and only if the intersection of any two elements of $\mathcal{C}$ is empty or an element of $\mathcal{C}$. A map is usually identified with its poset of vertices, edges, and faces, ordered by inclusion. If M is a polyhedral map, then this poset is a lattice when augmented by $\emptyset$ and S as smallest and largest elements. The dual lattice (obtained by reversing the order) again gives a polyhedral map, the ***dual map***, on the same 2-manifold S. Note that in the context of polyhedral maps, the qualification "polyhedral" does not mean that it can be realized as a polyhedral complex. However, a polyhedral 2-manifold can always be regarded as a polyhedral map.

An important problem is the following:

PROBLEM 20.1.1 General Embeddability Problem

When is a given finite poset isomorphic to the face poset of some polyhedral complex in a given space $\mathbb{R}^n$*? When can a ball complex be polyhedrally embedded or polyhedrally immersed in* $\mathbb{R}^n$*?*

These questions are different from the embeddability problems that are discussed in piecewise-linear topology, because simplicial subdivisions are excluded. A complete answer is available only for the face posets of spherical maps:

THEOREM 20.1.2 Steinitz's Theorem

Each polyhedral map M *on the* 2*-sphere is isomorphic to the boundary complex of a convex* 3*-polytope. Equivalently, a finite graph is the edge graph of a convex* 3*-polytope if and only if it is planar and* 3*-connected (it has at least* 4 *vertices and the removal of any* 2 *vertices leaves a connected graph).*

Very little is known about polyhedral embeddings of orientable polyhedral maps of positive genus g. There are some general necessary combinatorial conditions for the existence of polyhedral embeddings in n-space $\mathbb{R}^n$ [BGH91]. Given a simplicial polyhedral map of genus g it is generally difficult to decide whether or not it admits a polyhedral embedding in 3-space $\mathbb{R}^3$. For each $g \geq 6$, there are examples of simplicial polyhedral maps that cannot be embedded in $\mathbb{R}^3$ [BO 00]. Each nonorientable closed surface can be immersed but not embedded in $\mathbb{R}^3$. However, the Möbius strip and therefore each nonorientable surface can be triangulated in such a way that the resulting simplicial polyhedral map cannot be polyhedrally immersed in $\mathbb{R}^3$ [Brø83]. On the other hand, each triangulation of the torus can be polyhedrally embedded in $\mathbb{R}^3$ [ABM08], and each triangulation of the real projective plane $\mathbb{R}P^2$ can be polyhedrally embedded in $\mathbb{R}^4$ [BS95].

Another important type of problem asks for topological properties of the space of all polyhedral embeddings, or of all convex d-polytopes, with a given face lattice. This is the ***realization space*** for this face lattice. Every convex 3-polytope has an open ball as its realization space. However, the realization spaces of convex 4-polytopes can be arbitrarily complicated; see the "Universality Theorem" by Richter-Gebert [Ric96] in Chapter 15 of this Handbook.

For further embeddability results in higher dimensions, as well as for a discussion of some related problems such as the polytopality problems and isotopy problems, see [Zie95, BLS$^+$93, BW93]. For a computational approach to the embeddability problem in terms of oriented matroids, see Bokowski and Sturmfels [BS89], as well as Chapter 6 of this Handbook. We shall revisit the embeddability problem in Sections 20.2 and 20.5 for interesting special classes of polyhedral maps.

Many interesting maps M on compact surfaces S have a Schläfli symbol $\{p, q\}$; for examples, see Section 20.4. These maps can then be obtained from the regular tessellation $\{p, q\}$ of the 2-sphere, the Euclidean plane, or the hyperbolic plane by making identifications. Trivially, $qf_0 = 2f_1 = pf_2$, where f_0, f_1, f_2 are the numbers of vertices, edges, and faces of M, respectively. Also, if the Euler characteristic χ of S is negative and m denotes the number of flags (incident triples consisting of a vertex, an edge, and a face) of M, then

$$\chi = f_0 - f_1 + f_2 = \frac{m}{2}\left(\frac{1}{q} - \frac{1}{2} + \frac{1}{p}\right) \leq -\frac{m}{84}, \tag{20.1.1}$$

and equality holds on the right-hand side if and only if M is of type $\{3,7\}$ or $\{7,3\}$.

20.2 EXTREMAL PROPERTIES

There is a natural interest in polyhedral maps and polyhedra defined by certain minimality properties. For relations with the famous Map Color Theorem, which gives the minimum genus of a surface on which the complete graph K_n can be embedded, see Ringel [Rin74], Barnette [Bar83], Gross and Tucker [GT87], and Mohar and Thomassen [MT01]. See also Brehm and Wills [BW93].

GLOSSARY

***f*-vector:** For a map M, the vector $f(M) = (f_0, f_1, f_2)$, where f_0, f_1, f_2 are the numbers of vertices, edges, and faces of M, respectively.

Weakly neighborly: A polyhedral map is weakly neighborly (a ***wnp map***) if any two vertices lie in a common face.

Neighborly: A map is neighborly if any two vertices are joined by an edge.

Nonconvex vertex: A vertex x of a polyhedral 2-manifold M in $\mathbb{R}^3$ is a ***convex vertex*** if at least one of the two components into which M divides a small convex neighborhood of x in $\mathbb{R}^3$ is convex; otherwise, x is nonconvex.

Tight polyhedral 2-manifold: A polyhedral 2-manifold M embedded in $\mathbb{R}^3$ such that every hyperplane strictly supporting M locally at a point supports M globally.

BASIC RESULTS

THEOREM 20.2.1

Let M be a polyhedral map of Euler characteristic χ with f-vector (f_0, f_1, f_2). Then

$$f_0 \geq \lceil (7 + \sqrt{49 - 24\chi})/2 \rceil. \tag{20.2.1}$$

Here, $\lceil t \rceil$ denotes the smallest integer greater than or equal to t. This lower bound is known as the ***Heawood bound*** and is an easy consequence of Euler's formula $f_0 - f_1 + f_2 = \chi$ $(= 2 - 2g$ if M is orientable of genus $g)$.

THEOREM 20.2.2

Except for the nonorientable 2-manifolds with $\chi = 0$ (Klein bottle) or $\chi = -1$ and the orientable 2-manifold of genus $g = 2$ $(\chi = -2)$, each 2-manifold admits a triangulation for which the lower bound (20.2.1) *is attained.*

This is closely related to the Map Color Theorem. The same lower bound (20.2.1) holds for the number f_2 of faces of M, since the dual of M is a polyhedral map with the same Euler characteristic and with f-vector (f_2, f_1, f_0).

The exact minimum for the number f_1 of edges of a polyhedral map is known for only some manifolds. Let $E_+(\chi)$ or $E_-(\chi)$, respectively, denote the smallest number f_1 such that there is a polyhedral map with f_1 edges on the orientable 2-manifold, or on the nonorientable 2-manifold, respectively, of Euler characteristic χ. The known values of $E_+(\chi)$ and $E_-(\chi)$ are listed in Table 20.2.1; undecided cases are left blank. The polyhedral maps that attain the minimal values $E_+(2)$, $E_+(-8)$, $E_-(0)$, and $E_-(-6)$ are uniquely determined.

TABLE 20.2.1 The known values of $E_+(\chi)$ and $E_-(\chi)$.

χ	2	1	0	-1	-2	-3	-4	-5	-6	-7	-8	-26
$E_+(\chi)$	6	–	18	–	27	–	33	–	38	–	40	78
$E_-(\chi)$	–	15	18	23	26	30	33	35	36	40	42	

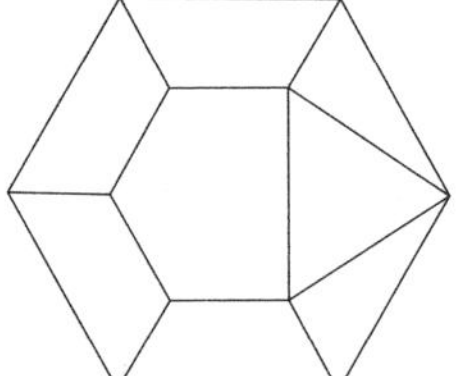

FIGURE 20.2.1
A self-dual polyhedral map on $\mathbb{RP}^2$ *with the minimum number* (15) *of edges.*

For a map on $\mathbb{RP}^2$ with 15 edges, see Figure 20.2.1. For the unique polyhedral map with 40 edges on the orientable 2-manifold of genus 5 ($\chi = -8$), see Figure 20.2.2 (and [Bre90a]). This map is weakly neighborly and self-dual, and has a cyclic group of automorphisms acting regularly on the set of vertices and on the set of faces.

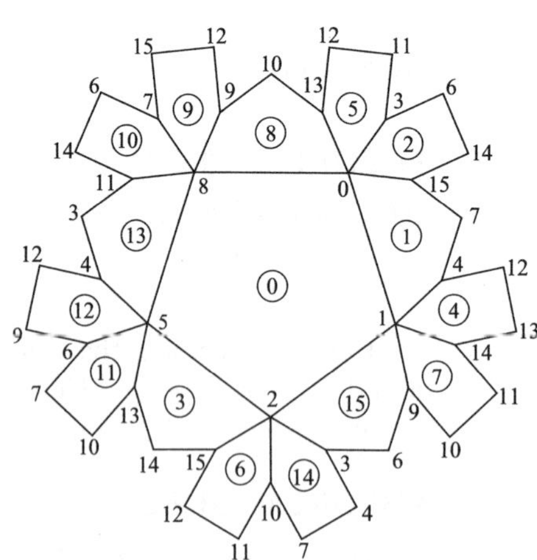

FIGURE 20.2.2
The unique polyhedral map of genus 5 *with the minimum number* (40) *of edges.*

A general bound for the number f_1 of edges is given by

THEOREM 20.2.3 [Bre90a]

$$f_1 \geq -\chi + \min\{y \in \mathbb{N} \mid y(\sqrt{2y} - 6) \geq -8\chi \text{ and } y \geq 8\},$$

where $\mathbb{N}$ *is the set of natural numbers.*

If M is a polyhedral map on a surface S, then a new polyhedral map M' on S can be obtained from M by the following operation, called ***face splitting***. A new edge xy is added across a face of M, where x and y are points on edges of M that are not contained in a common edge. The new vertices x and y of M' may be vertices of M, or one or both may be relative interior points of edges of M. The dual operation is called ***vertex splitting***. On the sphere S^2, the (boundary complex of the) tetrahedron is the only polyhedral map that is minimal with respect to face splitting. On the real projective plane $\mathbb{RP}^2$, there are exactly 16 polyhedral maps that are minimal with respect to face splitting [Bar91], and exactly 7 that are minimal with respect to both face splitting and vertex splitting. These are exactly the polyhedral maps on $\mathbb{RP}^2$ with 15 edges, which is the minimum number of edges for $\mathbb{RP}^2$. For an example, see Figure 20.2.1.

For neighborly polyhedral maps we always have equality in (20.2.1). Weakly neighborly polyhedral maps (wnp maps) are a generalization of neighborly polyhedral maps. On the 2-sphere, the only wnp maps are the (boundary complexes of the) pyramids and the triangular prism. Every other 2-manifold admits only finitely many combinatorially distinct wnp maps. Moreover,

$$\limsup_{\chi\to\infty} V_{\max}(\chi)\cdot|2\chi|^{-2/3} \le 1\,,$$

where $V_{\max}(\chi)$ denotes the maximum number of vertices of a wnp map of Euler characteristic χ; see [BA86], which also discusses further equalities and inequalities for general polyhedral maps. For several 2-manifolds, all wnp maps have been determined. For example, on the torus there are exactly five wnp maps, and three of them are geometrically realizable as polyhedra in $\mathbb{R}^3$.

In some instances, the combinatorial lower bound (20.2.1) can also be attained geometrically by (necessarily orientable) polyhedra in $\mathbb{R}^3$. Trivially, the tetrahedron minimizes f_0 $(= 4)$ for $g = 0$. For $g = 1$ there is a polyhedron with $f_0 = 7$ known as the ***Császár torus***; see Figure 20.2.3. A pair of congruent copies of the torus shown in Figure 20.2.3b can be linked (if the coordinates orthogonal to the plane of projection are sufficiently small). Polyhedra that have the minimum number of vertices have also been found for $g = 2$ (the exceptional case), 3, 4, or 5, with 10, 10, 11, or 12 vertices, respectively. For $g \le 4$ each triangulation with the minimum number of vertices admits a realization as a polyhedron in $\mathbb{R}^3$, but for $g = 5$ there are also minimal triangulations which do not admit such a realization. For $g = 6$, none of the 59 combinatorially different triangulations with 12 vertices admits a geometric realization as a polyhedron in $\mathbb{R}^3$ [Sch10].

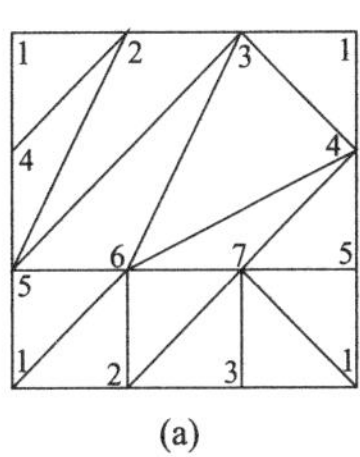

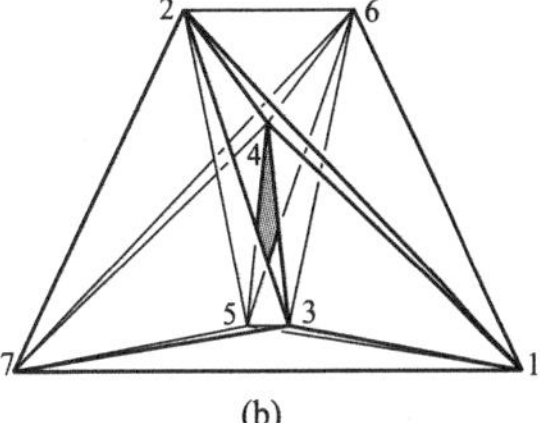

FIGURE 20.2.3
(a) *The unique 7-vertex triangulation of the torus and* (b) *a symmetric realization as a polyhedron.*

The minimum number of vertices for polyhedral maps that admit polyhedral immersions in $\mathbb{R}^3$ is 9 for the real projective plane $\mathbb{R}\mathrm{P}^2$ [Bre90b], the Klein bottle, and the surface of Euler characteristic $\chi = -1$. The minimum number is 10 for the surface with $\chi = -3$ [Leo09]. The lower bound for $\mathbb{R}\mathrm{P}^2$ follows directly from the fact that each immersion of $\mathbb{R}\mathrm{P}^2$ in $\mathbb{R}^3$ has (generically) a triple point (like the classical Boy surface).

There are also some surprising results for higher genus. For example, for each $q \geq 3$ there exists a polyhedral map M_q of type $\{4, q\}$ with $f_0 = 2^q$ and $g = 2^{q-3}(q-4)+1$ such that M_q and its dual have polyhedral embeddings in $\mathbb{R}^3$ [MSW83]. These polyhedra are combinatorially regular in the sense of Section 20.5. Note that $f_0 = O(g/\log g)$. Thus for sufficiently large genus, M_q has more handles than vertices, and its dual has more handles than faces.

Every polyhedral 2-manifold in $\mathbb{R}^3$ of genus $g \geq 1$ contains at least 5 nonconvex vertices. This bound is attained for each $g \geq 1$. For tight polyhedral 2-manifolds, the lower bound for the number of nonconvex vertices is larger and depends on g. For a survey on tight polyhedral submanifolds see [Küh95].

20.3 EBERHARD'S THEOREM AND RELATED RESULTS

Eberhard's theorem is one of the oldest nontrivial results about convex polyhedra. The standard reference is Grünbaum [Grü67, Grü70]. For recent developments see also Jendrol [Jen93].

GLOSSARY

p-sequence: For a polyhedral map M, the sequence $p(M) = (p_k(M))_{k\geq 3}$, where $p_k = p_k(M)$ is the number of k-gonal faces of M.

v-sequence: For a polyhedral map M, the sequence $v(M) = (v_k(M))_{k\geq 3}$, where $v_k = v_k(M)$ is the number of vertices of M of degree k.

EBERHARD-TYPE RESULTS

Significant results are known for the general problem of determining what kind of polygons, and how many of each kind, may be combined to form the faces of a polyhedral map M on an orientable surface of genus g. These refine results (for $d = 3$) about the boundary complex and the number of i-dimensional faces ($i = 0, \ldots, d-1$) of a convex d-polytope [Grü67, Zie95]; see Chapter 17.

If M is a polyhedral map of genus g with f-vector (f_0, f_1, f_2), then

$$\sum_{k\geq 3} p_k = f_2, \quad \sum_{k\geq 3} v_k = f_0, \quad \sum_{k\geq 3} k p_k = 2f_1 = \sum_{k\geq 3} k v_k \,. \tag{20.3.1}$$

Further, Euler's formula $f_0 - f_1 + f_2 = 2(1-g)$ implies the equations

$$\sum_{k\geq 3}(6-k)p_k + 2\sum_{k\geq 3}(3-k)v_k = 12(1-g) \tag{20.3.2}$$

and

$$\sum_{k\geq 3}(4-k)(p_k+v_k) = 8(1-g)\,. \tag{20.3.3}$$

These equations contain no information about p_6, v_3 and p_4, v_4, respectively.

Eberhard-type results deal with the problem of determining which pairs $(p_k)_{k\geq 3}$ and $(v_k)_{k\geq 3}$ of sequences of nonnegative integers can occur as p-sequences $p(M)$ and v-sequences $v(M)$ of polyhedral maps M of a given genus g. The above equations yield simple necessary conditions. As a consequence of Steinitz's theorem (Section 20.1), the problem for $g = 0$ is equivalent to a similar such problem for convex 3-polytopes [Grü67, Grü70]. The classical theorem of Eberhard says the following:

THEOREM 20.3.1 Eberhard's Theorem

For each sequence $(p_k \mid 3 \leq k \neq 6)$ of nonnegative integers satisfying

$$\sum_{k\geq 3}(6-k)p_k = 12,$$

there exists a value of p_6 such that the sequence $(p_k)_{k\geq 3}$ is the p-sequence of a spherical polyhedral map all of whose vertices have degree 3, or, equivalently, of a convex 3-polytope that is simple (has vertices only of degree 3).

This is the case $g = 0$ and $v_3 = f_0$, $v_k = 0$ $(k \geq 4)$.

More general results have been established [Jen93]. Given two sequences $p' = (p_k \mid 3 \leq k \neq 6)$ and $v' = (v_k \mid k > 3)$ of nonnegative integers such that the equation (20.3.2) is satisfied for a given genus g, let $E(p', v'; g)$ denote the set of integers $p_6 \geq 0$ such that $(p_k)_{k\geq 3}$ and $(v_k)_{k\geq 3}$, with $v_3 := (\sum_{k\geq 3} kp_k - \sum_{k\geq 4} kv_k)/3$ determined by (20.3.1), are the p-sequences and v-sequences, respectively, of a polyhedral map of genus g. For all but two admissible triples (p', v', g), the set $E(p', v'; g)$ is known up to a finite number of elements. For example, for $g = 0$, the set $E(p', v'; 0)$ is nonempty if and only if $\sum_{k\not\equiv 0\,(\text{mod}\,3)} v_k \neq 1$ or $p_k \neq 0$ for at least one odd k. In particular, for each such nonempty set, there exists a constant c depending on (p', v') such that $E(p', v'; 0) = \{j \mid c \leq j\}$, $\{j \mid c \leq j \equiv 0\,(\text{mod}\,2)\}$, or $\{j \mid c \leq j \equiv 1\,(\text{mod}\,2)\}$. Similarly, for each triple with $g \geq 2$, there is a constant c depending on (p', v', g) such that $E(p', v'; g) = \{j \mid c \leq j\}$. There are analogous results for sequences $(p_k \mid 3 \leq k \neq 4)$ and $(v_k \mid 3 \leq k \neq 4)$ that satisfy the equation (20.3.3) or other related equations.

For $g = 1$ there is also a more geometric Eberhard-type result available, which requires the polyhedral map M to be polyhedrally embedded in $\mathbb{R}^3$:

THEOREM 20.3.2 [Gri83]

Let s, p_k $(k \geq 3, k \neq 6)$ be nonnegative integers. Then there exists a toroidal polyhedral 2-manifold M in $\mathbb{R}^3$ with $p_k(M) = p_k$ $(k \neq 6)$ and $\sum_{k\geq 3}(k-3)v_k(M) = s$ if and only if $\sum_{k\geq 3}(6-k)p_k = 2s$ and $s \geq 6$.

Also, for toroidal polyhedral 2-manifolds in $\mathbb{R}^3$ (as well as for convex 3-polytopes), the exact range of possible f-vectors is known [Grü67, BW93].

THEOREM 20.3.3

A polyhedral embedding in $\mathbb{R}^3$ of some torus with f-vector (f_0, f_1, f_2) exists if and *only if* $f_0 - f_1 + f_2 = 0$, $f_2(11 - f_2)/2 \leq f_0 \leq 2f_2$, $f_0(11 - f_0)/2 \leq f_2 \leq 2f_0$, *and* $2f_1 - 3f_0 \geq 6$.

For generalizations of Eberhard's theorem to tilings of the Euclidean plane, see also [GS87].

20.4 REGULAR MAPS

Regular maps are topological analogues of the ordinary regular polyhedra and star-polyhedra on surfaces. Historically they became important in the context of transformations of algebraic equations and representations of algebraic curves in homogeneous complex variables. There is a large body of literature on regular maps and their groups. The classical text is Coxeter and Moser [CM80]. For more recent texts see McMullen and Schulte [MS02], and Conder, Jones, Siran, and Tucker [CJST].

GLOSSARY

(Combinatorial) automorphism: An incidence-preserving bijection (of the set of vertices, edges, and faces) of a map M on a surface S to itself. The (***combinatorial automorphism***) ***group*** $A(M)$ of M is the group of all such bijections. It can be "realized" by a group of homeomorphisms of S.

Regular map: A map M on S whose group $A(M)$ is transitive on the flags (incident triples consisting of a vertex, an edge, and a face) of M.

Chiral map: A map M on S whose group $A(M)$ has two orbits on the flags such that any two adjacent flags are in distinct orbits. Here two flags are ***adjacent*** if they differ in precisely one element: a vertex, an edge, or a face. (For a chiral map the underlying surface S must be orientable.)

GENERAL RESULTS

Each regular map M is of type $\{p, q\}$ for some finite p and q. Its group $A(M)$ is transitive on the vertices, the edges, and the faces of M. In general, the Schläfli symbol $\{p, q\}$ does not determine M uniquely. The group $A(M)$ is generated by involutions ρ_0, ρ_1, ρ_2 such that the ***standard relations***

$$\rho_0{}^2 = \rho_1{}^2 = \rho_2{}^2 = (\rho_0\rho_1)^p = (\rho_1\rho_2)^q = (\rho_0\rho_2)^2 = 1$$

hold, but in general there are also further independent relations. Any triangle in the "barycentric subdivision" (order complex) of M is a fundamental region for $A(M)$ on the underlying surface S; see Section 20.1. For any fixed such triangle, we can take for ρ_i the "combinatorial reflection" in its side opposite to the vertex that corresponds to an i-dimensional element of M. The set of standard relations gives a presentation for the symmetry group of the regular tessellation $\{p, q\}$ on the 2-sphere, in the Euclidean plane, or in the hyperbolic plane, whichever is the universal covering of M. See Figure 20.5.1 (a) for a conformal (hyperbolic) drawing of the Dyck map (shown also in Figure 20.1.1) with a fundamental region shaded. The identifications on the boundary of the drawing are indicated by letters.

The regular maps on orientable surfaces of genus $g \leq 301$ and on non-orientable surfaces of genus $h \leq 602$ have been enumerated by computer (see [Con12]). A similar enumeration is also known for chiral maps on orientable surfaces of genus $g \leq 301$. Up to isomorphism, if $g = 0$, there are just the Platonic solids (or regular spherical tessellations) $\{3,3\}$, $\{3,4\}$, $\{4,3\}$, $\{3,5\}$, and $\{5,3\}$. When $g = 1$ there are three infinite families of torus maps of type $\{3,6\}$, $\{6,3\}$, and $\{4,4\}$, each a quotient of the corresponding Euclidean universal covering tessellations $\{3,6\}$, $\{6,3\}$, and $\{4,4\}$, respectively. For $g \geq 2$, the universal covering tessellation $\{p,q\}$ is hyperbolic and there are only finitely many regular maps on a surface of genus g. The latter follows from the ***Hurwitz formula*** $|A(M)| \leq 84\,|\chi|$ (or from the inequality 20.1.1), where χ is the Euler characteristic of S. Each regular map on a nonorientable surface is doubly covered by a regular map of the same type on an orientable surface, and this covering map is unique [Wil78].

Generally speaking, given M, the topology of S is reflected in the relations that have to be added to the standard relations to obtain a presentation for $A(M)$. Conversely, many interesting regular maps can be constructed by adding certain kinds of extra relations for the group. Two examples are the regular maps $\{p,q\}_r$ and $\{p,q|r\}$ obtained by adding the extra relations $(\rho_0\rho_1\rho_2)^r = 1$ or $(\rho_0\rho_1\rho_2\rho_1)^r = 1$, respectively. Often these are "infinite maps" on noncompact surfaces, but there are also many (finite) maps on compact surfaces. The Dyck map $\{3,8\}_6$ and the famous ***Klein map*** $\{3,7\}_8$ (with group $PGL(2,7)$) are both of genus 3 and of the first kind, while the traditional regular skew polyhedra in Euclidean 3-space or 4-space are of the second kind. For more details and further interesting classes of regular maps, see [CM80, MS02, CJST] and Chapter 18 of this Handbook. In Section 20.5 we shall discuss polyhedral embeddings of regular maps in ordinary 3-space.

The rotation subgroup (orientation preserving subgroup) of the group of an orientable regular map (of type $\{3,7\}$ or $\{7,3\}$) that achieves equality in the Hurwitz formula is also called a ***Hurwitz group***. The Klein map is the regular map of smallest genus whose rotation subgroup is a Hurwitz group [Con90].

20.5 SYMMETRIC POLYHEDRA

Traditionally, much of the appeal of polyhedral 2-manifolds comes from their combinatorial or geometric symmetry properties. For surveys on symmetric polyhedra in $\mathbb{R}^3$ see Schulte and Wills [SW91, SW12], Bokowski and Wills [BW88], and Brehm and Wills [BW93].

GLOSSARY

Combinatorially regular: A polyhedral 2-manifold (or polyhedron) P is combinatorially regular if its combinatorial automorphism group $A(P)$ is flag-transitive (or, equivalently, if the underlying polyhedral map is a regular map).

Equivelar: A polyhedral 2-manifold (or polyhedron) P is equivelar of type $\{p,q\}$ if all its 2-faces are convex p-gons and all its vertices are q-valent.

GENERAL RESULTS

See Section 20.4 for results about regular maps. Up to isomorphism, the Platonic solids are the only combinatorially regular polyhedra of genus 0. For the torus, each regular map that is a polyhedral map also admits an embedding in $\mathbb{R}^3$ as a combinatorially regular polyhedron. Much less is known for maps of genus $g \geq 2$. Two infinite sequences of combinatorially regular polyhedra have been discovered, one consisting of polyhedra of type $\{4, q\}$ $(q \geq 3)$ and the other of their duals of type $\{q, 4\}$ (see [MSW83, RZ11]). These are polyhedral embeddings of the maps M_q and their duals mentioned in Section 20.2. Several famous regular maps have also been realized as polyhedra, including Klein's $\{3,7\}_8$, Dyck's $\{3,8\}_6$, the map $\{3,10\}_6$ (a relative of Dyck's map), and Coxeter's $\{4,6|3\}$, $\{6,4|3\}$, $\{4,8|3\}$, and $\{8,4|3\}$ [SW85, SW91, BS89, BL16]. It is conjectured that there are just eight regular maps in the genus range $2 \leq g \leq 6$ that can be realized as combinatorially regular polyhedra in $\mathbb{R}^3$ (see [SW12]). However, a complete classification of the regular maps of high genus that admit realizations as combinatorially regular polyhedra in $\mathbb{R}^3$ does not seem to be within reach at present. See Figure 20.5.1 for an illustration of a polyhedral realization of Dyck's regular map $\{3,8\}_6$ shown in Figure 20.1.1. Figure 20.5.1(a) shows a conformal drawing of the Dyck map, with a fundamental region shaded, while (b) shows a maximally symmetric polyhedral realization. For the Klein map $\{7,3\}_8$, the dual of $\{3,7\}_8$, there exists also a non-self-intersecting polyhedral realization with non-convex heptagonal faces [McC09].

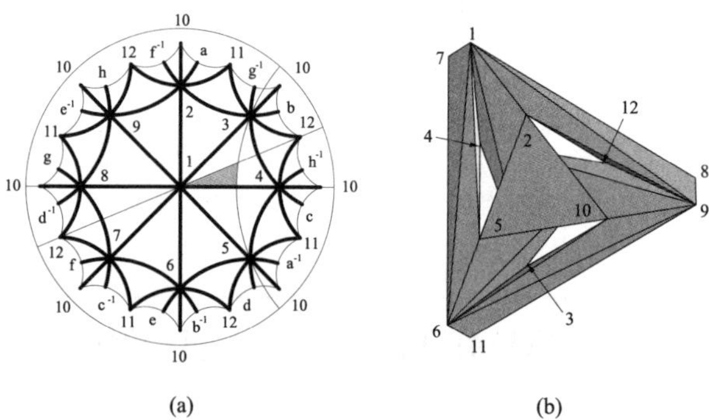

FIGURE 20.5.1
Dyck's regular map:
(a) *a conformal drawing, with fundamental region shaded;*
(b) *a symmetric polyhedral realization.*

For a more general concept of polyhedra in $\mathbb{R}^3$ or higher-dimensional spaces, as well as an enumeration of the corresponding regular polyhedra, see Chapter 18 of this Handbook. The latter also contains a depiction of the polyhedral realization of $\{4,8|3\}$.

Equivelarity is a local regularity condition. Each combinatorially regular polyhedron in $\mathbb{R}^3$ is equivelar. However, there are many other equivelar polyhedra. For sufficiently large genus g, for example, there are equivelar polyhedra for each of the types $\{3, q\}$ with $q = 7, 8, 9$; $\{4, q\}$ with $q = 5, 6$; and $\{q, 4\}$ with $q = 5, 6$ [BW93].

The symmetry group of a polyhedron is generally much smaller than the combinatorial automorphism group of the underlying polyhedral map. In particular, the five Platonic solids are the only polyhedra in $\mathbb{R}^3$ with a flag-transitive symmetry group. Note that even for higher genus (namely, for $g = 1, 3, 5, 7, 11$, and 19) there exist polyhedra with vertex-transitive symmetry groups. Such a polyhedral torus is shown in Figure 20.5.2.

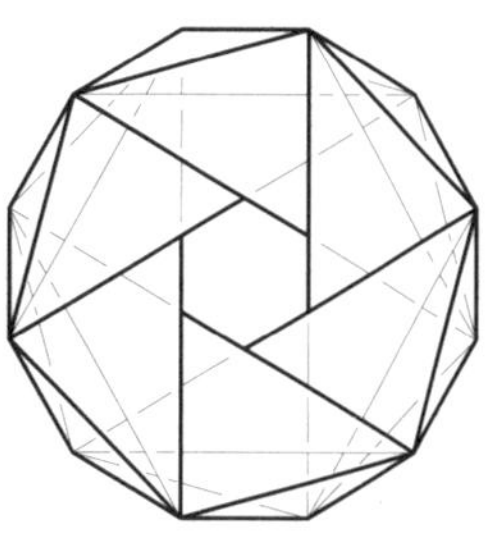

FIGURE 20.5.2
A vertex-transitive polyhedral torus.

Finally, if we relax the requirement that a polyhedron be free of self-intersections and allow more general "polyhedral realizations" of maps (for instance, polyhedral immersions), then there is much more flexibility in the construction of "polyhedra" with high symmetry properties. The most famous examples are the Kepler-Poinsot star-polyhedra, but there are also many others. For more details see [SW91, BW88, BW93, MS02] and Chapter 18 of this Handbook.

20.6 SOURCES AND RELATED MATERIAL

SURVEYS

[Bar83]: A text about colorings of maps and polyhedra.

[Bjö95]: A survey on topological methods in combinatorics.

[BLS$^+$93]: A monograph on oriented matroids.

[BS89]: A text about computational aspects of geometric realizability.

[BW93]: A survey on polyhedral manifolds in 2 and higher dimensions.

[Con90]: A survey on Hurwitz groups.

[CJST]: A monograph on regular maps on surfaces.

[Cox73]: A monograph on regular polytopes, regular tessellations, and reflection groups.

[CM80]: A monograph on discrete groups and their presentations.

[GT87]: A text about topological graph theory, in particular graph embeddings in surfaces.

[Grü67]: A monograph on convex polytopes.

[Grü70]: A survey on convex polytopes complementing the exposition in [Grü67].

[GS87]: A monograph on plane tilings and patterns.

[Küh95]: A survey on tight polyhedral manifolds.

[Moi77]: A text about geometric topology in low dimensions.

[MT01]: A text about embeddings of graphs in surfaces.

[MS02]: A monograph on abstract regular polytopes and their groups.

[Rin74]: A text about maps on surfaces and the Map Color Theorem.

[SW91]: A survey on combinatorially regular polyhedra in 3-space.

[Zie95]: A graduate textbook on convex polytopes.

RELATED CHAPTERS

Chapter 3: Tilings
Chapter 6: Oriented matroids
Chapter 15: Basic properties of convex polytopes
Chapter 16: Subdivisions and triangulations of polytopes
Chapter 17: Face numbers of polytopes and complexes
Chapter 18: Symmetry of polytopes and polyhedra

REFERENCES

[ABM08] D. Archdeacon, C.P. Bonnington, and J.A. Ellis-Monaghan. How to exhibit toroidal maps in space. *Discrete Comput. Geom.* 38:573–594, 2008.

[BA86] U. Brehm and A. Altshuler. On weakly neighborly polyhedral maps of arbitrary genus. *Israel J. Math.*, 53:137–157, 1986.

[Bar83] D.W. Barnette. *Map Coloring, Polyhedra, and the Four-Color Problem.* MAA, Washington, 1983.

[Bar91] D.W. Barnette. The minimal projective plane polyhedral maps. In P. Gritzmann and B. Sturmfels, editors, *Applied Geometry and Discrete Mathematics—The Victor Klee Festschrift*, vol. 4 of*DIMACS Ser. Discrete Math. Theoret. Comp. Sci.*, pages 63–70, AMS, Providence, 1991.

[BGH91] D.W. Barnette, P. Gritzmann, and R. Höhne On valences of polyhedra. *J. Combin. Theory Ser. A*, 58:279–300, 1991.

[Bjö95] A. Björner. Topological methods. In R.L. Graham, M. Grötschel, and L. Lovász, editors, *Handbook of Combinatorics*, pages 1819–1872, Elsevier, Amsterdam, 1995.

[BL16] U. Brehm and U. Leopold. Symmetric realizations of the regular map $\{3, 10\}_6$ of genus 6 with 15 vertices as a polyhedron. In preparation, 2016.

[BLS$^+$93] A. Björner, M. Las Vergnas, B. Sturmfels, N. White, and G.M. Ziegler. *Oriented Matroids.* Vol. 46 of *Encyclopedia Math. Appl.*, Cambridge University Press, 1993.

[BO 00] J. Bokowski and A.G. de Oliveira. On the generation of oriented matroids. *Discrete Comput. Geom.* 24:197–208, 2000.

[Bre83] U. Brehm. A nonpolyhedral triangulated Möbius strip. *Proc. Amer. Math. Soc.* 89:519–522, 1983.

[Bre90a] U. Brehm. Polyhedral maps with few edges. In R. Bodendiek and R. Henn, editors, *Topics in Combinatorics and Graph Theory*, pages 153–162, Physica-Verlag, Heidelberg, 1990.

[Bre90b] U. Brehm. How to build minimal polyhedral models of the Boy surface. *Math. Intelligencer* 12:51–56, 1990.

[BS89] J. Bokowski and B. Sturmfels. *Computational Synthetic Geometry.* Vol. 1355 of *Lecture Notes in Math.*, Springer-Verlag, Berlin, 1989.

[BS95] U. Brehm and G. Schild. Realizability of the torus and the projective plane in $\mathbb{R}^4$. *Israel J. Math.*, 91:249–251, 1995.

[BW88] J. Bokowki and J.M. Wills. Regular polyhedra with hidden symmetries. *Math. Intelligencer*, 10:27–32, 1988.

[BW93] U. Brehm and J.M. Wills. Polyhedral manifolds. In P.M. Gruber and J.M. Wills, editors, *Handbook of Convex Geometry,* Vol. A, pages 535–554, North-Holland, Amsterdam, 1993.

[CJST] M.D.E. Conder, G.A. Jones, J. Širáň, and T.W. Tucker. *Regular Maps.* In preparation.

[CM80] H.S.M. Coxeter and W.O.J. Moser. *Generators and Relations for Discrete Groups*, 4th edition. Springer-Verlag, Berlin, 1980.

[Con90] M. Conder. Hurwitz groups: A brief survey. *Bull. Amer. Math. Soc.*, 23:359–370, 1990.

[Con12] M.D.E. Conder. Lists of regular maps, hypermaps and polytopes, trivalent symmetric graphs, and surface actions. `www.math.auckland.ac.nz/~conder`.

[Cox73] H.S.M. Coxeter. *Regular Polytopes*, 3rd edition. Dover, New York, 1973.

[Gri83] P. Gritzmann. The toroidal analogue of Eberhard's theorem. *Mathematika*, 30:274–290, 1983.

[Grü67] B. Grünbaum. *Convex Polytopes.* Interscience, London, 1967. Revised edition (V. Kaibel, V. Klee, and G.M. Ziegler, editors), vol. 221 of *Grad. Texts in Math.*, Springer, New York, 2003.

[Grü70] B. Grünbaum. Polytopes, graphs, and complexes. *Bull. Amer. Math. Soc.*, 76:1131–1201, 1970.

[GS87] B. Grünbaum and G.C. Shephard. *Tilings and Patterns.* Freeman, New York, 1987.

[GT87] J.L. Gross and T.W. Tucker. *Topological Graph Theory.* Wiley, New York, 1987.

[Jen93] S. Jendrol. On face-vectors and vertex-vectors of polyhedral maps on orientable 2-manifolds. *Math. Slovaca*, 43:393–416, 1993.

[Küh95] W. Kühnel. *Tight Polyhedral Submanifolds and Tight Triangulations.* Vol. 1612 of *Lecture Notes in Math.*, Springer-Verlag, New York, 1995.

[Leo09] U. Leopold. *Polyhedral Embeddings and Immersions of Triangulated 2-Manifolds.* Diplomarbeit (Master's Thesis), Dresden, 2009.

[McC09] D. McCooey. A non-self-intersecting polyhedral realization of the all-heptagon Klein map. *Symmetry Cult. Sci.* 20:247–268, 2009.

[Moi77] E.E. Moise. *Geometric Topology in Dimensions 2 and 3.* Vol. 47 of *Graduate Texts in Math.*, Springer-Verlag, New York, 1977.

[MS02] P. McMullen and E. Schulte. *Abstract Regular Polytopes.* Cambridge University Press, 2002.

[MSW83] P. McMullen, C. Schulz, and J.M. Wills. Polyhedral manifolds in E^3 with unusually large genus. *Israel J. Math.*, 46:127–144, 1983.

[MT01] B. Mohar and C. Thomassen. *Graphs on Surfaces.* Johns Hopkins University Press, Baltimore, 2001.

[Ric96] J. Richter-Gebert. *Realization Spaces of Polytopes.* Vol. 1643 of *Lecture Notes in Math.*, Springer-Verlag, Berlin, 1996.

[Rin74] G. Ringel. *Map Color Theorem.* Springer-Verlag, Berlin, 1974.

[RZ11] T. Rörig and G.M. Ziegler. Polyhedral surfaces in wedge products. *Geom. Dedicata* 151:155–173, 2011.

[Sch10] L. Schewe. Nonrealizable minimal vertex triangulations of surfaces: showing nonrealizability using oriented matroids and satisfiability solvers. *Discrete Comput. Geom.* 43:289–302, 2010.

[SW12] E. Schulte and J.M. Wills. Convex-faced combinatorially regular polyhedra of small genus. *Symmetry* 4:1–14, 2012.

[SW85] E. Schulte and J.M. Wills. A polyhedral realization of Felix Klein's map $\{3,7\}_8$ on a Riemann surface of genus 3. *J. London Math. Soc.*, 32:539–547, 1985.

[SW91] E. Schulte and J.M. Wills. Combinatorially regular polyhedra in three-space. In K.H. Hofmann and R. Wille, editors, *Symmetry of Discrete Mathematical Structures and Their Symmetry Groups*, pages 49–88. Heldermann Verlag, Berlin, 1991.

[Wil78] S.E. Wilson. Non-orientable regular maps. *Ars Combin.*, 5:213–218, 1978.

[Zie95] G.M. Ziegler. *Lectures on Polytopes.* Vol. 152 of *Graduate Texts in Math.*, Springer-Verlag, New York, 1995.

Part III

COMBINATORIAL AND COMPUTATIONAL TOPOLOGY

21 TOPOLOGICAL METHODS IN DISCRETE GEOMETRY

Rade T. Živaljević

INTRODUCTION

A problem is solved or some other goal achieved by "topological methods" if in our arguments we appeal to the "form," the "shape," the "global" rather than "local" structure of the object or configuration space associated with the phenomenon we are interested in. This configuration space is typically a manifold or a simplicial complex. The global properties of the configuration space are usually expressed in terms of its homology and homotopy groups, which capture the idea of the higher (dis)connectivity of a geometric object and to some extent provide "an analysis properly geometric or linear that expresses location directly as algebra expresses magnitude."[1]

Thesis: *Any global effect that depends on the object as a whole and that cannot be localized is of homological nature, and should be amenable to topological methods.*

HOW IS TOPOLOGY APPLIED IN DISCRETE GEOMETRIC PROBLEMS?

In this chapter we put some emphasis on the role of *equivariant* topological methods in solving combinatorial or discrete geometric problems that have proven to be of relevance for computational geometry and topological combinatorics and with some impact on computational mathematics in general. The versatile *configuration space/test map* scheme (CS/TM) was developed in numerous research papers over the years and formally codified in [Živ98]. Its essential features are summarized as follows:

CS/TM-1: The problem is rephrased in topological terms.

The problem should give us a clue how to define a "natural" *configuration space* X and how to rephrase the question in terms of zeros or coincidences of the associated *test maps*. Alternatively the problem may be divided into several subproblems, in which case one is often led to the question of when the solution subsets of X corresponding to the various subproblems have *nonempty intersection.*

CS/TM-2: Standard topological techniques are used in the solution of the rephrased problem.

The topological technique that is most frequently and consistently used in problems of discrete geometry is based on various forms of *generalized Borsuk-Ulam theorems.* However many other tools (Lusternik-Schnirelmann category, cup product, cup-length, intersection homology, etc.) have also found important applications.

[1] A dream of G.W. Leibniz expressed in a letter to C. Huygens dated 1697; see [Bre93, Chap. 7].

21.1 THE CONFIGURATION SPACE/TEST MAP PARADIGM

GLOSSARY

Configuration space/test map scheme (CS/TM): A very useful and general scheme for proving combinatorial or geometric facts. The problem is reduced to the question of showing that there does not exist a G-equivariant map $f : X \to V \setminus Z$ (Section 21.5) where X is the configuration space, V the test space, and Z the test subspace associated with the problem, while G is a naturally arising group of symmetries.

Configuration space: In general, any topological space X that parameterizes a class of configurations of geometric objects (e.g., arrangements of points, lines, fans, flags, etc.) or combinatorial structures (trees, graphs, partitions, etc.). Given a problem $\mathcal{P}$, an associated configuration or ***candidate space*** $X_{\mathcal{P}}$ collects all geometric configurations that are (reasonable) candidates for a solution of $\mathcal{P}$.

Test map and ***test space:*** A map $t : X_{\mathcal{P}} \to V$ from the configuration space $X_{\mathcal{P}}$ into the so-called test space V that tests the validity of a candidate $p \in X_{\mathcal{P}}$ as a solution of $\mathcal{P}$. This is achieved by the introduction of a ***test subspace*** $Z \subset V$, where $p \in X$ is a solution to the problem if and only if $t(p) \in Z$. Usually $V \cong \mathbb{R}^d$ while Z is just the origin $\{0\} \subset V$ or more generally a linear subspace arrangement in V.

Equivariant map: The final ingredient in the CS/TM-scheme is a group G of symmetries that acts on both the configuration space $X_{\mathcal{P}}$ and the test space V (keeping the test subspace Z invariant). The test map t is always assumed G-equivariant, i.e., $t(g \cdot x) = g \cdot t(x)$ for each $g \in G$ and $x \in X_{\mathcal{P}}$. Some of the methods and tools of equivariant topology are outlined in Section 21.5.

EXAMPLE 21.1.1

Suppose that ρ is a metric on $\mathbb{R}^2$ that induces the same topology as the usual Euclidean metric. Let $\Gamma \subset \mathbb{R}^2$ be a compact subspace, for example Γ can be a finite union of arcs. The problem is to find equilateral triangles in Γ, i.e. the triples (a, b, c) of distinct points in Γ such that $\rho(a, b) = \rho(b, c) = \rho(c, a)$.

This is our first example that illustrates the CS/TM-scheme. The configuration space X should collect all candidates for the solution, so a first, "naive" choice is the space of all (ordered) triples $(x, y, z) \in \Gamma$. Of course we can immediately rule out some obvious nonsolutions, e.g., degenerate triangles (x, y, z) such that at least one of numbers $\rho(x, y), \rho(y, z), \rho(z, x)$ is zero. (This illustrates the fact that in general there may be several possible choices for a configuration space associated to the initial problem.) Our choice is $X := \Gamma^3 \setminus \Delta$ where $\Delta := \{(x, x, x) \mid x \in \Gamma\}$. A "triangle" $(x, y, z) \in X$ is ρ-equilateral if and only if $(\rho(x, y), \rho(y, z), \rho(z, x)) \in Z$, where $Z := \{(u, u, u) \in \mathbb{R}^3 \mid u \in \mathbb{R}\}$. Hence a test map $t : X \to \mathbb{R}^3$ is defined by $t(x, y, z) = (\rho(x, y), \rho(y, z), \rho(z, x))$, the test space is $V = \mathbb{R}^3$, and $Z \subset \mathbb{R}^3$ is the associated test subspace. A triangle $\{x, y, z\}$, viewed as a set of vertices, is in general labelled by six different triples in the configuration space X. This redundancy is a motivation for introducing the group of symmetries $G = S_3$, which acts on both the configuration space X and the test space V. The test map t is clearly S_3-equivariant. Summarizing, we observe that the set of all equilateral triangles $\{x, y, z\}$ in Γ is in one-to-one correspondence with the set $t^{-1}(Z)/S_3$ of all S_3-orbits of all elements in $t^{-1}(Z)$. If we are interested only in the existence of (non-degenerate) equilateral triangles in Γ, it is sufficient to show that $f^{-1}(Z) \neq \emptyset$ for each S_3-equivariant map $f : X \to \mathbb{R}^3$. This is a topological problem which may be reduced to the non-existence of a S_3-equivariant map $g : X \to S^1$, where $S^1 \subset Z^{\perp}$ is the (S_3-invariant) unit circle in the orthogonal complement to $Z \subset \mathbb{R}^3$.

Here is another example of how topology comes into play and proves useful in geometric and combinatorial problems. The *configuration space* associated to the next problem is a 2-dimensional torus $T^2 \cong S^1 \times S^1$. This time, however, the test map is not explicitly given. Instead, the problem is reduced to counting intersection points of two "test subspaces" in T^2.

EXAMPLE 21.1.2 *A watch with two equal hands*

A watch was manufactured with a defect so that both hands (minute and hour) are identical. Otherwise the watch works well and the question is to determine the number of ambiguous positions, i.e., the positions for which it is not possible to determine the exact time.

Each position of a hand is determined by an angle $\omega \in [0, 2\pi]$, so the configuration space of all positions of a hand is homeomorphic to the unit circle S^1. Two independent hands have the 2-dimensional torus $T^2 \cong S^1 \times S^1$ as their configuration space. If θ corresponds to the minute hand and ω is the coordinate of the hour hand, then the fact that the first hand is twelve times faster is recorded by the equation $\theta = 12\,\omega$. If the hands change places then the corresponding curve Γ_2 has equation $\omega = 12\,\theta$. The ambiguous positions are exactly the intersection points of these two curves (except those that belong to the diagonal $\Delta := \{(\theta, \omega) \mid \theta = \omega\}$, when it is still possible to tell the exact time without knowing which hand is for hours and which for minutes). The reader can now easily find the number of these intersection points and compute that there are 143 of them in the intersection $\Gamma_1 \cap \Gamma_2$, and 11 in the intersection $\Gamma_1 \cap \Gamma_2 \cap \Delta$, which shows that there are all together 132 ambiguous positions.

REMARK 21.1.3

Let us note that the "watch with equal hands" problem reduces to counting points or 0-dimensional manifolds in the intersection of two circles, viewed as 1-dimensional submanifolds of the 2-dimensional manifold T^2. More generally, one may be interested in how many points there are in the intersection of two or more submanifolds of a higher-dimensional ambient manifold. Topology gives us a versatile tool for computing this and much more, in terms of the so-called *intersection product* $\alpha \frown \beta$ *of homology classes* α *and* β in a manifold M. This intersection product is, via Poincaré duality, equivalent to the "cup" product, and has the usual properties [Bre93, Mun84]. In our Example 21.1.2, keeping in mind that $a \frown b = -\,b \frown a$ for all 1-dimensional classes, and in particular that $a \frown a = 0$ if dim $(a) = 1$, we have $[\Gamma_1] \frown [\Gamma_2] = ([\theta] + 12[\omega]) \frown ([\omega] + 12[\theta]) = [\theta] \frown [\omega] + 12([\omega] \frown [\omega]) + 12([\theta] \frown [\theta]) + 144([\omega] \frown [\theta]) = 143([\omega] \frown [\theta])$ and, taking the orientation into account, we conclude that the number of intersection points is 143.

CONFIGURATION SPACES

The selection of an appropriate configuration space is very often the crux of the application of the CS/TM-scheme. Their construction is often based on a variety of combinatorial and geometric ideas and the following examples serve as an illustration of fairly complex configuration spaces that have appeared in actual applications.

EXAMPLE 21.1.4 *Alon's "spaces of partitions"*

The proof of the necklace-splitting theorem (Theorem 21.4.3) provides a nice example of an application of the CS/TM scheme. A continuous model of a necklace is an interval $[0,1]$ together with k measurable subsets $A_1, \dots, A_k$ representing "beads" of different colors. It is elementary that the configuration space of all sequences $0 \leq x_1 \leq \dots \leq x_m \leq 1$ is an m-dimensional simplex, hence the totality of all m-cuts of a necklace is also identified as an m-dimensional simplex Δ^m. Given a cut $c \in \Delta^m$, the assembling of the resulting subintervals $I_0(c), \dots, I_m(c)$ of $[0,1]$ into r collections is determined by a function $f : [m+1] \to [r]$. Hence, a configuration space associated to the necklace-splitting problem is obtained by gluing together m-dimensional simplices Δ^m_f, one for each function $f \in \mathrm{Fun}([m+1],[r])$. The complex $\mathcal{N}_{m,r}$ obtained by this construction turns out to be an example of a complex obtained from a simplex by a *deleted join operation* [Mat08, Živ98].

EXAMPLE 21.1.5 *Multidimensional necklaces*

A model of a d-dimensional "necklace," used in the proof [LŽ08] of a multidimensional version of Theorem 21.4.3 is the cube $I^d = [0,1]^d$ together with a collection of k measurable subsets $A_1, \dots, A_k$ of I^d. The space of all partitions of I^d by axes-aligned hyperplanes (using m_i hyperplanes in the direction $i \in \{1, \dots, d\}$) is the product of simplices $Q = \Delta^{m_1} \times \dots \times \Delta^{m_d}$. A point c in the interior of Q describes a dissection of Q into d-dimensional parallelepipeds enumerated by the set $\Pi = [m_1+1] \times \dots \times [m_d+1]$. The configuration space Ω_Q associated to the d-dimensional "necklace-splitting problem" is obtained by gluing together polytopes $Q_f \cong Q$, one for each allocating function $f \in \mathrm{Fun}(\Pi,[r])$. There is an associated "moment map" $\mu : \Omega_Q \to Q$ which makes the configuration space Ω_Q a relative of *small covers, toric spaces, (generalized) moment-angle complexes* and other objects that appear in *toric topology* [BP15].

EXAMPLE 21.1.6 *Gromov's "spaces of partitions"*

Very interesting polyhedral partitions are introduced by Gromov in [Gro03]. His *spaces of partitions* [Gro03, Section 5] are defined as the configuration spaces of labelled binary trees T_d of height d, with $2^d - 1$ internal nodes N_d and 2^d external nodes L_d (leaves of the tree T_d). More explicitly a labelled binary tree $(T_d, \{H_\nu\}_{\nu \in N_d})$ has an oriented hyperplane H_ν associated to each of the internal nodes $\nu \in N_d$ of T_d. The left (respectively right) outgoing edge, emanating from $\nu \in N_d$ is associated the positive halfspace H_ν^+ (respectively the negative halfspace H_ν^-) determined by H_ν.

Each of the leaves $\lambda \in L_d$ is the end point of the unique maximal path π_λ in the tree T_d. Each of the maximal paths π_λ is associated a polyhedral region Q_λ defined as the intersection of all halfspaces associated to edges of the path

π_λ. The associated partition $\{Q_\lambda\}_{\lambda \in L_d}$ depends continuously on the chosen labels (hyperplanes) and defines an element of the associated "space of partitions."

These and related configuration spaces were used in [Gro03] for a proof of a general Borsuk-Ulam type theorem ($c_\bullet$-Corollary 5.3 on page 188) and utilized by Gromov for his proof of the *Waist of the Sphere Theorem.*

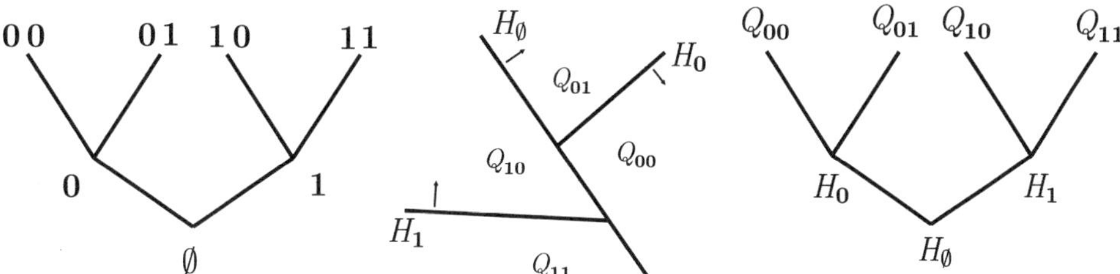

FIGURE 21.1.1
Labelled binary tree and the associated convex partition.

EXAMPLE 21.1.7 *Partitions via function-separating diagrams*

The classical configuration space $F_n(\mathbb{R}^d) = \{x \in (\mathbb{R}^d)^n \mid x_i \neq x_j \text{ for } i \neq j\}$ of labelled, distinct points in $\mathbb{R}^d$ has been used [KHA14] as a basis for constructions of convex (polyhedral) partitions of $\mathbb{R}^d$. These partitions typically arise as *generalized Voronoi partitions* (power diagrams), or more generally as "function-separating diagrams" of finite families $\mathcal{F} = \{f_1, \dots, f_n\}$ of real-valued functions with the domain $X \subset \mathbb{R}^d$. By definition the separating diagram of $\mathcal{F}$ is the collection of sets $\{X_i\}_{i=1}^n$ where $x \in X_i$ if and only if $f_i(x) = max_{j=1,\dots,n}\{f_j(x)\}$. For a point $z = (z_1, \dots, z_n) \in F_n(\mathbb{R}^d)$ and a weight vector $(r_i) \in \mathbb{R}^n$, the associated (generalized) Voronoi partition is the function-separating diagram of the family $\{\phi_i\}_{i=1}^n$, where $\phi_i(x) = r_i - \|x - z_i\|^2$. Note that essentially the same class of convex partitions is obtained if one uses linear (affine) functions defined on $\mathbb{R}^d$.

EXAMPLE 21.1.8 *Chessboard complexes and the chessboard transform*

The "chessboard transform" is a procedure of constructing a simplicial complex (a generalized chessboard complex) collecting all feasible candidates for the solution of a particular Tverberg-type problem (Section 21.4). Suppose that $S = \{x_i\}_{i=1}^m \subset \mathbb{R}^d$ is a finite set ("cloud") of points of cardinality m. This set may be structured or enriched by some (unspecified) structure $\mathcal{S}$ on S, for example the points may be colored in k-colors by a coloring map $\phi : [m] \to [k]$ where the cardinality m_i of each monochromatic subsets $C_i = \phi^{-1}(i)$ is prescribed in advance.

In a Tverberg-type problem, one looks for a collection $\{S_i\}_{i=1}^r$ of r disjoint subsets of S such that the intersection of all convex sets $\mathrm{conv}(S_i)$ is non-empty. Such a collection can be recorded (visualized) as a single set (simplex) $X \subset [m] \times [r]$ in the "chessboard" $[m] \times [r]$ where by definition $X = \{(i, j) \mid x_i \in S_j\}$. The set X may satisfy some additional condition expressing its compatibility with the structure $\mathcal{S}$. For example, if S is colored by a coloring function ϕ then the usual condition is that S_i is a "rainbow set" in the sense that it has at most one point in each color, $|S_i \cap C_j| \leq 1$ for each i and j. More generally, for a chosen simplicial complex $K \subset 2^{[m]}$, the structure $\mathcal{S}$ may arise from a simplicial map $f : K \to \mathbb{R}^d$

where $S = f(\text{Vert}(K))$ and the compatibility condition says that $S_i \in K$ for each $i = 1, \ldots, r$.

The ***chessboard transform*** of an $\mathcal{S}$-structured set S is the simplicial complex $\Delta_{m,q}[\mathcal{S}]$ which collects all $\mathcal{S}$-constrained ($\mathcal{S}$-compatible) subsets $X \subset [m] \times [r]$. If $\mathcal{S}$ is not structured, i.e., if there are no other constraints on X aside from the requirement that $S_i \cap S_j = \emptyset$ for $i \neq j$, we recover the complex $\mathcal{N}_{m,r}$ from Example 21.1.4. If $\mathcal{S}$ is the structure associated to a coloring function ϕ, then the complex $\Delta_{m\times q}[\mathcal{S}]$ is the join,

$$\Delta_{m,q}[\mathcal{S}] = \Delta_{m_1,q} * \ldots * \Delta_{m_k,q}$$

of standard chessboard complexes [BLVŽ94, VŽ11]. By definition the (standard) chessboard complex $\Delta_{p,q}$ is the complex of all non-attacking placements of rooks in a $(p \times q)$-chessboard (a placement is non-attacking (non-taking) if it is not allowed to have more than one rook in the same row or in the same column).

21.2 PARTITIONS OF MASS DISTRIBUTIONS

The problem of consensus division arises when two or more competitive or cooperative parties, each guided by their own preferences, divide an object according to some notion of fairness. There are many different mathematical reformulations of this problem depending on what kind of divisions are allowed, what kind of object is divided, whether the parties involved are cooperative or not, etc. Early examples of problems and results of this type are the "ham sandwich theorem" of Steinhaus and Banach, the "envy-free cake division problem" of Selfridge and Conway, and the "proportional cake-division problem" of Steinhaus, see [BT95]. The problems of partitioning mass distributions in Euclidean spaces, as a mathematical reformulation of the problem of fair division, form the first circle of discrete geometric problems where topological methods have been applied with great success.

GLOSSARY

Measure: A non-negative, countably additive (σ-additive) Borel measure, defined on the σ-algebra generated by all open (closed) sets in $\mathbb{R}^d$. A measure is finite if $\mu(\mathbb{R}^d) < +\infty$.

Measurable set: Any set in the domain of the function μ.

Mass distribution and density function: A density function is an integrable function $f : \mathbb{R}^d \to [0, +\infty)$ representing the density of a "mass distribution" (measure) on $\mathbb{R}^d$. The measure μ arising this way is defined by $\mu(A) := \int_A f\,dx$.

Counting measure: If $S \subset \mathbb{R}^d$ is a finite set, then $\mu(A) := |A \cap S|$ is the *counting measure* induced by the set S.

Halving hyperplane: A hyperplane that simultaneously bisects a family of measurable sets.

21.2.1 THE HAM SANDWICH THEOREM

Given a collection of d measurable sets (mass distributions, finite sets) in $\mathbb{R}^d$, the problem is to simultaneously bisect all of them by a single hyperplane.

THEOREM 21.2.1 *Ham Sandwich Theorem* [Bor33]

Let $\mu_1, \mu_2, \ldots, \mu_d$ be a collection of measures (mass distributions, measurable sets, finite sets). Then there exists a hyperplane H such that for all $i = 1, \ldots, d$, $\mu_i(H^+) \geq 1/2\, \mu_i(\mathbb{R}^d)$ and $\mu_i(H^-) \geq 1/2\, \mu_i(\mathbb{R}^d)$, where H^+ and H^- are the closed halfspaces associated with the hyperplane H.

In the special case where $\mu(H) = 0$, i.e., when the hyperplane itself has measure zero, H is called a *halving hyperplane* since $\mu_i(H^+) = \mu_i(H^-) = 1/2\, \mu_i(\mathbb{R}^d)$ for all i. A halving hyperplane H is also called a "ham sandwich cut," in agreement with the popular interpretation of the 3-dimensional case where a "sandwich" is cut into two equal parts by a straight cut of a knife.

TOPOLOGICAL BACKGROUND

The topological result lying behind the ham sandwich theorem is the Borsuk-Ulam theorem, [Ste85, Mat08]. The proof of the ham sandwich theorem historically marks one of the first applications of the CS/TM-scheme, with the $(d-1)$-sphere as the configuration space, $\mathbb{R}^d$ as the test space, and $G = \mathbb{Z}_2$ as the group of symmetries associated to the problem. Given a collection $\{A_i\}_{i=1}^d$ of d measurable sets, the test map $t : S^{d-1} \to \mathbb{R}^d$ is defined by $t(e) = (\alpha_1, \ldots, \alpha_d)$, with α_i determined by the condition that $H_i := \{x \in \mathbb{R}^d \mid \langle x, e \rangle = \alpha_i\}$ is a median halving hyperplane for the measurable set A_i. (The median halving hyperplane in any direction is the mid-hyperplane between the two extreme halving hyperplanes in that direction.) The test space is the diagonal $Z := \{(\alpha, \ldots, \alpha) \in \mathbb{R}^d \mid \alpha \in \mathbb{R}\}$. The test map t is obviously "odd," or $\mathbb{Z}_2$-equivariant, in the sense that $t(-e) = -t(e)$. The Borsuk-Ulam theorem is applied to the modified test map $\bar{t} = p \circ t : S^{n-1} \to Z^\perp$, where $Z^\perp \cong \mathbb{R}^{n-1}$ is the hyperplane orthogonal to Z and $p : \mathbb{R}^n \to Z^\perp$ is the associated orthogonal projection.

THEOREM 21.2.2 *Borsuk-Ulam Theorem* [Bor33]

For every continuous map $f : S^n \to \mathbb{R}^n$ from an n-dimensional sphere into n-dimensional Euclidean space, there exists a point $x \in S^n$ such that $f(x) = f(-x)$.

There is a different topological approach to the ham sandwich theorem closer to the earlier example about a watch with two indistinguishable hands. Here we mention only that the role of the torus T^2 is played by a manifold M representing all hyperplanes in $\mathbb{R}^d$ (the configuration space), while the curves Γ_1 and Γ_2 are replaced by suitable submanifolds N_i of M, one for each of the measures μ_i, $i = 1, \ldots, d$. N_i is defined as the space of all halving hyperplanes for the measurable set A_i.

21.2.2 THE CENTER POINT THEOREM

THEOREM 21.2.3 *Center Point Theorem* [Rad46]

Let μ be one of the measures described in Section 21.2.1. Then there exists a point $x \in \mathbb{R}^d$ such that for every closed halfspace $P \subset \mathbb{R}^d$, if $x \in P$ then

$$\mu(P) \geq \frac{\mu(\mathbb{R}^d)}{d+1}.$$

TOPOLOGICAL BACKGROUND

If the Borsuk-Ulam theorem is responsible for the ham sandwich theorem, then R. Rado's center point theorem can be seen as a consequence of another well-known topological result, *Brouwer's fixed point theorem* [Bro12].

In a nutshell, the center point theorem is, up to some technical details, deduced from Brouwer's theorem as follows. If x is not a center point of μ then there is a halfspace $P \ni x$ such that $\mu(P) < \mu(\mathbb{R}^d)/(d+1)$. The map $x \mapsto x + e$, where $e \in S^{d-1}$ is the outer normal unit vector of P, defines a vector field without zeros (contradicting the Brouwer's theorem).

Recall that the center point theorem was originally deduced (by R. Rado) from Helly's theorem about intersecting families of convex sets, which also has several topological relatives. For this reason, it is often viewed as a measure-theoretic equivalent of Helly's theorem.

APPLICATIONS AND RELATED RESULTS

As noted by Miller and Thurston (see [MTTV97, MTTV98]), the center point theorem and the Koebe theorem on the disk representation of planar graphs can be used to prove the existence of a small separator for a planar graph, a result proved originally (by Lipton and Tarjan) by different methods.

The ***regression depth*** $\mathrm{rd}_{\mathcal{P}}(H)$ of a hyperplane H relative to a collection $\mathcal{P}$ of n points in $\mathbb{R}^d$ is the minimum number of points that H must pass through in moving to the vertical position. Dually, given an arrangement $\mathcal{H}$ of n hyperplanes in $\mathbb{R}^d$, the regression depth $\mathrm{rd}_{\mathcal{H}}(x)$ of a point x relative to $\mathcal{H}$ is the smallest k such that x cannot escape to infinity without crossing (or moving parallel to) at least k hyperplanes. The problem of finding a point (resp. hyperplane) with maximum regression depth relative to $\mathcal{H}$ (resp. $\mathcal{P}$) is shown in [AET00] to be intimately connected with the problem of finding center points. The main result (confirming a conjecture of Rousseeuw and Hubert) is that there always exists a point with regression depth $\lceil n/(d+1) \rceil$; cf. Chapter 58 of this Handbook.

The "dual center point" theorem [Kar09] says that for each general position family of n hyperplanes in $\mathbb{R}^d$ there exists a point c such that any ray starting at c intersects at least $\lceil n/d + 1 \rceil$ hyperplanes. Both this result and the center point theorem are special cases of the "projective center point theorem" proved by Karasev and Matschke (see [KM14, Theorems 1.2 and 1.3]).

21.2.3 CENTER TRANSVERSAL THEOREM

THEOREM 21.2.4 *Center Transversal Theorem* [ŽV90]

Let $0 \leq k \leq d-1$ and suppose that $\mu_0, \mu_1, \ldots, \mu_k$ is a collection of finite, Borel measures on $\mathbb{R}^d$. Then there exists a k-dimensional affine subspace $D \subset \mathbb{R}^d$ such that for every closed halfspace $H(v, \alpha) := \{x \in \mathbb{R}^d \mid \langle x, v \rangle \leq \alpha\}$ and every $i \in \{0, 1, \ldots, k\}$,

$$D \subset H(v, \alpha) \Longrightarrow \mu_i(H(v, \alpha)) \geq \frac{\mu_i(\mathbb{R}^d)}{d - k + 1}.$$

TOPOLOGICAL BACKGROUND

The center transversal theorem contains the ham sandwich and center point theorems as two boundary cases [ŽV90]. The topological principle that is at the root of this result should be strong enough for this purpose. This result has several incarnations. One of them is a theorem of E. Fadell and S. Husseini [FH88] that claims the nonexistence of a $\mathbb{Z}_2^{\oplus k}$-equivariant map $f : V_{n,k} \to (\mathbb{R}^k)^{n-k} \setminus \{0\}$ from the Stiefel manifold of all orthonormal k-frames in $\mathbb{R}^n$ to the sum of $n-k$ copies of $\mathbb{R}^k$. The group $\mathbb{Z}_2^{\oplus k}$ can be identified with the group of all diagonal, orthogonal matrices and its action on $\mathbb{R}^k$ is induced by the obvious action of $O(k)$. A related result [FH88, ŽV90] is that the vector bundle $\xi_k^{\oplus(n-k)}$ does not admit a nonzero, continuous cross-section, where ξ_k is the tautological k-plane bundle over the Grassmann manifold $G_k(\mathbb{R}^n)$. For the introduction to vector bundles, classical Lie groups and associated manifolds the reader is referred to [Bre93, Chapters II and VII].

APPLICATIONS AND RELATED RESULTS

The following Helly-type transversal theorem, due to Dol'nikov, is a consequence of the same topological principle that is at the root of the center transversal theorem. Moreover, the center transversal theorem is related to Dol'nikov's result in the same way that the center point theorem is related to Helly's theorem.

THEOREM 21.2.5 [Dol93]

Let $\mathcal{K}_0, \dots, \mathcal{K}_k$ be families of compact convex sets. Suppose that for every i, and for each k-dimensional subspace $V \subset \mathbb{R}^d$, there exists a translate V_i of V intersecting every set in $\mathcal{K}_i$. Then there exists a common k-dimensional transversal of the family $\mathcal{K} := \bigcup_{i=0}^k \mathcal{K}_i$, i.e., there exists an affine k-dimensional subspace of $\mathbb{R}^d$ intersecting all the sets in $\mathcal{K}$.

Let $\mathcal{K} = \{K_0, \dots, K_k\}$ be a family of convex bodies in $\mathbb{R}^n$, $1 \le k \le n-1$. Then an affine l-plane $A \subset \mathbb{R}^n$ is called a ***common maximal l-transversal*** of $\mathcal{K}$ if $m(K_i \cap A) \ge m(K_i \cap (A+x))$ for each $i \in \{0, \dots, k\}$ and each $x \in \mathbb{R}^n$, where m is l-dimensional Lebesgue measure in A and $A+x$, respectively. It was shown in [MVŽ01] that, given a family $\mathcal{K} = \{K_i\}_{i=0}^k$ of convex bodies in R^n $(k < l)$, the set $C_l(\mathcal{K})$ of all common maximal l-transversals of $\mathcal{K}$ has to be "large" from both the measure-theoretic and the topological point of view.

21.2.4 EQUIPARTITION OF MASSES BY HYPERPLANES

Every measurable set $A \subset \mathbb{R}^3$ can be partitioned by three planes into 8 pieces of equal measure (H. Hadwiger). This is an instance of the general problem of characterizing all triples (d, j, k) such that for any j mass distributions (measurable sets) in $\mathbb{R}^d$, there exist k hyperplanes such that each of the 2^k "orthants" contains the fraction $1/2^k$ of each of the masses. Such a triple (d, j, k) will be called *admissible.* Let $\Delta(j, k)$ be the minimum dimension d such that the triple (d, j, k) is admissible. It is known (E. Ramos [Ram96]) that $d \ge j(2^k - 1)/k$ is a necessary condition and $d \ge j2^{k-1}$ a sufficient one for a triple (d, j, k) to be admissible. Ramos's method yields many interesting results in lower dimensions, including the admissibility of the triples $(9,3,3)$, $(9,5,2)$, and $(5,1,4)$, see [Živ15, BFHZ15, BFHZ16, VŽ15] for subsequent developments. The most interesting special case that is still out of reach is the triple $(4,1,4)$.

PROBLEM 21.2.6

Is it true that each measurable set $A \subset \mathbb{R}^4$ can be partitioned by four hyperplanes into sixteen parts of equal measure?

THEOREM 21.2.7 [Živ15] (see also [BFHZ15])

Each collection of $j = 4 \cdot 2^\nu + 1$ measures in $\mathbb{R}^d$ where $d = 6 \cdot 2^\nu + 2$ admits an equipartition by two hyperplanes. From here we deduce that,

$$\Delta(4 \cdot 2^\nu + 1, 2) = 6 \cdot 2^\nu + 2 \tag{21.2.1}$$

since following [Ram96] *the lower bound $\Delta(j,2) \geq 3j/2$ holds for each $j \geq 1$.*

APPLICATIONS AND RELATED RESULTS

According to [Mat08], an early interest of computer scientists in partitioning mass distributions by hyperplanes was stimulated in part by *geometric range searching*; cf. Chapter 40 of this Handbook. More general equipartitions involve not necessarily central hyperplane arrangements. An example is a result of S. Vrećica [Vre09, Theorem 3.1] which says that a mass distribution in $\mathbb{R}^d$ admits an equipartition by $4d-2$ polyhedral regions determined by a collection of $2d-2$ parallel hyperplanes and an additional transverse hyperplane. The existence of very general equipartitions made by iterated hyperplane cuts, where the directions of the hyperplanes were prescribed in advance, was established in [KRPS16]. This result can be also classified as a relative of the "splitting necklace theorem" (Theorem 21.4.3) since the equipartition is obtained after assembling the polyhedral regions into k collections (k is the number of parties involved in the consensus division).

21.2.5 PARTITIONS BY CONICAL, POLYHEDRAL FANS

An old result of R. Buck and E. Buck (*Math. Mag.*, 1949) says that for each continuous mass distribution in the plane, there exist three concurrent lines $l_1, l_2, l_3 \subset \mathbb{R}^2$ that partition $\mathbb{R}^2$ into six sectors of equal measure. It is natural to search for higher dimensional analogs of this result.

Suppose that $Q \subset \mathbb{R}^d$ is a convex polytope and assume that the origin $O \in \mathbb{R}^d$ belongs to the interior $\mathrm{int}(Q)$ of Q. Let $\{F_i\}_{i=1}^k$ be the collection of all facets of Q. Let $\mathcal{F} := \mathrm{fan}(Q)$ be the associated ***fan***, i.e., $\mathcal{F} = \{C_1, \dots, C_k\}$ where $C_i = \mathrm{cone}(F_i)$ is the convex closed cone with vertex O generated by F_i.

THEOREM 21.2.8 [Mak01]

Let Q be a regular dodecahedron with the origin $O \in \mathbb{R}^3$ as its barycenter. Then for any continuous mass distribution μ on $\mathbb{R}^3$, centrally symmetric with respect to O, there exists a linear map $L \in GL(3,\mathbb{R})$ such that

$$\mu(L(C_1)) = \mu(L(C_2)) = \ldots = \mu(L(C_{12})).$$

Makeev actually showed in [Mak01] that L can be found in the set of all matrices of the form $a \cdot t$, where t is an upper triangular matrix and $a \in GL(3, \mathbb{R})$ is a matrix given in advance. In an earlier paper (see [Mak98]) he showed that a radial partition by a fan determined by the facets of a cube always exists for an arbitrary measure in $\mathbb{R}^3$. Moreover, he shows in [Mak01] that a result analogous to Theorem 21.2.8 also holds for rhombic dodecahedra. Recall that the rhombic dodecahedron U_3 is the polytope bounded by twelve planes, each containing an edge of a cube and parallel to one of the great diagonal planes. A higher dimensional analogue of the rhombic dodecahedron is the polytope U_n in $\mathbb{R}^n$ described as the dual of the difference body of a regular simplex.

PROBLEM 21.2.9

Let $T \subset \mathbb{R}^n$ be a regular simplex and $Q := T - T$ the associated "difference polytope." Let $U_n := Q^\circ$ be the polytope polar to Q. Clearly U_n is a centrally symmetric polytope with $n^2 + n$ facets F_i, $i = 1, \ldots, n^2 + n$. Let $\{K_i\}_{i=1}^{n^2+n}$ be the associated conical dissection of $\mathbb{R}^n$, where $K_i := \text{cone}(F_i)$. Is it true that for any continuous mass distribution μ on $\mathbb{R}^n$ there exists a nondegenerate affine map $A : \mathbb{R}^n \to \mathbb{R}^n$ such that

$$\mu(A(K_1)) = \mu(A(K_2)) = \ldots = \mu(A(K_{n^2+n}))\,?$$

The center transversal theorem is a common generalization of the ham sandwich theorem and the center point theorem. There are other general statements of this type as illustrated by the following conjecture. For a given non-degenerate simplex $\Delta := \text{conv}(\{a_i\}_{i=0}^m) \subset \mathbb{R}^d$, let $\mathcal{D}(\Delta) = \{D_i\}_{i=0}^m$ be the associated dissection of $\mathbb{R}^d$ into $m+1$ wedgelike cones, where $D_i := P^\perp \oplus \text{cone}(\text{conv}(\{a_j\}_{j \neq i}))$ and $P := \text{aff}\,\Delta$.

CONJECTURE 21.2.10 [VŽ92]

Let $\mu_0, \ldots, \mu_k$ be a family of continuous mass distributions (measures), $0 \leq k \leq d-1$, defined on $\mathbb{R}^d$. Then there exists a $(d-k)$-dimensional regular simplex Δ such that for the corresponding dissection, $\mathcal{D}(\Delta)$, for some $x \in \mathbb{R}^d$, and for all i, j,

$$\mu_i(x + D_j) \geq \frac{\mu_i(\mathbb{R}^d)}{d - k + 1}.$$

Both $B(d, 0)$ and $B(d, d-1)$ are true ($B(d, d-1)$ is the ham sandwich theorem). The conjecture is also confirmed [VŽ92] in the case $B(d, d-2)$ for all d. Moreover, there exists a natural topological conjecture implying $B(d, k)$ that is closely related to the analogous statement needed for the center transversal theorem. This statement essentially claims that there is no $\mathbb{Z}_{k+1}$-equivariant map from the Stiefel manifold $V_k(\mathbb{R}^n)$ to the unit sphere $S(V)$ in an appropriate $\mathbb{Z}_{k+1}$-representation V.

21.2.6 PARTITIONS BY CONVEX SETS

THEOREM 21.2.11 [Sob12] [KHA14]

Let n and d be integers with $n, d \geq 2$. Assume that $\mu_1, \ldots, \mu_d$ are continuous mass distributions such that $\mu_1(\mathbb{R}^d) = \ldots = \mu_d(\mathbb{R}^d) = n$. Then there exists a partition of $\mathbb{R}^d$ into n sets $C_1, \ldots, C_n$ such that the interiors $\text{int}(C_i)$ are convex sets and $\mu_i(C_j) = 1$ for each $i = 1, \ldots, n$ and $j = 1, \ldots, n$.

The planar version of Theorem 21.2.11 was conjectured by Kaneko and Kano [KK99] and proved by a number of authors (see [BM01]) before it was established in full generality by P. Soberón [Sob12] and R. Karasev, A. Hubard, and B. Aronov [KHA14].

One of the key steps in the proof of Theorem 21.2.11 was the recognition of *weighted Voronoi diagrams* as elements of the configuration space adequate for this problem (Example 21.1.7). It was quickly recognized that the method has a potential for proving other results of similar nature (with equal or similar configuration space). Note however that some of these results are known to hold only under the assumption that n (the number of convex sets in the partition) is a prime power.

THEOREM 21.2.12 [KHA14, Corollary 1.1] (see also [BZ14, Zie15])
Given a convex body K in $\mathbb{R}^d$ and a prime power $n = p^k$ it is possible to partition K into n convex bodies with equal d-dimensional volumes and equal $(d-1)$-dimensional surface areas.

TOPOLOGICAL BACKGROUND

The ultimate topological statement responsible for all known results about convex equipartitions is the following far reaching theorem about maps equivariant with respect to the symmetric group Σ_n. Special cases of the result were originally proved by D. Fuchs (the case $d = 2, p = 2$) and V. Vassiliev (the case $d = 2$ and $p \geq 3$). The result was in full generality (following the general ideas of the proofs of Fuchs and Vassiliev) established by R. Karasev (see [KHA14]). A proof based on different ideas can be found in [BZ14], see also [Zie15] for an outline and some additional information.

THEOREM 21.2.13 (D. Fuchs, V. Vassiliev, R. Karasev) (see also [BZ14])
Let $F_n(\mathbb{R}^d)$ be the classical configuration space of all ordered n-tuples of distinct points in $\mathbb{R}^d$ where $n = p^k$ is a prime power. Suppose that $\mathrm{Mat}_0(d-1,n) \cong \mathbb{R}^{(d-1)(n-1)}$ is the vector space of all $(d-1) \times n$ matrices such that the entries in each of the $(d-1)$ rows add up to zero. Suppose that the symmetric group Σ_n acts by permuting the coordinates of $\overline{x} \in F_n(\mathbb{R}^d)$ and the columns of the matrix $m = (m_{ij}) \in \mathrm{Mat}_0(d-1,n)$. Then for each Σ_n-equivariant map $f : F_n(\mathbb{R}^d) \to \mathrm{Mat}_0(n, d-1)$ there exists a configuration $\overline{x}$ such that $f(\overline{x}) = 0$.

APPLICATIONS AND RELATED RESULTS

It is known from (first order) equivariant obstruction theory that the (non)existence of a Σ_n-equivariant map is closely related to the question of (non)existence of H-equivariant maps for different Sylow subgroups of the symmetric group Σ_n. If $n = 2^k$ then the 2-Sylow subgroup of Σ_n is precisely the group of all automorphisms of the binary tree described in Example 21.1.6. This observation establishes a link between Theorem 21.2.13 and Gromov's "Non-vanishing Lemma," [Gro03, p. 188].

21.2.7 PARTITIONS BY CONVEX SETS IN PRESCRIBED RATIOS

The conjecture of Kaneko and Kano (Section 21.2.6) motivated I. Bárány and J. Matoušek [BM01, BM02] to study general conical partitions of planar or spherical measures in prescribed ratios. Here we assume that all measures are continuous mass distributions.

An arrangement of k semilines in the Euclidean plane or on the 2-sphere is called a k-fan if all semilines start from the same point. A k-fan is an α-partition for a probability measure μ if $\mu(\sigma_i) = \alpha_i$ for each $i = 1, \ldots, k$, where $\{\sigma_i\}_{i=1}^{k}$ are conical sectors associated with the k-fan and $\alpha = (\alpha_1, \ldots, \alpha_k)$ is a given vector. The set of all $\alpha = (\alpha_1, \ldots, \alpha_m)$ such that for any collection of probability measures $\mu_1, \ldots, \mu_m$ there exists a common α-partition by a k-fan is denoted by $\mathcal{A}_{m,k}$. It was shown in [BM01] that the interesting cases of the problem of existence of α-partitions are $(k, m) = (2, 3), (3, 2), (4, 2)$.

CONJECTURE 21.2.14 [BM01, BM02]

Suppose that (k, m) *is equal to* $(2, 3), (3, 2)$ *or* $(4, 2)\}$. *Then* $\alpha \in \mathcal{A}_{k,m}$ *if and only if*

$$\alpha_1 + \ldots + \alpha_m = 1 \quad \text{and} \quad \alpha_i > 0 \quad \text{for each} \quad i = 1, \ldots m.$$

Bárány and Matoušek proposed a very nice approach to this problem [BM01, BM02] and showed that $(\frac{1}{4}, \frac{1}{4}, \frac{1}{4}, \frac{1}{4})$ and $(\frac{1}{5}, \frac{1}{5}, \frac{1}{5}, \frac{2}{5})$ are in $\mathcal{A}_{4,2}$. From here they deduced that $\{(\frac{1}{3}, \frac{1}{3}, \frac{1}{3}), (\frac{1}{2}, \frac{1}{4}, \frac{1}{4})\} \cup \{(\frac{p}{5}, \frac{q}{5}, \frac{r}{5}) \mid p, q, r \in N^{+}, p + q + r = 5\} \subset \mathcal{A}_{3,2}$.

21.3 THE PROBLEMS OF BORSUK AND KNASTER

The topological methods used in proofs of measure partition results are actually applicable to a much wider class of combinatorial and geometric problems. In fact quite different problems, which on the surface have very little in common may actually lead to the same or closely related configuration spaces and test maps. This in turn implies that such problems both follow from the same general topological principle and that they could, despite appearances, be classified as "relatives."

21.3.1 BORSUK'S PROBLEM

Borsuk's well-known problem [Bor33] about covering sets in $\mathbb{R}^n$ with sets of smaller diameter was solved in the negative by J. Kahn and G. Kalai [KK93] who proved that the size of a minimal cover is exponential in n; see Chapters 1 and 2 of this Handbook. This, however, gave a new impetus to the study of "Borsuk numbers" after the old exponential upper bounds suddenly became more plausible. This may be one of the reasons why results about "universal covers," originally used for these estimates, have received new attention.

The following result was proved originally by V. Makeev [Mak98]; see also [HMS02, Kup99]. Recall the rhombic dodecahedron U_3, the polytope bounded by twelve rhombic facets, which appeared in Section 21.2.5.

THEOREM 21.3.1 [Mak98] (see also [HMS02, Kup99])

A rhombic dodecahedron of width 1 *is a universal cover for all sets* $S \subset \mathbb{R}^3$ *of diameter* 1. *In other words, each set of diameter* 1 *in* 3*-space can be covered by a rhombic dodecahedron whose opposite faces are* 1 *unit apart.*

Let $\Sigma \subset \mathbb{R}^n$ be a regular simplex of edge-length 1, with vertices $v_1, \ldots, v_{n+1}$. Then the intersection of $n(n+1)/2$ parallel strips S_{ij} of width 1, where S_{ij} is bounded by the $(n-1)$-planes orthogonal to the segment $[v_i, v_j]$ passing through the vertices v_i and v_j $(i < j)$, is a higher dimensional analog of the rhombic dodecahedron. It is easy to see that this is just another description of the polytope U_n that we encountered in Problem 21.2.9.

CONJECTURE 21.3.2 *Makeev's conjecture* [Mak94]

The polytope U_n is a universal cover *in $\mathbb{R}^n$. In other words, for each set $S \subset \mathbb{R}^n$ of diameter* 1, *there exists an isometry $I : \mathbb{R}^n \to \mathbb{R}^n$ such that $S \subset I(U_n)$.*

The relevance of Makeev's conjecture for the general Borsuk problem is obvious since in low dimensions, $d = 2$ and $d = 3$, the solutions were based on the construction of suitable universal covers. (Note that the case $d = 4$ of the Borsuk partition problem is still open!) The following stronger conjecture is yet another example of a topological statement with potentially interesting consequences in discrete and computational geometry.

CONJECTURE 21.3.3 [HMS02]

Let $f : S^{n-1} \to \mathbb{R}$ be an odd function, and let $\Delta^n \subset \mathbb{R}^n$ be a regular simplex of edge-length 1, *with vertices $v_1, \ldots, v_{n+1}$. Then there exists an orthogonal linear map $A \in SO(n)$ such that the $n(n+1)/2$ hyperplanes H_{ij}, $1 \leq i < j \leq n+1$, are concurrent, where*

$$H_{ij} := \{x \in \mathbb{R}^n \mid \langle x, A(v_j - v_i) \rangle = f(A(v_j - v_i))\}.$$

G. Kuperberg showed in [Kup99] that, unlike the cases $n = 2$ and $n = 3$, for $n \geq 4$ there is homologically an even number of isometries $I : \mathbb{R}^n \to \mathbb{R}^n$ such that $S \subset I(U_n)$ for a given set S of constant width. Kuperberg showed that the Makeev conjecture can be reduced (essentially in the spirit of the CS/TM-scheme) to the question of the existence of a Γ-equivariant map $f : SO(n) \to V \setminus \{0\}$, where Γ is a group of symmetries of the root system of type A_n and the test space V is an $n(n-1)/2$-dimensional representation of Γ. The fact that such a map exists if and only if $n \geq 4$ may be an indication that the Makeev conjecture is false in higher dimensions.

21.3.2 KNASTER'S PROBLEM

Knaster's problem (Problem 4., *Colloq. Math.*, 1:30, 1947) is one of the old conjectures of discrete geometry with a distinct topological flavor. The conjecture is now known to be false in general, but the problem remains open in many interesting special cases.

PROBLEM 21.3.4 *Knaster's problem*

Given a finite subset $S = \{s_1, \ldots, s_k\} \subset S^n$ of the n-sphere, determine the conditions on k and n so that for each continuous map $f : S^n \to \mathbb{R}^m$ there will exist an isometry $O \in SO(n+1)$ with

$$f(O(s_1)) = f(O(s_2)) = \ldots = f(O(s_k)).$$

B. Knaster originally conjectured that such an isometry O always exists if $k \leq n-m+2$. Just as in the case of the Borsuk problem, the first counterexamples took a long time to appear. V. Makeev [Mak86, Mak90], and somewhat later K. Babenko and S. Bogatyi [BB89], showed that the condition $k \leq n-m+2$ is not sufficient, at least for some special values of n and m. New examples were discovered by W. Chen (*Topology*, 1998), B.S. Kashin and S.J. Szarek (*C. R. Math. Acad. Sci. Paris*, 2003), A. Hinrichs and C. Richter (*Israel J. Math*, 2005), see also [BK14] for an update and related results.

The fact that Knaster's conjecture is false in general does not rule out the possibility that for some special configurations $S \subset S^n$ the answer is still positive. The case where S is the set of vertices of a largest regular simplex inscribed in S^n is of special interest since it directly generalizes the Borsuk-Ulam theorem.

Questions closely related to Knaster's conjecture are the problems of inscribing or circumscribing polyhedra to convex bodies in $\mathbb{R}^n$; see [HMS02, Kup99]. G. Kuperberg observed that both the circumscription problem for constant-width bodies and Knaster's problem are special cases of the following problem.

PROBLEM 21.3.5 [Kup99]

Given a finite set T of points on S^{d-1} and a linear subspace L of the space of all functions from T to $\mathbb{R}^n$, decide if, for each continuous function $f : S^{d-1} \to \mathbb{R}^n$, there is an isometry O such that the restriction of $f \circ O$ to T is an element of L.

21.4 TVERBERG-TYPE THEOREMS

GLOSSARY

Tverberg-type problem: A problem in which a finite set $A \subset \mathbb{R}^d$ is to be partitioned into nonempty, disjoint pieces $A_1, \dots, A_r$, possibly subject to some constraints, so that the corresponding convex hulls $\{\text{conv}(A_i)\}_{i=1}^r$ intersect.

Colors: A set of $k+1$ colors is a collection $\mathcal{C} = \{C_0, \dots, C_k\}$ of disjoint subsets of $\mathbb{R}^d$. A set $B \subset \mathbb{R}^d$ is ***properly colored*** if it contains at most one point from each of the sets C_i; in this case $\text{conv}(B)$ is called a ***rainbow simplex*** (possibly degenerate).

Types A, B, and ***C***: Colored Tverberg problems are of type A, type B, or type C depending on whether $k = d, k < d$ or $k > d$ (resp.), where $k+1$ is the number of colors and d the dimension of the ambient space.

Tverberg numbers $T(r,d)$, $T(r,k,d)$: $T(r,k,d)$ is the minimal size of each of the colors C_i, $i = 0, \dots, k$, that guarantees that there always exist r intersecting, pairwise vertex-disjoint, rainbow simplices. $T(r,d) := T(r,d,d)$.

AN OVERVIEW AND FIRST EXAMPLES

"Tverberg problems" is a common name for a class of theorems and conjectures about finite sets of points (point clouds) in $\mathbb{R}^d$. The original Tverberg theorem (Theorem 21.4.1) claims that every set $K \subset \mathbb{R}^d$ with $(r-1)(d+1)+1$ elements can be partitioned $K = K_1 \cup \ldots \cup K_r$ into r nonempty, pairwise disjoint subsets $K_1, \ldots, K_r$ such that the corresponding convex hulls have a nonempty intersection:

$$\bigcap_{i=1}^{r} \operatorname{conv}(K_i) \neq \emptyset. \qquad (21.4.1)$$

This result can be reformulated as the statement that for each linear (affine) map $f : \Delta^D \xrightarrow{a} \mathbb{R}^d$ ($D = (r-1)(d+1)$) there exist r nonempty disjoint faces $\Delta_1, \ldots, \Delta_r$ of the simplex Δ^D such that $f(\Delta_1) \cap \ldots \cap f(\Delta_r) \neq \emptyset$. This form of Tverberg's result can be abbreviated as follows,

$$(\Delta^{(r-1)(d+1)} \xrightarrow{a} \mathbb{R}^d) \Rightarrow (r - \text{intersection}). \qquad (21.4.2)$$

Here we tacitly assume that the faces intersecting in the image are always vertex disjoint. The letter "a" over the arrow means that the map is affine and its absence indicates that it can be an arbitrary continuous map.

It is desirable to refine the original Tverberg theorem by specifying which simplicial complexes K can replace the full simplex $\Delta^{(r-1)(d+1)}$ in the proposition (21.4.2).

The following four statements are illustrative for results of "colored Tverberg type."

$$(K_{3,3} \longrightarrow \mathbb{R}^2) \Rightarrow (2 - \text{intersection}) \qquad (21.4.3)$$

$$(K_{3,3,3} \xrightarrow{a} \mathbb{R}^2) \Rightarrow (3 - \text{intersection}) \qquad (21.4.4)$$

$$(K_{5,5,5} \longrightarrow \mathbb{R}^3) \Rightarrow (3 - \text{intersection}) \qquad (21.4.5)$$

$$(K_{4,4,4,4} \longrightarrow \mathbb{R}^3) \Rightarrow (4 - \text{intersection}) \qquad (21.4.6)$$

$K_{t_1,t_2,\ldots,t_k} = [t_1] * [t_2] * \ldots * [t_k]$ is by definition the complete multipartite simplicial complex obtained as a join of 0-dimensional complexes (finite sets). By definition the vertices of this complex are naturally partitioned into groups of the same "color." For example $K_{p,q} = [p] * [q]$ is the complete bipartite graph obtained by connecting each of p "red vertices" with each of q "blue vertices." The simplices of $K_{t_1,t_2,\ldots,t_k}$ are often referred to as *rainbow simplices.*

The implication (21.4.3) says that for each continuous map $\phi : K_{3,3} \to \mathbb{R}^2$ there always exist two vertex-disjoint edges which intersect in the image. In light of the Hanani-Tutte theorem this statement is equivalent to the non-planarity of the complete bipartite graph $K_{3,3}$. The implication (21.4.4) is an instance of a result of Bárány and Larman [BL92]. It says that each collection of nine points in the plane, evenly colored by three colors, can be partitioned into three multicolored or "rainbow triangles" which have a common point. Note that a 9-element set $C \subset \mathbb{R}^2$ which is evenly colored by three colors, can be also described by a map $\alpha : [3] \sqcup [3] \sqcup [3] \to \mathbb{R}^2$ from a disjoint sum of three copies of [3]. In the same spirit an affine map $\phi : K_{3,3,3} \xrightarrow{a} \mathbb{R}^2$ parameterizes not only the colored set itself but takes into account from the beginning that some simplices (multicolored or rainbow simplices) play a special role.

A similar conclusion has statement (21.4.5) which is a formal analogue of the statement (21.4.3) in dimension 3. It is an instance of a result of Vrećica and Živaljević [VŽ94], which claims the existence of three intersecting, vertex-disjoint rainbow triangles in each constellation of 5 red, 5 blue, and 5 white stars in the

3-space. A non-linear version of this result is that $K_{5,5,5}$ is 3-non-embeddable in $\mathbb{R}^3$ in the sense that there always exists a triple point in the image.

The statement (21.4.6) is an instance of the result of Blagojević, Matschke, and Ziegler [BMZ15] saying that 4 intersecting, vertex disjoint rainbow tetrahedra in $\mathbb{R}^3$ will always appear if we are given sixteen points, evenly colored by four colors.

All statements (21.4.3)–(21.4.6) are instances of results of colored Tverberg type. They are respectively classified as the Type A ((21.4.4) and (21.4.6)), and the Type B ((21.4.3) and (21.4.5)) colored Tverberg results.

21.4.1 MONOCHROMATIC TVERBERG THEOREMS

THEOREM 21.4.1 *Affine Tverberg Theorem* [Tve66]

Every set $K = \{a_j\}_{j=0}^{(r-1)(d+1)} \subset \mathbb{R}^d$ with $(r-1)(d+1)+1$ elements can be partitioned into r nonempty, disjoint subsets $K_1, \dots, K_r$ so that the corresponding convex hulls have nonempty intersection:

$$\bigcap_{i=1}^{r} \operatorname{conv}(K_i) \neq \emptyset .$$

(The special case $q = 2$ is Radon's theorem; see Chapter 4.)

THEOREM 21.4.2 *Topological Tverberg Theorem* [BSS81, Öza87]

Assume that r is a prime ([BSS81]) *or a prime power* ([Öza87]). *Then for every continuous map $f : \Delta^{(r-1)(d+1)} \to \mathbb{R}^d$ there exist vertex-disjoint faces $\Delta_1, \dots, \Delta_r \subset \Delta^{(r-1)(d+1)}$ such that $\bigcap_{i=1}^{r} f(\Delta_i) \neq \emptyset$.*

APPLICATIONS AND RELATED RESULTS

The affine Tverberg theorem was proved by Helge Tverberg in 1966. The topological Tverberg theorem, proved by Bárány, Shlosman, and Szűcs in 1981 (and by Özaydin in 1987 in the prime power case), reduces to the affine version if f is an affine (simplicial) map. Some of the relevant references for these two theorems and their applications are [Bjö95, Sar92, Vol96a, Živ98, Mat02, Mat08].

Note that Theorem 21.4.1 is not a formal consequence of Theorem 21.4.2 since the latter needs an extra condition that r is a prime power. Surprisingly enough this condition turned out to be essential (see Section 21.4.5).

The following "necklace-splitting theorem" of Noga Alon is a very nice application of the continuous Tverberg theorem.

THEOREM 21.4.3 [Alo87]

Assume that an open necklace has ka_i beads of color i, $1 \leq i \leq t$, $k \geq 2$. Then it is possible to cut this necklace at $t(k-1)$ places and assemble the resulting intervals into k collections, each containing exactly a_i beads of color i.

The "Tverberg-Vrećica conjecture" is a statement that incorporates both the center transversal theorem (Theorem 21.2.4) and the (affine) Tverberg theorem (Theorem 21.4.1) in a single general statement.

CONJECTURE 21.4.4 [TV93]

Assume that $0 \leq k \leq d-1$ and let $S_0, S_1, \ldots, S_k$ be a collection of finite sets in $\mathbb{R}^d$ of given cardinalities $|S_i| = (r_i - 1)(d-k+1)+1$, $i = 0, 1, \ldots, k$. Then S_i can be split into r_i nonempty sets, $S_i^1, \ldots, S_i^{r_i}$, so that for some k-dimensional affine subspace $D \subset \mathbb{R}^d$, $D \cap \mathrm{conv}(S_i^j) \neq \emptyset$ for all i and j, $0 \leq i \leq k$, $1 \leq j \leq r_i$.

This conjecture was confirmed in [Živ99] for the case where both d and k are odd integers and $r_i = q$ for each i, where q is an odd prime number, and by Vrećica in the case $r_1 = \ldots = r_k = 2$ [Vre03]. Karasev [Kar07] extended this result to the case where q is a prime power (and arbitrary d and k). Further progress and update on the colored version of this problem can be found in [BMZ11].

The expository article [Kal01] is recommended as a source of additional information about Tverberg-type theorems and conjectures.

21.4.2 COLORED TVERBERG THEOREMS

Let $T(r,k,d)$ be the minimal number t so that for every collection of colors $\mathcal{C} = \{C_0, \ldots, C_k\}$ with the property $|C_i| \geq t$ for all $i = 0, \ldots, k$, there exist r *properly colored* sets $A_i = \{a_j^i\}_{j=0}^k$, $i = 1, \ldots, r$, that are pairwise disjoint but where the corresponding rainbow simplices $\sigma_i := \mathrm{conv}\, A_i$ have a nonempty intersection, $\bigcap_{i=1}^r \sigma_i \neq \emptyset$. A set X is "properly colored" if it does not contain more than one point of the same color.

The colored Tverberg problem is to establish the existence of, and then to evaluate or estimate, the integer $T = T(r,k,d)$. The cases $k = d$ and $k < d$ are related, but there is also an essential difference. In the case $k = d$, provided t is large enough, the number of intersecting rainbow simplices can be arbitrarily large. In the case $k < d$, for dimension reasons, one cannot expect more than $r \leq d/(d-k)$ intersecting k-dimensional rainbow simplices. This is the reason why colored Tverberg theorems are classified as type A or type B, depending on whether $k = d$ or $k < d$. The remaining case $k > d$ is classified as type C.

In the type A case, where $T(r,d,d)$ is abbreviated simply as $T(r,d)$, it is easy to see that a lower bound for this function is r. It is conjectured that this lower bound is attained:

CONJECTURE 21.4.5 (Type A) [BL92]

$T(r,d) = r$ for all r and d.

The colored Tverberg problem (type A) was originally conjectured and designed as a tool for solving important problems of computational geometry [ABFK92, BFL90, BL92]. The weak form of the conjecture, $T(r,d) < +\infty$ [BFL90], is already far from obvious.

Conjecture 21.4.5 is in [BL92] confirmed for $r = 2$ and for $d \leq 2$. Živaljević and Vrećica [ŽV92, Živ98, VŽ11] recognized the role of chessboard complexes (Example 21.1.8) as proper configuration spaces for colored Tverberg-type questions. Their central (type A) result [ŽV92] established the bound $T(r,d) \leq 2r-1$ if r is a prime power ($T(r,d) \leq 4r-3$ in the general case), providing the missing link in the solution of several opened problems in computational geometry (see the end of this section).

Blagojević, Matschke, and Ziegler [BMZ15] observed the importance of type C colored Tverberg questions and applied a similar CS/TM-scheme to settle Conjecture 21.4.5 in the affirmative if $r+1$ is a prime number.

THEOREM 21.4.6 (Type A) [BMZ15]

For every integer r such that $r+1$ is a prime number and every collection of $d+1$ disjoint sets ("colors") $C_0, C_1, \ldots, C_d$ in $\mathbb{R}^d$, each of cardinality at least r, there exist r disjoint, multicolored subsets $S_i \subset \bigcup_{i=0}^{d} C_i$ such that

$$\bigcap_{i=1}^{r} \operatorname{conv} S_i \neq \emptyset.$$

Recall that in the type B case of the general colored Tverberg problem it is assumed (as a necessary condition) that $r \leq d/(d-k)$. In light of the lower bound $T(r,k,d) \geq 2r-1$ [VŽ94] the following conjecture is quite natural.

CONJECTURE 21.4.7 (Type B)

$T(r,k,d) = 2r-1.$

Here is a theorem confirming Conjecture 21.4.7 if r is a prime power.

THEOREM 21.4.8 (Type B) [VŽ94, Živ98]

Let $C_0, \ldots, C_k$ be a collection of $k+1$ disjoint finite sets ("colors") in $\mathbb{R}^d$. Let r be a prime power such that $r \leq d/(d-k)$ and let $|C_i| = t \geq 2r-1$. Then there exist r properly colored k-dimensional simplices S_i, $i = 1, \ldots, r$, that are pairwise vertex-disjoint such that

$$\bigcap_{i=1}^{r} \operatorname{conv} S_i \neq \emptyset.$$

REMARK 21.4.9

Both Theorems 21.4.6 and 21.4.8 have their non-linear analogues since they are obtained by a topological argument (see the implications (21.4.8) and (21.4.7)). Moreover, as in the case of topological Tverberg theorem, the topological analogue of Theorem 21.4.8 is wrong if r is not a prime power [BFZ15, Remark 4.4].

The type C colored Tverberg results were originally conceived as a tool for the proof of Theorem 21.4.6, however they are certainly interesting in their own right.

THEOREM 21.4.10 (Type C) [BMZ15]

Let $r \geq 2$ be prime, $d \geq 1$, and $N = (r-1)(d+1)$. Let Δ^N be an N-dimensional simplex with a partition of its vertex set into $m+1 \geq d+2$ parts ("color classes"),

$$\operatorname{Vert}(\Delta^N) = C_0 \uplus C_1 \uplus \ldots \uplus C_m,$$

with $|C_i| \leq r-1$ for all i. Then for every continuous map $f : \Delta^N \to \mathbb{R}^d$, there is a collection $F_1, \ldots, F_r$ of disjoint faces of Δ^N such that,

(A) *$|C_i \cap F_j| \leq 1$ for each $i \in \{0, \ldots, m\}$ and $j \in \{1, \ldots, r\}$;*

(B) *$f(F_1) \cap \ldots \cap f(F_r) \neq \emptyset$.*

TOPOLOGICAL BACKGROUND

Suppose that $W_r = \{x \in \mathbb{R}^r \mid x_1 + \ldots + x_r = 0\}$ is the standard $(r-1)$-dimensional real representation of the cyclic group $\mathbb{Z}/r$. The $(r-1)d$-dimensional space $W_r^{\oplus d}$ can be described as the vector space of all real $(r \times d)$-matrices with column sums equal to zero. Let $K^{*r} = K * \ldots * K$ be the join of r copies of K.

Both Theorems 21.4.6 and 21.4.8 follow the CS/TS-scheme based on chessboard complexes $\Delta_{p,r}$ as configuration spaces (Example 21.1.8). The associated *test maps* are respectively (21.4.7) (for Theorem 21.4.8) and (21.4.8) (for Theorem 21.4.6).

$$(\Delta_{r,2r-1})^{*(k+1)} \xrightarrow{\mathbb{Z}/r} W_r^{\oplus d} \tag{21.4.7}$$

$$(\Delta_{r,r-1})^{*d} * [r] \xrightarrow{\mathbb{Z}/r} W_r^{\oplus d}. \tag{21.4.8}$$

Both theorems are consequences of the corresponding Borsuk-Ulam-type statements claiming that in the either case the $\mathbb{Z}/r$-equivariant map must have a zero if r is a prime number.

APPLICATIONS OF COLORED TVERBERG THEOREMS

The bound $T(d+1,d) \leq 4d+1$ established in [ŽV92] opened the possibility of proving several interesting results in discrete and computational geometry.

HALVING HYPERPLANES AND THE k-SET PROBLEM

The number $h_d(n)$ of halving hyperplanes of a set of size n in $\mathbb{R}^d$, i.e., the number of essentially distinct placements of a hyperplane that split the set in half, according to Bárány, Füredi, and Lovász [BFL90], satisfies

$$h_d(n) = O(n^{d-\epsilon_d}), \quad \text{where} \quad \epsilon_d = T(d+1,d)^{-(d+1)}.$$

POINT SELECTIONS AND WEAK ϵ-NETS

The equivalence of the following statements was established in [ABFK92] before the inequality $T(d+1,d) < +\infty$ was established in [ŽV92]. Considerable progress has since been made in this area [Mat02], and different combinatorial techniques for proving these statements have emerged in the meantime.

- Weak colored Tverberg theorem: $T(d+1,d)$ is finite.
- Point selection theorem: There exists a constant $s = s_d$, whose value depends on the bound for $T(d+1,d)$, such that any family $\mathcal{H}$ of $(d+1)$-element subsets of a set $X \subset \mathbb{R}^d$ of size $|\mathcal{H}| = p\binom{|X|}{d+1}$ contains a pierceable subfamily $\mathcal{H}'$ such that $|\mathcal{H}'| \gg p^s\binom{|X|}{d+1}$. ($\mathcal{H}'$ is ***pierceable*** if $\bigcap_{S \in \mathcal{H}'} \operatorname{conv} S \neq \emptyset$. $A \gg_d B$ if $A \geq c_1(d)B + c_2(d)$, where $c_1(d) > 0$ and $c_2(d)$ are constants depending only on the dimension d.)
- Weak ϵ-net theorem: For any $X \subset \mathbb{R}^d$ there exists a weak ϵ-net F for convex sets with $|F| \ll_d \epsilon^{(d+1)(1-1/s)}$, where $s = s_d$ is as above. (See Chapter 47 for the notion of ϵ-net; a *weak* ϵ-net is similar, except that it need not be part of X.)

- Hitting set theorem: For every $\eta > 0$ and every $X \subset \mathbb{R}^d$ there exists a set $E \subset \mathbb{R}^d$ that misses at most $\eta\binom{|X|}{d+1}$ simplices of X and has size $|E| \ll_d \eta^{1-s_d}$, where s_d is as above.

OTHER RELATED RESULTS

The configuration space that naturally arises via the CS/TM-scheme in proofs of Theorems 21.4.6 and 21.4.8 is the so-called ***chessboard complex*** $\Delta_{r,t}$, which owes its name to the fact that it can be described as the complex of all non-taking rook placements on an $r \times t$ chessboard. This is an interesting combinatorial object that arises independently as the coset complex of the symmetric group, as the complex of partial matchings in a complete bipartite graph, and as the complex of all partial injective functions. Due to the fact that the high connectivity of a configuration space is a property of central importance for applications (cf. Theorem 21.5.1), chessboard complexes have been studied from this point of view in numerous papers; see [Jon08, VŽ11, JVŽ17a, JVŽ17b] for more recent advances and references.

21.4.3 VAN KAMPEN-FLORES TYPE RESULTS

The classical Van Kampen-Flores theorem [Mat08, Theorem 5.1.1] says that the d-dimensional skeleton $K = (\Delta^{2d+2})^{\leq d}$ of a $(2d+2)$-dimensional simplex Δ^{2d+2} is *strongly non-embeddable* in $\mathbb{R}^{2d}$ in the sense that for each continuous map $f : (\Delta^{2d+2})^{\leq d} \to \mathbb{R}^{2d}$ there exist two vertex disjoint simplices $\sigma_1, \sigma_2 \in K$ such that $f(\sigma_1) \cap f(\sigma_2) \neq \emptyset$.

Both the Van Kampen-Flores theorem and its generalized version due to Sarkaria [Sar91], Volovikov [Vol96b], and Blagojević, Frick, and Ziegler [BFZ14], are relatives of the topological Tverberg theorem.

THEOREM 21.4.11 *Generalized Van Kampen-Flores Theorem*

Let $N = (r-1)(d+2)$ where $d \geq 1$ and r is a power of a prime. Let $k \geq \lceil \frac{r-1}{r} d \rceil$. Then for any continuous map $f : \Delta^N \to \mathbb{R}^d$ there are r pairwise disjoint faces $\sigma_1, \ldots, \sigma_r$ of Δ^N such that $\dim(\sigma_i) \leq k$ *for each i and $f(\sigma_1) \cap \cdots \cap f(\sigma_r) \neq \emptyset$.*

ADMISSIBLE AND PRESCRIBABLE TVERBERG PARTITIONS

In this section we discuss the problem whether each *admissible* r-tuple is *Tverberg prescribable* (or Van Kampan-Flores prescribable). This problem, as formulated in [BFZ14] (see also a related question of R. Bacher, Mathoverflow.net June 2011), will be referred to as the Tverberg A-P problem or the Tverberg A-P conjecture. Following [BFZ14, Definition 6.7] for $d \geq 1$ and $r \geq 2$, we say that an r-tuple $d = (d_1, \ldots, d_r)$ of integers is *admissible* if,

$$\lfloor d/2 \rfloor \leq d_i \leq d \qquad \text{and} \qquad \sum_{i=1}^{r} (d - d_i) \leq d. \tag{21.4.9}$$

An admissible r-tuple is *Tverberg prescribable* if there is an N such that for every continuous map $f : \Delta^N \to \mathbb{R}^d$ there is a Tverberg partition $\{\sigma_1, \ldots, \sigma_r\}$ for f with $\dim(\sigma_i) = d_i$.

PROBLEM 21.4.12 *Tverberg A-P problem* [BFZ14, Question 6.9.]

Is every admissible r-tuple Tverberg prescribable?

The "balanced case" of the Tverberg A-P conjecture is the case when the dimensions $d_1, \ldots, d_r$ satisfy the condition $|d_i - d_j| \leq 1$ for each i and j. In other words there exist $0 \leq s < r$ and k such that $d_1 = \ldots = d_s = k+1$ and $d_{s+1} = \ldots = d_r = k$. In this case the second admissibility condition in (21.4.9) reduces to the inequality $rk + s \geq (r-1)d$ while the first condition is redundant. The case when all dimensions are equal $d_1 = \ldots = d_r$ is answered by Theorem 21.4.11 (see for example [BFZ14, Theorem 6.5]). The following theorem of D. Jojić, S. Vrećica, and R. Živaljević covers the remaining cases.

THEOREM 21.4.13 *Balanced A-P theorem* [JVŽ17a]

Suppose that $r = p^\kappa$ is a prime power and let $\mathbf{d} = (d_1, \ldots, d_r)$ be a sequence of integers satisfying the condition $|d_i - d_j| \leq 1$ for each i and j. Then if the sequence $\mathbf{d}$ is admissible then it is Tverberg prescribable.

21.4.4 THE "CONSTRAINT METHOD"

The Gromov-Blagojević-Frick-Ziegler reduction, or the *"constraint method,"* is an elegant and powerful method for proving results of Tverberg-van Kampen-Flores type. In its basic form the method can be summarized as follows.

$$\begin{array}{ccc} K & \xrightarrow{f} & \mathbb{R}^d \\ {\scriptstyle e}\downarrow & & \downarrow{\scriptstyle i} \\ \Delta^N & \xrightarrow{F} & \mathbb{R}^{d+1} \end{array} \tag{21.4.10}$$

Suppose that the continuous Tverberg theorem holds for the triple $(\Delta^N, r, \mathbb{R}^{d+1})$ in the sense that for each continuous map $F : \Delta^N \to \mathbb{R}^{d+1}$ there exists a collection of r vertex disjoint faces $\Delta_1, \ldots, \Delta_r$ of Δ^N such that $f(\Delta_1) \cap \ldots \cap f(\Delta_r) \neq \emptyset$. For example, Theorem 21.4.2 says that this is the case if $r = p^k$ is a prime power and $N = (r-1)(d+2)$. Suppose that $K \subset \Delta^N$ is a simplicial complex which is *r-unavoidable* in the sense that if $A_1 \uplus \ldots \uplus A_r = [N+1]$ is a partition of the set $[N+1]$ (of vertices of Δ), then at least one of the faces $\Delta(A_i)$ of Δ^N (spanned by A_i) is in K. Then for each continuous map $f : K \to \mathbb{R}^d$ there exists vertex disjoint simplices $\sigma_1, \ldots, \sigma_r \in K$ such that $f(\sigma_1) \cap \ldots \cap f(\sigma_r) \neq \emptyset$.

Indeed, let $\bar{f}$ be an extension ($\bar{f} \circ e = f$) of the map f to Δ^N. Suppose that $\rho : \Delta^N \to \mathbb{R}$ is the function $\rho(x) := \mathrm{dist}(x, K)$, measuring the distance of the point $x \in \Delta^N$ from K. Define $F = (\bar{f}, \rho) : \Delta^N \to \mathbb{R}^{d+1}$ and assume that $\Delta_1, \ldots, \Delta_r$ is the associated family of vertex disjoint faces of Δ^N, such that $F(\Delta_1) \cap \ldots \cap F(\Delta_r) \neq \emptyset$. More explicitly, suppose that $x_i \in \Delta_i$ such that $F(x_i) = F(x_j)$ for each $i, j = 1, \ldots, r$. Since K is r-unavoidable, $\Delta_i \in K$ for some i. As a consequence $\rho(x_i) = 0$, and in turn $\rho(x_j) = 0$ for each $j = 1, \ldots, r$. If Δ_i' is the minimal face of Δ^N containing x_i then $\Delta_i' \in K$ for each $i = 1, \ldots, r$ and $f(\Delta_1') \cap \ldots \cap f(\Delta_r') \neq \emptyset$.

For a more complete exposition and numerous examples of applications of the "constraint method" the reader is referred to [BFZ14], see also [Gro10, Section 2.9(c)] and [Lon02, Proposition 2.5].

21.4.5 COUNTEREXAMPLES

In many results of Tverberg-van Kampen-Flores type there is a conspicuous condition that the number of intersecting simplices is a power of a prime $r = p^k$ (see Theorems 21.4.2, 21.4.8, 21.4.11). It is probably safe to say that a majority of specialists in the area believed that this condition is not essential, and that it will be eventually removed from these statements (note its absence in the formulation of the Affine Tverberg Theorem, Theorem 21.4.1).

It was a deep insight of Isaac Mabillard and Uli Wagner that the truth may be quite the opposite. Motivated by the intriguing absence of counterexamples in problems of Tverberg-Van Kampen type, they initiated in [MW14] the program of studying maps $f : K \to \mathbb{R}^d$ without triple, quadruple, or, more generally, global r-fold points (r-Tverberg points).

A necessary condition for such a map to exist is the existence of an S_r-equivariant map $F : K_\Delta^{\times r} \to S(W_r^{\oplus d})$ where S_r is the symmetric group, $W_r = \{x \in \mathbb{R}^d \mid x_1 + \ldots + x_r = 0\}$, and $K_\Delta^{\times r}$ is the associated r*-fold deleted product* of K defined as the union of all products $\sigma_1 \times \cdots \times \sigma_r$ of simplices such that $\sigma_i \in K$ for all i and $\sigma_i \cap \sigma_j \neq \emptyset$ for each $i \neq j$.

The following remarkable result was the first in a row of theorems which paved the way for long-awaited counterexamples in this area. The reader is referred to [MW14, MW15, MW16, AMSW17] for a complete exposition of the theory and subsequent developments.

THEOREM 21.4.14 (I. Mabillard and U. Wagner [MW14, MW15])

Suppose that $r \geq 2, k \geq 3$*, and let* K *be a simplicial complex of dimension* $(r-1)k$*. Then the following statements are equivalent:*

(i) *There exists an* S_r*-equivariant map* $F : K_\Delta^{\times r} \to S(W_r^{\oplus rk})$*.*

(ii) *There exists a continuous map* $f : K \to \mathbb{R}^{rk}$ *such that* $f(\sigma_1) \cap \cdots \cap f(\sigma_r) = \emptyset$ *for each collection of pairwise disjoint faces* $\sigma_1, \ldots, \sigma_r$ *of* K*.*

It has been known for quite some time that the statement (i) in Theorem 21.4.14 is satisfied if r is *not* a prime power (M. Özaydin [Öza87]). It follows that Theorem 21.4.14 alone is strong enough to provide counterexamples for van Kampen type questions in the non prime power case. For example it implies that Theorem 21.4.11 is in general false, unless r is a prime power.

As far as the topological Tverberg theorem (Theorem 21.4.2) is concerned Theorem 21.4.14 was not strong enough for this purpose. A new idea was needed and it came in the form of the Gromov-Blagojević-Frick-Ziegler reduction (Section 21.4.4). Surprisingly enough, this idea was introduced before the appearance of [MW14] and Theorem 21.4.14.

Gromov studied in [Gro10] the lower bounds on the topological complexity of the fibers $F^{-1}(y) \subset X$ of continuous maps $F : X \to Y$, in terms of combinatorial and topological invariants of spaces X and Y. The statement *"The topological Tverberg theorem, whenever available, implies the van Kampen-Flores theorem"* appears in Section 2.9(c), with a short proof.

Unaware of [Gro10], Blagojević, Frick, and Ziegler, developed in [BFZ14] their "constraint method" as a powerful tool for reducing various Tverberg type statements to the original topological Tverberg theorem.

The following result was announced in [Fri15], see also [BFZ15].

THEOREM 21.4.15 (Frick [Fri15], Blagojević, Frick, and Ziegler [BFZ15])
Assume that $r \geq 6$ is an integer that is not a prime power, let $k \geq 3$ and $N = (r-1)(rk+2)$. Then there exists a continuous map $f : \Delta^N \to \mathbb{R}^{rk+1}$ such that $f(\sigma_1) \cap \cdots \cap f(\sigma_r) = \emptyset$ for each collection of pairwise disjoint faces $\sigma_1, \ldots, \sigma_r$ of Δ^N.

For a more complete exposition, refinements and generalizations of these results the reader is referred to [MW15, BFZ15, AMSW17].

21.5 TOOLS FROM EQUIVARIANT TOPOLOGY

The method of equivariant maps is a versatile tool for proving results in discrete geometry and combinatorics. For many results these are the only proofs available. Equivariant maps are typically encountered at the final stage of application of the CS/TM-scheme (Section 21.1).

GLOSSARY

***G*-space *X*, *G*-action:** A group G acts on a space X if each element of G is a continuous transformation of X and multiplication in G corresponds to composition of transformations. Formally, a G-action α is a continuous map $\alpha : G \times X \to X$ such that $\alpha(g, \alpha(h, x)) = \alpha(gh, x)$. Then X is called a G-space and $\alpha(g, x)$ is often abbreviated as $g \cdot x$ or gx.

Free *G*-action: A G-action is free if $g \cdot x = x$ for some $x \in X$ implies $g = e$, where e is the unit element in G. An action is fixed-point free if for each $x \in X$ there exists $g \in G$ such that $g \cdot x \neq x$.

***G*-equivariant map:** A map $f : X \to Y$ of two G-spaces X and Y is equivariant if for all $g \in G$ and $x \in X$, $f(g \cdot x) = g \cdot f(x)$.

Borsuk-Ulam-type theorem: Any theorem establishing the nonexistence of a G-equivariant map between two G-spaces X and Y.

***n*-Connected space:** A path-connected and simply connected space with trivial homology in dimensions $1, 2, \ldots, n$. A path-connected space X is simply connected or 1-connected if every closed loop $\omega : S^1 \to X$ can be deformed to a point.

The following generalization of the Borsuk-Ulam theorem is the key result used in many proofs. Note that if $X = S^n$, $Y = S^{n-1}$, and $G = \mathbb{Z}_2$, it specializes to a statement equivalent to the standard version of the Borsuk-Ulam theorem (Theorem 21.2.2).

THEOREM 21.5.1 [Dol83]
Suppose X and Y are simplicial (more generally CW) complexes equipped with the free action of a finite group G, and that X is m-connected, where $m = \dim Y$. Then there does not exist a G-equivariant map $f : X \to Y$.

The following refinement of Theorem 21.5.1 (due to A.Yu. Volovikov) allows us to treat some interesting cases of non-free G-actions.

THEOREM 21.5.2 [Vol96a]

Let p be a prime number and $G = (\mathbb{Z}_p)^k$ an elementary abelian p-group. Suppose that X and Y are fixed-point free G-spaces such that $\widetilde{H}^i(X, \mathbb{Z}_p) \cong 0$ for all $i \leq n$ and Y is an n-dimensional cohomology sphere over $\mathbb{Z}_p$. Then there does not exist a G-equivariant map $f : X \to Y$.

A topological index theory is a complexity theory for G-spaces that allows us to conclude that there does not exist a G-equivariant map $f : X \to Y$ if the G-space X is of *larger complexity* than the G-space Y. A measure of complexity of a given G-space is the so-called *equivariant index* $\mathrm{Ind}_G(X)$. In general, an index function is defined on a class of G-spaces, say all finite G-CW complexes, and takes values in a suitable partially ordered set Ω. For example if $G = \mathbb{Z}_2$, an index function $\mathrm{Ind}_{\mathbb{Z}_2}(X)$ is defined as the minimum integer n such that there exists a $\mathbb{Z}_2$-equivariant map $f : X \to S^n$. In this case $\Omega := \mathbb{N}$ is the poset of nonnegative integers. Note that the Borsuk-Ulam theorem simply states that $\mathrm{Ind}_{\mathbb{Z}_2}(S^n) = n$.

PROPOSITION 21.5.3 [Mat08, Živ98]

For a nontrivial finite group G, there exists an integer-valued index function $\mathrm{Ind}_G(\cdot)$ *defined on the class of finite, G-simplicial complexes such that the following hold:*

(i) *If* $\mathrm{Ind}_G(X) > \mathrm{Ind}_G(Y)$, *then a G-equivariant map $f : X \to Y$ does not exist.*

(ii) *If X is $(n-1)$-connected then* $\mathrm{Ind}_G(X) \geq n$.

(iii) *If X is an n-dimensional, free G-complex then* $\mathrm{Ind}_G(X) \leq n$.

(iv) $\mathrm{Ind}_G(X * Y) \leq \mathrm{Ind}_G(X) + \mathrm{Ind}_G(Y) + 1$, *where $X * Y$ is the* join *of spaces.*

It is clear that the computation or good estimates of the complexity indices $\mathrm{Ind}_G(X)$ are essential for applications. Occasionally this can be done even if the details of construction of the index function are not known. Such a tool for finding the lower bounds for an index function described in Proposition 21.5.2 is provided by the following inequality.

PROPOSITION 21.5.4 *Sarkaria Inequality* [Mat08, Živ98]

Let L be a free G-complex and $L_0 \subset L$ a G-invariant, simplicial subcomplex. Let $\Delta(L \setminus L_0)$ be the order complex (cf. Chapter 20) *of the complementary poset $L \setminus L_0$. Then*

$$\mathrm{Ind}_G(L_0) \geq \mathrm{Ind}_G(L) - \mathrm{Ind}_G(\Delta(L \setminus L_0)) - 1.$$

The index function described Proposition 21.5.3 in its basic form relies on Proposition 21.5.1 and can be applied only to free group actions. A more general index function which has all the expected properties (including the Sarkaria's inequality) and which can be applied to some non-free group actions is described in [JVŽ16]. In some applications it is more natural, and sometimes essential, to use more sophisticated partially ordered sets of G-degrees of complexity. A notable example is the *ideal valued index theory* of S. Husseini and E. Fadell [FH88], which proved useful in establishing the existence of equilibrium points in incomplete markets (mathematical economics).

A BRIEF GUIDE TO THE LITERATURE

The book [Mat08] is an excellent introduction and a guide to other applications of topological methods, including the graph coloring problems (Kneser's conjecture). The monograph [Die87] provides a lot of foundational material and introduces the reader into the more advanced topics in the theory of transformation groups. The reader interested in applications to combinatorics and discrete geometry will find here both a detailed exposition of equivariant obstruction theory (Section II.3) and some very interesting applications (the Hopf classification theorem, the classification of equivariant maps between representation spheres, etc.).

The problem of the existence of G-equivariant maps $f : X \to Y$ (in the case when Y is a sphere) is closely related to the problem of finding a non-zero section of a vector bundle [Die87, Section I.7]. A far reaching "singularity approach" to this problem is described in the monograph [Kos81]. This approach can be quite effective, especially in the case when it is sufficient to calculate the first obstruction, see [Gro03, Section 5.1], [KHA14, Section 5.3], and [VŽ15, Section 2.4] for some illustrative examples.

RELATED CHAPTERS

Chapter 1: Finite point configurations
Chapter 4: Helly-type theorems and geometric transversals
Chapter 16: Subdivisions and triangulations of polytopes
Chapter 65: Applications to structural molecular biology
Chapter 66: Geometry and topology of genomic data

REFERENCES

[ABFK92] N. Alon, I. Bárány, Z. Füredi, and D. Kleitman. Point selections and weak ϵ-nets for convex hulls. *Combin. Probab. Comput.*, 1:189–200, 1992.

[AET00] N. Amenta, D. Eppstein, and S-H. Teng. Regression depth and center points. *Discrete Comput. Geom.*, 23:305–329, 2000.

[Alo87] N. Alon. Splitting necklaces. *Adv. Math.*, 63:247–253, 1987.

[AMSW17] S. Avvakumov, I. Mabillard, A. Skopenkov, and U. Wagner. Eliminating higher-multiplicity intersections, III. Codimension 2. Preprint, `arXiv:1511.03501`, version 4, 2017.

[BB89] I.K. Babenko and S.A. Bogatyi. On the mapping of a sphere into Euclidean space (Russian). *Mat. Zametki*, 46:3–8, 1989; translated in *Math. Notes*, 46:683–686, 1989.

[BFHZ15] P.V.M. Blagojević, F. Frick, A. Haase, and G.M. Ziegler. Topology of the Grünbaum-Hadwiger-Ramos hyperplane mass partition problem. Preprint, `arXiv:1502.02975`, 2015.

[BFHZ16] P.V.M. Blagojević, F. Frick, A. Haase, and G.M. Ziegler. Hyperplane mass partitions via relative equivariant obstruction theory. *Doc. Math.*, 21:735–771, 2016.

[BFL90] I. Bárány, Z. Füredi, and L. Lovász. On the number of halving planes. *Combinatorica*, 10:175–183, 1990.

[BFZ14] P.V.M. Blagojević, F. Frick, and G.M. Ziegler. Tverberg plus constraints. *B. London Math. Soc.*, 46:953–967, 2014.

[BFZ15] P.V.M. Blagojević, F. Frick, and G.M. Ziegler. Barycenters of polytope skeleta and counterexamples to the topological Tverberg conjecture, via constraints. Preprint, arXiv:1510.07984, 2015.

[Bjö95] A. Björner. Topological methods. In R. Graham, M. Grötschel, and L. Lovász, editors, *Handbook of Combinatorics*, pages 1819–1872, North-Holland, Amsterdam, 1995.

[BK14] B. Bukh and R.N. Karasev. Suborbits in Knaster's problem. *Bull. London Math. Soc.*, 46:269–278, 2014.

[BL92] I. Bárány and D.G. Larman. A colored version of Tverberg's theorem. *J. London Math. Soc.*, 45:314–320, 1992.

[BLVŽ94] A. Björner, L. Lovász, S.T. Vrećica, and R.T. Živaljević. Chessboard complexes and matching complexes. *J. London Math. Soc. (2)*, 49:25–39, 1994.

[BM01] I. Bárány and J. Matoušek. Simultaneous partitions of measures by k-fans, *Discrete Comput. Geom.*, 25:317–334, 2001.

[BM02] I. Bárány and J. Matoušek. Equipartitions of two measures by a 4-fan. *Discrete Comput. Geom.*, 27:293–301, 2002.

[BMZ11] P.V.M. Blagojević, B. Matschke, and G.M. Ziegler. Optimal bounds for a colorful Tverberg-Vrećica type problem. *Adv. Math.*, 226:5198–5215, 2011.

[BMZ15] P.V.M. Blagojević, B. Matschke, and G.M. Ziegler. Optimal bounds for the colored Tverberg problem. *J. Eur. Math. Soc.*, 17:739–754, 2015.

[Bor33] K. Borsuk. Drei Sätze über die n-dimensionale euklidische Sphäre. *Fund. Math.*, 20:177–190, 1933.

[BP15] V.M. Buchstaber and T.E. Panov, *Toric Topology*. Vol. 204 of *Math. Surveys Monog.*, AMS, Providence, 2015.

[Bre93] G.E. Bredon. *Topology and Geometry*, Vol. 139 of *Graduate Texts in Math.*, Springer, New York, 1993.

[Bro12] L.E.J. Brouwer. Über Abbildung von Mannigfaltigkeiten. *Math. Ann.*, 71:97–115, 1912.

[BSS81] I. Bárány, S.B. Shlosman, and A. Szűcs. On a topological generalization of a theorem of Tverberg. *J. London Math. Soc.*, 23:158–164, 1981.

[BT95] S.J. Brams and A.D. Taylor. An envy-free cake division protocol. *Amer. Math. Monthly*, 102:9–18, 1995.

[BZ14] P.V.M. Blagojević and G.M. Ziegler. Convex equipartitions via equivariant obstruction theory. *Israel J. Math.*, 200:49–77, 2014.

[Die87] T. tom Dieck. *Transformation Groups*. Vol. 8 of *Studies in Mathematics*, Walter de Gruyter, Berlin, 1987.

[Dol83] A. Dold. Simple proofs of some Borsuk-Ulam results. *Contemp. Math.*, 19:65–69, 1983.

[Dol93] V.L. Dol'nikov. Transversals of families of sets in $\mathbb{R}^n$ and a relationship between Helly and Borsuk theorems. *Mat. Sb.*, 184:111–131, 1993.

[FH88] E. Fadell and S. Husseini. An ideal-valued cohomological index theory with applications to Borsuk-Ulam and Bourgin-Yang theorems. *Ergodic Theory Dynam. Systems*, 8*:73–85, 1988.

[Fri15] F. Frick. Counterexamples to the topological Tverberg conjecture. *Oberwolfach Reports*, 12:318–321, 2015.

[Gro03] M. Gromov. Isoperimetry of waists and concentration of maps. *Geom. Funct. Anal.*, 13:178–215, 2003.

[Gro10] M. Gromov. Singularities, expanders and topology of maps. Part 2: From combinatorics to topology via algebraic isoperimetry. *Geom. Funct. Anal.*, 20:416–526, 2010.

[HMS02] T. Hausel, E. Makai, Jr., and A. Szűcs. Inscribing cubes and covering by rhombic dodecahedra via equivariant topology. *Mathematika*, 47:371–397, 2002.

[Jon08] J. Jonsson. *Simplicial Complexes of Graphs*. Vol. 1928 of *Lecture Notes in Math.*, Springer, Berlin, 2008.

[JVŽ16] D. Jojić, S. Vrećica, and R. Živaljević. Topology and combinatorics of 'unavoidable complexes.' Preprint, `arXiv:1603.08472`, 2016.

[JVŽ17a] D. Jojić, S. Vrećica, and R. Živaljević. Symmetric multiple chessboard complexes and a new theorem of Tverberg type. *J. Algebraic Combin.*, 46:15–31, 2017.

[JVŽ17b] D. Jojić, S. Vrećica, and R. Živaljević. Multiple chessboard complexes and the colored Tverberg problem. *J. Combin. Theory Ser. A*, 145:400–425, 2017.

[Kal01] G. Kalai. Combinatorics with a geometric flavor. In N. Alon, J. Bourgain, A. Connes, M. Gromov, and V. Milman, editors, *Visions in Mathematics: GAFA 2000 Special volume, Part II.*, pages 742–791, Birkhäuser, Basel, 2001.

[Kar07] R.N. Karasev. Tverberg's transversal conjecture and analogues of nonembeddability theorems for transversals. *Discrete Comput. Geom.*, 38:513–525, 2007.

[Kar09] R.N. Karasev. Dual theorems on central points and their generalizations. *Sbornik. Mathematics*, 199:1459–1479, 2008.

[KHA14] R.N. Karasev, A. Hubard, and B. Aronov. Convex equipartitions: the spicy chicken theorem. *Geom. Dedicata*, 170:263–279, 2014.

[KK93] J. Kahn and G. Kalai. A counterexample to Borsuk's conjecture. *Bull. Amer. Math. Soc.*, 29:60–62, 1993.

[KK99] A. Kaneko and M. Kano. Balanced partitions of two sets of points in the plane. *Comput. Geom.*, 13:253–261, 1999.

[KM14] R.N. Karasev and B. Matschke. Projective center point and Tverberg theorems. *Discrete Comput. Geom.*, 52:88–101, 2014.

[Kos81] U. Koschorke. *Vector Fields and Other Vector Bundle Morphisms: A Singularity Approach.* Vol. 847 of *Lecture Notes in Math.*, Springer, Heidelberg, 1981.

[KRPS16] R.N. Karasev, E. Roldán-Pensado, and P. Soberón. Measure partitions using hyperplanes with fixed directions. *Israel J. Math.*, 212:705–728, 2016.

[Kup99] G. Kuperberg. Circumscribing constant-width bodies with polytopes. *New York J. Math.* 5:91–100, 1999.

[Lon02] M. de Longueville. Notes on the topological Tverberg theorem, *Discrete Math.*, 241:207–233, 2001. Erratum: *Discrete Math.*, 247:271–297, 2002.

[LŽ08] M. de Longueville and R.T. Živaljević. Splitting multidimensional necklaces. *Adv. Math.*, 218:926–939, 2008.

[Mak86] V.V. Makeev. Some properties of continuous mappings of spheres and problems in combinatorial geometry. In L.D. Ivanov, editor, *Geometric Questions in the Theory of Functions and Sets* (Russian), pages 75–85, Kalinin State Univ., 1986.

[Mak90] V.V. Makeev. The Knaster problem on the continuous mappings from a sphere to a Euclidean space. *J. Soviet Math.*, 52:2854–2860, 1990.

[Mak94] V.V. Makeev. Inscribed and circumscribed polygons of a convex body. *Mat. Zametki*, 55:128–130, 1994; translated in *Math. Notes*, 55:423–425, 1994.

[Mak98] V.V. Makeev. Some special configurations of planes that are associated with convex compacta (Russian). *Zap. Nauchn. Sem. S.-Petersburg* (POMI), 252:165–174, 1998.

[Mak01] V.V. Makeev. Equipartition of a mass continuously distributed on a sphere and in space (Russian). *Zap. Nauchn. Sem. S.-Petersburg* (POMI), 279:187–196, 2001.

[Mat02] J. Matoušek. *Lectures on Discrete Geometry.* Volume 212 of *Graduate Texts in Math.*, Springer-Verlag, New York, 2002.

[Mat08] J. Matoušek. *Using the Borsuk-Ulam Theorem. Lectures on Topological Methods in Combinatorics and Geometry.* Second edition, Universitext, Springer, Heidelberg, 2008.

[MTTV97] G.L. Miller, S.-H. Teng, W. Thurston, and S.A. Vavasis. Separators for sphere-packings and nearest neighbor graphs. *J. ACM*, 44:1–29, 1997.

[MTTV98] G.L. Miller, S.-H. Teng, W. Thurston, S.A. Vavasis. Geometric separators for finite-element meshes. *SIAM J. Sci. Comput.*, 19:364–386, 1998.

[Mun84] J.R. Munkres. *Elements of Algebraic Topology.* Addison-Wesley, Menlo-Park, 1984.

[MVŽ01] E. Makai, S. Vrećica, and R. Živaljević. Plane sections of convex bodies of maximal volume. *Discrete Comput. Geom.*, 25:33–49, 2001.

[MW14] I. Mabillard and U. Wagner. Eliminating Tverberg points, I. An analogue of the Whitney trick. In *Proc. 30th Sympos. Comput. Geom.*, pages 171–180, ACM Press, 2014.

[MW15] I. Mabillard and U. Wagner. Eliminating higher-multiplicity intersections, I. A Whitney trick for Tverberg-type problems. Preprint, `arXiv:1508.02349`, 2015.

[MW16] I. Mabillard and U. Wagner. Eliminating higher-multiplicity intersections, II. The deleted product criterion in the r-metastable range. In *Proc. 32nd Sympos. Comput. Geom.*, article 51, vol. 51 of *LIPIcs*, Schloss Dagstuhl, 2016. Full version available at `arXiv:1601.00876`, 2016.

[Öza87] M. Özaydin. Equivariant maps for the symmetric group. Unpublished manuscript, available online at `http://minds.wisconsin.edu/handle/1793/63829`, 1987.

[Rad46] R. Rado. Theorem on general measure. *J. London Math. Soc.*, 21:291–300, 1946.

[Ram96] E.A. Ramos. Equipartitions of mass distributions by hyperplanes. *Discrete Comput. Geom.*, 15:147–167, 1996.

[Sar91] K.S. Sarkaria, A generalized van Kampen-Flores theorem. *Proc. Amer. Math. Soc.*, 11:559–565, 1991.

[Sar92] K.S. Sarkaria. Tverberg's theorem via number fields. *Israel J. Math.*, 79:317–320, 1992.

[Sob12] P. Soberón. Balanced convex partitions of measures in $\mathbb{R}^d$. *Mathematika*, 58:71-76, 2012.

[Ste85] H. Steinlein. Borsuk's antipodal theorem and its generalizations and applications: a survey. In *Topological Methods in Nonlinear Analysis*, volume 95 of *Sém. Math. Sup.*, pages 166–235, Presses de l'Université de Montréal, 1985.

[TV93] H. Tverberg and S. Vrećica. On generalizations of Radon's theorem and the ham sandwich theorem. *Europ. J. Combin.*, 14:259–264, 1993.

[Tve66] H. Tverberg. A generalization of Radon's theorem. *J. London Math. Soc.*, 41:123–128, 1966.

[Vol96a] A.Y. Volovikov. On a topological generalization of the Tverberg theorem. *Math. Notes*, 59:324–32, 1996.

[Vol96b] A.Y. Volovikov. On the van Kampen-Flores theorem, *Math. Notes*, 59:477–481, 1996.

[Vre03] S. Vrećica. Tverberg's conjecture. *Discrete Comput. Geom.*, 29:505–510, 2003.

[Vre09] S. Vrećica. Equipartitions of a mass in boxes. *J. Combin. Theory Ser. A*, 116:132–142, 2009.

[VŽ92] S. Vrećica and R. Živaljević. The ham sandwich theorem revisited. *Israel J. Math.*, 78:21–32, 1992.

[VŽ94] S. Vrećica and R. Živaljević. New cases of the colored Tverberg theorem. In H. Barcelo and G. Kalai, editors, *Jerusalem Combinatorics'93*, pages 325–334. Vol. 178 of *Contemp. Math.*, AMS, Providence, 1994.

[VŽ11] S. Vrećica and R. Živaljević. Chessboard complexes indomitable. *J. Combin. Theory Ser. A*, 118:2157–2166, 2011.

[VŽ15] S. Vrećica and R. Živaljević. Hyperplane mass equipartition problem and the shielding functions of Ramos. Preprint, `arXiv:1508.01552`, 2015.

[Zie15] G.M. Ziegler. Cannons at sparrows. *Eur. Math. Soc. Newsl.*, 95:25–31, 2015.

[Živ98] R. Živaljević. User's guide to equivariant methods in combinatorics, I and II. *Publ. Inst. Math. (Beograd) (N.S.)*, (I) 59:114–130, 1996 and (II) 64:107–132, 1998.

[Živ99] R. Živaljević. The Tverberg-Vrećica problem and the combinatorial geometry on vector bundles. *Israel. J. Math.*, 111:53–76, 1999.

[Živ15] R. Živaljević. Computational topology of equipartitions by hyperplanes. *Topol. Methods Nonlinear Anal.*, 45:63–90, 2015.

[ŽV90] R. Živaljević and S. Vrećica. An extension of the ham sandwich theorem. *Bull. London Math. Soc.*, 22:183–186, 1990.

[ŽV92] R.T. Živaljević and S.T. Vrećica. The colored Tverberg's problem and complexes of injective functions. *J. Combin. Theory Ser. A*, 61:309–318, 1992.

22 RANDOM SIMPLICIAL COMPLEXES

Matthew Kahle

INTRODUCTION

Random shapes arise naturally in many contexts. The topological and geometric structure of such objects is interesting for its own sake, and also for applications. In physics, for example, such objects arise naturally in quantum gravity, in material science, and in other settings. Stochastic topology may also be considered as a null hypothesis for topological data analysis.

In this chapter we overview combinatorial aspects of stochastic topology. We focus on the topological and geometric properties of random simplicial complexes. We introduce a few of the fundamental models in Section 22.1. We review high-dimensional expander-like properties of random complexes in Section 22.2. We discuss threshold behavior and phase transitions in Section 22.3, and Betti numbers and persistent homology in Section 22.4.

22.1 MODELS

We briefly introduce a few of the most commonly studied models.

22.1.1 ERDŐS–RÉNYI-INSPIRED MODELS

A few of the models that have been studied are high-dimensional analogues of the Erdős–Rényi random graph.

The Erdős–Rényi random graph

The Erdős–Rényi random graph $G(n,p)$ is the probability distribution on all graphs on vertex set $[n] = \{1, 2, \ldots, n\}$, where every edge is included with probability p jointly independently. Standard references include [Bol01] and [JŁR00].

One often thinks of p as a function of n and studies the asymptotic properties of $G(n,p)$ as $n \to \infty$. We say that an event happens *with high probability (w.h.p.)* if the probability approaches 1 as $n \to \infty$.

Erdős–Rényi showed that $\bar{p} = \log n/n$ is a sharp threshold for connectivity. In other words,: for every fixed $\epsilon > 0$, if $p \geq (1+\epsilon)\bar{p}$ then w.h.p. $G(n,p)$ is connected, and if $p \leq (1-\epsilon)\bar{p}$ then w.h.p. it is disconnected. A slightly sharper statement is given in the following section. Several thresholds for topological properties of $G(n,p)$ are summarized in Table 22.1.1.

TABLE 22.1.1 Topological thresholds for $G = G(n,p)$. The column SHARP indicates whether the threshold is sharp, coarse, or one-sided sharp. The column TIGHT indicates whether there is any room for improvement on the present bound.

PROPERTY	THRESHOLD	SHARP	TIGHT	SOURCE
G is not 0-collapsible	$1/n$	one-sided	yes	[Pit88]
$H_1(G) \neq 0$	$1/n$	one-sided	yes	[Pit88]
G is not planar	$1/n$	sharp	yes	[ŁPW94]
G contains arbitrary minors	$1/n$	sharp	yes	[AKS79]
G is pure 1-dimensional	$\log n/n$	sharp	yes	[ER59]
G is connected	$\log n/n$	sharp	yes	[ER59]

GLOSSARY

Threshold function: Let $\mathcal{P}$ be a graph property. We say that f is a threshold function for property $\mathcal{P}$ in the random graph $G \in G(n,p)$ if whenever $p = \omega(f)$, G has property $\mathcal{P}$ w.h.p. and whenever $p = o(f)$, G does not have property $\mathcal{P}$.

Sharp threshold: We say that f is a sharp threshold for graph property $\mathcal{P}$ if there exists a function $g = o(f)$ such that for $p < f - g$, $G \notin \mathcal{P}$ w.h.p. and if $p > f + g$, $G \in \mathcal{P}$ w.h.p.

Simplicial complex: A simplicial complex Δ is a collection of subsets of a set S, such that (1) if $U \subset V$ is nonempty and $V \in \Delta$ then $U \in \Delta$, and (2) $\{v\} \in \Delta$ for every $v \in S$. An element of Δ is called a face. Such a set system can be naturally associated a topological space by considering every set of size k in Δ to represent a $(k-1)$-dimensional simplex, homeomorphic to a closed Euclidean ball. This topological space is sometimes called the geometric realization of Δ, but we will slightly abuse notation and identify a simplicial complex with its geometric realization.

Link: Given a simplicial complex Δ and a face $\sigma \in \Delta$, the link of σ in Δ is defined by

$$\mathrm{lk}_\Delta(\sigma) = \{\tau \in \Delta \mid \tau \cap \sigma = \emptyset \text{ and } \tau \cup \sigma \in \Delta\}.$$

The link is itself a simplicial complex.

Homology: Associated with any simplicial complex X, abelian group G, and integer $i \geq 0$, $H_i(X,G)$ denotes the ith homology group of X with coefficients in G. If k is a field, then $H_i(X,k)$ is a vector space over k.

Homology is defined as "cycles modulo boundaries." Homology is invariant under homotopy deformations.

Betti numbers: If one considers homology with coefficients in $\mathbb{R}$, then $H_i(X,\mathbb{R})$ is a real vector space. The Betti numbers β_i are defined by $\beta_i = \dim H_i(X,\mathbb{R})$. The 0th Betti number β_0 counts the number of connected components of X, and in general the ith Betti number is said to count the number of i-dimensional holes in X.

The random 2-complex

Random hypergraphs have been well studied, but if we wish to study such objects topologically then random simplicial complexes is probably a more natural point of view.

Linial and Meshulam introduced the topological study of the random 2-complex $Y(n,p)$ in [LM06]. This model of random simplicial complex has n vertices, $\binom{n}{2}$ edges, and each of the $\binom{n}{3}$ possible 2-dimensional faces is included independently with probability p.

The random 2-complex is perhaps the most natural 2-dimensional analogue of $G(n,p)$. For example, the link of every vertex in $Y(n,p)$ has the same distribution as $G(n-1,p)$.

Several topological thresholds for $Y(n,p)$ discussed in the next section are described in Table 22.1.2.

TABLE 22.1.2 Topological thresholds for the random 2-complex $Y = Y(n,p)$. c.f. in the TIGHT column means that the bound is best possible up to a constant factor.

PROPERTY	THRESHOLD	SHARP	TIGHT	SOURCE
Y is not 1-collapsible	$2.455/n$	one-sided	yes	[CCFK12, ALŁM13, AL16]
$H_2(Y,\mathbb{R}) \neq 0$	$2.753/n$	one-sided	yes	[Koz10, LP14, AL15]
cdim $\pi_1(Y) = 2$	$\Theta(1/n)$	?	c.f.	[CF15a, New16]
Y is not embeddable in $\mathbb{R}^4$	$\Theta(1/n)$	?	c.f.	[Wag11]
Y is pure 2-dimensional	$2\log n/n$	sharp	yes	[LM06]
$H_1(Y,\mathbb{Z}/\ell\mathbb{Z}) = 0$	$2\log n/n$	sharp	yes	[LM06, MW09]
$H_1(Y,\mathbb{R}) = 0$	$2\log n/n$	sharp	yes	[LM06, HKP12]
$\pi_1(Y)$ has property (T)	$2\log n/n$	sharp	yes	[HKP12]
$H_1(Y,\mathbb{Z}) = 0$	$O(\log n/n)$	sharp	c.f.	[HKP17]
cdim $\pi_1(Y) = \infty$	$1/n^{3/5}$	coarse	yes	[CF13]
Y contains arbitrary subdivisions	$\theta\left(1/\sqrt{n}\right)$	?	c.f.	[GW16]
$\pi_1(Y) = 0$	$O(1/\sqrt{n})$	?	no	[BHK11, GW16, KPS16]

TABLE 22.1.3 Topological thresholds for $Y_d(n,p)$, $d > 2$. Definitions for the constants c_d and c_d^* are given in Section 22.3.2.

PROPERTY	THRESHOLD	SHARP	TIGHT	SOURCE
Y is not $(d-1)$-collapsible	c_d/n	one-sided	yes	[ALŁM13, AL16]
$H_d(Y,\mathbb{R}) \neq 0$	c_d^*/n	one-sided	yes	[Koz10, LP14, AL15]
Y is not embeddable in $\mathbb{R}^{2d}$	$\Theta(1/n)$	?	c.f.	[Wag11]
Y is pure d-dimensional	$d\log n/n$	sharp	yes	[MW09]
$H_{d-1}(Y,\mathbb{Z}/\ell\mathbb{Z}) = 0$	$d\log n/n$	sharp	yes	[MW09, HKP12]
$H_{d-1}(Y,\mathbb{R}) = 0$	$d\log n/n$	sharp	yes	[HKP12]
$H_{d-1}(Y,\mathbb{Z}) = 0$	$O(\log n/n)$	sharp	c.f.	[HKP17]
$\pi_{d-1}(Y) = 0$	$O(\log n/n)$	sharp	c.f.	[HKP17]

The random d-complex

The natural generalization to d-dimensional model was introduced by Meshulam and Wallach in [MW09]. For the random d-complex $Y_d(n,p)$, contains the complete $(d-1)$-skeleton of a simplex on n vertices, and every d-dimensional face appears independently with probability p. Some of the topological subtlety of the random 2-dimensional model collapses in higher dimensions: for $d \geq 3$, the complexes are $d-2$-connected, and in particular simply connected. By the Hurewicz theorem, $\pi_{d-1}(Y)$ is isomorphic to $H_{d-1}(Y, \mathbb{Z})$, so these groups have the same vanishing threshold.

The random clique complex

Another analogue of $G(n,p)$ in higher dimensions was introduced in [Kah09]. The random clique complex $X(n,p)$ is the *clique complex* of $G(n,p)$. It is the maximal simplicial complex compatible with a given graph. In other words, the faces of the clique complex $X(H)$ correspond to complete subgraphs of the graph H.

The random clique complex asymptotically puts a measure over a wide range of topologies. Indeed, every simplicial complex is homeomorphic to a clique complex, e.g., by barycentric subdivision.

There are several comparisons of this model to the random d-complex, but some important contrasts as well. One contrast to $Y_d(n,p)$ is that for every $k \geq 1$, $X(n,p)$ has not one but two phase transitions for kth homology, one where homology appears and one where it vanishes. In particular, higher homology is not monotone with respect to p. However, there are still comparisons to $Y_d(n,p)$—the appearance of homology $H_k(X)$ is analogous to the birth of top homology $H_d(Y)$. Similarly, the vanishing threshold for $H_k(X(n,p))$ is analogous to vanishing of $H_{d-1}(Y)$.

TABLE 22.1.4 Topological thresholds for $X(n,p)$

PROPERTY	THRESHOLD	SHARP	TIGHT	SOURCE
$\pi_1(X) = 0$	$p = 1/n^{1/3}$	?	no	[Kah09, Bab12, CFH15]
$H_k(X, \mathbb{R}) \neq 0$	$p = c/n^{1/k}$	one-sided	c.f.	[Kah09]
X^k is pure k-dimensional	$p = \left(\frac{(k/2+1)\log n}{n}\right)^{1/(k+1)}$	yes	yes	[Kah09]
$H_k(Y, \mathbb{R}) = 0$	$p = \left(\frac{(k/2+1)\log n}{n}\right)^{1/(k+1)}$	yes	yes	[Kah14a]

The multi-parameter model

There is a natural multi-parameter model which generalizes all of the models discussed so far. For every $i = 1, 2, \ldots$ let $p_i : \mathbb{N} \to [0,1]$. Then define the multiparameter random complex $X(n; p_1, p_2, \ldots)$ as follows. Start with n vertices. Insert every edge with probability p_1. Conditioned on the presence of all three boundary edges, insert a 2-face with probability p_2, etc.

The random d-complex $Y_d(n,p)$ with complete $(d-1)$-skeleton is equivalent to the case $p_1 = p_2 = \cdots = p_{d-1} = 1$, $p = p_d$, and $p_{d+1} = p_{d+2} = \cdots = 0$. The random clique complex $X(n,p)$ corresponds to $p_2 = p_3 = \cdots = 1$, and $p = p_1$. This multi-parameter model is first studied by Costa and Farber in [CF16].

22.1.2 RANDOM GEOMETRIC MODELS

The *random geometric graph* $\mathcal{G}(n,r)$ is a flexible model, defined as follows. Consider a probability distribution on $\mathbb{R}^d$ with a bounded, measurable, density function $f : \mathbb{R}^d \to \mathbb{R}$. Then one chooses n points independently and identically distributed (i.i.d.) according to this distribution. The n points are the vertices of the graph, and two vertices are adjacent if they are within distance r. Usually $r = r(n)$ and $n \to \infty$. The standard reference for random geometric graphs is Penrose's monograph [Pen03].

There are at least two commonly studied ways to build a simplicial complex on a geometric graph. The first is the Vietoris–Rips complex, which is the same construction as the clique complex above—one fills in all possible faces, i.e., the faces of the Vietoris–Rips complex correspond to the cliques of the graph. The second is the Čech complex, where one considers the higher intersections of the balls of radius $r/2$. This leads to two natural models for random geometric complexes $VR(n,r)$ and $C(n,r)$.

TABLE 22.1.5 Topological thresholds for $C(n,r)$

PROPERTY	THRESHOLD	SHARP	TIGHT	SOURCE
$H_k(X) \neq 0$	$nr^d = n^{-(k+2)/(k+1)}$	coarse	yes	[Kah11]
$H_k(Y) = 0$	$nr^d = \log n + \theta(\log\log n)$	sharp	essentially	[BW17]

TABLE 22.1.6 Topological thresholds for $VR(n,r)$

PROPERTY	THRESHOLD	SHARP	TIGHT	SOURCE
$H_k(X) \neq 0$	$nr^d = n^{-(2k+2)/(2k+1)}$	coarse	yes	[Kah11]
$H_k(Y) = 0$	$nr^d = \theta(\log n)$	sharp	c.f.	[Kah11]

22.2 HIGH-DIMENSIONAL EXPANDERS

GLOSSARY

Cheeger number: The normalized Cheeger number of a graph $h(G)$ with vertex set V is defined by

$$h(G) = \min_{\emptyset \subsetneq A \subsetneq V} \frac{\#E(A,\bar{A})}{\min\{\operatorname{vol} A, \operatorname{vol} \bar{A}\}},$$

where

$$\operatorname{vol} A = \sum_{v \in A} \deg(v)$$

and $\bar{A}$ is the complement of A in V.

Laplacian: For a connected graph H, the normalized graph Laplacian $L = L[H]$ is defined by

$$L = I - D^{-1/2}AD^{-1/2}.$$

Here A is the adjacency matrix, and D is the diagonal matrix with vertex degrees along the diagonal.

Spectral gap: The eigenvalues of the normalized graph Laplacian of a connected graph satisfy

$$0 = \lambda_1 < \lambda_2 \leq \cdots \leq \lambda_n \leq 2.$$

The smallest positive eigenvalue $\lambda_2[H]$ is of particular importance, and is sometimes called the spectral gap of H.

Expander family: Let $\{G_i\}$ be an infinite sequence of graphs where the number of vertices tends to infinity. We say that $\{G_i\}$ is an expander family if

$$\liminf \lambda_2[G_i] > 0.$$

Expander graphs are of fundamental importance for their applications in computer science and mathematics [HLW06]. It is natural to seek their various higher-dimensional generalizations. See [Lub14] for a survey of recent progress on higher-dimensional expanders, particularly Ramanujan complexes which generalize Ramanujan graphs.

Gromov suggested that one property that higher-dimensional expanders should have is geometric or topological overlap. A sequence of d-dimensional simplicial complexes $\Delta_1, \Delta_2, \ldots,$ is said to have the *geometric overlap property* if for every geometric map (affine-linear on each face) $f : \Delta_i \to \mathbb{R}^d$, there exists a point $p \in \mathbb{R}^d$ such that $f^{-1}(p)$ intersects the interior of a constant fraction of the d-dimensional faces. The sequence is said to have the stronger *topological overlap property* if this holds even for continuous maps.

One way to define a higher-dimensional expander is via *coboundary expansion*, which generalizes the Cheeger number of a graph. Following Linial and Meshulam's coisoperimetric ideas, Dotterrer and Kahle [DK12] pointed out that d-dimensional random simplicial complexes are coboundary expanders. By a theorem of Gromov [Gro09, Gro10] (see the note [DKW15] for a self-contained proof of Gromov's theorem), this implies that random complexes have the topological overlap property.

Lubotzky and Meshulam introduced a new model of random 2-complex, based on random Latin squares, in [LM15]. The main result is the existence of coboundary expanders with bounded edge-degree, answering a question asked implicitly in [Gro10] and explicitly in [DK12].

Another way to define higher-dimensional expanders is via the *spectral gap* of various Laplacian operators. Hoffman, Kahle, and Paquette studied the spectral gap of random graphs [HKP12], and applied Garland's method to prove homology-vanishing theorems. Gundert and Wagner extended this to study higher-order spectral gaps of these complexes [GW16]. Parzanchevski, Rosenthal, and Tessler showed that this implies the geometric overlap property [PRT15].

22.3 PHASE TRANSITIONS

There has been a lot of interest in identifying thresholds for various topological properties, such as vanishing of homology. As some parameter varies, the topology passes a *phase transition* where some property suddenly emerges. In this section we review a few of the most well-studied topological phase transitions.

22.3.1 HOMOLOGY-VANISHING THEOREMS

The following theorem describing the connectivity threshold for the random graph $G(n,p)$ is the archetypal homology-vanishing theorem.

THEOREM 22.3.1 *Erdős–Rényi theorem* [ER59]

If

$$p \geq \frac{\log n + \omega(1)}{n}$$

then w.h.p. $G(n,p)$ is connected, and if

$$p \leq \frac{\log n - \omega(1)}{n}$$

then w.h.p. $G(n,p)$ is disconnected. Here $\omega(1)$ is any function so that $\omega(1) \to \infty$ as $n \to \infty$.

We say that it is a homology-vanishing theorem because path connectivity of a topological space X is equivalent to $\widetilde{H}_0(X,G) = 0$ with any coefficient group G.

It is also a cohomology-vanishing theorem, since $\widetilde{H}^0(X,G) = 0$ is also equivalent to path-connectivity. In many ways it is better to think of it as a cohomology theorem, since the standard proof, for example in [Bol01][Chapter 10], is really a cohomological one. This perspective helps when understanding the proof of the Linial–Meshulam theorem.

The following cohomological analogue of the Erdős–Rényi theorem was the first nontrivial result for the topology of random simplicial complexes.

THEOREM 22.3.2 *Linial–Meshulam theorem* [LM06]

Let $Y = Y(n,p)$. If

$$p \geq \frac{2\log n + \omega(1)}{n}$$

then w.h.p. $H_1(Y, \mathbb{Z}/2\mathbb{Z}) = 0$, and if

$$p \leq \frac{2\log n - \omega(1)}{n}$$

then w.h.p. $H_1(Y, \mathbb{Z}/2\mathbb{Z}) \neq 0$.

One of the main tools introduced in [LM06] is a new co-isoperimetric inequality for the simplex, which was discovered independently by Gromov. These co-isoperimetric inequalities were combined by Linial and Meshulam with intricate cocycle-counting combinatorics to get a sharp threshold.

See [DKW15] for a comparison of various definitions co-isoperimetry, and a clean statement and self-contained proof of Gromov's theorem.

Theorem 22.3.2 was generalized further by Meshulam and Wallach.

THEOREM 22.3.3 [MW09]

Fix $d \geq 1$, and let $Y = Y_d(n,p)$. Let G be any finite abelian group. If

$$p \geq \frac{d \log n + \omega(1)}{n}$$

then w.h.p. $H_{d-1}(Y,G) = 0$, and if

$$p \leq \frac{d \log n - \omega(1)}{n}$$

then w.h.p. $H_{d-1}(Y,G) \neq 0$.

Theorem 22.3.3 generalizes Theorem 22.3.2 in two ways: by letting the dimension $d \geq 2$ be arbitrary, and also by letting coefficients be in an arbitrary finite abelian group G.

Spectral gaps and Garland's method

There is another approach to homology-vanishing theorems for simplicial complexes, via Garland's method [Gar73]. The following refinement of Garland's theorem is due to Ballman and Światkowski.

THEOREM 22.3.4 [Gar73, BŚ97]

If Δ is a finite, pure d-dimensional, simplicial complex, such that

$$\lambda_2[lk_\Delta(\sigma)] > 1 - \frac{1}{d}$$

for every $(d-2)$-dimensional face $\sigma \in \Delta$, then $H_{d-1}(\Delta, \mathbb{R}) = 0$.

This leads to a new proof of Theorem 22.3.3, at least over a field of characteristic zero.

THEOREM 22.3.5 [HKP12]

Fix $d \geq 1$, and let $Y = Y_d(n,p)$. If

$$p \geq \frac{d \log n + \omega(1)}{n}$$

then w.h.p. $H_{d-1}(Y, \mathbb{R}) = 0$, and if

$$p \leq \frac{d \log n - \omega(1)}{n}$$

then w.h.p. $H_{d-1}(Y, \mathbb{R}) \neq 0$. Here $\omega(1)$ is any function that tends to infinity as $n \to \infty$.

This is slightly weaker than the Meshulam–Wallach theorem topologically speaking, since $H_i(Y,G) = 0$ for any finite group G implies that $H_i(Y,\mathbb{R}) = 0$ by the universal coefficient theorem, but generally the converse is false. However, the proof via Garland's method avoids some of the combinatorial complications of co-cycle counting.

Garland's method also provides proofs of theorems which have so far eluded other methods. For example, we have the following homology-vanishing threshold in the random clique complex model. Note that $k = 0$ again corresponds to the Erdős–Rényi theorem.

THEOREM 22.3.6 [Kah14a]

Fix $k \geq 1$ and let $X = X(n,p)$. Let $\omega(1)$ denote a function that tends to ∞ arbitrarily slowly. If

$$p \geq \left(\frac{\left(\frac{k}{2}+1\right)\log n + \left(\frac{k}{2}\right)\log\log n + \omega(1)}{n} \right)^{1/(k+1)}$$

then w.h.p. $H_k(X,\mathbb{R}) = 0$, and if

$$\frac{1}{n^k} \leq p \leq \left(\frac{\left(\frac{k}{2}+1\right)\log n + \left(\frac{k}{2}\right)\log\log n - \omega(1)}{n} \right)^{1/(k+1)}$$

then w.h.p. $H_k(X,\mathbb{R}) \neq 0$.

It may be that this theorem holds with $\mathbb{R}$ coefficients replaced by a finite group G or even with $\mathbb{Z}$, but for the most part this remains an open problem. The only other case that seems to be known is the case $k = 1$ and $G = \mathbb{Z}/2$ by DeMarco, Hamm, and Kahn [DHK13], where a similarly sharp threshold is obtained.

The applications of Garland's method depends on new results on the spectral gap of random graphs.

THEOREM 22.3.7 [HKP12]

Fix $k \geq 0$. Let $\lambda_1, \lambda_2, \ldots$ denote the eigenvalues of the normalized graph Laplacian of the random graph $G(n,p)$. If

$$p \geq \frac{(k+1)\log n + \omega(1)}{n}$$

then

$$1 - \sqrt{\frac{C}{np}} \leq \lambda_2 \leq \cdots \leq \lambda_n \leq 1 + \sqrt{\frac{C}{np}}$$

with probability at least $1 - o\left(n^{-k}\right)$. Here $C > 0$ is a universal constant.

Theorem 22.3.6, combined with some earlier results [Kah09], has the following corollary.

THEOREM 22.3.8 [Kah14a]

Let $k \geq 3$ and $\epsilon > 0$ be fixed. If

$$\left(\frac{(C_k + \epsilon)\log n}{n} \right)^{1/k} \leq p \leq \frac{1}{n^{1/(k+1)+\epsilon}},$$

where $C_3 = 3$ and $C_k = k/2 + 1$ for $k > 3$, then w.h.p. X is rationally homotopy equivalent to a bouquet of k-dimensional spheres.

The main remaining conjecture for the topology of random clique complexes is that these rational homotopy equivalences are actually homotopy equivalences.

CONJECTURE 22.3.9 *The bouquet-of-spheres conjecture.*
Let $k \geq 3$ and $\epsilon > 0$ be fixed. If

$$\frac{n^{\epsilon}}{n^{1/k}} \leq p \leq \frac{n^{-\epsilon}}{n^{1/(k+1)}}$$

then w.h.p. X is homotopy equivalent to a bouquet of k-spheres.

Given earlier results, this is equivalent to showing that $H_k(X, \mathbb{Z})$ is torsion free. So far, integer homology for $X(n,p)$ is not very well understood. Some progress has been made for $Y_d(n,p)$, described in the following.

Integer homology

Unfortunately, neither method discussed above (the cocycle-counting methods pioneered by Linial and Meshulam or the spectral methods of Garland), seems to handle integral homology. There is a slight subtlety here—if one knows for some simplicial complex Σ that $H_i(\Sigma, G) = 0$ for every finite abelian group G then $H_i(\Sigma, \mathbb{Z}) = 0$ by the universal coefficient theorem. See, for example, Hatcher [Hat02][Chapter 2].

So it might seem that the Theorem 22.3.3 will also handle $\mathbb{Z}$ coefficients, but the proof uses cocycle counting methods which require G to be fixed, or at least for the order of the coefficient group $|G|$ to be growing sufficiently slowly. Cocycle counting does not seem to work, for example, when $|G|$ is growing exponentially fast. The following gives an upper bound on the vanishing threshold for integer homology.

THEOREM 22.3.10 [HKP12]
Fix $d \geq 2$, and let $Y = Y_d(n,p)$. If

$$p \geq \frac{80d \log n}{n},$$

then w.h.p. $H_{d-1}(Y, \mathbb{Z}) = 0$.

The author suspects that the true threshold for homology with $\mathbb{Z}$ coefficients is the same as for field coefficients: $d \log n / n$.

CONJECTURE 22.3.11 *A sharp threshold for $\mathbb{Z}$ homology.*
If

$$p \geq \frac{d \log n + \omega(1)}{n}$$

then w.h.p. $H_{d-1}(Y, \mathbb{Z}) = 0$, and if

$$p \leq \frac{d \log n - \omega(1)}{n}$$

then w.h.p. $H_{d-1}(Y, \mathbb{Z}) \neq 0$.

22.3.2 THE BIRTH OF CYCLES AND COLLAPSIBILITY

$G(n,p)$ in the $p = 1/n$ regime

There is a remarkable phase transition in structure of the random graph $G(n,p)$ at the threshold $p = 1/n$. A "giant" component, on a constant fraction of the vertices, suddenly emerges. This is considered an analogue of percolation on an infinite lattice, where an infinite component appears with probability 1.

THEOREM 22.3.12 [ER59]

Let $p = c/n$ for some $c > 0$ fixed, and $G = G(n,p)$.

- *If $c < 1$ then w.h.p. all components are of order $O(\log n)$.*
- *If $c > 1$ then w.h.p. there is a unique giant component, of order $\Omega(n)$.*

An overview of this remarkable phase transition can be found in Chapter 11 of Alon and Spencer [AS08].

In random graphs, the appearance of cycles with high probability has the same threshold $1/n$.

THEOREM 22.3.13 [Pit88]

Suppose $p = c/n$ where $c > 0$ is constant.

- *If $c \geq 1$ then w.h.p. G contains at least one cycle, i.e.,*

$$\mathbb{P}\left[H_1(G) \neq 0\right] \to 1.$$

- *If $c < 1$ then*

$$\mathbb{P}\left[H_1(G) = 0\right] \to \sqrt{1-c}\,\exp(c/2 + c^2/4).$$

The analogy in higher dimensions is only just beginning to be understood.

The birth of cycles

Kozlov first studied the vanishing threshold for top homology in [Koz10].

THEOREM 22.3.14 [Koz10]

Let $Y = Y_d(n,p)$, and G be any abelian group.

(1) *If $p = o(1/n)$ then w.h.p. $H_d(Y,G) = 0$.*

(2) *If $p = \omega(1/n)$ then w.h.p. $H_d(Y,G) \neq 0$.*

Part (1) of this theorem cannot be improved. Indeed, let S be the number of subcomplexes isomorphic to the boundary of a $(d+1)$-dimensional simplex. If $p = c/n$ for some constant $c > 0$, then

$$\mathbb{E}[S] \to c^{d+2}/(d+2)!,$$

as $n \to \infty$. Moreover, S converges in law to a Poisson distribution with this mean in the limit, so

$$\mathbb{P}[H_d(Y,G) \neq 0] \geq \mathbb{P}[S \neq 0] \to 1 - \exp(-c^{d+2}/(d+2)!).$$

In particular, for $p = c/n$ and $c > 0$, $\mathbb{P}[H_d(Y,G) \neq 0]$ is bounded away from zero.

On the other hand, part (2) can be improved. Indeed straightforward computation shows that if $p \geq c/n$ and $c > d+1$ then w.h.p. the number of d-dimensional faces is greater than the number of $(d-1)$-dimensional faces. Simply by dimensional considerations, we conclude that $H_d(Y,G) \neq 0$.

This can be improved more though. Aronshtam and Linial found the best possible constant factor c_d^*, defined for $d \geq 2$ as follows.

Let $x \in (0,1)$ be the unique root to the equation

$$(d+1)(1-x) + (1+dx)\log x = 0,$$

and then set

$$c_d^* = \frac{-\log x}{(1-x)^d}.$$

THEOREM 22.3.15 [AL15]

Let $Y = Y_d(n,p)$. If $p \geq c/n$ where $c > c_d^*$, then w.h.p. $H_d(Y,G) \neq 0$.

In the other direction, Linial and Peled showed that this result is tight, at least in the case of $\mathbb{R}$ coefficients.

THEOREM 22.3.16 [LP14]

If $p \leq c/n$ where $c < c_d^*$ then w.h.p. $H_d(Y,G)$ is generated by simplex boundaries. So

$$\mathbb{P}\left[H_d(Y,\mathbb{R}) = 0\right] \to \exp(-c^{d+2}/(d+2)!).$$

Linial and Peled also showed the birth of a *giant (homological) shadow* at the same point. This is introduced and defined in [LP14], and it is discussed there as a higher-dimensional analogue of the birth of the giant component in $G(n,p)$.

The threshold for d-collapsibility

In a d-dimensional simplicial complex, an *elementary collapse* is an operation that deletes a pair of faces (σ, τ) such that

- τ is a d-dimensional face,
- σ is a $(d-1)$-dimensional face contained in τ, and
- σ is not contained in any other d-dimensional faces.

An elementary collapse results in a homotopy equivalent simplicial complex.

If a simplicial complex can be reduced to a $(d-1)$-dimensional complex by a series of elementary collapses, we say that it is *d-collapsible*.

For a graph, 1-collapsible is equivalent to being a forest. In other words, a graph G is 1-collapsible if and only if $H_1(G) = 0$. This homological criterion does not hold in higher dimensions. In fact, somewhat surprisingly, d-collapsibility and $H_d \neq 0$ have distinct thresholds for random complexes.

Let $d \geq 2$ and set

$$g_d(x) = (d+1)(x+1)e^{-x} + x(1-e^{-x})^{d+1}.$$

Define c_d to be the unique solution $x > 0$ of $g_d(x) = d + 1$.

THEOREM 22.3.17 [ALŁM13, AL16]
Let $Y = Y_d(n, p)$.

- *If $p \geq c/n$ where $c > c_d$ then w.h.p. Y is not d-collapsible, and*
- *if $p \leq c/n$ where $c < c_d$ then Y is d-collapsible with probability bounded away from zero.*

So again, this is a one-sided sharp threshold. Regarding collapsibility in the random clique complex model, Malen showed in his PhD thesis [Mal16] that if $p \ll n^{-1/(k+1)}$, then w.h.p. $X(n, p)$ is k-collapsible.

Embeddability

Every d-dimensional simplicial complex is embeddable in $\mathbb{R}^{2d+1}$, but not necessarily in $\mathbb{R}^{2d}$. Wagner studied the threshold for non-embeddability of random d-complexes in $\mathbb{R}^{2d}$, and showed the following for $Y = Y_d(n, p)$.

THEOREM 22.3.18 [Wag11]
There exist constants $c_1, c_2 > 0$ depending only on the dimension d such that:

- *if $p < c_1/n$ then w.h.p. Y is embeddable in $\mathbb{R}^{2d}$, and*
- *if $p > c_2/n$ then w.h.p. Y is not embeddable in $\mathbb{R}^{2d}$.*

There is a folklore conjecture that a d-dimensional simplicial complex on n vertices embeddable in $\mathbb{R}^{2d}$ can have at most $O(n^d)$ faces [Dey93, Kal91]. See, for example, the discussion in the expository book chapter [Wag13]. The $d = 1$ case is equivalent to showing that a planar graph may only have linearly many edges, which follows immediately from the Euler formula, but the conjecture is open for every $d \geq 2$. Theorem 22.3.18 shows that it holds generically.

22.3.3 PHASE TRANSITIONS FOR HOMOLOGY IN RANDOM GEOMETRIC COMPLEXES

Penrose described sharp thresholds for connectivity of random geometric graphs, analogous to the Erdős–Rényi theorem. In the case of a uniform distribution on the unit cube $[0, 1]^d$ or a standard multivariate distribution, these results are tight [Pen03].

Thresholds for homology in random geometric complexes was first studied in [Kah11]. A homology vanishing threshold for random geometric complexes is obtained in [Kah11], which is tight up to a constant factor, but recently a much sharper result was obtained by Bobrowski and Weinberger.

THEOREM 22.3.19 [BW17]
Fix $1 \geq k \geq d - 1$. If

$$nr^d \geq \log n + k \log \log n + \omega(1),$$

then w.h.p. $\beta_k = 0$, *and if*

$$nr^d \leq \log n + (k-2)\log\log n - \omega(\log\log\log n),$$

then w.h.p. $\beta_k \to \infty$.

22.3.4 RANDOM FUNDAMENTAL GROUPS

GLOSSARY

Fundamental group: In a path-connected topological space X, choose an arbitrary base point p. Then the homotopy classes of loops in X based at p, i.e., continuous functions $f : [0,1] \to X$ with $f(0) = f(1) = p$ may be endowed with the structure of a group, where the group operation is a concatenation of two loops at double speed. This is called the fundamental group $\pi_1(X)$, and up to isomorphism it does not depend on the choice of base point p. If $\pi_1(X) = 0$ then X is said to be simply connected. The first homology group $H_1(X, \mathbb{Z})$ is isomorphic to the abelianization of $\pi_1(X)$.

A chain of implications: The following implications hold for an arbitrary simplicial complex X.

$$\pi_1(X) = 0 \implies H_1(X,\mathbb{Z}) = 0 \implies H_1(X,\mathbb{Z}/q\mathbb{Z}) = 0 \implies H_1(X,\mathbb{R}) = 0.$$

Here q is any prime. This is a standard application of the universal coefficient theorem for homology [Hat02].

A partial converse to one of the implications is the following. If $H_1(X, \mathbb{Z}/q\mathbb{Z}) = 0$ for every prime q, then $H_1(X, \mathbb{Z}) = 0$.

Hyperbolic group: A finitely presented group is said to be word hyperbolic if it can be equipped with a word metric satisfying certain characteristics of hyperbolic geometry [Gro87].

Kazhdan's property (T): A group G is said to have property (T) if the trivial representation is an isolated point in the unitary dual equipped with the Fell topology. Equivalently, if a representation has almost invariant vectors then it has invariant vectors.

Group cohomology: Associated with a finitely presented group G is a contractible CW complex EG on which G acts freely. The quotient BG is the classifying space for principle G bundles. The group cohomology of G is equivalent to the cohomology of BG.

Cohomological dimension: The cohomological dimension of a group G, denoted cdim G, is the largest dimension k such that $H^k(G, R) \neq 0$ for some coefficient ring R.

The random fundamental group $\pi_1(Y(n,p))$ may fruitfully be compared to other models of random group studied earlier, such as Gromov's density model [Oll05]. The techniques and flavor of the subject owes as much to geometric group theory as to combinatorics.

The vanishing threshold and hyperbolicity

Babson, Hoffman, and Kahle showed that the vanishing threshold for simple connectivity is much larger than the homology-vanishing threshold.

THEOREM 22.3.20 [BHK11]
Let $\epsilon > 0$ be fixed and $Y = Y(n, p)$. If

$$p \geq \frac{n^{\epsilon}}{\sqrt{n}}$$

then w.h.p. $\pi_1(Y) = 0$, and if

$$p \leq \frac{n^{-\epsilon}}{\sqrt{n}}$$

then w.h.p. $\pi_1(Y)$ is a nontrivial hyperbolic group.

Most of the work in proving Theorem 22.3.20 is showing that, on the sparse side of the threshold, π_1 is hyperbolic. This in turn depends on a local-to-global principle for hyperbolicity due to Gromov [Gro87].

Gundert and Wagner showed that it suffices to assume that

$$p \geq \frac{C}{\sqrt{n}}$$

for some constant $C > 0$ to show that w.h.p. $\pi_1(Y) = 0$ [GW16]. Korándi, Peled, and Sudakov showed that it suffices to take $C = 1/2$ [KPS16].

The author suspects that there is a sharp threshold for simple connectivity at $C/\sqrt{n}$ for some $C > 0$.

CONJECTURE 22.3.21 *A sharp vanishing threshold for $\pi_1(Y)$.*
There exists some constant $C > 0$ such that if

$$p \geq \frac{C + \epsilon}{\sqrt{n}},$$

with high probability, $\pi_1(Y) = 0$; and if

$$p \leq \frac{C - \epsilon}{\sqrt{n}},$$

with high probability, $\pi_1(Y) \neq 0$.

Kazhdan's property (T)

One of the most important properties studied in geometric group theory is property (T). Loosely speaking, a group is (T) if it does not have many unitary representations. Property (T) is also closely related to the study of expander graphs. For a comprehensive overview of the subject, see the monograph [BHV08].

Inspired by Garland's method, Żuk gave a spectral condition sufficient to imply (T). Hoffman, Kahle, and Paquette applied Żuk's condition, together with Theorem 22.3.7 to show that the threshold for $\pi_1(Y)$ to be (T) coincides with the Linial–Meshulam homology-vanishing threshold.

THEOREM 22.3.22 [HKP12]

Let $Y = Y(n,p)$.

- *If*
$$p \geq \frac{2\log n + \omega(1)}{n}$$
then w.h.p. $\pi_1(Y)$ is (T), and
- *if*
$$p \leq \frac{2\log n - \omega(1)}{n}$$
then w.h.p. $\pi_1(Y)$ is not (T).

Cohomological dimension

Costa and Farber [CCFK12] studied the cohomological dimension of the random fundamental group in [CF13]. Their main findings are that there are regimes when the cohomological dimension is 1, 2, and ∞, before the collapse of the group at $p = 1/\sqrt{n}$.

THEOREM 22.3.23 [HKP12]

Let $Y = Y(n,p)$.

- *If*
$$p \ll \frac{1}{n}$$
then w.h.p. cdim $\pi_1(Y) = 1$ [CCFK12].
- *If*
$$\frac{3}{n} \leq p \ll n^{-3/5}$$
then w.h.p. cdim $\pi_1(Y) = 2$ [CF15a] .
- *if*
$$n^{-3/5} \ll p \leq n^{-1/2-\epsilon}$$
then w.h.p. cdim $\pi_1(Y) = \infty$ [CF13].

Here we use $f \ll g$ to mean $f = o(g)$, i.e.,
$$\lim_{n\to\infty} f/g = 0.$$

Newman recently refined part of this picture [New16], showing that if $p < 2.455/n$ w.h.p. cdim $\pi_1(Y) = 1$, and if $p > 2.754/n$ then w.h.p. cdim $\pi_1(Y) = 2$. The precise constants are c_2 and c_2^*, defined in Section 22.3.2.

The fundamental group of the clique complex

Babson showed that $p = n^{-1/3}$ is the vanishing threshold for $\pi_1(X(n,p))$ in [Bab12]. An independent and self-contained proof, including more refined results regarding torsion and cohomological dimension, was given by Costa, Farber, and Horak in [CFH15].

Finite quotients

Meshulam studied finite quotients of the random fundamental group, and showed that if they exist then the index must be large—the index must tend to infinity with n. His technique is a version of the cocycle-counting arguments in [LM06] and [MW09], for non-abelian cohomology.

THEOREM 22.3.24 Meshulam, [Mes13]

Let $c > 0$ be fixed. If $p \geq \frac{(6+7c)\log n}{n}$ then w.h.p. $\pi_1(Y)$ has no finite quotients with index less than n^c. Moreover, if H is any fixed *finite group and $p \geq \frac{(2+c)\log n}{n}$ then w.h.p. there are no nontrivial maps to H.*

22.3.5 PHASE TRANSITIONS IN THE MULTI-PARAMETER MODEL

Applying Garland's method, Fowler described the homology-vanishing phase transition in the multi-parameter model in [Fow15].

THEOREM 22.3.25 Fowler [Fow15]

Let $X = X(n, p_1, p_2, \dots)$ with $p_i = n^{-\alpha_i}$ and $\alpha_i \geq 0$ for all i. If

$$\sum_{i=1}^{k} \alpha_i \binom{k}{i} < 1,$$

then w.h.p. $H^{k-1}(X, \mathbb{Q}) = 0$. If

$$\sum_{i=1}^{k} \alpha_i \binom{k}{i} \geq 1$$

and

$$\sum_{i=1}^{k-1} \alpha_i \binom{k-1}{i} < 1$$

then w.h.p. $H^{k-1}(X, \mathbb{Q}) \neq 0$.

22.4 BETTI NUMBERS AND PERSISTENT HOMOLOGY

22.4.1 BETTI NUMBERS

The random clique complex

In the random clique complex $X(n, p)$, it was noted in [Kah09] that if

$$1/n^{1/k} \ll p \ll 1/n^{1/(k+1)},$$

then

$$\mathbb{E}[\beta_k] = (1 - o(1)) \binom{n}{k+1} p^{\binom{k+1}{2}}.$$

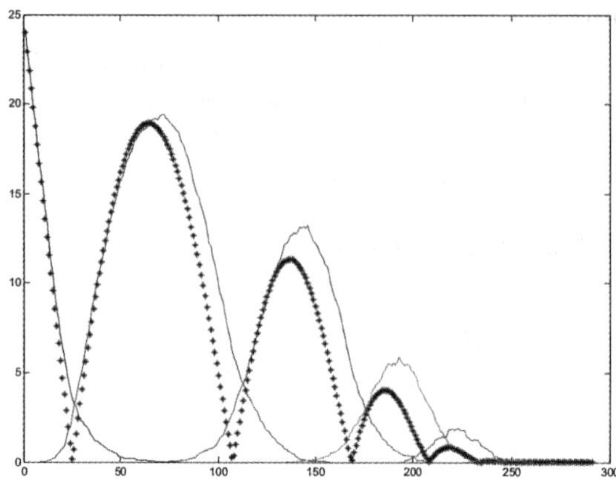

FIGURE 22.4.1
$|\mathbb{E}[\chi]|$ *plotted in blue, against the Betti numbers of the random flag complex* $X(n,p)$*. Here* $n = 25$ *and* p *varies from* 0 *to* 1*. The horizontal axis is the number of edges. Computation and image courtesy of Vidit Nanda.*

A more refined estimate may be obtained along the following lines.

We have the Euler relation

$$\chi = f_0 - f_1 + f_2 - \cdots = \beta_0 - \beta_1 + \beta_2 - \ldots,$$

where f_i denotes the number of i-dimensional faces. The expected number of i-dimensional faces is easy to compute—by linearity of expectation we have

$$\mathbb{E}[f_i] = \binom{n}{i+1} p^{\binom{i+1}{2}}.$$

If we make the simplifying assumption that only one Betti number β_i is nonzero, then we have

$$|\chi| = |f_0 - f_1 + f_2 - \ldots| = \beta_i.$$

So we obtain a plot of all of the Betti numbers by plotting the single function

$$\begin{aligned} |\chi| &= |f_0 - f_1 + f_2 - \ldots| \\ &= \left| \binom{n}{1} - \binom{n}{2} p^1 + \binom{n}{3} p^2 - \binom{n}{4} p^6 + \ldots \right|. \end{aligned}$$

This seems to work well in practice. See for example Figure 22.4.1. It is interesting that even though all of the theorems we have discussed are asymptotic as $n \to \infty$, the above heuristic gives a reasonable prediction of the shape of the Betti number curves, even for $n = 25$ and $p \leq 0.6$.

Homological domination in the multi-parameter model

In [CF15b, CF17], Costa and Farber show that for many choices of parameter (an open, dense subset of the set of allowable vectors of exponents) in the multi-parameter model, the homology is dominated in one degree.

Random geometric complexes

Betti numbers of random geometric complexes were first studied by Robins in [Rob06].

Betti numbers of random geometric complexes are also studied in [Kah11]. Estimates are obtained for the Betti numbers in the subcritical regime $nr^d \to 0$. In this regime the Vietoris–Rips complex and Čech complex have small connected components (bounded in size), so all the topology is local.

For the following theorems, we assume that n points are chosen i.i.d. according to a probability measure on $\mathbb{R}^d$ with a bounded measurable density function f. So the assumptions on the underlying probability distribution are fairly mild.

The following describes the expectation of the Betti numbers of the Vietoris–Rips complex in the subcritical regime. In this regime the homology of $VR(n,r)$ is dominated by subcomplexes combinatorially isomorphic to the boundary of the cross-polytope.

THEOREM 22.4.1 [Kah11]

Fix $d \geq 2$ and $k \geq 1$. If $nr^d \to 0$ then

$$\mathbb{E}\left[\beta_k[VR(n,r)]\right] / \left(n^{2k+2} r^{d(2k+1)}\right) \to C_k,$$

as $n \to \infty$ where D_k is a constant which depends on k, d, and the function f.

The analogous story for the Čech complex is the following. Here the homology is dominated by simplex boundaries.

THEOREM 22.4.2 [Kah11]

Fix $d \geq 2$ and $1 \leq k \leq d-1$. If $nr^d \to 0$ then

$$\mathbb{E}\left[\beta_k[C(n,r)]\right] / \left(n^{k+2} r^{d(k+1)}\right) \to C_k,$$

as $n \to \infty$ where D_k is a constant which depends on k, d, and the function f.

In the thermodynamic limit

The thermodynamic limit, or critical regime, is when $nr^d \to C$ for some constant $C > 0$. In [Kah11], it is shown that for every $1 \leq k \leq d-1$, we have $\beta_k = \Theta(n)$. Yogeshwaran, Subag, and Adler obtained the strongest results so far for Betti numbers in the thermodynamic limit, including strong laws of large numbers [YSA15], in particular that $\beta_k/n \to C_k$.

Limit theorems

Kahle and Meckes computed variance of the Betti numbers, and proved Poisson and normal limiting distributions for Betti numbers in the subcritical regime $r = o\left(n^{-1/d}\right)$ in [KM13].

More general point processes

Yogeshwaran and Adler obtained similar results for Betti numbers in a much more general setting of stationary point processes [YA15].

22.4.2 PERSISTENT HOMOLOGY

Bubenik and Kim studied persistent homology for i.i.d. random points on the circle, in the context of a larger discussion about foundations for topological statistics [BK07].

Bobrowski, Kahle, and Skraba studied maximally persistent cycles in $VR(n,r)$ and $C(n,r)$. They defined the *persistence* of a cycle as $p(\sigma) = d(\sigma)/b(\sigma)$, and found a law of the iterated logarithm for maximal persistence.

THEOREM 22.4.3 [BKS15]

Fix $d \geq 2$, and $1 \leq i \leq d-1$. Choose n points i.i.d. uniformly randomly in the cube $[0,1]^d$. With high probability, the maximally persistent cycle has persistence

$$\max_{\sigma} p(\sigma) = \Theta\left(\frac{\log n}{\log\log n}\right)^{1/i}.$$

CONJECTURE 22.4.4 *A law of large numbers for persistent homology.*

$$\max_{\sigma} p(\sigma) / \left(\frac{\log n}{\log\log n}\right)^{1/i} \to C,$$

for some constant $C = C_{d,i}$.

OTHER RESOURCES

For an earlier survey of Erdős–Rényi based models with a focus on the cohomology-vanishing phase transition, see also [Kah14b]. For a more comprehensive overview of random geometric complexes, see [BK14].

Several other models of random topological space have been studied. Ollivier's survey [Oll05] provides a comprehensive introduction to random groups, especially to Gromov's density random groups and the triangular model. Dunfield and Thurston introduced a new model of random 3-manifold [DT06] which has been well studied since then.

RELATED CHAPTERS

Chapter 21: Topological methods in discrete geometry
Chapter 24: Persistent homology
Chapter 25: High-dimensional topological data analysis

REFERENCES

[AKS79] M. Ajtai, J. Komlós, and E. Szemerédi. Topological complete subgraphs in random graphs. *Studia Sci. Math. Hungar.*, 14:293–297, 1979.

[AL15] L. Aronshtam and N. Linial. When does the top homology of a random simplicial complex vanish? *Random Structures Algorithms*, 46:26–35, 2015.

[AL16] L. Aronshtam and N. Linial. The threshold for collapsibility in random complexes. *Random Structures Algorithms*, 48:260–269, 2016.

[ALŁM13] L. Aronshtam, N. Linial, T. Łuczak, and R. Meshulam. Collapsibility and vanishing of top homology in random simplicial complexes. *Discrete Comput. Geom.*, 49:317–334, 2013.

[AS08] N. Alon and J.H. Spencer. *The probabilistic method.* Wiley-Interscience Series in Discrete Mathematics and Optimization, Wiley, Hoboken, 3rd edition, 2008.

[Bab12] E. Babson. Fundamental groups of random clique complexes. Preprint, `arXiv:1207.5028`, 2012.

[BHV08] B. Bekka, P. de la Harpe, and A. Valette. *Kazhdan's property (T)*, vol. 11 of *New Mathematical Monographs.* Cambridge University Press, 2008.

[BHK11] E. Babson, C. Hoffman, and M. Kahle. The fundamental group of random 2-complexes. *J. Amer. Math. Soc.*, 24:1–28, 2011.

[BK07] P. Bubenik and P.T. Kim. A statistical approach to persistent homology. *Homology, Homotopy Appl.*, 9:337–362, 2007.

[BK14] O. Bobrowski and M. Kahle. Topology of random geometric complexes: a survey. In *Topology in Statistical Inference, Proc. Sympos. Appl. Math.*, AMS, Providence, to appear. Preprint, `arXiv:1409.4734`, 2014.

[BKS15] O. Bobrowski, M. Kahle, and P. Skraba. Maximally persistent cycles in random geometric complexes. Preprint, `arXiv:1509.04347`, 2015.

[Bol01] B. Bollobás. *Random graphs*, 2nd edition. Vol. 73 of *Cambridge Studies in Advanced Mathematics*, Cambridge University Press, 2001.

[BŚ97] W. Ballmann and J. Świątkowski. On L^2-cohomology and property (T) for automorphism groups of polyhedral cell complexes. *Geom. Funct. Anal.*, 7:615–645, 1997.

[BW17] O. Bobrowski and S. Weinberger. On the vanishing of homology in random Čech complexes. *Random Structures Algorithms*, 51:14–51, 2017.

[CCFK12] D. Cohen, A. Costa, M. Farber, and T. Kappeler. Topology of random 2-complexes. *Discrete Comput. Geom.*, 47:117–149, 2012.

[CF13] A. Costa and M. Farber. Geometry and topology of random 2-complexes. *Israel J. Math.*, 209:883–927, 2015.

[CF16] A. Costa and M. Farber. Random simplicial complexes. In *Configuration Spaces: Geometry, Topology and Representation Theory*, pages 129–153, Springer, Cham, 2016.

[CF15a] A. Costa and M. Farber. The asphericity of random 2-dimensional complexes. *Random Structures Algorithms*, 46:261–273, 2015.

[CF15b] A. Costa and M. Farber. Homological domination in large random simplicial complexes. Preprint, `arXiv:1503.03253`, 2015.

[CF17] A. Costa and M. Farber. Large random simplicial complexes, III the critical dimension. *J. Knot Theory Ramifications*, 26:1740010, 2017.

[CFH15] A. Costa, M. Farber, and D. Horak. Fundamental groups of clique complexes of random graphs. *Trans. London Math. Soc.*, 2:1–32, 2015.

[Dey93] T.K. Dey. On counting triangulations in d dimensions. *Comput. Geom.*, 3:315–325, 1993.

[DHK13] B. DeMarco, A. Hamm, and J. Kahn. On the triangle space of a random graph. *J. Combin.*, 4:229–249, 2013.

[DK12] D. Dotterrer and M. Kahle. Coboundary expanders. *J. Topol. Anal.*, 4:499–514, 2012.

[DKW15] D. Dotterrer, T. Kaufman, and U. Wagner. On expansion and topological overlap. In *Proc. 32nd Sympos. Comp. Geom.*, vol. 51 of *LIPIcs*, pages 35:1–35:10, Dagstuhl, 2016.

[DT06] N.M. Dunfield and W.P. Thurston. Finite covers of random 3-manifolds. *Invent. Math.*, 166:457–521, 2006.

[ER59] P. Erdős and A. Rényi. On random graphs. I. *Publ. Math. Debrecen*, 6:290–297, 1959.

[Fow15] C. Fowler. Generalized random simplicial complexes. Preprint, `arXiv:1503.01831`, 2015.

[Gar73] H. Garland. p-adic curvature and the cohomology of discrete subgroups of p-adic groups. *Ann. of Math. (2)*, 97:375–423, 1973.

[Gro87] M. Gromov. Hyperbolic groups. In *Essays in Group Theory*, vol. 8 of *Math. Sci. Res. Inst. Publ.*, pages 75–263, Springer, New York, 1987.

[Gro09] M. Gromov. Singularities, expanders and topology of maps. I. Homology versus volume in the spaces of cycles. *Geom. Funct. Anal.*, 19:743–841, 2009.

[Gro10] M. Gromov. Singularities, expanders and topology of maps. Part 2: From combinatorics to topology via algebraic isoperimetry. *Geom. Funct. Anal.*, 2:416–526, 2010.

[GW16] A. Gundert and U. Wagner. On topological minors in random simplicial complexes. *Proc. Amer. Math. Soc.*, 144:1815–1828, 2016.

[GW16] A. Gundert and U. Wagner. On eigenvalues of random complexes. *Israel J. Math*, 216:545–582, 2016.

[Hat02] A. Hatcher. *Algebraic topology.* Cambridge University Press, Cambridge, 2002.

[HKP12] C. Hoffman, M. Kahle, and E. Paquette. Spectral gaps of random graphs and applications to random topology. Preprint, `arXiv:1201.0425`, 2012.

[HKP17] C. Hoffman, M. Kahle, and E. Paquette. The threshold for integer homology in random d-complexes. *Discrete Comput. Geom.*, 57:810—823, 2017.

[HLW06] S. Hoory, N. Linial, and A. Wigderson. Expander graphs and their applications. *Bull. Amer. Math. Soc. (N.S.)*, 43:439–561, 2006.

[JŁR00] S. Janson, T Łuczak, and A. Rucinski. *Random Graphs.* Wiley-Interscience Series in Discrete Mathematics and Optimization. Wiley, New York, 2000.

[Kah09] M. Kahle. Topology of random clique complexes. *Discrete Math.*, 309:1658–1671, 2009.

[Kah11] M. Kahle. Random geometric complexes. *Discrete Comput. Geom.*, 45:553–573, 2011.

[Kah14a] M. Kahle. Sharp vanishing thresholds for cohomology of random flag complexes. *Ann. of Math. (2)*, 179:1085–1107, 2014.

[Kah14b] M. Kahle. Topology of random simplicial complexes: a survey. In *Algebraic topology: applications and new directions*, vol. 620 of *Contemp. Math.*, pages 201–221, AMS, Providence, 2014.

[Kal91] G. Kalai. The diameter of graphs of convex polytopes and f-vector theory. In *Applied geometry and discrete mathematics*, vol. 4 of *DIMACS Ser. Discrete Math. Theoret. Comput. Sci.*, pages 387–411, AMS, Providence, 1991.

[KM13] M. Kahle and E. Meckes. Limit theorems for Betti numbers of random simplicial complexes. *Homology Homotopy Appl.*, 15:343–374, 2013.

[Koz10] D.N. Kozlov. The threshold function for vanishing of the top homology group of random d-complexes. *Proc. Amer. Math. Soc.*, 138:4517–4527, 2010.

[KPS16] D. Korándi, Y. Peled, and B. Sudakov. A random triadic process. *SIAM J. Discrete Math.*, 30:1–19, 2016.

[LM06] N. Linial and R. Meshulam. Homological connectivity of random 2-complexes. *Combinatorica*, 26:475–487, 2006.

[LM15] A. Lubotzky and R. Meshulam. Random Latin squares and 2-dimensional expanders. *Adv. Math.*, 272:743–760, 2015.

[LP14] N. Linial and Y. Peled. On the phase transition in random simplicial complexes. *Ann. of Math.*, 184:745–773, 2016.

[ŁPW94] T. Łuczak, B. Pittel, and J.C. Wierman. The structure of a random graph at the point of the phase transition. *Trans. Amer. Math. Soc.*, 341:721–748, 1994.

[Lub14] A. Lubotzky. Ramanujan complexes and high dimensional expanders. *Jpn. J. Math.*, 9:137–169, 2014.

[Mal16] G. Malen. *The Topology of Random Flag and Graph Homomorphism Complexes.* Ph.D. thesis, The Ohio State University, 2016.

[Mes13] R. Meshulam. Bounded quotients of the fundamental group of a random 2-complex. Preprint, `arXiv:1308.3769`, 2013.

[MW09] R. Meshulam and N. Wallach. Homological connectivity of random k-dimensional complexes. *Random Structures Algorithms*, 34:408–417, 2009.

[New16] A. Newman. On freeness of the random fundamental group. Preprint, `arXiv:1601.07520`, 2016.

[Oll05] Y. Ollivier. *A January 2005 invitation to random groups*, vol. 10 of *Ensaios Matemáticos [Mathematical Surveys].* Sociedade Brasileira de Matemática, Rio de Janeiro, 2005.

[Pen03] M. Penrose. *Random Geometric Graphs.* Vol. 5 of *Oxford Studies in Probability*, Oxford University Press, 2003.

[Pit88] B. Pittel. A random graph with a subcritical number of edges. *Trans. Amer. Math. Soc.*, 309:51–75, 1988.

[PRT15] O. Parzanchevski, R. Rosenthal, and R.J. Tessler. Isoperimetric inequalities in simplicial complexes. *Combinatorica*, 35:1–33, 2015.

[Rob06] V. Robins. Betti number signatures of homogeneous Poisson point processes. *Physical Review E*, 74:061107, 2006.

[Wag11] U. Wagner. Minors in random and expanding hypergraphs. In *Proc. 27th Sympos. Comput. Geom.*, pages 351–360, ACM Press, 2011.

[Wag13] U. Wagner. Minors, embeddability, and extremal problems for hypergraphs. In J. Pach, editor, *Thirty Essays on Geometric Graph Theory*, pages 569–607, Springer, New York, 2013.

[YA15] D. Yogeshwaran and R.J. Adler. On the topology of random complexes built over stationary point processes. *Ann. Appl. Probab.*, 25:3338–3380, 2015.

[YSA15] D. Yogeshwaran, E. Subag, and R.J. Adler. Random geometric complexes in the thermodynamic regime. *Probab. Theory and Related Fields*, 1–36, 2015.

23 COMPUTATIONAL TOPOLOGY OF GRAPHS ON SURFACES

Éric Colin de Verdière

INTRODUCTION

This chapter surveys computational topology results in the special, low-dimensional case where the ambient space is a surface. Surface topology is very well-understood and comparably simpler than the higher-dimensional counterparts; many computational problems that are undecidable in general (e.g., homotopy questions) can be solved efficiently on surfaces. This leads to a distinct flavor of computational topology and to dedicated techniques for revisiting topological problems on surfaces from a computational viewpoint.

Topological surfaces and graphs drawn on them appear in various fields of mathematics and computer science, and these aspects are not surveyed here:

- in *topology* of three-dimensional manifolds, also in connection to the recent resolution of the Poincaré conjecture, combinatorial and algebraic structures defined on surfaces are often relevant, e.g., via the study of mapping class groups and Teichmüller spaces [FM11];
- in *topological graph theory*, a branch of structural graph theory, graphs on surfaces are studied from a combinatorial point of view, also in relation to the theory of Robertson and Seymour on graph minors; for example, colorability questions of graphs on surfaces, generalizing the four-color theorem for planar graphs, are well-studied [MT01];
- in *enumerative combinatorics*, a natural problem is to count (exactly or asymptotically) maps with given properties in the plane or on surfaces, with the help of generating series; moreover, typical properties of random maps are investigated [Mie09, Bet12, LZ04];
- various *applications* involve surface meshes, in particular in geometry processing and computer graphics, for approximation [CDP04], topological simplification [GW01, WHDS04], compression [AG05], and parameterization [GY03]. Techniques for general surfaces apply also to subsets of the plane, and are thus relevant in VLSI design [LM85] and map simplification [BKS98].

This chapter is organized as follows. We first review the basic concepts and properties of topological surfaces and graphs embedded on them (Sections 23.1 and 23.2). Then we consider three categories of topological problems, mostly from a computational perspective: drawing an abstract input graph on a surface (Section 23.3), homotopy questions and variations (Section 23.4), and optimization of curves and graphs on surfaces, also from a homological point of view (Section 23.5). Then we survey techniques that allow us to solve general graph problems faster in the case where the input graph is embedded on a fixed surface (Section 23.6). Finally, we collect other miscellaneous results (Section 23.7).

23.1 SURFACES

Surfaces are considered from a topological point of view: Two homeomorphic surfaces are regarded as equivalent. Surfaces such as the sphere or the disk are topologically uninteresting; our focus is on surfaces in which some closed curves are non-contractible (they cannot be deformed to a point by a continuous motion on the surface).

GLOSSARY

Homeomorphism: Given two topological spaces X and X', a map $h\colon X \to X'$ is a homeomorphism if h is bijective and both h and its inverse are continuous.

Surface (topological definition): In this chapter, a surface S is a *compact* two-dimensional manifold possibly with boundary. Equivalently, S is a compact topological space that is Hausdorff (any two distinct points have disjoint neighborhoods) and such that every point has a neighborhood homeomorphic to the plane or the closed half-plane. The set of points of a surface S that have no neighborhood homeomorphic to the plane is the ***boundary*** of S.

Surface (combinatorial definition): Equivalently, a surface S is a topological space obtained from finitely many disjoint triangles by identifying some pairs of edges of the triangles (by the quotient topology). The ***boundary*** of S is the union of the edges that are not identified with any other edge.

Path: A path on S is a continuous map $p\colon [0,1] \to S$. Its two ***endpoints*** are $p(0)$ and $p(1)$.

Connectedness: A surface is ***connected*** if any two points of the surface are the endpoints of some path. The inclusionwise maximal connected subsets of a surface form its ***connected components***.

Orientability: A surface is ***non-orientable*** if some subset of it (with the induced topology) is homeomorphic to the Möbius strip (defined in Figure 23.1.1). Otherwise, it is ***orientable***.

PROPERTIES: CLASSIFICATION OF SURFACES

Every connected surface is homeomorphic to exactly one of the following surfaces:

- the orientable surface of ***genus*** $g \geq 0$ with $b \geq 0$ ***boundary components*** (or, more concisely, ***boundaries***), obtained from the sphere by removing g disjoint open disks, attaching a handle (defined in Figure 23.1.1) to each of the resulting g circles, and finally removing b open disks with disjoint closures;
- the non-orientable surface of ***genus*** $g \geq 1$ with $b \geq 0$ ***boundary components*** (or ***boundaries***), obtained from the sphere by removing g disjoint open disks, attaching a Möbius strip (defined in Figure 23.1.1) to each of the resulting g circles, and finally removing b open disks with disjoint closures.

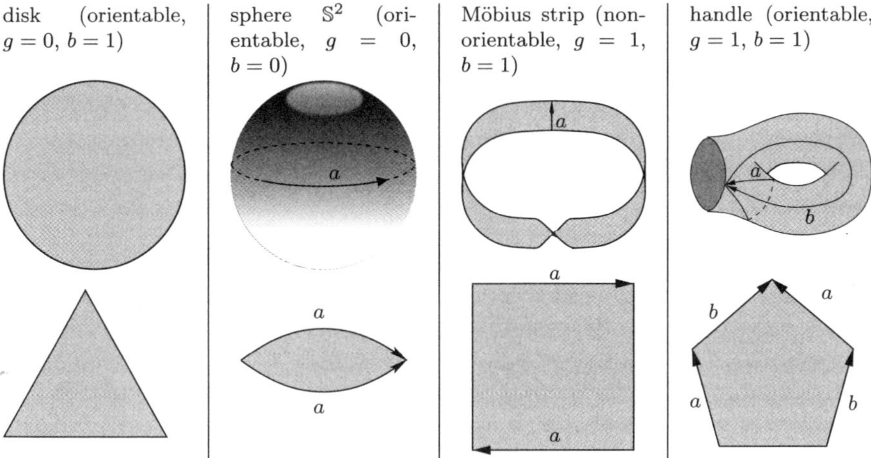

FIGURE 23.1.1
Examples of surfaces. Each surface (top row) comes with a polygonal schema (bottom row), a polygon with some labeled and directed edges; the surface can be obtained by identifying the pairs of edges with the same labels, respecting their direction. The genus g and number of boundary components b are specified, as well as whether the surface is orientable.

Every surface can be obtained by identifying pairs of edges of disjoint triangles. More concisely, every surface can be defined by a ***polygonal schema***, a polygon with labels and directions on some of the edges specifying how they must be identified. In particular, one can define a ***canonical*** polygonal schema for every connected surface without boundary:

- The canonical polygonal schema of the orientable surface of genus $g \geq 1$ is a $4g$-gon whose successive edges are labeled $a_1, b_1, \bar{a}_1, \bar{b}_1, \ldots, a_g, b_g, \bar{a}_g, \bar{b}_g$, and where edge x is directed clockwise, edge $\bar{x}$ is directed counterclockwise. Identifying edge x with $\bar{x}$, as indicated by their directions, gives the orientable surface of genus g. See Figure 23.1.2.
- Similarly, the canonical polygonal schema of the non-orientable surface of genus $g \geq 1$ is a $2g$-gon whose successive edges are labeled $a_1, a_1, \ldots, a_g, a_g$, and where all edges are directed clockwise.

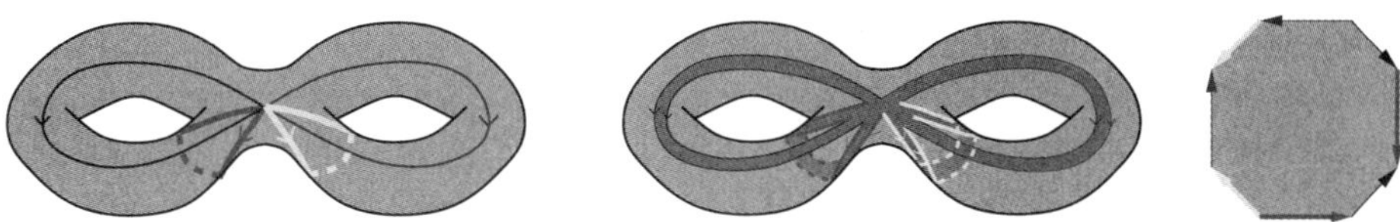

FIGURE 23.1.2
A double torus with a system of loops (left); the surface cut along the loops (middle) is a disk, shown in the form of a (canonical) polygonal schema (right).

EXAMPLES

Table 23.1.1 lists some common connected surfaces. See also Figures 23.1.1 and 23.1.2.

TABLE 23.1.1 Some common surfaces.

Surface	Orientable?	Genus	# of boundary components
Sphere	Yes	0	0
Disk	Yes	0	1
Annulus = cylinder	Yes	0	2
Pair of pants	Yes	0	3
Torus	Yes	1	0
Handle	Yes	1	1
Double torus	Yes	2	0
Projective plane	No	1	0
Möbius strip	No	1	1
Klein bottle	No	2	0

23.2 GRAPHS ON SURFACES

GLOSSARY

Let S be a surface.

Loop: A loop is a path whose two endpoints are equal to a single point, called the ***basepoint*** of the loop.

Closed curve: A closed curve on S is a continuous map from the unit circle $\mathbb{S}^1$ to S. This is almost the same as a loop, except that a closed curve has no distinguished basepoint. A closed curve is sometimes called a ***cycle***, although, contrary to the standard terminology in graph theory, here a cycle may self-intersect.

Curve: A curve is either a path or a closed curve. For most purposes, the parameterization is unimportant; for example, a path p could be regarded as equivalent to $p \circ \varphi$, where $\varphi : [0,1] \to [0,1]$ is bijective and increasing.

Simplicity: A path or a closed curve is ***simple*** if it is injective. A loop $\ell\colon [0,1] \to S$ is ***simple*** if its restriction to $[0,1)$ is injective.

Graph: In this chapter, unless specified otherwise, graphs are finite, undirected, and may have loops and multiple edges.

Curve (in a graph): A ***curve*** in a graph G (also called ***walk*** in the terminology of graph theory) is a sequence of directed edges $e_1, \ldots, e_k$ of G where the target of e_i equals the source of e_{i+1}. Repetitions of vertices and edges are allowed. The ***endpoints*** of the curve are the source of e_1 and the target of e_k. If they are equal, the curve is ***closed***.

Graph embedding (topological definition): A graph G naturally leads to a topological space $\hat{G}$, defined as follows: One considers a disjoint set of segments, one per edge of G, and identifies the endpoints that correspond to the same vertex of G. This gives a topological space, from which $\hat{G}$ is obtained by adding one isolated point per isolated vertex of G. (As a special case, if G has no loop and no multiple edge, then G is a one-dimensional simplicial complex, and $\hat{G}$

is the associated topological space.) An ***embedding*** of G is a continuous map from $\hat{G}$ into S that is a homeomorphism from $\hat{G}$ onto its image.

Graph embedding (concrete definition): Equivalently, an embedding of G on S is a "crossing-free" drawing of G: It maps the vertices of G to distinct points of S, and its edges to paths of S whose endpoints are the images of their incident vertices; the image of an edge can self-intersect, or intersect the image of another edge or vertex, only at its endpoints. When no confusion arises, we identify G with its embedding on S, or with the image of that embedding.

Face: The faces of an embedded graph G are the connected components of the complement of the image of G.

Degree: The degree of a vertex v is the number of edges incident to v, counted with multiplicity (if an edge is a loop). The degree of a face f is the number of edges incident to f, counted with multiplicity (if an edge has the same face on both sides).

Cellular embedding: A graph embedding is ***cellular*** if its faces are homeomorphic to open disks.

Triangulation: A graph embedding is a triangulation if it is cellular and its faces have degree three. The triangulation may fail to be a simplicial complex: A triangle is not necessarily incident to three distinct vertices, or even to three distinct edges.

Cutting: Given an embedded graph G on S without isolated vertex, the operation of cutting S along G results in a (possibly disconnected) surface with boundary, denoted $S \setminus\!\!\setminus G$ (or sometimes S✂G or similar); each connected component of $S \setminus\!\!\setminus G$ corresponds to a face of G on S, and by identifying pieces of the boundaries of these components in the obvious way, one recovers the surface S. Similarly, one can cut along a set of disjoint, simple closed curves. (Technically, if S has non-empty boundary, an additional condition is needed: The intersection of an edge with the boundary of S can be either the entire edge, its two endpoints, or one of its endpoints.)

Planarity: A graph is planar if it has an embedding to the plane (or equivalently the sphere).

Dual graph: A dual graph of a cellularly embedded graph G on S (assumed without boundary) is a graph G^* embedded on S with one vertex f^* inside each face f of S, and with an edge e^* for each edge e of G, such that e^* crosses e and no other edge of G. A dual graph is cellularly embedded. Its combinatorial map (see below) is uniquely determined by the combinatorial map of G.

Euler genus: The Euler genus $\bar{g}$ of a connected surface S with genus g equals $2g$ if S is orientable, and g if S is non-orientable.

Euler characteristic: The Euler characteristic of a cellularly embedded graph G equals $\chi(G) := v - e + f$, where v, e, and f are its number of vertices, edges, and faces, respectively.

PROPERTIES: EULER'S FORMULA AND CONSEQUENCES

1. ***Euler's formula:*** If G is *cellularly* embedded on a connected surface S of Euler genus $\bar{g}$ with b boundary components, then $\chi(G) = 2 - \bar{g} - b$. In particular,

$\chi(G)$ does not depend on G, only on S, and is consequently called the ***Euler characteristic*** of S.

2. The number of vertices and faces of a graph G cellularly embedded on a connected surface is at most linear in the number of its edges. In particular, the combinatorial complexity of G is linear in its number of edges.
3. Conversely, let G be a (not necessarily cellular) graph embedding on a connected surface with Euler genus $\bar{g}$ and b boundaries. Assume that G has no face of degree one or two that is an open disk. Then the numbers e of edges and v of vertices of G satisfy $e = O(v + \bar{g} + b)$.

DATA STRUCTURES

In all the problems we shall consider, the exact embedding of a graph on a surface is irrelevant; only the actual combinatorial data associated to the embedding is meaningful. If G is a graph cellularly embedded on a surface S without boundary, we only need the information of G together with the *facial walks*, namely, the closed walks in G encountered when walking along the boundary of the faces of G. This information is called the *combinatorial map* of G, and allows us to reconstruct the surface, by attaching disks to every facial walk. (Some conditions on the walks are needed to ensure that the resulting space is indeed a surface.) If S has boundaries, one can specify the corresponding faces of G. If S is orientable and G has no loop edge, instead of the facial walks one could as well specify the cyclic ordering of the edges incident to each vertex.

However, more complicated data structures are needed to perform basic operations efficiently. For example, one should be able to compute the degree of a face in time linear in the degree; to count the number of faces of G in linear time; to determine whether the surface is orientable in linear time; etc. (The last two operations, together with counting the number of vertices and edges, allow us to identify the topology of the surface in linear time using Euler's formula; this can also be done in logarithmic space [BEKT16].)

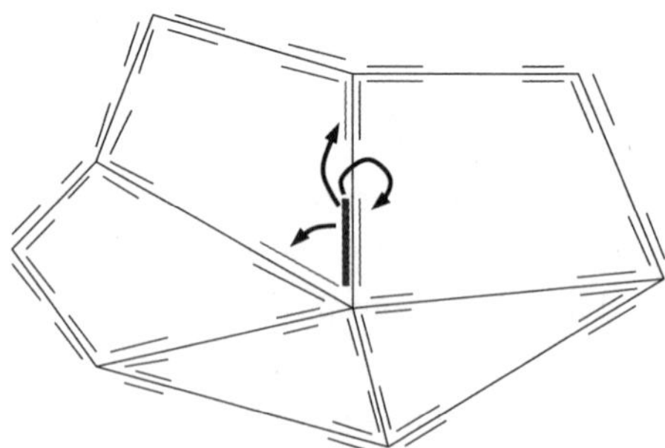

FIGURE 23.2.1
The graph-encoded map data structure. Each edge bears four flags, drawn parallel to it. Three operations allow us to move from a flag to a nearby flag.

One such data structure, the *graph-encoded map* or *gem representation* [Lin82], uses flags (quarter-edges, or, equivalently, incidences between a vertex, an edge, and a face) of G, see Figure 23.2.1; three involutive operations can be applied to a flag to move to an incident flag. Alternative data structures have been designed for more general situations (e.g., to allow surfaces with boundaries) or to take advantage of special situations (e.g., in the case where G is a triangulation, or where S is orientable); see the survey [Ket99]. However, the choice of the data structure is irrelevant for the theoretical design and asymptotic analysis of the algorithms.

CONVENTIONS FOR THIS CHAPTER

Henceforth, **we assume all surfaces to be connected.**

In several works mentioned in the following, only orientable surfaces are considered. In some cases, non-orientable surfaces are just as easy to handle, but sometimes they lead to additional difficulties. **We refer to the original articles to determine whether the results hold on non-orientable surfaces.**

Also, in most problems studied in this chapter, surfaces with boundaries are no harder to handle than surfaces without boundary: Any algorithm for surfaces without boundary immediately implies an algorithm for surfaces with boundary (with the same running time, or by replacing g by $g+b$ in the complexity, where g and b are the genus and the number of boundary components). For this reason, **we mostly focus on computational problems for surfaces without boundary.**

Finally, when we consider cellularly embedded graphs for algorithmic problems, **we implicitly assume that they are specified in the form of a data structure as described above** (e.g., a graph-encoded map).

23.3 EMBEDDING AND DRAWING GRAPHS ON SURFACES

Being able to build embeddings of a graph on a surface with small genus is important; almost all algorithms for graphs embeddable on a fixed surface require an embedding of the input graph (there are a few exceptions [ES14, Kel06, MS12]). We discuss algorithmic results related to the problem of embedding a graph on a surface, and then consider more general drawings where crossings are allowed.

EMBEDDING GRAPHS ON SURFACES

Let G be an abstract graph (not embedded on any surface), given, e.g., by the (unordered) list of the edges incident to every vertex. We assume that G is connected. Let n denote the combinatorial complexity of G, that is, the total number of vertices and edges of G.

1. *General facts:* An embedding on an orientable surface with minimum possible genus is cellular. If G is embeddable on an orientable (resp., non-orientable) surface of genus g, then it is embeddable on an orientable (resp., non-orientable) surface of genus g', for every $g' \geq g$.
2. *General bound:* G can be cellularly embedded on some orientable surface with genus $O(n)$.
3. *Planar case:* There is an $O(n)$-time algorithm for deciding embeddability in the sphere (equivalently, in the plane) [HT74]; also in $O(n)$ time, the graph can

be embedded with straight-line segments in the plane [Sch90a] (see also [NR04, Ch. 4]), if it has no loop or multiple edge. See Chapter 55 for more results on graph drawing.

4. *Time complexity:* Given a graph G and a surface S, specified by its Euler genus $\bar{g}$ and by whether it is orientable, determining whether G embeds on S is NP-hard [Tho89], but can be done in $2^{\text{poly}(\bar{g})} \cdot n$ time [KMR08, Moh99] (where poly($\bar{g}$) is a polynomial in $\bar{g}$), which is linear if $\bar{g}$ is fixed. Such an embedding can be computed in the same amount of time if it exists.
5. *Space complexity:* For every *fixed* $\bar{g}$, determining whether an input graph G embeds on some surface (orientable or not) of Euler genus at most $\bar{g}$ can be done in space logarithmic in the input size [EK14].
6. *Approximation:* Given as input a graph G and an integer $\bar{g}$, one can in polynomial time either correctly report that G embeds on no surface of Euler genus $\bar{g}$, or compute an embedding on some surface of Euler genus $\bar{g}^{O(1)}$ [KS15].

Except for the planar case, these algorithms are rather complicated, and implementing them is a real challenge. For example, there seems to be no available implementation of a polynomial-time algorithm for testing embeddability in the torus, and no publicly available implementation of any algorithm to decide whether a graph embeds on the double torus; attempts of implementing some known embedding algorithms, even in the simplest cases, have unveiled some difficulties [MK11]. On the other hand, a recent approach is promising in practice for graphs of moderate size, using integer linear programming or Boolean satisfiability reformulations [BCHK16].

In contrast, determining the *maximum* genus of an orientable surface without boundary on which a graph can be *cellularly* embedded can be done in polynomial time [FGM88]. There are also results on the embeddability of two-dimensional simplicial complexes on surfaces [Moh97].

On a less algorithmic side, in the field of topological graph theory, a lot more is known about the embeddability of some classes of graphs on some surfaces; see, e.g., [Arc96, Sect. 4.2] and references therein.

GLOSSARY ON DRAWINGS

Let G be a graph and S be a surface.

Drawing: Drawings are more general than embeddings in that they allow a finite set of crossing points, where exactly two pieces of edges intersect and actually cross. Formally, recall that G has an associated topological space $\hat{G}$. A (topological) ***drawing*** of G on S is a continuous map from $\hat{G}$ into S such that the preimage of every point in S has cardinality zero or one, except for a finite set of points ("crossings"), whose preimages have cardinality two; moreover, each such crossing point has a disk neighborhood that contains exactly the images of two pieces of edges of $\hat{G}$, which form, up to homeomorphism, two crossing straight lines.

Arrangement: Let D be a drawing of G on S. The ***arrangement*** of D on S is the graph G' embedded on S that has the same image as D and is obtained from D by inserting a vertex of degree four at each crossing in D and subdividing the edges of G accordingly. Similarly, one can consider the arrangement of a set of curves drawn on S.

Crossing number: The crossing number of G with respect to S is the minimum number of crossings that G has in any drawing of G on S.

Pair crossing number: The pair crossing number of G with respect to S is the minimum number of pairs of edges of G that cross, over all drawings of G on S.

Odd crossing number: The odd crossing number of G with respect to S is the minimum number of pairs of edges of G that cross an odd number of times, over all drawings of G on S.

DRAWING GRAPHS ON SURFACES WITH FEW CROSSINGS

1. *Crossing numbers:* Computing the planar crossing number of a graph is NP-hard (even in very special cases, such as that of a planar graph with a single additional edge [CM13]), and there exists no polynomial-time algorithm with approximation guarantee better than a certain constant [Cab13]. However, for every fixed k, one can, in linear time, determine whether an input graph has planar crossing number at most k [KR07], although the problem admits no polynomial kernel [HD16]. Some approximation algorithms for the planar crossing number are known in restricted cases, such as bounded maximum degree [HC10, Chu11].
2. *Variations on crossing numbers:* The relations between the various notions of crossing numbers are not fully understood. Let c, p, and o denote the planar crossing number, planar pair crossing number, and planar odd crossing number, respectively, of some graph G. It is clear that $o \leq p \leq c$, and it is known that the left inequality can be strict [PSŠ08]. It is widely believed that $p = c$, but the best bound known so far is $c = O(p^{3/2} \log^2 p)$ (this follows essentially from [Tót12]). See, e.g., [Mat14] for more details, and [Sch13a] for a wide survey on the various notions of crossing numbers.
3. *Hanani–Tutte theorem:* The (weak) Hanani–Tutte theorem [Han34, Tut70], however, states that if $o = 0$ then $c = 0$. Furthermore it holds not only for the plane, but for arbitrary surfaces [CN00, PSŠ09b]: If a graph G can be drawn on a surface S in a way that every pair of edges crosses an even number of times, then G can be embedded on S. In the planar case, it actually suffices to assume that every pair of *independent* edges (which do not share any endpoints) crosses an even number of times, but whether this generalizes to arbitrary surfaces is open, except for the projective plane [PSS09a, CVK$^+$16]. We refer to surveys [Sch13b, Sch14] for more details.

23.4 HOMOTOPY AND ISOTOPY

Most works in computational topology for surfaces do not take as input a given abstract graph, as in the previous section; instead, they consider an already embedded graph, given by its combinatorial map.

GLOSSARY

Let S be a surface.

Reversal: The reversal of a path $p\colon [0,1] \to S$ is the path $p^{-1}\colon [0,1] \to S$ defined by $p^{-1}(t) = p(1-t)$.

Concatenation: The concatenation of two paths $p, q\colon [0,1] \to S$ with $p(1) = q(0)$ is the path $p \cdot q$ defined by $(p \cdot q)(t) = p(2t)$ if $t \leq 1/2$ and $(p \cdot q)(t) = q(2t-1)$ if $t \geq 1/2$.

Homotopy for paths: Given two paths $p, q\colon [0,1] \to S$, a ***homotopy*** between p and q is a continuous deformation between p and q that keeps the endpoints fixed. More formally, it is a continuous map $h\colon [0,1] \times [0,1] \to S$ such that $h(0,\cdot) = p$, $h(1,\cdot) = q$, and both $h(\cdot,0)$ and $h(\cdot,1)$ are constant maps (equal, respectively, to $p(0) = q(0)$ and to $p(1) = q(1)$). The paths p and q are ***homotopic***. Being homotopic is an equivalence relation, partitioning the paths with given endpoints into ***homotopy classes***.

Fundamental group: The homotopy classes of loops with a given basepoint form a group, where concatenation of loops accounts for the multiplication and reversal accounts for the inverse operation: if $[p]$ denotes the homotopy class of path p, then we have $[p \cdot q] = [p] \cdot [q]$ and $[p^{-1}] = [p]^{-1}$.

Homotopy for closed curves (also called ***free homotopy***): Given two closed curves $\gamma, \delta\colon \mathbb{S}^1 \to S$, a ***homotopy*** between γ and δ is a continuous deformation between them, namely, a continuous map $h\colon [0,1] \times \mathbb{S}^1 \to S$ such that $h(0,\cdot) = \gamma$ and $h(1,\cdot) = \delta$.

Contractibility: A loop or closed curve is ***contractible*** if it is homotopic to a constant loop or closed curve.

Isotopy: An isotopy between two *simple* paths, loops, or closed curves is a homotopy h that does not create self-intersections: for each t, $h(t,\cdot)$ is a simple path, loop, or closed curve. An isotopy of a graph G is a continuous family of embeddings of G (the vertices and edges move continuously).

Ambient isotopy: An ambient isotopy of a surface S is a continuous map i : $[0,1] \times S \to S$ such that for each $t \in [0,1]$, $i(t,\cdot)$ is a homeomorphism.

Minimally crossing: A family of closed curves $\Gamma = (\gamma_1, \ldots, \gamma_k)$ is minimally crossing if for every family of closed curves $\Gamma' = (\gamma'_1, \ldots, \gamma'_k)$ with γ_i and γ'_i homotopic for each i, the number of intersections and self-intersections in Γ is no larger than in Γ'.

Covering space: Let $\tilde{S}$ be a possibly non-compact connected surface. A continuous map $\pi\colon \tilde{S} \to S$ is a ***covering map*** if every point $x \in S$ has a connected neighborhood U such that $\pi^{-1}(U)$ is a disjoint union of open sets $(U_i)_{i \in I}$ and $\pi|_{U_i}\colon U_i \to U$ is a homeomorphism for each i. We say that $(\tilde{S}, \pi)$ is a ***covering space*** of S. A ***lift*** of a path p is a path $\tilde{p}$ on $\tilde{S}$ such that $\pi \circ \tilde{p} = p$. Finally, if each loop in $\tilde{S}$ is contractible, then $(\tilde{S}, \pi)$ is a ***universal covering space*** of S, which is essentially unique (precisely: if $(\tilde{S}, \pi)$ and $(\tilde{S}', \pi')$ are universal covering spaces, then there is a homeomorphism $\tau : \tilde{S} \to \tilde{S}'$ such that $\pi = \pi' \circ \tau$).

BASIC PROPERTIES

1. Two paths p and q are homotopic if and only if $p \cdot q^{-1}$ is a (well-defined and) contractible loop.
2. Two loops p and q with the same basepoint are freely homotopic (viewed as closed curves without basepoint) if the homotopy classes of the loops p and q are conjugates in the fundamental group.
3. The fundamental group of a surface S without boundary of genus g is best understood by looking at a canonical polygonal schema of the surface: If S is orientable, it is the group generated by $2g$ generators $a_1, b_1, \ldots, a_g, b_g$ and with a single relation, $a_1 b_1 a_1^{-1} b_1^{-1} \ldots a_g b_g a_g^{-1} b_g^{-1}$, corresponding to the boundary of the polygonal schema. Similarly, if S is non-orientable, it is the group generated by g generators $a_1, \ldots, a_g$ and with a single relation, $a_1 a_1 \ldots a_g a_g$.
4. The fundamental group of a surface with at least one boundary component is a free group (because such a surface has the homotopy type of a graph).
5. Let $(\tilde{S}, \pi)$ be a covering space of S. Every path p on S admits lifts on $\tilde{S}$; moreover, if $\tilde{x}$ is a lift of $p(0)$, then p has a unique lift $\tilde{p}$ such that $\tilde{p}(0) = \tilde{x}$. Two paths are homotopic on S if and only if they have homotopic lifts on $\tilde{S}$. In particular, two paths are homotopic if they admit lifts with the same endpoints in the universal covering space.

DECIDING HOMOTOPY AND ISOTOPY

1. *Homotopy:* One of the first and most studied problems regarding curves on surfaces is concerned with homotopy tests: (1) The *contractibility problem*: Is a given closed curve (or, equivalently here, loop) contractible? (2) The *free homotopy problem*: Are two given closed curves (freely) homotopic? These problems translate to central problems from group theory, in the special case of fundamental groups of surfaces: Given a finitely generated group, presented in the form of generators and relations, (1) does a given word in the generators represent the trivial element of the group (the *word problem*)? Do two given words in the generators represent conjugate elements in the group (the *conjugacy problem*)?

   In computational geometry, these problems are studied in the following context: The input is a cellularly embedded graph G and one or two closed curves in G, represented as closed walks in G. There exist linear-time (and thus optimal) algorithms for both the contractibility and the free homotopy problems [LR12, EW13]. (An earlier article [DG99] claims the same results, but it is reported [LR12] that the algorithm for free homotopy in that article has a subtle flaw.) The approaches rely on the construction of a part of the universal covering space, or on results from small cancellation theory in group theory [GS90]. We remark that Dehn's algorithm [Deh12] can be implemented in linear time, but assuming that the surface is fixed and that the graph has a single face, which the other algorithms mentioned above do not require.
2. *Isotopy:* Deciding whether two simple closed curves are isotopic can also be done in linear time, because this equivalence relation is a simple refinement of

homotopy for simple closed curves [Eps66]. Deciding isotopy of graph embeddings is more complicated, but can also be done efficiently, since it essentially reduces to homotopy tests for closed curves [CVM14].

3. *Minimum-cost homotopies:* Often, when it is known that two curves are homotopic, one would like to compute a "reasonable" homotopy. Relevant questions include finding a homotopy that sweeps the minimum possible area (in a discretized sense) [CW13], or has the minimum possible number of "steps"; a homotopy in which the maximum length of the intermediate curves is minimal ("height" of the homotopy) [CL09]; a homotopy in which the maximum distance traveled by a point from the first to the second curve is minimal ("width" of the homotopy—this is related to the *homotopic Fréchet distance*) [HPN+16]; etc. Several of these questions have been studied only in the case of the plane, and extensions to surfaces are still open.

ELEMENTARY MOVES AND UNCROSSING

FIGURE 23.4.1
The four Reidemeister moves, up to ambient isotopy. The pictures represent the intersection of the union of the curves with a small disk on S; in particular, in these pictures, the regions bounded by the curves are homeomorphic to disks, and no other parts of curves intersect the parts of the curves shown.

1. *Elementary moves:* Every family of closed curves in general position can be made minimally crossing by a finite sequence of Reidemeister moves, described in Figure 23.4.1. If a closed curve has k self-crossings, $\Omega(k^2)$ Reidemeister moves can be needed; this is tight if the curve is homotopic to a simple curve, but in general no subexponential upper bound seems to be known [CE16]. Actually, one can deform a family of curves continuously to make it minimally crossing without increasing the total number of crossings at any step, and moreover, in a minimally crossing family, each curve is itself minimally self-crossing, and each pair of curves is minimally crossing [GS97] (see also [HS94a]). There are other characterizations of curves not in minimally crossing position [HS85].
2. *Making curves simple:* Let G be a graph cellularly embedded on a surface S. One can decide whether an input curve, represented by a closed walk in G, is homotopic to a simple closed curve in S in near-linear time. More generally, one can compute the minimum number of self-intersections of a curve in S homotopic to an input closed walk in G, and the minimum number of intersections between two curves in S respectively homotopic to two input closed walks in G, in quadratic time [DL17].
3. *Untangling curves by a homeomorphism:* Given two families of disjoint, simple curves, one can try to minimize the number of crossings between them by changing one of them by a homeomorphism of the surface; some bounds are known on the number of crossings that one can achieve [MSTW16].

4. *Simultaneous graph drawing:* This also relates to the problem of embedding two input graphs on the same surface in a way that the embeddings cross each other few times. Here also some results are known [Neg01, RS05, HKMT16]; one can also require both combinatorial maps to be fixed.
5. *Number of homotopy classes:* How many simple closed curves in different homotopy classes can one draw such that they pairwise cross at most k times, for a given integer k? On orientable surfaces of genus $g \geq 2$ without boundary and $k = 0$, the answer is $3g - 2$ (a pants decomposition, see below, together with a contractible closed curve). The problem is more interesting for larger values of k; it was recently proved that, for fixed k, the number of curves one can draw is polynomial in the genus [Prz15].

23.5 OPTIMIZATION: SHORTEST CURVES AND GRAPHS

The problem of computing shortest curves and graphs satisfying certain topological properties on surfaces has been widely considered. This leads to problems with a flavor of combinatorial optimization.

For these problems to be meaningful, a metric must be provided. In computational geometry, one could naturally consider piecewise linear surfaces in some Euclidean space (perhaps $\mathbb{R}^3$); however, efficient algorithms for computing shortest paths in such surfaces [MMP87, CH96] need additional assumptions because distances involve square roots, which leads to deep and unrelated questions on the complexity of comparing sums of square roots [Blö91]. Furthermore, in the context of graph problems in the specific case of surface-embedded graphs (Section 23.7 below), that model would be insufficient. The notions of combinatorial and cross-metric surfaces, defined below, have been developed to avoid these technical distractions, and are suitable in various settings. On the other hand, with an oracle for shortest path computations, several of the results in this section extend to more geometric settings, for example piecewise linear surfaces in some Euclidean space (see, e.g., [EW05, Sect. 3.6]).

GLOSSARY

Discrete metrics on surfaces

Combinatorial surface: A combinatorial surface is the data of a cellular graph embedding G, with positive weights on the edges. The only allowed curves are walks in G; the length of a curve is the sum of the weights of the edges of G traversed by the curve, counted with multiplicity. Algorithmically, curves are stored as closed walks in G. The complexity of the combinatorial surface is the complexity of the embedding G (asymptotically, its number of edges).

Cross-metric surface: A cross-metric surface [CVE10] is also the data of a cellular graph embedding G on some surface S, with positive weights on the edges. However, in contrast to the combinatorial surface model, here the curves are drawn on the surface S in *general position* with respect to G; the length of a curve is the sum of the weights of the edges of G crossed by the curve, counted with multiplicity. Algorithmically, a family of curves (or a graph) on a cross-metric surface is stored by the combinatorial map of the arrangement of that family of curves (or graph) together with G. The complexity of the cross-metric surface is the complexity of the embedding G (asymptotically, its number of edges).

Without loss of generality, one could draw the curves in a neighborhood of the dual graph G^* of G. Pushing them completely onto G^* would transform them into curves on the combinatorial surface defined by G^*. However, the cross-metric surface defined by G retains more information than the combinatorial surface defined by G^*: In the latter case, when curves share edges of G^*, they automatically overlap; the cross-metric model allows us to make them disjoint except at some well-defined crossing points. (We should point out that it is still possible to define the notion of crossing between two curves in a combinatorial surface, but this is still insufficient for some of the algorithms described below.)

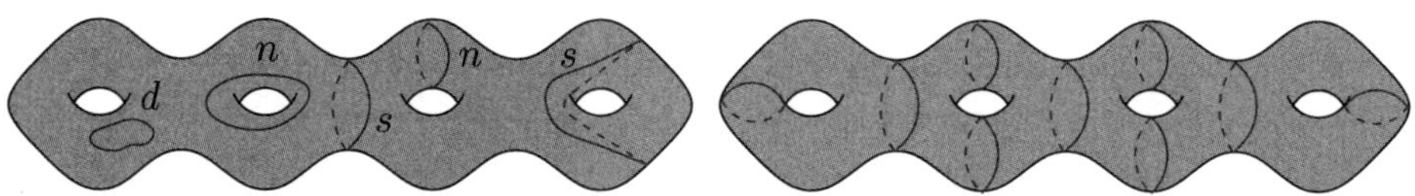

FIGURE 23.5.1
Left: Some closed curves on surfaces, (d) disk-bounding, (n) non-separating, (s) splitting. Right: A pants decomposition of a surface.

Types of simple closed curves

Let γ be a simple closed curve in the interior of a surface S. See Figure 23.5.1.

Disk-bounding curve: γ is disk-bounding if the surface S cut along γ (denoted by $S \backslash\!\backslash \gamma$) has two connected components, one of which is homeomorphic to the disk.

Separating curve: γ is separating if $S \backslash\!\backslash \gamma$ has two connected components.

Splitting curve: γ is splitting if γ is separating but not disk-bounding.

Essential curve: γ is essential if no component of $S \backslash\!\backslash \gamma$ is a disk or an annulus.

Topological decompositions

Cut graph: A cut graph is a graph G embedded on a surface S such that $S \backslash\!\backslash G$ is homeomorphic to a closed disk.

System of loops: A system of loops on a surface without boundary is a cut graph with a single vertex. See Figure 23.1.2.

Canonical system of loops: A system of loops G on a surface without boundary S is canonical if the edges of the polygon $S \backslash\!\backslash G$ appear in the same order as in a canonical polygonal schema (see Section 23.1)

Pants decomposition: A pants decomposition of an orientable surface S is a family Γ of simple, disjoint closed curves on S such that $S \backslash\!\backslash \Gamma$ is a disjoint union of pairs of pants. See Figure 23.5.1.

Octagonal decomposition: An octagonal decomposition of an orientable surface S without boundary is a family Γ of closed curves on S such that each (self-)intersection point in Γ is a crossing between exactly two closed curves, and each face of the arrangement of Γ on S is an octagon (a disk with eight sides).

Homology

In the context of graphs on surfaces, *one-dimensional homology on surfaces over the field* $\mathbb{Z}/2\mathbb{Z}$ is used; it can be described somewhat more concisely than more general homology theories. Let S be a surface. Here we assume graph embeddings to be piecewise linear (with respect to a fixed triangulation of S).

Homological sum: By the previous assumption, the closure of the symmetric difference of the images of two graph embeddings G and G' is the image of some graph embedding G'', called the homological sum of G and G'. (G'' is defined up to subdivision of edges with degree-two vertices, insertion of isolated vertices, and the reverse operations; here, graph embeddings are considered up to such operations.)

Homology cycle: A graph G embedded on S is a homology cycle if every vertex of G has even degree. The set of homology cycles forms a vector space over the field $\mathbb{Z}/2\mathbb{Z}$: The empty graph is the trivial element and addition is the homological sum.

Homology boundary: A graph G embedded on S is a homology boundary if the faces of G can be colored in two colors, say black and white, such that G is the "boundary" between the two colors: Exactly one side of each edge of G is incident to a black face. The set of homology boundaries forms a vector space over $\mathbb{Z}/2\mathbb{Z}$. Every homology boundary is a homology cycle.

Homology group: It is the $\mathbb{Z}/2\mathbb{Z}$-vector space, denoted by $H_1(S)$, that is the quotient of the homology cycles by the homology boundaries. A graph embedding is ***homologically trivial*** if it is a homology boundary.

The homology of sets of loops or closed curves can be defined similarly, because these loops and closed curves are the images of some graph embedding. Using the more advanced theory of *singular homology* one can remove the restriction of dealing with piecewise-linear graph embeddings.

BASIC PROPERTIES

1. A simple closed curve is disk-bounding if and only if it is contractible.
2. A simple closed curve is separating if and only if it is homologically trivial.
3. The homology group of a surface S without boundary has dimension $\bar{g}$, the Euler genus of S, and is generated by the loops appearing on the boundary of a canonical polygonal schema.

SHORTEST CURVES

Deciding whether a simple closed curve in a cross-metric (or combinatorial) surface is separating or disk-bounding can be done in time linear in the size of the data structure used to store the cellular graph and the curve; this boils down to determining whether some graph is connected, or whether some surface is a disk (which is easy using Euler's formula). Here we consider the optimization version, by looking for shortest curves with a given topological type in a combinatorial or cross-metric surface. Non-disk-bounding or non-separating curves are of particular interest, because cutting along such a curve simplifies the topology of a surface. Below we use *non-trivial* as a shorthand for either non-disk-bounding or non-separating.

TABLE 23.5.1 Algorithms for shortest non-trivial closed curves on surfaces without boundary, depending on whether the graph is weighted and whether it is directed. "Non-sep" and "non-db" mean non-separating and non-disk-bounding, respectively; k is the size of the output. The best complexities known to date are in bold (there can be several of them in each category due to the tradeoff between g, n, and k). Of course, the undirected case reduces to the directed case, and the unweighted case reduces to the weighted case; in each cell, we do not repeat the algorithms that are available for more general scenarios.

	UNDIRECTED	DIRECTED
WEIGHTED	$\boldsymbol{O(n^2 \log n)}$ [EHP04] $O(g^{3/2}n^{3/2} \log n)$ non-sep, $g^{O(g)}n^{3/2}$ non-db [CM07] $g^{O(g)}n \log n$ [Kut06] $O(g^3 n \log n)$ [CC07] $\boldsymbol{O(g^2 n \log n)}$ [CCE13] $g^{O(g)}n \log\log n$ [INS+11] $\boldsymbol{2^{O(g)}n \log\log n}$ [Fox13] $\boldsymbol{O(gn \log n)}$ for 2-approx. [EHP04]	$O(n^2 \log n)$ [CCVL16] $\boldsymbol{O(g^{1/2}n^{3/2} \log n)}$ [CCVL16] $2^{O(g)}n \log n$ non-sep [EN11b] $\boldsymbol{O(g^2 n \log n)}$ non-sep, $g^{O(g)}n \log n$ non-db [Eri11] $\boldsymbol{O(g^3 n \log n)}$ non-db [Fox13]
UNWEIGHTED	$O(n^3)$ [Tho90] (see [MT01]) $\boldsymbol{O(n^2)}$ [CCVL12] $\boldsymbol{O(gnk)}$ [CCVL12] $\boldsymbol{O(gn/\varepsilon)}$ for $(1+\varepsilon)$-approx. [CCVL12]	$\boldsymbol{O(n^2)}$ [CCVL16] $\boldsymbol{O(gnk)}$ [CCVL16]

1. *Structural properties:* In a combinatorial surface, a shortest noncontractible or non-null-homologous loop based at a vertex x is made of two shortest paths from x and of a single edge (this is the so-called 3-path condition [Tho90]). It follows that the globally shortest non-contractible and non-null-homologous closed curves do not repeat vertices and edges, and are also shortest non-disk-bounding and non-separating closed curves. More generally, in the algorithms mentioned below, a typical tool is to prove a bound on the number of crossings between the (unknown) shortest curve and any shortest path.

2. *Different scenarios for shortest non-trivial curves:* Table 23.5.1 summarizes the running times of the known algorithms. In such problems, it is relevant to look for more efficient algorithms in the case where the genus g is smaller compared to the complexity n of the graph defining the surface. The standard scenario, which is the only one considered elsewhere in this chapter, is that of a combinatorial (or equivalently, cross-metric) surface (the undirected,

weighted case, in the upper left corner in Table 23.5.1). One can also aim for faster algorithms in the *unweighted* case (unit weights). Finally, one can extend the techniques to the case of *directed* graphs, where the edges of the combinatorial surface are directed and can only be used in a specified direction (equivalently, the edges of the cross-metric surface can only be crossed in a specific direction).

3. *Other topological types:* Shortest simple closed curves of other topological types have been investigated as well (in the following, n denotes the complexity of the cross-metric surface): shortest splitting curves [CCV+08] (NP-hard, but computable in $O(n \log n)$ time for fixed genus); shortest essential curves [EW10] ($O(n^2 \log n)$ time, or $O(n \log n)$ for fixed genus and number of boundaries—in this case, surfaces with boundary require more sophisticated techniques); and non-separating curves which are shortest in their (unspecified) homotopy class [CDEM10] ($O(n \log n)$).
4. *Shortest homotopic curves:* A slightly different problem is that of computing a shortest curve homotopic to a given curve (either a path or a closed curve); this is also doable in small polynomial time, using octagonal decompositions to build a part of the universal covering space [CVE10] (earlier algorithms dealt with simple curves only, with an iterated shortening process that leads to a global optimum [CVL05, CVL07]).
5. *Shortest paths:* All these algorithms rely on shortest path computations on combinatorial (or cross-metric) surfaces, which can be done in $O(n \log n)$ time using Dijkstra's algorithm [Dij59] classically speeded up with Fibonacci heaps [FT87] in the primal (or dual) graph. This actually computes the shortest paths from a single source to all other vertices of the combinatorial surface. Other algorithms are available for computing multiple shortest paths quickly under some conditions on the locations of the endpoints [CCE13].

SHORTEST DECOMPOSITIONS

Decompositions of surfaces are central in topology; for example, the standard proof of the classification theorem transforms an arbitrary cut graph into a canonical system of loops. Many algorithms described in the previous subsection rely on topological decompositions and their properties.

1. *Shortest cut graph:* The problem of computing a shortest cut graph on a cross-metric surface has been extensively studied. Computing the shortest cut graph is NP-hard, but there is an $O(\log^2 g)$-approximation algorithm that runs in $O(g^2 n \log n)$ time [EHP04]. Moreover, for every $\varepsilon > 0$ one can compute a $(1+\varepsilon)$-approximation in $f(\varepsilon, g) \cdot n^3$ time, for some function f [CAM15]. If one is looking for a shortest cut graph with a specified vertex set P (for example, a shortest system of loops with given basepoint [EW05]), then there is an algorithm with running time $O(n \log n + gn + |P|)$ [CV10]. At the root of several of these articles lies the *tree-cotree property* [Epp03]: If G is a cellular graph embedding, there exists a partition (T, C, X) of the edges of G such that T is a spanning tree of G and the edges dual to C form a spanning tree of the dual graph G^*. Contracting T and deleting C transforms G into a system of loops, each loop corresponding to an element of X.

2. *Other topological decompositions:* Some canonical system of loops (for orientable surfaces without boundary) can be computed in $O(gn)$ time [LPVV01]. An octagonal decomposition or a pants decomposition made of closed curves which are as short as possible in their respective homotopy classes can be computed in $O(gn \log n)$ time [CVE10]. But in general the complexity of computing shortest such decompositions is open. On the other hand, there are bounds on the maximum length of some decompositions, assuming that the combinatorial surface is an unweighted triangulation, or, dually, that the cross-metric surface is unweighted and each vertex has degree three [CVHM15].
3. *Stretch:* Let S be a cross-metric surface, and let G be the associated embedded graph. The stretch of S is the minimum of the *product* of the lengths of γ and δ, over all closed curves γ and δ crossing exactly once. This quantity is related to the planar crossing number and the size of a largest toroidal grid minor of G^* [HC10], and can be computed in small polynomial time [CCH14].

HOMOLOGY AND ITS RELATION TO CUTS AND FLOWS

As hinted above, homology is useful because a simple closed curve is separating if and only if it is null-homologous; the algorithms for computing shortest non-separating closed curves actually compute shortest non-null-homologous closed curves, which turn out to be simple.

Homology is a natural concept; in particular, it is interesting to look for a family of closed curves, of minimum total length, the homology classes of which generate the homology group. Some efficient algorithms have been given for this purpose [EW05], also in connection with an algorithm to compute a minimum cycle basis of a surface-embedded graph [BCFN16].

Another reason for the importance of homology is its relation to cuts: Given a graph G cellularly embedded on a surface S without boundary, the (s,t)-cuts in G are dual to the subgraphs of G^* in some fixed homology class on the surface obtained from S by removing the faces of G^* containing s and t. Thus, computing minimum cuts amounts to computing shortest homologous subgraphs. This property has been exploited to study general graph problems, where better algorithms can be designed in the specific case of graphs embedded on a fixed surface, to:

1. compute minimum (s,t)-cuts in near-linear time [CEN09, EN11b]. The best algorithm runs in $2^{O(g)} n \log n$ time, where g is the genus [EN11b], and relies on the *homology cover*, a particular type of covering space;
2. compute maximum (s,t)-flows faster, by exploiting further the duality between flows and cuts [CEN12, BEN$^+$16];
3. count and sample minimum (s,t)-cuts efficiently [CFN14];
4. compute global minimum cuts efficiently (without fixing s and t) [EFN12];
5. deal with other problems, e.g., to compute the edge expansion and other connectivity measures [Pat13] or to bound the space complexity of bipartite matching [DGKT12].

23.6 ALGORITHMS FOR GRAPHS EMBEDDED ON A FIXED SURFACE

Some general graph problems can be solved faster in the special case of graphs embedded on a fixed surface. Examples include cut and flow problems (see previous section), multicommodity problems, domination and independence problems, connectivity problems (Steiner tree, traveling salesman problem, etc.), disjoint paths problems, shortest paths problems, subgraph problems, and more.

Sometimes the problems are solvable in polynomial-time on arbitrary graphs, and the goal is to obtain faster algorithms for surface-embedded graphs. But in many cases, the problems considered are NP-hard on arbitrary graphs, and polynomial-time algorithms are obtained for graphs embeddable on a fixed surface (occasionally by fixing some other parameters of the problem). Typically, optimization problems are considered, in which case it is relevant to look for approximation algorithms.

The methods involved usually combine topological aspects (as described above) with techniques from structural and algorithmic graph theory.

GLOSSARY

Minor: A graph H is a minor of another graph G if H can be obtained from G by removing edges and isolated vertices, and contracting edges.

Minor-closed family: A family $\mathscr{F}$ of graphs is minor-closed if every minor of a graph in $\mathscr{F}$ is also in $\mathscr{F}$.

Tree decomposition: A tree decomposition of a graph $G = (V, E)$ is a tree T in which each node is labeled by a subset of V, such that:

- for each $v \in V$, the set of nodes in T whose labels contain v induces a non-empty connected subtree of T, and
- if G has an edge connecting vertices u and v, then the label of at least one node of T contains both u and v.

Width: The width of a tree decomposition is the maximum cardinality of the labels minus one.

Treewidth: The treewidth of a graph G is the minimum width of a tree decomposition of G.

SURVEY OF TECHNIQUES

Central to algorithmic and structural graph theory is the study of minor-closed families of graphs; by a deep result of Robertson and Seymour [RS04], for each such family $\mathscr{F}$, there is a *finite* set $X_{\mathscr{F}}$ of graphs such that $G \in \mathscr{F}$ if and only if no graph in $X_{\mathscr{F}}$ is a minor of G. We refer to [KM07] for a survey on these structural aspects.

The graphs embeddable on a fixed surface form a minor-closed family, and have the benefit that they can be studied using topological techniques. Robertson and Seymour provide a decomposition theorem for minor-closed families of graphs involving graphs embeddable on a fixed surface [RS03]; efficient algorithms for surface-embedded graphs are sometimes extended to minor-closed families of graphs (different from the family of all graphs).

It is impossible to list all results in algorithms for surface-embedded graphs here, so we focus on general methods. Several algorithms are based on topological techniques described in the previous sections (in particular, shortest non-trivial curves or shortest decompositions), in several cases with advanced algorithmic techniques [EN11a, KKS11, ES14, PPSL14]. Sometimes the same techniques have led to new results for planar graphs [Eri10, EN11c, CV17b]. Methods applicable to several algorithmic problems have also emerged, in many cases extending previous ones invented for planar graphs:

1. *Graph separators and treewidth:* Let G be a graph with n vertices embedded on a surface with genus g. In linear time, one can compute a balanced separator of size $O(\sqrt{gn})$, namely, a set of $O(\sqrt{gn})$ vertices whose removal leaves a graph without connected component of more than $2n/3$ vertices [GHT84, Epp03]. Also, the treewidth of G is $O(\sqrt{gn})$.
2. *Dynamic programming:* Small treewidth implies efficient algorithms using dynamic programming in arbitrary graphs. When the graph is embedded, one can exploit this fact to obtain algorithms with smaller dependence on the treewidth for some problems [Bon12, RST13, RST14].
3. *Irrelevant vertex technique:* Several graph problems enjoy the following property [Thi12]: If the input graph has large treewidth, there exists an irrelevant vertex, whose removal creates an equivalent instance of the problem (e.g., a vertex at the center of a large grid minor). This property is widely used in structural graph theory and has been exploited several times in the context of algorithms for surface-embedded graphs [KR07, KT12, RS12].
4. *Polynomial-time approximation schemes (PTASs):* Baker [Bak94] has introduced a technique for designing approximation schemes for some optimization problems with local constraints in planar graphs: She has showed that one can delete a small part of the input graph without changing too much the value of the solution and such that the resulting graph has small treewidth. The technique has been extended to graphs embeddable on a fixed surface [Epp00], to graphs that can be drawn on a fixed surface with a bounded number of crossings per edge [GB07], and to more general *contraction-closed* problems where contraction instead of deletion must be used [Kle05, DHM10]. A crucial step in making the latter technique effective is the construction of a *spanner*: In the case of a minimization problem, this is a subgraph of the input graph containing a near-optimal solution and whose weight is linear in that of the optimal solution. *Brick decomposition* is a technique that builds spanners for some problems, originally in planar graphs, but also sometimes in graphs on surfaces [BDT14].
5. *Bidimensionality:* This theory [Thi15, DFHT05, DHT06, DH08] applies to minimization problems on unweighted graphs where contracting an edge of the graph does not increase the value of the solution, and where the value of the solution in grid graphs (and generalizations) is large. It leads to output-sensitive algorithms for graphs embeddable on a fixed surface with running time of the form $2^{O(\sqrt{k})} \cdot n^{O(1)}$, where k is the value of the solution and n is the input

size. This also provides PTASs in some cases [DH05]. For the problems where bidimensionality applies, PTASs can sometimes also be obtained in weighted graphs using a different framework [CACV+16].

6. *Stochastic embeddings:* Let $G = (V, E)$ and $G' = (V', E')$ be positively edge-weighted graphs. A *non-contracting metric embedding* f from G to G' is a mapping from V to V' such that $d'(f(x), f(y)) \geq d(x, y)$ for each $x, y \in V$, where d and d' represent shortest path distances in G and G', respectively. The *distortion* of f is the maximum of $d'(f(x), f(y))/d(x, y)$ over all $x \neq y \in V$ (see Chapter 8). Every graph G embeddable on an orientable surface S of genus g admits a probability distribution of non-contracting metric embeddings into planar graphs such that for each $x, y \in V$, one has $\mathbb{E}[d'(f(x), f(y))] \leq O(\log g) \cdot d(x, y)$, where the expectation is over all f in the distribution [Sid10]. This reduces several optimization problems on graphs on S to the same problem in planar graphs, up to the loss of an $O(\log g)$ factor. Actually, such a distribution can be computed in polynomial time even if no embedding of G on S is known [MS12].

23.7 OTHER MODELS

A rather large number of results relate to the concepts described in this chapter, and it would be impossible to cover them all. Below, we provide a selection of miscellaneous results that consider other models for representing graphs on surfaces.

COMPUTATIONAL TOPOLOGY IN THE PLANE WITH OBSTACLES

The plane minus finitely many points or polygons ("obstacles") forms a (non-compact) surface S. Taking any cellular graph embedding on S makes S a combinatorial (or cross-metric) surface, so most of the topological algorithms above apply. However, it is much more natural to consider arbitrary piecewise-linear curves in S, whose length is defined by the Euclidean metric. In this model, S is defined by the obstacles (a finite set of disjoint simple polygons, for simplicity of exposition); curves are arbitrary polygonal lines avoiding the interior of the obstacles. Some of the problems defined in the previous sections and related problems have been studied in this model:

1. *Homotopy and isotopy tests:* There are efficient algorithms to test whether two curves are (freely) homotopic [CLMS04], or whether two graphs are isotopic [CVM14].
2. *Shortest homotopic paths* can be computed efficiently as well [HS94b, Bes03, EKL06]; see also Section 31.2. A variant where several simple and disjoint paths must be shortened while preserving their homotopy class and keeping their neighborhoods simple and disjoint (i.e., the paths are "thick") has also been investigated [GJK+88].
3. *Shortest disjoint paths:* Here the goal is to compute disjoint paths with minimum total length (or, more precisely, non-crossing paths, since in the limit case, the solution may consist of overlapping paths). If the endpoints lie on the boundary of a bounded number of obstacles, the problem is solvable in polynomial time [EN11c].

4. *Other results* include a constant-factor approximation algorithm for the shortest pants decomposition in the case where the obstacles are points [Epp09] and an algorithm for computing the homotopic Fréchet distance, a measure of similarity between curves that takes the obstacles into account topologically [CCV+10, HPN+16].

SIMPLE AND DISJOINT CURVES IN GRAPHS

In the cross-metric model, defined by a cellularly embedded graph G, one can think of curves as being drawn in a neighborhood of G^*. So, intuitively, curves are drawn in G^*, but they can share vertices and edges of G^* while being simple and pairwise disjoint.

It is very natural, especially in topological graph theory, to forbid such overlaps: A set of disjoint simple curves cannot repeat any vertex or edge of G^*. Many of the problems mentioned in the previous sections make sense in this setup, which turns out to be generally more difficult to handle than the cross-metric model. In this model, the following results are known (here by *circuit* we mean a closed curve in the graph without repeated vertex, and containing at least one edge):

1. Determining whether there exists a separating (resp., splitting) circuit is NP-complete [CCVL11].
2. Determining some contractible (resp., non-contractible, resp., non-separating) circuit, if such a circuit exists, is possible in linear time, even if one requires the circuit to pass through a given vertex [CCVL11].
3. Computing a *shortest* contractible circuit is possible in polynomial time, but if one requires the circuit to pass through a given vertex, the problem becomes NP-hard [Cab10].
4. Computing a *shortest* separating circuit is NP-hard [Cab10].
5. There is a combinatorial characterization on whether curves can be made simple and disjoint in the graph by a homotopy on the surface [Sch91]. In the case of a planar surface with boundaries, this leads to a polynomial-time algorithm [Sch90b, Th. 31], which in turn has some algorithmic consequences on the problem of computing vertex-disjoint paths in planar graphs [Sch90b, Th. 34]. See also [Sch03, Ch. 76].

NORMAL CURVES ON SURFACES

Let Γ be a family of disjoint simple closed curves on a surface S in general position with respect to a triangulation T of S. A natural way to represent Γ, as described in the previous sections, is by its arrangement with T. *Normal curves* are a more economical representation, at the price of a mild condition: For every triangle t of T, the intersection of the image of Γ with t must be a set of (disjoint simple) paths, called *normal arcs*, connecting *different* sides of t. For such a Γ, and for each triangle t of T, one stores three integers recording the number of normal arcs connecting each of the three pairs of sides of t. Overall, Γ is described by $3n$ non-negative integers, where n is the number of triangles in T. Conversely, given a vector of $3n$ non-negative integers, one can unambiguously reconstruct Γ

up to *normal isotopy*, that is, up to an ambient isotopy that leaves the edges of T globally unchanged.

To store the vector of normal coordinates, $O(n \log(X/n))$ bits are needed, where X is the number of crossing points of Γ with T. In contrast, representing these curves on a cross-metric surface requires at least to store a constant amount of information per vertex of the arrangement, which is $\Theta(n+X)$ in total. So the normal curve representation can be exponentially compressed compared to the "naïve" one. Despite this, in time polynomial in the input size one can:

1. count the number of connected components of a normal curve (note that a "normal curve" does not have to be connected), and partition these components according to their (normal or not) isotopy classes, given by their multiplicities and the normal coordinates of a representative [SSŠ02, EN13];
2. decide whether two normal curves are isotopic [SSŠ02, EN13];
3. compute the algebraic [SSŠ02, EN13] or the geometric [SSŠ08] intersection number of two normal curves. (The algebraic intersection number of γ and δ is the sum, over all crossings between γ and δ, of the sign of the crossing, which is $+1$ if γ crosses δ from left to right at that crossing point and -1 otherwise; this is well-defined if the surface is orientable, since it is invariant by isotopy. The geometric intersection number of γ and δ is the minimum number of crossings between curves γ' and δ' isotopic to γ and δ.)

These problems have been initially studied using *straight-line programs*, a concise encoding of words over a finite alphabet; many algorithms on words can be solved efficiently using the straight-line program representation, in particular because straight-line programs can represent exponentially long words; this leads to efficient algorithms for normal curves [SSŠ02, SSŠ08]. The same and other problems have been revisited using more topological techniques [EN13]. Normal curves are the lower-dimensional analog of *normal surfaces*, widely used in three-dimensional topology.

23.8 OTHER RESOURCES

Books. Graphs on surfaces from a combinatorial viewpoint are treated in detail in [MT01]; see also [GT87]. For basic surface topology, we recommend [Arm83, Sti93, Hen94].

Survey. [Eri12] surveys optimization problems for surface-embedded graphs, providing more details on a large fraction of Section 23.5.

Course notes and unpublished material. [Eri13] provides some notes in computational topology with a strong emphasis on graphs on surfaces. [CV12, CV17a] survey some algorithms for optimization of graphs and curves on surfaces. [DMST11] emphasizes graph algorithms for surface-embedded graphs.

RELATED CHAPTERS

Chapter 20: Polyhedral maps
Chapter 31: Shortest paths and networks
Chapter 55: Graph drawing
Chapter 66: Geometry and topology of genomic data

REFERENCES

[AG05] P. Alliez and C. Gotsman. Recent advances in compression of 3D meshes. In N.A. Dodgson, M.S. Floater, and M.A. Sabin, editors, *Advances in Multiresolution for Geometric Modelling*, pages 3–26, Springer-Verlag, Berlin, 2005.

[Arc96] D. Archdeacon. Topological graph theory. A survey. *Congr. Numer.*, 115:5–54, 1996.

[Arm83] M.A. Armstrong. *Basic Topology.* Undergraduate Texts in Mathematics, Springer-Verlag, Berlin, 1983.

[Bak94] B.S. Baker. Approximation algorithms for NP-complete problems on planar graphs. *J. ACM*, 41:153–180, 1994.

[BCFN16] G. Borradaile, E.W. Chambers, K. Fox, and A. Nayyeri. Minimum cycle and homology bases of surface embedded graphs. In *Proc. 32nd Sympos. Comput. Geom.*, vol 51 of *LIPIcs*, article 23, Schloss Dagstuhl, 2016.

[BCHK16] S. Beyer, M. Chimani, I. Hedtke, and M. Kotrbčík. A practical method for the minimum genus of a graph: Models and experiments. In *Proc. 15th Sympos. Experimental Algorithms*, vol. 9685 of *LNCS*, pages 75–88, Springer, Cham, 2016.

[BDT14] G. Borradaile, E.D. Demaine, and S. Tazari. Polynomial-time approximation schemes for subset-connectivity problems in bounded-genus graphs. *Algorithmica*, 68:287–311, 2014.

[BEKT16] B.A. Burton, M. Elder, A. Kalka, and S. Tillmann. 2-manifold recognition is in logspace. *J. Comput. Geom.*, 7:70–85, 2016.

[BEN+16] G. Borradaile, D. Eppstein, A. Nayyeri, and C. Wulff-Nilsen. All-pairs minimum cuts in near-linear time for surface-embedded graphs. In *Proc. 32nd Sympos. Comput. Geom.*, vol. 51 of *LIPIcs*, article 22, Schloss Dagstuhl, 2016.

[Bes03] S. Bespamyatnikh. Computing homotopic shortest paths in the plane. *J. Algorithms*, 49:284–303, 2003.

[Bet12] J. Bettinelli. The topology of scaling limits of positive genus random quadrangulations. *Ann. Probab.*, 40:1897–1944, 2012.

[BKS98] M. de Berg, M. van Kreveld, and S. Schirra. Topologically correct subdivision simplification using the bandwidth criterion. *Cartography and GIS*, 25:243–257, 1998.

[Blö91] J. Blömer. Computing sums of radicals in polynomial time. In *Proc. 32nd IEEE Sympos. Found. Comp. Sci.*, pages 670–677, 1991.

[Bon12] P. Bonsma. Surface split decompositions and subgraph isomorphism in graphs on surfaces. In *Proc. 29th Sympos. Theoret. Aspects Comp. Sci.*, vol. 14 of *LIPIcs*, pages 531–542, Schloss Dagstuhl, 2012.

[Cab10] S. Cabello. Finding shortest contractible and shortest separating cycles in embedded graphs. *ACM Trans. Algorithms*, 6:24, 2010.

[Cab13] S. Cabello. Hardness of approximation for crossing number. *Discrete Comput. Geom.*, 49:348–358, 2013.

[CACV+16] V. Cohen-Addad, É. Colin de Verdière, P.N. Klein, C. Mathieu, and D. Meierfrankenfeld. Approximating connectivity domination in weighted bounded-genus graphs. In *Proc. 48th ACM Sympos. Theory Comput.*, pages 584–597, 2016.

[CAM15] V. Cohen-Addad and A. de Mesmay. A fixed parameter tractable approximation scheme for the optimal cut graph of a surface. In *Proc. 23rd European Sympos. Algorithms*, vol. 9294 of *LNCS*, pages 386–398, Springer, Berlin, 2015.

[CC07] S. Cabello and E.W. Chambers. Multiple source shortest paths in a genus g graph. In *Proc. 18th ACM-SIAM Sympos. Discrete Algorithms*, pages 89–97, 2007.

[CCE13] S. Cabello, E.W. Chambers, and J. Erickson. Multiple-source shortest paths in embedded graphs. *SIAM J. Comput.*, 42:1542–1571, 2013.

[CCH14] S. Cabello, M. Chimani, and P. Hliněný. Computing the stretch of an embedded graph. *SIAM J. Discrete Math.*, 28:1391–1401, 2014.

[CCV+08] E.W. Chambers, É. Colin de Verdière, J. Erickson, F. Lazarus, and K. Whittlesey. Splitting (complicated) surfaces is hard. *Comput. Geom.*, 41:94–110, 2008.

[CCV+10] E.W. Chambers, É. Colin de Verdière, J. Erickson, S. Lazard, F. Lazarus, and S. Thite. Homotopic Fréchet distance between curves — or, walking your dog in the woods in polynomial time. *Comput. Geom.*, 43:295–311, 2010.

[CCVL11] S. Cabello, É. Colin de Verdière, and F. Lazarus. Finding cycles with topological properties in embedded graphs. *SIAM J. Discete Math.*, 25:1600–1614, 2011.

[CCVL12] S. Cabello, É. Colin de Verdière, and F. Lazarus. Algorithms for the edge-width of an embedded graph. *Comput. Geom.*, 45:215–224, 2012.

[CCVL16] S. Cabello, É. Colin de Verdière, and F. Lazarus. Finding shortest non-trivial cycles in directed graphs on surfaces. *J. Comput. Geom.*, 7:123–148, 2016.

[CDEM10] S. Cabello, M. DeVos, J. Erickson, and B. Mohar. Finding one tight cycle. *ACM Trans. Algorithms*, 6:61, 2010.

[CDP04] S.-W. Cheng, T.K. Dey, and S.-H. Poon. Hierarchy of surface models and irreducible triangulations. *Comput. Geom.*, 27:135–150, 2004.

[CE16] H.-C. Chang and J. Erickson. Untangling planar curves. In *Proc. 32nd Sympos. Comput. Geom.*, vol. 51 of *LIPIcs*, article 29, Schloss Dagstuhl, 2016.

[CEN09] E.W. Chambers, J. Erickson, and A. Nayyeri. Minimum cuts and shortest homologous cycles. In *Proc. 25th Sympos. Comput. Geom.*, pages 377–385, ACM Press, 2009.

[CEN12] E.W. Chambers, J. Erickson, and A. Nayyeri. Homology flows, cohomology cuts. *SIAM J. Comput.*, 41:1605–1634, 2012.

[CFN14] E.W. Chambers, K. Fox, and A. Nayyeri. Counting and sampling minimum cuts in genus g graphs. *Discrete Comput. Geom.*, 52:450–475, 2014.

[CH96] J. Chen and Y. Han. Shortest paths on a polyhedron. *Internat. J. Comput. Geom. Appl.*, 6:127–144, 1996.

[Chu11] J. Chuzhoy. An algorithm for the graph crossing number problem. In *Proc. 43rd ACM Sympos. Theory Comput.*, pages 303–312, 2011.

[CL09] E.W. Chambers and D. Letscher. On the height of a homotopy. In *Proc. 21st Canad. Conf. Comput. Geom.*, pages 103–106, 2009. See also the erratum at `mathcs.slu.edu/~chambers/papers/hherratum.pdf`.

[CLMS04] S. Cabello, Y. Liu, A. Mantler, and J. Snoeyink. Testing homotopy for paths in the plane. *Discrete Comput. Geom.*, 31:61–81, 2004.

[CM07] S. Cabello and B. Mohar. Finding shortest non-separating and non-contractible cycles for topologically embedded graphs. *Discrete Comput. Geom.*, 37:213–235, 2007.

[CM13] S. Cabello and B. Mohar. Adding one edge to planar graphs makes crossing number hard. *SIAM J. Comput.*, 42:1803–1829, 2013.

[CN00] G. Cairns and Y. Nicolayevsky. Bounds for generalized thrackles. *Discrete Comput. Geom.*, 23:191–206, 2000.

[CV10] É. Colin de Verdière. Shortest cut graph of a surface with prescribed vertex set. In *Proc. 18th European Sympos. Algorithms, part 2*, vol. 6347 of *LNCS*, pages 100–111, Springer, Berlin, 2010.

[CV12] É. Colin de Verdière. *Topological algorithms for graphs on surfaces.* Habilitation thesis, École normale supérieure, Paris, 2012. Available at `http://monge.univ-mlv.fr/~colinde/pub/12hdr.pdf`.

[CV17a] É. Colin de Verdière. *Algorithms for embedded graphs.* Course notes, `http://monge.univ-mlv.fr/~colinde/cours/all-algo-embedded-graphs.pdf`, 2017.

[CV17b] É. Colin de Verdière. Multicuts in planar and bounded-genus graphs with bounded number of terminals. *Algorithmica*, 78:1206–1224, 2017.

[CVE10] É. Colin de Verdière and J. Erickson. Tightening nonsimple paths and cycles on surfaces. *SIAM J. Comput.*, 39:3784–3813, 2010.

[CVHM15] É. Colin de Verdière, A. Hubard, and A. de Mesmay. Discrete systolic inequalities and decompositions of triangulated surfaces. *Discrete Comput. Geom.*, 53:587–620, 2015.

[CVK+16] É. Colin de Verdière, V. Kaluža, P. Paták, Z. Patáková, and M. Tancer. A direct proof of the strong Hanani–Tutte theorem on the projective plane. In *Proc. 24th Sympos. Graph Drawing Network Visualization*, pages 454–467, vol. 9801 of *LNCS*, Springer, Cham, 2016.

[CVL05] É. Colin de Verdière and F. Lazarus. Optimal system of loops on an orientable surface. *Discrete Comput. Geom.*, 33:507–534, 2005.

[CVL07] É. Colin de Verdière and F. Lazarus. Optimal pants decompositions and shortest homotopic cycles on an orientable surface. *J. ACM*, 54:18, 2007.

[CVM14] É. Colin de Verdière and A. de Mesmay. Testing graph isotopy on surfaces. *Discrete Comput. Geom.*, 51:171–206, 2014.

[CW13] E.W. Chambers and Y. Wang. Measuring similarity between curves on 2-manifolds via homotopy area. In *Proc. 29th Sympos. Comput. Geom.*, pages 425–434, ACM Press, 2013.

[Deh12] M. Dehn. Transformation der Kurven auf zweiseitigen Flächen. *Math. Ann.*, 72:413–421, 1912.

[DFHT05] E.D. Demaine, F.V. Fomin, M.T. Hajiaghayi, and D.M. Thilikos. Subexponential parameterized algorithms on bounded-genus graphs and H-minor-free graphs. *J. ACM*, 52:866–893, 2005.

[DG99] T.K. Dey and S. Guha. Transforming curves on surfaces. *J. Comput. System Sci.*, 58:297–325, 1999.

[DGKT12] S. Datta, A. Gopalan, R. Kulkarni, and R. Tewari. Improved bounds for bipartite matching on surfaces. In *Proc. Sympos. Theoret. Aspects Comp. Sci.*, vol. 14 of *LIPIcs*, pages 254–265, Schloss Dagstuhl, 2012.

[DH05] E.D. Demaine and M. Hajiaghayi. Bidimensionality: new connections between FPT algorithms and PTASs. In *Proc. 16th ACM-SIAM Sympos. Discrete Algorithms*, pages 590–601, 2005.

[DH08] E.D. Demaine and M. Hajiaghayi. The bidimensionality theory and its algorithmic applications. *The Computer Journal*, 51:292–302, 2008.

[DHM10] E.D. Demaine, M. Hajiaghayi, and B. Mohar. Approximation algorithms via contraction decomposition. *Combinatorica*, 30:533–552, 2010.

[DHT06] E.D. Demaine, M. Hajiaghayi, and D.M. Thilikos. The bidimensional theory of bounded-genus graphs. *SIAM J. Discrete Math.*, 20:357–371, 2006.

[Dij59] E.W. Dijkstra. A note on two problems in connexion with graphs. *NumerMath.*, 1:269–271, 1959.

[DL17] V. Despré and F. Lazarus. Computing the geometric intersection number of curves. In *Proc. 33rd Sympos. Comput. Geom.*, vol. 77 of *LIPIcs*, article 35, Schloss Dagstuhl, 2017.

[DMST11] E. Demaine, S. Mozes, C. Sommer, and S. Tazari. *Algorithms for Planar Graphs and Beyond.* Course notes, `http://courses.csail.mit.edu/6.889/fall11/lectures/`, 2011.

[EFN12] J. Erickson, K. Fox, and A. Nayyeri. Global minimum cuts in surface embedded graphs. In *Proc. 23rd ACM-SIAM Sympos. Discrete Algorithms*, pages 1309–1318, 2012.

[EHP04] J. Erickson and S. Har-Peled. Optimally cutting a surface into a disk. *Discrete Comput. Geom.*, 31:37–59, 2004.

[EK14] M. Elberfeld and K. Kawarabayashi. Embedding and canonizing graphs of bounded genus in logspace. In *Proc. 46th ACM Sympos. Theory of Computing*, pages 383–392, 2014.

[EKL06] A. Efrat, S.G. Kobourov, and A. Lubiw. Computing homotopic shortest paths efficiently. *Comput. Geom.*, 35:162–172, 2006.

[EN11a] J. Erickson and A. Nayyeri. Computing replacement paths in surface-embedded graphs. In *Proc. 22nd ACM-SIAM Sympos. Discrete Algorithms*, pages 1347–1354, 2011.

[EN11b] J. Erickson and A. Nayyeri. Minimum cuts and shortest non-separating cycles via homology covers. In *Proc. 22nd ACM-SIAM Sympos. Discrete Algorithms*, pages 1166–1176, 2011.

[EN11c] J. Erickson and A. Nayyeri. Shortest non-crossing walks in the plane. In *Proc. 22nd ACM-SIAM Sympos. Discrete Algorithms*, pages 297–308, 2011.

[EN13] J. Erickson and A. Nayyeri. Tracing compressed curves in triangulated surfaces. *Discrete Comput. Geom.*, 49:823–863, 2013.

[Epp00] D. Eppstein. Diameter and treewidth in minor-closed graph families. *Algorithmica*, 27:275–291, 2000.

[Epp03] D. Eppstein. Dynamic generators of topologically embedded graphs. In *Proc. 14th ACM-SIAM Sympos. Discrete Algorithms*, pages 599–608, 2003.

[Epp09] D. Eppstein. Squarepants in a tree: sum of subtree clustering and hyperbolic pants decomposition. *ACM Trans. Algorithms*, 5, 2009.

[Eps66] D.B.A. Epstein. Curves on 2-manifolds and isotopies. *Acta Math.*, 115:83–107, 1966.

[Eri10] J. Erickson. Maximum flows and parametric shortest paths in planar graphs. In *Proc. 21st ACM-SIAM Sympos. Discrete Algorithms*, pages 794–804, 2010.

[Eri11] J. Erickson. Shortest non-trivial cycles in directed surface graphs. In *Proc. 27th Sympos. Comput. Geom.*, pages 236–243, ACM Press, 2011.

[Eri12] J. Erickson. Combinatorial optimization of cycles and bases. In A. Zomorodian, editor, *Advances in Applied and Computational Topology*, vol. 70 of *Proc. Sympos. Appl. Math.*, pages 195–228, AMS, Providence, 2012.

[Eri13] J. Erickson. *Computational Topology.* Course notes, `http://compgeom.cs.uiuc.edu/~jeffe/teaching/comptop/`, 2013.

[ES14] J. Erickson and A. Sidiropoulos. A near-optimal approximation algorithm for asymmetric TSP on embedded graphs. In *Proc. 13th Sympos. Comput. Geom.*, pages 130–135, ACM Press, 2014.

[EW05] J. Erickson and K. Whittlesey. Greedy optimal homotopy and homology generators. In *Proc. 16th ACM-SIAM Sympos. Discrete Algorithms*, pages 1038–1046, 2005.

[EW10] J. Erickson and P. Worah. Computing the shortest essential cycle. *Discrete Comput. Geom.*, 44:912–930, 2010.

[EW13] J. Erickson and K. Whittlesey. Transforming curves on surfaces redux. In *Proc. 24th ACM-SIAM Sympos. Discrete Algorithms*, pages 1646–1655, 2013.

[FGM88] M.L. Furst, J.L. Gross, and L.A. McGeoch. Finding a maximum-genus graph imbedding. *J. ACM*, 35:523–534, 1988.

[FM11] B. Farb and D. Margalit. *A Primer on Mapping Class Groups.* Princeton University Press, 2011.

[Fox13] K. Fox. Shortest non-trivial cycles in directed and undirected surface graphs. In *Proc. 24th ACM-SIAM Sympos. Discrete Algorithms*, pages 352–364, 2013.

[FT87] M.L. Fredman and R.E. Tarjan. Fibonacci heaps and their uses in improved network optimization algorithms. *J. ACM*, 34:596–615, 1987.

[GB07] A. Grigoriev and H.L. Bodlaender. Algorithms for graphs embeddable with few crossings per edge. *Algorithmica*, 49:1–11, 2007.

[GHT84] J.R. Gilbert, J.P. Hutchinson, and R.E. Tarjan. A separator theorem for graphs of bounded genus. *J. Algorithms*, 5:391–407, 1984.

[GJK$^+$88] S. Gao, M. Jerrum, M. Kaufmann, K. Mehlhorn, W. Rülling, and C. Storb. On continuous homotopic one layer routing. In *Proc. 4th Sympos. Comput. Geom.*, pages 392–402, ACM Press, 1988.

[GS90] S.M. Gersten and H.B. Short. Small cancellation theory and automatic groups. *Invent. Math.*, 102:305–334, 1990.

[GS97] M. de Graaf and A. Schrijver. Making curves minimally crossing by Reidemeister moves. *J. Combin. Theory Ser. B*, 70:134–156, 1997.

[GT87] J.L. Gross and T.W. Tucker. *Topological Graph Theory.* Wiley, New York, 1987.

[GW01] I. Guskov and Z.J. Wood. Topological noise removal. In *Proc. Graphics Interface*, pages 19–26, Canad. Inf. Process. Soc., Toronto, 2001.

[GY03] X. Gu and S.-T. Yau. Global conformal surface parameterization. In *Proc. Eurographics/ACM Sympos. Geom. Processing*, pages 127–137, 2003.

[Han34] C. Chojnacki (H. Hanani). Über wesentlich unplättbare Kurven im dreidimensionalen Raume. *Fund. Math.*, 23:135–142, 1934.

[HC10] P. Hliněný and M. Chimani. Approximating the crossing number of graphs embeddable in any orientable surface. In *Proc. ACM-SIAM Sympos. Discrete Algorithms*, pages 918–927, 2010.

[HD16] P. Hliněný and M. Derňár. Crossing number is hard for kernelization. In *Proc. 32nd Sympos. Comput. Geom.*, vol. 51 of *LIPIcs*, article 42, Schloss Dagstuhl, 2016.

[Hen94] M. Henle. *A Combinatorial Introduction to Topology*. Dover Publications, Mineola, 1994.

[HKMT16] A. Hubard, V. Kaluža, A. de Mesmay, and M. Tancer. Shortest path embeddings of graphs on surfaces. In *Proc. 32nd Sympos. Comput. Geom.*, vol. 51 of *LIPIcs*, article 43, Schloss Dagstuhl, 2016.

[HPN$^+$16] S. Har-Peled, A. Nayyeri, M. Salavatipour, and A. Sidiropoulos. How to walk your dog in the mountains with no magic leash. *Discrete Comput. Geom.*, 55:39–73, 2016.

[HS85] J. Hass and P. Scott. Intersections of curves on surfaces. *Israel J. Math.*, 51:90–120, 1985.

[HS94a] J. Hass and P. Scott. Shortening curves on surfaces. *Topology*, 33:25–43, 1994.

[HS94b] J. Hershberger and J. Snoeyink. Computing minimum length paths of a given homotopy class. *Comput. Geom.*, 4:63–98, 1994.

[HT74] J. Hopcroft and R. Tarjan. Efficient planarity testing. *J. ACM*, 21:549–568, 1974.

[INS$^+$11] G.F. Italiano, Y. Nussbaum, P. Sankowski, and C. Wulff-Nilsen. Improved algorithms for min cut and max flow in undirected planar graphs. In *Proc. 43rd ACM Sympos. Theory of Computing*, pages 313–322, 2011.

[Kel06] J.A. Kelner. Spectral partitioning, eigenvalue bounds, and circle packings for graphs of bounded genus. *SIAM J. Comput.*, 35:882–902, 2006.

[Ket99] L. Kettner. Using generic programming for designing a data structure for polyhedral surfaces. *Comput. Geom.*, 13:65–90, 1999.

[KKS11] K. Kawarabayashi, P.N. Klein, and C. Sommer. Linear-space approximate distance oracles for planar, bounded-genus and minor-free graphs. In *Proc. Int. Coll. Automata, Languages and Progr., part 1*, vol. 6755 of *LNCS*, pages 135–146, Springer, Berlin, 2011.

[Kle05] P.N. Klein. A linear-time approximation scheme for planar weighted TSP. In *Proc. 46th IEEE Sympos. Found. Comp. Sci.*, pages 647–657, 2005.

[KM07] K. Kawarabayashi and B. Mohar. Some recent progress and applications in graph minor theory. *Graphs Combin.*, 23:1–46, 2007.

[KMR08] K. Kawarabayashi, B. Mohar, and B. Reed. A simpler linear time algorithm for embedding graphs into an arbitrary surface and the genus of graphs of bounded tree-width. In *Proc. 49th IEEE Sympos. Found. Comp. Sci.*, pages 771–780, 2008.

[KR07] K. Kawarabayashi and B. Reed. Computing crossing number in linear time. In *Proc. 39th ACM Sympos. Theory of Computing*, pages 382–390, 2007.

[KS15] K. Kawarabayashi and A. Sidiropoulos. Beyond the Euler characteristic: approximating the genus of general graphs. In *Proc. 47th ACM Sympos. Theory of Computing*, pages 675–682, 2015.

[KT12] M. Kamiński and D.M. Thilikos. Contraction checking in graphs on surfaces. In *Proc. 29th Sympos. Theoret. Aspects Comp. Sci.*, vol. 14 of *LIPIcs*, pages 182–193, Schloss Dagstuhl, 2012.

[Kut06] M. Kutz. Computing shortest non-trivial cycles on orientable surfaces of bounded genus in almost linear time. In *Proc. 22nd Sympos. Comput. Geom.*, pages 430–438, ACM Press, 2006.

[Lin82] S. Lins. Graph-encoded maps. *J. Combin. Theory Ser. B*, 32:171–181, 1982.

[LM85] C.E. Leiserson and F.M. Maley. Algorithms for routing and testing routability of planar VLSI layouts. In *Proc. 17th ACM Sympos. Theory of Computing*, pages 69–78, 1985.

[LPVV01] F. Lazarus, M. Pocchiola, G. Vegter, and A. Verroust. Computing a canonical polygonal schema of an orientable triangulated surface. In *Proc. 17th Sympos. Comput. Geom.*, pages 80–89, ACM Press, 2001.

[LR12] F. Lazarus and J. Rivaud. On the homotopy test on surfaces. In *Proc. 53rd IEEE Sympos. Found. Comp. Sci.*, pages 440–449, 2012.

[LZ04] S.K. Lando and A.K. Zvonkin. *Graphs on Surfaces and Their Applications.* Springer-Verlag, Berlin, 2004.

[Mat14] J. Matoušek. String graphs and separators. In J. Nešetřil and M. Pellegrini, *Geometry, Structure and Randomness in Combinatorics*, pages 61–97, Scuola Normale Superiore, Pisa, 2014.

[Mie09] G. Miermont. Tessellations of random maps of arbitrary genus. *Annales Scientifiques de l'École normale supérieure, Quatrième série*, 42:725–781, 2009.

[MK11] W. Myrvold and W. Kocay. Errors in graph embedding algorithms. *J. Comput. System Sci.*, 77:430–438, 2011.

[MMP87] J.S.B. Mitchell, D.M. Mount, and C.H. Papadimitriou. The discrete geodesic problem. *SIAM J. Comput.*, 16:647–668, 1987.

[Moh97] B. Mohar. On the minimal genus of 2-complexes. *J. Graph Theory*, 24:281–290, 1997.

[Moh99] B. Mohar. A linear time algorithm for embedding graphs in an arbitrary surface. *SIAM J. Discete Math.*, 12:6–26, 1999.

[MS12] Y. Makarychev and A. Sidiropoulos. Planarizing an unknown surface. In *Proc. 15th Workshop on Approximation*, vol. 7408 of *LNCS*, pages 266–275, Springer, Berlin, 2012.

[MSTW16] J. Matoušek, E. Sedgwick, M. Tancer, and U. Wagner. Untangling two systems of noncrossing curves. *Israel J. Math.*, 212:37–79, 2016.

[MT01] B. Mohar and C. Thomassen. *Graphs on Surfaces.* Johns Hopkins University Press, 2001.

[Neg01] S. Negami. Crossing numbers of graph embedding pairs on closed surfaces. *J. Graph Theory*, 36:8–23, 2001.

[NR04] T. Nishizeki and M.S. Rahman. *Planar Graph Drawing.* World Scientific, Singapore, 2004.

[Pat13] V. Patel. Determining edge expansion and other connectivity measures of graphs of bounded genus. *SIAM J. Comput.*, 42:1113–1131, 2013.

[PPSL14] Ma. Pilipczuk, Mi. Pilipczuk, P. Sankowski, and E.J. van Leeuwen. Network sparsification for Steiner problems on planar and bounded-genus graphs. In *Proc. 55th IEEE Sympos. Found. Comp. Sci.*, pages 186–195, 2014.

[Prz15] P. Przytycki. Arcs intersecting at most once. *Geom. Funct. Anal.*, 25:658–670, 2015.

[PSŠ08] M.J. Pelsmajer, M. Schaefer, and D. Štefankovič. Odd crossing number and crossing number are not the same. *Discrete Comput. Geom.*, 39:442–454, 2008.

[PSS09a] M.J. Pelsmajer, M. Schaefer, and D. Stasi. Strong Hanani–Tutte on the projective plane. *SIAM J. Discrete Math.*, 23:1317–1323, 2009.

[PSŠ09b] M.J. Pelsmajer, M. Schaefer, and D. Štefankovič. Removing even crossings on surfaces. *European J. Combin.*, 30:1704–1717, 2009.

[RS03] N. Robertson and P.D. Seymour. Graph minors. XVI. Excluding a non-planar graph. *J. Combin. Theory Ser. B*, 89(1):43–76, 2003.

[RS04] N. Robertson and P.D. Seymour. Graph minors. XX. Wagner's conjecture. *J. Combin. Theory Ser. B*, 92:325–357, 2004.

[RS05] R.B. Richter and G. Salazar. Two maps with large representativity on one surface. *J. Graph Theory*, 50:234–245, 2005.

[RS12] N. Robertson and P.D. Seymour. Graph minors. XXII. Irrelevant vertices in linkage problems. *J. Combin. Theory Ser. B*, 102:530–563, 2012.

[RST13] J. Rué, I. Sau, and D.M. Thilikos. Asymptotic enumeration of non-crossing partitions on surfaces. *Discrete Math.*, 313:635–649, 2013.

[RST14] J. Rué, I. Sau, and D.M. Thilikos. Dynamic programming for graphs on surfaces. *ACM Trans. Algorithms*, 10:8, 2014.

[Sch90a] W. Schnyder. Embedding planar graphs on the grid. In *Proc. 1st ACM-SIAM Sympos. Discrete Algorithms*, pages 138–148, 1990.

[Sch90b] A. Schrijver. Homotopic routing methods. In B. Korte, L. Lovász, H.J. Prömel, and A. Schrijver, editors, *Paths, Flows, and VLSI-layout*, pages 329–371. Springer-Verlag, Berlin, 1990.

[Sch91] A. Schrijver. Disjoint circuits of prescribed homotopies in a graph on a compact surface. *J. Combin. Theory Ser. B*, 51:127–159, 1991.

[Sch03] A. Schrijver. *Combinatorial optimization: Polyhedra and efficiency.* Vol. 24 of *Algorithms and Combinatorics*, Springer-Verlag, Berlin, 2003.

[Sch13a] M. Schaefer. The graph crossing number and its variants: A survey. *Electron. J. Combin.*, Dynamic Surveys, article 21, 2013. Updated in 2014.

[Sch13b] M. Schaefer. Toward a theory of planarity: Hanani–Tutte and planarity variants. *J. Graph. Alg. Appl.*, 17:367–440, 2013.

[Sch14] M. Schaefer. Hanani–Tutte and related results. In I. Bárány, K.J. Böröczky, and L. Fejes Tóth, editors, *Geometry—Intuitive, Discrete, and Convex: A tribute to László Fejes Tóth*, vol. 24 of *Bolyai Society Math. Studies*, pages 259–299, Springer, Berlin, 2014.

[Sid10] A. Sidiropoulos. Optimal stochastic planarization. In *Proc. 51st IEEE Sympos. Found. Comp. Sci.*, pages 163–170, 2010.

[SSŠ02] M. Schaefer, E. Sedgwick, and D. Štefankovič. Algorithms for normal curves and surfaces. In *Proc. 8th Conf. Computing and Combinatorics*, vol. 2387 of *LNCS*, pages 370–380, Springer, Berlin, 2002.

[SSŠ08] M. Schaefer, E. Sedgwick, and D. Štefankovič. Computing Dehn twists and geometric intersection numbers in polynomial time. In *Proc. 20th Canad. Conf. Comput. Geom.*, pages 111–114, 2008.

[Sti93] J. Stillwell. *Classical Topology and Combinatorial Group Theory*, 2nd edition. Springer-Verlag, New York, 1993.

[Thi12] D.M. Thilikos. Graph minors and parameterized algorithm design. In H.L. Bodlaender, R.G. Downey, F.V. Fomin, and D. Marx, editors, *The Multivariate Algorithmic Revolution and Beyond*, vol. 7370 of *LNCS*, pages 228–256, Springer, Berlin, 2012.

[Thi15] D.M. Thilikos. Bidimensionality and parameterized algorithms. In *Proc. 10th Sympos. Parameterized and Exact Computation*, vol. 43 of *LIPIcs*, pages 1–16, Schloss Dagstuhl, 2015.

[Tho89] C. Thomassen. The graph genus problem is NP-complete. *J. Algorithms*, 10:568–576, 1989.

[Tho90] C. Thomassen. Embeddings of graphs with no short noncontractible cycles. *J. Combin. Theory Ser. B*, 48:155–177, 1990.

[Tót12] G. Tóth. A better bound for pair-crossing number. In J. Pach, editor, *Thirty Essays on Geometric Graph Theory*, pages 563–567, Springer, New York, 2012.

[Tut70] W.T. Tutte. Toward a theory of crossing numbers. *J. Combin. Theory*, 8:45–53, 1970.

[WHDS04] Z.J. Wood, H. Hoppe, M. Desbrun, and P. Schröder. Removing excess topology from isosurfaces. *ACM Trans. Graph.*, 23:190–208, 2004.

24 PERSISTENT HOMOLOGY

Herbert Edelsbrunner and Dmitriy Morozov

INTRODUCTION

Persistent homology was introduced in [ELZ00] and quickly developed into the most influential method in computational topology. There were independent developments of this idea preceding this paper, and we mention the little known paper by Marston Morse [Mor40], the spectral sequences introduced by Jean Leray in [Ler46], the notion of prominence in mountaineering [Mun53], the size function introduced by Frosini [Fro90], and the study of fractal sets by Robins [Rob99]. Perhaps the fast algorithm described in [ELZ00] triggered the explosion of interest we currently observe because its availability as software facilitates the application to a broad collection of problems and datasets.

From the mathematical perspective, persistent homology is part of Morse theory [Mil63]. Functions come naturally, which explains the affinity to applications and to data. Persistent homology quantifies the critical points, which we illustrate with the resolution of a basic question in mountaineering: *What is a mountain?* We can identify a mountain with its peak, but clearly not every local maximum on Earth qualifies as a mountain. The summit of Mt. Everest is prominent but its South summit is not, despite being higher than any other mountain in the world. To identify peaks that are prominent, climbers measure how far they have to descend before they can climb to an even higher peak. This measurement, called the *topographic prominence*, assigns a significance to every local maximum of the elevation function on planet Earth, and peaks with prominence above 500 meters qualify as separate mountains. Persistence addresses the multi-scale aspects of natural phenomena by generalizing topographic prominence to more general functions and to higher-dimensional features and holes. It also applies to abstract and high-dimensional data, such as sound, documents, DNA sequences, and languages by equipping the data with a possibly discrete metric.

Section 24.1 explains persistence algebraically, as an extension of the classical notion of homology. Section 24.2 approaches persistence geometrically, focusing on the functions and complexes whose persistence we compute. Section 24.3 is analytic in nature, studying the space of persistence diagrams and the stability of persistence. Section 24.4 focuses on algorithms, explaining the connection to matrix reduction and how the computation can be made fast in important cases. Section 24.5 returns to the algebraic foundations, describing zigzag persistence as a unifying as well as generalizing concept. Section 24.6 illustrates the application of persistent homology by considering two mathematical questions.

24.1 ALGEBRA

We recall basic concepts from algebraic topology to explain homology as well as its recent extension, persistent homology. The first four of these concepts can also be found in Chapter 17, and we repeat the definitions to be consistent with the notation adopted in this chapter.

GLOSSARY

Abstract simplex: A non-empty finite subset of a universal vertex set, $\alpha \subseteq \mathbf{U}$. Its *dimension* is one less than its cardinality, $\dim \alpha = \text{card}\alpha - 1$. Setting $p = \dim \alpha$, we call α an *(abstract) p-simplex.* A *face* is a non-empty subset of α. An *ordering* of α is a sequence of its $p+1$ vertices, denoted $\alpha = [v_0, v_1, \ldots, v_p]$. An *orientation* is a class of orderings that differ by an even number of transpositions. Every simplex has two orientations, except for a 0-simplex, which has only one.

Abstract simplicial complex: A finite collection of abstract simplices, A, that is closed under the face relation: $\beta \in A$ and $\alpha \subseteq \beta$ implies $\alpha \in A$. Its *dimension* is the maximum dimension of any of its simplices. A *subcomplex* is an abstract simplicial complex $B \subseteq A$. The *p-skeleton* is the largest subcomplex whose dimension is p. For example, the 1-skeleton of A is a graph.

Geometric simplex: The convex hull of a non-empty and affinely independent set of points in $\mathbb{R}^n$. Assuming $\mathbf{U}$ is a set of points in $\mathbb{R}^n$, we define *dimension, (geometric) p-simplex, face, ordering,* and *orientation* as in the abstract case.

Geometric simplicial complex: A finite collection of geometric simplices, K, that is closed under the face relation such that $\sigma, \tau \in K$ implies $\sigma \cap \tau$ is either empty or a face of both simplices. We define *dimension* and *subcomplex* as in the abstract case. The *underlying space* of K, denoted $|K|$, is the set of points in $\mathbb{R}^n$ contained in simplices of K together with the Euclidean topology inherited from $\mathbb{R}^n$. While K is a combinatorial object, $|K|$ is a topological space.

Chain: A formal sum of ordered simplices of the same dimension, $c = \sum_i a_i \sigma_i$. The coefficients a_i are elements of an abelian group. If all simplices have dimension p, then we call c a *p-chain.* We *add* p-chains like polynomials. The set of p-chains of a simplicial complex, K, is the *p-chain group* of K, denoted $\mathsf{C}_p = \mathsf{C}_p(K)$. If the coefficients are elements of a field, then C_p is a vector space.

Boundary map: The linear map $\partial_p \colon \mathsf{C}_p \to \mathsf{C}_{p-1}$ defined by mapping an ordered p-simplex to the alternating sum of its $(p-1)$-dimensional faces. For $\sigma = [v_0, v_1, \ldots, v_p]$, we get $\partial_p \sigma = \sum_{i=0}^{p} (-1)^i \hat{\sigma}_i$, in which $\hat{\sigma}_i$ is σ without vertex v_i. Note that $\partial_0 [v_i] = 0$ for every i. It is not difficult to verify that $\partial_{p-1} \circ \partial_p = 0$ for all p.

Chain complex: An infinite sequence of chain groups connected by boundary maps, $\ldots \stackrel{\partial_{p+1}}{\to} \mathsf{C}_p \stackrel{\partial_p}{\to} \mathsf{C}_{p-1} \stackrel{\partial_{p-1}}{\to} \ldots$. The *p-cycle group* is the kernel of the p-th boundary map, $\mathsf{Z}_p = \ker \partial_p$. It is the subgroup of p-chains with zero boundary. The *p-boundary group* is the image of the $(p+1)$-st boundary map, $\mathsf{B}_p = \text{img}\partial_{p+1}$. It is the subgroup of p-cycles that are the boundary of a $(p+1)$-chain.

Homology group: The quotient of the cycle group over the boundary group. Call two cycles *homologous* if they differ only by a boundary. A *homology class* is the maximal collection of homologous cycles in a complex. Fixing p, the set of homology classes of p-cycles is the *p-th homology group*, $\mathsf{H}_p = \mathsf{Z}_p / \mathsf{B}_p$. If the coefficients are elements in a field, then H_p is a vector space.

Relative homology: The generalization of homology to pairs $K_0 \subseteq K$. Here we distinguish two chains only if they differ within $K \setminus K_0$: $\mathsf{C}_p(K, K_0) = \mathsf{C}_p(K)/\mathsf{C}_p(K_0)$ for all p. The boundary maps $\partial_p \colon \mathsf{C}_p(K, K_0) \to \mathsf{C}_{p-1}(K, K_0)$ are well defined, and we set $\mathsf{Z}_p(K, K_0) = \ker \partial_p$, $\mathsf{B}_p(K, K_0) = \mathrm{img} \partial_{p+1}$, and $\mathsf{H}_p(K, K_0) = \mathsf{Z}_p(K, K_0)/\mathsf{B}_p(K, K_0)$.

Exact sequence: A sequence of abelian groups connected by homomorphisms, $\ldots \to V_i \to V_{i-1} \to \ldots$, in which the image of each map is equal to the kernel of the next map. In the case of vector spaces, the homomorphisms are linear maps. An exact sequence of five abelian groups that begins and ends with zero, $0 \to V_3 \to V_2 \to V_1 \to 0$, is called a *short exact sequence.*

Filtered simplicial complex: A simplicial complex, K, together with a function $f \colon K \to \mathbb{R}$ such that $f(\sigma) \leq f(\tau)$ whenever σ is a face of τ. The *sublevel set* at a value $r \in \mathbb{R}$ is $f^{-1}(-\infty, r]$, which is a subcomplex of K. Letting $r_0 < r_1 < \ldots < r_m$ be the values of the simplices and writing $K_i = f^{-1}(-\infty, r_i]$, we call $K_0 \subseteq K_1 \subseteq \ldots \subseteq K_m$ the *sublevel set filtration* of f.

Persistence module: A sequence of vector spaces connected by linear maps, $\mathbf{f}_r^s \colon U_r \to U_s$, for every pair of values $r \leq s$, such that $\mathbf{f}_r^r = \mathrm{id}$ and $\mathbf{f}_r^t = \mathbf{f}_s^t \circ \mathbf{f}_r^s$ for all $r \leq s \leq t$. We denote such a module by $\mathcal{U} = (U_r, \mathbf{f}_r^s)$. For example, the persistence module of the above filtered simplicial complex is $\mathcal{H}(K) = (\mathsf{H}(K_i), \mathbf{f}_i^j)$ in which $\mathsf{H}(K_i)$ is the direct sum of the p-th homology groups of K_i, over all p, and $\mathbf{f}_i^j \colon \mathsf{H}(K_i) \to \mathsf{H}(K_j)$ is induced by the inclusion $K_i \subseteq K_j$.

HOMOLOGY

While we restricted the above definitions to simplicial complexes, homology groups can be defined in much greater generality. For example, we may use *singular simplices* (continuous maps of simplices) to construct the *singular homology groups* of a topological space. An example of such a space is $\mathbb{X} = |K|$, and importantly, the singular homology groups of $\mathbb{X}$ are isomorphic to the simplicial homology groups of K. We refer to the axiomatization of Eilenberg and Steenrod [ES52] for the main tool to reach a unified view of the many different constructions of homology in the literature.

Homology groups, Betti numbers, and Euler characteristic. Assuming coefficients in a field, $\mathbb{F}$, each homology group is a vector space, $\mathbb{F}^\beta$. Here, β is a non-negative integer, namely the *dimension* of the vector space or the *rank* of the group. Writing $\mathsf{H}_p(\mathbb{X}) = \mathbb{F}^{\beta_p}$, we call $\beta_p = \beta_p(\mathbb{X})$ the *p-th Betti number* of $\mathbb{X}$. Historically, Betti numbers preceded homology groups. In turn, the *Euler characteristic* of $\mathbb{X}$, which we can define as $\chi(\mathbb{X}) = \sum_{p \geq 0} (-1)^p \beta_p(\mathbb{X})$, preceded the Betti numbers. For example, it was known already to Leonhard Euler that χ of the boundary of any convex polytope in $\mathbb{R}^3$—then defined as the alternating sum of face numbers—is 2.

Relative and local homology. Instead of defining homology for a space that is partially open, it is more convenient to define it for a pair of closed spaces, $K_0 \subseteq K$. One motivation for introducing this concept is the desire to define homology locally, at a point $x \in \mathbb{X}$. We may capture the homology within an open neighborhood N of x by constructing $\mathsf{H}_p(\mathbb{X}, \mathbb{X} \setminus N)$. Shrinking the neighborhood towards x, the *local homology* of $\mathbb{X}$ at x is the limit of the relative homology.

Snake Lemma. As one of the main achievements of algebraic topology, this lemma is a recipe for the construction of long exact sequences. Specifically, if $0 \to \mathsf{C} \to \mathsf{D} \to \mathsf{E} \to 0$ is a short exact sequence of chain complexes, then

$$\ldots \to \mathsf{H}_{p+1}(\mathsf{E}) \to \mathsf{H}_p(\mathsf{C}) \to \mathsf{H}_p(\mathsf{D}) \to \mathsf{H}_p(\mathsf{E}) \to \mathsf{H}_{p-1}(\mathsf{C}) \to \ldots$$

is a long exact sequence of homology groups. This is useful because exact sequences provide a powerful language to compactly encode relationships between homology groups. We illustrate this with two examples.

- ***Long exact sequence of a pair.*** Letting $K_0 \subseteq K$ be two simplicial complexes, we consider $0 \to \mathsf{C}_p(K_0) \to \mathsf{C}_p(K) \to \mathsf{C}_p(K, K_0) \to 0$, for every integer $p \geq 0$, in which the middle two maps are inclusions. It is not difficult to see that the kernel of every map is the image of the preceding map. By the Snake Lemma,

  $$\ldots \to \mathsf{H}_{p+1}(K, K_0) \to \mathsf{H}_p(K_0) \to \mathsf{H}_p(K) \to \mathsf{H}_p(K, K_0) \to \mathsf{H}_{p-1}(K_0) \to \ldots$$

  is a long exact sequence.
- ***Mayer–Vietoris long exact sequence.*** Letting $K = A \cup B$ be three simplicial complexes, we consider $0 \to \mathsf{C}_p(A \cap B) \to \mathsf{C}_p(A) \oplus \mathsf{C}_p(B) \to \mathsf{C}_p(A \cup B) \to 0$, for every integer $p \geq 0$, in which the second map is the direct sum of two inclusions and the third map is the sum of two inclusions. By the Snake Lemma,

  $$\ldots \to \mathsf{H}_{p+1}(K) \to \mathsf{H}_p(A \cap B) \to \mathsf{H}_p(A) \oplus \mathsf{H}_p(B) \to \mathsf{H}_p(K) \to \mathsf{H}_{p-1}(A \cap B) \to \ldots$$

  is a long exact sequence. It has many applications within algebraic topology, including the construction of the zigzag pyramid as discussed in Section 24.5.

PERSISTENCE

While homology defines holes and counts them, persistent homology also measures them. This additional feature opened a floodgate of applications to the sciences and beyond and has established computational topology as a viable new field of mathematical inquiry; see e.g., Edelsbrunner and Harer [EH10]. The idea itself has several independent roots in the mathematical literature, the earliest of which is a little known paper by Marston Morse [Mor40]. The description in [ELZ00] together with the algorithms for computing persistence diagrams have initiated the current interest in the subject.

Persistence module of a function. Let $\mathbb{X}$ be a topological space and $f \colon \mathbb{X} \to \mathbb{R}$ a real-valued function. Writing $\mathbb{X}_r = f^{-1}(-\infty, r]$, the sublevel sets form a filtration of the topological space, $\mathbb{X}_r \subseteq \mathbb{X}_s$ for all $r \leq s$. Taking the homology of every sublevel set, we get a persistence module for each dimension, $\mathcal{H}_p(f) = (\mathsf{H}_p(\mathbb{X}_r), \mathbf{p}_r^s)$, in which the maps $\mathbf{p}_r^s$ are induced by the inclusions $\mathbb{X}_r \subseteq \mathbb{X}_s$. We often simplify the notation by taking direct sums of homology groups and maps, $\mathsf{H}(\mathbb{X}_r) = \bigoplus_{p \geq 0} \mathsf{H}_p(\mathbb{X}_r)$ and $\mathbf{f}_r^s = \bigoplus_{p \geq 0} \mathbf{p}_r^s$, and consider the persistence module $\mathcal{H}(f) = (\mathsf{H}(\mathbb{X}_r), \mathbf{f}_r^s)$ that simultaneously captures all dimensions.

Decomposition into summands. Under mild assumptions, a persistence module decomposes uniquely into elementary pieces. To describe this, we call a persistence module, $\mathcal{U} = (U_r, \mathbf{f}_r^s)$, *q-tame* ("$q$" for quadrant), if the rank of $\mathbf{f}_r^s$ is finite for all $r < s$. Given an interval $[b, d)$, the corresponding *interval module* is $\mathcal{I} = (I_r, \mathbf{i}_r^s)$, with vector spaces $I_r = \mathbb{F}$ whenever $b \leq r < d$ and $I_r = 0$ otherwise, as well as maps $\mathbf{i}_r^s = \text{id}$ whenever $b \leq r \leq s < d$ and $\mathbf{i}_r^s = 0$, otherwise. Every q-tame persistence module decomposes uniquely as a direct sum of interval modules, $\mathcal{U} = \bigoplus_j \mathcal{I}_j$. The interval modules are *indecomposable summands.* There is an alternative way to describe this decomposition: select elements u_i^r in each vector space U_r, such that the nonzero elements form a basis of U_r, and the maps diagonalize with respect to these bases, i.e., $\mathbf{f}_r^s(u_i^r) = u_i^s$ whenever $u_i^r \neq 0$. Note that the u_i^r are not unique, but the intervals they define are. If the decomposition of module $\mathcal{U}$ contains a module for interval $[b, d)$, we say that a class is born at U_b and dies entering U_d.

Persistence diagram. The information contained in a persistence module has intuitive combinatorial representations. The *persistence diagram* associated with $\mathcal{U} = \bigoplus_j \mathcal{I}_j$, denoted $\text{Dgm}(\mathcal{U})$, is the multi-set of points (b_j, d_j) with $\mathcal{I}_j$ defined by $[b_j, d_j)$. For technical reasons that will become clear when we discuss the stability of persistence diagrams, we usually add infinitely many copies of the points (x, x) on the diagonal to the persistence diagram. When $\mathcal{U} = \mathcal{H}(f)$ is defined by the sublevel sets of a function, we abbreviate the notation to $\text{Dgm}(f) = \text{Dgm}(\mathcal{H}(f))$, or $\text{Dgm}_p(f) = \text{Dgm}(\mathcal{H}_p(f))$ if we wish to restrict the information to a single dimension p. Sometimes $\mathcal{U}$ is represented by the corresponding multi-set of intervals, $[b_j, d_j)$, which is referred to as the *barcode* of $\mathcal{U}$.

Equivalence of persistence modules. Two persistence modules, $\mathcal{U} = (U_r, \mathbf{f}_r^s)$ and $\mathcal{V} = (V_r, \mathbf{g}_r^s)$, are *isomorphic* if the vector spaces are pairwise isomorphic, $U_r \simeq V_r$ for all r, and these isomorphisms commute with the maps $\mathbf{f}_r^s$ and $\mathbf{g}_r^s$ in the modules. This situation can be graphically represented by the squares

$$\begin{array}{ccc} U_r & \xrightarrow{\mathbf{f}_r^s} & U_s \\ \simeq\downarrow & & \downarrow\simeq \\ V_r & \xrightarrow{\mathbf{g}_r^s} & V_s, \end{array} \tag{24.1.1}$$

which are required to commute for all $r \leq s$.

THEOREM 24.1.1 Persistence Equivalence Theorem

Isomorphic persistence modules imply identical persistence diagrams, $\text{Dgm}(\mathcal{U}) = \text{Dgm}(\mathcal{V})$.

Extended persistence. Some applications of persistent homology require an extension of the filtration defining the module; see e.g., [AEHW06, EP16]. We explain this for the filtration of sublevel sets of a function $f\colon \mathbb{X} \to \mathbb{R}$ for which the extension consists of the pairs $(\mathbb{X}, \mathbb{X}^r)$, with $\mathbb{X}^r = f^{-1}[r, \infty)$ the *superlevel set* of f at r. Applying homology, we get vector spaces of the form $\mathsf{H}(\mathbb{X}, \mathbb{X}^r)$, and a persistence module with maps $\mathbf{f}_s^r\colon \mathsf{H}(\mathbb{X}, \mathbb{X}^s) \to \mathsf{H}(\mathbb{X}, \mathbb{X}^r)$ for all $r < s$. Since $\mathsf{H}(\mathbb{X}) = \mathsf{H}(\mathbb{X}, \emptyset)$, we can append the new persistence module to $\mathcal{H}(f)$ and get the *extended persistence module*, $\mathcal{H}^{\text{ext}}(f)$, and the *extended persistence diagram*, $\text{Dgm}(\mathcal{H}^{\text{ext}}(f))$.

24.2 GEOMETRY

In many applications of persistent homology, the essential geometric information is encoded in the filtration. While this is not necessary, it is a convenient vehicle for measuring geometry with topology.

GLOSSARY

Star of a simplex: The set of simplices in a simplicial complex that contain the given simplex, $\mathrm{St}\sigma = \{\tau \in K \mid \sigma \subseteq \tau\}$. The *closed star* also contains the faces of the simplices that contain the given simplex, $\overline{\mathrm{St}}\,\sigma = \{\upsilon \in K \mid \upsilon \subseteq \tau \in \mathrm{St}\sigma\}$. The *link* contains all simplices in the closed star that avoid the given simplex, $\mathrm{Lk}\sigma = \{\upsilon \in \overline{\mathrm{St}}\,\sigma \mid \upsilon \cap \sigma = \emptyset\}$.

Piecewise-linear function: A real-valued function $f\colon |K| \to \mathbb{R}$ that is specified on the vertices of K and is interpolated linearly on the simplices. To stress the point that the complex is not refined before interpolation, f is sometimes referred to as a *simplexwise-linear function.*

Lower star of a vertex: The subset of the star in which the vertices of every simplex have function values smaller than or equal to the value at the given vertex, $\mathrm{St}_{-}u = \{\tau \in \mathrm{St}u \mid f(v) \leq f(u) \text{ for all } v \in \tau\}$. Similarly, the *lower link* of the vertex is $\mathrm{Lk}_{-}u = \{\tau \in \mathrm{Lk}u \mid f(v) \leq f(u) \text{ for all } v \in \tau\}$.

Barycentric subdivision: An abstract simplicial complex, $\mathrm{Sd}K$, which has the simplices $\sigma_i \in K$ as its vertices, and which has a simplex $\tau = [\sigma_0, \sigma_1, \ldots, \sigma_p]$ iff its vertices form a chain of faces in K, i.e., $\sigma_0 \subseteq \sigma_1 \subseteq \ldots \subseteq \sigma_p$.

Homotopy equivalence: An equivalence relation between topological spaces. To define it, we call a continuous map $h\colon \mathbb{X} \times [0,1] \to \mathbb{Y}$ a *homotopy* between the maps $a, b\colon \mathbb{X} \to \mathbb{Y}$ that satisfy $a(x) = h(x,0)$ and $b(x) = h(x,1)$ for all $x \in \mathbb{X}$. Now $\mathbb{X}$ and $\mathbb{Y}$ are *homotopy equivalent*, or they have the same *homotopy type*, denoted $\mathbb{X} \simeq \mathbb{Y}$, if there are maps $f\colon \mathbb{X} \to \mathbb{Y}$ and $g\colon \mathbb{Y} \to \mathbb{X}$ such that there is a homotopy between $g \circ f$ and $\mathrm{id}_{\mathbb{X}}$ as well as between $f \circ g$ and $\mathrm{id}_{\mathbb{Y}}$. A *contractible* set is homotopy equivalent to a point.

Deformation retraction from $\mathbb{X}$ to $\mathbb{Y}$: A continuous map $D\colon \mathbb{X} \times [0,1] \to \mathbb{X}$ such that $D(x,0) = x$, $D(x,1) \in \mathbb{Y}$, and $D(y,t) = y$ for all $x \in \mathbb{X}$, $y \in \mathbb{Y}$, and $t \in [0,1]$. If such a D exists, then $\mathbb{Y}$ is a *deformation retract* of $\mathbb{X}$. Note that $\mathbb{Y}$ is then homotopy equivalent to $\mathbb{X}$. Indeed, setting $f\colon \mathbb{X} \to \mathbb{Y}$ and $g\colon \mathbb{Y} \to \mathbb{X}$ defined by $f(x) = D(x,1)$ and $g(y) = y$, we get D as a homotopy between $g \circ f$ and $\mathrm{id}_{\mathbb{X}}$, and we get the identity as a homotopy between $f \circ g$ and $\mathrm{id}_{\mathbb{Y}}$.

Hausdorff distance: An extension of a distance between points to a distance between point sets,

$$d_H(X,Y) = \max\{\sup_{x \in X} \inf_{y \in Y} \|x - y\|, \sup_{y \in Y} \inf_{x \in X} \|x - y\|\}. \tag{24.2.1}$$

Writing $X^r = X + B(0,r)$ for the Minkowski sum with the closed ball of radius $r \geq 0$, $d_H(X,Y)$ is the infimum radius r such that $X \subseteq Y^r$ and $Y \subseteq X^r$.

Nerve: A simplicial complex associated to a collection of sets. The sets are the vertices of the complex, and a simplex belongs to the complex iff its vertices have a non-empty intersection, $\text{Nrv}S = \{\alpha \subseteq S \mid \bigcap_{A \in \alpha} A \neq \emptyset\}$. If the sets in S are convex, then the nerve is homotopy equivalent to the union of these sets. This result is known as the Nerve Theorem [Bor48, Ler46], and it generalizes to the case in which all non-empty common intersections are contractible.

Čech complex: The nerve of the ball neighborhoods of a set of points $X \subseteq \mathbb{R}^n$. Writing $B(x, r)$ for the closed ball of radius $r \geq 0$ centered at x, the *Čech complex* of X for radius r is $\text{Čech}_r(X) = \text{Nrv}\{B(x, r) \mid x \in X\}$.

Alpha complex: The nerve of the clipped ball neighborhoods of a set of points $X \subseteq \mathbb{R}^n$. Here we clip $B(x, r)$ with the *Voronoi domain* of x, which consists of all points $a \in \mathbb{R}^n$ that are at least as close to x as to any other point in X. Writing $V(x) = \{a \in \mathbb{R}^n \mid \|a - x\| \leq \|a - y\| \text{ for all } y \in X\}$, the *alpha complex* of X for radius r is $\text{Alpha}_r(X) = \text{Nrv}\{B(x, r) \cap V(x) \mid x \in X\}$. To stress the connection to the dual of the Voronoi domains, $\text{Alpha}_r(X)$ is sometimes referred to as the *Delaunay complex* of X for radius r. It is a subcomplex of the *Delaunay triangulation*, $\text{Alpha}_\infty(X)$, which is the dual of the Voronoi domains.

Vietoris–Rips complex: The largest simplicial complex whose 1-skeleton is also the 1-skeleton of the Čech complex. Given a graph $G = (V, E)$, the *clique complex* contains all cliques of G, namely all simplices $\alpha \subseteq V$ for which E contains all edges of α. With this notation, the *Vietoris–Rips complex* of X for radius r is the clique complex of the 1-skeleton of the Čech complex of X and r, $\text{Rips}_r(X) = \{\alpha \subseteq X \mid \|u - v\| \leq 2r \text{ for all } u, v \in \alpha\}$.

PIECEWISE-LINEAR FUNCTIONS

Functions generally do not have finite descriptions, so we have to work with approximations. For example, we may fix the values at a finite set of points and make up the information in between by interpolation. Due to its simplicity, the description in which the interpolation is piecewise linear is popular in applications.

Sublevel sets and lower stars. Given a piecewise-linear function $f \colon |K| \to \mathbb{R}$ and a value $r \in \mathbb{R}$, we recall that the corresponding sublevel set is $|K|_r = f^{-1}(-\infty, r]$. It is not necessarily the underlying space of a subcomplex of K, but it is homotopy equivalent to one. Specifically, let $K_r \subseteq K$ contain all simplices whose vertices have values smaller than or equal to r. Equivalently, K_r is the union of the lower stars of all vertices with function values at most r. Clearly $|K_r| \subseteq |K|_r$, and it is not difficult to see that there is a deformation retraction from $|K|_r$ to $|K_r|$, implying that the two are homotopy equivalent.

Lower star filtration. Starting with the piecewise-linear function $f \colon |K| \to \mathbb{R}$, we construct a function $g \colon K \to \mathbb{R}$ that maps every simplex to the maximum function value of its vertices. Observe that K together with g is a filtered complex, and that $K_r = g^{-1}(-\infty, r]$ for every r. The resulting nested sequence of complexes is usually referred to as the *lower star filtration* of f or of g.

Hierarchy of equivalences between spaces. Recall that a *homeomorphism* $h \colon \mathbb{X} \to \mathbb{Y}$ is a continuous bijection whose inverse is continuous. If such an h exists, then $\mathbb{X}$ and $\mathbb{Y}$ are *topologically equivalent* or they have the same *topology type*. Topological equivalence is stronger than homotopy equivalence, which in turn is stronger than homology. In other words, if $\mathbb{X}$ and $\mathbb{Y}$ are homeomorphic, then they are homotopy equivalent, and if they are homotopy equivalent, then they have isomorphic homology groups. These implications cannot be reversed.

Equivalent persistence modules. Passing to homology, we get two persistence modules, $\mathcal{F} = (\mathsf{H}(|K|_r), \mathbf{f}_r^s)$ and $\mathcal{G} = (\mathsf{H}(K_r), \mathbf{g}_r^s)$. For any two values $r \leq s$, we get a square,

$$\begin{array}{ccc} \mathsf{H}(|K|_a) & \longrightarrow & \mathsf{H}(|K|_b) \\ \uparrow & & \uparrow \\ \mathsf{H}(K_a) & \longrightarrow & \mathsf{H}(K_b), \end{array} \tag{24.2.2}$$

in which all four maps are induced by inclusion. It follows that the square commutes. Since $|K_r| \simeq |K|_r$, for all r, the vertical maps are isomorphisms. As explained earlier, this implies that the two modules have the same persistence diagram, $\mathrm{Dgm}(f) = \mathrm{Dgm}(g)$.

Filtration as a piecewise-linear function. Given a filtered simplicial complex, K with $f\colon K \to \mathbb{R}$, we define a piecewise-linear function on its barycentric subdivision, $g\colon |\mathrm{Sd}K| \to \mathbb{R}$, by setting the value at the vertices to $g(\sigma_i) = f(\sigma_i)$, for all $\sigma_i \in K$. The two persistence modules, $\mathcal{H}(f)$ and $\mathcal{H}(g)$, are isomorphic because the sublevel sets of g deformation retract onto the sublevel sets of f.

DISTANCE FUNCTIONS

Besides piecewise-linear functions, distance functions on the ambient space of given data are most popular in applications of persistent homology. Given $X \subseteq \mathbb{R}^n$, it is the function $d_X\colon \mathbb{R}^n \to \mathbb{R}$ defined by mapping every $a \in \mathbb{R}^n$ to the distance to X, $d_X(a) = \inf_{x \in X} \|a - x\|$. The sublevel set of d_X for $r \geq 0$ is the union of balls of radius r centered at the points of X.

Distance functions and Hausdorff distance. If two sets $X, Y \subseteq \mathbb{R}^n$ are close in Hausdorff distance, then their distance functions are close:

$$\sup_{a \in \mathbb{R}^n} |d_X(a) - d_Y(a)| \leq d_H(X, Y). \tag{24.2.3}$$

This motivates us to recover properties of a shape that are stable under Hausdorff perturbations from point samples of the shape. Indeed, we will see in the next section how the homology of a shape can be recovered from the homology of a point sample. Meanwhile, we study three filtered complexes that are commonly used to represent the sublevel sets of a distance function.

Persistence and nerves. Let S be a collection of convex sets and recall the Nerve Theorem, which asserts that $\mathrm{Nrv}S \simeq \bigcup S$.

THEOREM 24.2.1 Persistence Nerve Theorem

Let T be a second collection of convex sets and $b\colon S \to T$ a bijection such that $s \subseteq b(s)$ for every set $s \in S$. Then the inclusion between the two nerves induced by b and the inclusion between the two unions commute with the homotopy equivalences,

$$\begin{array}{ccc} \bigcup S & \longrightarrow & \bigcup T \\ \uparrow & & \uparrow \\ \mathrm{Nrv}S & \longrightarrow & \mathrm{Nrv}T. \end{array} \tag{24.2.4}$$

Čech and alpha complexes are homotopy equivalent. Indeed, both complexes are homotopy equivalent to the corresponding union of balls. Moreover, the filtrations of Čech complexes, of alpha complexes, and of sublevel sets of the distance function have the same persistence diagram. For the Čech complexes and the sublevel sets, this statement follows from the Persistence Nerve Theorem 24.2.1 and the Persistence Equivalence Theorem 24.1.1. For the alpha complexes, we observe in addition that the union of balls and the union of clipped balls are the same.

Čech and Vietoris–Rips complexes are interleaved. By construction, the Vietoris–Rips complex of a set $X \subseteq \mathbb{R}^n$ for radius $r \geq 0$ contains the Čech complex of X for r. However, the holes left by $p+1$ balls that have pairwise non-empty intersections cannot be large. Indeed, if we grow each ball to twice its initial radius, each ball contains all $p+1$ centers. It follows that the $p+1$ balls of radius $2r$ have a non-empty common intersection. In summary, $\text{Čech}_r(X) \subseteq \text{Rips}_r(X) \subseteq \text{Čech}_{2r}(X)$. For the special case of Euclidean distance, growing the balls to radius $\sqrt{2}r$ suffices to guarantee a non-empty common intersection, giving $\text{Rips}_r(X) \subseteq \text{Čech}_{\sqrt{2}r}(X)$.

24.3 ANALYSIS

The step from functions to homology is a temporary excursion to algebra, and we are right back to analysis when we reason about the persistence diagrams that summarize the features of the function.

GLOSSARY

Interleaving distance: A notion of distance between persistence modules. Specifically, $\mathcal{U} = (U_r, \mathbf{f}_r^s)$ and $\mathcal{V} = (V_r, \mathbf{g}_r^s)$ are *ε-interleaved* if there are maps $\varphi_r \colon U_r \to V_{r+\varepsilon}$ and $\psi_r \colon V_r \to U_{r+\varepsilon}$, for all $r \in \mathbb{R}$, that commute with the maps inside the modules. We get the *interleaving distance* between $\mathcal{U}$ and $\mathcal{V}$ by taking the infimum of the $\varepsilon \geq 0$ for which the modules are ε-interleaved.

Bottleneck distance: A notion of distance between persistence diagrams. Finding a bijection, $\gamma \colon \text{Dgm}(\mathcal{U}) \to \text{Dgm}(\mathcal{V})$, we quantify it by taking the maximum L_∞-distance of any two corresponding points. The *bottleneck distance* between the persistence diagrams of $\mathcal{U}$ and $\mathcal{V}$ is the infimum over all bijections:

$$\text{W}_\infty(\text{Dgm}(\mathcal{U}), \text{Dgm}(\mathcal{V})) = \inf_\gamma \sup_{x \in \text{Dgm}(\mathcal{U})} \|x - \gamma(x)\|_\infty . \tag{24.3.1}$$

Wasserstein distances: A 1-parameter family of distances between persistence diagrams. Fixing a real number $q \geq 1$, we quantify a bijection $\gamma \colon \text{Dgm}(\mathcal{U}) \to \text{Dgm}(\mathcal{V})$ by taking the sum of the q-th powers of the L_∞-distances between corresponding points. The *q-Wasserstein distance* between the persistence diagrams of $\mathcal{U}$ and $\mathcal{V}$ is the infimum of the q-th roots of these sums over all bijections:

$$\text{W}_q(\text{Dgm}(\mathcal{U}), \text{Dgm}(\mathcal{V})) = \inf_\gamma \left(\sum_{x \in \text{Dgm}(\mathcal{U})} \|x - \gamma(x)\|_\infty^q \right)^{1/q} . \tag{24.3.2}$$

It approaches the bottleneck distance as q goes to infinity.

Moments: A 1-parameter family of summaries of a persistence diagram. Fixing a real number $q \geq 0$, the *q-th moment* of $\mathcal{U}$ is the sum of the q-th powers of the persistences:

$$M_q(\mathcal{U}) = \sum_{(b,d)\in \mathrm{Dgm}(\mathcal{U})} |d-b|^q. \tag{24.3.3}$$

It is sometimes referred to as the *q-th total persistence* of $\mathcal{U}$. There are variants, such as the *level set moment*, defined if $\mathcal{U}$ is the direct sum of the $\mathcal{U}_p$, for $p \geq 0$, which is the alternating sum of persistences:

$$M_\chi(\mathcal{U}) = \sum_{p\geq 0}(-1)^p \sum_{(b,d)\in \mathrm{Dgm}_p(\mathcal{U})} (d-b). \tag{24.3.4}$$

We note that $M_\chi(\mathcal{U})$ is equal to the integral of the Euler characteristic of $f^{-1}(r)$, over all $r \in \mathbb{R}$, provided $\mathcal{U} = \mathcal{H}^{\mathrm{ext}}(f)$. Indeed, each point in the extended persistence diagram contributes its persistence to the level set moment, which is the contribution of the represented cycle to the integral of the Euler characteristic. We note that points below the diagonal contribute negative values because they represent level set homology classes of one dimension lower.

Local feature size at a point: The infimum distance of $x \in \mathbb{X}$ from a point $a \in \mathbb{R}^n$ that has two or more nearest points in $\mathbb{X} \subseteq \mathbb{R}^n$, denoted $\mathrm{lfs}(x)$. The closure of the set of points $a \in \mathbb{R}^n$ with two or more nearest points in $\mathbb{X}$ is commonly referred to as the *medial axis* of $\mathbb{X}$. The global version of the local feature size is the *reach* of $\mathbb{X}$ defined as the infimum local feature size over all points of $\mathbb{X}$, $\mathrm{reach}(\mathbb{X}) = \inf_{x\in\mathbb{X}} \mathrm{lfs}(x)$. For example, if $\mathbb{X}$ is a smoothly embedded compact manifold, then the reach is positive. The reach is at most one over the maximum curvature but it can be smaller.

Homological critical value of a real-valued function: Any value $a \in \mathbb{R}$ such that the homology of the sublevel sets of the function changes at a, i.e., the map $\mathsf{H}\left(f^{-1}(-\infty, a-\varepsilon]\right) \to \mathsf{H}\left(f^{-1}(-\infty, a]\right)$, induced on homology by inclusion, is not an isomorphism for any sufficiently small ε.

Weak feature size of a set: The infimum positive homological critical value of the distance function, $d_{\mathbb{X}} \colon \mathbb{R}^n \to \mathbb{R}$, denoted $\mathrm{wfs}(\mathbb{X})$. It is larger than or equal to the reach.

STABILITY

Persistence diagrams are stable under perturbations of the function. This is perhaps the most important result in the theory. It was first proved for functions using the bottleneck distance between persistence diagrams by Cohen-Steiner, Edelsbrunner, and Harer [CEH07]. Under some restrictions, stability holds also using Wasserstein distances, as proved in [CEHM10]. The stability under the bottleneck distance has been strengthened to a statement about persistence modules in [BL14, CDGO16].

Bottleneck stability for functions. Recall the bottleneck distance between two persistence diagrams.

THEOREM 24.3.1 Bottleneck Stability Theorem

Given $f, g\colon \mathbb{X} \to \mathbb{R}$, the bottleneck distance between their persistence diagrams does not exceed the L_∞-difference between the functions:

$$\mathrm{W}_\infty\left(\mathrm{Dgm}(f), \mathrm{Dgm}(g)\right) \le \|f - g\|_\infty. \tag{24.3.5}$$

We refer to Figure 24.3.1 for an illustration of this inequality, which is true in great generality. For example, $\mathbb{X}$ is not constrained to manifolds or other special classes of topological spaces. Also, the requirement on f and g are mild: they need to be *tame*, which means that they have only finitely many homological critical values and every sublevel set has finite rank homology.

FIGURE 24.3.1
Left: *the graphs of two scalar functions on $\mathbb{X} = \mathbb{R}$.* Right: *the persistence diagrams of the two functions. Each point marks the birth of a component at a local minimum and its death at a local maximum.*

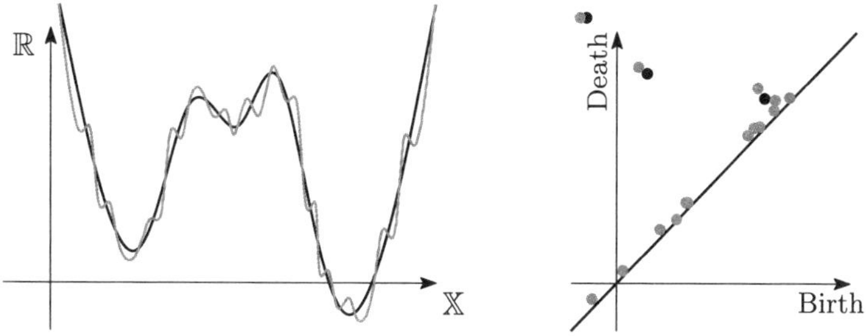

The inequality can be strengthened by composing g with a homeomorphism $h\colon \mathbb{X} \to \mathbb{X}$. In other words, the bottleneck distance between the two diagrams is bounded from above by $\inf_h \|f - g \circ h\|_\infty$.

Wasserstein stability for functions. For finite values of q, we get stability under the q-Wasserstein distance only if we impose conditions on the space and the function.

THEOREM 24.3.2 Wasserstein Stability Theorem

Let $\mathbb{X}$ be a triangulable compact metric space, assume there exists a constant C such that the k-th moment of the persistence diagram of any function with Lipschitz constant 1 is bounded from above by C, and let $f, g\colon \mathbb{X} \to \mathbb{R}$ be two tame functions with Lipschitz constant 1. Then

$$\mathrm{W}_q\left(\mathrm{Dgm}(f), \mathrm{Dgm}(g)\right) \le C^{\frac{1}{q}} \|f - g\|_\infty^{1-\frac{k}{q}}, \tag{24.3.6}$$

for all $q \ge k$.

Provided the exponent is positive, $1 - \frac{k}{q} > 0$, the right-hand side vanishes as the L_∞-difference between f and g goes to zero. Hence, we have stability for $q > k$.

Interleaving isometry for persistence modules. The interleaving distance between persistence modules can be inserted between the two sides of the inequality in the Bottleneck Stability Theorem 24.3.1. Doing so, we get an equality on the left-hand side. But this equality holds more generally, namely also for persistence modules that are not necessarily defined by functions:

$$\mathrm{W}_\infty(\mathrm{Dgm}(\mathcal{U}), \mathrm{Dgm}(\mathcal{V})) = \inf\{\varepsilon \mid \mathcal{U} \text{ and } \mathcal{V} \text{ are } \varepsilon\text{-interleaved}\}. \tag{24.3.7}$$

Traditionally, the upper bound implied by the equality is referred to as stability. For its proof it suffices to use the following two commuting diagrams and their symmetric versions:

$$\begin{array}{ccc} & U_r \xrightarrow{\mathbf{f}_r^s} U_s & \\ {\scriptstyle \psi_{r-\varepsilon}} \nearrow & & \searrow {\scriptstyle \varphi_s} \\ V_{r-\varepsilon} & \xrightarrow{\mathbf{g}_{r-\varepsilon}^{s+\varepsilon}} & V_{s+\varepsilon} \end{array} \qquad \begin{array}{ccc} U_{r+\varepsilon} & \xrightarrow{\mathbf{f}_{r+\varepsilon}^{s+\varepsilon}} & U_{s+\varepsilon} \\ {\scriptstyle \psi_r} \nearrow & & {\scriptstyle \psi_s} \nearrow \\ V_r & \xrightarrow{\mathbf{g}_r^s} & V_s \end{array} \tag{24.3.8}$$

Interleaving on log-scale. The interleaving of Čech and Vietoris–Rips complexes mentioned in the previous section suggests we draw the persistence diagrams for the two filtrations in log-scale. All critical values are non-negative, which we preserve by mapping $r \geq 0$ to $\log_2(r+1)$. Drawing the diagrams in log-scale is therefore the same as drawing the persistence diagrams for the Čech and Vietoris–Rips filtrations using $f = \log_2(d_{\mathbb{X}} + 1)$ instead of the distance function. The corresponding persistence modules are ε-interleaved for $\varepsilon = 1$, which implies $\mathrm{W}_\infty(\mathrm{Dgm}(\text{Čech}), \mathrm{Dgm}(\mathrm{Rips})) \leq 1$ for general metrics. The bound on the right-hand side improves to $\frac{1}{2}$ when we specialize to the Euclidean metric.

INFERENCE

The stability of persistent homology can be exploited to inferring the homology of a space from a finite point sample. We follow [CEH07] in how this is done and compare it with the inference result obtained without stability [NSW08].

Homology inference using stability. Given $\mathbb{X} \subseteq \mathbb{R}^n$, we recall that $d_{\mathbb{X}}\colon \mathbb{R}^n \to \mathbb{R}$ maps every point to its distance from the closest point in $\mathbb{X}$. Under mild conditions, it is possible to infer the homology of $\mathbb{X}$ from the distance function defined by a finite point set.

THEOREM 24.3.3 Homology Inference Theorem

Let $\mathbb{X} \subseteq \mathbb{R}^n$ be compact, with weak feature size $\mathrm{wfs}(\mathbb{X}) = 4\varepsilon > 0$, *and let $X \subseteq \mathbb{R}^n$ be finite with Hausdorff distance $d_H(X, \mathbb{X}) \leq \varepsilon$ from $\mathbb{X}$. Then*

$$\mathrm{rk}\mathsf{H}(\mathbb{X}) = \mathrm{rk}\left[\mathsf{H}(d_X^{-1}[0,\varepsilon]) \to \mathsf{H}(d_X^{-1}[0,3\varepsilon])\right]. \tag{24.3.9}$$

In words, every homology class of $\mathbb{X}$ can already be seen in the sublevel set at ε, and it can still be seen in the sublevel set at 3ε. Furthermore, under the given assumptions, no other homology classes can be seen in both these sublevel sets.

Homology inference without using stability. Here we let $\mathbb{X}$ be a manifold that is smoothly embedded in $\mathbb{R}^n$. Its reach is positive, and we let $X \subseteq \mathbb{X}$ have small Hausdorff distance, $d_H(X, \mathbb{X}) \leq \frac{\varepsilon}{2}$. Then for any $\varepsilon < \sqrt{3/5}\mathrm{reach}(\mathbb{X})$, the homology of the sublevel set of $d_X\colon \mathbb{R}^n \to \mathbb{R}$ for ε is the homology of the manifold:

$$\mathrm{rk}\mathsf{H}(\mathbb{X}) = \mathrm{rk}\mathsf{H}(d_X^{-1}[0,\varepsilon]). \tag{24.3.10}$$

This is so because the union of balls of radius ε centered at the points in X do not have any holes that are not also present in the manifold. Indeed, $d_X^{-1}[0,\varepsilon]$ is homotopy equivalent to $\mathbb{X}$, which is stronger than the claim about homology.

Comparison. The above two results differ in their assumptions on the space $\mathbb{X}$ whose homology is inferred. The first result is sensitive to the weak feature size, while the second result needs positive reach. We have $\text{reach}(\mathbb{X}) \leq \text{wfs}(\mathbb{X})$, which favors homology inference with stability. For example, if $\mathbb{X}$ is a manifold in $\mathbb{R}^n$ but not smoothly embedded, then its reach is zero—which prevents homology inference without using stability—while the weak feature size may very well be positive.

24.4 ALGORITHMS

Persistent homology owes a great deal of its popularity to the development of efficient algorithms. Edelsbrunner, Letscher, and Zomorodian [ELZ00] launched the current line of research with the introduction of a fast algorithm, which we review in this section together with various shortcuts, including optimizations of Chen and Kerber [CK11], fast updates for time-varying persistence [CEM06], and a sample of dualities in persistent homology [DMVJ11b].

GLOSSARY

Boundary matrix: A matrix representation of the boundary map. Given a simplicial complex with m_p p-simplices, the *p-th boundary matrix* is denoted $D_p[1..m_{p-1}, 1..m_p]$, with $D_p[i,j] = 0$ if σ_i is not a face of σ_j, and $D_p[i,j] = (-1)^k$ if $\sigma_j = [u_0, u_1, \ldots, u_p]$ and σ_i is σ_j with u_k removed.

Column operation: Adding a multiple of one column to another. Similarly, a *row operation* adds a multiple of one row to another. Given a matrix, M, we write $M[i,\cdot]$ for its i-th row and $M[\cdot,j]$ for its j-th column.

Pivot: The lowest non-zero element in a column, $M[\cdot,j]$, denoted $\text{pvt}M[\cdot,j]$. The row of the pivot is denoted $\text{low}M[\cdot,j]$, with $\text{low}M[\cdot,j] = 0$ if the entire columns is zero.

Reduced matrix: A matrix M in which the pivots are in distinct rows; that is: $\text{low}M[\cdot,j] \neq \text{low}M[\cdot,k]$ or $\text{low}M[\cdot,j] = \text{low}M[\cdot,k] = 0$ whenever $j \neq k$.

REDUCTION ALGORITHM

As described in Munkres [Mun84], the homology groups of a simplicial complex can be computed by reduction of the boundary matrices. Similarly, we can compute the persistent homology by matrix reduction, but there are differences. In persistence, the ordering of the simplices is essential, and we do the reduction so it preserves order. It is therefore convenient to work with a single matrix that represents the boundary maps in all dimensions.

Filtered boundary matrix. Given a filtration, $K_0 \subseteq K_1 \subseteq \ldots \subseteq K_m$ of a simplicial complex, $K = K_m$, in which consecutive complexes differ by a single simplex, $K_i = K_{i-1} \cup \{\sigma_i\}$, we order the rows and the columns of the boundary matrix of K by the index of their first appearance. We call the resulting matrix D, dropping the dimension subscript since the matrix combines all dimensions.

Column algorithm. Similar to Gaussian elimination, the following greedy algorithm reduces the filtered boundary matrix D: it computes a decomposition $R = DV$, in which the matrix R is reduced and the matrix V is invertible upper-triangular.

$R = D$; $V = I$;
for each column $R[\cdot, j]$ from 1 to m **do**
  while $\text{low}R[\cdot, j] = \text{low}R[\cdot, k] \neq 0$, with $k < j$ **do**
    $c = \text{pvt}R[\cdot, j]/\text{pvt}R[\cdot, k]$;
    $R[\cdot, j] = R[\cdot, j] - cR[\cdot, k]$;
    $V[\cdot, j] = V[\cdot, j] - cV[\cdot, k]$.

Observe that the algorithm works exclusively with left-to-right column operations. Each such operation is a loop, so we have three nested loops, which explains why the algorithm takes a constant times m^3 operations in the worst case.

Persistence pairing. The reduced matrix, R, contains the persistence information we are interested in. Specifically, the persistence module $\mathsf{H}(K_0) \to \mathsf{H}(K_1) \to \ldots \to \mathsf{H}(K_m)$ contains an indecomposable summand $\mathcal{I}[i, j)$ iff $\text{low}R[\cdot, j] = i$. It contains a summand $\mathcal{I}[i, \infty)$ iff $R[\cdot, i] = 0$ and there is no column j with $\text{low}R[\cdot, j] = i$.

Homology generators. While the indecomposable summands are encoded in the reduced matrix, the corresponding generators of the homology groups are sometimes stored in V and sometimes in R. Note that the algorithm maintains that $R[\cdot, j]$ is the boundary of the chain $V[\cdot, j]$, for every j. If $R[\cdot, i] = 0$, then $V[\cdot, i]$ is a cycle that first appears in complex K_i. If there is no column $R[\cdot, j]$ with $\text{low}R[\cdot, j] = i$, then $V[\cdot, i]$ is a generator of homology in the final complex K. In contrast, if $\text{low}R[\cdot, j] = i$, then $R[\cdot, j]$ records a cycle that appears in K_i and becomes a boundary in K_j. In other words, it is a homology generator in groups $\mathsf{H}(K_i)$ through $\mathsf{H}(K_{j-1})$.

Uniqueness of pairing. Given D, the decomposition $R = DV$ such that R is reduced and V is invertible upper-triangular is not unique. However, any two such decompositions have the same map, low, from the columns to the rows, including 0. When reducing the boundary matrix, we can therefore perform column operations in any order, as long as columns are added from left to right. Once the matrix is reduced, it gives the correct persistence pairing.

Row algorithm. Since the columns can be processed in any order, we can use an alternative reduction algorithm that processes the matrix row-by-row, from the bottom up. Despite its name, the algorithm reduces the matrix using column operations.

$R = D$; $V = I$;
for each row $R[i, \cdot]$ from m back up to 1 **do**
  $C = \{j \mid \text{low}R[\cdot, j] = i\}$; $\mathit{leftmost} = \min C$;
  for $j \in C \setminus \{\mathit{leftmost}\}$ **do**
    $c = \text{pvt}R[\cdot, j]/\text{pvt}R[\cdot, \mathit{leftmost}]$;
    $R[\cdot, j] = R[\cdot, j] - cR[\cdot, \mathit{leftmost}]$;
    $V[\cdot, j] = V[\cdot, j] - cV[\cdot, \mathit{leftmost}]$.

Unlike the column algorithm, which produces the pairs in the order of deaths (from earliest to latest), the row algorithm produces the pairs in the order of births (from latest to earliest). If we are only interested in the persistence pairing, the row algorithm can discard columns once they have been used for the reduction. In other words, column $R[\cdot, \textit{leftmost}]$ can be dropped after completing its inner for-loop.

SHORTCUTS

For large complexes, even storing the full boundary matrix would be prohibitively expensive. To cope, we tacitly assume a sparse matrix representation that focuses on the non-zero elements. With this understanding, the above matrix reduction algorithms are surprisingly efficient, namely much faster than the worst case, which is a constant times m^3 operations. Needless to say that fast is never fast enough, and there is still much to be saved if we are clever about operations or exploit special properties that sometimes present themselves.

Compression optimization. If column $R[\cdot, j] \neq 0$ after reduction, then row $R[j, \cdot]$ can be set to zero. Indeed, the non-zero column witnesses that σ_j destroys a cycle, therefore it cannot create one. In the column algorithm, it is convenient to do this update on the fly: when processing column $R[\cdot, i]$, we can remove all those elements whose (already reduced) columns in R are not zero.

Clearing optimization. If $\text{low}R[\cdot, j] = i \neq 0$ after reduction of the column, then $R[\cdot, i]$ is necessarily 0. This is helpful in speeding up the row algorithm by setting $R[\cdot, i] = 0$ immediately after processing $R[i, \cdot]$ and finishing with a non-zero row. This optimization produces a significant speed-up in practice, but it sacrifices matrix V. In other words, the optimized algorithm no longer computes the decomposition $R = DV$, but only matrix R, which is sufficient to recover the persistence pairing.

Fast algorithm for 0-dimensional persistence. Recall that both the row algorithm and the column algorithm require a constant times m^3 operations in the worst case, in which m is the number of simplices in K. If we are interested only in 0-dimensional persistence, we can take advantage of a faster algorithm that runs in time at most a constant times $m \log m$. The algorithm takes advantage of the standard union-find data structure, which maintains a collection of disjoint sets supporting the operation $\text{F}\textsc{ind}(i)$, which finds the lowest-value representative of the set containing vertex i, and the operation $\text{U}\textsc{nite}(i, j)$, which unites the sets represented by i and j and, assuming $i < j$, makes i the new representative. The initial sorting dominates the running time.

Updates after a transposition. A filtration that changes continuously over time can be modeled as a sequence of transpositions of consecutive simplices. If we switch the order of simplices σ_i and σ_{i+1}, the boundary matrix changes by a transposition of the rows i and $i+1$ and of the columns i and $i+1$. Letting P be the corresponding permutation matrix, the new matrix $D' = PDP$. Performing the same transposition in matrices R and V, we get matrices $R' = PRP$ and $V' = PVP$, with $R' = D'V'$. But R' is not necessarily reduced and V' is not necessarily invertible upper-triangular. The latter condition can fail only if $V[i, i+1] \neq 0$. In this case, we can subtract a multiple of column $V[\cdot, i]$ from $V[\cdot, i+1]$, before the transposition, to ensure that $V[i, i+1] = 0$. The former condition can fail if (1) $\text{low}R[\cdot, i] = \text{low}R[\cdot, i+1]$ because of the update to matrix V, or (2) there are two columns with $\text{low}R[\cdot, k] = \text{low}R[\cdot, \ell] = i+1$. Both cases can be fixed by subtracting the lower-index column from the higher-index column. In all cases, we perform only a constant number of column operations, which implies that a single transposition takes time at most a constant times m.

Relative homology pairs and generators. Let $K_0 \subseteq K_1 \subseteq \ldots \subseteq K_m$ be a filtration of $K = K_m$. Mapping every complex, K_i, to the relative homology of the pair (K, K_i), we get a persistence module,

$$\mathsf{H}(K, K_0) \to \mathsf{H}(K, K_1) \to \ldots \to \mathsf{H}(K, K_{m-1}) \to \mathsf{H}(K, K_m), \tag{24.4.1}$$

which we denote as $\mathcal{H}^{\text{rel}}(K)$. Its decomposition into interval summands is closely related to the decomposition of the module $\mathcal{H}(K)$. The intervals can be recovered from the decomposition $R = DV$ computed by either the row algorithm or the column algorithm. Specifically, $\mathcal{H}^{\text{rel}}(K)$ has a summand $\mathcal{I}[i, j)$ in $(p+1)$-dimensional homology iff $\mathcal{H}(K)$ has the same summand in p-dimensional homology, which happens iff low$R[\cdot, j] = i$. In this case, $V[\cdot, j]$ is the generator of the relative homology class. Module $\mathcal{H}^{\text{rel}}(K)$ has a summand $\mathcal{I}[0, i)$ in p-dimensional homology iff module $\mathcal{H}(K)$ has a summand $\mathcal{I}[i, \infty)$ in p-dimensional homology, which happens iff $R[\cdot, i] = 0$ and there is no j with low$R[\cdot, j] = i$. In this case, $V[\cdot, i]$ is the generator of the relative homology class.

24.5 ZIGZAG PERSISTENCE

Persistent homology fits naturally in the theory of quiver representations, where we view a discrete persistence module as a representation of the so-called A_n quiver, a path with n vertices. The theory implies that it is not important that all linear maps point in the same direction. Instead, they may alternate, and the resulting sequence still decomposes into interval summands. Carlsson and de Silva [CD10] introduced the connection between persistence modules and quiver representations to the computational topology community under the name of zigzag persistence. For technical reasons, we work with Steenrod homology throughout this section, which is a homology theory equipped with an axiom that ensures the Mayer–Vietoris sequences are exact for arbitrary spaces. Furthermore, we assume that all homology groups are finite, and that all functions have finitely many critical values. In the case of finite simplicial complexes, Steenrod homology agrees with simplicial homology, and all the piecewise-linear functions have a finite number of critical values.

GLOSSARY

Zigzag filtration: A discrete sequence of topological spaces, $(\mathbb{X}_i)_{i>0}$, such that $\mathbb{X}_i \subseteq \mathbb{X}_{i+1}$ or $\mathbb{X}_i \supseteq \mathbb{X}_{i+1}$ for every i. Its *type* is the sequence of symbols $(\tau_i)_{i>0}$, with $\tau_i \in \{\to, \leftarrow\}$, such that $\tau_i = \to$ implies $\mathbb{X}_i \subseteq \mathbb{X}_{i+1}$ and $\tau_i = \leftarrow$ implies $\mathbb{X}_i \supseteq \mathbb{X}_{i+1}$.

Zigzag persistence module: A sequence of vector spaces over a field $\mathbb{F}$ connected by linear maps $\mathbf{f}_i^{i+1} \colon U_i \to U_{i+1}$ if $\tau_i = \to$ and $\mathbf{f}_{i+1}^{i} \colon U_{i+1} \to U_i$ if $\tau_i = \leftarrow$. We denote such a module by $\mathcal{U} = (U_i, \mathbf{f})$, always assuming that the type is clear from the context. We call $\mathcal{U}$ a *zigzag interval module* if there exists an interval $[k, \ell]$ such that $U_i = 0$ whenever $i < k$ or $\ell < i$, $U_i = \mathbb{F}$ whenever $k \leq i \leq \ell$, and a linear map is zero if either the source or the target are zero, and it is the identity otherwise. We denote this zigzag interval module as $\mathcal{I}[k, \ell]$.

Zigzag interval decomposition: A zigzag persistence module decomposes as a direct sum of zigzag interval modules of the matching type, $\mathcal{U} = \oplus_j \mathcal{I}[k_j, \ell_j]$. As with ordinary persistence modules, this means we can select elements u_j^r in each vector space U_r, such that the nonzero elements form a basis of U_r, and the maps diagonalize with respect to the bases, i.e., $\mathbf{f}_r^s(u_j^r) = u_j^s$ whenever $u_j^r \neq 0$.

Relative Mayer–Vietoris long exact sequence: A long exact sequence associated to the relative homology groups of two nested pairs of spaces. Specifically, given two pairs of spaces, $(\mathbb{U}, \mathbb{A})$ and $(\mathbb{V}, \mathbb{B})$, with $\mathbb{A} \subseteq \mathbb{U}$ and $\mathbb{B} \subseteq \mathbb{V}$, the sequence

$$\begin{aligned} \cdots \to \mathsf{H}_{p+1}(\mathbb{U} \cup \mathbb{V}, \mathbb{A} \cup \mathbb{B}) \to \mathsf{H}_p(\mathbb{U} \cap \mathbb{V}, \mathbb{A} \cap \mathbb{B}) \to \mathsf{H}_p(\mathbb{U}, \mathbb{A}) \oplus \mathsf{H}_p(\mathbb{V}, \mathbb{B}) \to \\ \mathsf{H}_p(\mathbb{U} \cup \mathbb{V}, \mathbb{A} \cup \mathbb{B}) \to \mathsf{H}_{p-1}(\mathbb{U} \cap \mathbb{V}, \mathbb{A} \cap \mathbb{B}) \to \ldots, \end{aligned} \tag{24.5.1}$$

in which the middle two maps are induced by inclusions, is long exact.

LEVEL SET ZIGZAG

There is a natural zigzag associated to a real-valued function. One can view it as sweeping the level sets of the function from bottom to top. The construction was first introduced and studied by Carlsson et al. [CDM09].

Level set zigzag. Given a real-valued function $f\colon \mathbb{X} \to \mathbb{R}$ with critical values $s_0 < s_1 < \ldots < s_{m-1}$ and an interleaved sequence of regular values $-\infty = r_0 < s_0 < r_1 < \ldots < s_{m-1} < r_m = \infty$, we denote by $\mathbb{X}_i^{i+1} = f^{-1}[r_i, r_{i+1}]$ the pre-image of the interval $[r_i, r_{i+1}]$. We call the strictly alternating zigzag filtration,

$$\mathbb{X}_0^1 \supseteq \mathbb{X}_1^1 \subseteq \mathbb{X}_1^2 \supseteq \ldots \subseteq \mathbb{X}_{m-2}^{m-1} \supseteq \mathbb{X}_{m-1}^{m-1} \subseteq \mathbb{X}_{m-1}^m, \tag{24.5.2}$$

a *level set zigzag filtration* of f. Its type is $\leftarrow \to \leftarrow \to \ldots \leftarrow \to \leftarrow \to$. Passing to homology, we get the corresponding *level set zigzag persistence module*,

$$\mathsf{H}(\mathbb{X}_0^1) \leftarrow \mathsf{H}(\mathbb{X}_1^1) \to \ldots \leftarrow \mathsf{H}(\mathbb{X}_{m-1}^{m-1}) \to \mathsf{H}(\mathbb{X}_{m-1}^m), \tag{24.5.3}$$

which is unique up to isomorphism.

Mayer–Vietoris pyramid. We can arrange pairs of pre-images of the function in a grid, as shown in Figure 24.5.1, called the *Mayer–Vietoris pyramid*. The nodes of the grid are pairs of spaces: $(\mathbb{X}_k^\ell, \emptyset)$ in the bottom quadrant, $(\mathbb{X}_0^\ell, \mathbb{X}_0^k)$ in the left quadrant, $(\mathbb{X}_k^m, \mathbb{X}_\ell^m)$ in the right quadrant, and $(\mathbb{X}, \mathbb{X}_0^k \cup \mathbb{X}_\ell^m)$ in the top quadrant.

Distinguished paths. Several paths through the Mayer–Vietoris pyramid have natural interpretations. The zigzag filtration along the bottom edge is the level set zigzag (24.5.2). The major diagonal, running from the bottom-left to the upper-right corner, and the minor diagonal, running from the bottom-right to the upper-left corner, have the following sequences of spaces:

$$\emptyset \subseteq \mathbb{X}_0^1 \subseteq \ldots \subseteq \mathbb{X}_0^m = \mathbb{X} \subseteq (\mathbb{X}, \mathbb{X}_{m-1}^m) \subseteq \ldots \subseteq (\mathbb{X}, \mathbb{X}_0^m) = (\mathbb{X}, \mathbb{X}), \tag{24.5.4}$$

$$\emptyset \subseteq \mathbb{X}_{m-1}^m \subseteq \ldots \subseteq \mathbb{X}_0^m = \mathbb{X} \subseteq (\mathbb{X}, \mathbb{X}_0^1) \subseteq \ldots \subseteq (\mathbb{X}, \mathbb{X}_0^m) = (\mathbb{X}, \mathbb{X}). \tag{24.5.5}$$

These are the extended filtrations of f and $-f$.

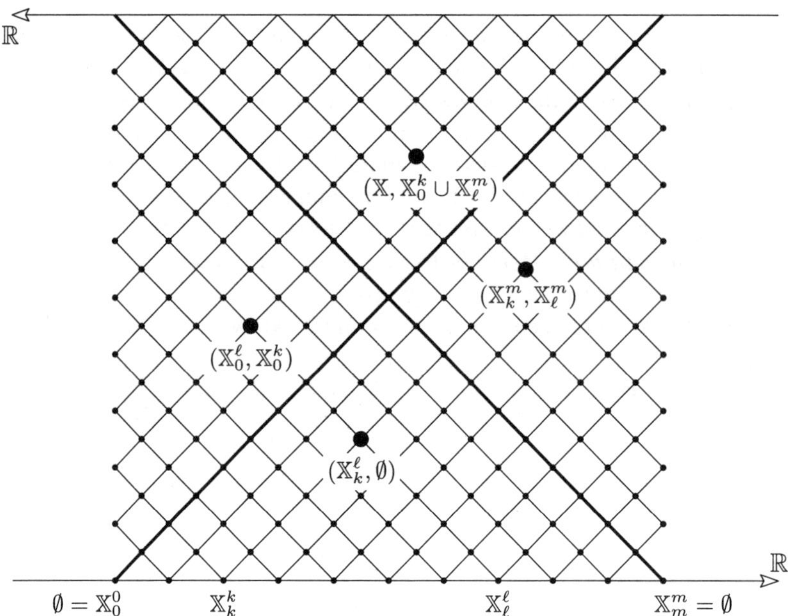

FIGURE 24.5.1
The Mayer–Vietoris pyramid. The two diagonals carry the extended persistence of f and $-f$. The pairs at the corners of every diamond satisfy the relative Mayer–Vietoris long exact sequence. The boundary map from the top row to the bottom row extends the square and turns the entire diagram into a Möbius strip.

Sub-diagrams of extended persistence. We distinguish between three sub-diagrams of the extended persistence diagram. $\mathrm{Ord}_p(f)$ comprises the p-dimensional classes whose birth and death occur in absolute homology; $\mathrm{Rel}_p(f)$ comprises the classes whose birth and death occur in relative homology; and $\mathrm{Ext}_p(f)$ comprises the classes born in absolute but dying in relative homology. They are referred to as the *ordinary*, the *relative*, and the *extended sub-diagrams* of $\mathrm{Dgm}(\mathcal{H}^{\mathrm{ext}}(f))$.

Mayer–Vietoris diamond. For every two pairs connected by a monotonically rising path in the grid, the lower pair includes into the upper pair. Moreover, any four pairs at the corners of a rectangle have a special property which we now explain; see Figure 24.5.1. The intersection of the left and right corners (as pairs of spaces) gives the bottom corner, and their union gives the top corner. For the spaces chosen in the figure, we have $(\mathbb{X}_k^\ell, \emptyset) = (\mathbb{X}_0^\ell \cap \mathbb{X}_k^m, \mathbb{X}_0^k \cap \mathbb{X}_\ell^m)$ and $(\mathbb{X}, \mathbb{X}_0^k \cup \mathbb{X}_\ell^m) = (\mathbb{X}_0^\ell \cup \mathbb{X}_k^m, \mathbb{X}_0^k \cup \mathbb{X}_\ell^m)$. This is true for all rectangles, independent of whether their corners lie in different quadrants or not. This property ensures that the homology groups of any four such spaces can be arranged in the relative Mayer–Vietoris long exact sequence, which lends its name to the pyramid.

Infinite strip. Passing to homology, the pyramid of spaces unrolls into an (infinite) strip of homology groups. The top edge of the pyramid in homological dimension p connects to the bottom edge in dimension $p-1$ via the boundary map, $\mathsf{H}_p(\mathbb{X}, \mathbb{X}_0^i \cup \mathbb{X}_{i+1}^m) \to \mathsf{H}_{p-1}(\mathbb{X}_i^i) \oplus \mathsf{H}_{p-1}(\mathbb{X}_{i+1}^{i+1})$.

DECOMPOSITION

An important property of the Mayer–Vietoris pyramid of homology groups is its decomposition into one-dimensional summands; see [CDM09, BEMP13]. It implies that the full pyramid contains exactly the same information as the level set zigzag or extended persistence, and we can infer the homology of any interlevel set, as well as the rank of any map between any pair of interlevel sets, from the decomposition of the level set zigzag.

Flush diamonds. We distinguish a special shape inside the pyramid unrolled into an infinite strip, which we refer to as a *flush diamond*. It consists of all grid nodes inside a rectangle with sides running at 45° angles, and whose left and right corners are flush with pyramid sides. A *flush diamond indecomposable* is a pyramid that consists of zero vector spaces outside a flush diamond and of one-dimensional vector spaces, connected by identity maps, inside the diamond. Figure 24.5.2 illustrates the supports of four such indecomposables in the pyramid.

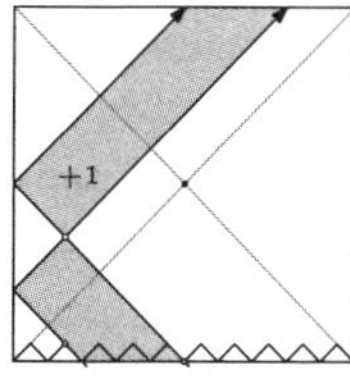

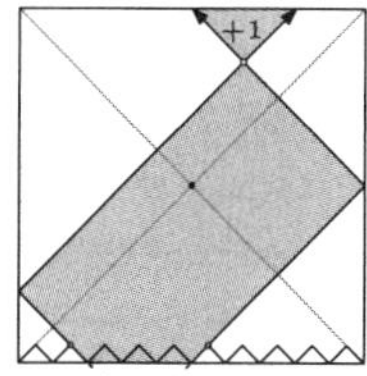

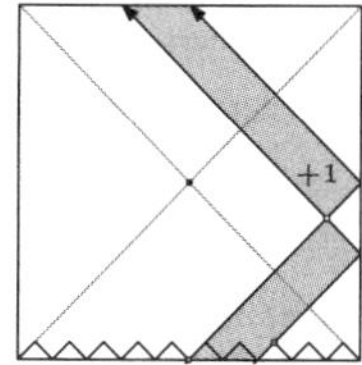

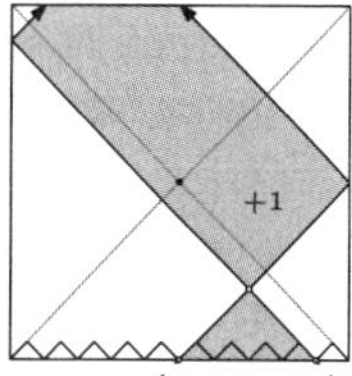

FIGURE 24.5.2
Flush diamonds in the Mayer–Vietoris pyramid that correspond to the four different types of intervals in the levelset zigzag.

Pyramid decomposition and restrictions. Any pyramid of homology groups with the Mayer–Vietoris structure decomposes as a direct sum of flush diamond indecomposables. Because flush diamond indecomposables restrict uniquely to the levelset zigzag, we can recover the decomposition of the entire pyramid from the decomposition of the zigzag. Therefore, we can read off the dimension of any (absolute or relative) homology group of the pre-image of any interval from the decomposition of the levelset zigzag. Furthermore, we can read off the rank of any map between any two such groups from the same decomposition. All zigzags that span the entire pyramid have this property, so we can recover all of the above structures from the decomposition of any other path through the pyramid—for example, from the extended persistence.

Symmetry of extended persistence. Because the two diagonals of the pyramid correspond to the extended persistence of functions f and $-f$, the decomposition of the pyramid into diamonds relates the indecomposables of the two functions. The following theorem relates the three sub-diagrams for f and $-f$; its statement and proof follow from Figure 24.5.2.

THEOREM 24.5.1 Symmetry Theorem

Given a function $f\colon \mathbb{X} \to \mathbb{R}$, the ordinary, relative, and extended persistence sub-

diagrams of f and $-f$ are related as follows:

$$\mathrm{Ord}_p(f) = \mathrm{Rel}^0_{p+1}(-f), \tag{24.5.6}$$

$$\mathrm{Ext}_p(f) = \mathrm{Ext}^R_p(-f), \tag{24.5.7}$$

$$\mathrm{Rel}_p(f) = \mathrm{Ord}^0_{p-1}(-f), \tag{24.5.8}$$

in which the superscript 0 denotes the central reflection through the origin, $(x, y) \mapsto (-x, -y)$, in the diagram as drawn in Figure 24.3.1. Similarly, the superscript R denotes the reflection across the minor diagonal, $(x, y) \mapsto (-y, -x)$.

24.6 APPLICATIONS

Persistent homology is extraordinarily versatile, contributing to numerous questions in a variety of directions. In this section, we showcase the application of persistent homology to two particular topics. Following [DMVJ11a], we use persistence to write data in circular coordinates. Sketching results in [EP16], we show how persistent homology can be used to obtain converging Crofton-type formulas for the intrinsic volume of not necessarily convex shapes.

CIRCLE-VALUED COORDINATES

Machine learning offers a variety of methods to understand high-dimensional data. A popular class of techniques reduces the dimension by mapping a point set from high- to low-dimensional space, e.g., given $X \subseteq \mathbb{R}^n$, find $f\colon X \to \mathbb{R}^d$ with $d = 2$ or 3. How can we exploit topological constraints, such as the persistent homology of X in $\mathbb{R}^n$? Consider the group of homotopy classes of continuous maps from a space to the circle, denoted $[\mathbb{X}, \mathbb{S}^1]$. A classical equation in homotopy theory relates this group to the 1-dimensional cohomology group of the space, a dual of the 1-dimensional homology group, computed with integer coefficients:

$$[\mathbb{X}, \mathbb{S}^1] = \mathsf{H}^1(\mathbb{X}, \mathbb{Z}). \tag{24.6.1}$$

If we detect prominent 1-dimensional (co)cycles in the data, we can turn them into circle-maps. This idea is due to de Silva et al. [DMVJ11a], who propose to compute the persistence diagram of a filtration built on the input point sample, to select a persistent cohomology class, and to turn it into a circle-valued map. A distinctive feature of this method is that it can detect different homotopy classes of such maps and find the smoothest representative within any class. Specifically, de Silva et al. propose the following algorithm:

1. Build a Vietoris–Rips filtration, Rips(X), of the data. Let k be prime, use persistent cohomology to find a significant cohomology class in the filtration, and select its generator in a particular complex $[\alpha_k] \in \mathsf{H}^1(\mathrm{Rips}_r(X), \mathbb{Z}_k)$.
2. Lift $[\alpha_k]$ to a cohomology class with integer coefficients, $[\alpha] \in \mathsf{H}^1(\mathrm{Rips}_r(X), \mathbb{Z})$.
3. Smooth the integer cocycle to a harmonic cocycle in the same cohomology class, $\bar{\alpha} \in \mathsf{C}^1(\mathrm{Rips}_r(X), \mathbb{R})$.
4. Integrate the harmonic cocycle $\bar{\alpha}$ to a circle-valued function, $g\colon \mathrm{Rips}_r(X) \to \mathbb{S}^1$, which restricts to a circle-valued function on the data, $f\colon X \to \mathbb{S}^1$.

FIGURE 24.6.1
Two circle-valued parameterizations of a sample of a figure-8. The hue of the points is set to the circle coordinate. The arrows indicate the circle along which the coordinate varies smoothly.

Figure 24.6.1 shows the two maps for a point sample of a figure-8.

INTRINSIC VOLUMES

We begin with a discussion of the convex case, which is well understood and contains some of the most beautiful theorems in geometry. Given a convex body, $\mathbb{K} \subseteq \mathbb{R}^n$, we write $\mathbb{K}^r$ for the Minkowski sum with the ball of radius $r \geq 0$, and we recall the *Steiner polynomial* of degree n, which gives the n-dimensional volume of the thickened body:

$$\mathrm{Vol}(\mathbb{K}^r) = \sum_{p=0}^{n} b_p V_{n-p}(\mathbb{K}) \cdot r^p. \tag{24.6.2}$$

There are $n+1$ coefficients, each the product of the p-dimensional volume of the unit ball in $\mathbb{R}^p$, b_p, and the $(n-p)$*-th intrinsic volume* of $\mathbb{K}$, $V_{n-p}(\mathbb{K})$. In $\mathbb{R}^3$, the intrinsic volumes are a constant times the volume, surface area, total mean curvature, and total Gaussian curvature. Hadwiger's Characterization Theorem asserts that every measure on convex sets that is invariant under rigid motions, additive, and continuous is a linear combination of the intrinsic volumes [Had52]. The Crofton Formula provides an integral geometric representation of the intrinsic volume. Writing $\mathcal{L}_p^n \subseteq \mathcal{E}_p^n$ for the *linear* and *affine Grassmannians*, namely the p-dimensional linear and affine subspaces of $\mathbb{R}^n$, it asserts that

$$V_{n-p}(\mathbb{K}) = c_{p,n} \int_{E \in \mathcal{E}_p^n} \chi(\mathbb{K} \cap E) \, \mathrm{d}E, \tag{24.6.3}$$

for $0 \leq p \leq n$, in which $\chi(\mathbb{K} \cap E)$ is the Euler characteristic of the intersection, and $c_{p,n} = \binom{n}{p} \frac{b_n}{b_p b_{n-p}}$. The right-hand side of Crofton's Formula makes sense also for non-convex sets and can thus be used to generalize the concept of intrinsic volumes beyond convex bodies. Indeed, even the Steiner polynomial extends, albeit only to bodies with positive reach, for which it gives the correct volume for sufficiently small values of r; see e.g., the tube formulas of Weyl [Wey39].

While Crofton's Formula is invariant under rigid motion and additive also for non-convex bodies, it is not necessarily continuous. To see this, let $\mathbb{X} = B(0,1)$ be the unit disk in $\mathbb{R}^2$ and recall that the length of its boundary is 2π. Writing $r\mathbb{Z}^2$

for the scaled integer grid in the plane, we approximate $\mathbb{X}$ by the union of squares centered at grid points inside $\mathbb{X}$:

$$\mathbb{X}_r = \bigcup_{(ri,rj)\in\mathbb{X}} [ri - \tfrac{r}{2}, ri + \tfrac{r}{2}] \times [rj - \tfrac{r}{2}, rj + \tfrac{r}{2}]. \tag{24.6.4}$$

The total length of the left-facing edges bounding $\mathbb{X}_r$ varies between $2-r$ and $2+r$, and so do total lengths of the right-facing, up-facing, and down-facing edges. It follows that the length of the boundary of $\mathbb{X}_r$ converges to 8 as r goes to 0. But 8 is not 2π. While the Crofton Formula does not converge to the correct value, we can modify it to do so. To this end, we rewrite the first intrinsic volume as an integral over level set moments:

$$V_1(\mathbb{X}) = c_{n-1,n} \int_{E\in\mathcal{E}^n_{n-1}} \chi(\mathbb{X}\cap E)\,\mathrm{d}E \tag{24.6.5}$$

$$= c_{n-1,n} \int_{L\in\mathcal{L}^n_{n-1}} \int_{y=-\infty}^{\infty} \chi(f_L^{-1}(y))\,\mathrm{d}y\,\mathrm{d}L \tag{24.6.6}$$

$$= c_{n-1,n} \int_{L\in\mathcal{L}^n_{n-1}} M_\chi(\mathcal{F}_L)\,\mathrm{d}L, \tag{24.6.7}$$

in which $f_L\colon \mathbb{X} \to \mathbb{R}$ is the height function in direction normal to L, $\mathcal{F}_L = \mathcal{H}^{\text{ext}}(f_L)$, and $M_\chi(\mathcal{F}_L)$ is the level set moment of $\mathcal{F}_L$. For $\varepsilon \geq 0$, we define $M_\chi(\mathcal{F}_L, \varepsilon)$ by ignoring contributions of persistence ε or less. Finally, we define the *modified first intrinsic volume* of $\mathbb{X} \subseteq \mathbb{R}^n$ and $\varepsilon \geq 0$ as

$$V_1(\mathbb{X}, \varepsilon) = c_{n-1,n} \int_{L\in\mathcal{L}^n_{n-1}} M_\chi(\mathcal{F}_L, \varepsilon)\,\mathrm{d}L. \tag{24.6.8}$$

Choosing ε equal to the length of the cube diagonal, we can prove that the thus defined modified first intrinsic volume of the approximating body, $V_1(\mathbb{X}_r, r\sqrt{n})$, differs from $V_1(\mathbb{X})$ by at most some constant times r and therefore converges to the correct first intrinsic volume for compact bodies $\mathbb{X}$ whose boundaries are smoothly embedded $(n-1)$-manifolds in $\mathbb{R}^n$; see [EP16]. At the time of writing this article, such a convergence result was not known for all intrinsic volumes. The first open case is the surface area of bodies in $\mathbb{R}^3$, that is: $n = 3$ and $p = 1$.

24.7 SOURCES AND RELATED MATERIAL

The literature directly or indirectly related to persistent homology is extensive, and we had to make choices. Since this chapter does not focus on applications—which are numerous—we have touched the body of applied literature only tangentially.

BOOKS AND SURVEY ARTICLES

There are many textbooks in algebraic topology available, including Hatcher [Hat02] just to mention one. Of these, very few say anything about computations. In contrast, there are only three books in total that say anything about persistent homology, and the first two are heavily computational.

[Zom05]: a slightly modified version of the author's doctoral thesis.

[EH10]: an introductory text in computational topology, with a heavy focus on persistent homology.

[Ghr14]: a text on applied topology, which includes a few sections on persistent homology.

[Oud15]: a monograph on persistence, covering foundations as well as applications.

There are five survey articles on the topic of persistent homology available to aid in the introduction of non-experts into the field.

[Ghr07]: discusses the concept of the topology of data, focusing on the barcode as a concrete expression thereof.

[EH08]: surveys the state-of-the-art in 2008, giving weight to algorithms computing persistent homology.

[Car09]: takes a more abstract approach and conveys his vision of topology and data.

[Wei11]: gives a brief introduction to persistent homology addressing the mathematics community at large.

[EM12]: surveys the field in 2012, stressing the dichotomy and interplay between theory and practice.

RELATED CHAPTERS

Chapter 21: Topological methods in discrete geometry
Chapter 22: Random simplicial complexes
Chapter 23: Computational topology of graphs on surfaces
Chapter 25: High-dimensional topological data analysis
Chapter 66: Geometry and topology of genomic data

REFERENCES

[AEHW06] P.K. Agarwal, H. Edelsbrunner, J. Harer, and Y. Wang. Extreme elevation on a 2-manifold. *Discrete Comput. Geom.*, 36:553–572, 2006.

[BL14] U. Bauer and M. Lesnick. Induced matchings of barcodes and the algebraic stability of persistence. In *Proc. 30th Sympos. Comput. Geom.*, pages 355–364, ACM Press, 2014.

[BEMP13] P. Bendich, H. Edelsbrunner, D. Morozov, and A. Patel. Homology and robustness of level and interlevel sets. *Homology Homotopy Appl.*, 15:51–72, 2013.

[Bor48] K. Borsuk. On the imbedding of systems of compacta in simplicial complexes. *Fund. Math.*, 35:217–234, 1948.

[Car09] G. Carlsson. Topology and data. *Bull. Amer. Math. Soc.*, 46:255–308, 2009.

[CD10] G. Carlsson and V. de Silva. Zigzag persistence. *Found. Comput. Math.*, 10:367–405, 2010.

[CDGO16] F. Chazal, V. de Silva, M. Glisse, and S. Oudot. *The Sstructure and Stability of Persistence Modules.* Springer, Berlin, 2016.

[CDM09] G. Carlsson, V. de Silva, and D. Morozov. Zigzag persistent homology and real-valued functions. In *Proc. 25th Sympos. Comput. Geom.*, pages 227–236, ACM Press, 2009.

[CEH07] D. Cohen-Steiner, H. Edelsbrunner, and J. Harer. Stability of persistence diagrams. *Discrete Comput. Geom.*, 37:103–120, 2007.

[CEHM10] D. Cohen-Steiner, H. Edelsbrunner, J. Harer, and Y. Mileyko. Lipschitz functions have L_p-stable persistence. *Found. Comput. Math.*, 10:127–139, 2010.

[CEM06] D. Cohen-Steiner, H. Edelsbrunner, and D. Morozov. Vines and vineyards by updating persistence in linear time. In *Proc. 22nd Sympos. Comput. Geom.*, pages 119–126, ACM Press, 2006.

[CK11] C. Chen and M. Kerber. Persistent homology computation with a twist. In *Proc. 27th European Workshop Comput. Geom.*, Morschach, 2011.

[DMVJ11a] V. de Silva, D. Morozov, and M. Vejdemo-Johansson. Persistent cohomology and circular coordinates. *Discrete Comput. Geom.*, 45:737–759, 2011.

[DMVJ11b] V. de Silva, D. Morozov. and M. Vejdemo-Johansson. Dualities in persistent (co)homology. *Inverse Problems*, 27:124003, 2011.

[EH08] H. Edelsbrunner and J. Harer. Persistent homology—a survey. In J.E. Goodman, J. Pach, and R. Pollack, editors, *Surveys on Discrete and Computational Geometry: Twenty Years Later*, vol. 453 of *Contemp. Math.*, pages 257–282, AMS, Providence, 2008.

[EH10] H. Edelsbrunner and J. Harer. *Computational Topology: An Introduction.* AMS, Providence, 2010.

[ELZ00] H. Edelsbrunner, D. Letscher, and A. Zomorodian. Topological persistence and simplification. *Discrete Comput. Geom.*, 28:511–533, 2002.

[EM12] H. Edelsbrunner and D. Morozov. Persistent homology: theory and practice. In *Proc. Europ. Congress Math.*, pages 31–50, 2012.

[EP16] H. Edelsbrunner and F. Pausinger. Approximation and convergence of the intrinsic volume. *Adv. Math.*, 287:674–703, 2016.

[ES52] S. Eilenberg and N. Steenrod. *Foundations of Algebraic Topology.* Princeton University Press, 1952.

[Fro90] P. Frosini. A distance for similarity classes of submanifolds of a Euclidean space. *Bull. Australian Math. Soc.*, 42:407–416, 1990.

[Ghr07] R. Ghrist. Barcodes: the persistent topology of data. *Bull. Amer. Math. Soc.*, 45:61–75, 2007.

[Ghr14] R. Ghrist. *Elementary Applied Topology.* CreateSpace, Scotts Valley, 2014.

[Had52] H. Hadwiger. Additive Funktionale k-dimensionaler Eikörper I. *Arch. Math.*, 3:470–478, 1952.

[Hat02] A. Hatcher. *Algebraic Topology.* Cambridge University Press, 2002.

[Ler46] J. Leray. L'anneau d'homologie d'une représentation. *Les Comptes rendus de l'Académie des sciences*, 222:1366–1368, 1946.

[Mil63] J. Milnor. *Morse Theory.* Princeton University Press, 1963

[Mor40] M. Morse. Rank and span in functional topology. *Ann. Math.*, 41:419–454, 1940.

[Mun53] H.T. Munro et al. Munro's tables of the 3000-feet mountains of Scotland; and other tables of lesser heights. Scottish Mountaineering Club, 1953.

[Mun84] J.R. Munkres. *Elements of Algebraic Topology.* Perseus, Cambridge, 1984.

[NSW08] P. Niyogi, S. Smale, and S. Weinberger. Finding the homology of submanifolds with high confidence from random samples. *Discrete Comput. Geom.*, 39:419–441, 2008.

[Oud15] S.Y. Oudot. *Persistence Theory: From Quiver Representations to Data Analysis.* AMS, Providence, 2015.

[Rob99] V. Robins. Toward computing homology from finite approximations. *Topology Proc.*, 24:503–532, 1999.

[Wei11] S. Weinberger. What is ... persistent homology? *Notices Amer. Math. Soc.*, 58:36–39, 2011.

[Wey39] H. Weyl. On the volume of tubes. *Amer. J. Math.*, 61:461–472, 1939.

[Zom05] A. Zomorodian. *Topology for Computing.* Cambridge University Press, 2005.

25 HIGH-DIMENSIONAL TOPOLOGICAL DATA ANALYSIS

Frédéric Chazal

INTRODUCTION

Modern data often come as point clouds embedded in high-dimensional Euclidean spaces, or possibly more general metric spaces. They are usually not distributed uniformly, but lie around some highly nonlinear geometric structures with nontrivial topology. *Topological data analysis* (TDA) is an emerging field whose goal is to provide mathematical and algorithmic tools to understand the topological and geometric structure of data. This chapter provides a short introduction to this new field through a few selected topics. The focus is deliberately put on the mathematical foundations rather than specific applications, with a particular attention to stability results asserting the relevance of the topological information inferred from data.

The chapter is organized in four sections. Section 25.1 is dedicated to distance-based approaches that establish the link between TDA and curve and surface reconstruction in computational geometry. Section 25.2 considers homology inference problems and introduces the idea of interleaving of spaces and filtrations, a fundamental notion in TDA. Section 25.3 is dedicated to the use of persistent homology and its stability properties to design robust topological estimators in TDA. Section 25.4 briefly presents a few other settings and applications of TDA, including dimensionality reduction, visualization and simplification of data.

25.1 GEOMETRIC INFERENCE AND RECONSTRUCTION

Topologically correct reconstruction of geometric shapes from point clouds is a classical problem in computational geometry. The case of smooth curve and surface reconstruction in $\mathbb{R}^3$ has been widely studied over the last two decades and has given rise to a wide range of efficient tools and results that are specific to dimensions 2 and 3; see Chapter 35. Geometric structures underlying data often appear to be of higher dimension and much more complex than smooth manifolds. This section presents a set of techniques based on the study of distance-like functions leading to general reconstruction and geometric inference results in any dimension.

GLOSSARY

Homotopy equivalence: Given two topological spaces X and Y, two maps $f_0, f_1 : X \to Y$ are *homotopic* if there exists a continuous map $H : [0,1] \times X \to Y$ such that for all $x \in X$, $H(0,x) = f_0(x)$ and $H(1,x) = f_1(x)$. The two spaces X

and Y are said to be *homotopy equivalent*, or to *have the same homotopy type* if there exist two continuous maps $f : X \to Y$ and $g : Y \to X$ such that $g \circ f$ is homotopic to the identity map in X and $f \circ g$ is homotopic to the identity map in Y.

Isotopy: Given $X, Y \subseteq \mathbb{R}^d$, an *(ambient) isotopy* between X and Y is a continuous map $F : \mathbb{R}^d \times [0,1] \to \mathbb{R}^d$ such that $F(.,0)$ is the identity map on $\mathbb{R}^d$, $F(X,1) = Y$ and for any $t \in [0,1]$, $F(.,t)$ is an homeomorphism of $\mathbb{R}^d$.

Probability measure: A *probability measure* μ on $\mathbb{R}^d$ is a function mapping every (Borel) subset B of $\mathbb{R}^d$ to a nonnegative number $\mu(B)$ such that whenever $(B_i)_{i\in I}$ is a countable family of disjoint (Borel) subsets, then $\mu(\cup_{i\in I} B_i) = \sum_{i\in I} \mu(B_i)$, and $\mu(\mathbb{R}^d) = 1$. The *support* of μ is the smallest closed set S such that $\mu(\mathbb{R}^d \backslash S) = 0$. Probability measures are similarly defined on metric spaces.

Hausdorff distance: Given a compact subset $K \subset \mathbb{R}^d$, the distance function from K, $d_K : \mathbb{R}^d \to [0,+\infty)$, is defined by $d_K(x) = \inf_{y\in K} d(x,y)$. The Hausdorff distance between two compact subsets $K, K' \subset \mathbb{R}^d$ is defined by $d_H(K,K') = \|d_K - d_{K'}\|_\infty = \sup_{x\in\mathbb{R}^d} |d_K(x) - d_{K'}(x)|$.

DISTANCE-BASED APPROACHES AND GEOMETRIC INFERENCE

The general problem of geometric inference can be stated in the following way: given an *approximation* P (e.g., a point cloud) of a *geometric object* K in $\mathbb{R}^d$, is it possible to reliably and efficiently estimate the topological and geometric properties of K? Obviously, it needs to be instantiated in precise frameworks by defining the class of geometric objects that are considered and the notion of distance between these objects. The idea of distance-based inference is to associate to each object a real-valued function defined on $\mathbb{R}^d$ such that the sublevel sets of this function carry some geometric information about the object. Then, proving geometric inference results boils down to the study of the stability of the sublevel sets of these functions under perturbations of the objects.

Distance to compact sets and distance-like functions. A natural and classical example is to consider the set of compact subsets of $\mathbb{R}^d$, which includes both continuous shapes and point clouds. The space of compact sets is endowed with the Hausdorff distance and to each compact set $K \subset \mathbb{R}^d$ is associated its distance function $d_K : \mathbb{R}^d \to [0,+\infty)$ defined by $d_K(x) = \inf_{y\in K} d(x,y)$. The properties of the r***-offsets*** $K^r = d_K^{-1}([0,r])$ of K (i.e., the union of the balls of radius r centered on K) can then be used to compare and relate the topology of the offsets of compact sets that are close to each other with respect to the Hausdorff distance. When the compact K is a smooth submanifold, this leads to basic methods for the estimation of the homology and homotopy type of K from an approximate point cloud P, under mild sampling conditions [NSW08, CL08]. This approach extends to a larger class of nonsmooth compact sets K and leads to stronger results on the inference of the isotopy type of the offsets of K [CCSL09a]. It also leads to results on the estimation of other geometric and differential quantities such as normals [CCSL09b], curvatures [CCSLT09] or boundary measures [CCSM10] from shapes sampled with a moderate amount of noise (with respect to Hausdorff distance).

These results mainly rely on the ***stability*** of the map associating to a compact set K its distance function d_K (i.e., $\|d_K - d_{K'}\|_\infty = d_H(K,K')$ for any compact sets $K, K' \subset \mathbb{R}^d$) and on the *1-semiconcavity* of the squared distance function d_K^2 (i.e., the convexity of the map $x \to \|x\|^2 - d_K^2(x))$ motivating the following definition.

Definition: A nonnegative function $\varphi : \mathbb{R}^d \to \mathbb{R}_+$ is *distance-like* if it is proper (the pre-image of any compact in $\mathbb{R}$ is a compact in $\mathbb{R}^d$) and $x \to \|x\|^2 - \varphi^2(x)$ is convex.

The 1-semiconcavity property of a distance-like function φ allows us to define its gradient vector field $\nabla\varphi : \mathbb{R}^d \to \mathbb{R}^d$. Although not continuous, this gradient vector field can be integrated [Pet06] into a continuous flow that is used to compare the geometry of the sublevel sets of two close distance functions. In particular, the topology of the sublevel sets of a distance-like function φ can only change at levels corresponding to critical points, i.e., points x such that $\|\nabla_x\varphi\| = 0$:

LEMMA 25.1.1 Isotopy Lemma [Gro93, Proposition 1.8]

Let φ be a distance-like function and $r_1 < r_2$ be two positive numbers such that φ has no critical point in the subset $\varphi^{-1}([r_1, r_2])$. Then all the sublevel sets $\varphi^{-1}([0, r])$ are isotopic for $r \in [r_1, r_2]$.

This result suggests the following definitions.

Definition: Let φ be a distance-like function. We denote by $\varphi^r = \varphi^{-1}([0, r])$ the r sublevel set of φ.

- A point $x \in \mathbb{R}^d$ is called *α-critical* if $\|\nabla_x\varphi\| \leq \alpha$.
- The *weak feature size* of φ at r is the minimum $r' > 0$ such that φ doesn't have any critical value between r and $r + r'$. We denote it by $\mathrm{wfs}_\varphi(r)$. For any $0 < \alpha < 1$, the *α-reach* of φ is the maximum r such that $\varphi^{-1}((0, r])$ does not contain any α-critical point.

Note that the isotopy lemma implies that all the sublevel sets of φ between r and $r + \mathrm{wfs}_\varphi(r)$ have the same topology. Comparing two close distance-like functions, if φ and ψ are two distance-like functions, such that $\|\varphi - \psi_\infty\| \leq \varepsilon$ and $\mathrm{wfs}_\varphi(r) > 2\varepsilon$, $\mathrm{wfs}_\psi(r) > 2\varepsilon$, then for every $0 < \eta \leq 2\varepsilon$, $\varphi^{r+\eta}$ and $\psi^{r+\eta}$ have the same homotopy type. An improvement of this result leads to the following reconstruction theorem from [CCSM11].

THEOREM 25.1.2 Reconstruction Theorem

Let φ, ψ be two distance-like functions such that $\|\varphi - \psi\|_\infty < \varepsilon$, with $\mathrm{reach}_\alpha(\varphi) \geq R$ for some positive ε and α. Then, for every $r \in [4\varepsilon/\alpha^2, R - 3\varepsilon]$ and every $\eta \in (0, R)$, the sublevel sets ψ^r and φ^η are homotopy equivalent when

$$\varepsilon \leq \frac{R}{5 + 4/\alpha^2}.$$

Under similar but slightly more technical conditions the Reconstruction theorem can be extended to prove that the sublevel sets are indeed homeomorphic and even isotopic, and that their normals and curvatures can be compared [CCSL09b, CCSLT09].

As an example, distance functions from compact sets are obviously distance-like and the above reconstruction result gives the following result.

THEOREM 25.1.3 *Let $K \subset \mathbb{R}^d$ be a compact set and let $\alpha \in (0, 1]$ be such that $r_\alpha = \mathrm{reach}_\alpha(d_K) > 0$. If $P \subset \mathbb{R}^d$ such that $d_H(K, P) \leq \kappa\alpha$ with*

$\kappa < \alpha^2/(5\alpha^2+12)$, then the offsets K^r and $P^{r'}$ are homotopy equivalent when

$$0 < r < r_\alpha \quad \text{and} \quad \frac{4\mathrm{d_H}(\mathrm{P},\mathrm{K})}{\alpha^2} \leq \mathrm{r}' \leq \mathrm{r}_\alpha - 3\mathrm{d_H}(\mathrm{P},\mathrm{K}).$$

In particular, if K is a smooth submanifold of $\mathbb{R}^d$, then $r_1 > 0$ and $P^{r'}$ is homotopy equivalent to K.

It is interesting to notice that indeed, distance-like functions are closely related to distance functions from compact sets: any distance-like function $\varphi : \mathbb{R}^d \to \mathbb{R}^+$ is the restriction to a hyperplane of the distance function from a compact set in $\mathbb{R}^{d+1}$ [CCSM11, Prop.3.1].

DTM AND KERNEL DISTANCES: THE MEASURE POINT OF VIEW

The major drawback of the geometric inference framework derived from the Hausdorff distance and distances between compact sets is its instability in the presence of outliers in the approximate data (i.e., points that are not close to the underlying geometric object). One way to circumvent this problem is to consider the approximate data as an empirical measure (i.e., a weighted sum of Dirac measures centered on the data points) rather than a point cloud, and to consider the probability measures on $\mathbb{R}^d$ instead of the compact subsets of $\mathbb{R}^d$ as the new class of geometric objects.

As the distance between a point $x \in \mathbb{R}^d$ and a compact set K is defined as the radius of the smallest ball centered at x and containing a point of K, a basic and natural idea to associate a distance-like function to a probability measure is to mimic this definition in the following way: given a probability measure μ and a parameter $0 \leq l < 1$, define the function $\delta_{\mu,l} : \mathbb{R}^d \to \mathbb{R}_+$ by

$$\delta_{\mu,l} : x \in \mathbb{R}^d \mapsto \inf\{r > 0 : \mu(\bar{B}(x,r)) > l\}$$

where $\bar{B}(x,r)$ is the closed Euclidean ball of radius r with center x. Unfortunately, the map $\mu \to \delta_{\mu,l}$ turns out to be, in general, not continuous for standard metrics on the space of probability measures. This continuity issue is fixed by averaging over the parameter l.

Definition: Let μ be a probability measure on $\mathbb{R}^d$, and $m \in (0,1]$ be a positive parameter. The function defined by

$$d^2_{\mu,m} : \mathbb{R}^d \to \mathbb{R}_+, \; x \mapsto \frac{1}{m}\int_0^m \delta_{\mu,l}(x)^2 dl$$

is called the *distance-to-measure (DTM) function to μ with parameter m.*

From a practical point of view, if $P \subset \mathbb{R}^d$ is a finite point cloud and $\mu = \frac{1}{|P|}\sum_{x\in P}\delta_x$ is the uniform measure on P, then for any x the function $l \to \delta_{\mu,l}(x)$ is constant on the intervals $(k/|P|,(k+1)/|P|)$ and equal to the distance between x and its kth nearest neighbor in P. As an immediate consequence for $m = k/|P|$,

$$d^2_{\mu,m}(x) = \frac{1}{k}\sum_{i=1}^{k}\|x - X_{(i)}(x)\|^2$$

where $X_{(i)}(x)$ is the ith nearest neighbor of x in P. In other words, $d^2_{\mu,m}(x)$ is just the average of the squared distances from x to its first k nearest neighbors.

Distance-to-measure functions turn out to be distance-like; see Theorem 25.1.4 for distance-to-measures below. The application of Theorem 25.1.2 of the previous section to DTM functions require stability properties relying on a well-chosen metric on the space of measures. For this reason, the space of probability measures is equipped with a so-called *Wasserstein distance* W_p $(p \geq 1)$ whose definition relies on the notion of a transport plan between measures, which is strongly related to the theory of optimal transport [Vil03].

A *transport plan* between two probability measures μ and ν on $\mathbb{R}^d$ is a probability measure π on $\mathbb{R}^d \times \mathbb{R}^d$ such that for every $A, B \subseteq \mathbb{R}^d$ $\pi(A \times \mathbb{R}^d) = \mu(A)$ and $\pi(\mathbb{R}^d \times B) = \nu(B)$. Intuitively $\pi(A \times B)$ corresponds to the amount of mass of μ contained in A that will be transported to B by the transport plan. Given $p \geq 1$, the p-cost of such a transport plan π is given by

$$\mathcal{C}_p(\pi) = \left(\int_{\mathbb{R}^d \times \mathbb{R}^d} \|x - y\|^p \, d\pi(x, y) \right)^{1/p}.$$

This cost is finite when the measures μ and ν both have finite *p-moments*, i.e., $\int_{\mathbb{R}^d} \|x\|^p \, d\mu(x) < +\infty$ and $\int_{\mathbb{R}^d} \|x\|^p \, d\nu(x) < +\infty$. The set of probability measures on $\mathbb{R}^d$ with finite p-moment includes all probability measures with compact support, such as, e.g., empirical measures. The *Wasserstein distance* of order p between two probability measures μ and ν on $\mathbb{R}^d$ with finite p-moment is the minimum p-cost $\mathcal{C}_p(\pi)$ of a transport plan π between μ and ν. It is denoted by $W_p(\mu, \nu)$.

For geometric inference, the interest in Wasserstein distance comes from its weak sensibility to the presence of a small number of outliers. For example, consider a reference point cloud P with N points, and define a noisy version P' by replacing n points in P by points $o_1, \ldots, o_n$ such that $d_P(o_i) \geq R$ for some $R > 0$. Considering the cost of the transport plan between P' and P that moves the outliers back to their original position, and keeps the other points fixed, we get $W_p(\mu_P, \mu_{P'}) \leq \frac{n}{N}(R + \mathrm{diam}(P))$ while the Hausdorff distance between P and P' is at least R. Hence, if the number of outliers is small, i.e., $n \ll N$, the Wasserstein distance between μ_P and $\mu_{P'}$ remains small. Moreover, if the N points of P are independently drawn from a common measure μ, then μ_P converges almost surely to μ in the Wasserstein metric W_p (see [BGV07] for precise statements).

THEOREM 25.1.4 Stability of distance-to-measures [CCSM11]

For any probability measure μ in $\mathbb{R}^d$ and $m \in (0, 1)$ the function $d_{\mu,m}$ is distance-like. Moreover, if ν is another probability measure on $\mathbb{R}^d$ and $m > 0$, then

$$\|d_{\mu,m} - d_{\nu,m}\|_\infty \leq \frac{1}{\sqrt{m}} \mathrm{W}_2(\mu, \nu).$$

This theorem allows us to apply the reconstruction theorem (Theorem 25.1.2) to recover topological and geometric information of compact shapes from noisy data containing outliers [CCSM11, Cor. 4.11].

More recently, a new family of distance-like functions associated to probability measures, called *kernel distances*, has been introduced in [PWZ15] that are closely related to classical kernel-based density estimators. They offer similar, but complementary, properties as the DTM functions and come with stability properties ensuring the same topological guarantees for topological and geometric inference.

Probabilistic and statistical considerations. The distance-based approach is well-suited to explore reconstruction and geometric inference from a statistical perspective, in particular when data are assumed to be randomly sampled. The problem of approximation of smooth manifolds with respect to the Hausdorff distance from random samples under different models of noise has been studied in [GPP⁺12a, GPP⁺12b]. The statistical analysis of DTM and kernel distances remains largely unexplored despite a few recent preliminary results [CMM15, CFL⁺14]; see also the open problems below.

Some open problems. Here are a few general problems related to the distance-based approach that remain open or partly open.

1. The computation of the DTM at a given point only require us to compute nearest neighbors but the efficient global computation of the DTM, e.g., to obtain its sublevel sets or its persistent homology, turns out to have prohibitive complexity as it is closely related to the computation of higher-order Voronoi diagrams. The difficulty of efficiently approximating the DTM function is still rather badly understood despite a few results in this direction [BCOS16, GMM13, Mér13]; see also Chapter 27.
2. The dependence of DTM functions on the parameter m raises the problem of the choice of this parameter. The same problem also occurs with the kernel distances that depend on a bandwidth parameter. Very little is known about the dependency of DTM on m (the situation is slightly better for the kernel distances) and data-driven methods to choose these parameters still need to be developed. Preliminary results in this direction have recently been obtained in [CMM15, CFL⁺14].

RECONSTRUCTION IN HIGH DIMENSION

Although the above-mentioned approaches provide general frameworks for geometric inference in any dimension, they do not directly lead to efficient reconstruction algorithms. Here, a reconstruction algorithm is meant to be an algorithm that:

- takes as input a finite set of points P sampled from an unknown shape K,
- outputs a triangulation or a simplicial complex that approximates K, and
- provides a topologically correct reconstruction (i.e., homeomorphic or isotopic to K) when certain sampling conditions quantifying the quality of the approximation of K by P are satisfied.

Efficient algorithms with such guarantees are possible if we restrict ourselves to specific classes of shapes to reconstruct.

- **Low-dimensional smooth manifolds in high dimension:** except for the case of curve and surface reconstruction in $\mathbb{R}^2$ and $\mathbb{R}^3$; see Chapter 35. The attempts to develop effective reconstruction algorithms for smooth manifolds in arbitrary dimension remain quite limited. Extending smooth manifold reconstruction algorithms in $\mathbb{R}^3$ to $\mathbb{R}^d$, $d > 3$, raises several major difficulties. In particular, important topological properties of restricted Delaunay triangulations used for curve and surface reconstruction no longer hold in higher dimensions, preventing direct generalization of the existing low-dimensional algorithms. Moreover, classical data structures involved in reconstruction algorithms, such as the Delaunay triangulation, are global and their complexity depends exponentially on the ambient dimension, which make them almost intractable in dimensions larger than 3. However, a few attempts have been made to overcome these issues. In [BGO09], using the so-called ***witness complex*** [SC04], the authors design a reconstruction algorithm whose complexity scales up with the intrinsic dimension of the submanifold. More recently, a new data structure, the *tangential Delaunay complex*, has been introduced and used to design effective reconstruction algorithms for smooth low-dimensional submanifolds of $\mathbb{R}^d$ [BG13].

- **Filamentary structures and stratified spaces:** 1-dimensional filamentary structures appear in many domains (road networks, network of blood vessels, astronomy, etc.) and can be modeled as 1-dimensional stratified sets, or (geometric) graphs. Various methods, motivated and driven by specific applications, have been developed to reconstruct such structures from point cloud data. From a general perspective, the (relatively) simple structure of graphs allows to propose new approaches to design metric graph reconstruction algorithms with various topological guarantees, e.g., homeomorphy or homotopy type and closeness in the Gromov-Hausdorff metric [GSBW11, ACC$^+$12, CHS15]. Despite a few attempts [BCSE$^+$07, BWM12], reconstruction of stratified sets of higher dimension turns out to be a much more difficult problem that remains largely open.

25.2 HOMOLOGY INFERENCE

The results on geometric inference from the previous section provide a general theoretical framework to "reconstruct" unknown shapes from approximate data. However, it is not always desirable to fully reconstruct a geometric object to infer some relevant topological properties from data. This is illustrated in this section by two examples. First, we consider a weaker version of the reconstruction paradigm where the goal is to infer topological invariants, more precisely homology and Betti numbers. Second, we consider coverage problems in sensor networks that can be answered using homology computations. Both examples rely on the idea that relevant topological information cannot always be directly inferred from the data at a given scale, but by considering how topological features relate to each other across different scales. This fundamental idea raises the notion of interleaving between spaces and filtrations and leads to persistence-based methods in TDA that are considered in the next section.

GLOSSARY

Abstract simplicial complex: Given a set X, an abstract simplicial complex C with vertex set X is a set of finite subsets of X, the simplices, such that the elements of X belong to C and if $\sigma \in C$ and $\tau \subset \sigma$, then $\tau \in C$.

Homology: Intuitively, homology (with coefficient in a field) associates to any topological space X, a family of vector spaces, the so-called homology groups $H_k(X)$, $k = 0, 1, \ldots$, each of them encoding k-dimensional topological features of X. A fundamental property of homology is that any continuous function $f : X \to Y$ induces a linear map $f_* : H_k(X) \to H_k(Y)$ between homology groups that encodes the way the topological features of X are mapped to the topological features of Y by f. This linear map is an isomorphism when f is a homeomorphism or a homotopy equivalence (homology is thus a homotopy invariant). See [Hat01] or Chapter 22 for a formal definition.

Betti numbers: The kth Betti number of X, denoted $\beta_k(X)$, is the rank of $H_k(X)$ and represents the number of "independent" k-dimensional features of X: for example, $\beta_0(X)$ is the number of connected components of M, $\beta_1(X)$ the number of independent cycles or tunnels, $\beta_2(X)$ the number of cavities, etc.

Čech complex: Given P, a subset of a metric space X and $r > 0$, the Čech complex $\check{\mathrm{C}}\mathrm{ech}(P, r)$ built on top of P, with parameter r is the abstract simplicial complex defined as follows: (i) the vertices of $\check{\mathrm{C}}\mathrm{ech}(P, r)$ are the points of P and (ii) $\sigma = [p_0, \dots, p_k] \in \check{\mathrm{C}}\mathrm{ech}(P, r)$ if and only if the intersection of balls of radius r and centered at the p_i's have nonempty intersection.

Vietoris-Rips complex: Given a metric space (X, d_X) and $r \geq 0$, the Vietoris-Rips complex $\mathrm{Rips}(X, r)$ is the (abstract) simplicial complex defined by i) the vertices of $\mathrm{Rips}(X, r)$ are the points of X and, ii) $\sigma = [x_0, \dots, x_k] \in \mathrm{Rips}(X, r)$ if and only if $d_X(x_i, x_j) \leq r$ for any $i, j \in \{0, \dots, k\}$.

ČECH COMPLEX, VIETORIS-RIPS COMPLEX, AND HOMOLOGY INFERENCE

An important advantage of simplicial complexes is that they are not only combinatorial objects but they can also be seen as topological spaces. Let C be a finite simplicial complex with vertex set $X = \{x_1, \dots, x_n\}$. Identifying each x_i with the point e_i of $\mathbb{R}^n$ all of whose coordinates are 0 except the ith which is equal to 1, one can identify each simplex $\sigma = [x_{i_0}, \cdot x_{i_k}] \in C$ with the convex hull of the points $e_{i_0}, \cdot e_{i_k}$. The union of these sets inherits a topology as a subset of $\mathbb{R}^n$ and is called the geometric realization of C in $\mathbb{R}^n$. In the following, the topology or the homotopy type of a simplicial complex refers to the ones of its geometric realization.

Thanks to this double nature, simplicial complexes play a fundamental role to bridge the gap between continuous shapes and their discrete representations. In particular, the classical nerve theorem [Hat01][Corollary 4G3] is fundamental in TDA to relate continuous representation of shapes to discrete description of their topology through simplicial complexes.

Definition: Let X be a topological space and let $\mathcal{U} = \{U_i\}_{i \in I}$ be an open cover of X, i.e., a family of open subsets such that $X = \cup_{i \in I} U_i$. The *nerve of* $\mathcal{U}$, denoted $N(\mathcal{U})$, is the (abstract) simplicial complex defined by the following:

(i) the vertices of $N(\mathcal{U})$ are the U_i's, and
(ii) $\sigma = [U_{i_0}, \dots, U_{i_k}] \in N(\mathcal{U})$ if and only if $\bigcap_{j=0}^k U_{i_j} \neq \emptyset$.

THEOREM 25.2.1 Nerve Theorem

Let $\mathcal{U} = \{U_i\}_{i \in I}$ be an open cover of a paracompact topological space X. If any nonempty intersection of finitely many sets in $\mathcal{U}$ is contractible, then X and $N(\mathcal{U})$ are homotopy equivalent. In particular, their homology groups are isomorphic.

An immediate consequence of the Nerve Theorem is that under the assumption of Theorem 25.1.2, the computation of the homology of a smooth submanifold $K \subset \mathbb{R}^d$ approximated by a finite point cloud P boils down to the computation of the homology of the Čech complex $\check{\mathrm{C}}\mathrm{ech}(P, r)$ for some well-chosen radius r. However, this direct approach suffers from several drawbacks: first, computing the nerve of a union of balls requires the extensive use of an awkward predicate testing the nonemptiness of the intersection of finite sets of balls; second, the suitable choice of the radius r relies on the knowledge of the reach of K and of the Hausdorff distance between K and P that are usually not available. Moreover, the assumption that the underlying shape K is a smooth manifold is often too restrictive in practical applications. To overcome this latter restriction, [CL07, CSEH07] consider the linear map $H_k(P^\varepsilon) \to H_k(P^{3\varepsilon})$ induced by the inclusion $P^\varepsilon \hookrightarrow P^{3\varepsilon}$ of small offsets of P and prove that its rank is equal to the kth Betti number of K^δ when $d_H(K, P) < \varepsilon < \mathrm{wfs}(K)/4$ and $0 < \delta < \mathrm{wfs}(K)$, where $\mathrm{wfs}(K) = \mathrm{wfs}_{d_k}(0)$ is the infimum of the positive critical values of d_K. The idea of using nested pairs of offsets to infer the homology of compact sets was initially introduced in [Rob99] for the study of attractors in dynamical systems. Beyond homology, the inclusion $P^\varepsilon \hookrightarrow P^{3\varepsilon}$ also induces group morphisms between the homotopy groups of theses offsets whose images are isomorphic to the homotopy groups of K^δ [CL07]. Homotopy inference and the use of homotopy information in TDA raise deep theoretical and algorithmic problems and remains rather unexplored despite a few attempts such as, e.g., [BM13].

The homotopy equivalences between P^ε, $P^{3\varepsilon}$ and $\check{\mathrm{C}}\mathrm{ech}(P^\varepsilon)$, $\check{\mathrm{C}}\mathrm{ech}(P^{3\varepsilon})$, respectively, given by the nerve theorem can be chosen in such a way that they commute with the inclusion $P^\varepsilon \hookrightarrow P^{3\varepsilon}$, leading to an algorithm for homology inference based upon the Čech complex. To overcome the difficulty raised by the computation of the Čech complex, [CO08] proposes to replace it by the Vietoris-Rips complex. Using the elementary interleaving relation

$$\check{\mathrm{C}}\mathrm{ech}(P, r/2) \subseteq \mathrm{Rips}(P, r/2) \subset \check{\mathrm{C}}\mathrm{ech}(P, r),$$

one easily obtains that, for any integer $k = 0, 1 \ldots$, the rank of the linear map $H_k(\mathrm{Rips}(P, \varepsilon)) \to H_k(\mathrm{Rips}(P, 4\varepsilon))$ is equal to that of $H_k(K^\delta)$ when $2d_H(P, K) < \varepsilon < (\mathrm{wfs}(K) - d_H(P, K))/4$ and $0 < \delta < \mathrm{wfs}(K)$. A similar result also holds for witness complexes built on top of the input data P. To overcome the problem of the choice of the Vietoris-Rips parameter ε, a greedy algorithm is proposed in [CO08] that maintains a nested sequence of Vietoris-Rips complexes and eventually computes the Betti numbers of the offsets K^δ for various relevant scales δ. When K is an m-dimensional smooth submanifold of $\mathbb{R}^d$ this algorithm recovers the Betti numbers of K in times at most $c(m)n^5$ where $n = |P|$ and $c(m)$ is a constant depending exponentially on m and linearly on d. Precise information about the complexity of the existing homology inference algorithms is available in [Oud15, Chapter 4].

From a statistical perspective, when K is a smooth submanifold and P is a random sample, the estimation of the homology has been considered in [NSW08, NSW11] while [BRS+12] provides minimax rates of convergence.

COVERAGE PROBLEMS IN SENSOR NETWORKS

Given that sensors located at a set of nodes $P = \{p_1, \ldots, p_n\} \subset \mathbb{R}^d$ spread out in a bounded region $D \subset \mathbb{R}^d$, assume that each sensor can sense its environment within a disc of fixed *covering radius* $r_c > 0$. Basic coverage problems in sensor networks address the question of the full coverage of D by the sensing areas covered by the sensor. When the exact position of the nodes is not known but only the graph connecting sensors within distance less than some *communication radius* $r_c > 0$ from each other, Vietoris-Rips complexes appear as a natural tool to infer topological information about the covered domain. Following this idea, [SG07a, SG07b] propose to use the homology of nested pairs of such simplicial complexes to certify that the domain D is covered by the union of the covering discs in various settings.

More precisely, assume that each node can detect and communicate with other nodes via a strong signal within radius $r_s > 0$ and via a weak signal within a radius $r_s > 0$, respectively, such that $r_c \geq r_s/\sqrt{2}$ and $r_w \geq r_s\sqrt{10}$. Assume moreover that the nodes can detect the presence of the boundary ∂D within a *fence detection radius* r_f and denote by $F \subset P$ the set of nodes that are at distance at most r_f from ∂D. Regarding the domain D, assume that $D \backslash (\partial D)^{r_f + r_s/\sqrt{2}}$ is connected and the injectivity radius of the hypersurface $d_{\partial D}^{-1}(r_f)$ is larger than r_s. Then [SG07a] introduces the following criterion involving the relative homology of the pairs of Vietoris-Rips complexes built on top of F and P.

THEOREM 25.2.2 Coverage criterion

If the morphism between relative homology groups

$$i_\star : H_d(\text{Rips}(P, F, r_s)) \to H_d(\text{Rips}(P, F, r_w))$$

induced by the inclusion of the pairs of complexes

$$i : (\text{Rips}(P, r_s), \text{Rips}(F, r_s)) \hookrightarrow (\text{Rips}(P, r_w), \text{Rips}(F, r_w))$$

is nonzero, then $D \setminus (\partial D)^{r_f + r_s/\sqrt{2}}$ *is contained in the union of the balls of radius* r_c *centered at the points of* P.

This result has given rise to a large literature on topological methods in sensor networks. In particular, regarding the robustness of this criterion, its stability under perturbations of the networks is studied in [HK14]. Similar ideas, combined with zigzag persistent homology, have also been used to address other problems such as, e.g., the detection of evasion paths in mobile sensor networks [AC15].

25.3 PERSISTENCE-BASED INFERENCE

Beyond homology, persistent homology (see Chapter 22) plays a central role in topological data analysis. It is usually used in two different ways. It may be applied to functions defined on data in order to estimate topological features of these functions (number and relevance of local extrema, homology of sublevel sets, etc.). Persistent homology may also be applied to geometric filtrations built on top of the data in order to infer topological information about the global structure of data. These two ways give rise to two main persistence-based pipelines that are presented in the next two sections and illustrated in Figure 25.3.1. The resulting persistence diagrams are then used to reveal and characterize topological features for further data analysis tasks (classification, clustering, learning, etc.). From a theoretical perspective, the stability properties of persistent homology allow us to establish the stability and thus the relevance of these features.

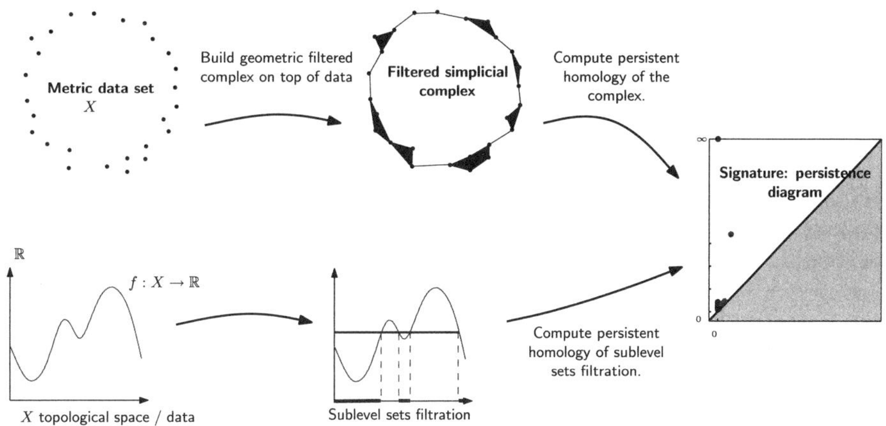

FIGURE 25.3.1
The classical pipelines for persistence in TDA.

GLOSSARY

Filtered simplicial complex: Given a simplicial complex C and a finite or infinite subset $A \subset \mathbb{R}$, a *filtration of* C is a family $(C_\alpha)_{\alpha \in A}$ of subcomplexes of C such that for any $\alpha \leq \alpha'$, $C_\alpha \subseteq C_{\alpha'}$ and $C = \cup_{\alpha \in A} C_\alpha$.

Sublevel set filtration: Given a topological space X and a function $f : X \to \mathbb{R}$, the *sublevel set filtration of* f is the nested family of sublevel sets of f: $(f^{-1}((-\infty, \alpha]))_{\alpha \in \mathbb{R}}$.

Metric space: A metric space is a pair (X, d_X) where X is a set and $d_X : X \times X \to \mathbb{R}_+$ is a nonnegative map such that for any $x, y, z \in X$, $d_X(x, y) = d_X(y, x)$, $d_X(x, y) = 0$ if and only if $x = y$ and $d_X(x, z) \leq d_X(x, y) + d_X(y, z)$.

Gromov-Hausdorff distance: The Gromov-Hausdorff distance extends the notion of Hausdorff distance between compact subsets of the same metric spaces to general spaces. More precisely, given two compact metric spaces (X, d_X) and (Y, d_Y) and a third metric space (Z, d_Z), a map $\varphi : X \to Z$ (resp., $\psi : Y \to Z$) is an isometric embedding if for any $x, x' \in X$, $d_Z(\varphi(x), \varphi(x')) = d_X(x, x')$ (resp., any $y, y' \in Y$, $d_Z(\psi(y), \psi(y')) = d_Y(y, y')$). The *Gromov-Hausdorff distance* $d_{GH}(X, Y)$ between X and Y is defined as the infimum of the Hausdorff distances $d_H(\varphi(X), \psi(Y))$ where the infimum is taken over all the metric spaces (Z, d_Z) and all the isometric embeddings $\varphi : X \to Z$ and $\psi : Y \to Z$.

Persistent homology: Persistent homology provides a framework and efficient algorithms to encode the evolution of the homology of families of nested topological spaces (filtrations) indexed by a set of real numbers, such as the sublevel sets filtration of a function, a filtered complex, etc. These indices may often be seen as scales, as for example in the case of the Vietoris-Rips filtration where the index is the radius of the balls used to build the complex. Given a filtration

$(F_\alpha)_{\alpha \in A}$, its homology changes as α increases: new connected components can appear, existing connected components can merge, cycles and cavities can appear or be filled, etc. Persistent homology tracks these changes, identifies features and associates, with each of them, an interval or lifetime from α_{birth} to α_{death}. For instance, a connected component is a feature that is born at the smallest α such that the component is present in F_α, and dies when it merges with an older connected component. Intuitively, the longer a feature persists, the more relevant it is. The set of intervals representing the lifetime of the identified features is called the barcode of the filtration. As an interval can also be represented as a point in the plane with coordinates $(\alpha_{birth}, \alpha_{death})$, the set of points (with multiplicity) representing the intervals is called the persistence diagram of the filtration. See Chapter 22 for formal definitions.

Bottleneck distance: Given two persistence diagrams, D and D', the *bottleneck distance* $d_B(D, D')$ is defined as the infimum of $\delta \geq 0$ for which there exists a matching between the diagrams, such that two points can only be matched if their distance is less than δ and all points at distance more than δ from the diagonal must be matched. See Chapter 22, for more details.

PERSISTENCE OF SUBLEVEL SET FILTRATIONS

Persistent homology of sublevel set filtration of functions may be used from two different perspectives in TDA.

Collections of complex objects. When data are collections in which each element is already a "complex" geometric object such as, e.g., an image or shape, functions defined on each data element may be used to highlight some of their features. The persistence diagrams of the sublevel set filtrations of such functions can be used for comparison and classification of the data elements. The bottleneck distance between the diagrams is then used as a measure of dissimilarity between the elements. The idea of using persistence of functions defined on images and shapes was first introduced in the setting of *size theory* where it was used for shape analysis [VUFF93]; see also [FL99] for a survey. These ideas are not restricted to images and shapes and can also be applied to other "geometric" data such as, for example, textures or hand gesture data [LOC14, RHBK15]. In practical applications the main difficulty of this approach is in the design of functions whose persistent homology provides sufficiently informative and discriminative features for further classification or learning tasks.

Scalar field analysis. Another problem arising in TDA is the estimation of the persistent homology of a function defined on a possibly unknown manifold, from a finite approximation. As an example, assume that we are given a collection of sensors spread out in some region and that these sensors measure some physical quantity, such as temperature or humidity. Assuming that the nodes do not know their geographic location but that they can detect which other nodes lie in their vicinity, the problem is then to recover global topological information about the physical quantity through the estimation of its persistence diagrams. Another example is the estimation of the persistence diagrams of a probability density function f defined on some domain from a finite set of points sampled according to f. The persistence diagram of f may be used to provide information about the modes (peaks) of f and their shape and prominence. More formally, the problem can be stated in the following way: given an unknown metric space X and a function $f : X \to \mathbb{R}$ whose values are known only at a finite set of sample points $P \subset X$, can we reliably estimate the persistence diagrams of the sublevel set filtration of f?

When X is a compact Riemannian manifold and f is a Lipschitz function, [CGOS11] provide an algorithm computing a persistence diagram whose bottleneck distance to the diagram of f is upper bounded by a function depending on the Lipschitz constant of f and on $d_H(P, X)$ when this latter quantity is smaller than some geometric quantity, namely the so-called *convexity radius* of the manifold X. Applied to the case where f is a density estimate, this result has led to new clustering algorithms on the Riemannian manifold where persistence is used to identify and characterize relevant clusters [CGOS13]. Applied to curvature-like functions on surfaces, it has also been used for shape segmentation [SOCG10]. As already noted in Section 25.1, the dependence of the quality of the estimated persistence diagrams on the Hausdorff distance $d_H(P, X)$ makes this approach very sensitive to data corrupted by outliers. Some recent attempts have been made to overcome this issue [BCD$^+$15] but the existing results apply in only very restrictive settings and the problem remains largely open.

PERSISTENCE-BASED SIGNATURES

Relevant multiscale topological signatures of data can be defined using the persistent homology of filtered simplicial complexes built on top of the data. Formally, given a metric space (Y, d_Y), the data, approximate a (possibly unknown) metric space (X, d_X). The idea is to build a filtered simplicial complex on top of Y whose homotopy type, homology, or persistent homology is related to the one of X. Considering the Vietoris-Rips filtration $\mathbb{R}\text{ips}(X)$, it was proven in [Hau95] that if X is a closed Riemannian manifold, then for any sufficiently small $\alpha > 0$, $\text{Rips}(X, \alpha)$ is homotopy equivalent to X. This result was later generalized to prove that if (Y, d_Y) is close enough to (X, d_X) with respect to the Gromov-Hausdorff distance, then there exists $\alpha > 0$ such that $\text{Rips}(Y, \alpha)$ is homotopy equivalent to X [Lat01]. Quantitative variants of this result were obtained in [ALS13] for a class of compact subsets of ε^d. Considering the whole filtration and its persistent homology allows us to relax the assumptions made on X. For the Čech and Vietoris-Rips complexes, the following stability result holds in any compact metric space [CSO14].

THEOREM 25.3.1 Stability of persistence-based signatures

Let (X, d_X) and (Y, d_Y) be two compact metric spaces. Then

$$\mathrm{d_b}(\mathsf{dgm}(\mathrm{H}(\check{\mathbb{C}}\mathrm{ech}(X))), \mathsf{dgm}(\mathrm{H}(\check{\mathbb{C}}\mathrm{ech}(Y)))) \leq 2\mathrm{d_{GH}}(X, Y),$$
$$\mathrm{d_b}(\mathsf{dgm}(\mathrm{H}(\mathbb{R}\mathrm{ips}(X))), \mathsf{dgm}(\mathrm{H}(\mathbb{R}\mathrm{ips}(Y)))) \leq 2\mathrm{d_{GH}}(X, Y)$$

where $\mathsf{dgm}(\mathrm{H}(\check{\mathbb{C}}\mathrm{ech}(X)))$ *(resp.,* $\mathsf{dgm}(\mathrm{H}(\mathbb{R}\mathrm{ips}(X)))$*) denotes the persistence diagrams of the Čech (resp., Vietoris-Rips) filtrations built on top of X and* $\mathrm{d_b}(.,.)$ *is the bottleneck distance.*

This result indeed holds for larger families of geometric complexes built on top of metric spaces, in particular for the so-called witness complexes [SC04], and also extends to spaces endowed with a dissimilarity measure (no need of the triangle inequality). Computing persistent homology of geometric filtrations built on top of data is a classical strategy in TDA; see for example [CISZ08] for a "historical" application.

A first version of Theorem 25.3.1, restricted to the case of finite metric spaces, is given in [CCSG+09] where it is applied to shape comparison and classification. From a practical perspective, the computation of the Gromov-Hausdorff distance between two metric spaces is in general out of reach, even for finite metric spaces with relatively small cardinality. The computation of persistence diagrams of geometric filtrations built on top of metric spaces thus provides a tractable way to compare them. It is however important to notice that the size of the k-dimensional skeleton of geometric filtrations, such as the Rips-Vietoris or Čech complexes, built on top of n data points is $O(n^k)$, leading to severe practical restriction for their use. Various approaches have been proposed to circumvent this problem. From an algorithmic point of view, new data structures have been proposed to efficiently represent geometric filtrations [BM14] and compute their persistence; see Chapter 22. Other lighter filtrations have also been proposed, such as the graph induced complex [DFW15] or the sparse Rips complex [She13]. From a statistical point of view, subsampling and bootstrap methods have been proposed to avoid the prohibitive computation of the persistent homology on filtrations built on the whole data; see the next paragraph. Despite these recent attempts, the practical computation of persistent homology of geometric filtrations built on top of a large data set remains a severe issue.

STATISTICAL ANALYSIS OF PERSISTENCE-BASED SIGNATURES

In the context of data analysis, where data usually carries some noise and outliers, the study of persistent homology from a statistical perspective has recently attracted some interest. Assuming that the data $X_n = \{x_1, \ldots, x_n\}$ is an i.i.d. sample from some probability measure μ supported on a compact metric space (M, d_M), the persistence diagram of geometric filtrations built on top of X_n becomes a random variable distributed according a probability measure in the space of persistence diagrams endowed with the bottleneck distance. Recent efforts have been made to understand and exploit the statistical properties of these distributions of diagrams. For example, building on the stability result for persistence-based signatures, [CGLM15] established convergence rates for the diagrams built on top of X_n to the diagrams built on top of M as $n \to +\infty$. In the same direction, considering subsamples of fixed size m, [BGMP14] and [CFL+15a] prove stability results for the associated distributions of diagrams under perturbations of the probability measure μ in the Gromov-Prohorov and Wasserstein metrics respectively. The latter results provide new promising methods for inferring persistence-based topological information that are resilient to the presence of noise and outliers in the data and that turn out to be practically efficient (persistent homology being computed on filtrations built on top of small fixed-size subsamples).

More generally, a main difficulty in the use of persistent homology in statistical settings hinges on the fact that the space of persistence diagrams is highly nonlinear. This makes the definition and computation of basic statistical quantities such as, e.g., means, nonobvious. Despite this difficulty it has been shown that several standard statistical notions and tools can still be defined and used with persistent diagrams, such as Fréchet means [MMH11], confidence sets [FLR+14], or bootstrap techniques [CFL+15b], etc. Attempts have also been made to find new representations of persistence diagrams as elements of linear spaces in which statistical tools are easier to handle. A particularly interesting contribution in this direction is the introduction of the notion of *persistence landscape*, a representation of persistence diagrams as a family of piecewise linear functions on the real line [Bub15].

25.4 OTHER APPLICATIONS OF TOPOLOGICAL METHODS IN DATA ANALYSIS

Topological Data Analysis has known an important development during the last decade and it now includes a broad spectrum of tools, methods, and applications that go beyond the mathematical results presented in the first three sections of this paper. In this section, we present other directions in which TDA has been developed or applied.

VISUALIZATION AND DIMENSIONALITY REDUCTION

Beyond mathematical and statistical relevance, the efficient and easy-to-understand visualization of the topological and geometric structure of data is an important task in data analysis. The TDA toolbox proposes a few methods to represent and visualize some topological features of data.

Data visualization using Mapper. *Mapper* is a method to visualize high-dimensional and complex data using simplicial complexes. Introduced in [SMC07], it relies on the idea that local and partial clustering of the data leads to a cover of the whole data whose nerve provides a simplified representation of the global structure. Given a data set X, a function $f : X \to \mathbb{R}$, and a finite cover $(I_i)_{i=1,\dots,n}$ of $f(X) \subset \mathbb{R}$ by a family of intervals, the Mapper method first clusterizes each preimage $f^{-1}(I_i)$, of the interval I_i to obtain a (finite) cover $U_1, \dots, U_{k_i}$ of $f^{-1}(I_i)$. The union of the obtained clusters for all the intervals I_i's is a cover of X and Mapper outputs a graph, the 1-skeleton of this cover. The method is very flexible as it leaves the choice of the function f, the cover $(I_i)_{i=1,\dots,n}$, and the clustering methods to the user. The output graph provides an easy-to-visualize representation of the structure of the data driven by the function f. The Mapper algorithm has been popularized and is widely used as a visualization tool to explore and discover hidden insights in high-dimensional data sets; see, e.g., [Car09, LSL$^+$13] for a precise description and a discussion of the Mapper algorithm. When the length of the intervals I_i's is small, the output of Mapper can be seen as a discrete version of the Reeb graph of the function f. However, despite a few recent results, the theoretical analysis of the Mapper method and its formal connection with the Reeb graph remain an open research area.

Morse theory. Other topological methods, including in particular Morse theory, are also successfully used for data visualization, but in a rather different perspective than Mapper. The interested reader is referred to the following collection of books providing a good survey on the topic: [PTHT11, PHCF12, BHPP14].

Circular coordinates and dimensionality reduction. Nonlinear dimensionality reduction (NLDR) includes a set of techniques whose aim is to represent high-dimensional data in low-dimensional spaces while preserving the intrinsic structure of the data. Classical NLDR methods map the data in a low-dimensional Euclidean space $\mathbb{R}^k$ assuming that real-valued coordinates are sufficient to correctly and efficiently parametrize the underlying structure M (which is assumed to be a manifold) of the data. More precisely, NLDR methods intend to infer a set of functions $f_1, \ldots, f_k : M \to \mathbb{R}$ such that the map $F = (f_1, \ldots, f_k) : M \to \mathbb{R}^k$ is an embedding preserving the geometric structure of M. As a consequence, the theoretical guarantees of NLDR methods require M to have a very simple geometry. For example, ISOMAP [TSL00] assumes M to be isometric to a convex open subset of $\mathbb{R}^k$. To enrich the class of functions used to parametrize the data, [SMVJ11] introduces a persistence-based method to detect and construct circular coordinates, i.e., functions $f : M \to \mathbb{S}^1$ where $\mathbb{S}^1$ is the unit circle. The approach relies on the classical property that $\mathbb{S}^1$ is the classifying space of the first cohomology group (with integer coefficients) $H^1(M, \mathbb{Z})$, i.e., $H^1(M, \mathbb{Z})$ is equal to the set of equivalence classes of maps from M to $\mathbb{S}^1$, where two maps are equivalent if they are homotopic [Hat01]. The method consists first in building a filtered simplicial complex on top of the data and using persistent cohomology to identify relevant, i.e., persistent, cohomology classes. Then a smooth (harmonic) cocycle is chosen in each of these classes and integrated to give a circular function on the data.

This approach opens the door to new NLDR methods combining real-valued and circle-valued coordinates. Using time-delay embedding of time series and time-dependent data [Tak81], the circular coordinates approach also opens the door to new topological approaches in time series analysis [PH13, Rob14].

TOPOLOGICAL DATA ANALYSIS IN SCIENCES

Despite its youth, TDA has already led to promising applications and results in various domains of science and the TDA toolbox is now used with many different kinds of data. The following list provides a short and nonexhaustive selection of domains where topological approaches appear to be particularly promising.

- ***Biology:*** Biology is currently probably the largest field of application of TDA. There already exists a vast literature using persistent homology and the Mapper algorithm to analyze various types of biological data; see, e.g., [DCCW+10, NLC11] for an application to breast cancer data.
- ***Networks analysis:*** Beyond sensor network problems, the use of topological data analysis tools to understand and analyze the structure of networks has recently attracted some interest. A basic idea is to build filtered simplicial complexes on top of weighted networks and to compute their persistent homology. Despite a few existing preliminary experimental results, this remains a widely unexplored research direction.
- ***Material science:*** Persistent homology recently found some promising applications in the study of structure of materials, such as for example granular media [KGKM13] or amorphous materials [NHH+15].
- ***Shape analysis:*** The geometric nature of 2D and 3D shapes makes topological methods particularly relevant to design shape descriptors for various tasks such as classification and segmentation of registration; see, for example, [CZCG05, FL12, FL11, COO15].

25.5 FURTHER READINGS

[Car09, Ghr08]: two survey papers that present various aspects of TDA addressing a large audience.

[Oud15]: a recent book that offers a very good introduction.

Although not discussed in this chapter, (discrete) Morse theory, Reeb graphs [DW13] and, more recently, category and sheaf theory are among the mathematical tools used in TDA. An introduction to these topics from a computational and applied perspective can be found in the recent books [EH10, Ghr14].

RELATED CHAPTERS

Chapter 22: Random simplicial complexes
Chapter 24: Persistent homology
Chapter 35: Curve and surface reconstruction
Chapter 43: Nearest neighbors in high-dimensional spaces
Chapter 66: Geometry and topology of genomic data

REFERENCES

[AC15] H. Adams and G. Carlsson. Evasion paths in mobile sensor networks. *Internat. J. Robotics Research*, 34:90–104, 2015.

[ACC$^+$12] M. Aanjaneya, F. Chazal, D. Chen, M. Glisse, L.J. Guibas, and D. Morozov. Metric graph reconstruction from noisy data. *Internat. J. Comput. Geom. Appl.*, 22:305–325, 2012.

[ALS13] D. Attali, A. Lieutier, and D. Salinas. VietorisRips complexes also provide topologically correct reconstructions of sampled shapes. *Comput. Geom.*, 46:448–465, 2013.

[BCD$^+$15] M. Buchet, F. Chazal, T.K. Dey, F. Fan, S.Y. Oudot, and Y. Wang. Topological analysis of scalar fields with outliers. In *Proc. 31st Sympos. Comput. Geom.*, pages 827–841, ACM Press, 2015.

[BCOS16] M. Buchet, F. Chazal, S.Y. Oudot, and D.R. Sheehy. Efficient and robust persistent homology for measures. *Comput. Geom.*, 58:70–96, 2016.

[BCSE$^+$07] P. Bendich, D. Cohen-Steiner, H. Edelsbrunner, J. Harer, and D. Morozov. Inferring local homology from sampled stratified spaces. In *Proc. 48th IEEE Sympos. Found. Comp. Sci.*, pages 536–546, 2007.

[BG13] J.-D. Boissonnat and A. Ghosh. Manifold reconstruction using tangential Delaunay complexes. *Discrete Comput. Geom.*, 51, 2013.

[BGMP14] A.J. Blumberg, I. Gal, M.A. Mandell, and M. Pancia. Robust statistics, hypothesis testing, and confidence intervals for persistent homology on metric measure spaces. *Found. Comput. Math.*, 14:745–789, 2014.

[BGO09] J.-D. Boissonnat, L.J. Guibas, and S.Y. Oudot. Manifold reconstruction in arbitrary dimensions using witness complexes. *Discrete Comput. Geom.*, 42:37–70, 2009.

[BGV07] F. Bolley, A. Guillin, and C. Villani. Quantitative concentration inequalities for empirical measures on non-compact spaces. *Probab. Theory Rel.*, 137:541–593, 2007.

[BHPP14] P.-T. Bremer, I. Hotz, V. Pascucci, and R. Peikert, editors. *Topological Methods in Data Analysis and Visualization III: Theory, Algorithms, and Applications.* Mathematics and Visualization, Springer, Berlin, 2014.

[BM13] A.J. Blumberg and M.A. Mandell. Quantitative homotopy theory in topological data analysis. *Found. Comput. Math.*, 13:885–911, 2013.

[BM14] J.-D. Boissonnat and C. Maria. The simplex tree: An efficient data structure for general simplicial complexes. *Algorithmica*, 70:406–427, 2014.

[BRS+12] S. Balakrishnan, A. Rinaldo, D. Sheehy, A. Singh, and L.A. Wasserman. Minimax rates for homology inference. In *Proc. 15th Conf. Artif. Intell. Stats.*, pages 64–72, JMLR W&CP, 2012.

[Bub15] P. Bubenik. Statistical topological data analysis using persistence landscapes. *J. Mach. Learn. Res.*, 16:77–102, 2015.

[BWM12] P. Bendich, B. Wang, and S. Mukherjee. Local homology transfer and stratification learning. In *Proc. 23rd ACM-SIAM Sympos. Discrete Algorithms*, pages 1355–1370, 2012.

[Car09] G. Carlsson. Topology and data. *Bull. Amer. Math. Soc.*, 46:255–308, 2009.

[CCSG+09] F. Chazal, D. Cohen-Steiner, L.J. Guibas, F. Mémoli, and S.Y. Oudot. Gromov-Hausdorff stable signatures for shapes using persistence. *Computer Graphics Forum*, 28:1393–1403, 2009.

[CCSL09a] F. Chazal, D. Cohen-Steiner, and A. Lieutier. A sampling theory for compact sets in Euclidean space. *Discete Comput. Geom.*, 41:461–479, 2009.

[CCSL09b] F. Chazal, D. Cohen-Steiner, and A. Lieutier. Normal cone approximation and offset shape isotopy. *Comput. Geom.*, 42:566–581, 2009.

[CCSLT09] F. Chazal, D. Cohen-Steiner, A. Lieutier, and B. Thibert. Stability of curvature measures. *Computer Graphics Forum*, 28:1485–1496, 2009.

[CCSM10] F. Chazal, D. Cohen-Steiner, and Q. Mérigot. Boundary measures for geometric inference. *Found. Comput. Math.*, 10:221–240, 2010.

[CCSM11] F. Chazal, D. Cohen-Steiner, and Q. Mérigot. Geometric inference for probability measures. *Found. Comput. Math.*, 11:733–751, 2011.

[CFL+14] F. Chazal, B.T. Fasy, F. Lecci, B. Michel, A. Rinaldo, and L. Wasserman. Robust topological inference: Distance to a measure and kernel distance. Preprint, `arXiv:1412.7197`, 2014.

[CFL+15a] F. Chazal, B. Fasy, F. Lecci, B. Michel, A. Rinaldo, and L. Wasserman. Subsampling methods for persistent homology. In *Proc. 32nd Internat. Conf. Machine Learning (ICML)*, pages 2143–2151, JMLR W&CP, 2015.

[CFL+15b] F. Chazal, B.T. Fasy, F. Lecci, A. Rinaldo, and L. Wasserman. Stochastic convergence of persistence landscapes and silhouettes. *J. Comput. Geom.*, 6:140–161, 2015.

[CGLM15] F. Chazal, M. Glisse, C. Labruère, and B. Michel. Convergence rates for persistence diagram estimation in topological data analysis. *J. Machine Learning Research*, 16:3603–3635, 2015.

[CGOS11] F. Chazal, L.J. Guibas, S.Y. Oudot, and P. Skraba. Scalar field analysis over point cloud data. *Discrete Comput. Geom.*, 46:743–775, 2011.

[CGOS13] F. Chazal, L.J. Guibas, S.Y. Oudot, and P. Skraba. Persistence-based clustering in Riemannian manifolds. *J. ACM*, 60, 2013.

[CHS15] F. Chazal, R. Huang, and J. Sun. Gromov-Hausdorff approximation of filamentary structures using Reeb-type graphs. *Discrete Comput. Geom.*, 53:621–649, 2015.

[CISZ08] G. Carlsson, T. Ishkhanov, V. de Silva, and A. Zomorodian. On the local behavior of spaces of natural images. *Internat. J. Computer Vision*, 76:1–12, 2008.

[CL07] F. Chazal and A. Lieutier. Stability and computation of topological invariants of solids in $\mathbb{R}^n$. *Discrete Comput. Geom.*, 37:601–617, 2007.

[CL08] F. Chazal and A. Lieutier. Smooth manifold reconstruction from noisy and non-uniform approximation with guarantees. *Comput. Geom.*, 40:156–170, 2008.

[CMM15] F. Chazal, P. Massart, and B. Michel. Rates of convergence for robust geometric inference. *Electronic J. Stat.*, 10:2243–2286, 2016.

[CO08] F. Chazal and S.Y. Oudot. Towards persistence-based reconstruction in Euclidean spaces. In *Proc. 24th Sympos. Comput. Geom.*, pages 232–241, ACM Press, 2008.

[COO15] M. Carrière, S.Y Oudot, and M. Ovsjanikov. Stable topological signatures for points on 3d shapes. *Computer Graphics Forum*, 34:1–12, 2015.

[CSEH07] D. Cohen-Steiner, H. Edelsbrunner, and J. Harer. Stability of persistence diagrams. *Discrete Comput. Geom.*, 37:103–120, 2007.

[CSO14] F. Chazal, V. de Silva, and S.Y. Oudot. Persistence stability for geometric complexes. *Geom. Dedicata*, 173:193–214, 2014.

[CZCG05] G. Carlsson, A. Zomorodian, A. Collins, and L.J. Guibas. Persistence barcodes for shapes. *Internat. J. Shape Model*, 11, 2005.

[DCCW+10] D. DeWoskin, J. Climent, I. Cruz-White, M. Vazquez, C. Park, and J. Arsuaga. Applications of computational homology to the analysis of treatment response in breast cancer patients. *Topology Appl.*, 157:157–164, 2010.

[DFW15] T.K. Dey, F. Fan, and Y. Wang. Graph induced complex on point data. *Comput. Geom.*, 48:575–588, 2015.

[DW13] T.K. Dey and Y. Wang. Reeb graphs: approximation and persistence. *Discrete Comput. Geom.*, 49:46–73, 2013.

[EH10] H. Edelsbrunner and J.L. Harer. *Computational Topology: An Introduction.* AMS, Providence, 2010.

[FL99] P. Frosini and C. Landi. Size theory as a topological tool for computer vision. *Pattern Recognit. Image Anal.*, 9:596–603, 1999.

[FL11] B. di Fabio and C. Landi. A Mayer-Vietoris formula for persistent homology with an application to shape recognition in the presence of occlusions. *Found. Comput. Math.*, 11:499–527, 2011.

[FL12] B. di Fabio and C. Landi. Persistent homology and partial similarity of shapes. *Pattern Recog. Lett.*, 33:1445–1450, 2012.

[FLR+14] B.T. Fasy, F. Lecci, A. Rinaldo, L. Wasserman, S. Balakrishnan, A. Singh, et al. Confidence sets for persistence diagrams. *Ann. Statist.*, 42:2301–2339, 2014.

[Ghr08] R. Ghrist. Barcodes: The persistent topology of data. *Bull. Amer. Math. Soc.*, 45:61–75, 2008.

[Ghr14] R. Ghrist. *Elementary Applied Topology.* CreateSpace, 2014.

[GMM13] L. Guibas, D. Morozov, and Q. Mérigot. Witnessed k-distance. *Discrete Comput. Geom.*, 49:22–45, 2013.

[GPP+12a] C.R. Genovese, M. Perone-Pacifico, I. Verdinelli, and L. Wasserman. Manifold estimation and singular deconvolution under Hausdorff loss. *Ann. Statist.*, 40:941–963, 2012.

[GPP+12b] C.R. Genovese, M. Perone-Pacifico, I. Verdinelli, and L. Wasserman. Minimax manifold estimation. *J. Machine Learning Research*, 13:1263–1291, 2012.

[Gro93] K. Grove. Critical point theory for distance functions. In *Proc. Sympos. Pure Math.*, vol. 54, 1993.

[GSBW11] X. Ge, I. Safa, M. Belkin, and Y. Wang. Data skeletonization via Reeb graphs. In *Proc. 24th Int. Conf. Neural Information Processing Systems*, pages 837–845, 2011.

[Hat01] A. Hatcher. *Algebraic Topology.* Cambridge University Press, 2001.

[Hau95] J.-C. Hausmann. On the Vietoris-Rips complexes and a cohomology theory for metric spaces. *Ann. Math. Stud.*, 138:175–188, 1995.

[HK14] Y. Hiraoka and G. Kusano. Coverage criterion in sensor networks stable under perturbation. Preprint, `arXiv:1409.7483`, 2014.

[KGKM13] M. Kramar, A. Goullet, L. Kondic, and K. Mischaikow. Persistence of force networks in compressed granular media. *Physical Review E*, 87, 2013.

[Lat01] J. Latschev. Vietoris-Rips complexes of metric spaces near a closed Riemannian manifold. *Arch. Math.*, 77:522–528, 2001.

[LOC14] C. Li, M. Ovsjanikov, and F. Chazal. Persistence-based structural recognition. In *Proc. IEEE Conf. Comp. Vis. Pattern Recogn.*, pages 2003–2010, 2014.

[LSL+13] P.Y. Lum, G. Singh, A. Lehman, T. Ishkanov, M. Vejdemo-Johansson, M. Alagappan, J. Carlsson, and G. Carlsson. Extracting insights from the shape of complex data using topology. *Scientific Reports*, 3:1236, 2013.

[Mér13] Q. Mérigot. Lower bounds for k-distance approximation. In *Proc. 29th Sympos. Comput. Geom.*, pages 435–440, ACM Press, 2013.

[MMH11] Y. Mileyko, S. Mukherjee, and J. Harer. Probability measures on the space of persistence diagrams. *Inverse Problems*, 27, 2011.

[NHH+15] T. Nakamura, Y. Hiraoka, A. Hirata, E.G. Escolar, and Y. Nishiura. Persistent homology and many-body atomic structure for medium-range order in the glass. *Nanotechnology*, 26, 2015.

[NLC11] M. Nicolau, A.J. Levine, and G. Carlsson. Topology based data analysis identifies a subgroup of breast cancers with a unique mutational profile and excellent survival. *Proc. Natl. Acad. Sci. USA*, 108:7265–7270, 2011.

[NSW08] P. Niyogi, S. Smale, and S. Weinberger. Finding the homology of submanifolds with high confidence from random samples. *Discrete Comput. Geom.*, 39:419–441, 2008.

[NSW11] P. Niyogi, S. Smale, and S. Weinberger. A topological view of unsupervised learning from noisy data. *SIAM J. Comput.*, 40:646–663, 2011.

[Oud15] S.Y. Oudot. *Persistence Theory: From Quiver Representations to Data Analysis.* Vol. 209 of *Math. Surv. Monogr.*, AMS, Providence, 2015.

[Pet06] A. Petrunin. Semiconcave functions in Alexandrov's geometry. In *Surveys in Differential Geometry*, vol. 11, pages 137–201. International Press, Somerville, 2006.

[PH13] J.A. Perea and J. Harer. Sliding windows and persistence: An application of topological methods to signal analysis. *Found. Comput. Math.*, 15:799–838, 2013.

[PHCF12] R. Peikert, H. Hauser, H. Carr, and R. Fuchs, editors. *Topological Methods in Data Analysis and Visualization II: Theory, Algorithms, and Applications.* Mathematics and Visualization, Springer, Berlin, 2012.

[PTHT11] V. Pascucci, X. Tricoche, H. Hagen, and J. Tierny. *Topological Methods in Data Analysis and Visualization: Theory, Algorithms, and Applications*, 1st edition. Springer, Berlin, 2011.

[PWZ15] J.M. Phillips, B. Wang, and Y. Zheng. Geometric inference on kernel density estimates. In *Proc. 31st Sympos. Comput. Geom.*, pages 857–871, ACM Press, 2015.

[RHBK15] J. Reininghaus, S. Huber, U. Bauer, and R. Kwitt. A stable multi-scale kernel for topological machine learning. In *Proc. IEEE Conf. Comp. Vis. Pattern Recogn.*, pages 4741–4748, 2015.

[Rob99] V. Robins. Towards computing homology from finite approximations. In *Topology Proceedings*, vol. 24, pages 503–532, 1999.

[Rob14] M. Robinson. *Topological Signal Processing.* Springer, Berlin, 2014.

[SC04] V. de Silva and G. Carlsson. Topological estimation using witness complexes. In *Proc. 1st Eurographics Conf. on Point-Based Graphics*, pages 157–166, 2004.

[SG07a] V. de Silva and R. Ghrist. Coverage in sensor networks via persistent homology. *Algebraic & Geometric Topology*, 7:339–358, 2007.

[SG07b] V. de Silva and R. Ghrist. Homological sensor networks. *Notices Amer. Math. Soc.*, 54, 2007.

[She13] D.R. Sheehy. Linear-size approximations to the Vietoris-Rips filtration. *Discrete Comput. Geom.*, 49(4):778–796, 2013.

[SMC07] G. Singh, F. Mémoli, and G. Carlsson. Topological methods for the analysis of high dimensional data sets and 3D object recognition. In *Proc. Eurographics Sympos. on Point-Based Graphics (SPBG)*, pages 91–100. Eurographics, 2007.

[SMVJ11] V. de Silva, D. Morozov, and M. Vejdemo-Johansson. Persistent cohomology and circular coordinates. *Discrete Comput. Geom.*, 45:737–759, 2011.

[SOCG10] P. Skraba, M. Ovsjanikov, F. Chazal, and L. Guibas. Persistence-based segmentation of deformable shapes. In *Proc. IEEE Conf. Comp. Vis. Pattern Recogn.*, pages 45–52, 2010.

[Tak81] F. Takens. *Detecting Strange Attractors in Turbulence.* Springer, Berlin, 1981.

[TSL00] J.B. Tenenbaum, V. De Silva, and J.C. Langford. A global geometric framework for nonlinear dimensionality reduction. *Science*, 290:2319–2323, 2000.

[Vil03] C. Villani. *Topics in Optimal Transportation*, AMS, Providence, 2003.

[VUFF93] A. Verri, C. Uras, P. Frosini, and M. Ferri. On the use of size functions for shape analysis. *Biological Cybernetics*, 70:99–107, 1993.